Dr. John E. Bertie
Department of Chemistry
The University of Alberta
Edmonton, Alberta, Canada T6G 2G2

Theory and Methods of Calculation of Molecular Spectra

Theory and Methods of Calculation of Molecular Spectra

L. A. Gribov

VI Vernadsky Institute of Geochemistry and Analytical Chemistry, USSR Academy of Sciences, Moscow

W. J. Orville-Thomas

Department of Chemistry and Applied Chemistry University of Salford

JOHN WILEY & SONS

Chichester · New York · Brisbane · Toronto · Singapore

Library of Congress Cataloging in Publication Data:

Gribov, L. A. (Lev Aleksandrovich)
Theory and methods of calculation of molecular spectra.
1. Molecular spectra. I. Orville-Thomas, W. J., 1921– . II. Title.
QD96.M65G737 1988 543.'0858 88-181

ISBN 0 471 91882 2

British Library Cataloguing in Publication Data:

Gribov, L. A.
Theory and methods of calculation of molecular spectra.
1. Molecular spectroscopy
I. Title II. Orville-Thomas, W. J.
535.8'4 QC454.M6

ISBN 0 471 91882 2

Phototypesetting by Thomson Press (India) Limited, New Delhi
Printed and bound in Great Britain by Anchor Brendon Ltd, Tiptree, Essex

Contents

Preface

This book gives an account of how computers can be used to supplement, and in some cases to replace, experimental spectroscopic studies in the identification of molecular substances and to provide detailed information on their geometrical and electronic structures.

The traditional way of identifying an unknown chemical substance was to study it experimentally in an analytical laboratory. Over the course of years the methods of identification changed from classical analytical procedures to those based on physico-chemical, mainly spectroscopic, techniques. The standard method of identifying unknown molecular structures was to compare the experimental spectra (UV, IR, NMR, etc.) with the spectra of known substances.

The experimental approach led to a wealth of data such as characteristic bands, bond lengths and other geometrical parameters, and spin–spin coupling constants, etc. In the early days tables of characteristic bands were constructed which facilitated the identification of molecular substances. Also molecular constants derived from spectral analysis and diffraction studies were collected and used in discussions on molecular structure. Nowadays such information is fed into data banks for quick information retrieval.

The non-destructive spectroscopic methods, allied to developments in chemical theory, led to a revolution in academic and industrial laboratories so far as the identification of unknown substances was concerned.

The availability of a large quantity of experimental data led to the rapid development of a semi-empirical approach to structural identification. This approach reached its peak in the books of L. J. Bellamy.

The experimental approach demands considerable effort, a great deal of instrumental time and skilled operators. The current increase in the work load on spectroscopic services has led to a crisis in industrial research and development where increasingly it is becoming more difficult to meet the analytical demands being made. Financial and other considerations, such as the finite supply of highly qualified and experienced spectroscopists, militate against the obvious solution of increasing the number of spectrometers available and improving the semi-empirical databank basis for structural identification.

This impasse can be solved by adding a theoretical approach to the two-pronged attack consisting of a spectrometer plus operator. Since about 1950 a new revolution of great magnitude has occurred in the scientific world. This has been the advent of computers which daily have an overwhelming impact on society and the quality of life. As with the Industrial Revolution chemists are playing a dominant role in this new 'Computer Revolution'.

The ready accessibility of computers has led to the replacement of the dual spectrometer–operator approach by a three-pronged spectrometer–computer–operator attach. This change of emphasis merely acknowledges the tremendous advances made in computer calculations of molecular properties which nowadays rival and sometimes are more accurate than the experimental values. It is true to say that sophisticated computer software now forms an essential ingredient of laboratory equipment and ranks alongside an adequate supply of chemicals and spectrometers.

The experimental–empirical structural correlation approach represented a half-way house to the completely theoretical approach where a 'theoretical' spectrum is generated by solving the appropriate quantum mechanical equations. For large molecules the theoretical approach is impossibly difficult and so recourse has to be made to more approximate methods. Following this route so much progress has been made that we are fast approaching the point where the semi-empirical approach based on databanks will soon be replaced by an artificial intelligence system which will generate theoretical spectra for direct comparison with those obtained experimentally.

The central problem in molecular spectroscopy is not only to identify a structural formula for a molecule but to obtain detailed information on its geometrical and electronic structure—ideally in the ground and excited states.

From the mathematical point of view the investigation of polyatomic molecules reduces to the determination of sets of molecular parameters associated with different models for the molecule. For example to describe a static three-dimensional model, one would need to know atomic charges, bond lengths and bond angles. It is the solution of the inverse spectral problem which gives these sets of parameters for molecular models. To obtain the most accurate solution for the inverse spectral problem it is necessary to consider a series of models which increase in complexity as the study proceeds. Once a particular model has been obtained it is used as the starting point to simulate a theoretical spectrum corresponding to the chosen model. This theoretical spectrum is then compared with that obtained experimentally. This comparison makes it clear whether the chosen model needs to be changed or not. If necessary the model is improved until the theoretical and experimental spectra coincide satisfactorily. Once this has been achieved, a more complex model is set up and the iterative process repeated. This procedure is repeated until the most accurate possible molecular model is obtained. The final product of this iterative process of solving the inverse spectral problem is a refined set of molecular parameters which gives the most accurate model for the molecule in terms of summarizing its geometrical and electronic structure. To accomplish this task in a reasonably short period of time it is necessary to use powerful computers.

In going from a simple to a more complicated model it is necessary to increase the input information, i.e. the new set of parameters is larger and more complicated than the previous set. But, of course, when the corresponding inverse spectral problem is solved one gets more valuable information about the molecular structure.

For maximum efficiency the whole artificial intelligence system should operate with the minimum of help from the operator. Hence one has to build into the system the possibility of generating automatically the description of the molecular models and the corresponding theoretical spectral display. These displays are then compared directly with the experimental spectra and any necessary changes made in the set of model parameters to ensure greater coincidence between the theoretical and experimental spectra.

In summary, then, the current theoretical approach can be described as follows. The molecular system is described by a series of models of increasing complexity. For example one could proceed from an empirical formula CH_4O to a structural model CH_3OH; this in turn is replaced by a three-dimensional representation

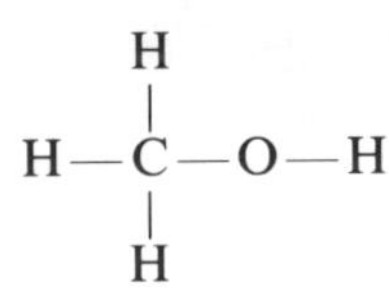

in which the bond lengths and angles have definite values. One could then introduce information on the force field and bond moments, etc.

Once a particular model has been chosen the methods of computational chemistry are then used to calculate a simulated spectrum which can then be compared with that obtained experimentally. In order to improve the fit, the parameters of the assumed model are varied by an iterative process and the corresponding theoretical spectrum calculated at each stage until a satisfactory degree of concordance is achieved between the experimental and theoretical spectra. This amounts to carrying out a 'structural' investigation on a computer. During the course of this procedure the computer generates a large number of simulated spectra which are simultaneously compared with those obtained experimentally.

We hope that the book will be found helpful by the reader who will accept it with goodwill. We will be grateful for any criticism which might be made, for we fully realize that the book is not free from shortcomings. We apologize for any errors and await corrections from readers.

The text of this book was first written in Russian, the major portion of which was then translated into English by Richard Hainsworth who at that time was working in the USSR with 'Mir' Publishers. The authors would like to acknowledge their sincere gratitude to him for this preliminary draft.

Lev A. Gribov (Moscow)
W. J. Orville-Thomas (Salford)

Introduction

Molecular spectroscopy as a method for investigating the structure and properties of polyatomic molecules, molecular associates, polymers, crystals, etc. is now going through a period of rebirth, accompanied by a rapid increase in its potential value. This can be attributed to:

(1) A substantial improvement in instrumentation and techniques. The discovery of many new methods enables the measurement of the optical spectral characteristics of matter of the most varied nature and in different states;
(2) The advance in theory and computational algorithms, which enables us to calculate these characteristics quantitatively. The models chosen for the trial calculations differ only in the details and accuracy of the structural information used.

In order to clarify the role and significance of the theory and computation technique let us consider the basics of methodology of spectral analysis. First, observations can include the frequencies of band centres or lines in the absorption, emission, scattering and reflection spectra, band intensities (integral and peak values, absolute and relative) and characteristics of band envelopes and lines (halfwidths, moments of various orders, etc.). All these features are characterized by numbers which form a set of experimentally defined quantities, and this set characterizes the entity being investigated under the conditions of measurement. However, it is only capable of implicitly characterizing a molecular entity. No matter how accurate the measurements of the experimental data become, these sets of values *per se* in no way provide the researcher with an accurate structural model.

Indeed, when referring to a molecular entity we can represent this, for example, as an empirical or structural formula or a geometrical spatial model with distributed charges. It is this type of information which every researcher tries to obtain in order to study a particular substance. In principle, any investigation of polyatomic molecules, molecular associates, polymers, etc., reduces to the construction of various models in a certain hierarchical succession: from the simplest model in the form of an empirical formula to much more complicated ones such as those involving parameters of the potential surfaces of the ground and excited states and charge distribution. Each of these models can be further characterized by a collection of numerical parameters: number and type (mass or charge) of atoms, bond lengths and valence angles, coefficients in the potential functions, charges at the atoms and bonds, etc. These parameters form the second set of features to be studied with the aid of spectral methods, and represents the essence of spectral analysis of the structure of a substance. In principle, such an analysis, in turn, becomes feasible in those cases alone where we have managed to develop a theory powerful enough to establish a one-to-one relationship between the two sets of parameters.

From the mathematical point of view the logic of spectral analysis is determined by that of solving inverse spectral problems, and this is done as follows. Assume that we possess some initial ideas concerning the model of a substance being studied. On the basis of these ideas let us presuppose a collection of features which comprehensively characterize a relevant model. Now let us assign a certain parameter to each of these features which can take on, depending upon the type of the model, either discrete or continuous values (for example, atomic mass, bond length, valence angle, etc.).

Then, by varying the values of these parameters, starting from initial estimates and thus assuming the theoretical spectrum of a model, we will approximate the numerical values characterizing the set of experimental features to those of similar features, also produced theoretically, until the matching of the values of these two sets is seen to be adequate. In these circumstances the model under investigation can be considered as established, provided that, in addition, it meets certain reasonable physical requirements which are initially imposed upon it and which must be observed throughout the entire process of finding a solution.

In general, then, to undertake the solution of the inverse problem (or, to put it another way, to carry out spectrochemical investigation) it is necessary, first, to be able to specify the models of the physical entities in adequate detail, and second, to know how to construct the sets of spectral features adequate for these models. Furthermore, it is clear that the theoretical spectra of the models must vary in a deterministic way in response to changes, either discrete or continuous, in any of the parameters specifying a given model. This inevitably implies the availability of adequately developed theory and computational algorithms which enable us to compute quantitatively the major spectral features of complicated molecular models with an accuracy adequate for comparison with experiment. Without these, it would be impossible to make a detailed spectrochemical analysis, no matter how advanced are the experimental technique and tools.

Therefore experiment and computation constitute an indissoluble dialectical unity which underlies all the spectral methods of studying the structure and properties of matter. In the ideal case these two aspects are developed in parallel, as any priority in elaborating one aspect is in no way capable of compensating for inadequate development of the other. However, in a number of cases development of experimental methods has overtaken the construction of relevant theories and computational techniques. In the majority of cases this has resulted in the impossibility of extracting all the information on the structure of the entity being studied. An example of how inadequate progress in computational techniques may produce an adverse effect upon the general level of applied spectral research is the great difference between the potential of spectroscopy in the field of organic and organometallic compounds, on the one hand, and of inorganic compounds on the other, particularly those comprising heavy atoms. Although the same experimental methods can be applied to both types of compounds, at present we have no proper computational capabilities applicable to inorganic compounds, which might considered equal to those applied to organic compounds. This accounts for the much wider use of spectral analysis in the field of organic and organometallic compounds.

Despite this lag of a theoretical and, in particular, of a computational basis compared with experimental capabilities, marked progress has now been made in this direction during the last two decades. For complex molecules it is now possible to calculate, at the harmonic level, values for vibrational frequencies which compare well with experimental spectra. In computing accurate values for band intensities the position is less satisfactory, but crucial advances have been made in the past ten years.

The use of straightforward vibrational variational methods has, for the first time, enabled us to solve anharmonic vibrational problems for complex molecules in the high-energy range up to the boundary of quasi-continuum and dissociation. This, in turn, provides the necessary computation basis for constructing an up-to-date theory of overtone spectroscopy. Methods of quantum chemistry, which have had an unusual growth during recent years, underlie the design of a self-consistent theory of computing algorithms for the purpose of analysis of purely electronic spectra. Straightforward quantum calculations of the potential surface parameters are already becoming common; we are witnessing the arrival of similar types of calculations of intensities of absorption and emission bands. Substantial progress has been achieved in computations in an adiabatic approximation of a vibrational structure of electronic bands in the absorption and emission spectra, so that in the near future we can expect a shift towards a mass utilization of this technique. The machinery for solving inverse spectral problems has been appreciably improved. Considerable progress has also been made in the theory and methods for computing vibro-rotational spectra.

A new class of problems has been introduced into theoretical spectroscopy. These are concerned with the processing of information expressed in qualitative terms in the form of propositions, with the result that, for the first time, it has become possible to present, in the form of algorithms, the whole process of discovering information on a substance from spectral measurements, beginning with its experimentally determined quantities and ending with the construction of detailed models. All these successful steps have led towards a new phase which consists of creating a computational spectroscopic procedure, in which a complete treatment of spectra will be carried out concurrently with obtaining experimental results.

For more than two decades one of the authors of this book and his co-workers have been concerned with the elaboration of a self-consistent theory and computational methods of molecular spectroscopy, mainly of vibrational and electron–vibrational spectroscopy.

The results obtained have been summarized in a number of studies published in the USSR. These consist of the following (which are cited not in chronological order but according to the increasing complexity of the models): *Molecular Spectral Analysis and the Computer* (M. E. Elyashberg, L. A. Gribov and V. V. Serov, Nauka, Moscow, 1980); *Intensity Theory for Infrared Spectra of Polyatomic Molecules* (L. A. Gribov, USSR Academy of Science, Moscow, 1963); *Molecular Vibrations* (M. V. Wolkenstein, L. A. Gribov, M. A. Eliashevich and B. I. Stepanov, Main Editorial Board for Physico-Mathematical Literature, Moscow, 1972); *Computational Methods and Algorithms in the Theory of Molecular Vibrational Spectra* (L. A. Gribov and V. A. Dement'ev, Nauka, Moscow, 1961); *Methods of Computing Electronic Vibrational Spectra of Polyatomic Molecules* (L. A. Gribov, V. I. Baranov and B. K. Novosadov, Nauka, Moscow, 1984); and *Theory of Infrared Spectra of Polymers* (L. A. Gribov, Main Editorial Board for Physico-Mathematical Literature, Moscow, 1977). In this book an attempt is made to present the material included in the works listed above. In addition, a large number of original papers by the same authors are dealt with, and this explains the few references to non-Russian literature. However, in spite of its length, this volume does not represent a comprehensive survey. It has been designed to emphasize the major contribution of the Moscow school directed by one of us (L. A. G.).

In writing this book the only computational methods considered by the authors are, first, those that have been developed to a considerable extent, which can thus recommended as a basis for carrying out large-scale computations using available computer software, and second, those that have been thoroughly tested by L. A. G. and his research team. All the computational case studies given in this book originated in our own work. Of course, no claim for uniqueness is made concerning development of the theory and methods of computing molecular spectra. As is well known, these have been developed by Hartree, Fock, Roothan, Pariser, Parr, Pople, Herzberg, Slater, Wilson, Crawford, Eliashevich, Stepanov, Wolkenstein and many other scientists who have contributed much to the new stage in the development of computational molecular spectroscopy.

Correct attributions of priority to particular scientists have not been attempted in this book. For this reason, we apologize in advance to those research workers who may consider that their names have been incorrectly omitted. Classical studies can be found in any branch of science, and their contribution is well known without recourse to any special references. We have, of course, widely used the results obtained. Our own contribution is much more modest, being confined primarily to a further elaboration of practically applicable computing techniques. In this direction our research proceeded fairly independently of the efforts of other scholars, and in a number of cases this can be considered as pioneering work. If similar results have been obtained by others (perhaps even earlier than we did), this was done independently, and is another reason why relatively few references are found in this book.

It was our intention to write this book in such a way as to give an orderly and self-consistent presentation of the methods for computing electronic, electronic–vibrational and vibrational spectra of polyatomic molecules, thus enabling the reader to use it as a basis for further research without resorting at each step to other sources. In this regard, the book is fairly comprehensive and self-contained.

A general treatment of the interaction between light and matter has not been included, since many excellent accounts are readily available. This book is therefore addressed mainly to those skilled in the art of theoretical and applied spectroscopy of polyatomics and of quantum chemistry, and who are well aware of the fact that any computation of spectra reduces to the following two problems: (1) calculation of energy levels and (2) estimation of the probabilities of optical transitions, the latter usually being dipole transitions. The probabilities of these transitions are determined, in the case of single-quantum processes of emission and absorption of electromagnetic radiations, by the squares of the respective matrix elements. In the case of multi-quantum processes, including the simplest two-quantum Raman scattering, the probabilities depend on the components of tensors of rank two and higher orders. These tensor components are expressed by single or multiple sums of the terms, each of which is represented by a factorized product of the Cartesian projections of the dipole transition vectors on the axes of a laboratory frame of reference, when the transitions take place between the initial, intermediate and final energy states of a molecule.

To derive and analyse general formulae for all the above processes we use quantum electrodynamics. This branch of science does not make any distinctions between an atom, diatomic or small polyatomic molecules or complex systems. The problems to be solved by the special theory of molecular spectra of complex systems begin with the investigation of a number of general properties, and an elaboration of the methods for computing energy levels and matrix elements of dipole transitions (quadrupolar and other types of transitions play a very minor role). These are the topics dealt with in this book. In a sense, the methods elaborated here apply equally to the calculation of the processes of simple absorption, emission or Raman scattering, and also to the analysis of non-linear effects. However, the lack of expertise in computing spectral manifestations of non-linear effects and, most important, the absence of a parametric approach prevents a claim that a theoretical basis for specifying these effects has already been established. The gap between experimental and theoretical capabilities in this field is so great today that it seems likely that it will take a very long time for the most recent achievements in non-linear spectroscopy to lead to the emergence of new standard methods of studying the structure of matter.

Yet another important comment should be made. Despite the fact that many of the methods presented in this book are applicable to the computation of the spectra of diatomics and small polyatomics, as a whole, they are intended mainly for handling large molecules. It is not by chance that such methods are typically associated with parametric approaches which are not applicable to small molecules, where *ab initio* calculations are preferable.

Chapter 1

General Problems of Methodology of Spectrum Analysis of Complex Molecules

In this chapter we shall develop the concepts mentioned in the Introduction and emphasize how practical conclusions follow from those rather general reflections. We consider this to be necessary, the more so as theorists in molecular spectroscopy, intent on overcoming mathematical difficulties in solving their particular problems, forget that the most important aim of any theory is to apply it in practice, and therefore to work out an adequate strategy for practical applications. This is especially true for complex molecular systems, which have many different properties. If the main aim of the study is an 'object' (a molecule, molecular associate or crystal), then spectral analysis is a means of understanding rather than an end in itself. It is important that a researcher who studies specific aspects of, for example, a complex molecular system be clearly aware that the aim is to build up a model concept of the object. A model is the only means of providing a physical representation and gives a researcher grounds for discussing an object's existence at various points in time and its appearance (or structure) in different situations.

As is well known, molecular structural studies employ a great number of molecular models. The different ideas which characterize particular models form a pictorial language, the use of which is to provide an understanding of the mysteries of Nature and also to make it possible to forecast the properties of molecular systems on the basis of general concepts. The use of molecular models, which are described mathematically as precisely as possible, is thus not only helpful but also necessary for the complete understanding at the molecular level of the world which surrounds us. This is especially so when a computer, which cannot operate on inexact concepts, is involved in the process of comprehension.

To understand the methodology of modern spectral chemical analysis it is useful to list the basic molecular models used in various fields of chemistry and physics and to be aware of how these models are described mathematically.

The most general model which reflects the concept of a molecule as a micro-system comprising atomic nuclei and electrons is the empirical formula of the substance. To fully characterize this model it is sufficient to indicate only the number, charges and mass numbers of the atomic nuclei occurring in this molecular system. Assuming electrical neutrality, the number of electrons is determined automatically. Should the concept inherent in the empirical formula be supported by the hypothesis that the charged particles which make up the system interact according to the Coulomb law only and that their motion follows the quantum laws, the general Schroedinger equation may be written. This equation, given in all texts on quantum chemistry and molecular spectrum theory, is

$$(\hat{T}_{\mathrm{n}} + \hat{T}_{\mathrm{e}} + V_{\mathrm{en}} + V_{\mathrm{nn}} + V_{\mathrm{ee}})\Psi_{\mathrm{en}} = E_{\mathrm{en}}\Psi_{\mathrm{en}} \tag{1.1}$$

where $\hat{T}_{\mathrm{n}}$ and $\hat{T}_{\mathrm{e}}$ are the kinetic energy operators for nuclei and electrons; V_{en}, V_{nn}, V_{ee} are, respectively, the operators of the Coulomb attraction of electrons and nuclei and nucleus–nucleus and electron–electron repulsions; Ψ_{en} is the general wave function of the system; and E_{en} is the total energy of the system.

Although the molecular representations by empirical formulae and Schroedinger equations of the type (1.1) are attractive because of their high generality and strictness (such molecular models may be correlated with experimental data with a great degree of accuracy), they are practically useless for polyatomic molecules and carry little information since, for example, they do not discriminate benzene from prismane, which are very different systems in every way.

Undoubtedly, an all-embracing intellect would be able to appreciate immediately all the consequences compatible with equation (1.1), including the existence of benzene and prismone together with all their properties. The human intellect is, however, limited in its resources and is not able to reason on such a general basis. Therefore to specify the object we should use a more detailed model based on a knowledge of particular atomic groups. As a result, the number of possible topological isomers which can be constructed from a given set of atoms according to general chemical rules drops sharply.

A more detailed molecular model is provided by a molecular structural formula which describes a definite isomer. The next possibility, at a more highly detailed level, is a model in the form of rigid geometrical figures with fixed bond lengths, valence and torsion angles, etc. More complex molecular models include elastic variations of the geometric characteristics leading to inversions, internal rotations, etc.; the limits are set so as to exclude a change of one topological isomer to another. Next, we introduce the charge characteristics either in the form of definitely spaced charges or as continuous distributions of electronic density which are clearly seen on relief electronic density maps.

As polyatomic molecules may exist in excited as well as in ground electronic states the models discussed above are also applicable for studying excited states. We can then discuss a change in molecular geometrical form when electron excitation, charge transfer, etc. take place.

Two principal factors are important for further discussion. First, each molecular model can be characterized exactly, unambiguously and with mathematical strictness by a discrete series of parameters. Second, each model corresponds to a definite spectrum which may be generated to a satisfactory degree of accuracy on the basis of the given set of parametric values.

As stated earlier, a molecular model in the form of an empirical formula is unambiguously characterized by a set of numbers representing the nuclear charges and mass. The number of atoms of a given kind can also be characterized by repeating the same pairs of numbers.

The existence of one or several fixed structural groups may be indicated by the so-called contiguity or incidence matrices, in which the lines and columns are numbered according to the atoms of the groupings. The matrix elements lying at the intersections of the lines and columns connecting the bound atoms are 1, 2 or 3, depending on the number of similar bonds. The remaining elements are zero.

A molecular structural formula may be characterized by the full contiguity matrix, and these matrices unambiguously characterize the chemical structures of atomic groups and molecules.

To describe fully the rigorous spatial molecular model we need, in addition to the parameters and contiguity matrix mentioned above, a set of numbers which characterize definite values for bond lengths, valence angles, etc. In order to describe 'soft' or 'elastic' spatial molecular models a set of potential parameters must also be introduced to indicate, for example, the bond elasticity, inversion potentials and internal rotational potentials.

The elastic properties of a simple molecule with one sufficiently deep minimum on the potential energy surface may be characterized by the coefficients of the second- or higher-degree polynomial according to the co-ordinates with the help of which possible deformations of the molecule are described. Usually, such 'natural' co-ordinates are represented by the variations of bond lengths, valence and other angles as compared with their equilibrium values. In addition, the charges on atoms, bonds, etc. may be characterized with the help of a certain definite set of numbers. Even continuous distributions of electronic densities in molecules can be described in terms of discrete sets of values which are the coefficients in the linear combinations of atomic orbitals representing molecular orbitals.

Arguments of this kind may be extended, but what has been said is sufficient to substantiate the statement that it is possible to find an adequate mathematical representation of any known molecular model.

Before discussing spectra corresponding to different models let us consider how information on the properties of a particular system increases as we go from one model to another in the hierarchy of models listed above. The information available is determined by the set of parameters introduced for the description of the model, and it is assumed that these may be found experimentally. This set of values for the molecular parameters can then be used to generate a model spectrum which can be compared with that obtained experimentally.

Molecular models which were first introduced on purely empirical grounds may now be justified as a result of a series of successive approximations of the general quantum theory of systems of complex structures. The most important approximation of theory is the adiabatic one, where the motion of electrons and nuclei is dealt with separately and the concepts of potential surfaces, equilibrium geometry, etc. are introduced.

Let us briefly discuss one more general issue. The construction of a pictorial representation of a phenomenon by introducing various models inevitably means omitting many details and, therefore, mathematical accuracy is not obtained. This shows, in particular, in the fact that only the solution of the Schroedinger equation expressed as equation (1.1) may result in almost complete agreement with experiment, the error being insignificant. However, a very non-specific idea of a molecule as a set of bound charged particles is adequate for this equation. Other more pictorial models cannot be strictly correlated with experimental data, but they can produce the complex and pictorial representations of molecules that are most useful in present-day science. Here we deal with a peculiar manifestation of Bohr's principle of complementarity, which may be formulated as an attempt to achieve a strict mathematical description of a phenomenon hinders the most clear understanding. It was not by accident that Bohr, when asked the question 'What is truth supplemented with?' answered 'Clarity'.

Let us now discuss the theoretical spectrum corresponding to various molecular models, beginning with those which are more or less specific. The existence of characteristic vibrations and atomic chromophores which form the basis of spectra structure correlation tables makes it possible to construct the theoretical spectrum for a molecular model which is given in terms of an empirical formula containing individual structural groups, or of models in structural formulae expressed by a set of spectral bands and, in particular cases, by band intensities which are designated as high, medium or low intensity. Today the physical conditions causing the existence of characteristic vibrations or electronic transitions of chromophores are well known. The existence of spectral bands within known limits in which the absorption or emission band rather than a fixed-frequency band may appear (provided the group is comparatively closed in structure) is related to the fact that such atomic groups are influenced to various extents by their immediate molecular surroundings. If the complete structural formula is given and, hence, the surroundings of any characteristic structural unit are known, the construction of a spectral representation in the form of possible absorption or emission bands becomes more complete and the position of the bands more precise. Undoubtedly, more detailed information on spectra–structure correlations would be valuable for the accurate construction of such 'theoretical' spectra. This information may be obtained either experimentally (and the data used to improve spectra structure correlation tables) or by way of approximate calculations based on general spectral theory. Note that the limits within which spectral bands appear may be confined within boundaries by means of inequalities or more flexibly with the aid of the so-called 'penalty functions'; these are in better harmony with natural phenomena and make it possible to construct spectral representations. In this case, band limits are not fixed but are, to some extent, variable, but when the band extremities are approached the penalty starts rising very quickly. This is indicative of a low probability for the corresponding situation.

Let us now consider the next higher level of complexity, i.e. the rigid geometric model of a polyatomic molecule. By definition, this model prohibits relative motions of the molecular nuclei and only electronic motions and molecular rotations are permissible. Such a model results in purely electronic and rotational spectra, and in the latter case it is true only if the molecule has a static dipole moment.

To obtain the corresponding theoretical spectra it is necessary to solve a problem of energy levels and their probable transitions for a system of electrons moving in the field of motionless Coulomb centres (i.e. the atomic nuclear charges) and the problem of a dipole system rotating in space. The solution of the first problem is discussed in detail below, whilst the second is outside the scope of this book. It is important to note that a problem of electronic motion in a nuclear field can be solved numerically in a very strict way.

The electronic spectrum corresponding to such a rigid geometric model will be a set of narrow absorption or emission lines with very definite intensities. Such a spectral representation does not include a most important element, i.e. the existence of the permitted or overlapped vibrational structure from which experimentally complex band contours are obtained in the electronic spectra of polyatomic molecules. Intermolecular interactions are excluded. This model also does not produce a purely vibrational spectrum which manifests itself in the form of numerous bands and lines occurring in infra-red and Raman spectra. Nevertheless, it is most important, and, as experience shows, permits us to detect the main features observed in electronic absorption and luminescence spectra.

The vibrational structure of electronic spectra will appear in the theoretical spectra of the next 'soft' or 'elastic' model. This model includes relative motions of the nuclei in the molecule, and so a vibrational spectrum will appear but band intensities are not predicted. It is also possible to obtain information on the Raman scattering spectrum.

When such a model is improved by the inclusion of information on the charge distribution a more realistic picture is obtained, particularly when changes in electrical charges induced by molecular deformations are taken into account. In this case the infra-red spectrum may be predicted in much greater detail, i.e. not only the frequencies of absorption bands but also their absolute and relative intensities may be obtained. The construction of such theoretical spectra in practice requires the solution of the respective Schroedinger equations and the calculation of transition probabilities between energy levels. The techniques of making such calculations for complex systems are given later.

Further improvements in this type of model are possible. For instance, a 'soft' spatial molecular model can be improved if we consider that the molecular deformations follow a more complex law than that of Hooke. Mathematically, this means that instead of using a simple potential function, which is quadratic with respect to deformation co-ordinates, a more complex one is chosen. This could be in the form of the higher-degree polynomial or a linear combination of Morse potentials. When such a potential is chosen a theoretical spectrum can be calculated which reproduces correctly the main regularities in the spectra, including overtone and combination bands. When more detailed models are used, the calculated spectrum is an even closer fit to that obtained experimentally.

Without dwelling any longer on these questions, note that the possibility exists of calculating a theoretical spectrum for a molecular model to a good degree of precision, and this is the subject of this book. In practice, we can transfer experimental values and parameters used in the model to the set used for the calculation of the theoretical spectrum. Hence, a strict mathematical correspondence links two sets of values. It is this circumstance, unknown two decades ago, that makes it possible to raise spectroscopy as a research technique for elucidating the structure of complex molecular systems to a new qualitative level, which can be called a 'computing spectroscopic' level and which has promising possibilities.

The possibility of setting up with mathematical accuracy a correspondence between the parameters of the molecular models and those leading to the theoretical spectrum allows us to formulate strictly the solution of the so-called 'inverse spectral problems' which are the mathematical representations of general logic for any spectral chemical studies. As stated in the Introduction, the essence of the inverse spectral problem lies in a reconstruction of the values of molecular model parameters from various experimentally measured spectra. We shall discuss some of the most important features of solving inverse spectral problems in Chapter 15. Here we shall indicate only the solution sequence.

Let us assume that we have initial information on the system under study which permits us to set up approximate values of the parameters characteristic of a particular molecular model. Let us then construct corresponding theoretical spectra for these models and compare them with the experimental spectra. In general, the theoretical spectrum will differ from that obtained experimentally. Now since we have an unambiguous correspondence between the experimental set of parameters and the theoretical set we are able to modify the theoretical spectrum corresponding to the model by a direct variation of the parametric values by mathematical means. In this way they approach experimental values. When agreement is achieved between theory and practice, support is provided for the molecular model and the corresponding set of parameters conforms to the natural system. In practice, the convergence between theory and experiment requires a series of iterations in each of which the direct spectral problems are solved to yield the spectrum corresponding to the given set of model parameters. A number of important aspects of the solution of the inverse spectral problem and the resulting general knowledge questions of spectral-chemical studies will be dealt with below.

The concepts contained in molecular models of different levels of sophistication form the basis of a glossary of scientific language used by all researchers in studying the structure of a substance at microscopic level.

However, it is not expedient either to introduce a great variety of model types or to decrease the models to two or three general types. The first leads to unnecessary verbosity and the second to the loss of expressive potentials of the 'language'.

Apart from these general considerations, the important fact is that only by the use of models is it possible to compute the desired spectrum characteristic of large systems in real time. Indeed, the requirements of widespread computation of complex molecular spectra imposed by present-day practice cannot be met if the computation for a particular molecule is not related from the very beginning to other molecules, and to the information available on the properties of particular structural elements. This information, in the form of values for a variety of molecular properties, is stored in specific databanks which are now being set up.

In order to use the model approach in molecular spectroscopy it is essential that all the known models are characterized mathematically by specific sets of parameters. Two requirements must be met by the parametric systems. First, the parameters should be strictly introduced at the stage of transition from the general case to a more precise example. Meeting this requirement guarantees that the parameters introduced are valid and physically reasonable. The set of parameters is chosen by studying individual molecular groups. Second, this set (based on individual molecular components) should predict correctly the features found experimentally for the complex molecular systems under discussion.

At this stage we will consider the problems of the selection of models and sets of parameters for the calculation of the optical characteristics of complex polyatomic molecules. The basic molecular models are introduced at the adiabatic approximation based on the concepts of potential energy surfaces for the ground and excited electronic states. This leads, first, to rigid and dynamic geometric models of complex compounds. The sets of parameters associated with these models consist of values for bond lengths, valence angles (these are fixed for a rigid model and may vary somewhat with respect to equilibrium values for a dynamic model), atomic mass and potential function constants such as force constants. All these quantities are directly related to fundamental concepts, and may be estimated by means of well-developed quantum-chemical methods. They also possess the required local properties.

The rigid and dynamic spatial models make it possible to calculate the frequencies of purely electronic, electronic–vibrational and vibrational single- and multiquantum optical transitions whose spectral signs appear to be unambiguous algorithmic functions of molecular parameters, whose values are very stable in a series of molecules constructed from similar fragments.

Simple rigid molecules have been well studied and will not be discussed further. The problem is more complicated for those models which permit the calculation not only of frequencies but also of the band intensities of the transitions. In this case, rigid models must be supplemented by a new set of parameters which characterize the static and dynamic properties of the distribution of charges in a molecule. However, for a number of reasons, it is necessary to introduce different sets of parameters to describe the intensities in different types of spectra.

It is common knowledge that the intensities of the absorption and emission bands occurring in such processes are determined by the second powers of the dipole transition matrix elements. To obtain representations of spectra containing information on the intensities of absorption bands or lines in Raman scattering spectra we need to know the charge distribution and use the so-called 'electro-optical' parameters. These questions are considered in detail in the corresponding chapters of this book.

Let us now return to general considerations. As stated above, the process of understanding what happens in spectroscopic investigations of polyatomic molecules lies in constructing more complex molecular models and is reduced mathematically to the solution of the corresponding inverse spectral problem using computers. The solution of inverse problems at each iteration cycle calls for a comparison of the theoretical representation of the spectrum corresponding to a particular molecular model with that obtained experimentally.

As stated previously, the only mathematical model of a molecule which is in very good agreement with Nature is a model conforming to Schroedinger's equation in the very general form of equation (1.1) but, unfortunately, it carries little structural information.

Although all the remaining models used for spectral computation lead to a theoretical representation of the spectrum which, in principle, contains the same features as experimental spectra, they are by no means accurate. Therefore, if we have a certain molecular model and have constructed the corresponding theoretical spectral representation by solving the adequate Schroedinger equation with as much mathematical accuracy as possible, there will not in general, be good agreement with experimental data. Certainly, we may try to obtain almost exact agreement with the experimental data by varying the set of parameters inherent to the model. However, it is not obvious that this is possible when we consider the spectrum over a sufficiently wide range. In addition, the variations necessary may lead to values of the molecular parameters which violate basic chemical concepts. Thus, in order to obtain something resembling the truth we must sometimes be able to govern not only the model itself but the computing procedure, as is true for, say, semi-empirical methods for solving a purely electronic spectrum problem, where parameters are substituted for some of the accurately computed integrals. In other words, similar constants should be introduced in the description of the model. Therefore a transition to a semi-empirical method in studying the theory of polyatomic molecular spectra is not only helpful in simplifying the computation problems but it is also necessary for the understanding process at this stage of solving spectral problems. Of course, these parameters may be chosen in different ways. For example, they can be related to atomic properties. This procedure leads to a decrease in the number of parameters but to difficulties in dealing with molecular fragments when computations on very large systems are attempted. Alternatively, we can introduce the parameters related to fragments. These are greater in number but, in turn, there is a possibility of raising the accuracy of computation of spectra of really large systems. These statements are true for the spectral representations of other molecular models, in particular, for those which are intended for computation of band intensities. In this case, the need to simplify the computation procedure by introducing internal parameters is not always necessary.

It is clear that the process of describing molecular models with the aid of a set of parameters cannot be completely objective in principle since it depends, to a certain extent, upon the researcher's views and expertise. To make this point clear, let us take as an example the construction of the theoretical spectral representation corresponding to a vibrational spectrum for an 'elastic' model in the harmonic approximation.

Let us assume that we have found the position of minimum energy on a potential surface and its curvature with respect to all the internal co-ordinates of the system for the molecule, having solved a purely electronic problem and overcome all mathematical difficulties. If independent co-ordinates are used, it is an unambiguous procedure which can be (and has been) carried out repeatedly. This makes it possible to assign unambiguously numerical values to the force constants governing the atomic vibrations in the molecule. Using these force constants, frequencies for the vibrational bands can be calculated. Comparison of these values with a real infra-red spectrum shows that discrepancies of up to $100\,\mathrm{cm}^{-1}$ or more occur even when the molecular geometry is well known. That is, the comparison of the calculated and experimental spectrum is quite incorrect.

Evidently, the model has to be corrected, and this is usually done by rejecting small-value force constants, those of larger value being corrected by means of so-called 'scaling multipliers'. This procedure cannot be formalized and an element of subjectivity is introduced.

Let us now consider the principles involved in the solution of inverse spectral problems. In all cases it is necessary to vary the parameters of the model so that the spectral representations corresponding to the molecular model and the experimental spectrum converge. Usually, the decrease of a function of the form given below serves as a criterion:

$$F(\nu) = \sum_k (\nu_{\mathrm{ex}}^{(k)} - \nu_{\mathrm{th}}^{(k)})^2; \quad Q(I) = \sum_k (I_{\mathrm{ex}}^{(k)} - I_{\mathrm{th}}^{(k)})^2 \tag{1.2}$$

where ν_{ex} and ν_{th} are the experimental and theoretical computed values of frequencies and I_{ex} and I_{th} are the experimental and computed values of band intensities.

This type of equation representing the root-mean-square deviations was introduced by Hauss. There are several mathematical procedures for obtaining the minimum point and how it is approached (for example, linear and non-linear variants of the least squares method, Monte Carlo method and A. N. Tikhonov's regularization method).

Depending on the nature of a problem and its method of solution, we can find the increments in the initial values of the model parameters which lead to the minimum point of equation (1.2). It is also possible to ensure that the new set of molecular parameters have the least deviation from the initial set.

However, in general, except for a few small molecules, a unique solution for the inverse problem cannot be obtained due to the principal features introduced subjectively into the mathematical problem. To maximize accuracy it is necessary that limits for possible variation in the values for parameters which ensures physical plausibility are introduced from the beginning, either by inequalities in the form of allowed ranges or with the help of penalty functions. The total numerical result depends not only on the way the limitations are imposed but frequently on the selection of the initial approximate values of all the parameters used in computation. However, these limitations and the values are assigned by a human being, and are therefore subjective, thus negating the possibility of achieving the unique 'correct' solution.

This subjectivity then causes a certain unavoidable error in the final result, and this differs in its nature from the usual errors caused by errors in measurement. The 'subjective' error is introduced during the process of searching for sets of parameters and constructing models, even though the problem is mainly formulated within a strictly mathematical framework.

Two questions are raised: Do we not have an exclusive circle? Are the sceptics right in saying 'We will get what we want'? Fortunately not. The molecular models proper, which may be strictly derived from first principles, are objective. The strict theoretical spectral representations corresponding to these models, though they differ from experimental spectra, do not exhibit serious inconsistencies. Even for a crude model, the desired agreement is not bad. Thus, subjectivity may only be important when the set of parameter values is being quantitatively refined. It is pointless, therefore, to include values for force constants to a third place of decimals when a harmonic model is used for a vibrating molecule, even if a computer has produced ten decimal places. Again, we should not be delighted with the fact that we can reproduce a theoretical infra-red spectrum which differs from the experimental one by $\pm 0.1\,\text{cm}^{-1}$.

It is also pointless to expect that the set of parameters corresponding to a particular model, and chosen as described above, should be as close as possible in their numerical values to those obtained from successive iterative quantum computations. For instance, the values of bond dipole moments determined for specific molecular groupings and used to predict vibrational spectra of polyatomic molecules may appear to be far from the values of similar quantities computed by *ab initio* methods or found on the basis of other experiments. However, we should not give way to despair. It should be remembered that the use of a particular model and the choice of a parametric system depends on the specific problem which is to be solved by a researcher.

There are no all-purpose unique solutions in this type of work. Thus, despite progress in theory, calculation techniques and computers, the process of finding molecular parameters useful for spectral chemical studies remains a creative process. It cannot be reduced to a formal operation but is based, on the one hand, on accurate mathematical knowledge and, on the other, on common sense. There exists a constant danger of crossing an invisible border which separates 'physically reasonable' results from pseudo-scientific speculation. This should be remembered not only by experts in chemical spectroscopy but also by those who use their results. This danger exists in an acute form especially where complex mathematical formalism and the use of computers mesmerize and create an illusion of absolute infallibility. There is nothing to be done here—truth is supplemented by clarity. There is nothing for it but to console ourselves that, as Freud stated, the main virtue of a scientific way of thinking is the ability to be satisfied with incomplete knowledge and to continue scientific research, despite the absence of a complete confirmation.

It would be a mistake, however, to conclude that we should therefore retreat to the pre-computer period of molecular spectroscopy. The opposite is true—it is the developing mathematical procedures and computers that promise qualitative changes of spectral chemical studies in the near future. A researcher will be able to consider several models quickly, checking them for agreement with experiment, and at the same time determine on the basis of experience how subjectivity influences the final conclusions. This development, called 'computing spectroscopy', represents a natural growth arising from the previous trends in the development of molecular spectroscopic studies. The first and significant steps in this direction have already been made. However, many complex problems remain to be solved so as to increase efficiency at the scientist/computer interface. These include the development of more efficient control of the presentation of initial data and in methods of solution. In addition, it is necessary to improve the estimates of computed data and to set up specialized databanks, etc.

In conclusion, it can be stated that molecular spectroscopic theory has travelled a long way. Its successes have been spectacular during the last two decades, and in many cases it is now possible to arrive at a quantitative description and prediction of spectra from qualitative considerations.

The use of models makes necessary the setting up of parameter banks corresponding to different models and types of spectra. This could be achieved by an improvement in the mathematical techniques of solution of generalized inverse spectral problems. It could take the form of including not only experimental data on frequencies but also band intensities for a large number of molecules which could then be processed simultaneously. Certain semi-empirical approaches could also be improved. In addition, the possibility of predicting, say, the vibrational structure of electronic–vibrational spectra of large molecules on a semi-empirical basis also needs to be verified. Overcoming these problems is the price we inevitably have to pay for the visual clarity of using molecular models.

The stage of development of methods of calculating theoretical representations of spectra corresponding to molecular models (solution of direct spectral problems), which is the main topic of this book, has now reached a satisfactory degree of development. The widespread solution of inverse spectral problems and the setting up of specialized databanks are now on the way. Only after this second stage has been completed successfully will the entire field of molecular spectroscopy rise to a higher level of sophistication.

Chapter 2

General Principles of Calculating the Electronic and Vibration Energy Levels in Polyatomic Molecules

2.1 Separating Electronic and Nuclear Motions and the Adiabatic Approximation

Before considering how to separate electronic and nuclear motions we must make some preliminary points concerning the separation of different types of motions in general and the use of the frame of reference in particular.

A polyatomic molecule can both move through space and rotate as a whole. In addition, the particles that make up the molecule move relative to each other (internal motion). How these types of motion can be separated has often been discussed in the literature[1-3]. Therefore we shall only give some general considerations here.

The initial Schroedinger equation, written in a laboratory frame of reference, is

$$(\hat{T}_e + \hat{T}_n + V_{ee} + V_{en} + V_{nn})\Psi_{en} = E\Psi_{en} \tag{2.1}$$

where $\hat{T}_e$ and $\hat{T}_n$ are the kinetic energy operators of the electrons and nuclei and V_{en}, V_{ee} and V_{nn} are their Coulomb interaction operators (we shall only consider this case here). The potential energy terms contain only the moduli of the differences of the radii-vectors of interacting particles. These terms are therefore invariant with respect to the translational and rotational motion of the system as a whole. Hence the problem of separating the variables and choosing the corresponding equations only arises when the kinetic part is being analysed.

We shall show in Section 6.1 of Chapter 6 that this part of the Hamiltonian function can be written in separated variables with the term that depends solely on the relative motions being isolated. This can be done by choosing an appropriate system of relative co-ordinates and by expressing their rate of change in terms of Cartesian co-ordinates based on an internal system of axes. Without touching upon whether such internal co-ordinates can in fact be chosen, we shall simply assume that they have been, and we find, in matrix notation, that

$$\dot{q} = \vec{B}\dot{\vec{r}}$$

where $\dot{q}$ is the generalized velocity, $\dot{r}$ is the Cartesian velocity and the matrix $\vec{B}$ describes the transformation of one into the other. The relation between the generalized and Cartesian momenta can thus be written as

$$\vec{p} = \tilde{\vec{B}}p$$

The $\vec{B}$ matrix's vectors are given in terms of the laboratory system of co-ordinates. In the generalized co-ordinate system the kinetic energy matrix will be $T = \vec{B}E\tilde{\vec{B}}$, and it will not now depend on any translation or rotation of the system as a whole. This is because the vectors constituting the $\vec{B}$ and $\tilde{\vec{B}}$ matrices are multiplied as scalars during the matrix multiplications, and since the E matrix, being a diagonal matrix containing the reciprocals of the atoms' masses, is also scalar.

The nuclear vibrations (and internal rotations) can be separated in curvilinear co-ordinates (see Chapter 6). However, the same method cannot be used to describe the electronic motions: we can solve the pure electronic problem in Cartesian co-ordinates only.

We will therefore proceed as follows. The T matrix is reduced to the identity matrix. This can always be done, and entails making a transformation to new co-ordinates Q, which also describe relative motions. The relation between the generalized and Cartesian momenta can then be written as

$$\vec{p} = \tilde{\vec{B}}L_p P$$

If this expression is left-multiplied by the E matrix (bearing in mind that $P = \dot{Q}$ for generalized momenta corresponding to an identity matrix), we obtain

$$\dot{\vec{r}} = E\tilde{\vec{B}}L_p\dot{Q} \qquad (2.2)$$

whence

$$\mathrm{d}\vec{r} = E\tilde{\vec{B}}L_p\mathrm{d}Q \qquad (2.3)$$

Here, matrix L_p reduces the T matrix to an identity one.

A displacement of the particles in the chosen Cartesian laboratory co-ordinates which arise from an arbitrary change in the generalized co-ordinates is given by

$$\Delta\vec{r} = E\sum_k \tilde{\vec{B}}_k L_{pk}\delta Q_k \rightarrow E\int \tilde{\vec{B}}L_p\mathrm{d}Q \qquad (2.4)$$

During practical calculations the $\vec{B}$ and L_p matrices should be recalculated using equation (2.4), since the matrices and small atom displacements made during the previous step are known.

An important result for subsequent sections (for example, Section 8.3) is that the particle displacements found in this way (i.e. displacements from some initial state defined by the original choice of matrix vectors in the laboratory system of co-ordinates) will obey the condition that the momentum and angular momentum of the system equal zero. This means that the displacements are associated with purely relative motions. Note, then, that since the masses of the electrons are much smaller than those of atomic nuclei, it is the motion of the nuclei that will be the main factor when separating the internal and external motions in a polyatomic molecule.

Thus if equation (2.4) is used only for the nuclear co-ordinates, then the displacements of the nuclei from their initial positions that are obtained as a result will, with a good degree of accuracy, be the same as those obtained when the external motions are separated out.

In other words, if we accept that every displacement of the atoms in the molecule obeys equation (2.4) and that this condition is always followed, then we can assume that the molecule does not move as a whole with respect to the laboratory frame of reference or (what amounts to the same thing) the laboratory frame moves with the molecule as a whole. The motions of the electrons can then be considered in terms of the same Cartesian co-ordinates as movements with respect to the nuclei (or rather in a field of nuclei whose distribution in terms of an arbitrary Cartesian reference frame changes strictly according to equation (2.4)).

This argument has a corollary. If the electrons in the nuclei field can be considered to move in a Cartesian reference frame (which is especially true when the nuclei are assumed to be immobile (see below) and is convenient at the first stage of solving the whole problem), then we can choose an appropriate relative (internal) set of curvilinear co-ordinates to describe the inherent motions of the nuclei and solve the equations of motion in these co-ordinates.

We shall always assume later that the co-ordinate system being used for the sake of argument is stationary with respect to the molecule and does not rotate. This enables us to study the electronic and vibrational spectra whilst neglecting rotational energy and all the interactions between the electron and vibrational motions and the rotational motions. This approximation is very satisfactory for describing the external features of the electron and vibrational spectra of a polyatomic molecule. The appearance of rotational structures in these spectra is only really observed for very small molecules in the gaseous phase. We shall mainly be interested in molecules with five to ten atoms, mostly in liquids or solids, where rotation can, as a rule, be neglected and in which the influence of the environment is small.

These stipulations enable us to write the Schroedinger equation as

$$
\begin{aligned}
&-\frac{\hbar^2}{2m_e}\sum_i\left(\frac{\partial^2\Psi_{ev}}{\partial x_i^2}+\frac{\partial^2\Psi_{ev}}{\partial y_i^2}+\frac{\partial^2\Psi_{ev}}{\partial z_i^2}\right)\\
&-\frac{\hbar^2}{2}\sum_\alpha\frac{1}{m_\alpha}\left(\frac{\partial^2\Psi_{ev}}{\partial x_\alpha^2}+\frac{\partial^2\Psi_{ev}}{\partial y_\alpha^2}+\frac{\partial^2\Psi_{ev}}{\partial z_\alpha^2}\right)\\
&+\left(\sum_{\substack{i,j\\ i>j}}\frac{1}{r_{ij}}-\sum_{i,\alpha}\frac{z_\alpha}{r_{i\alpha}}+\sum_{\substack{\alpha,\beta\\ \alpha>\beta}}\frac{z_\alpha z_\beta}{r_{\alpha\beta}}\right)\Psi_{ev}=E_{ev}\Psi_{ev}
\end{aligned}
\tag{2.5}
$$

where m_e and m_α are the respective masses of an electron and a nucleus; Ψ_{ev} is the wave function of the system; r_{ij} is the distance between the ith and jth electrons; $r_{i\alpha}$ and $r_{\alpha\beta}$ are, respectively, the separations between the ith electron and the αth nucleus and between the αth and βth nuclei; and z_α and z_β are the charges on the nuclei in units of the charge on an electron. The first term corresponds to the kinetic energy of the electrons, to which we shall from now assign the subscripts i and j. The summation is over all electrons. The second term describes the kinetic energy of the nuclei. We shall denote the nuclei with subscripts α and β. The summation is over all nuclei. The third term characterizes the potential of electron–electron Coulomb repulsions (the first component in the brackets), the Coulomb nucleus–electron attractions (the second component) and the nucleus–nucleus repulsions (the third term). The summation is over every possible distinct permutation of index pairs.

Equation (2.5) is written for the Coulomb approximation in which the weaker magnetic interactions are neglected. This approximation is completely satisfactory for solving most spectroscopic problems. Since spins are not explicitly included in equation (2.5), then when solving it we need only find the space part of the complete wave function. The spin constitutent arises from other considerations dealt with later (Chapter 4).

An exact solution of equation (2.5) is not possible, hence more assumptions must be made. First, we note that since $m_\alpha \gg m_e$ and if the Ψ_{ev} function is changed less during nuclear co-ordinate changes than during those of electronic ones, then, in many cases, the influence of the second term must be less than the first. We can then consider the following new approximation:

$$-\frac{\hbar^2}{2m_e}\sum_i\left(\frac{\partial^2\Psi_{ev}}{\partial x_i^2}+\frac{\partial^2\Psi_{ev}}{\partial y_i^2}+\frac{\partial^2\Psi_{ev}}{\partial z_i^2}\right)$$
$$+\left(\sum_{\substack{i,j\\ i>j}}\frac{1}{r_{ij}}-\sum_{i,\alpha}\frac{z_\alpha}{r_{i\alpha}}+\sum_{\substack{\alpha,\beta\\ \alpha>\beta}}\frac{z_\alpha z_\beta}{r_{\alpha\beta}}\right)\Psi_{ev}=E_{ev}\Psi_{ev} \tag{2.6a}$$

Neglecting the second term in equation (2.5) (i.e. the kinetic energy of the nuclei) means that the nuclei are assumed to be completely immobile (the Born–Oppenheimer approximation). The task clearly then reduces to one of determining the motion of the electrons within a field of stationary charged centres (the nuclei). It is obvious that, given these conditions, we need only find that part of the Ψ_{ev} function corresponding to the purely electronic motion. Hence equation (2.6a) can be rewritten as

$$-\frac{\hbar^2}{2m}\sum_i\left(\frac{\partial^2\Psi_e}{\partial x_i^2}+\frac{\partial^2\Psi_e}{\partial y_i^2}+\frac{\partial^2\Psi_e}{\partial z_i^2}\right)$$
$$+\left(\sum_{i,j}\frac{1}{r_{ij}}-\sum_{i,\alpha}\frac{z_\alpha}{r_{i\alpha}}\right)\Psi_e=E_e\Psi_e \tag{2.6b}$$

The

$$\sum_{\alpha,\beta}\frac{z_\alpha z_\beta}{r_{\alpha\beta}}$$

term is omitted because it is a constant and can simply be added to each of the E_e values after equation (2.6b) has been solved.

Equation (2.6b) is the basic equation for calculating the electronic part of the E_{ev} function. We write it as

$$[\hat{T}_e(r) + V(r_{ij}) + V(r_{i\alpha})]\Psi_e = E_e\Psi_e \tag{2.6c}$$

where $\hat{T}_e(r)$ is the electron kinetic energy operator and $V(r_{ij})$ and $V(r_{i\alpha})$ are potential energy operators. The forms they take follow from a comparison of equations (2.6a) and (2.6b). The $V(r_{i\alpha})$ term in equation (2.6c) contains nuclear co-ordinates, and its contribution to the equation will be different for different distributions of the nuclei, so we shall obtain different solutions for E_e and Ψ_e. It is clear that the differences will only appear when the set of nuclear co-ordinates is obtained as a result of molecular deformations, and not as a result of translation or rotation of the molecule as a whole.

Every displacement of the atomic nuclei can be described in terms of the reference frame associated with the molecule using what are called 'normal' co-ordinates (Q_k). This means that the solutions of equation (2.6) must depend on these normal co-ordinates, i.e. $E_e = E_e(Q_k)$ and $\Psi_e = \Psi_e(r_i, Q_k)$. Continuous changes in the Q_k will make E_e describe a surface (a curve for either a diatomic molecule or, if changes in only one Q_k co-ordinate are being considered, for a polyatomic molecule). This is called the electronic potential surface of a polyatomic molecule, the potential curve of a diatomic molecule or the section of the potential surface of a polyatomic molecule along the axis of the single co-ordinate. The potential surface is a very important concept for developing the theory of electron shells and spectra of polyatomic molecules. Examples of potential energy curves and a very simple potential surface are shown in Figures 2.1 and 2.2.

Usually the potential surface is not considered as the function $E_e = E_e(Q)$, but as

$$E'_e = E_e(Q) + V[r_{\alpha\beta}(Q)] \tag{2.7}$$

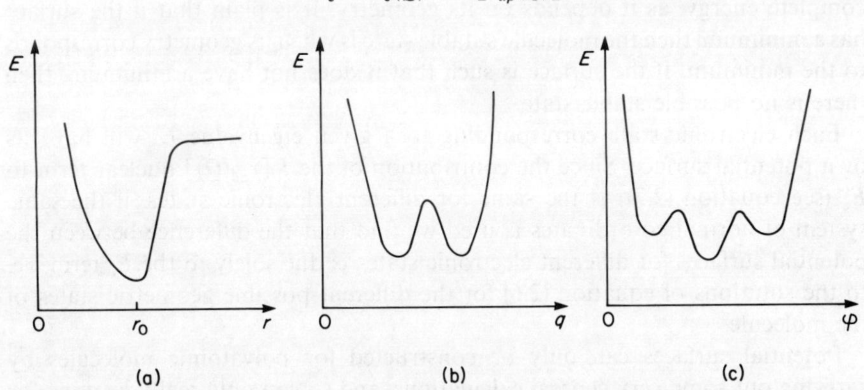

Figure 2.1. Examples of potential energy curves. (a) diatomic molecule; (b) potential of inversion; (c) potential for internal rotation

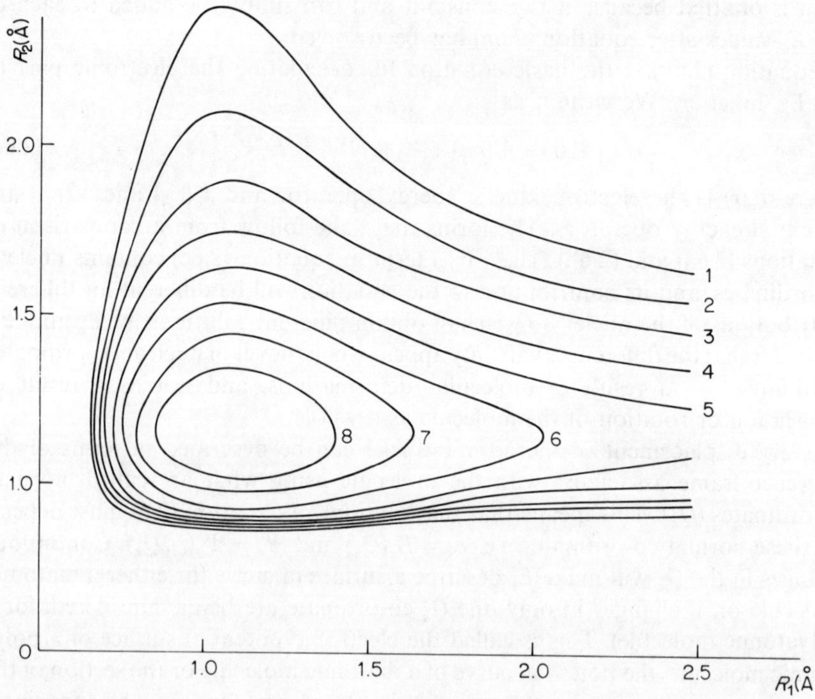

Figure 2.2. The map of potential surface for a HCN molecule; linear distortions; R_1 = HC distance, R_2 = CN distance. Each curve (1, 2...N) correspond to one value of energy: (1) $8.10^4\ cm^{-1}$, (2) $7.10^4\ cm^{-1}$, (3) $6.10^4\ cm^{-1}$,...(8) $1.10^4\ cm^{-1}$

where $V[r_{\alpha\beta}(Q)]$ is the nuclear Coulomb repulsion energy, and this clearly depends only on the relative dispositions of the nuclei and is consequently a function of the normal co-ordinates. The quantity E'_e describes the molecule's complete energy, as it depends on its geometry. It is plain that if the surface has a minimum, then the molecule's stable state is when its geometry corresponds to the minimum. If the surface is such that it does not have a minimum, then there is no possible stable state.

Each electronic state corresponding to a given eigenvalue E_e will have its own potential surface. Since the contribution of the $V[r_{\alpha\beta}(Q)]$ nuclear term to E'_e (see equation (2.7)) is the same for different electronic states, if the same system of normal co-ordinates is used we find that the difference between the potential surfaces for different electronic states is due solely to the E_e term, i.e. to the solutions of equation (2.6) for the different possible geometric states of the molecule.

Potential surfaces can only be constructed for polyatomic molecules by carrying out some very general calculations, and so can only really be done for not very large molecules.

It is timely now to consider a series of points that are important for setting

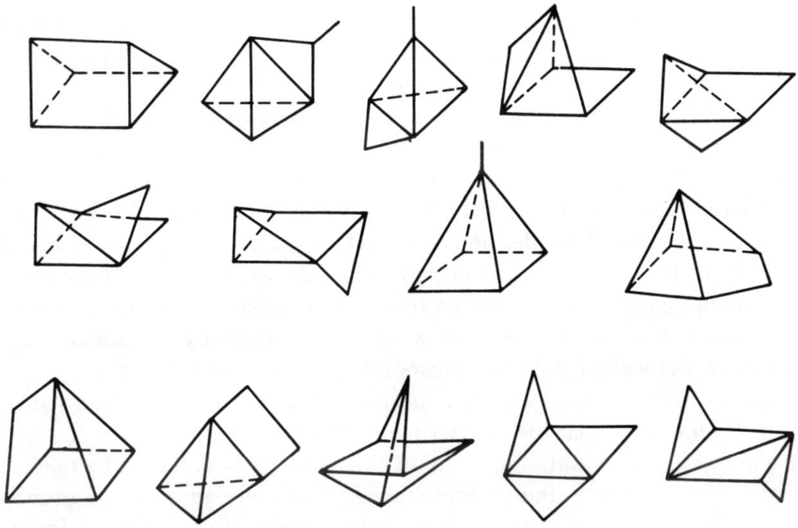

Figure 2.3. Examples of isomers for C_6H_6

up the problem. The potential surface of even quite small molecules may (indeed must) have not just one but several minima, even for the ground electronic state. If each minimum has its own molecular geometry, i.e. its own set of interatomic separations, angles between the vectors connecting atom pairs, dihedral angles, etc., then the molecule is said to have geometrical or topological isomers. For example, the C_6H_6 system of atoms has, together with the usual configuration of benzene, another 216 isomers which have different geometrical structures (for example, prismatic) (see Figure 2.3). This variety of configurations for the same set of atoms, each of which is constructed using the classical rules of valency, are found in most cases either chemically or spectrally as different molecular systems. This means that transitions between some are very rare and the energy barrier between the respective minima very high. These minima must, in turn, be deep enough for the potential well to contain sufficient energy levels to allow for nuclear movements.

In this way we can speak of a stable geometry of the molecule as being the geometry in which the arrangement of the material points (atomic nuclei) corresponds to a minimum on the multidimensional potential surface. This has an important practical consequence in that it is pointless to try to solve a complicated molecule's Schroedinger equation in its most general form (equation (2.1)) because of the unimaginable number of possible solutions. Hence, from the beginning the area in which a solution is being sought must be based on a definite choice of the initial geometry for the molecule and its permissible variations, and be localized in the neighbourhood of a deep minimum (or close group of minima) which can be treated as a single molecule, even though the group of minima may correspond to a group of topological isomers.

In future we shall always have in mind that this additional restriction applies and that the potential surface being considered does not apply to every domain it is defined for but is somewhat localized.

An important class of geometric isomers is formed by rotational isomers for which the potential surface changes periodically along one or more internal (relative) co-ordinate axes. As we shall see below, we can consider this case fairly strictly. However, it should be borne in mind that, as a rule, the relevant minima are distributed on the potential surface quite close to each other and are not deep. Hence insufficient energy levels involving nuclear motions are 'placed' within each minimum for us to speak of 'separate' molecular systems. The geometry of a rotation isomer is thus determined by the square of the maximum of the wave function corresponding to nuclear motion.

It may happen that the potential surface in the region in question has two or more minima such that the geometrical parameters (bond lengths or valency angles) are the same but another form of the molecule is obtained by reflecting the first in a plane or by a similar operation. The left and right forms of ammonia are examples of this (see Figure 2.4). We can then speak of 'inversion' or 'inverted isomers'.

The potential surfaces corresponding to different electronic states may intersect. However, a strict analysis that also takes vibronic interactions into account shows that the potential curves for a diatomic molecule and for states belonging to the same symmetry type do not intersect. This is known as the 'non-intersection rule'. This rule may be broken in the general case for polyatomic molecules. At points where potential surfaces intersect, the energies of the electronic states E_e' coincide and random degeneracy occurs.

We need only deal with the effects associated with intersecting potential surfaces in a few special cases—for example, molecular dissociations or radiationless transitions. The regions of intersections, as a rule, correspond to large distortions from the equilibrium configurations of the molecule. For this

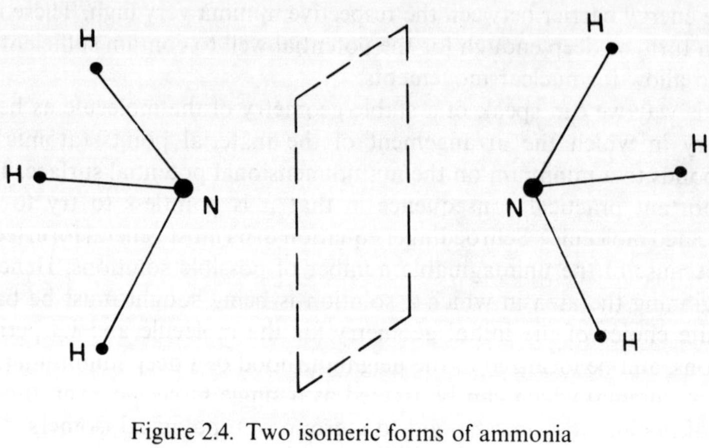

Figure 2.4. Two isomeric forms of ammonia

reason, we shall pay no attention to this effect and shall consider that all the potential surfaces are distinct and do not intersect.

Let us assume that the solutions of equation (2.6) have been found. Since the electronic part of the complete function Ψ_{ev} can be separated out, naturally we can search for solutions of the form

$$\Psi_{ev} = \Psi_{ev}(r, Q) = \Psi_e(r, Q)\Psi_v(Q) \tag{2.8}$$

Here $\Psi_v(Q)$ is the vibrational constituent of the complete wave function of the molecule Ψ_{ev} and depends only on the relative arrangements of the nuclei or normal co-ordinates.

We shall now establish which equation the $\Psi_v(Q)$ function must satisfy. In order to do this, we left-multiply the original equation of equation (2.5) by the Ψ_{ev}^* of equation (2.8) and integrate the result over the electronic variable (r) and the nuclear variable (Q). In order to simplify the notation we use a form analogous to equation (2.6b) by introducing a nucleus–nucleus repulsion operator $V(r_{\alpha\beta})$ and a nuclear kinetic operator $T_v(Q)$. Hence we obtain

$$\int \Psi_e^*(r, Q)\Psi_v^*(Q)[\hat{T}_e(r) + \hat{T}_v(Q) + V(r_{ij}) + V(r_{i\alpha}) + V(r_{\alpha\beta})]\Psi_e(r, Q)\Psi_v(Q)\,\mathrm{d}V\,\mathrm{d}Q$$

$$= E\int \Psi_e^2(r, Q)\Psi_v^2(Q)\,\mathrm{d}V\mathrm{d}Q$$

or

$$\begin{aligned}
&\int \Psi_e^*(r, Q)\hat{T}_e(r)\Psi_e(r, Q)\,\mathrm{d}V\,\Psi_v^2(Q)\,\mathrm{d}Q \\
&\quad + \int \Psi_e^*(r, Q)\Psi_v^*(Q)\hat{T}_v(Q)\Psi_e(r, Q)\Psi_v(Q)\,\mathrm{d}V\mathrm{d}Q \\
&\quad + \int \psi_e^2(r, Q)V(r_{ij})\mathrm{d}V\,\Psi_v^2(Q)\mathrm{d}Q \\
&\quad + \int \Psi_e^2(r, Q)V(r_{i\alpha})\mathrm{d}V\,\Psi_v^2(Q)\,\mathrm{d}Q \\
&\quad + \int \Psi_e^2(r, Q)\mathrm{d}V\,\Psi_v^2(Q)V(r_{\alpha\beta})\,\mathrm{d}Q = E
\end{aligned} \tag{2.9}$$

(since, by normalizing, $\int \Psi_e^2(r, Q)\Psi_v^2(Q)\,\mathrm{d}V\mathrm{d}Q = 1$).

It is often a good approximation to assume that the $\Psi_e(r, Q)$ function only depends weakly on the Q co-ordinates close to the molecule's equilibrium state, i.e. it changes very little for molecular deformations around equilibrium. Hence we can consider the $\hat{T}_v(Q)$ operator not to influence the $\Psi_e(r, Q)$ function, and can rewrite the second integral in equation (2.9) as

$$\int \Psi_e^2(r, Q)\mathrm{d}V\,\Psi_v^*(Q)\hat{T}_v(Q)\Psi_v(Q)\,\mathrm{d}Q$$

Then by integrating all the integrals with respect to the electronic co-ordinate (r) and allowing for the fact that the $\Psi_e(r,Q)$ function is a normalized solution of equation (2.6) by definition, we find, having collected like terms, that equation (2.9) can be written thus:

$$\int \{\Psi_e^*(r,Q)\hat{T}_e(r)\Psi_e(r,Q)\,dV + \Psi_e^2(r,Q)V(r_{ij})\,dV + \Psi_e^2(r,Q)V(r_{i\alpha})\,dV\}\Psi_v^2(Q)\,dQ$$

$$+ \int \Psi_v^*(Q)\hat{T}_v(Q)\Psi_v(Q)dQ + \int V(r_{\alpha\beta})\psi_v^2(Q)\,dQ = \int E_e(Q)\Psi_v^2(Q)\,dQ$$

$$+ \int \Psi_v^*(Q)\hat{T}_v(Q)\Psi_v(Q)\,dQ + \int V(r_{\alpha\beta})\Psi_v^2(Q)\,dQ = E \qquad (2.10)$$

The $\int E_e(Q)\Psi_v^2(Q)\,dQ$ integral's physical sense is that it is the average electronic energy of an electronic state when the nuclei are 'smeared' as described by the function $\Psi_v(Q)$. This quantity and the integral $\int V(r_{\alpha\beta})\Psi_v^2(Q)\,dQ$, which is the average value of the energy of the Coulomb repulsions of nuclei for 'smearing', make up the average value of the system's complete energy, and can be written as

$$E'' = E_e(0) + V[r_{\alpha\beta}(0)] + \Delta E_e(Q) + \Delta V[r_{\alpha\beta}(Q)]$$

where the $E_e(0)$ and $V[r_{\alpha\beta}(0)]$ terms are, respectively, the electron energy at the potential surfaces's minimum, if it exists, and the energy of the Coulomb repulsions of the nuclei in the configuration that corresponds to the minimum.

Using this expression we can rewrite equation (2.10) as

$$\{E_e(0) + V[r_{\alpha\beta}(0)]\}\int \Psi_v^2(Q)\,dQ$$

$$+ \int \Psi_v^*(Q)\{\hat{T}_v(Q) + \Delta E_e(Q) + \Delta V[r_{\alpha\beta}(Q)]\}\Psi_v(Q)\,dQ = E \qquad (2.11)$$

The equality is fulfilled if $E = E_e(0) + V[r_{\alpha\beta}(0)] + E_v$ and the E_v and $\Psi_v(Q)$, moreover, satisfy the relation

$$\hat{T}_v(Q)\Psi_v(Q) + \Delta\{E_e(Q) + V[r_{\alpha\beta}(Q)]\}\Psi_v(Q) = E_v\Psi_e(Q) \qquad (2.12)$$

Equation (2.12) contains only nuclear co-ordinates. This means that it is the natural equation to use to obtain the nuclear term $\Psi_v(Q)$ of the complete eigenfunction $\Psi_{ev} = \Psi_e(r,Q)\Psi_v(Q)$. The role of potential energy in this equation is played by the quantity $\Delta\{E_e(Q) + V[r_{\alpha\beta}(Q)]\}$, i.e. the difference between the energy of the 'well bottom' and that due to the deformation of nuclei.

Following the general laws of quantum mechanics, equation (2.12) will have a set of discrete solutions if the molecule's potential surface has a sufficiently deep minimum. Since the nuclear masses are significantly larger than the electronic mass, then clearly $E_v \ll E_e$. This yields the distribution of the electronic and vibrational energy levels shown in Figure 2.5.

These considerations give rise to the following method for approximating

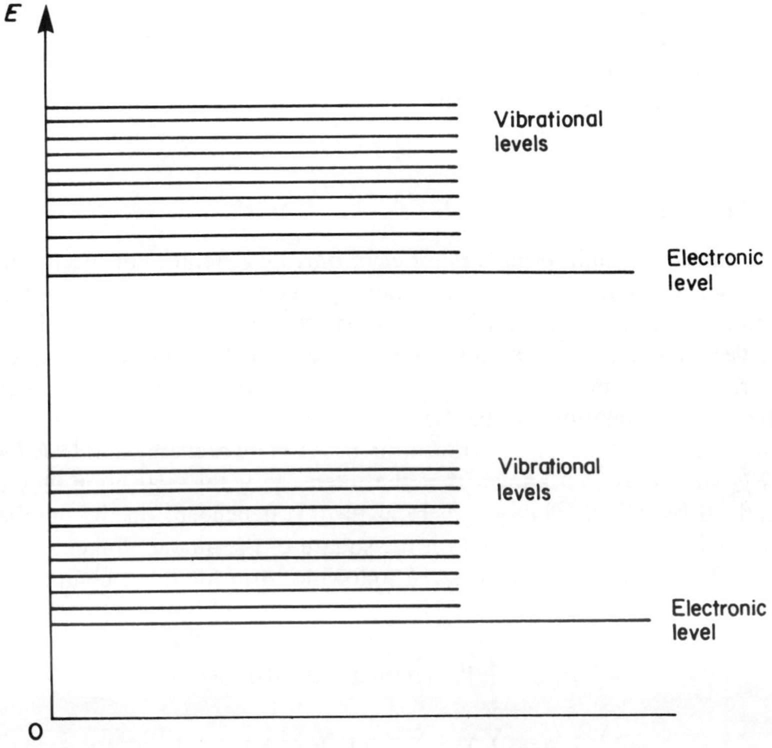

Figure 2.5. Schematic diagram of electronic–vibrational levels

the energy levels of a polyatomic molecule that correspond to electronic and relative nuclear (vibrational) motions.

The problem is to compile and solve an equation like equation (2.6) and to construct a series of potential surfaces. Then once these surfaces are known and cast, say, in the form of a series in Q co-ordinates for each electronic state, an equation similar to equation (2.12) must be compiled and solved. A set of vibrational states is then obtained for each electronic state. If a potential surface has a minimum, then a discrete set of vibrational states can be obtained, otherwise a continuous spectrum will arise for the vibrational states. It is obvious that the potential surface for the ground state, and those close to it, will always have a minimum, and consequently equation (2.12) will always yield a set of discrete vibrational energy levels.

Obviously, such a programme for estimating energy levels is not practicable for many states even for fairly simple molecules because of the very large amount of calculation. This can only be accomplished for individual special cases, and even then only approximately. However, experience has shown that the introduction of approximate theoretical considerations helps to solve some important problems.

Dividing the electronic and vibrational wave function due to the adiabaticity of electronic and vibratory motion within a molecule gives rise to several characteristic features when describing and interpreting the electronic and vibrational spectra. We shall discuss this topic in more detail later and shall only note the main features here.

2.2 The Franck–Condon Principle

This is one of the main principles for electronic and vibrational spectra and is very important when studying large non-symmetric molecules. According to the principle, an electron transition takes place between two states whilst the molecule's geometry remains unchanged. The Franck–Condon principle was first formulated from an analysis of experimental evidence and shows that the adiabatic approximation can be used.

Consider an optical dipole transition between two energy levels $E'_{\text{ev}} = E_m$ and $E''_{\text{ev}} = E_n$. It is well known that the intensity of this transition would be determined by its probability, which, in turn, is dependent on the number of molecules in the initial state and on the square of the dipole moment's matrix element. This matrix element has the following form for the x component of the vibronic transition:

$$\mu_{mnx} = \int \Psi''_{\text{ev}}(r, Q)\mu_x \Psi'_{\text{ev}}(r, Q)\,\mathrm{d}v\,\mathrm{d}Q$$

$$\mu_x = \sum_i x_i + \sum_\alpha z_\alpha x_\alpha(Q) \qquad (2.13)$$

where x_i and x_α are the co-ordinates of an electron and a nucleus, respectively. Moreover the x_α are all functions of the normal co-ordinates. Analogous formulae are obtained for the y- and z components.

Assuming that

$$\Psi_m = \Psi'_{\text{ev}} = \Psi'_{\text{e}}(r, Q)\Psi'_{\text{v}}(Q)$$

and

$$\Psi_n = \Psi''_{\text{ev}} = \Psi''_{\text{ev}}(r, Q)\Psi''_{\text{v}}(Q)$$

we obtain

$$\begin{aligned}
\mu_{mnx} &= \sum_i \int \Psi''_{\text{e}}(r, Q)\Psi''_{\text{v}}(Q)x_i\Psi'_{\text{e}}(r, Q)\Psi'_{\text{v}}(Q)\,\mathrm{d}v\,\mathrm{d}Q \\
&\quad + \sum_\alpha z_\alpha \int \Psi''_{\text{e}}(r, Q)\Psi''_{\text{v}}(Q)x_\alpha\Psi'_{\text{e}}(r, Q)\Psi'_{\text{v}}(Q)\,\mathrm{d}v\,\mathrm{d}Q \\
&= \sum_i \int \Psi''_{\text{e}}(r, Q)x_i\Psi'_{\text{e}}(r, Q)\,\mathrm{d}v\,\Psi''_{\text{v}}(Q)\Psi'_{\text{v}}(Q)\,\mathrm{d}Q \\
&\quad + \sum_\alpha \int \Psi''_{\text{e}}(r, Q)\Psi'_{\text{e}}(r, Q)\,\mathrm{d}v\,\Psi''_{\text{v}}(Q)x_\alpha\Psi'_{\text{v}}(Q)\,\mathrm{d}Q
\end{aligned}$$

If the second integral is integrated with respect to the electronic co-ordinates (along r) and taking into account that $f(Q) = \int \Psi_e''(r,Q)\Psi_e'(r,Q)\,dv = 0$ for any given Q and for different electronic states because of the orthogonality of the eigenfunctions, we then obtain

$$\mu_{mnx} = \int \{\Psi_v''(Q)\Psi_v'(Q)\,dQ \sum_i \Psi_e''(r,Q)x_i\Psi_e'(r,Q)\,dv\} \tag{2.14}$$

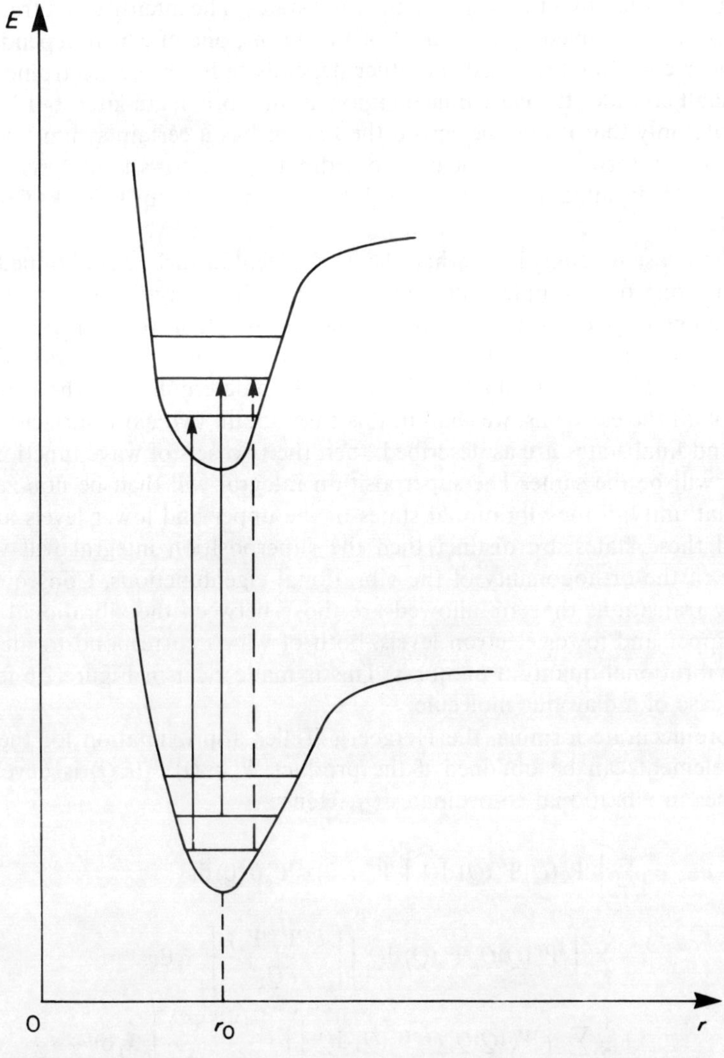

Figure 2.6. The permitted (solid lines) and prohibited (broken line) transitions between two electronic or vibrational states that correspond to similar potential surfaces which are only shifted up the energy axis

In most cases $\Psi_e(r,Q)$ does not depend very much on Q. Hence we shall not obtain too inaccurate a result if we set $\Psi_e(r,Q) \approx \Psi_e(r,0)$ in equation (2.14), from which

$$\mu_{mnx} = \int \Psi''_e(Q)\Psi'_v(Q)\,\mathrm{d}Q \cdot \sum_i \int \Psi''_e(r,0)x_i\Psi'_e(r,0)\,\mathrm{d}v \tag{2.15}$$

The zero in the electronic constituent of the eigenfunction means that the values of the electronic functions are taken in the neighbourhood of the potential minima corresponding to the initial and final states. The intensity of a transition will thus be determined by a product of two terms, one of which depends only on nuclear co-ordinates whilst the other depends only on the electronic ones.

We shall consider the electronic component in more detail later, but here we shall note only that it may be zero if the system has a certain symmetry.

We now return to the nuclear co-ordinates, i.e. to an analysis of the $\int \Psi''_v(Q)\Psi'_v(Q)\,\mathrm{d}Q$ integral, which is called the superposition (or Franck–Condon) integral.

The first case we consider is when the final potential surface is obtained from the initial one by a simple shift upwards along the energy axis. Strictly, this never happens. However, the approximation is often true in polyatomic molecules with many electrons when the states only differ by the transition of one electron. Since the geometry of a molecule is determined by the collective action of all the electrons, we shall in this case obtain two close surfaces. If the initial and final states are as described, then the two sets of wave functions Ψ''_v and Ψ'_v will be the same. The superposition integral will then be non-zero (it will equal unity) if the vibrational states of the upper and lower levels are the same. If these states are distinct then the superposition integral will vanish because of the orthogonality of the vibrational eigenfunctions. Consequently, the only transitions that are allowed are those between the vibrational levels of the upper and lower electron levels, both of which correspond to identical sets of vibrational quantum numbers. This is made clear in Figure 2.6 for the special case of a diatomic molecule.

A more accurate formula, the Herzberg–Teller approximation for the μ_{mnx} matrix element can be obtained if the product $\Psi''_e(r,Q)\Psi'_e(r,Q)$ is developed as a series in vibrational co-ordinates Q_k. Hence

$$\begin{aligned}\mu_{mnx} = {} & \sum_i \int \Psi''_v(Q)\Psi'_v(Q)\,\mathrm{d}Q \int \Psi''_e(r,0)x_i\Psi'_e(r,0)\,\mathrm{d}v \\ & + \sum_{i,k} \int \Psi''_v(Q)Q_k\Psi'_v(Q)\,\mathrm{d}Q \int \left[\frac{\partial(\Psi''_e\Psi'_e)}{\partial Q_k}\right]_0 x_i\,\mathrm{d}v \\ & + \frac{1}{2}\sum_{i,k,l} \int \Psi''_v(Q)Q_kQ_l\Psi'_v(Q)\,\mathrm{d}Q \int \left[\frac{\partial^2(\Psi''_e\Psi'_e)}{\partial Q_k \partial Q_l}\right]_0 x_i\,\mathrm{d}v + \cdots\end{aligned} \tag{2.16}$$

The zero subscript on a derivative means that it is evaluated at the equilibrium position of the co-ordinates of the ground state. This more complex integral with respect to vibrational co-ordinates is called the Herzberg–Teller integral.

If we again assume that the combining electronic surfaces are similar (simple shifts along the energy axis), then the first term in equation (2.16) is non-zero only for transitions for which the vibrational state is unchanged. The second and third terms in equation (2.16) will then be zero. For electronic transitions in which one of the normal vibrational co-ordinates is additionally and simultaneously excited, the second term in equation (2.16) is non-zero whilst the third and first terms vanish. If two normal co-ordinates (or the square of one) are excited simultaneously with the electronic transition, then only the third term will be non-zero.

Transitions without changes in the vibrational states all have the same frequency corresponding to the difference between the energies of the potential surfaces and their minima. A transition with an excitation (or loss of excitation) of one of the normal co-ordinates will have a frequency that differs from that of the pure electronic transition by a quantity that equals the frequency of normal oscillator. Transition frequencies when pairs of normal co-ordinates or their squares are excited (or when they are lost or fluoresce) differ from the purely electronic transition frequencies by quantities corresponding to overtone or combination vibrational frequencies. That is, a vibrational structure is observed in the electronic absorption (or emission) bands. The greatest intensity, however, will be observed for the purely electronic transitions.

The situation is more complicated when the potential surfaces of the two electronic states are not similar. The vibrational eigenfunctions of the upper and lower states are then distinct. The difference is larger, the more the forms of the potential surfaces of the combining electron states differ and the positions of their minima move apart. Note, too, that the relative distributions of the nuclei (the molecule's geometry) for both combining states can be described by the same set of generalized co-ordinates. These co-ordinates can be either the normal ones of one of the states with shifts along the relevant axes or the natural co-ordinates, which are increments in bond lengths, valency angles, etc. with respect to some fixed values taken origins for as the calculation.

In any case, the moduli of the functions Ψ'_v and Ψ''_v will be at a maximum in some places and vanish at others when the co-ordinates are varied. Clearly, the superposition integral may be non-zero (and consequently the transition is not allowed) when the generalized co-ordinates take on a set of values such that the moduli of both the Ψ''_v and Ψ'_v functions are at a maximum simultaneously. None of the regions in co-ordinate-change space, in which the modulus of the vibrational function of one state attains a maximum while that of the other state is zero, will contribute noticeably to the overall value of the superposition integral. This result may be used as a qualitative conclusion about the probability of electronic and vibrational transitions, especially when the potential surfaces are like this.

Since the Ψ''_v and Ψ'_v functions have well-defined domains where they differ significantly from zero, if the minima of the potential surfaces are displaced so that the Ψ''_v function can be reduced practically to zero at a point where the

Ψ'_v function only starts to appear (and this is true for most vibration levels) we find that the superposition integrals are zero. Hence if the minima of the potential curves of the combining electronic states are widely separated then the corresponding electronic transitions are not allowed.

As the index of the vibrational state grows, the number of domains in which the vibrational functions vanish increases. Moreover, domains may appear in which different products of the functions Ψ''_v and Ψ'_v are similar in size but different in sign. Both will lead to a reduction (sometimes very large) in the overall value of the superposition integral when integrated over all space. We

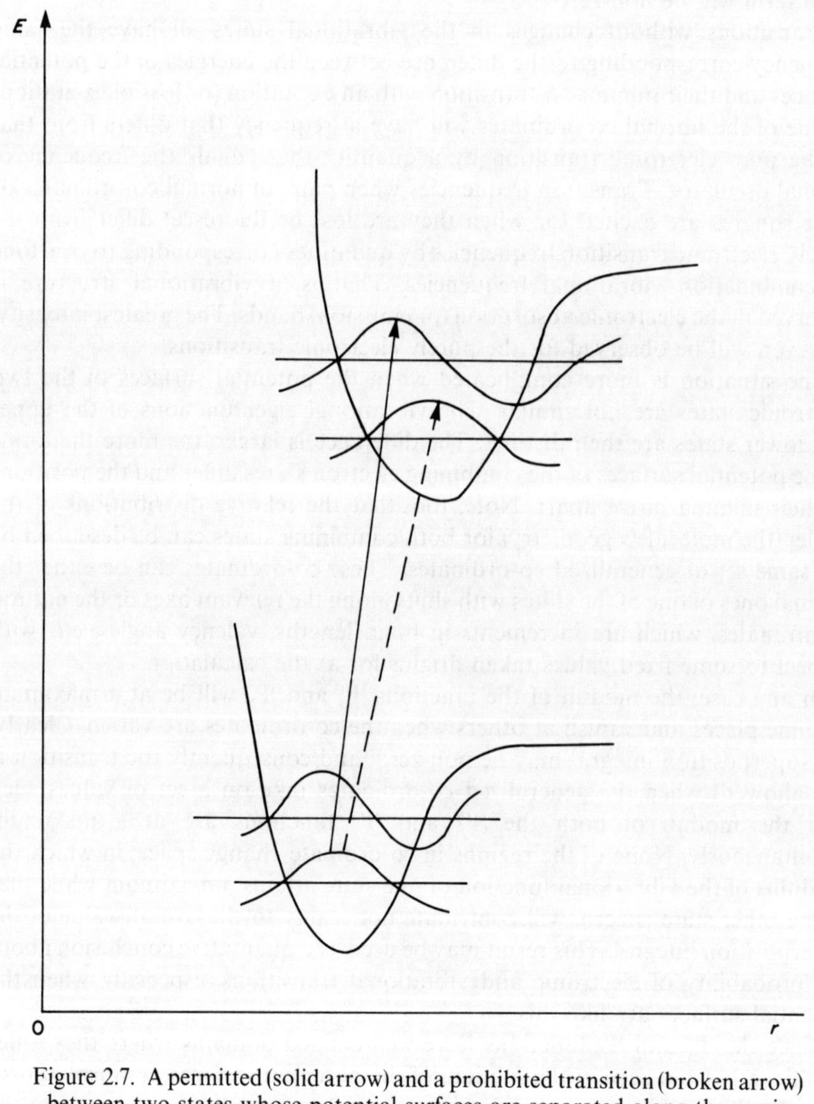

Figure 2.7. A permitted (solid arrow) and a prohibited transition (broken arrow) between two states whose potential surfaces are separated along the r-axis

can therefore conclude that in general the probability of transitions allowed by the Franck–Condon principle will fall off as the difference in the quantum number of the vibrational states grows.

Both a possible and an impossible transition are illustrated in Figure 2.7 for a diatomic molecule. Note that the probability of a transition from a lower state to an upper one, leading to an absorption band, will be related to the number of molecules in the lower vibrational state. This number diminishes following Boltzmann's law, i.e.

$$N_v \sim \exp[-E_v/kT]$$

where N_v is the number of molecules in a given vibrational state, E_v is the energy of the state, k is Boltzmann's constant and T is the absolute temperature. Thus the intensity of the relevant transitions, as shown in Figure 2.5, will diminish as the vibrational state's index increases.

Given the situation shown in Figure 2.7, transitions from a low vibrational state of a low electronic level to a low vibrational state of a higher electronic level will be prohibited, whereas transitions to higher vibrational states in the higher electronic level may be allowed (see the second curve of the lower state in Figure 2.7). Since the higher vibrational levels of the lower states are filled less often, the corresponding transition probabilities will be lower. This is reflected in the relatively low intensities of their absorption bands.

Distinct transitions corresponding to combinations of vibrational states during electronic and vibrational transitions in a complicated molecule may lie quite close together. The absorption bands obtained thus overlap to form a broad common band. The contour of this band will depend on the relative intensities of all its constituents. It will consequently be different for the cases shown in Figures 2.6 and 2.7.

Under normal conditions the majority of molecules are in a lower (ground) vibrational state of the ground electronic state. When these molecules are irradiated with light of the appropriate wavelength, transitions will be observed from the ground vibrational state of the ground electron state to a vibrational level of a higher electron state. The probability of this transition will be determined by the various integrals in equation (2.12). If the potential surfaces are similar, then the most intense transition will be one without excited vibrations. The absorption band of longest wavelength will correspond to this transition. The transitions during which vibrations are excited will be less intense and will correspond to shorter wavelengths.

As a result, the characteristic asymmetric absorption band contour is obtained with a sharp fall in intensity at the long wavelength end and a decrease in intensity towards the shorter wavelengths. If the potential surfaces are not the same then the greatest intensity will correspond to transitions with excitations of one or more normal vibrations. The absorption contour will thus be changed. It will depend first, on the form and mutual positioning of the upper and lower potential surfaces, and second, on the temperature, which determines the populations in the vibrational levels in the lower electron states.

In order to calculate integrals such as $\int \Psi_v''(Q)\Psi_v'(Q)\,dQ$ the vibrational eigenfunction of the upper and lower electron states must be expressed in a single co-ordinate system. The system of the normal co-ordinates of the lower (ground) state can be taken. The deviations in bond length, valency angles and other geometric parameters from their equilibrium values, when the geometries of the ground and excited electron states are not the same, will clearly be different. It is not difficult to obtain the relationships between the not-so-great deviations of the geometric parameters (from their equilibrium values) for the combining electron states.

Consider a state of a molecule that is described by a set of bond lengths s_1, s_2, etc. and valency angles α_1, α_2, etc. If the bond lengths and valency angles of the equilibrium configurations of the lower electron state are designated s'_{01}, s'_{02}, etc. and α'_{01}, α'_{02}, etc., respectively, and those of the equilibrium configuration of the upper state by s''_{01}, s''_{02}, etc., $\alpha''_{01}, \alpha''_{02}$, etc., respectively, then we can write the following system:

$$s_1 = s'_{01} + \Delta s'_1 = s''_{01} + \Delta s''_1$$
$$s_2 = s'_{02} + \Delta s'_2 = s''_{02} + \Delta s''_2$$
$$\cdots\cdots\cdots\cdots\cdots\cdots$$
$$\alpha_1 = \alpha'_{01} + \Delta\alpha'_1 = \alpha''_{01} + \Delta\alpha''_1$$
$$\alpha_2 = \alpha'_{02} + \Delta\alpha'_2 = \alpha''_{02} + \Delta\alpha''_2$$

Here, $\Delta s'_1$ indicates the increment to the appropriate geometrical parameter that must be introduced with respect to its equilibrium value to obtain the parameter's value for the selected configuration of the molecule.

Now we introduce the notation R'_0, R''_0, q' and q'', where R'_0 and R''_0 are column matrices for the equilibrium parameters of the two electron states and q' and q'' are the respective increments. We can then rewrite the above system in matrix notation as

$$R''_0 + q'' = R'_0 + q'$$

or

$$q'' = q' + R'_0 - R''_0 = q' + \Delta R_0$$

Increments to the equilibrium bonds and angles in the theory of the vibrations of polyatomic molecules form the system of natural co-ordinates and are a special case of generalized co-ordinates. This means that the expression $q'' = q' + \Delta R_0$ reflects the relationship between the natural vibration co-ordinates of the combining electronic states.

In turn, the natural vibrational co-ordinates can be expressed in terms of the normal co-ordinates using the matrix equations

$$q' = L'_q Q', \quad q'' = L''_q Q''$$

(see Chapter 7). Hence we can clearly write

$$L''_q Q'' = L'_q Q' + \Delta R_0$$

Using independent co-ordinates the L'' matrix has an inverse. By left-multiplying the last equation by L''^{-1}, we finally obtain the relation between the normal co-ordinates of the combining electronic states, i.e.

$$Q'' = L_q''^{-1} L_q' Q' + L_q''^{-1} \Delta R_0 \tag{2.17}$$

Calculation of the L_q'' and L_q' matrices will be considered later. This equation is called Dushinsky's relation.[4] The vibrational eigenfunctions Ψ_v'' and Ψ_v' have their simplest forms in the Q'' and Q' co-ordinates systems, respectively, i.e. in the normal co-ordinates of the respective electron states. Since a single system of normal co-ordinates (for example, $Q' = Q$) must be used to calculate the superposition integrals, then the co-ordinates must be changed in the $\Psi_v''(Q)$ function from the Q'' co-ordinates to the $Q' = Q$ co-ordinates. We shal investigate this operation in Chapter 10.

If the vibronic interactions and the anharmonic vibrations are neglected, then we will obtain the following expression for the energy levels for an electron state:

$$E = E_e + \sum_k \nu_k (v_k + \tfrac{1}{2})$$

where E_e is the energy of the electrons taking nuclear–nuclear repulsions into account at the potential surface's minimum, ν_k is a frequency of a normal vibration in appropriate units and v_k is the vibrational quantum number and can take the values $0, 1, 2, \ldots$. The summation runs over all of the molecule's degrees of freedom. Omitting the constant, we obtain the following expression for the frequency of a vibronic transition from a low state (single prime) to an upper one (two primes):

$$\omega_{ev} = E_e'' - E_e' + \sum_k (\nu_k'' v_k'' - \nu_k' v_k') + \tfrac{1}{2} \sum_k (\nu_k'' - \nu_k') \tag{2.18}$$

When the potential surfaces of the upper and lower states are close we obtain $\nu_k'' \approx \nu_k'$ and

$$\omega_{ev} \approx E_e'' - E_e' + \sum_k \nu_k (v_k'' - v_k') \tag{2.19}$$

According to the Franck–Condon principle in this approximation, the only transitions that are allowed are those for which the vibrational quantum numbers do not change. Finally we find that $\omega_{ev} = E_e'' - E_e'$, i.e. the frequencies correspond to purely electronic changes in energy. When we are calculating how the electronic wave functions depend on the nuclear co-ordinates, other frequencies corresponding to the general form of equation (2.18) will be observed, as shown above. The second most intense bands, after those of the purely electronic transitions, will be those for which the expression $[v_k'' - v_k'] = 1$, $v_l'' = v_l'$ holds true for all l, except $l = k$.

In real situations the selection rules are not strictly fulfilled and, moreover, the potential surfaces of the combining states may be very different. As a result, some transitions will often be observed, corresponding to different sets of vibrational quantum numbers for the combining electronic states.

The transitions from one vibrational level in the initial electronic state to all the different vibrational levels in the final electronic state form a 'progression'. The most intense band in a complicated molecule in which the potential surfaces are similar is that corresponding to the transition between the vibrational levels whose vibrational quantum numbers are close in value.

2.3 Types of Electronic Spectra and Some Important Patterns

The electronic spectra of a polyatomic molecule are divided into absorption and emission spectra. When a molecule absorbs energy it goes from a lower level to one of its possible excited states. The transition is associated with the absorption of a quantum of electromagnetic energy of the appropriate frequency. Since every stable electronic state is characterized by its own potential surface and set of vibrational and rotational energy levels, we find that during a transition from one electron state to another the molecule goes from one electronic–vibration–rotational level to another. Usually the transition in complicated molecules that are associated with changes in rotational states are difficult to classify, and hence it is fairly accurate to speak of 'vibronic' transitions, especially for polyatomic molecules.

Transitions between the closest electronic states have the greatest probability. These states must have the same spin. Inter-combinational transitions, during which the spin in the molecule is changed (particularly single–triplet transitions), are prohibited (see Chapter 4).

If the molecule has a small number of vibrational degrees of freedom, then the vibrational energy levels make up a discrete series which are fairly well separated. We shall thus observe, in either the absorption or emission spectra, bands which correspond to the electronic transitions separated by individual maxima that correspond to transitions between different combinations of vibrational states in the upper and lower energy levels.

If, however, the molecule has a large number of vibrational degrees of freedom, then the vibrational levels will be so close together that individual transitions corresponding to the changes in the vibrational states of the upper and lower electronic levels become indistinguishable experimentally. In other words, the vibrational structure of the electron transition will be blurred, a general envelope will be observed and the absorption band will become quite broad.

It is useful to distinguish between simple molecules whose vibronic structure is discrete and complex ones with band envelope spectra in which discrete vibrational structure cannot be observed.

Under certain conditions (i.e. very low temperatures) very complicated discrete structures can be observed in the absorption and emission spectra of molecular crystals such as benzene, naphthalene, anthracene and some other aromatic compounds. At low temperatures very sharp discrete spectra with distinct sequences of lines (Shpolsky's effect) can be observed in the emission spectra of the frozen solutions of koronene and other aromatic hydrocarbons. This line system is the result of vibrations in these molecules which appear to be due to the effect of the solvent. This effect does not disrupt the discreteness of spectra.

In many cases there is a simple relationship between absorption and emission spectra which has been called the 'mirror symmetry law'. This means if an absorption band has a particular contour in the absorption spectrum then a band will be observed in the emission spectrum whose contour is the mirror image of the one in the absorption spectrum (Figure 2.8). The place where the absorption and emission contours intersect corresponds to a purely electronic transition, i.e. to the transition from the zero vibrational level in the lower state to the similar level in the upper one.

The mirror symmetry law can be explained quite naturally if it is assumed that for a polyatomic molecule the potential surfaces of the combining electron states are very close. This can happen, as mentioned above, when we are dealing with a polyatomic molecule with many electrons and when a transition is associated with a change in state of only one of these electrons.

If the upper and lower states have almost identical potential surfaces, then we also obtain an identical set of vibrational levels for these states. The absorption and emission band contours are envelopes of the vibronic transitions bands. Each of these transitions, and its band's intensity, is related to two factors: first, to the probability of the transition between the two levels involved, and second, to the population of the initial level using the adiabatic approximation the transition probability is defined by the matrix element of the dipole moment, i.e.

$$\vec{\mu}_{nm} = \int \Psi''_{e}\Psi''_{v}\vec{\mu}\Psi'_{e}\Psi'_{v}\,dv\,dQ \tag{2.20}$$

This value of this matrix element does not depend on whether the transition

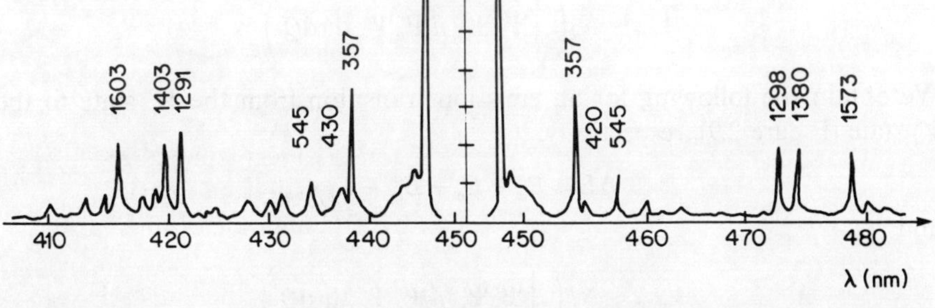

Figure 2.8. The absorption (a) and fluorescence (δ) spectra of perylene in n-hexane in relative units, $T = 4{,}2$ K. (By Bolotnikova *et al.*, with permission of the authors)

is from the lower level to the upper level or vice versa, since the value of equation (2.20) is independent of a reversal of the indices of the eigenfunctions of the two states.

The number of molecules at each vibrational level of the states involved follows Boltzmann's distribution under ordinary conditions, i.e. it satisfies $N(E) \sim \exp(-E/kT)$. At relatively low temperatures we can assume that before absorption takes place all molecules are Boltzmann distributed with respect to the vibrational levels of the lower electronic state, and transitions occur to the upper electronic state from each of the possible lower electronic ones. The intensity of each vibronic transition will be defined by the probability of the transition and the number of molecules in the original lower level.

Let us assume that an electron is excited and that all the molecules are transferred to the upper electronic state. If the lifespan of the molecule in the excited state is long enough then they will be Boltzmann distributed with respect to all the possible vibrational energy levels in the excited state. If the potential surface of the upper state is not substantially different from that of the lower one then we shall obtain a distribution with respect to energy levels that corresponds exactly with the distribution that occurred in the lower state. When light is emitted there will be transitions from the upper state into the now-empty lower one.

We shall now consider two vibrational levels of each of the upper and lower electronic states, denoting the two for the lower level by v'_1 and v'_j and the two for the upper one v''_i and v''_j. These energetic states will correspond to the eigenfunctions $(\Psi'_e\Psi'_{v_i})$, $(\Psi'_e\Psi'_{v_j})$, $(\Psi''_e\Psi''_{v_i})$ and $(\Psi''_e\Psi''_{v_j})$, respectively. Let us assume that the v'_i (and v''_i) vibration level lies below the v'_j (and v''_j) level in every electronic state. The number of molecules at these energy levels are, respectively, N'_i, N'_j, N''_i and N''_j.

Now let us consider a transition from the v'_i level of the lower state to the v''_j level of the upper one (an absorption). The frequency of this transition depends on the difference in the energies of the initial and final states, i.e. on

$$\Delta E = E''_e - E'_e + E''_{v_j} - E'_{v_i}$$

The intensity of the transition is proportional to

$$A_{abs} \sim N'_i \left(\int \Psi''_e \Psi''_{v_j} \vec{\mu} \Psi'_e \Psi'_{v_i} \, dv \, dQ \right)^2$$

We obtain the following for an emission transition from the V''_i state to the V'_j state (Figure 2.9), respectively:

$$\Delta E = E''_e - E'_e + E''_{v_i} - E'_{v_j}$$

and

$$A_{emis} \sim N''_i \left(\int \Psi'_e \Psi'_{v_j} \vec{\mu} \Psi''_e \Psi''_{v_i} \, dv \, dQ \right)^2$$

If the potential surfaces of the combining states are similar and the distri-

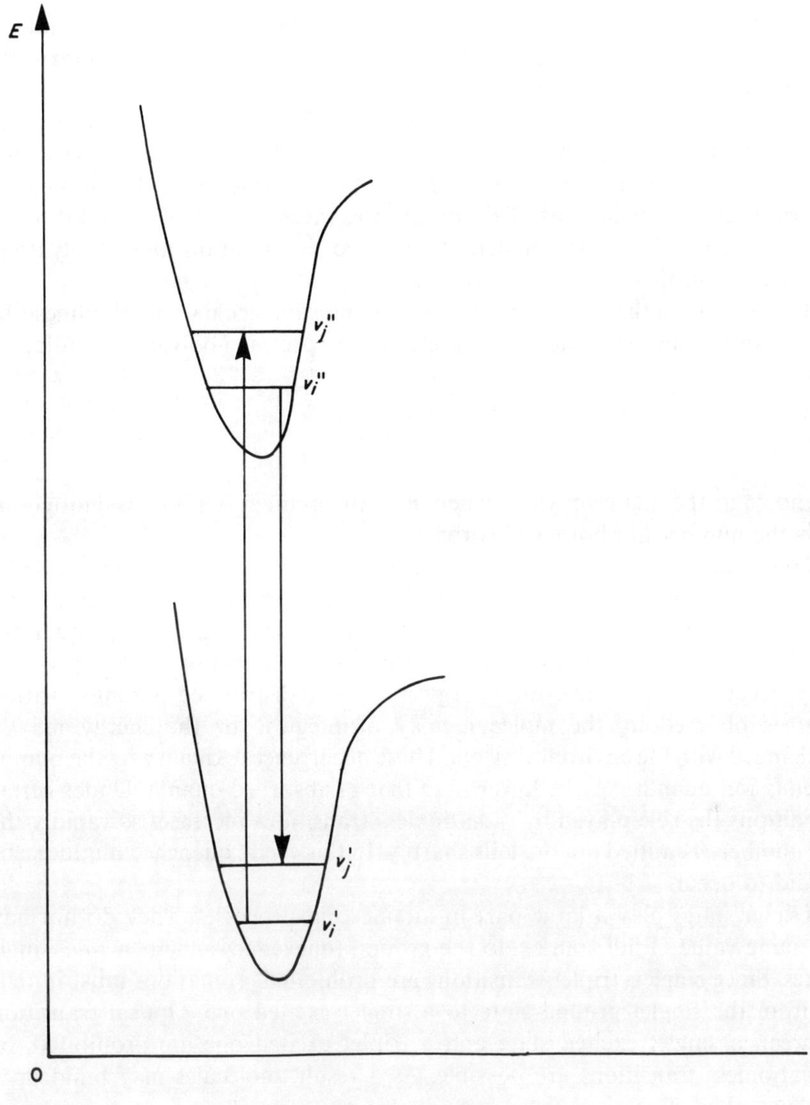

Figure 2.9. Sketch of mirror symmetry transitions

butions of the molecules with respect to the vibration levels are also similar, then because $N_i' = N_i''$, $\Psi'_{v_i} = \Psi''_{v_i}$, $\Psi'_{v_j} = \Psi''_{v_j}$, $E_{v_i} = E''_{v_i}$, and $E_{v_j} = E''_{v_j}$ we find that $A_{\text{absorption}} = A_{\text{emission}}$ and the emission and absorption frequencies will be symmetric to one another with respect to the quantity $E_e'' - E_e'$, which corresponds to the purely electronic transition.

A similar argument can be made for any pair of vibrational energy levels in the upper and lower states and this demonstrates the mirror symmetry law.

The situation is more complicated in real molecules, and by no means all molecules transfer to the excited electronic state. Moreover, not every molecule has similar potential surfaces for the electronic states involved. This means that the mirror symmetry law is often broken.

Since the shape of the absorption band is related to the population of the energy levels of the combining electronic states, it is clear that the shape must be sensitive both to temperature and to the influence of different kinds of intermolecular interactions. This makes the investigation of the contours of absorption bands a very important source of information about polyatomic molecules and their environments.

The concept of the quantum yield of a complex molecule's photoluminescence is very important to the theory of the electronic spectra of polyatomic molecules. The ratio

$$\kappa = \frac{n_e}{n_a} \leqslant 1 \tag{2.21}$$

is known as the quantum yield, where n_e is the number of photons emitted and n_a is the number of photons absorbed.

To explain why the concept is needed, first note that after the molecule has been excited into the upper electronic state it stays there for a certain finite length of time and then descends to a lower level by emitting an energy quantum. Whilst the molecule is in the upper electron state a number of different processes may occur. For example, the energy may be redistributed amongst various degrees of freedom, the molecule may disintegrate or the energy may be exchanged with the environment, etc. These occurrences explain why the number of emission quanta may be lower than that of absorbed quanta. Under certain conditions the role played by radiationless transitions increases so rapidly that the number of emitted quanta falls sharply. In this event, quenched luminescence is said to occur.

Triplet states play a large part in luminescent processes. They do not have the same value of full spin as do the ground (unexcited) and, as a rule, singlet states. Since singlet–triplet transitions are prohibited, excitations must initially be from the singlet ground state to a singlet excited one. Optical transitions between a singlet excited state and a triplet excited one are prohibited, but radiationless transitions are possible. As a result, molecules may build up in triplet excited states.

Transitions from a triplet state to a lower one are prohibited. However, we must bear in mind that these prohibition rules are not absolute and are only strictly obeyed given a whole series of approximations. Hence optical transition with electromagnetic quanta being emitted must be possible from excited triplet states to a lower unexcited (ground) singlet one, even though this transition may not be highly probable. Since the probability is low, the fluorescence process is hindered, and consequently the initial excited state of the molecule persists for a long time. Since triplet excited levels are lower than singlet excited ones, according to Hund's rule, this persistence has a longer wavelength than the emission wavelength from the absorbing state.

The second important characteristic of electronic transitions related to luminescence is the polarization of the emitted light. The appearance of polarization is due to the fact that luminescent emission, like that of the phenomenon of Raman scattering, is a two-quantum process. The polarization of luminescence yields valuable information on intermolecular interactions in particular.

To conclude, we note that if an excited state has a potential well then there will be discrete absorption and emission spectra, otherwise continuous spectra will be produced.

2.4 Symmetry of Wave Functions and Selection Rules by Symmetry

If a molecule has a symmetry in an electronic state in the neighbourhood of a minimum in its potential surface this means that the space of the molecule can be divided into regions that are symmetric with respect to the symmetry elements possessed by the molecule. Symmetry is apparent when there is a definite relative arrangement of the molecule's atoms and when the distribution of the electronic density in the symmetrical parts of the molecule's space is the same.

If the molecules only have second-rank symmetry elements, i.e. a centre of symmetry, a plane or a second-order axis, then a wave function of a molecule may only be symmetric or anti-symmetric with respect to these symmetry elements. Each of the symmetry elements divide the molecule's space into two regions in which, as we mentioned above, the electronic density distributions must be identical. Since this density is given by the square of the wave function at different points in the molecule's space it follows that the wave function at different points in the molecule's space must remain unchanged (symmetric functions) or only change sign (anti-symmetric functions) from one of the symmetry regions to another.

Now assume that the wave functions of the combining electron states (i.e. $\Psi_e'(r,0)$ and $\Psi_e''(r,0)$) are symmetric with respect to the same plane of symmetry, and consider the integral $\int \Psi_e' x \Psi_e'' \mathrm{d}V$, i.e. the matrix component element of a dipole transition. We can assume that the x axis is perpendicular to the symmetry plane and the origin lies on the plane without limiting generality. The symmetry plane divides the domain over which the integral is made into two parts. It is evident that we shall only obtain a non-zero value when the integral for one side of the symmetry plane equals the integral on the other. Since the values of x change sign as the plane is passed, the integral will only be non-zero if the Ψ_e' and Ψ_e'' functions possess different symmetries with respect to the plane of symmetry. The integrand in our case does not change sign as it goes from one of the symmetry subspaces to the other, i.e. it is symmetric. If the molecule has a centre of symmetry in the ground and excited states then when we take into account that symmetric points satisfy the equation

$$\Psi(x, y, z) = \pm \Psi(-x, -y, -z)$$

we find that the components of the co-ordinate axes of the matrix element of the dipole transition are zero if Ψ_e' and Ψ_e'' have the same parity. Consequently the transitions are only allowed if the different states have different parities with respect to inversions.

This rule is obeyed for observations in both polarized and unpolarized light. This means that the x component of the dipole transition's matrix element is non-zero only when the eigenfunctions of the combining states have different symmetries (parities) with respect to the plane of symmetry, otherwise the x component is zero. It can be easily verified that the two other components will also be non-zero for eigenfunctions with the same parity. If a molecule has a definite orientation in space (for example, in molecular crystals), then when its absorption spectrum is viewed in polarized light with its electrical vector pointed along the x axis a transitions's intensity will only be non-zero when Ψ_e' and Ψ_e'' have different parities. Odd–odd and even–even transitions are thus prohibited. By contrast, when the absorption of light is observed with E_y and E_z polarization, the transitions between states with identical symmetry will be active. A similar argument can be made for centres and second-order axes of symmetry.

Thus the presence of symmetry in the wave functions can be detected by the absorption of polarized light for orientated molecules.

We now consider the superposition integral. Since this only contains products of the vibrational functions of the upper and lower electronic states, we find that when there are general symmetry elements it will be non-zero if the products of the wave functions do not change sign from one symmetry subspace to the other. This can only occur when both the combining wave functions have the same symmetry with respect to the symmetry elements common to both combining electronic states.

Given high-order symmetry elements, i.e. C_p and S_p with $p \geqslant 3$, the argument is somewhat more complicated. However, even in these cases there is a rule which states that a vibronic transition is only allowed when the integrand in its dipole moment's matrix element and that in its superposition integral are fully symmetric with respect to all of the symmetry elements that are common to the combining electronic states.

Having analysed the integrands, using the selection rules by symmetry, and by knowing the symmetry of a molecule in the combining electronic states, it is therefore possible to establish whether the vibronic transition in question is prohibited or allowed under different observational conditions.

2.5 Solving the Non-adiabatic Vibronic Problem

Let us return to the general solution of the vibronic problem (equation (2.1)):

$$(\hat{T}_e + \hat{T}_n + V_{en} + V_{ee} + V_{nn})\Psi(\vec{r}, Q) = E\Psi(\vec{r}, Q)$$

which can be derived by using the variational method. In the variational procedure it is expedient to choose the basis functions so that they are as close as possible to exact functions obeying equation (2.1). This leads to a reduction in the dimensionality of the basis and, hence, simplifies the computations. For this reason, it is necessary to take into account physical peculiarities of electron and nuclear motions in the molecule so as to reflect them in a chosen basis. It is evident, therefore, that electronic wave functions $\Psi_{ek}(r, Q)$, representing the solutions of the pure electronic problem

$$(\hat{T}_e + V_{en} + V_{ee} + V_{nn})\Psi_{ek}(r, Q) = E_{ek}(Q)\Psi_{ek}(r, Q) \tag{2.22}$$

provide a suitable basis. The potential energy of internuclear interaction, V_{nn}, is included in the adiabatic electron Hamiltonian, for it gives only an additive increment to the electron energy $E_{ek}(Q)$ without affecting the electron wave function $\Psi_{ek}(r, Q)$.

The total solution of the vibronic problem (2.1) will sought in the form

$$\Psi(r, Q) = \sum_k c_k(Q)\Psi_{ek}(r, Q) \tag{2.23}$$

The variational coefficients, $c_k(Q)$, depend on the nuclear co-ordinates and a set of equations for calculating these coefficients is required. Let

$$\Psi_i(r, Q) = \sum_k c_{ik}(Q)\Psi_{ek}(r, Q)$$

and

$$\Psi_j(r, Q) = \sum_k c_{jk}(Q)\Psi_{ek}(r, Q)$$

be the ith and jth solutions of the complete vibronic problem (2.1). Substituting the ith solution in equation (2.1) gives

$$\begin{aligned}\hat{H}\Psi_i(r, Q) &= E_i\Psi_i(r, Q); (\hat{T}_e + V_{en} + V_{ee} + V_{nn})\Psi_i(r, Q) + \hat{T}_n\Psi_i(r, Q)\\ &= E_i\Psi_i(r, Q); \sum_k E_{ek}c_{ik}(Q)\Psi_{ek}(r, Q) + \hat{T}_n\Psi_i(r, Q)\\ &= E_i\Psi_i(r, Q)\end{aligned} \tag{2.24}$$

Multiplying the left-hand side of equation (2.24) by $\Psi_j^*(r, Q)$, then integrating over electron co-ordinates gives

$$\sum_k c_{jl}^*(Q)\left\{\sum_k [(E_{ek} - E_i)\delta_{kl} + \hat{t}_{lk}(Q)]c_{ik}(Q)\right\} = 0 \tag{2.25a}$$

where the operator

$$\hat{t}_{lk}(Q) = \int \Psi_{el}^*(r, Q)\hat{T}_n(Q)\Psi_{ek}(r, Q)\,dv \tag{2.25b}$$

Equation (2.25a) must be obeyed for any i and j, therefore the term in braces in equation (2.25a) must equal zero for any i. This gives the following set of differential equations for determining $c_k(Q)$ and E (the subscript i is omitted here):

$$\begin{aligned}(E_{e1}(Q) + \hat{t}_{11}(Q) - E)c_1(Q) + \hat{t}_{12}(Q)c_2(Q) + \cdots + \hat{t}_{1N}(Q)c_N(Q) = 0\\ \hat{t}_{21}(Q)c_1(Q) + (E_{e2}(Q) + \hat{t}_{22}(Q) - E)c_2(Q) + \cdots + \hat{t}_{2N}(Q)c_N(Q) = 0\\ \hat{t}_{N1}(Q)c_1(Q) + \hat{t}_{N2}(Q)c_2(Q) + \cdots + (E_{eN}(Q) + \hat{t}_{NN}(Q) - E)c_N(Q) = 0\end{aligned} \tag{2.26}$$

Operators $\hat{t}_{lk}(Q)$ in equation (2.25) and (2.26) represent vibronic interactions in the molecule. When $l \neq k$, $\hat{t}_{lk}(Q)$ describes interaction between the lth and kth pure electronic terms caused by nuclear motions, while $\hat{t}_{kk}(Q)$ gives corrections to the pure electronic and vibrational wave functions. Note that the off-diagonal operator $\hat{t}_{lk}(Q)$ is a first-order differential operator, whereas $\hat{t}_{kk}(Q)$ is a second-order differential one, being the sum of the second-order term $\hat{T}_{\mathrm{n}}(Q)$ and the sum $\hat{t}'_{kk}(Q)$ of zero- and first-order terms determined by vibronic interaction.

It seems, at best, extremely difficult to solve set (2.26) directly, since, on the one hand, it involves complex differential operators $\hat{t}_{lk}(Q)$ and, on the other, the order (determined by the number of basis functions) is so high. It is, therefore, necessary to simplify the set by making physically reasonable assumptions.

First, we can neglect, at a certain stage, those operator-coefficients in set (2.26) which couple the equations, i.e. operators $\hat{t}_{lk}(Q)$ $(l \neq k)$. With this assumption, set (2.26) splits into a system of uncoupled equations, and the problem reduces to solving the second-order differential equations:

$$(E_{\mathrm{e}k}(Q) + \hat{t}_{kk}(Q) - E)c_k(Q) = 0 \tag{2.27}$$

Bearing in mind the form of the operators $\hat{t}_{kk}$, equation (2.27) can be rewritten as

$$(\hat{T}_{\mathrm{n}}(Q) + E_{\mathrm{e}k}(Q) + \hat{t}'_{kk}(Q))c_k(Q) = Ec_k(Q) \tag{2.28}$$

thus separating terms $\hat{t}'_{kk}$, which determine vibronic interactions, i.e. zero- and first-order differential terms.

If the terms for vibronic interaction, i.e. $\hat{t}'_{kk}(Q)$, are neglected, equation (2.28) becomes

$$(\hat{T}_{\mathrm{n}}(Q) + E_{\mathrm{e}k}(Q))c_k(Q) = Ec_k(Q) \tag{2.29}$$

The physical meaning of equation (2.29) is that it describes nuclear motions in the potential field represented by the potential function $E_{\mathrm{e}k}(Q)$ and it can be solved by the methods generally used in molecular vibration computations. Potential functions $E_{\mathrm{e}k}(Q)$ describe the potential surface of a molecule in the kth pure electronic state. Operators $\hat{t}_{lk}(Q)$ and $\hat{t}_{kk}(Q)$ arise, as mentioned above, from the coupling between electronic and nuclear motions. However, the neglect of them in set (2.26) certainly does not imply complete disregard for vibronic interaction, since parametrically the interaction is included in the potential function $E_{\mathrm{e}k}(Q)$, determining the possible occurrence of nuclear vibrations in polyatomics.

It is evident from the form of basis functions (2.23) and the uncoupled set of equations (2.29) that the above assumptions result in separation of the electron and nuclear variables and, thereby, represent the so-called adiabatic approximation.

Naturally, questions arise as to the applicability of the approximation and on how large (or small) are the errors caused by neglecting the operators $\hat{t}_{lk}(Q)$ and $\hat{t}'_{kk}(Q)$.

First, note that the eigenvalue spectrum for equation (2.29) may, in principle, be both discrete and continuous. We shall confine ourselves to the case of a discrete spectrum, which is equivalent to the assumption that atoms in the molecule are involved only in finite motions.

This, in turn, implies that the potential surfaces of the molecule, $E_{ek}(Q)$, have minima in the vicinity of which these vibrational motions occur. Therefore, for the functions $c_k(Q)$ and, consequently, for $\Psi_{ek}(r, Q)$, reference points for nuclear co-ordinates can be identified for each k with the minimum point of the respective kth potential surface. Physically these reference points may be interpreted in the adiabatic approximation as representing the equilibrium molecular geometry in particular electron states.

In this case the adiabatic electronic wave function $\Psi_{ek}(r, Q)$ may be written as

$$\Psi_{ek}(r, Q) = \Psi_{ek}(r, 0) + \Delta\Psi_{ek}(r, Q) \tag{2.30}$$

where $\Psi_{ek}(r, 0)$ corresponds to the equilibrium geometry and the function $\Delta\Psi_{ek}(r, Q)$ describes variations of the total electron wave function when deviations from equilibrium molecular geometry occur. This function can be explained into a power series in vibrational co-ordinates, Q:

$$\Delta\Psi_{ek}(r, Q) = \left(\frac{\partial \Psi_{ek}(r, Q)}{\partial Q}\right)_0 Q + \left(\frac{\partial^2 \psi_{ek}(r, Q)}{\partial Q^2}\right)_0 Q^2 + \cdots \tag{2.31}$$

It should be borne in mind that Q in equation (2.31) represents the set of relative co-ordinates whose number is equal to the number of internal degrees of freedom of the molecule.

It is clear from equations (2.25a) and (2.30) that neglecting the terms $\hat{t}_{lk}(Q)$ and $\hat{t}'_{kk}(Q)$ is equivalent to neglecting the dependence of the electronic wave function $\Psi_{ek}(r, Q)$ on nuclear co-ordinates, namely the function $\Delta\Psi_{ek}(r, Q)$. This approximation (Franck–Condon or 'crude' adiabatic) appears, in a number of cases, to be adequate and, therefore, is widely used in analysis and interpretation of vibronic spectra of polyatomic molecules.

Note that in order to uncouple set (2.26) so as to separate the electron and nuclear variables it is sufficient to neglect only the off-diagonal operators $\hat{t}_{lk}(Q)$. The diagonal vibronic interaction operators $\hat{t}'_{kk}(Q)$ correct the wave functions only through the vibronic interaction within a given kth adiabatic electron term.

In this connection it is expedient first to estimate the relative smallness of elements $\hat{t}_{lk}$ and $\hat{t}'_{kk}$. Assume vibronic interaction to be weak and suppose that variations of the electron–nuclear interaction energy, V_{en}, caused by nuclear motions only, slightly affect the electron wave functions, i.e. that the function $\Psi_{ek}(r, Q)$ is close to the Frank–Condon function $\Psi_{ek}(r, 0)$, which represents the solution of the problem with the nuclei fixed at their equilibrium positions. Vibronic coupling can then be treated in terms of perturbation theory:

$$\Psi_{ek}(r, Q) = \Psi_{ek}(r, 0) + \sum_{i \neq k} \sum_{k} \xi_{kki} Q_k \Psi_{ei}(r, 0) + \cdots \tag{2.32}$$

where only terms of zero and first order are written explicitly. The quantities

ξ_{kki} represent matrix elements of the vibronic interaction operator, their explicit form not being essential. It is important only that they are of the order of ε, the perturbation parameter describing the weakness of vibronic interaction. Terms of the second order and above are of the order ε^2 and higher.

Substituting expression (2.32) for Ψ_{el} and Ψ_{ek} into equation (2.25a) for matrix elements $\hat{\tau}_{lk}$ and neglecting the terms higher than first order in ε, we obtain

$$\hat{\tau}_{lk} = \sum_{k} \xi_{kkl} \hat{\tau}_k^0 \tag{2.33}$$

$$\hat{\tau}'_{kk} = \sum_{i \neq k} \sum_{k,k'} \xi_{kki} \xi_{k'ki} Q_{k'} \hat{\tau}_{k'}^0 \tag{2.34}$$

where $\hat{\tau}_k^0 = \partial/\partial Q_k$ is the operator of first-order differentiation with respect to the nuclear parameter Q_k, i.e. with an accuracy of factor, momentum conjugate to co-ordinate Q_k.

Thus it follows from equations (2.33) and (2.34) that $\hat{\tau}_{lk} \sim \varepsilon$ and $\hat{\tau}'_{kk} \sim \varepsilon^2$ (i.e. the diagonal operator $\hat{\tau}'_{kk}$) is of the second order with respect to the off-diagonal operator and, hence, vibronic interaction 'within the adiabatic electron term' is relatively weak compared with the interaction between different electron states caused by nuclear vibrations. It should be borne in mind, therefore, that neglecting the off-diagonal matrix elements, $\hat{\tau}_{lk}$, inherently implies that terms $\hat{\tau}'_{kk}$ also have to be neglected.

2.6 The Herzberg–Teller Approximation

In this context the question arises as to the validity of the Herzberg–Teller approximation that partially takes into account vibronic interaction in the adiabatic approximation. Actually, in the Herzberg–Teller approach vibronic wave functions are believed to be the product of the electronic wave function, as presented in the form of equation (2.32), and the vibrational wave function obtained as a solution of the adiabatic vibration equation (2.29). Thus, the functions include the electron component depending on nuclear co-ordinates as parameters, but are of the product type as in the Franck–Condon approximation.

The slight dependence of the electronic function $\Psi_{ek}(r, Q)$ on nuclear co-ordinates leads to off-diagonal differential operators $\hat{\tau}_{lk}(Q)$ of the same approximate magnitude. Consequently, vibrational functions $c_k(Q)$ representing the solutions of set (2.26) will have corrections of the same order and, therefore, these corrections must be included in the complete wave function. In this sense the Herzberg–Teller approximation seems to be incorrect.

In an experimental study mean values of certain physical quantities, not wave functions, are measured. Therefore when assessing the validity of a particular approximation it is necessary to estimate not only corrections to wave functions but also their influence on the measurable quantities. In spectroscopic studies these quantities are the transition intensity and its related probability.

Consider the probability of the dipole transition between the ith and jth electron–vibration terms. The appropriate wave functions have the form

$$\Psi_i(r,Q)=\sum_k c_{ik}(Q)\Psi_{ek}(r,Q) \tag{2.35}$$

$$\Psi_j(r,Q)=\sum_l c_{jl}(Q)\Psi_{el}(r,Q) \tag{2.36}$$

when electron and vibration components, to the first order in ε, may be presented as

$$\Psi_{ek}(r,Q)=\Psi_{ek}(r,0)+\varepsilon\Psi'_{ek}(r,Q) \tag{2.37}$$

$$c_{ik}(Q)=c_{ik}^{0}(Q)\delta_{ik}+\varepsilon c'_{ik}(Q) \tag{2.38}$$

The particular form of the functions Ψ'_{ek} and c'_{ik} entering equations (2.37) and (2.38) is not essential here, this being important only in that these functions are of zero order in ε. The Kroneker symbol, δ_{ik}, reflects the fact that electron and nuclear motions can be separated in the absence of the off-diagonal interactions in set (2.26), thus reducing the sums in equations (2.35) and (2.36) to only one ith (for equation (2.35)) or jth (for equation (2.36)) term.

For the probability of an $i\leftrightarrow j$ dipole transition we obtain

$$\begin{aligned} w_{ij}^{1/2} &\sim \left|\int \Psi_i^*(r,Q)\hat{\mu}\Psi_j(r,Q)\,\mathrm{d}v\,\mathrm{d}Q\right| \\ &= \left|\left(\int \Psi_{ei}^*(r,0)\hat{\mu}\Psi_{ej}(r,0)\mathrm{d}v\right)\left(\int c_{ii}^{0}(Q)c_{jj}^{0}(Q)\mathrm{d}Q\right)\right. \\ &\quad + \int c_{ii}^{0}(Q)\left(\int \Psi_{ei}^*(r,0)\hat{\mu}\varepsilon\Psi'_{ej}(r,Q)\,\mathrm{d}v\right)c_{jj}^{0}(Q)\,\mathrm{d}Q \\ &\quad \left. + \sum_l\left(\int \Psi_{ei}^*(r,0)\hat{\mu}\Psi_{el}(r,0)\mathrm{d}v\right)\left(\int c_{ii}^{0}(Q)\varepsilon c'_{jl}(Q)\,\mathrm{d}Q\right)+\cdots\right| \end{aligned} \tag{2.39}$$

where only terms corresponding to the Franck–Condon approximation (first term) and those resulting from vibronic corrections to the function $\Psi_j(r,Q)$, covered by equations (2.37) and (2.38), are written explicitly. The terms above first order in ε are neglected, the others being similar to the second and third terms.

The total probability of the dipole transition is composed of the Franck–Condon (first) term, the Herzberg–Teller (second, and not written but similar to this one) terms and the third term, not taken into account in the Herzberg–Teller approach, although it is of the same order. An essential factor determining the applicability of the Herzberg–Teller approximation is that the integral over electron co-ordinates has the same form as the corresponding Franck–Condon integral, representing the square root of the probability of the appropriate pure electron transition as calculated in the Franck–Condon approximation. It is evident that the main contribution to the third term in equation (2.39) is due to the term with $l=j$, i.e. corresponding to $\varepsilon c'_{jl}=\varepsilon c'_{jj}$, and other terms can be classified as being of higher order, following from the very form of set (2.26).

Hence, if the transition between the ith and jth electron terms is symmetry prohibited in the Franck–Condon approximation, then the Herzberg–Teller approach can be used to estimate the vibronic effects. In all other cases the Herzberg–Teller approximation cannot, strictly speaking, be applied and it is necessary to take into account the contribution resulting from vibronic perturbation of vibrational wave functions, i.e. the third term in equation (2.39). If, however, the Franck–Condon transition is symmetry prohibited but appears to be of low intensity, the Herzberg–Teller approach may be used if the integral $\int \Psi_{ei}^*(r,0)\hat{\mu}\varepsilon\Psi'_{ej}(r,Q)\,dv$ is considerably greater than $\int \Psi_{ei}^*(r,0)\hat{\mu}\Psi_{ej}(r,0)\,dv$ which determines, in the Franck–Condon approximation, the probability of electron transition.

It must be borne in mind that it seems incorrect and inappropriate to retain in the Herzberg–Teller approximation terms in the wave function (2.32) which are second order in ε, since their contribution to the transition probability is of the same order as the terms in equation (2.39), which have already been neglected.

2.7 Algorithm for Solving the Vibronic Problem

A correct solution of the non-adiabatic vibronic problem requires consistent treatment of the vibronic interaction. The resulting solution should involve not only the parametric dependence of the electron wave function on nuclear co-ordinates, derived particularly in terms of perturbation theory, but also appropriate modification of the vibration functions and their 'mixing' with the electron functions which makes it necessary to solve the set of equations (2.26).

Finding the total solution of the electronic vibrational problem in the form of equation (2.23) is convenient only for determination of the main physical features of the solution. In practice, this will necessitate certain simplifications, consisting of specifying the form of the functional dependences of the basis functions $\Psi_{ek}(r,Q)$ and variational coefficients $c_{ik}(Q)$.

An analytical solution of the adiabatic electron problem (2.22) is not possible because of the complexity of the functional dependence of V_{en} on nuclear co-ordinates (if only the vibronic interaction is not assumed to be weak, then it can be considered in terms of perturbation theory). The variational method can also be used in this case by choosing $\Psi_{ek}^0(r) = \Psi_{ek}(r,0)$ as the basis functions. Here, $\Psi_{ek}(r,0)$ is a solution of equation (2.22) obtained in the Franck–Condon approximation, i.e. at some given nuclear geometry $Q = Q_0$, which also serves as a reference point, and therefore it is accepted that $Q_0 = 0$. If the general solution is represented by

$$\Psi_{ek}(r,Q) = \sum_p b_{kp}(Q)\Psi_{ep}^0(r) \tag{2.40}$$

then the variational set of algebraic equations determining $b_{kp}(Q)$ and $E_{ek}(Q)$ takes the form

$$\sum_p b_{kp}Q[(E_{ep}^0 - E_{ek}(Q))\delta_{pq} + \theta_{qp}(Q)] = 0 \tag{2.41}$$

where $q = 1, 2, \ldots, N$; $\theta_{qp}(Q) = \int \Psi^*_{eq}(r) \Delta V_{en} \Psi^0_{ep}(r)\,dv$;

$$\Delta V_{en} = V_{en}(Q) - V_{en}(0); \quad \text{and} \quad E^0_{ep} = E_{ep}(0).$$

It is evident, however, that solving the two variational problems (2.41) and (2.26) can be reduced to solving one general variational problem with Ψ^0_{ek} as basis functions.

Substituting equation (2.40) into equation (2.23) gives

$$\Psi_i(r, Q) = \sum_k c_{ik}(Q)\Psi_{ek}(r, Q) = \sum_k c_{ik}(Q) \sum_p b_{kp}(Q)\psi^0_{ep}(r)$$

$$= \sum_p \zeta_{ip}(Q)\Psi^0_{ep}(r); \; \zeta_{ip}(Q) = \sum_k c_{ik}(Q) b_{kp}(Q) \tag{2.42}$$

The variational coefficients $\zeta_{ip}(Q)$ can be found from the set of algebraic equations formally coinciding with set (2.26), differing only in the sense of the coefficients

$$\sum_p [(E^0_{ep}(Q) - E_i)\delta_{qp} + \hat{\tau}_{qp}(Q)]\zeta_{ip}(Q) = 0$$

$$E^0_{ep}(Q) = E^0_{ep} + V_{nn}, \quad \hat{\tau}_{qp}(Q) = \theta_{qp}(Q) + \hat{T}_n \delta_{qp} \tag{2.43}$$

Thus, in the basis of electronic functions, $\Psi^0_{ep}(r)$ the problem reduces to solving the set of equations (2.43) for variational coefficients $\zeta_{ip}(Q)$. As can easily be seen, for the case of weak electronic vibrational interaction set (2.43) can be uncoupled (by complete analogy with equation (2.26)) and the coefficients $\zeta_{ip}(Q)$ satisfy the usual adiabatic vibrational equations with $(E^0_{ep}(Q) + \theta_{pp}(Q))$ as a potential function.

Two ways of solving set (2.43) are possible. In the first method, simplification of the form of the operators $\hat{\tau}_{qp}(Q)$ allows direct solution of set (2.43) (for example, by methods of perturbation theory). Alternatively, a suitable matrix representation is used to describe the electronic vibrational motions in the molecule, then the matrices are analysed and simplified to permit their easy diagonalization.

The advantages of the second approach are obvious, namely the clarity of the results obtained and the absence of any additional restrictions on the strength of vibronic interactions. Although the variational approach requires diagonalization of matrices of very high order, the choice of suitable representations, accurately reflecting the peculiarities of the electronic and nuclear motions in the molecule, allows the problem to be reduced to the transformation of matrices of considerably lower order (<500). This is well within the capabilities of modern techniques, as will now be discussed in detail.

Solution of set (2.43) can be obtained easily by the variational method if the vibrational functions $\zeta_{ip}(Q)$ are presented as a linear combination of basis functions $c^0_k(Q)$ (adiabatic vibrational wave functions) representing the solution of equation (2.29). The problem is thus reduced to searching for the solution of a set of adiabatic equations, as in the case of the usual Ritz method. Such a problem is extremely unwieldy, but the solution is not impossible in principle.

An inherent feature of the vibronic problem is that the number of vibrational degrees of freedom of polyatomic molecules is large (~ 50 for a medium-sized molecule) and, if the basis functions are not at least nearly coincident with the eigenfunctions, a large basis of vibrational functions is required. This, of course, considerably complicates the problem, because matrices of very high order have to be diagonalized. Thus, as mentioned above, the choice of the basis functions is of key importance for the practical application of the variational approach.

Consider one possible variational solution of the vibronic problem, representing a further improvement of the conceptions presented in reference 5. When selecting basis functions we shall try, from the beginning, to include the important feature of vibronic motions that, in the adiabatic approximation, molecular vibrations are executed near the appropriate potential surface minimum. This means that the arguments of the adiabatic vibration wave functions for different electron terms are generally shifted relative to each other along the axes of the natural (or normal) co-ordinates. In addition, vibration wave functions for different electron states are distinguished because of mixing of the normal modes (Dushinsky effect) and the difference between vibrational frequencies for different electron terms (frequency effect). It follows from the analysis of the vibronic spectra of molecules that the shift of the normal co-ordinates is the main (or first-order) effect determining the inherent features of the variations of molecular structure. Such an effect is evidently of major importance when choosing the basis functions.

One possible way of taking into account these features of the vibronic interaction is to use a modified basis of vibronic wave functions. Let us assume that the adiabatic electron equation is to be solved by some suitable quantum-mechanical method, and that the wave functions and the energies of electron states, i.e. the potential energy surfaces of the molecule, are to be found. Then it is expedient to choose, as the basis for the non-adiabatic vibronic problem, those functions for which the electron part is a solution of the adiabatic electron equation at a given equilibrium geometry. However, the vibrational components consist of the functions representing the solutions of the vibration equation for the ground adiabatic electron state with shifted argument. This shift of the argument is equivalent to a shift in the minima of the corresponding potential surfaces with respect to the equilibrium geometry. Thus, the vibrational components of electronic vibrational basis functions are associated with the minima of potential surfaces.

The problem of what molecular geometry to choose for calculating the electron functions must be considered. As a rule, solutions of the vibronic problem, close in energy to the minima of the adiabatic potential surfaces of the ground and nearest excited states, are of interest. Therefore both vibrational and electronic components of the basis functions should be chosen so that they correspond to the geometry close to that at the appropriate potential surface minimum. This results in a reduction of the required vibrational basis and, also, in diminution of the values of the off-diagonal matrix elements. Therefore, in electron-state calculations using this method it is appropriate to choose the molecular geometry as close as possible (smallest root-mean-square deviation) to the geometries of the minima of the most interesting lower electron terms.

Let the basis electronic–vibrational functions have the form

$$\Psi_{kv}(r,Q) = \Psi_k(r,0)\phi_{kv}(Q) \tag{2.44}$$

where

$$\psi_k(r,0) \equiv \Psi^0_{ek}(r) = \sum_i \tau_{ik}\Phi_i$$

is the wave function of the kth adiabatic state, Φ_i are electron configurations built on the Hartree–Fock (see Chapter 4) solution for the equilibrium molecular geometry discussed above, τ_{ik} are elements of the configuration interaction matrix and $\phi_{kv}(Q)$ are vibrational basis functions. The last are determined as follows. $\phi_{1v}(Q)$ is the solution of the problem with Hamiltonian $\hat{H} = \hat{T}_n + W$, where W is (or is close to) the potential function of the Born–Oppenheimer operator in the ground electron state; $\phi_{kv}(Q)$ $(k \neq 1)$ coincide in form with $\phi_{1v}(Q)$ but their arguments are shifted by as much as the potential surfaces.

It is essential that basis functions Ψ_{kv} of equation (2.44) are orthonormal, since functions ψ_k are orthonormal and ϕ_{kv} are orthonormal for each k. The overlap matrix is therefore the unit matrix and the variational procedure in such a modified basis is no more complicated than in the usual case.

The matrix elements, $a_{k,v,k'v'}$, of the electron–vibration Hamiltonian

$$\hat{H} = \hat{T}_e + \hat{T}_n + V_{ee} + V_{en} + V_{nn}$$

are

$$a_{kv,k'v'} = E_k\delta_{kk'} + \theta_{k,vv'} + \omega_{kv,k'v'} \tag{2.45}$$

where E_k is the energy of the kth adiabatic electron term:

$$\begin{aligned}\theta_{k,vv'} &= \langle v_k | \hat{H}_n - W + V_{nn} | v'_{k'}\rangle \\ &+ \left\langle v_k \left| \sum_i \tau_{ki}\tau_{k'i}\left(\sum_{p,m}\left\langle \lambda_p^{(i)} \left| \left(\frac{\partial V_{en}}{\partial Q_m}\right)_0 \right| \lambda_p^{(i)}\right\rangle Q_m + \cdots\right) \right| v'_k\right\rangle \\ &+ \left\langle v_k \left| \sum_{\substack{i,j\\ i\neq j}} \tau_{ik}\tau_{jk}\left(\sum_m\left\langle \lambda_p^{(i)} \left| \left(\frac{\partial V_{en}}{\partial Q_m}\right)_0 \right| \lambda_q^{(j)}\right\rangle Q_m + \cdots\right) \right| v'_{k'}\right\rangle\end{aligned} \tag{2.46}$$

$$\begin{aligned}\omega_{kv,k'v'} = (1-\delta_{kk'})\Bigg[&\left\langle v_k \left| \sum_i \tau_{ik}\tau_{ik'}\left(\sum_{p,m}\left\langle \lambda_p^{(i)} \left| \left(\frac{\partial V_{en}}{\partial Q_m}\right)_0 \right| \lambda_p^{(i)}\right\rangle Q_m + \cdots\right) \right| v'_{k'}\right\rangle \\ &+ \left\langle v_k \left| \sum_{\substack{i,j\\ i\neq j}} \tau_{ik}\tau_{jk'}\left(\sum_m\left\langle \lambda_p^{(i)} \left| \left(\frac{\partial V_{en}}{\partial Q_m}\right)_0 \right| \lambda_q^{(j)}\right\rangle Q_m + \cdots\right) \right| v'_{k'}\right\rangle\Bigg]\end{aligned} \tag{2.47}$$

and λ_p are one-electron molecular spin–orbitals (see below).

It is seen that matrix A, composed of matrix elements $a_{kv,k'v'}$ equation (2.45) has a specific form, i.e. it can be represented as the sum of three matrices: the diagonal matrix of the energies of adiabatic electron terms, E_k; matrix θ diagonal in numbers of electron functions; and matrix ω whose elements diagonal in k equal zero. That is, matrix θ describes vibronic coupling within a given kth electron term and its diagonalization is actually equivalent to solving the vibrational adiabatic problem for each kth electron state with partial regard to vibronic interaction. The matrix ω is determined by vibronic interaction between different adiabatic electron terms.

According, diagonalization of the complete variational matrix A will be carried out in two steps. Consider, first, the part diagonal in k, i.e. $(E+\theta)$.

Let us expand V_{nn} in the series in vibrational co-ordinates, represent potential W by a quadric quantic and take into account that $\langle v_k|\hat{H}_n|v_k'\rangle = \delta_{1k}E_v\delta_{vv'} + (1-\delta_{1k})E_{vv'}^k$ where E_v is the vibrational energy of the ground electronic state. Then the following expression can be obtained for $\theta_{k,vv'}$:

$$\begin{aligned}\theta_{k,vv'} &= \delta_{vv'}(\delta_{1k}E_v + V_{\text{nn}}^0) + (1-\delta_{1k})E_{vv'}^k \\ &+ \sum_m \left[\left(\frac{\partial V_{\text{nn}}}{\partial Q_m}\right)_0 + \sum_i \tau_{ik}\tau_{ik}\sum_p \left\langle \lambda_p^{(i)}\middle|\left(\frac{\partial V_{\text{en}}}{\partial Q_m}\right)_0\middle|\lambda_p^{(i)}\right\rangle\right]\langle v_k|Q_m|v_k'\rangle \\ &+ \sum_{m,m'}\left[-w_{mm'} + \sum_i \tau_{ik}\tau_{ik}\sum_p \left\langle \lambda_p^{(i)}\middle|\left(\frac{\partial^2 V_{\text{en}}}{\partial Q_m \partial Q_{m'}}\right)_0\middle|\lambda_p^{(i)}\right\rangle\right] \\ &\times \langle v_k|Q_m Q_{m'}|v_k'\rangle + \cdots \\ &+ \sum_m \left[\sum_{i,j\neq i}\tau_{ik}\tau_{jk}\left\langle \lambda_p^{(i)}\middle|\left(\frac{\partial V_{\text{en}}}{\partial Q_m}\right)_0\middle|\lambda_q^{(j)}\right\rangle\right]\langle v_k|Q_m|v_k'\rangle \\ &+ \sum_{m,m'}\left[\sum_{i,j\neq i}\tau_{ik}\tau_{jk}\left\langle \lambda_p^{(i)}\middle|\left(\frac{\partial^2 V_{\text{en}}}{\partial Q_m \partial Q_{m'}}\right)_0\middle|\lambda_q^{(j)}\right\rangle\right]\langle v_k|Q_m Q_{m'}|v_k'\rangle + \cdots\end{aligned} \tag{2.48}$$

To diagonalize the matrix $(E+\theta)$ it is necessary to diagonalize the individual blocks corresponding to different k, i.e. to different electron states, E_k, appearing in each block as an additive term. This is actually equivalent to solving the vibrational problem for different electron states with partial regard to vibronic interactions.

The dimensions of each individual block of matrix θ are determined by the number of vibration wave functions included in the basis. If the basis involves vibration functions corresponding to normal, overtone and compound vibrations (with one- and twofold excitations of no more than two normal modes simultaneously), then the order of blocks is $N_v = 2(M!/(M-2)!) + M(v_0+1)+1$, where M is the number of degrees of freedom of the molecule and v_0 is the maximum quantum number of the overtone vibration considered. For $M=5$ and $v_0=2$, $N_v = 6276$. If compound vibrations are neglected, then $N_v = 1376$. Methods are available which allow matrices of such an order to be diagonalized. The procedure can be simplified because we are really interested in the energies of, at most, a few hundred of the lowest levels rather than all the eigenvalues of the variational matrix.

It should also be borne in mind that as the problem is solved in the harmonic approximation, then there is no need to obtain the complete solution of the vibrational problem in terms of combining electron states. It is sufficient to determine just the normal vibrations, since only these are important in non-adiabatic (vibronic) interaction, and the overtone and combinational vibrations can be neglected. It is then sufficient to perform so-called 'quasi-diagonalization' of the blocks of matrix θ, i.e. to reduce each block to a form in which all elements in the rows and columns corresponding to normal vibrations (except for diagonal elements) equal zero, the values of other elements being arbitrary. Since the basis was built of vibrational functions associated with the potential surface minima, i.e. close to the real vibrational functions, then the diagonal elements in the diagonal parts of the blocks of matrix θ will determine the corresponding vibrational frequencies. Hence, by neglecting less important interactions between overtone and compound vibrations, the technique of quasi-diagonalization allows a considerable mathematical simplification of the problem.

Quasi-diagonalization of such large matrices for molecules with several dozen normal modes (~ 50–100) is about as complex as the complete diagonalization of a matrix of the order of a few hundreds (~ 100–300), as it presents no special problems. Since, in practice, vibronic interaction between only a few (< 5) of the lowest adiabatic electron terms need to be considered, quasi-diagonalization of no more than five such matrices is required.

After quasi-diagonalization matrix A takes the form

$$A' = \tilde{T}'AT' = 1 + \theta' + \omega' = 1 + \beta + B + \omega' \tag{2.49}$$

where matrix T' has a block form in k and β and B are diagonal and off-diagonal parts of matrix θ', respectively.

To obtain the total solution of the non-adiabatic problem, vibronic interaction between different electron terms must also be considered. As noted above, the main contribution to this interaction is due to the normal vibrations of the molecule. Therefore, component B and the corresponding elements of matrix ω' in the off-diagonal part of matrix A' (equation (2.49)) may be neglected, and matrix A' takes the form

$$A'' = 1 + \beta + \omega'' \tag{2.50}$$

where ω'' involves elements of matrix ω' with quantum numbers $v, v' \leqslant 1$ and, consequently, the order of matrix A'' is $N_0 = k_0 M$, where k_0 is the number of interacting electron terms ($k_0 \leqslant 5$). The problem thus reduces to diagonalizing a matrix of order no higher than 500.

Final solution of the non-adiabatic problem can be obtained after diagonalizing matrix A'' (equation (2.50)) by using the appropriate transformation matrix T'', the resulting diagonal elements being equal to the energies of the vibronic states. Corresponding wave functions are

$$\Psi_{kv} = \sum_{k'v'} \tau''_{kv,k'v'} \left(\sum_{k''} \tau_{k'k''} \Phi_{k''} \right) \left(\sum_{v''} \tau'_{k',v'v''} \phi_{k'v''} \right)$$

where τ' and τ'' are the elements of the transformation matrices T' and T'', respectively.

Thus solving the complete electronic vibrational problem by the variational Ritz method in a modified basis of electron–vibration states involves two main stages. The mathematical transformations required in each stage reflect certain physical features of the electronic vibrational motions in the molecule and, therefore, the results obtained have a clear physical meaning.

The above discussion shows that, in spite of the unwieldiness of the method, it is possible in principle (and, moreover, quite feasible using modern computers) to obtain an exact solution of the non-adiabatic problem, since the solution reduces to the diagonalization of a few ($\lesssim 5$) matrices of order no higher than several hundreds ($\lesssim 500$).

Attention should be paid to a characteristic feature distinguishing adiabatic and non-adiabatic solutions of the electronic–vibrational equation. In the adiabatic approximation electronic and vibrational motions can be separated and, therefore, the vibronic function can be presented as the product of electron and vibration parts, each independently satisfying their own equations.* This means that electron and vibration states can be separated and the concepts of electronic and vibrational energy levels introduced. The term 'electronic level', for example, indicates the energy of the electronic motion corresponding to the minimum (or some other point) of the potential surface of a molecule. Moreover, the very concept of the potential surface of a molecule may be introduced for a given adiabatic electronic state, and for this reason it is meaningful to speak of an equilibrium molecular geometry corresponding to the minimum of a potential surface.

In the general non-adiabatic case these concepts do not work and cannot, in principle, be introduced, as the electronic and vibrational motions cannot be separated. We can then speak only of average values of nuclear co-ordinates and the most probable molecular geometry in a given vibronic state.

This certainly complicates the analysis of the vibronic states of the molecule, making it impossible to use the familiar interpretation of states which is valid in the case of the adiabatic approximation. The variational method of solution allows a non-adiabatic electronic state to be related to adiabatic states and to determine, from the maximum values of the variational coefficients, those states which make the greatest contributions. Thus the variational solution provides a physically lucid interpretation of vibronic states which is of special importance in the case of weak vibronic interaction, where a few adiabatic vibronic states can be specified as being represented in the total wave function with very high weighting coefficients.

Additional information about problems considered above can be found in reference 6–20.

*In the adiabatic approximation there is an additive dependence between the electronic and vibrational equations.

References

1. M. A. Elyashevich. Rotational–vibrational energy of polyatomic molecules. *Trans. State Optic Inst.*, **12**, 106 (1938) (in Russian).
2. G. A. Natanzon and M. N. Adamov. *Bull. Leningrad State Univ.*, **4**, 28 (1974); **10**, 24 (1974) (in Russian).
3. J. D. Louck. *J. Molec. Spectr.*, **61**, 107 (1976).
4. F. Dushinsky. *Acta Physicochim. URSS*, **7**, 551 (1937).
5. L. A. Gribov and N. I. Prokofieva. *Zh. Prikl. Spectr.*, **31**, 1054 (1979); **35**, 827 (1981) (in Russian).
6. J. C. Slater. *Quantum Theory of Matter*, McGraw-Hill, New York, (1968).
7. R. G. Parr. *The Quantum Theory of Molecular Electronic Structure*, W. A. Benjamin, New York/Amsterdam (1963).
8. G. M. Barrow. *Introduction to Molecular Spectroscopy*, McGraw-Hill, New York (1962).
9. H. H. Jaffe and M. Orchin. *Theory and Applications of Ultraviolet Spectroscopy*, John Wiley, New York (1962).
10. J. N. Murrell. *The Theory of Electronic State of Organic Molecules*, Wiley, London (1964).
11. C. W. King. *Spectroscopy and Molecular Structure*, Holt, Rinehart and Winston, New York (1964).
12. M. A. Elyashevich. *Atomic and Molecular Spectroscopy Nauka*, Physico-Mathematical Literature, Moscow (1962) (in Russian).
13. I. B. Bersuker. *Opt. Spectr.*, **II**, 319 (1961) (in Russian).
14. I. B. Bersuker. *J. Exp. Theor. Phys.*, **43**, 1315 (1962) (in Russian).
15. I. B. Bersuker and V. Z. Polinger. *Vibronic Interactions in Molecules and Cristalls*, Nauka, Physico-Mathematical Literature, Moscow (1983) (in Russian).
16. W. Moffit and A. D. Liehr. *Phys. Rev.* **106**, 1195 (1956).
17. W. Moffit and W. Thorson. *Phys. Rev.*, **108**, 1251 (1957).
18. L. A. Gribov, V. I. Baranov and Yu. Nefedov. *J. Mol. Struct.* (Theochem), **148**, 1 (1986)
19. L. A. Gribov. *Introduction to Molecular Spectroscopy*, Nauka, Fis.-Math. Lit., Moscow (1976) (in Russian).
20. V. I. Baranov and L. A. Gribov. *J. Molec. Struct.*, **104**, 267 (1983).

Chapter 3

Non-interacting Electron Approximation

3.1 The Variational Method for Solving Schroedinger's Equation

Before solving a problem with many electrons in a polyatomic molecule we need to consider some general principles for solving problems concerning the motion of a single electron in a field of many positively charged centres. These then allow us, using Pauli's principle, to approximate the energy levels of a molecule with many electrons, assuming at first that they do not interact.

All molecular problems are so complicated that they are usually solved by using numerical approximation methods, the variational method being a major technique in this respect. This method is based on the fact that the differential equation $\hat{H}\psi = E\psi$ can be approximately solved for limiting values of a functional of the form $\bar{H} = \int \psi \hat{H} \psi \, dV$*, where ψ is an orthogonal test functions.

If $\psi = \psi_n$ is a solution of Schroedinger's equation with this operator $\hat{H}$, then $\bar{H}_n = E_n$, i.e. it is equal to one of the system's possible energetic states (energy levels). It is therefore also possible to deal with the function $\bar{H} = \int \psi \hat{H} \psi \, dV$, even when $\bar{H} \neq E_n$, and treat it as an average value of the system's energy with the operator $\hat{H}$ when the set of particles, whose state it describes, is characterized by the wave function ψ. If ψ is not a solution of the Schroedinger equation then $\bar{H} \neq E_n$. What we try to do when approximating a solution of the equation is to choose a function ψ so that the equality $\bar{H} = E_n$ is approximately satisfied. Thus ψ is varied in any way so long as the wave function is kept normalized. Two cases are possible.

First, the values of E_n are known from previous experiments. Hence ψ functions are sought so that the integral

$$\bar{H} = \int \psi \hat{H} \psi \, dV$$

is equal or close to the experimental values of E_n.

The second case is the most interesting practically, and is when the E_n values are not known experimentally. The functions can also be found in this case by starting from the expression for the average energy of $\bar{H}$, but some supplementary physical requirement must be imposed. One such requirement can be introduced from the general physical law governing the system's minimum

*Here and below we shall assume that the eigenfunctions and the variational functions are real.

energy. This states that all physical systems try to adopt the state which corresponds to the minimum overall energy.

We have already mentioned that the values of E_n are the limiting values of the integral $\bar{H} = \int \psi \hat{H} \psi \, \mathrm{d}V$ if the ψ functions are being varied over the whole of the domain. One of these values will correspond to the minimum possible energetic state, and is called the ground state. This condition indicates a way of determining E_0 and selecting ψ_0, i.e. a way of solving Schroedinger's equation. That is, the test function ψ must be varied to achieve as low a value of $\bar{H}$ as possible.

If the variation is over a wide enough range of values then we would hope that the minimum possible energetic state will be close to the true value of the ground state. If the search for extremum values of $\bar{H} = \int \psi \hat{H} \psi \, \mathrm{d}V$ is continued then other permissible energy levels and their eigenfunctions can be estimated, remembering that these eigenfunctions are orthonormal.

In reality, in order to vary $\bar{H}$, an initial representation of the test function ψ must be chosen so that it includes a series of parameters which can be varied so as to change the value of the functions over a wide range. Although, generally speaking, there is no limitation on the method of parameterizing the function, mathematically the simplest procedure of selecting a solution arises when the function being varied depends linearly on its constituent parameters. This means that the Ψ function should be looked for in the form

$$\Psi = \sum_k c_k \phi_k \tag{3.1}$$

Here, c_k are the parameters being varied, considered always to be real, while the ϕ_k are the unchanged functions whose choice depends on the problem in question.

In this case, $\bar{H}$ takes the form

$$\bar{H} = \sum_{k,m} c_k c_m \left(\int \phi_k \hat{H} \phi_m \, \mathrm{d}V \right) \tag{3.2}$$

Thus $\bar{H}$ has been transformed into a quadratic form with respect to the c_k variables whose coefficients equal $\int \phi_k \hat{H} \phi_m \, \mathrm{d}V$. Since only the c_k coefficients are changed as $\bar{H}$ is being varied while the ϕ_k functions remain unchanged, the integrals $\int \phi_k \hat{H} \phi_m \, \mathrm{d}V$ also remain unchanged. The search for the limiting values of this reduces to one for the extremum values of the quadratic form in equation (3.2). The normalization requirement of the eigenfunctions results in an expression of the form

$$\int \psi^2 \, \mathrm{d}V = \sum_{k,m} c_k c_m \int \phi_k \phi_m \, \mathrm{d}V = 1 \tag{3.3}$$

This means that we must search for the extremum values of the quadratic form on an ellipsoid described by the expression in equation (3.3). This problem is analogous to reducing two quadratic form to sums of squares, or to consistently diagonalizing the two matrices whose elements are

$$h_{km} = \int \phi_k \hat{H} \phi_m \, \mathrm{d}V$$

and

$$s_{km} = \int \phi_k \phi_m \, \mathrm{d}V$$

This modification of the variational procedure is called Ritz's method.

In a special case we may have $s_{km} = 1$ when $k = m$ and $s_{km} = 0$ when $k \neq m$. This means that the ϕ_k functions are orthonormal.

The matrix of the quadratic form must be symmetric. We shall show that this requirement is satisfied. With respect to the integrals $s_{km} = \int \phi_k \phi_m \, \mathrm{d}V$ this is clear because the integration result does not depend on the order of the two multiplicands. Now what happens with the integrals $h_{km} = \int \phi_k \hat{H} \phi_m \, \mathrm{d}V$? For simplicity, we shall consider that the problem has been formulated in Cartesian co-ordinates and we shall limit ourselves to one variable. Thus for one particle we have

$$\hat{H} = -\frac{\hbar^2}{2m} \frac{\mathrm{d}^2}{\mathrm{d}x^2} + V(x) \quad \text{and} \quad h_{km} = \int \phi_k \hat{H} \phi_m \, \mathrm{d}x$$

$$= -\frac{\hbar^2}{2m} \int \phi_k \frac{\mathrm{d}^2}{\mathrm{d}x^2} \phi_m \, \mathrm{d}x + \int \phi_k \, V(x) \phi_m \, \mathrm{d}x$$

The second term does not change when ϕ_k and ϕ_m are switched because the integration is simply over the product of three functions. Let us consider the first integral rewritten as

$$\int \phi_k \frac{\mathrm{d}}{\mathrm{d}x} \left(\frac{\mathrm{d}\phi_m}{\mathrm{d}x} \right) \mathrm{d}x$$

Integrating by parts, the integral in question becomes

$$\int_a^b \phi_k \frac{\mathrm{d}}{\mathrm{d}x} \left(\frac{\mathrm{d}\phi_m}{\mathrm{d}x} \right) \mathrm{d}x = \phi_k \frac{\mathrm{d}\phi_m}{\mathrm{d}x} \bigg|_a^b - \int_a^b \frac{\mathrm{d}\phi_m}{\mathrm{d}x} \frac{\mathrm{d}\phi_k}{\mathrm{d}x} \, \mathrm{d}x$$

where a and b are the bounds of the intervals within which the functions ϕ_k and ϕ_m are given.

The product

$$\phi_k(x) \frac{\mathrm{d}\phi_m(x)}{\mathrm{d}x}$$

vanishes if the functions and their derivatives vanish at the boundaries of the interval in which the functions are defined. The

$$\int_a^b \frac{\mathrm{d}\phi_m(x)}{\mathrm{d}x} \frac{\mathrm{d}\phi_k(x)}{\mathrm{d}x} \, \mathrm{d}x$$

integral is clearly independent of the order of the subscripts ϕ_k and ϕ_m.

By using a similar approach it is easy to show that in the multidimensional case $\int \phi_k \hat{H} \phi_m \, dV = \int \phi_m \hat{H} \phi_k \, dV$ if at the bounds of the integration the functions ϕ_k and ϕ_m and their first partial derivatives with respect to every argument all vanish. As is known, this property of the h_{km} matrix elements being independent of the order of the subscripts is called Hermitian, and $\hat{H}$ itself is called a Hermitian operator. Consequently we must require that when choosing the ϕ_k functions they are zero at the boundaries of the domain of significance.

We shall now consider one important aspect of a search for an approximate solution. Let us assume that we have found a function ψ_{approx} which is found to be quite close to the real function of the ground state ψ_0. We shall show that the average value of the $\bar{H}$ energy obtained from this approximate function is then always larger than E_0. In other words, the approximate method of solution only allows us to find an upper estimate of the energy of the ground state. In order to do this, note that any approximate function may be represented as a linear combination of a series of accurate eigenfunctions, i.e.

$$\psi = c_0 \psi_0 + \sum_{k=1}^{\infty} c_k \psi_k \tag{3.4}$$

If the approximate function ψ_{A} is normalized then we must have

$$c_0^2 + \sum_{k=1}^{\infty} c_k^2 = 1 \tag{3.5}$$

This relationship can easily be obtained if expression (3.1) is substituted into the normalization integral, and it is realized that the ψ_0 and ψ_k functions are orthornormal.

We shall calculate the integral

$$\bar{H} = \int \psi \bar{H} \psi \, dV$$

By substituting equation (3.4) and by taking equation (3.5) into account, we obtain

$$\bar{H} = c_0^2 E_0 + \sum_{k=1}^{\infty} c_k^2 E_k = E_0 - \sum_{k=1}^{\infty} c_k^2 E_0 + \sum_{k=1}^{\infty} c_k^2 E_k$$

$$= E_0 + \sum_{k=1}^{\infty} (E_k - E_0) c_k^2 > E_0$$

since $E_k > E_0$ because E_0 corresponds to the ground state.

Thus a search for a solution as described may only give an upper approximate value for the ground state. This solution is closer to the truth, the more flexible the test function, i.e. the more functions there are incorporated into the test function the better they reflect the nature of the problem in question. The flexibility of a function can be increased if the ϕ_k functions are also varied. However, the values of the h_{km} and s_{km} integrals have then to be recalculated at each stage of the solution, since the calculation of these integrals is, as a rule, the most difficult aspect. Varying the ϕ_k is usually not done for polyatomic calculations.

A second limitation inherent in this method and in the test function is associated with the finiteness of our choice of solutions (approximate energy levels and eigenfunctions). The number of these values is equal to the number of functions in the expression for the test function or the dimension of the basis. It is clear that it cannot be too large because a very large matrix would have to be diagonalized and, moreover, the number of h_{km} and s_{km} integrals that must be determined grows rapidly.

Real operators have an infinite number of eigenvalues and eigenfunctions. Thus we must bear in mind that whenever calculating the energy levels of a system with many particles (electronic spectra in particular) we are compelled to limit ourselves to a considerably wider task from the start. Thus we must, for example study the lowest or the highest energy levels and their corresponding possible optical transitions. We shall run into this situation repeatedly in the specialized sections later.

To complete this discussion we wish to dwell on one more method of constructing approximate solutions which specifically allows us a series of cases to choose the form of the test function. This method is to divide the space in which the problem is to be solved into sections so that the whole problem can be solved for each section separately. This approach is convenient when the problem in each section reduces to one that is either already understood or can be solved simply.

The disadvantage of this method, which is often inapplicable in practice, is that the solutions from the separate sections must be 'fastened together'. This 'fastening' has to be done so that the wave function and its derivatives do not lose their continuity at any of the boundaries between the sections. If this is attainable then the solution obtained is exact.

A good approximation to the solution can be achieved if the functions found for each of the sections of space vanish rapidly at the boundaries of their section. The better this condition is fulfilled, the closer the approximate solution is to the truth.

An argument of this kind leads in particular to the choice of test function in the form of linear combinations of atomic orbitals when electronic energy level problems are being solved.

Linear combinations of atomic orbitals (LCAOs) are compiled in the following way. An atomic function is introduced for each atom in the molecule based on the atom's electronic configuration in the free state. The centre of an atomic function is placed at the nucleus of the atom in the molecule. The atomic functions belonging to all the nuclei and chosen in this way form the variational basis which, obviously, is not orthogonal. Note that the introduction of atomic functions that correspond to the ground states of the atoms in their free state, in fact, means that the molecule is formed from unperturbed atoms. It is clear that in a number of cases the concept is not correct, and hence we can only, as a rule, obtain solutions on this basis for the ground and first excited states of the molecule. The atomic basis of the variational problem in which only the atomic orbitals of the ground valency states of the atoms appear is called the 'minimum' basis. As a rule (and in most calculations) this basis is used even though in some cases the atomic orbitals of excited atomic states are used in order to obtain greater accuracy. Such a basis is called an 'extended' one.

In principle, the usual hydrogen-like atomic functions can be used as the atomic orbitals. However, in order to simplify subsequent calculations of matrix elements it is convenient to use, instead, what are called Slater's functions. These differ from the hydrogen-like functions in that their radial part is simplified. They have the form $F_n(r) = r^{n-1}e^{-\xi r}$, where n is the major quantum number and ξ is Slater's parameter and is related to the effective atomic charge. This change in the atomic basis set does not lead to any substantial error in the final result if the basis is large enough. In addition. Gaussian functions have quite commonly been used; these have exponents which depend on the square of the distance between the centres of the functions. The calculation of the matrix element is then even simpler, though the number of Gaussian functions that must be introduced into the calculation is increased. Methods of calculations of different matrix elements are considered in the Appendix.

The choice of a test function for variation in the form of an LCAO can be justified qualitatively as follows. Let us consider a set of atoms that make up a molecule and let us surround each atom with a sphere large enough for the atomic functions of its ground state to be practically zero at the boundary of the sphere. We distribute all the atoms as they are distributed within the molecule, but in such a way that their spheres only touch each other. The quantum-mechanical problem for such a molecule can be solved by dividing all the space into sections—in this case into the spheres surrounding the atoms and the space between them. Then in each region the solution will correspond to a set of atomic orbitals and the general solution will simply be the sum of all the atomic orbitals, taking into account the distribution of their centres.

If the atoms are brought closer so that the interatomic distances approach their true values in the molecule, then the spheres surrounding the atoms, and consequently their atomic wave functions, begin to overlap. If the overlap is not large, then at first we can take the overall wave function as being the sum of the atomic orbitals. The error that arises from this can be reduced by varying the amplitude of the atomic functions, i.e. by multiplying the atomic functions of an appropriate numerical coefficient. As a result, we arrive at a function in the form of an LCAO.

It is clear that from the method of constructing a function in the form of an LCAO it will reproduce the system's true wave function increasingly worse as the overlap of the atomic orbitals of neighbouring atoms increases. Calculations have shown that the approximate nature of a test function itself when in the form of an LCAO is the reason for one of the basic errors that arise when calculating the electron energy levels of polyatomic molecules.

As mentioned earlier, the accuracy of the test function and of the calculations can be improved by introducing a large number of atomic orbitals for each atom. This is first, because when the number of atomic orbitals in the LCAO is increased the test function becomes more flexible, and second, because as the atoms come closer together the mixing of their excited states begins to be an important factor. Unfortunately, in practice the number of atomic orbitals in an LCAO is seldom increased, i.e. a minimum basis set is used because of the large increase in the amount of calculation.

Despite this deficiency, test functions in the form of LCAOs in many cases yield very satisfactory results, and so the method based on them is now widespread.

LCAO functions have a number of important properties, including clarity, transparent physical sense for all the matrix elements that are obtained and the ability therefore to use different semi-empirical methods based on reasonable parameterizations. This very important situation makes calculation methods based on LCAOs very convenient for solving the large number of applied problems encountered in spectrochemistry.

We have considered above the usual method of solving the problem associated with the motion of a single electron in a field of many static nuclei. A new, more exact method for this very important problem has been developed by Novosadov[1,2], who used the Schroedinger equation in a momentum space.

3.2 The Approximation of Non-interacting Electrons

We shall now consider how to calculate the electron states by neglecting the Coulomb repulsion of the electrons. In this case all the electrons are distributed according to Pauli's principle in levels which corresponds to the motion of an unit electron in a field of many positively charged nuclei[3–5]. The Schroedinger equation is

$$\hat{H}\psi = [\hat{T}_e + V_{en}(r_{i\alpha})]\psi = E\psi \tag{3.6}$$

The problem is solved assuming that all the nuclei are in fixed positions corresponding to the equilibrium configuration.

We look for a solution by striving to find limiting values of the average overall energy of $\bar{H}$ with an LCAO–MO test function in the form

$$\psi = \sum_k c_k \chi_k \tag{3.7}$$

Here, χ_k is an atomic wave function (atomic orbital). By substituting equation (3.7) into the expression for $\bar{H}$, we obtain

$$\bar{H} = \sum_k h_{kk} c_k^2 + \sum_{k,l} h_{kl} c_k c_l \tag{3.8}$$

Here

$$h_{kk} = \int \chi_k \hat{H} \chi_k \, \mathrm{d}V$$

$$h_{kl} = \int \chi_k \hat{H} \chi_l \, \mathrm{d}V$$

Hence $\bar{H}$ is a quadratic form with respect to the c_k variables having constant coefficients h_{kk} and h_{kl} called the Coulomb and resonance integrals, respectively.

We introduce a normalizing condition on the test function (3.7) as an additional condition when searching for the limiting values of $\bar{H}$. We obtain

$$1 = \int \psi^2 \, dV = \sum_k c_k^2 \int \chi_k^2 \, dV + \sum_{k,l} c_k c_l \int \chi_k \chi_l \, dV$$

$$= \sum_k c_k^2 + \sum_{k,l}^{k} c_k c_l s_{kl} \tag{3.9}$$

The $s_{kl} = \int \chi_k \chi_l \, dV$ coefficients are called the 'overlap' integrals. If the χ_k functions are orthonormalized, then for each atom $s_{kk} = 1$ while $s_{kl} = 0$.

The normalization condition (3.9) is, like $\bar{H}$, quadratic in form. Consequently the problem reduces to a search for the limiting values of a quadratic form (3.8) on the ellipsoid of equation (3.9), or, what amounts to the same thing, to the simultaneous transformation of these two quadratic forms to a sum of squares, or to the simultaneous diagonalization of two matrices with the elements h_{kl} and s_{kl}. These matrices are symmetric, i.e. $s_{kl} = s_{lk}$ and $h_{kl} = h_{lk}$. The first is clear from the form of the s_{kl} integral, while the second can be seen from the Hermiticity of the $\hat{H}$ operator and the disappearance of the χ_k function at infinity (see Section 3.1). In order for this pair of matrices to be diagonalizable, all the eigenvalues of the S matrix (whose elements are s_{kl}) must be positive. When the test function is chosen in the form of an LCAO, this is always true.

As a result of diagonalizing the H and S matrices we find approximate values of the energy levels E_n of equation (3.6) and approximate value of the $c_k^{(n)}$ coefficients in the LCAO–MO of equation (3.7). The ψ_n functions will also be orthogonal for different sets of $c_k^{(n)}$.

We shall now consider some important properties of the matrix elements s_{kl} and h_{kl}. All the s_{kl} and h_{kl} integrals die away as the distances between the atomic orbital centres grow, and this is shown in Figures 3.1 and 3.2. The relatively rapid fall-off in the s_{kl} integrals with an increase in the distances between the centres of the atomic orbitals means that in approximate methods all the elements of S matrix are set equal to zero with the exception of the diagonal elements, which are set equal to unity. This makes it possible to restrict the diagonalization to only one matrix.

We now turn to an analysis of the physical sense of the Coulomb and resonance integrals. Consider the Coulomb integral, h_{kk}, in more detail, i.e.

$$h_{kk} = \int \chi_k \hat{H} \chi_k \, dV = \int \chi_k \left(\hat{T}_e - \sum_\alpha \frac{Z_\alpha}{r_{i\alpha}} \right) \chi_k \, dV$$

$$= \int \chi_k \left(\hat{T}_e - \frac{Z_{\alpha_k}}{r_{i\alpha_k}} \right) \chi_k \, dV - \sum_{\alpha \neq \alpha_k} \int \chi_k^2 \frac{Z_\alpha}{r_{i\alpha}} \, dV$$

The first term in the sum contains the operator $(\hat{T}_e - Z_{ak}/r_{iak})$ in the integrand and is encountered in the hydrogen-like atom problem. The α_k subscript means

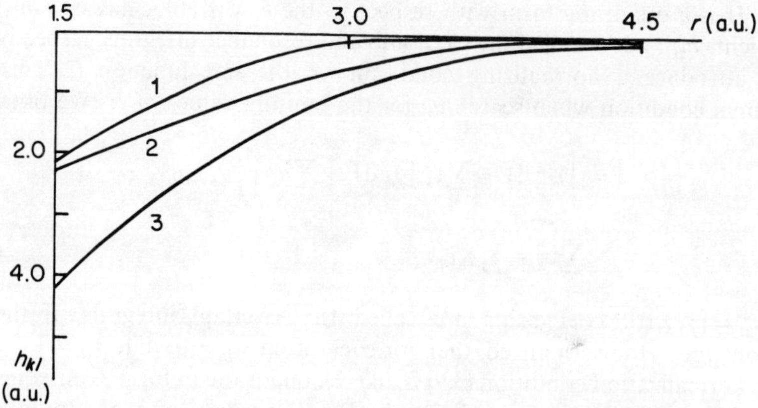

Figure 3.1. Examples of the way bicentric resonance integrals depend on the distance between two atomic orbitals' centres. The integrals were calculated using a Slater parameter equal to 1.62. The charge on the centres is taken to be 4. Curve 1 refers to two $2p_y$ functions; curve 2 refers to a $2s$ and a $2p_x$ function; and curve 3 refers to a two $2s$ function. The x-axis is directed along the straight line connecting the two atomic orbitals' centres

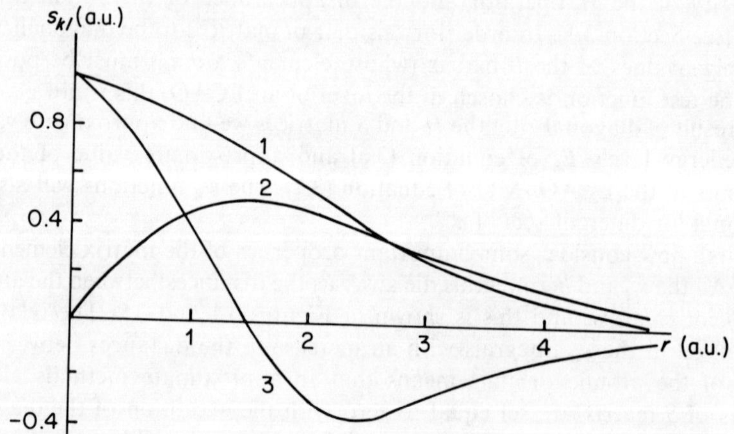

Figure 3.2. Examples of the way the overlap integrals depend on the distance between the centres of atomic orbitals. The integrals were calculated for Slater parameters equal to 1.62. Curve 1 refers to a choice of two $2p_y$ functions; curve 2 to a $2s$ and a $2p_x$ function; and curve 3 to a $2p_x$ and $2p_x$ function. The x-axis is directed along the straight line connecting the two orbitals

that the nucleus is chosen at which the χ_k function is centred. Thus if the kth atomic orbital (χ_k) is chosen as a hydrogen-like orbital then

$$\int \chi_k \left(\hat{T}_e - \frac{Z_{\alpha_k}}{r_{i\alpha_k}} \right) \chi_k \, dV = E_k$$

where E_k is that atomic energy of the electron in the kth state. If χ_k is a Slater orbital then the integral yields the average value of the overall energy of the electron that is distributed about the given nucleus according to the law given by the χ_k function. The integrals that we meet here and below yield the average values of the operators that depend on the normalizing of the wave function. Since this normalization is not essential to later discussions, we can ignore it. Hence the E_k is purely an atomic parameter and is constant in any molecule, but only if the form of χ_k remains unchanged for a given type of atom.

The integral $\int \chi_k^2 Z_a/r_{ia}\,\mathrm{d}V$ has the sense of an average potential energy of the electrostatic attraction of the negative charge, which is continuously distributed in the space about the α_k atom according to the value of the function χ_k^2, for the α nucleus whose charge is Z_α. Clearly, when the distance between the main nucleus and another is large, this integral must simply equal the quantity $Z_\alpha/r_{\alpha_k\alpha}$, where $r_{\alpha_k\alpha}$ is the distance between the two nuclei. This can easily be estimated if it is realized that the integration in fact only contributes to the final result in the region where χ_k^2 is substantially non-zero. If this region is small with respect to the distance to the αth nucleus, then we can set $r_{i\alpha} \approx r_{\alpha_k\alpha}$, i.e.

$$\int \chi_k^2 \frac{Z_\alpha}{i_{i\alpha}}\,\mathrm{d}V \approx \frac{Z_\alpha}{r_{\alpha_k\alpha}} \int \chi_k^2\,\mathrm{d}V = \frac{Z_\alpha}{i_{\alpha_k\alpha}}$$

for atomic orbitals χ_k normalized to unity.

Consequently in complicated molecules we have

$$h_{kk} = E_k - \sum_{\alpha \neq \alpha_k} \int \chi_k^2 \frac{Z_\alpha}{r_{i\alpha}}\,\mathrm{d}V \approx E_k - \sum_{\alpha \neq \alpha_k} \frac{Z_\alpha}{r_{\alpha_k\alpha}} \tag{3.10}$$

It is clear that h_{kk} contains a quantity E_k that is unchanged for any molecule (if the atomic orbital and sort of atom remains the same) and a part that depends on the type of molecule, i.e.

$$\sum_{\alpha \neq \alpha_k} \int \chi_k^2 \frac{Z_\alpha}{r_{i\alpha}}\,\mathrm{d}V \approx \sum_{\alpha \neq \alpha_k} \frac{Z_\alpha}{r_{\alpha_k\alpha}}$$

If we consider a series of molecules that all contain the same atomic grouping, then all the integrals contained in h_{kk} (equation (3.10)) will be the same within the limits of the grouping and will depend relatively little on the surroundings. This opens the way for using a semi-empirical theory for the electronic spectra of molecules which belong to the same class.

We turn now to an analysis of the resonance integrals $h_{kl} = \int \chi_k \hat{H} \chi_l\,\mathrm{d}V$. There are two types of resonance integrals: those for which the χ_k and χ_l are centred on the same nucleus and those for which they are centred on different nuclei.

The first type can be simplified if we use hydrogen-like atomic orbitals. Thus by separating out of the $\hat{H}$ operator the part that corresponds to the hydrogen-like atomic term (i.e. $(\hat{T}_\mathrm{e} - Z_{\alpha_{kl}}/r_{i\alpha_{kl}})$, where α_{kl} is the atom on which the χ_k

and χ_l are centred), we can write

$$h_{kl}=\int\chi_k\left(\hat{T}_e-\frac{Z_{\alpha_{kl}}}{r_{i\alpha_{kl}}}\right)\chi_l\,\mathrm{d}v-\sum_{\alpha\neq\alpha_{kl}}\int\chi_k\chi_l\frac{Z_\alpha}{r_{i\alpha}}\,\mathrm{d}v$$

$$=\int\chi_k E_l\chi_l\,\mathrm{d}v-\sum_{\alpha=\alpha_{kl}}\int\chi_k\chi_l\frac{Z_\alpha}{r_{i\alpha}}\,\mathrm{d}v=-\sum_{\alpha\neq\alpha_k}\int\chi_k\chi_l\frac{Z_\alpha}{r_{i\alpha}}\,\mathrm{d}v \tag{3.11}$$

Here we have taken into account that hydrogen-like atomic orbitals of the same type are orthogonal and hence the first term cancels out.

The $\int\chi_k\chi_l Z_\alpha/r_{i\alpha}$ integral is like a part of a Coulomb integral, which we studied earlier, and represents the electrostatic attraction of the negative charge, which distributed about the α_{kl} nucleus according to the function $\chi_k\chi_l$, for all the remaining nuclei. However, it differs from the Coulomb integral case in that this integral is non-zero only when the distance from the α_{kl} nucleus to another nucleus is small compared with the region in which the $(\chi_k\chi_l)$ product is non-zero. If, however, this condition is not met, then, because the χ_k and χ_l functions are orthogonal, we can write

$$\int\chi_k\chi_l\frac{Z_\alpha}{r_{i\alpha}}\,\mathrm{d}v\approx\frac{Z_\alpha}{r_{\alpha_{kl}\alpha}}\int\chi_k\chi_l\,\mathrm{d}v=0$$

Taking into account what has been said about the resonance integrals of the first type, we can, in approximate calculations, assume them to be zero.

We now consider the second type of resonance integrals, i.e. those which contain atomic orbitals centred on different nuclei. In order to do this we introduce an auxiliary integral, i.e.

$$h'_{kl}=\int(\chi_k+\chi_l)\hat{H}(\chi_k+\chi_l)\,\mathrm{d}v$$

Hence for the h_{kl} resonance integral we obtain

$$h_{kl}=\tfrac{1}{2}(h'_{kl}-h_{kk}-h_{ll})=h_{lk} \tag{3.12}$$

The form and significance of the h_{kk} and h_{ll} integrals we already know, and we rewrite the h_{kl} integral in more detail, separating out of the $\hat{H}$ operator the part containing the α_k and α_l atomic terms, i.e.

$$h'_{kl}=\int(\chi_k+\chi_l)\left(\hat{T}_l-\frac{Z_{\alpha_k}}{r_{i\alpha_k}}-\frac{Z_{\alpha_l}}{r_{i\alpha_l}}\right)(\chi_k+\chi_l)\,\mathrm{d}v$$

$$-\sum_{\alpha\neq\alpha_k\neq\alpha_l}\int(\chi_k+\chi_l)^2\frac{Z_\alpha}{r_{i\alpha}}\,\mathrm{d}v \tag{3.13}$$

The first term in this sum represents the average overall energy of an electron in the $(\chi_k+\chi_l)$ state in a field with two centres, indexed k and l. If the χ_k and χ_l are atomic wave functions then the sum of χ_k and χ_l can be no other than an approximate reflection of one of the possible molecular orbitals of a diatomic molecule or a bond, whilst the integral must be the energy E_{kl} of the electron corresponding to this orbital.

The second term in equation (3.13) yields the sum of the energies of the electrostatic attractions of the negative charge distributed according to the function $(\chi_k + \chi_l)^2$ for the rest of the nuclei. For a nucleus to which the distance is greater than the region in which the $(\chi_k + \chi_l)^2$ function is localized, the electrostatic integral is approximately

$$\frac{(1 + s_{kl})2Z_\alpha}{r_{kl,\alpha}}$$

where $r_{kl,\alpha}$ is the distance from the centre of gravity of the charge $(\chi_k + \chi_l)^2$ to the atom. This can in most cases be considered as being halfway between the k and l nuclei. Thus we have

$$\begin{aligned} h'_{kl} &= E_{kl} - \sum_{\alpha \neq \alpha_k \neq \alpha_l} \int (\chi_k + \chi_l)^2 \frac{Z_\alpha}{r_{i\alpha}} \mathrm{d}V \\ &\approx E_{kl} - \sum_{\alpha \neq \alpha_k \neq \alpha_l} \frac{(1 + s_{kl})2Z_\alpha}{r_{kl,\alpha}} \end{aligned} \tag{3.14}$$

Hence we have (see equation (3.12))

$$h_{kl} \approx \frac{1}{2}\left(E_{kl} - E_k - E_l + \int \chi_k^2 \frac{Z_{\alpha l}}{r_{i\alpha_l}} \mathrm{d}v + \int \chi_l^2 \frac{Z_{\alpha_k}}{r_{i\alpha_k}} \mathrm{d}V - 2s_{kl} \sum_{\alpha \neq \alpha_k \neq \alpha_l} \frac{Z_\alpha}{r_{i\alpha}} \right) \tag{3.15}$$

We have taken into account here that whenever the distance of the α atom is greater than the region in which the $\chi_k\chi_l$ product is localized, then the quantity Z_α/r_{ia} can be taken outside the integral sign.

We mentioned above that the overlap integral s_{kl} dies away rapidly for different centres as the distance between the centres grow. Thus the final term of equation (3.15) vanishes for most resonance integrals.

Equation (3.15) shows that a constant part can be separated out of a h_{kl} integral when a series of molecules retain the α_k and α_l atoms and the same atomic orbitals are used. Thus this constant part can be used as a parameter in a semi-empirical theory. It is interesting that since $s_{kl} < 1$, the influence of the environment on the h_{kl} integral is even less than that on the h_{kk} integral, i.e. its value is even more constant over a series of molecules than that of a Coulomb integral.

The electrostatic parts of the h_{kk} and h_k integrals do not play any important roles. Hence when the values of h_{kk} and h_{kl} are compared they can be ignored, so we have

$$2h_{kl} \approx E_{kl} - E_k - E_l$$

Note that a state with the function $(\chi_k + \chi_l)$, which increases the probability of an electron residing in an internuclear region, is more favourable when the electron moves through a field of two centres than a state for which the electron is localized about one or other of the nuclei. This has been confirmed by quantitative calculations. Thus we must have

$$|E_{kl}| \geqslant |E_k + E_l|$$

and, moreover, as usual, E_k, E_l, $E_{kl} < 0$. It is because of this that the appearance of resonance integrals reduces the overall energy of the molecule compared with that of a system with non-interacting atoms. We indicated at the beginning of this section that a set of limiting values E_n can be found by transforming the quadratic form for $\bar{H}$ (see equation (3.8)) into a sum of squares. The number of these values will equal the number of atomic orbitals in the LCAO. Each of these limiting values corresponds to one of the approximate solutions to equation (3.6) for a given disposition of nuclei. Usually this disposition is taken as being the equilibrium configuration of the molecule in the ground electron state. We obtain the following for each E_n:

$$E_n = \sum_k c_k^{(n)2} h_{kk} + \sum_{k,l} c_k^{(n)} c_l^{(n)} h_{kl} \tag{3.16}$$

In matrix notation equation (3.16) can be written as:

$$E_n = \tilde{c}^{(n)} H c^{(n)} \tag{3.17}$$

where H is a symmetric matrix containing the elements h_{kk} and h_{kl}, and $\tilde{c}^{(n)}$ and $c^{(n)}$ are row and column matrices, respectively, containing the elements $c_k^{(n)}$.

Up to this point we have only dealt with one particle. We shall now consider the problems associated with the presence of many particles in the system but we shall limit ourselves to electrons.

We note first, that formally each electron is assigned a definite number and set of inherent co-ordinates. We can rewrite the Schroedinger equation taking this into account, i.e. an approach to some degree analogous with the classical one. However, it will later be seen that all observable phenomena are independent of such a numeration.

We introduce at this stage the concept of electron spin. The quantum property of an electron which can be interpreted, by analogy with classical mechanics, as its angular momentum is called 'spin'. Any projection of it onto any chosen direction can take one of two values, i.e. $+1/2$ or $-1/2$ (in units of $\hbar$). The state of an electron is characterized by its special co-ordinates and spin. Correspondingly, the wave function of each electron is written as either

$$\psi = \varphi(x, y, z)\alpha$$

or

$$\psi = \varphi(x, y, z)\beta \tag{3.18}$$

Here, $\varphi(x, y, z)$ is the wave function which describes the spatial distribution of the electron while α $(+1/2)$ and β $(-1/2)$ are the alternative spin constituents of the overall wave function. Spin can be considered to be an additional quantum number. The function α and β can be represented as the column matrices

$$\begin{bmatrix} 1 \\ 0 \end{bmatrix} \text{ and } \begin{bmatrix} 0 \\ 1 \end{bmatrix}$$

and are orthogonal.

If the state of many electrons is described with each electron being attributed its own wave function, then such a description is called a 'single-electron' configuration. The whole of the subsequent discussion will generally be conducted on this basis.

An approximate description of the behaviour of many electrons is called the 'independent electrons approximation' if none of the possible interactions between the electrons are taken into account. When we were dealing with one particle then no question about its possible distribution in a given state (energy level) could arise. However, when many particles are being considered, this must obviously arise.

The distribution of electrons among the energy levels follows Pauli's principle. This states that only one electron can occupy a state with a given full set of quantum numbers. This means that no more than two electrons, which must have opposite spins, can occupy a state with a given spatial function (a given energy level). Obviously, this situation must be taken into account when an eigenfunction ψ that describes a group of electrons is constructed.

Pauli's principle is the result of the antisymmetry of a multi-electron function ψ with respect to permutations of electrons. If the order in which a pair of electrons are placed in the multi-electron function ψ is reversed, then the function must also change sign. Thus we have

$$\psi(x_1, y_1, z_1; x_2, y_2, z_2; x_3, y_3, z_3; \ldots) = -\psi(x_2, y_2, z_2; x_1, y_1, z_1; x_3, y_3, z_3; \ldots) \tag{3.19}$$

Here, x_1, y_1, z_1; x_2, y_2, z_2 and x_3, y_3, z_3, etc. are the co-ordinates of the first, second and third, etc. electrons.

We shall use the rules we introduced above and find an expression for a multi-electron wave function, limiting ourselves to the independent electron approximation. In this approximation a molecule's Hamiltonian must clearly not contain terms containing the co-ordinates of two or more electrons simultaneously, or be the sums of separate terms each of which contains the co-ordinates of one electron.

We can therefore write the Hamiltonian of a system with many electrons as

$$\hat{H} = \hat{H}(1) + \hat{H}(2) + \hat{H}(3) + \cdots \tag{3.20}$$

where the parentheses contain the indices of the electrons. The overall eigenfunctions must in this case be a product of the eigenfunctions of the system, i.e. $\psi = \psi_1(1)\psi_2(2)\psi_3(3), \ldots$, where $\psi_1(1)$ is one of the possible solutions of the equation $\hat{H}(1)\psi(1) = E\psi(1)$; $\psi_2(2)$ is one of the possible solutions of the equation $\hat{H}(2)\psi(2) = E\psi(2)$, etc. We choose one of the possible solutions of the equation

$\hat{H}(1)\psi(1) = E\psi(1)$ and designate it $\psi_1(1)$. We do the same for the equation $\hat{H}(2)\psi(2) = E\psi(2)$ calling it $\psi_2(2)$, etc. Thus the function

$$\psi = \psi_1(1)\psi_2(2)\psi_2(3)\cdots \tag{3.21}$$

will yield one of the possible solutions to the equation with the Hamiltonian of equation (3.20).

We note also that the attribution of a number to each electron is a formal operation and that physically the electrons all remain indistinguishable. The physical result must be independent of whether we put the first electron into the ψ_1 state and the second into the ψ_2 state, or vice versa. Thus the overall energies of the system in the states

$$\psi = \psi_1(1)\psi_2(2)\psi_3(3)\cdots \quad \text{and} \quad \psi = \psi_1(2)\psi_2(1)\psi_3(3)\cdots$$

must be the same. In other words, we are dealing with degeneracy with respect to all the possible permutations of the indices of the electrons.

Let us take into account that when the eigenfunction is constructed as a linear combination of all the eigenfunctions which correspond to degenerate energetic states, it will be a solution of the equation whose Hamiltonian is similar to equation (3.20). It only remains for us to determine the coefficients of the linear combination. These can be found if we take advantage of the anti-symmetry requirement for the overall eigenfunction with the same requirement for the overall eigenfunction with respect to switches of the co-ordinates (or indices) of a pair of the electrons. Hence we must write, for example, the following for a system of two electrons:

$$\Psi = \psi_1(1)\psi_2(2) - \psi_1(2)\psi_2(1) \tag{3.22}$$

For a system of three electrons we obtain, respectively:

$$\begin{aligned}\Psi = \psi_1(1)\psi_2(2)\psi_3(3) - \psi_1(2)\psi_2(1)\psi_3(3) + \psi_1(2)\psi_2(3)\psi_3(1)\\ - \psi_1(3)\psi_2(2)\psi_3(1) + \psi_1(3)\psi_2(1)\psi_3(2) - \psi_1(1)\psi_2(3)\psi_3(2)\end{aligned} \tag{3.23}$$

Thus the coefficients in front of the products, which differ by permutations of the electronic indices, are ± 1. It is easy to see that changing the place of a pair of electrons in a function such as equations (3.22) or (3.23) will change the sign of the overall function.

Looking at equations (3.22) and (3.23) we see that they are expansions of the determinants of the matrices

$$\begin{bmatrix} \psi_1(1) & \psi_1(2) \\ \psi_2(1) & \psi_2(2) \end{bmatrix}$$

and

$$\begin{bmatrix} \psi_1(1) & \psi_1(2) & \psi_1(3) \\ \psi_2(1) & \psi_2(2) & \psi_2(3) \\ \psi_3(1) & \psi_3(2) & \psi_3(3) \end{bmatrix}$$

Thus we conclude that the overall multi-electron function for the case of independent electrons must be the determinant of the matrix:

$$\begin{bmatrix} \psi_1(1) & \psi_1(2) & \cdots & \psi_1(n) \\ \psi_2(1) & \psi_2(2) & \cdots & \psi_2(n) \\ \cdots & \cdots & \cdots & \cdots \\ \psi_n(1) & \psi_n(2) & \cdots & \psi_n(n) \end{bmatrix} \tag{3.24}$$

The choice of the functions $\psi_1, \psi_2, \psi_3, \ldots, \psi_n$, etc. determines the electronic configuration of the system.

It is easy to see that this kind of function satisfies Pauli's principle. That is, if two electrons happen to occupy the same state (the spin also being taken into account), then we must have $\psi_1 = \psi_2$ and so two rows of matrix (3.24) must be the same, and thus the determinant is zero. Hence such a situation is prohibited.

The determinant is also zero if there is any dependency between the ψ_k functions. Switching two electrons means that the corresponding columns of the matrix must be switched and the determinant then changes sign.

Thus a function in the form of the determinant of a matrix such as equation (3.24) meets all the requirements imposed on a system of many electrons. It now only remains for us to normalize it. We shall consider this operation for the example of a two-electron function.

We must calculate the integral

$$\begin{aligned} \int \Psi^2 \,\mathrm{d}V &= N^2 \int [\psi_1(1)\psi_2(2) - \psi_1(2)\psi_2(1)]^2 \,\mathrm{d}v_1 \,\mathrm{d}v_2 \\ &= N^2 \left[\int \psi_1^2(1)\psi_2^2(2) \,\mathrm{d}v_1 \,\mathrm{d}v_2 + \int \psi_1^2(2)\psi_2^2(1) \,\mathrm{d}v_1 \,\mathrm{d}v_2 \right. \\ &\quad \left. - 2 \int \psi_1(1)\psi_2(2)\psi_1(2)\psi_2(1) \,\mathrm{d}v_1 \,\mathrm{d}v_2 \right] = 1 \end{aligned}$$

where N is the normalizing factor. The indices within the parentheses ($\psi_1(1)$, for instance) indicate that the function only contains the co-ordinates of that electron (i.e. electron 1). The integration variables can therefore be separated, and we may write

$$\begin{aligned} \int \psi^2 \,\mathrm{d}V = N^2 \left[\int \psi_1^2(1) \,\mathrm{d}v_1 \int \psi_2^2(2) \,\mathrm{d}v_2 + \int \psi_1^2(2) \,\mathrm{d}v_2 \int \psi_2^2(1) \,\mathrm{d}v_1 \right. \\ \left. - 2 \int \psi_1(1)\psi_2(1) \,\mathrm{d}v_1 \int \psi_1(2)\psi_2(2) \,\mathrm{d}v_2 \right] = 1 \end{aligned}$$

If we choose the ψ_1 and ψ_2 functions to be normalized and orthogonal, then the first two integrals become unity and the third vanishes. As a result we obtain $\int \psi^2 \,\mathrm{d}V = 2N^2 = 1$, from which $N = 1/\sqrt{2}$.

In general, when an arbitrary $\int \psi^2 \,\mathrm{d}V$ for which the function is a determinant is calculated we obtain a sum of integrals containing squares of permutations (which all become unity when the ψ_k functions are orthonormalized) and products of two different permutations (which vanish). Hence we conclude that the normalizing multiplier must be $1/\sqrt{n!}$, where n is the number of states among which the electrons are distributed. This implies that all the functions in the determinant are orthonormal. These functions may have the same spatial parts but different spin.

Finally, the wave function for a multi-electron system for the selected electronic configuration has the form

$$\psi = \frac{1}{\sqrt{n!}} \mathrm{Det} \begin{bmatrix} \psi_1(1)\psi_1(2) & \cdots & \psi_1(n) \\ \psi_2(1)\psi_2(2) & \cdots & \psi_2(n) \\ \cdots\cdots & \cdots & \cdots\cdots \\ \psi_n(1)\psi_n(2) & \cdots & \psi_n(n) \end{bmatrix} \tag{3.25}$$

It is easy to get the total energy $\mathscr{E}^{(m)}$ for a selected electronic configuration (determinal function) as a sum of the energies of individual molecular orbitals (MO) taking into account the number of electrons which occupy one MO:

$$\mathscr{E}^{(m)} = \sum_n \sigma_n^{(m)} E_n = \sum_n \sigma_n^{(m)} \tilde{c}^{(n)} H c^{(n)} \tag{3.26}$$

where $\sigma_n^{(m)}$ are the occupation numbers for a chosen electronic configuration (m); this can be 0, 1 and 2, corresponding to the fact that a particular MO can contain 0, 1 or 2 electrons (with opposite spins) at the given one-electron level with function φ_n. The difference between two multi-electron states will be defined only by sets $\sigma_n^{(m)}$, which therefore play the role of original quantum numbers.

Equation (3.26) can be written in a matrix form, convenient for practical calculations:

$$Sp(\Sigma^{(m)}\tilde{C}HC) = Sp(C\Sigma^{(m)}\tilde{C}H) = Sp(P^{(m)}H)$$

where Sp is the trace of matrices product, $\Sigma^{(m)}$ is a diagonal matrix with elements $\sigma_n^{(m)}$, $P^{(m)} = C\Sigma^{(m)}\tilde{C}$ is the density matrix of the chosen multi-electron state, C is the matrix in which the nth column possesses elements $c_k^{(n)}$ in order of enumeration of the AOs and the C matrix represents an eigenvectoral matrix of a pair of matrices H and S.

In order to make clear the physical meaning of the density matrix $P^{(m)}$ let us consider the density of probability distribution of a one-electron state φ_n:

$$\rho_n = \varphi_n^2 = \sum_{k,l} c_k^{(n)} c_l^{(n)} \chi_k \chi_l$$

The total density of many electrons will be equal to the sum of densities of all electrons occupying one of the electronic states, i.e.

$$\rho^{(m)} = \sum_n \sigma_n^{(m)} \rho_n$$

Using matrix symbols this equation can be written in the form

$$\rho^{(m)} = Sp(\Sigma^{(m)}\tilde{C}AC) = Sp(C\Sigma^{(m)}\tilde{C}A) = Sp(P^{(m)}A) \tag{3.27}$$

where A = symmetric matrix with elements $a_{kl} = \chi_k \chi_l$. We can see that $\rho^{(m)}$ is defined only by $P^{(m)}$ because the elements of the A matrix do not depend on the determinantal function selected (electronic configuration).

For non-interacting electrons all electronic configurations which differ only by the spin components of spin–orbitals will have the same energy value. Such a case is called spin degeneracy.

There are two types of electronic configurations. The first has $\sigma_n^{(m)} = 2$ for all occupied MO. For the second type some MO will have $\sigma_n^{(m)} = 1$. These electronic configurations correspond to occupied electronic shells and open ones.

Configurations with $\sigma_n^{(m)} = 2$ do not have spin degeneracy. For the open shells some degree of degeneracy will exist, depending on spin direction and the number of single occupied MO. For example, if there is only one singly occupied MO, two configurations with spin projections 1/2 and $-1/2$ are possible. If there are two such MOs four configurations are possible. For interactions between electrons omitting spin degeneracy partially or completely see Chapter 4.

The formula for a matrix element for a dipole transition between the ψ' and ψ'' states will now be derived. To do so, wave functions for selected electronic states in the form of corresponding determinants should be substituted into the integral

$$\vec{\mu}^{\prime,\prime\prime} = \int \psi' \sum_i \vec{r}_i \psi'' \, \mathrm{d}V$$

If we have spin–orbitals as MO for determinantal functions then the summation over the spin components of spin–orbitals should be done during matrix elements calculations. Such a summation converts to zero all matrix elements for dipole transitions between states with different total spins. This means that an electron should move from one energy level to another without a change in spin.

In the expression for the integral $\vec{\mu}^{\prime,\prime\prime}$ all terms with differring permutations of indices of electrons are equal to zero. Therefore only terms corresponding to the products of the same permutation of indices of electrons of combining functions ψ' and ψ'' will be non-zero. Evidently there will be $n!$ terms for n electrons. Let us take one of them:

$$\int \varphi_{n''_1}(1)\varphi_{n''_2}(2)\cdots \sum_i \vec{r}_i \varphi_{n'_1}(1)\varphi_{n'_2}(2)\cdots \mathrm{d}v_1\,\mathrm{d}v_2\cdots$$

$$= \sum_i \int \varphi_{n''_1}(1)\varphi_{n''_2}(2)\cdots \vec{r}_i \varphi_{n'_1}(1)\varphi_{n'_2}(2)\cdots \mathrm{d}v_1\,\mathrm{d}v_2\cdots \qquad (3.28)$$

Each integral in equation (3.28) for a given index i differs from zero only if functions $\varphi_{n'_j}$ and $\varphi_{n''_j}$, depending on the co-ordinates of electrons with indices not equal to i, will be the same. In the opposite case such integrals become zero because of the orthogonality of the functions φ_n. Hence such a transition is only possible when the space and energy states of a single electron is changed.

In other words, multi-electronic transitions, even between configurations with the same spin, are prohibited.

Because of the indistinguishability of electrons there will be $n!$ identical terms in $\int \varphi_{n''}(1)\vec{r}_1\varphi_{n'}(1)\mathrm{d}v_1$ or, if using a normalizing factor for the product of determinantal functions, we obtain

$$\vec{\mu}^{\prime,\prime\prime} = \int \varphi_{n'}\vec{r}\varphi_{n''}\,\mathrm{d}v \tag{3.29}$$

References

1. B. K. Novosadov. *Optica and Spectr.*, **41**, 832 (1976) (in Russian).
2. B. K. Novosadov, *J. Molec. Struct.*, **52**, 119 (1979).
3. J. C. Slater. *Electronic Structure of Molecules*, McGraw-Hill, New York (1963).
4. T. E. Peacock. *Electronic Properties of Aromatic and Heterocyclic Molecules*, Academic Press, London/New York (1965).
5. R. McWeeny and B. T. Sutcliffe. *Methods of Molecular Quantum Mechanics*, Academic Press, London/New York (1969).

Chapter 4

General Solution of the Polyelectronic Problem

4.1 The Method of Configuration Interaction

Until now we have only considered interactions between electrons and nuclei when we studied the motion of the electrons in a polyatomic molecule, and we discarded that part of the Schroedinger equation which corresponded to elctron–electron repulsions. In this chapter we shall present the principles for solving the electronic problems when these repulsions are taken into account. We start by considering the method that is based on searching for a solution in the form of a linear combination of determinantal functions (configurations)[1].

We shall consider a number of determinantal functions which contain a given selection of single-electron wave functions. One electron occupies each single-electron state. We shall, for completeness, use a spin–orbital, that is, a function which contains a spatial and spin part. To shorten the notation, however, we shall not write out the spatial and spin parts separately, and shall use the single symbol, ϕ_m. Each of these determinants is an eigenfunction of the Schroedinger equation written without taking any account of the electron–electron interacions. If we compile these determinants for all possible, or a very large number of the possible eigenfunctions, which must be of the same species, then the solution to the equation that does account for electron–electron interactions may be obtained in the form of a linear combination of the original determinantal functions:

$$\Psi = \sum_k c_k \phi_k \tag{4.1}$$

where ϕ_k are the determinant functions (electronic configurations).

Each of these determinantal functions involves the problem of searching for the limiting values of a quadratic form

$$\bar{H} = \sum_{k,n} c_k c_n \int \phi_k \hat{H} \phi_n \, \mathrm{d}V \tag{4.2}$$

in a sphere with unit radius.

An overlap integral, $\int \phi_k \phi_n \, \mathrm{d}V$, is zero if different determinantal functions are contained within the integral and unity if the determinantal functions are the same. Until now we have assumed that each determinantal function is one of the eigenfunctions of the multi-electron problem without the electron–electron repulsions being allowed for, this follows directly from the orthogonality of all the possible determinantal functions. This automatically occurs if the determinantal functions are made up of functions that are eigenfunctions of the problem associated with the motion of a single electron in a field of many static nuclei.

Thus the search for the limiting values of the quadratic equation (4.2) reduces to the diagonalization of the H matrix, which contains the elements

$$h_{kn} = \int \phi_k \hat{H} \phi_n \,\mathrm{d}V = \int \phi_k \hat{h} \phi_n \,\mathrm{d}V + \int \phi_k \hat{g} \phi_n \,\mathrm{d}V$$

In this equation the symbol $\hat{h}$ stands for the single-electron part, i.e. that part of the complete Hamiltonian of the molecule which corresponds to the motion of a single electron in a field of nuclei. In discrete form this expression is

$$\hat{h} = \sum_i \hat{T}_{\mathrm{e}}(i) + \sum_{i,\alpha} V(r_{i\alpha}) = \sum_i \hat{h}(i)$$

The symbol $\hat{g}$ stands for the sum

$$\hat{g} = \frac{1}{2}\sum_{i,j}\frac{1}{r_{ij}} = \sum_{i,j} V(r_{ij})$$

which corresponds to the Coulomb repulsion of electrons.

We shall start by considering those matrix elements which only contain operators corresponding to the kinetic energies of the electrons and to the attractions of the electrons to the nuclei, i.e. integrals of the form $\int \phi_k \hat{h} \phi_n \,\mathrm{d}V$. If the determinantal function is composed of solutions of the single-electron problems which allows for the electrons to be distributed among the states, then the function will itself be a solution to the problem with the operator $\hat{h}$. Consequently $\hat{h}\phi_n = E_n \phi_n$ and $\int \phi_k \hat{h} \phi_n \,\mathrm{d}V = E_n \int \phi_k \phi_n \,\mathrm{d}V = 0$ when $k \neq n$ and E_n when $k = n$. Thus, due to the orthogonality of the determinant eigenfunctions, all the single-electron parts in the non-diagonal matrix elements will be zero.

Before we continue we should look at the role that spin part plays in the spin–orbitals that form the determinantal functions.

If the determinant is written out in detail then each of the terms is a product of spin–orbitals. Since a spin–orbital is itself a product of a spatial part and a spin part, by summation of the spin and spatial parts we obtain products of spin functions and spatial functions for each term in the determinant. That is, we shall obtain products of the form $\prod_k \varphi_k(i) \prod_k s_k(i)$, where φ_k and s_k are the spatial and spin parts, respectively, of a spin–orbital. The components of the determinantal function differ by permutations of the index of the i electrons. If we look at the matrix elements between any two determinant functions of an operator $\hat{A}$ which acts solely upon the spatial part of the eigenfunctions (and all the operators we shall be dealing with belong to this category, e.g. $\hat{H}$, $\vec{\mu}$, etc.), then these matrix elements will be linear combinations of terms of the form

$$\int \prod_k \varphi_k''(i) \hat{A} \prod_k \varphi_k'(i) \prod_k s_k''(i) s_k'(i) \, \mathrm{d}V \mathrm{d}s \tag{4.3}$$

The prime and double primes refer to the different combining determinantal functions. When working out the matrix elements, the spatial parts of the spin–orbitals are integrated, whilst spin variables with the same index are summed. If the spin states of two electronic configurations are different then one of the products of the spin constituents will vanish for any switch in the indices of the electrons in the spin parts of the spin–orbitals. Two single-electron states differing by the projections of their spins are orthogonal. This also means that formally they will vanish in the sum taken over the spin variables in equation (4.3).

Consequently all the elements of the H matrix which correspond to pairs of combining determinant eigenfunctions with different overall spins will, in the configuration interaction method, be zero. It follows that when solving a problem using this method (for configurations in the form of linear combinations of determinants), only configurations with the same overall spin must be used.

This rule, governing the disappearance of matrix elements between determinantal functions, is retained unchanged even when the determinants are constructed from self-consistent orbitals (see Section 4.3).

The overall task of calculating the limiting values of the quadratic form $\bar{H}$ breaks down, as a result, into a series of problems for each of the possible spin states. The number of determinantal functions in the linear combination is correspondingly reduced.

We turn now to the calculation of matrix elements between determinantal functions for the two-electron terms of the Schroedinger operator, i.e. matrix elements of the form

$$\int \phi' \hat{g} \phi'' \, \mathrm{d}V = \frac{1}{2} \sum_{i,j} \int \phi' \frac{1}{r_{ij}} \phi'' \, \mathrm{d}V \tag{4.4}$$

where ϕ' and ϕ'' are determinantal functions constructed of orthonormal spin–orbitals. In order to make clear the various manipulations, we shall consider the special case of three electrons which are distributed among three spin–orbitals, corresponding to two different electron configurations. In order to abbreviate the notation, we shall designate these spin–orbitals a', b' and c', for the first configuration, and a'', b'' and c'' for this second.

Thus, we obtain the following matrix element for the interaction of the two configurations

$$\begin{aligned} \int \phi' \hat{g} \phi'' \, \mathrm{d}V = \frac{1}{6} \int & [a'(1)b'(2)c'(3) + c'(1)a'(2)b'(3) \\ & + b'(1)c'(2)a'(3) - b'(1)a'(2)c'(3) - c'(1)b'(2)a'(3) \\ & - a'(1)c'(2)b'(3)] \left(\frac{1}{r_{12}} + \frac{1}{r_{13}} + \frac{1}{r_{23}} \right) \\ & \times [a''(1)b''(2)c''(3) + c''(1)a''(2)b''(3) + \cdots] \, \mathrm{d}V_1 \, \mathrm{d}V_2 \, \mathrm{d}V_3 \end{aligned} \tag{4.5}$$

The factor 1/6 arises due to the normalization of the determinantal functions ($N = 1\sqrt{n!}$, where n is the number of spin–orbitals). We separate out of equation (4.5) the term containing $1/r_{12}$ and group the terms in the determinants thus:

$$\begin{aligned}\{[a'(1)b'(2) - b'(1)a'(2)]c'(3) + [c'(1)a'(2) - a'(1)c'(2)]b'(3) \\ + [b'(1)c'(2) - c'(1)b'(2)]a'(3)\} \\ = \{d'(1,2)c'(3) + e'(1,2)b'(3) + f'(1,2)a'(3)\}\end{aligned} \tag{4.6}$$

and similarly for the second term. Thus we find that the first term containing $1/r_{12}$ in equation (4.5) is

$$\begin{aligned}\frac{1}{6}\Bigg\{ &\int d'(1,2)d''(1,2)\frac{1}{r_{12}}\mathrm{d}V_1\,\mathrm{d}V_2 \int c'(3)c''(3)\,\mathrm{d}V_3 \\ &+ \int e'(1,2)e''(1,2)\frac{1}{r_{12}}\mathrm{d}V_1\,\mathrm{d}V_2 \int b'(3)b''(3)\,\mathrm{d}V_3 \\ &+ \int f'(1,2)f''(1,2)\frac{1}{r_{12}}\mathrm{d}V_1\,\mathrm{d}V_2 \int a'(3)a''(3)\,\mathrm{d}V_3 \\ &+ \int d'(1,2)e''(1,2)\frac{1}{r_{12}}\mathrm{d}V_1\,\mathrm{d}V_2 \int c'(3)b''(3)\,\mathrm{d}V_3 \\ &+ \int e'(1,2)d''(1,2)\frac{1}{r_{12}}\mathrm{d}V_1\,\mathrm{d}V_2 \int b'(3)c''(3)\,\mathrm{d}V_3 \\ &+ \int d'(1,2)f''(1,2)\frac{1}{r_{12}}\mathrm{d}V_1\,\mathrm{d}V_2 \int c'(3)a''(3)\,\mathrm{d}V_3 \\ &+ \int f'(1,2)d''(1,2)\frac{1}{r_{12}}\mathrm{d}V_1\,\mathrm{d}V_2 \int a'(3)c''(3)\,\mathrm{d}V_3 \\ &+ \int e'(1,2)f''(1,2)\frac{1}{r_{12}}\mathrm{d}V_1\,\mathrm{d}V_2 \int b'(3)a''(3)\,\mathrm{d}V_3 \\ &+ \int f'(1,2)e''(1,2)\frac{1}{r_{12}}\mathrm{d}V_1\,\mathrm{d}V_2 \int a'(3)b''(3)\,\mathrm{d}V_3 \Bigg\}\end{aligned} \tag{4.7}$$

If the combining electronic states contain different and orthogonal orbitals (the orthogonality is automatically guaranteed if the eigenfunctions of the single-electron problem are used), then equation (4.7) vanishes.

This conclusion is true even in the more general case, and enables us to formulate a rule which states that matrix elements of configuration interaction will be non-zero when they differ from each other by no more than two spin–orbitals. If $a' = a''$, but $b' \neq b''$ and $c' \neq c''$ (and, of course, $a' \neq b' \neq c'$ and $a'' \neq b'' \neq c''$), then equation (4.7) will simplify to

$$\frac{1}{6}\int f'(1,2)\,f''(1,2)\frac{1}{r_{12}}\,\mathrm{d}V_1\,\mathrm{d}V_2$$
$$=\frac{1}{6}\int [b'(1)c'(2)-c'(1)b'(2)]\frac{1}{r_{12}}[b''(1)c''(2)-c''(1)b''(2)]\,\mathrm{d}V_1\,\mathrm{d}V_2$$
$$=\frac{1}{6}\int [b'(1)b''(1)c'(2)c''(2)+c'(1)c''(1)b'(2)b''(2)]\frac{1}{r_{12}}\,\mathrm{d}V_1\,\mathrm{d}V_2 \qquad (4.8)$$
$$-\frac{1}{6}\int [b'(1)c'(2)c''(1)b''(2)+c'(1)b''(1)c''(2)b'(2)]\frac{1}{r_{12}}\,\mathrm{d}V_1\,\mathrm{d}V_2$$
$$=\frac{1}{3}\int [b'(1)b''(1)c'(2)c''(2)-b'(1)c''(1)c'(2)b''(2)]\frac{1}{r_{12}}\,\mathrm{d}V_1\,\mathrm{d}V_2$$

Here, we have taken into account that the products of the spin–orbitals in the square brackets only differ by permutations of electronic indices and that upon integration they yield similarly valued integrals.

If $a'=a$ and $b'=b''$, but $c'\neq c''$, then equation (4.7) takes the form:

$$\frac{1}{6}\left\{\int e'(1,2)e''(1,2)\frac{1}{r_{12}}\,\mathrm{d}V_1\,\mathrm{d}V_2+\int f'(1,2)\,f''(1,2)\frac{1}{r_{12}}\,\mathrm{d}V_1\,\mathrm{d}V_2\right\}$$
$$=\frac{1}{3}\int [c'(1)c''(1)a'(2)a''(2)-c'(1)a''(1)a'(2)c''(2)]\frac{1}{r_{12}}\,\mathrm{d}V_1\,\mathrm{d}V_2 \qquad (4.9)$$
$$+\frac{1}{3}\int [b'(1)b''(1)c'(2)c''(2)-b'(1)c''(1)c'(2)b''(2)]\frac{1}{r_{12}}\,\mathrm{d}V_1\,\mathrm{d}V_2$$

When all the spin–orbitals in the two determinants are pairwise the same, i.e. $a'=a''=a$, $b'=b''=b$ and $c'=c''=c$, then equation (4.7) for a diagonal element of H becomes

$$\frac{1}{6}\left\{\int d^2(1,2)\frac{1}{r_{12}}\,\mathrm{d}V_1\,\mathrm{d}V_2+\int e^2(1,2)\frac{1}{r_{12}}\,\mathrm{d}V_1\,\mathrm{d}V_2+\int f^2(1,2)\frac{1}{r_{12}}\,\mathrm{d}V_1\,\mathrm{d}V_2\right\}$$
$$=\frac{1}{3}\left\{\int [a^2(1)b^2(2)-a(1)b(1)a(2)b(2)]\frac{1}{r_{12}}\,\mathrm{d}V_1\,\mathrm{d}V_2\right.$$
$$+\int [a^2(1)c^2(2)-a(1)c(1)a(2)c(2)]\frac{1}{r_{12}}\,\mathrm{d}V_1\,\mathrm{d}V_2 \qquad (4.10)$$
$$\left.+\int [b^2(1)c^2(2)-b(1)c(1)b(2)c(2)]\frac{1}{r_{12}}\,\mathrm{d}V_1\,\mathrm{d}V_2\right\}$$

The first integrals in each term of equation (4.10) are called Coulomb two-electron integrals. They correspond to the Coulomb energy of repulsion of two like charges distributed in space with densities given by the functions a^2, b^2 and c^2, respectively. The second integral in each term are exchange integrals and arise due to the required antisymmetry of the overall electronic function.

Since it is necessary to sum the electron-spin projections, in addition to the integration, we find that if the spin–orbitals in the exchange integrals have opposite spins, then the exchange integral must vanish.

We have calculated one component in the overall matrix element, the one corresponding to $1/r_{12}$. The other two terms containing $1/r_{13}$ and $1/r_{23}$ are calculated in the same way, by combining the terms in the determinants so that functions containing the indices of the second electron (if the integral contains the $1/r_{13}$) or the first electron (if the integral contains the $1/r_{23}$) are separated out. For example, for the term containing the $1/r_{13}$ we write the determinant as

$$\{[a'(1)c'(3) - c'(1)a'(3)]b'(2) + [c'(1)b'(3) - b'(1)c'(3)]a'(2) \\ + [b'(1)a'(3) - a'(1)b'(3)]c'(2)\} \tag{4.11}$$

Equation (4.11) is exactly the same as equation (4.6) if the electron indices are changed. Since this operation can have no effect on the values of the relevant integrals, we must obtain constituent matrix elements that are identical to those found for the constituents with $1/r_{12}$. This also remains true for the $1/r_{23}$ constituents.

Summing three identical expressions leads to the disappearance of the factor 1/3 and finally from equations (4.8), (4.9) and (4.10) we obtain the following expressions for the two-electron parts of the non-zero matrix elements between two determinants:

(1)
$$\int\left[b'(1)c'(2)\frac{1}{r_{12}}b''(1)c''(2) - b'(1)c'(2)\frac{1}{r_{12}}c''(1)b''(2)\right]\mathrm{d}V_1\,\mathrm{d}V_2 \tag{4.12}$$

if $a' = a'' = a$, $b' \neq b''$ and $c' \neq c''$;

(2)
$$\int\left[a(1)c'(2)\frac{1}{r_{12}}a(1)c''(2) - a(1)c'(2)\frac{1}{r_{12}}a(2)c''(1)\right]\mathrm{d}V_1\,\mathrm{d}V_2 \\ + \int\left[b(1)c'(2)\frac{1}{r_{12}}b(1)c''(2) - b(1)c'(2)\frac{1}{r_{12}}b(2)c''(1)\right]\mathrm{d}V_1\,\mathrm{d}V_2 \tag{4.13}$$

if $a' = a'' = a$, $b' = b'' = b$ but $c' \neq c''$;

(3)
$$\int\left[a^2(1)\frac{1}{r_{12}}b^2(2) - a(1)b(1)\frac{1}{r_{12}}a(2)b(2)\right]\mathrm{d}V_1\,\mathrm{d}V_2 \\ + \int\left[a^2(1)\frac{1}{r_{12}}c^2(2) - a(1)c(1)\frac{1}{r_{12}}a(2)c(2)\right]\mathrm{d}V_1\,\mathrm{d}V_2 \\ + \int\left[b^2(1)\frac{1}{r_{12}}c^2(2) - b(1)c(1)\frac{1}{r_{12}}b(2)c(2)\right]\mathrm{d}V_1\,\mathrm{d}V_2 \tag{4.14}$$

if $a' = a'' = a$, $b' = b'' = b$ and $c' = c'' = c$.

Expressions such as equations (4.12), (4.13) and (4.14) remain true for the general case of arbitrary order determinants. By using these formulae, we can find matrix elements for interacting determinants (configurations) for any concrete case.

The larger the number of terms in the linear combination of determinants the more accurate will be the solution. We have here an apparent analogy with the linear combinations of atomic orbitals, but we should bear in mind that atomic orbitals are not solutions of the Schroedinger equation of the molecule whilst the determinants are solutions, albeit for the special case of no electron–electron interaction. For this reason, we can expect that by considerably increasing the number of determinantal functions in the linear combinations we can obtain a very close approximation to the exact result. Obviously, in practical calculations the number of terms in the linear combinations of determinants must be limited, and so we shall only obtain an approximate solution.

Each individual determinant corresponds to one definite energy of the non-interacting electrons or to a definite energy of the electron configuration. Accounting for electron–electron interactions will yield new values for the energy which are comparatively close to the values of the energy of the electron configurations that do not allow for these interactions. Thus, if, using the configuration interaction method, we hope to obtain accurate values of the energies and eigenvalues for a molecular state, then we must take a linear combination of the determinantal functions of the initial approximation (i.e. the one that does not allow for electron–electron interactions) which are close in value to the real state.

It can be shown that if functions corresponding to energetic states far from the one being sought are contained in the linear combination of functions, then their contribution is small. This then makes it possible in practice to truncate the length of the linear combination of the determinantal functions if the task is only to calculate the energies and eigenfunctions of individual energetic states. For a number of reasons, which we shall go into below, such calculations can only usually be done for the low-energy states. In these calculations only configurations which possess the same overall spin are included in the linear combination of determinants.

4.2 Spin-degeneracy Multiplets

Let us assume that all the inner orbitals of a molecule are occupied by electrons in pairs with opposite spins, whilst the two outer orbitals each have one electron. This state corresponds to the excitation of one of the electrons. Thus, depending on the direction of spin, we can obtain four possible electronic configurations, which are conventionally as shown in Figures 4.1. The overall spin's projections of the first and second configurations are, respectively, plus and minus unity, while they are zero for the third and fourth.

If we do not take the electron–electron interactions into account, then the energies of the molecule in all these configurations are the same. Thus we have fourfold spin degeneracy. Let us consider the case of four electrons distributed among three energy states.

In order to solve the problem taking the electron–electron interactions into account we compile a linear combination of all four determinantal functions for the configurations given in Figure 4.1. The relevant spin–orbitals have the form

$$\varphi_1(1)\alpha(1);\quad \varphi_1(2)\beta(2);\quad \varphi_2(3)\alpha(3);\quad \varphi_3(4)\alpha(4);$$

for the first configuration;

$$\varphi_1(1)\alpha(1);\quad \varphi_1(2)\beta(2);\quad \varphi_2(3)\beta(3);\quad \varphi_3(4)\beta(4);$$

for the second;

$$\varphi_1(1)\alpha(1);\quad \varphi_1(2)\beta(2);\quad \varphi_2(3)\alpha(3);\quad \varphi_3(4)\beta(4);$$

for the third; and

$$\varphi_1(1)\alpha(1);\quad \varphi_1(2)\beta(2);\quad \varphi_2(3)\beta(3);\quad \varphi_3(4)\alpha(4);$$

for the fourth. Here, φ represents the space component of a spin–orbital and α and β are the spin components (α corresponding to the spin projection of $+1/2$ and β to the spin projection of $-1/2$).

The indices in the parentheses are those of the electrons for the permutation. The first two configurations have overall spin projections of $+1$ and -1, respectively. The third and fourth configurations have zero overall spin projections.

We obtain the following for the two-electron parts of the diagonal elements of the H-matrix when using linear combinations of the determinantal functions based on equation (4.14):

$$\begin{aligned} g_{11} = g_{22} = & \int \varphi_1^2(1)\varphi_1^2(2)\frac{1}{r_{12}}\,\mathrm{d}V_1\,\mathrm{d}V_2 + \int \varphi_1^2(1)\varphi_2^2(3)\frac{1}{r_{13}}\,\mathrm{d}V_1\,\mathrm{d}V_3 \\ & + \int \varphi_1^2(1)\varphi_3^2(4)\frac{1}{r_{14}}\,\mathrm{d}V_1\,\mathrm{d}V_4 + \int \varphi_1^2(2)\varphi_2^2(3)\frac{1}{r_{23}}\,\mathrm{d}V_2\,\mathrm{d}V_3 \\ & + \int \varphi_1^2(2)\varphi_3^2(4)\frac{1}{r_{24}}\,\mathrm{d}V_2\,\mathrm{d}V_4 + \int \varphi_2^2(3)\varphi_3^2(4)\frac{1}{r_{34}}\,\mathrm{d}V_3\,\mathrm{d}V_4 \qquad (4.15) \\ & - \int \varphi_1(1)\varphi_2(3)\varphi_1(3)\varphi_2(1)\frac{1}{r_{13}}\,\mathrm{d}V_1\,\mathrm{d}V_3 \\ & - \int \varphi_1(1)\varphi_3(4)\varphi_1(4)\varphi_3(1)\frac{1}{r_{14}}\,\mathrm{d}V_1\,\mathrm{d}V_4 \\ & - \int \varphi_2(3)\varphi_3(4)\varphi_2(4)\varphi_3(3)\frac{1}{r_{34}}\,\mathrm{d}V_3\,\mathrm{d}V_4 \\ = & J_{11} + 2J_{12} + 2J_{13} + J_{23} - K_{12} - K_{13} - K_{23} \end{aligned}$$

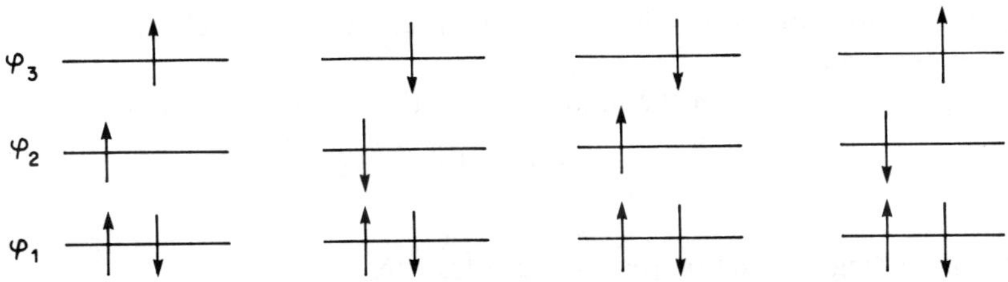

Figure 4.1. Spin distributions for four electrons

We have introduced here the notation

$$J_{nm} = \int \varphi_n^2(1)\varphi_m^2(2)\frac{1}{r_{12}}\,\mathrm{d}V_1\,\mathrm{d}V_2$$

for the Coulomb integrals and

$$K_{nm} = \int \varphi_n(1)\varphi_m(1)\varphi_n(2)\varphi_m(2)\frac{1}{r_{12}}\,\mathrm{d}V_1\,\mathrm{d}V_2$$

for the exchange integrals, and have taken into account that the values of these integrals do not depend on the indices of the electrons.

In order to abbreviate the notation we designate

$$J_{11} + 2J_{12} + 2J_{13} + J_{23} - K_{12} - K_{13} = J - K$$

Hence

$$g_{11} = J - K - K_{23}$$

For the g_{33} and g_{44} elements we obtain

$$g_{33} = g_{44} = J_{11} + 2J_{12} + 2J_{13} + J_{23} - K_{12} - K_{13} = J - K \tag{4.16}$$

Since the projections of the overall spins of the first, second and third (fourth) configurations are different, the non-diagonal elements

$$g_{12} = g_{13} = g_{14} = g_{23} = g_{24} = 0$$

The third and fourth configurations differ by two spin–orbitals (we have the states $\varphi_2\alpha$ and $\varphi_3\beta$ for the third configuration and $\varphi_2\beta$ and $\varphi_3\alpha$ for the fourth). Thus the matrix element h_{34} must be calculated according to equation (4.12) and equals

$$h_{34} = g_{34} = -\int \varphi_2(1)\varphi_3(1)\varphi_2(2)\varphi_3(2)\frac{1}{r_{12}}\,\mathrm{d}V_1\,\mathrm{d}V_2 = -K_{23} \tag{4.17}$$

In this way, the H-matrix for the interaction case of four configurations has the form

$$\mathrm{H} = \begin{bmatrix} E + J - K - K_{23} & 0 & 0 & 0 \\ 0 & E + J - K - K_{23} & 0 & 0 \\ 0 & 0 & E + J - K & -K_{23} \\ 0 & 0 & -K_{23} & E + J - K \end{bmatrix} \tag{4.18}$$

E is the energy of the configuration without taking the electron–electron interaction into account.

The block in the box in the matrix of equation (4.18) is symmetric,

$$\begin{bmatrix} a & b \\ b & a \end{bmatrix}$$

and can be diagonalized by the standard transformation:

$$C = (1/\sqrt{2})\begin{bmatrix} 1 & -1 \\ 1 & 1 \end{bmatrix}$$

(see Chapter 5), which changes its form from

$$\begin{bmatrix} a & b \\ b & a \end{bmatrix}$$

to the matrix

$$\begin{bmatrix} a+b & 0 \\ 0 & a-b \end{bmatrix}$$

By using this transformation we find the following expressions for the energies and eigenfunctions of the four states taking electron–electron interactions into account:

$$\begin{aligned} &E_1 = E + J - K - K_{23}, \quad &&\Psi_1 = \Phi_1, \quad &&S_z^{(1)} = 1; \\ &E_2 = E + J - K - K_{23}, \quad &&\Psi_2 = \Phi_2, \quad &&S_z^{(2)} = -1; \\ &E_3 = E + J - K - K_{23}, \quad &&\Psi_3 = \tfrac{1}{2}(\Phi_3 + \Phi_4) \quad &&S_z^{(3)} = 0; \\ &E_4 = E + J - K + K_{23}, \quad &&\Psi_4 = \tfrac{1}{2}(\Phi_3 - \Phi_4) \quad &&S_z^{(4)} = 0 \end{aligned} \tag{4.19}$$

Here, the Φ_n are the determinantal functions of the initial approximation (without the electron–electron interaction).

It can be seen that $E_1 = E_2 = E_3 = E_{\text{trip}}$ and $E_4 = E_{\text{singlet}} > E_{\text{trip}}$. A triply degenerate state (triplet) and a non-degenerate state (singlet) are obtained. The eigenfunctions of the triplet have projections $S_z = \pm 1$ and 0, whilst the eigenfunction has a projection of the overall spin equal to zero. The result $E_{\text{sing}} > E_{\text{trip}}$ is a special case of Houdin's rule. This states that the electronic state with the greatest spin has lower energy than the state with the least.*

We have considered a special case which corresponds to partially filled shells. Such unfilled shells are obtained when one electron in a previously full shell is excited and jumps to the next possible energy level. Consequently, the singlet and triplet in this case are excited states in our system.

Usually it is assumed that the pairs of electrons are distributed in all the spin–orbitals with opposite spins in the ground state of a neutral molecule. Any redistribution of the electrons to higher spin–orbitals corresponds to their

*Note that the ground state in molecules is very often a singlet state. Thus Hund's rule only applies to excited states.

being excited. Singlet and triplet states are most often encountered because they are obtained when only one electron is excited, thus not requiring too great an energy. Given a large amount of spin interaction in the molecule or repeated excitation, more complicated distributions of electrons among the states are possible. The degeneracy with respect to spin then grows and states with larger degrees of degeneracy (multiplet states) appear in addition to triplet states.

4.3 The Matrix Method of Constructing Spin Multiplets

Multiplet distributions can be studied and the necessary configurations can be derived for more complicated cases by considering their corresponding matrices[2]. In this section we shall look at this method in some detail.

The arrangement of N electrons over m orbitals following Pauli's principle with either one or two electrons in each energy level for a given overall spin for the molecule results in a very large number of configurations. A complete enumeration of these configurations is a very time-consuming task, and the use of a considerable number of configurations is rendered impossible due to the limitations imposed by the array dimensions of a computer and the time available for calculating the matrix elements.

At the same time, formally enumerating all the configurations suffers from an excess of information which is not really needed to construct a molecule's spectrum. As a rule, this spectrum corresponds to transitions between levels having spin projections of $S_z = 0, \pm 1/2, \pm 1$, for multiplets with overall spins of

$$S^2 \leqslant \frac{N}{2}\left(\frac{N}{2}+1\right)$$

Thus in practice the necessity of considering overall spin projections greater in modulus than unity falls away. Moreover, configurations in which the lower orbitals are empty whilst the upper ones are occupied, and the differences between them corresponding to X-ray quanta energies, clearly need not be considered because they do not correspond to any physically excited molecule in the optical waveband. Accordingly, we can restrict ourselves to modelling molecular electronic shells consisting of m lower doubly filled orbitals and n singly filled orbitals such that $N = 2m + n$.

We shall compile the linear combination of configurations (LCC) by a direct matrix method leading to results obtainable traditionally using the group representation theory.

All the results will be consequences of the properties of the system of matrices in question in which the permutation symmetry of the electronic system is reflected. Owing to this property, it becomes partially possible to diagonalize the Hamiltonian matrix, thus dividing the full matrix into blocks that correspond to definite projections of the overall spin of the molecule's electrons. This procedure gives the process of searching for an LCC particularly clarity obvious.

4.3.1 The general scheme for singly filled orbitals

Consider a molecule with N electrons which populate N different single-electron orbitals. The molecule's Hamiltonian in the purely electronic approximation includes single-electron operators of the interactions of the electrons with the molecule's nuclei and the Coulomb electron–electron repulsions.

As we have already shown, the wave function Ψ is constructed in the form of an LCC by the configuration interaction method, i.e.

$$\Psi = \sum_k C_k \phi_k$$

where ϕ_k is a Slater determinant that is complied from N spin–orbitals and the C_k are the coefficients of the LCC which should be determined. The ϕ_k determinantal functions are orthogonal with respect to spin.

Using either the variational procedure or the Bubnov–Galerkin method[2] we can obtain an algebraic system of equations for the C_k coefficients. The matrix elements of the basis functions ϕ_k can be calculated using the rules we have already considered.

The basis configurations ϕ_k for a given set of N spin–orbitals correspond to states of the molecule that are degenerate with respect to energy when electronic interactions are ignored. Allowing for the two-spin directions of each electron, we have 2^N configurations in all. Since the number of configurations grows very rapidly as the number of electrons increases, it is possible to show the electron configurations graphically as a tree, and the way the branches divide can be understood from Figure 4.2, which is for the case of $N = 4$.

An individual configuration corresponds to the graph that connects a series of edges upwards. The position of an edge in a given layer of the graph corresponds to a single-electron level occupied by an electron with the given spin projection. In this way, all the possible vertical graphs of a given tree reflect all the electronic configurations of the system. An upward arrow corresponds to spin 1/2 and a downward one to spin $-1/2$. The sum of the projections of

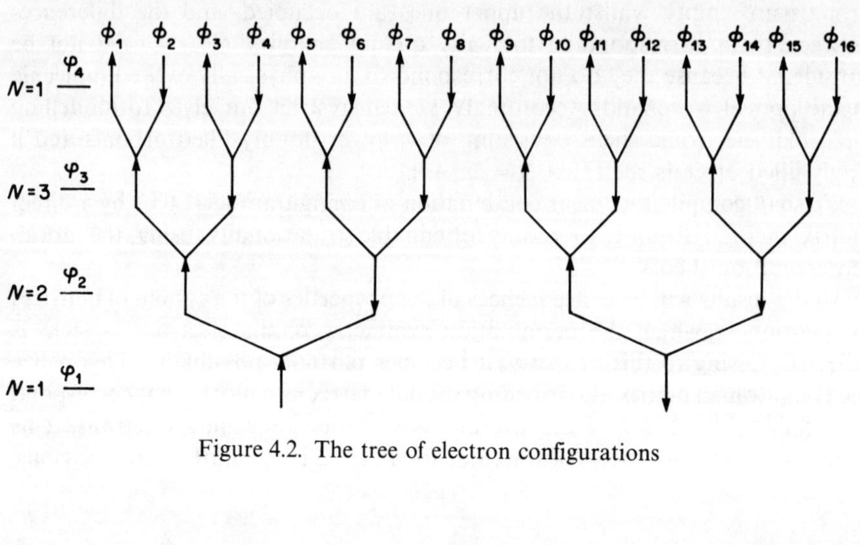

Figure 4.2. The tree of electron configurations

the spins for a graph is the projection of the complete spin of the system on the z-axis. It follows from Figure 4.2 that the extreme left and right graphs give the maximum modulus projections of the overall spin of the system.

These projections comprise part of the spin multiplet with the greatest possible multiplicity. In this way, we only have one multiplet with the maximum overall spin for N orbitals, the square of which is

$$S^2 = \frac{N}{2}\left(\frac{N}{2}+1\right)$$

It is not difficult to count how many remaining multiplets there are for which

$$S^2 < \frac{N}{2}\left(\frac{N}{2}+1\right)$$

Thus we have the following numbers of configurations for the case of $N=4$ shown in Figure 4.2 corresponding, respectively, to the projections of overall spins of $S_z = -2, -1, 0, 1$ and 2. One configuration for $S_z = -2$, four for $S_z = -1$, six for $S_z = 0$, four for $S_z = 1$ and one for $S_z = 2$. Note that the sequence of numbers of configurations corresponds to the coefficients C_0^4, C_1^4, C_2^4, C_3^4, C_4^4.

For the case of N electrons we have the following distribution of numbers of configurations among projections with an overall spin of

$$S^2 \leqslant \frac{N}{2}\left(\frac{N}{2}+1\right)$$

(we shall write the number relating to its configurations in parentheses next to the projection of the spin S_z):

$$-\frac{N}{2}(C_0^N),\quad -\frac{N}{2}+1(C_1^N),\quad -\frac{N}{2}+2(C_2^N),\ldots,\frac{N}{2}-2(C_{N-2}^N),$$
$$\frac{N}{2}-1(C_{N-1}^N),\quad \frac{N}{2}(C_N^N)$$

It follows that the multiplicities of the electronic states for which the greatest modulus of the projections of the overall spin onto the z-axis are the sequence

$$N/2,\quad N/2-1,\quad N/2-2,\ldots,$$

which equals the differences between the binomial sequence coefficients, i.e.

$$C_0^N,\quad C_1^N - C_0^N,\quad C_2^N - C_1^N,\ldots$$

Thus we have for $N=4$: one multiplet with $\max|S_z| = 2$, three triplets with $\max|S_z| = 1$ and two singlets with $S_z = 0$

In an arbitrary case of N electrons we have one multiplet with $\max|S_z| = N/2$, $C_1^N - 1$ multiplets with $\max|S_z| = N/2-1$ and $C_2^N - C_1^N$ multiplets with $\max|S_z| = N/2-2$, etc. The total number of multiplets is equal to the largest binomial coefficient $C_{N/2}^N$ for N even and $C_{(N+1)/2}^N$ for N odd. The number of configurations with the least spin is equal to $C^N_{N/2} - C^N_{(N/2)-1}$ for even N and $C^N_{(N-1)/2} - C^N_{(N-1)/2-1}$ for odd N. The number C^N_K, where $k = 1, 2, \ldots, N-1$, defines the order of the energy matrix in the configuration interaction method, the configurations describing the electronic terms of the system with projections of the overall spin of

$$S_z = -\frac{N}{2} + k$$

Thus given $N = 4$, we have fourth-order secular equations for $S_z = \pm 1$ and a sixth-order equation for $S_z = 0$. Moreover, after diagonalizing the corresponding matrices, the eigenvalues and LCCs corresponding to them must be recorded by grouping similar eigenvalues. Thus an LCC can be attributed to a multiplet with a given value of S^2, and this completes the construction of an LCC which describes the electronic states with given values of the square of the overall spin and the projection onto the z-axis.

To sum up, the construction of electronic multiplets reduces to a calculation of the matrices of the configuration interaction method and their consequent diagonalization. The dimension and structure of one of these matrices are defined by the permutation symmetry of the electronic system, and in a given basis of electronic configurations they can be determined using trees of the electronic configurations (Figure 4.2). Since these matrices have a special structure they will be considered in more detail. In a series of special cases the matrices of the configuration interaction method can be diagonalized analytically, and therefore we shall conside these cases and obtain some formulae that have important applications.

Consider all the configurations leading to the same projection of overall spin S_z. This means we should select from the tree of electronic configurations all the graphs whose number of 'up' and 'down' arrows is the same. We call the layers of the tree 'molecular orbitals' and designate them φ_i, where $i = 1, \ldots, N$, so that φ_1 corresponds to the lowest layer.

In Figure 4.2 the horizontal lines to the left correspond to the orbital energy levels. For the case when $N = 4$ in Figure 4.2, all sixteen electron configurations ϕ_k are numbered upwards for a set of four molecular orbitals φ_1, φ_2, φ_3 and φ_4.

We shall compile the LCC belonging to a given projection of the overall spin. To calculate the energy E of the system by the configuration interaction method we have a system of linear equations with respect to the coefficients C_k, i.e.

$$\sum_k C_k \left(\int \phi_l \hat{H} \phi_k \, \mathrm{d}V - E\delta_{lk} \right) = 0 \quad l, k = 1, \ldots, N \tag{4.20}$$

where $\hat{H}$ is the Hamiltonian of the atom or molecule in the Born–Oppenheimer approximation, $\mathrm{d}V$ is an element of volume in the co-ordinate space of all the system's electrons and δ_{lk} is the Kronecker delta.

Calculating the matrix elements from the configurations leads to the following result. The integrals of the single-electron operators make similar contributions to the diagonal elements with respect to the configurations. The diagonal two-electron integrals ($k = l$) from the configurations are equal to sum of Coulomb two-electron integrals J_{ij}, which correspond to all the possible pairs of orbitals φ_i and φ_j, i.e.

$$J = \sum_{i>j=1} J_{ij}$$

minus the exchange integrals $K_{i'j'}$, in which the orbitals $\varphi_{i'}$ and $\varphi_{j'}$ are chosen so that in the kth configuration every combination of two orbitals in the subset of orbitals populated by spins with the same direction is considered. The non-diagonal elements ($k \neq l$) are equal to the exchange integrals taken with the minus sign ($-K_{ij}$) between the orbitals φ_i and φ_j, the numbers i and l being chosen so that in the kth and lth configuration we strike out the orbitals whose spin directions are the same, to leave two orbitals which are the required i and j. If, in addition, the number of such pairs is larger than two, then the exchange integrals will be zero. These rules follow from the requirement that the matrix elements vanish due to the orthogonality of spin–orbitals with respect to spin.

Thus the summed contribution of the single- and two-electron Coulomb integrals to the diagonal elements will be similar for all the configurations, and if we deduct them from the energy E, then the remaining matrix will only contain exchange integrals, and, moreover, their sum in any row and in any column of the matrix will be the same. A matrix with such a property is called 'convex'.

We know that a convex matrix has an eigenvalue equal to the sum of the elements in a row which corresponds to an eigenvector of the form

$$\vec{X} = \{n^{-1/2}, n^{-1/2}, \ldots, n^{-1/2}\}$$

where n is the order of the matrix. An LCC corresponding to a given eigenvalue is contained in the multiplet with the greatest value of S^2.

Thus for any number of electrons the multiplet with the greatest system spin value can be compiled, explicitly and avoiding a matrix diagonalization.

All the matrices for the configuration interaction method for a given tree of configurations of N electrons contain the same set of $N(N-1)/2$ exchange integrals.

In the next sub-section we shall consider in some detail the cases for $N = 2, 3$ and 4, and find the LCCs corresponding to all the permissible multiplets.

4.3.2 *Model of closed shells*

We shall now consider the following model for electron shells of atoms or molecules. Let the m lower orbitals φ_i be occupied by pairs of electrons with opposite spins and the upper N-$2m$ orbitals contain one electron. Since the m lower orbitals are doubly degenerate, the spin of the two electrons may not, according to Pauli's principle, point in the same direction, forming thereby a complete shell.

Thus the LCC of a complete shell must be constructed only from those multiplets which correspond to the summed projections $S_z = 0$ for the lower orbitals φ_i, and not from the whole set of multiplets with $S_z = 0$. In addition, the most plausible model is one in which the overall spin of the shell is zero.

An LCC corresponding to a zero overall shell spin can be written in an explicit form. Consider again Figure 4.2; for the $N = 4$ case, $\varphi_1 = \varphi_2$ and $\varphi_3 = \varphi_4$. We should choose $\phi_6, \phi_7, \phi_{10}$ and ϕ_{11} as the basis configurations. If we form the LCC as

$$\Psi_1 = \frac{1}{\sqrt{2}}(\phi_6 - \phi_7), \quad \Psi_2 = \frac{1}{\sqrt{2}}(\phi_{10} - \phi_{11}) \tag{4.21}$$

then they will correspond to zero overall spin of the pair of electron in $\varphi_3 = \varphi_4$.

Clearly, any linear combination of Ψ_1 and Ψ_2 will not change the overall spin of this pair of electrons. Hence by forming the LCC as

$$\Psi = \frac{1}{\sqrt{2}}(\Psi_1 - \Psi_2) \tag{4.22}$$

we have obtained a zero overall spin state for the remaining electron pairs in the orbitals $\varphi_1 = \varphi_2$.

Hence the LCC we are looking for has the form

$$\Psi = \tfrac{1}{2}(\phi_6 - \phi_7 - \phi_{10} + \phi_{11}) \tag{4.23}$$

This LCC corresponds to $S^2 = S_z = 0$.

The LCC with $S^2 = S_z = 0$ can be easily constructed in the general case of $2m$ shell electrons if the following electronic configuration tree is considered (see Figure 4.3).

Here the layers of the tree first refer to the electron orbitals. However, each orbital is now occupied by two electrons with opposite spin. We begin to construct the LCC by combining the configurations of the upper layer. We take the differences of the configurations, $\phi_1 - \phi_2, \phi_3 - \phi_4, \ldots, \phi_{2m-1} - \phi_{2m}$ for pairs of neighbouring graphs, beginning with, say, the extreme left-hand pair. The zero overall spin is obtained for the pair of electrons occupying the mth orbital.

Now we descend to the $(m-1)$th layer of the tree and take the differences of the combinations obtained, which correspond to neighbouring graphs on

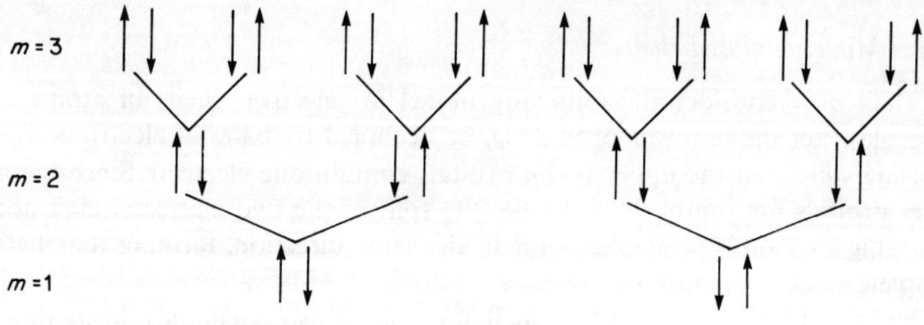

Figure 4.3. The tree of electron configurations of a closed core

the $(m-1)$th layer of the tree. We have

$$\phi_1-\phi_2-(\phi_3-\phi_4), \phi_5-\phi_6-(\phi_7-\phi_8), \ldots$$
$$\phi_{2^m-3}-\phi_{2^m-2}-(\phi_{2^m-1}-\phi_{2^m})$$

We continue until we reach the first layer of the tree, then, leaving the LCC corresponding to the left-hand half of the tree and deducting the LCC corresponding to the right-hand half, we obtain an LCC with a zero overall spin shell.

A general rule for alternating the plus and minus signs before the ϕ_1, $\phi_2, \ldots$, ϕ_{2^m} in the LCC can be obtained by considering sequentially the LCCs for $m = 1, 2, \ldots, .$

The normalizing coefficients in the LCC obtained should be equal to $2^{-m/2}$. Note that it is enough to find the signs for the left (or right) sub-tree of the one shown in Figure 4.3. Note also that there are no other LCCs of the shell which correspond to $S^2 = S_z = 0$, $\phi_1 + \phi_2 = 0$ since the orbitals for pairs of electrons are the same, and anti-symmetrically combining them yieds zero. The LCCs for all the electrons in the system, including both the shell and valency electrons, must be constructed starting from only those obtained in which the shell has a full spin of zero.

The system's full spin is determined by the valency electrons, and if we require that there are no more than four of them, then the LCCs of a given spin with a modulus not greater than two can be constructed analytically.

Our next task is to combine the LCCs for given S^2 and S_z, which correspond to energies of the system that are reasonably close to the energy of the state we must consider. This problem reduces to numerical solution of a secular equation in which the matrix elements are calculated from the LCC for the given S^2 and S_z, and not from the ϕ_i configurations. We call an LCC for given S^2 and S_z a spin configuration (SC). Thus the wave function of an atom or molecule constructed using this method will be a linear combination of spin configurations (LCSC). The functional type of an LCSC will depend on the representation of S^2 and S_z.

Having solved numerically the equations from the LCSC method we shall obtain a series of terms that describe the state of the system with certain values of S^2 and S_z. Transitions between the terms can be classified in terms of the series, in which $\Delta S = 0, \pm 1$, and so on.

By allowing for the spatial symmetry of the system we can classify both the terms and the spectra in the forms of the irreducible representations of the corresponding symmetry groups. As a result, the matrix of the LCSC method for certain values of S^2 and S_z will break down into blocks according to the symmetry groups.

4.3.3 Calculation of the spin configurations for $N = 2, 3$ and 4

The case for $N = 2$ was considered above, but it is useful to repeat it here in order to develop the approach. The electronic configurations have the form illustrated in Figure 4.4. The configurations ϕ_1 and ϕ_2 are components of the

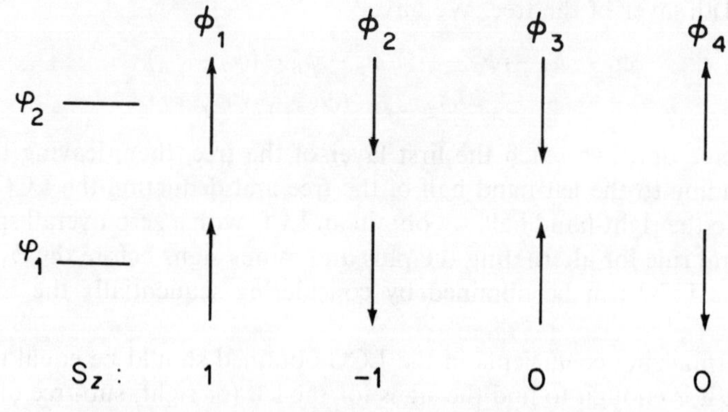

Figure 4.4. Electron configurations for two electrons

multiplet for spin $S=1$, which correspond to the projections $S_z=+1$ and $S_z=-1$, respectively.

The energies of these configurations have the same value which equals $E+J_{12}-K_{12}$, where E is the energy of the configuration when not allowing for electron–electron interactions, J_{12} is the Coulomb two-electron integral and $K_{1,2}$ is the exchange two-electron integral.

The other two configurations ϕ_3 and ϕ_4 reduce to a second-order matrix:

$$\begin{bmatrix} E+J_{12} & -K_{12} \\ -K_{12} & E+J_{12} \end{bmatrix} \tag{4.24}$$

which has the eigenvalues $E+J_{12}-K_{12}$ and $E+J_{12}+K_{12}$ and the eigenvectors $(2^{-1/2}, 2^{-1/2})$ and $(2^{-1/2}, -2^{-1/2})$, respectively.

These two LCCs correspond to the energies $E+J_{12}-K_{12}$ and $E+J_{12}+K_{12}$. Therefore, the first eigenvalue refers to a triplet and the second must refer to a singlet, and both have the same spin projection of $S_z=0$.

We thus obtain the following expressions for the energies and their corresponding spin configurations:

$$\begin{aligned} E &= E+J_{12}-K_{12}, \quad \Psi_1=\phi_1, \quad S_z=1 \\ E_2 &= E+J_{12}-K_{12}, \quad \Psi_2=\phi_2, \quad S_z=1 \\ E_3 &= E+J_{12}-K_{12}, \quad \Psi_3=\frac{1}{\sqrt{2}}(\phi_3+\phi_4), \quad S_z=0 \\ E_4 &= E+J_{12}+K_{12}, \quad \Psi_4=\frac{1}{\sqrt{2}}(\phi_3-\phi_4), \quad S_z=0 \end{aligned} \tag{4.25}$$

E_1, E_2 and E_3 comprise the spin triplet, whilst E_4 is the spin singlet.

For $N=3$ we shall consider the term tree upto the third layer of branches. We have the following configurations. It follows from Figure 4.5 that a system of three electrons has one quartet and two doublets. By compiling the LCCs we obtain two third-order secular equations with identically structured matrices.

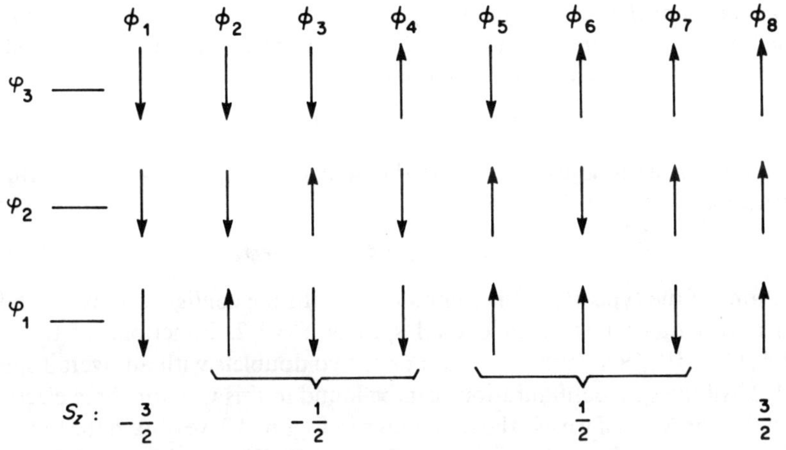

Figure 4.5. Electron configurations for three electrons

These have the form:

$$A = \begin{bmatrix} J-K_{12}; & -K_{23}; & -K_{13} \\ -K_{23}; & J-K_{13}; & -K_{12} \\ -K_{13}; & -K_{12}; & J-K_{23} \end{bmatrix} = \begin{bmatrix} J & 0 & 0 \\ 0 & J & 0 \\ 0 & 0 & J \end{bmatrix} - \begin{bmatrix} K_{12} & K_{23} & K_{13} \\ K_{23} & K_{13} & K_{12} \\ K_{13} & K_{12} & K_{23} \end{bmatrix} \tag{4.26}$$

where $J = J_{12} + J_{13} + J_{23}$.

The special form of these matrices means we can immediately find their eigenvalues. They are:

$$\begin{aligned} E_1 &= J - K_{12} - K_{13} - K_{23} \\ E_2 &= J - [K_{12}^2 + K_{13}^2 + K_{23}^2 - K_{12}K_{13} - K_{12}K_{23} - K_{13}K_{23}]^{1/2} \\ E_3 &= J + [K_{12}^2 + K_{13}^2 + K_{23}^2 - K_{12}K_{13} - K_{12}K_{23} - K_{13}K_{23}]^{1/2} \end{aligned} \tag{4.27}$$

Their corresponding eigenvectors are:

$$\begin{aligned} \vec{C}_1 &= \frac{1}{\sqrt{3}}\{1, 1, 1\} \\ \vec{C}_2 &= \frac{1}{\sqrt{6}}\{2e_1, -e_1 + \sqrt{3}e_2, -e_1 - \sqrt{3}e_2\} \\ \vec{C}_3 &= \frac{1}{\sqrt{6}}\{2e_3, -e_3 + \sqrt{3}e_4, -e_3 - \sqrt{3}e_4\} \end{aligned} \tag{4.28}$$

where

$$\begin{aligned} &e_1 = (t + \lambda^{1/2})[u^2 + (t + \lambda^{1/2})^2]^{-1/2}, \quad e_2 = u[u^2 + (t + \lambda^{1/2})^2]^{-1/2} \\ &e_3 = (t - \lambda^{1/2})[u^2 + (t - \lambda^{1/2})^2]^{-1/2}, \quad e_4 = u[u^2 + (t - \lambda^{1/2})^2]^{-1/2} \\ &t = K_{12} - 0.5(K_{23} + K_{13}), \quad u = 0.5\sqrt{3}(K_{23} - K_{13}) \\ &\lambda = K_{12}^2 + K_{13}^2 + K_{23}^2 - K_{12}K_{13} - K_{12}K_{23} - K_{13}K_{23} \end{aligned}$$

Starting from the basis function ϕ_2, ϕ_3, ϕ_4, the eigenfunctions $\Psi_i^{(1/2)}$ for the projection of the overall spin $S_z = 1/2$ are constructed as the scalar products of the eigenvectors $\vec{C}_i$ with the vectors $\vec{\phi} = \{\phi_2, \phi_3, \phi_4\}$, i.e.

$$\Psi_i^{(1/2)} = \vec{C}_i \vec{\phi}, \quad i = 1, 2, 3$$

We obtain an analogous results for the spin projection $S_z = -1/2$, with the substitutions

$$\phi_2 \to \phi_7, \quad \phi_3 \to \phi_6, \quad \phi_4 \to \phi_5$$

Functions of the type $\Psi_1^{(\pm 1/2)}$ in combination with the configurations ϕ_1 and ϕ_8 belong to a quartet with an overall spin of $S = 3/2$. Functions of the form $\{\Psi_2^{(1/2)}, \Psi_2^{(-1/2)}\}$, $\{\Psi_3^{(1/2)}, \Psi_3^{(-1/2)}\}$ comprise two doublets with an overall spin of $S = 1/2$. All the spin configurations can be found in this way for three electrons.

Finally, for $N = 4$ if we use the tree shown in Figure 4.2 we obtain the following configurations for four electrons (see Figure 4.6). We have one quintet, three triplets and two singlets. Configurations ϕ_1 and ϕ_{16} belong to the quintet.

For $S_z = 1$ and $S_z = -1$ we have two identical fourth-order matrices with the form:

$$A_4(S_z = 1, -1)$$
$$= \begin{bmatrix} J - K_{12} - K_{13} - K_{23} & -K_{34} & -K_{24} & -K_{14} \\ -K_{34} & J - K_{12} - K_{14} - K_{24} & -K_{23} & -K_{13} \\ -K_{24} & -K_{23} & J - K_{13} - K_{14} - K_{34} & -K_{12} \\ -K_{14} & -K_{13} & -K_{12} & J - K_{23} - K_{24} - K_{34} \end{bmatrix} \tag{4.29}$$

where

$$J = \sum_{j>i=1}^{4} J_{ij}$$

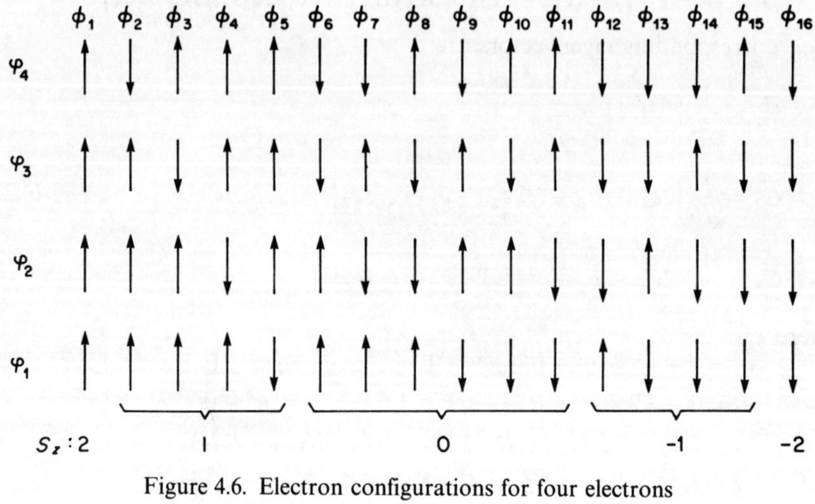

Figure 4.6. Electron configurations for four electrons

The rows and column of this matrix correspond to $\phi_2, \phi_3, \phi_4, \phi_5$, respectively, or $\phi_{12}, \phi_{13}, \phi_{14}, \phi_{15}$, respectively.

Using the orthogonal transformation

$$C_4 = \frac{1}{2}\begin{bmatrix} 1 & 1 & 1 & 1 \\ 1 & 1 & -1 & -1 \\ 1 & -1 & -1 & 1 \\ 1 & -1 & 1 & -1 \end{bmatrix} \tag{4.30}$$

the matrix of equation (4.29) is turned into the block form

$$\tilde{C}_4 A_4 C_4 = \begin{bmatrix} J-K & 0 & 0 & 0 \\ 0 & J-K_{12}-K_{34} & K_{24}-K_{13} & K_{14}-K_{23} \\ 0 & K_{24}-K_{13} & J-K_{23}-K_{14} & K_{34}-K_{12} \\ 0 & K_{14}-K_{23} & K_{34}-K_{12} & J-K_{13}-K_{24} \end{bmatrix} \tag{4.31}$$

where

$$K = \sum_{j>i=1}^{4} K_{ij}$$

The eigenvalue $J-K$ belongs to a quintet, and its eigenvector is $\{\frac{1}{2}, \frac{1}{2}, \frac{1}{2}, \frac{1}{2}\}$. Diagonalizing the remaining third-order block of equation (4.31) yields eigenvectors corresponding to three triplets, which have different energies.

In order to calculate the LCC for $S_z = 0$, we have the sixth-order matrix

$$A_6(S_z=0) = \begin{bmatrix} J-K_{12}-K_{34} & -K_{23} & -K_{24} & -K_{13} & -K_{14} & 0 \\ -K_{23} & J-K_{13}-K_{24} & -K_{34} & -K_{12} & 0 & -K_{14} \\ -K_{24} & -K_{34} & J-K_{14}-K_{23} & 0 & -K_{12} & -K_{13} \\ -K_{13} & -K_{12} & 0 & J-K_{14}-K_{23} & -K_{34} & -K_{24} \\ -K_{14} & 0 & -K_{12} & -K_{34} & J-K_{13}-K_{24} & -K_{23} \\ 0 & -K_{14} & -K_{13} & -K_{24} & -K_{23} & J-K_{12}-K_{34} \end{bmatrix} \tag{4.32}$$

The rows and columns correspond to the configurations $\phi_6, \phi_7, \phi_8, \phi_9, \phi_{10}$ and ϕ_{11}. Using the orthogonal transformation,

$$C_6 = \left(\frac{1}{2}\right)^{1/2}\begin{bmatrix} 1 & 0 & 0 & 1 & 0 & 0 \\ 0 & 1 & 0 & 0 & 0 & 1 \\ 0 & 0 & 1 & 0 & -1 & 0 \\ 0 & 0 & 1 & 0 & 1 & 0 \\ 0 & 1 & 0 & 0 & 0 & -1 \\ 0 & 0 & 0 & -1 & 0 & 0 \end{bmatrix} \tag{4.33}$$

the matrix is transformed into the block form matrix

$$\tilde{C}_6 A_6 C_6 = \begin{bmatrix} J-K_{12}-K_{34} & -K_{23}-K_{14} & -K_{24}-K_{13} & & & \\ -K_{23}-K_{14} & J-K_{13}-K_{24} & -K_{34}-K_{12} & & 0 & \\ -K_{24}-K_{13} & -K_{34}-K_{12} & J-K_{14}-K_{23} & & & \\ & & & J-K_{12}-K_{34} & K_{24}-K_{13} & K_{14}-K_{23} \\ & 0 & & K_{24}-K_{13} & J-K_{14}-K_{23} & K_{34}-K_{12} \\ & & & K_{14}-K_{23} & K_{34}-K_{12} & J-K_{13}-K_{24} \end{bmatrix} \quad (4.34)$$

This matrix has been broken down into two third-order blocks. The lower block is the same as the lower one of matrix (4.31) and thus describes the component of three triplets which correspond to a projection of the overall spin $S_z = 0$.

The upper block of the matrix in equation (4.34) has the same structure as matrix (4.26). Hence it can be diagonalized in the same way and we obtain formulae similar to equations (4.27) and (4.28) for the eigenvalues and eigenvectors.

The eigenvalue

$$E_1 = J - \sum_{j>i=1}^{4} K_{ij}$$

and the eigenvector $\vec{C}_1 = \{3^{-1/2}, 3^{-1/2}, 3^{-1/2}\}$ are the components of a quintuplet for which $S_z = 0$. The remaining two-spin configurations of this block describe two singlets with $S^2 = 0$ and $S_z = 0$.

In the general case of N electrons it is also possible to break down the secular equations into blocks, each equation corresponding to a particular projection of the overall spin S_z. However, the smallest blocks have an order of $N-1$. The structures of the resultant large-order matrices can be simplified further by neglecting small exchange integrals.

4.4 The Self-consistent Field Method

In Section 4.3 we considered the general principles for studying the states of molecules with many electrons by starting from sets of determinantal functions. These are constructed from orbitals that are themselves solutions of the problem concerning a single electron moving in a field of many charged centres. Naturally, when these orbitals were constructed we did not take into account the influence of other electrons. Thus these orbitals were, so to say, not very well defined. We would like to know whether it is possible to improve the overall accuracy by retaining the solution in the form of one determinant and the molecular orbital in the form of an LCAO, and nevertheless partially account for the interelectron interactions.

It is in fact possible to do so by using the method of self-consistent fields. In this method the Schroedinger equation is substituted by Hartree–Fock's equation.

Self-consistent field equations can be obtained in the most general form using Bubnov and Galerkin's projection method[3]. This is a consequence of the linearity of the original Schroedinger equation. Essentially the Bubnov–Galerkin method tries to approximate the solution of a given differential equation in the form of a linear combination of basis functions, the coefficients of which are defined by an orthogonality requirement on the errors of the equation to the basis functions. The equivalence of Ritz's method and the Bubnov–Galerkin method for linear differential and integral equations enables us to derive the self-consistent field equations by projection methods, avoiding the procedure of varying the single-particle functions.

Note that projection methods are of general utility because they do not use limiting properties of equations. They can be used to obtain the self-consistent field equations very simply both for the ground and for the excited states of a multiparticle system.

The task for many particles reduces to a solution of a multidimensional differential or integral equation. The Schroedinger equation for a system of molecular electrons in a field of fixed atomic nuclei is an example.

The basis functions must be multidimensional functions of the co-ordinates of the particles and must satisfy the quantum boundary conditions (for example, quadratic integrability with respect to all variables). In a projection method we try to select basis elements that will automatically satisfy the boundary conditions. The approximate equations are then obtained when equations to the multidimensional basis elements are orthogonalized. The orthogonalization is carried out in a space with small dimensionality so that the differential or integral equation is obtained with respect to the remaining variables in a space which has far fewer dimensions than the original equation.

In a system of electrons it is essential to allow for yet another limiting condition, i.e. the anti-symmetry of the wave equation with respect to the interchange of particles. This condition results in an approximate function that is constructed as one of a linear combination of Slater determinants with single-particle functions taken from an orthogonal basis of the complete set of elements which satisfy the quantum boundary condition of quadratic intergrability. We wish to emphasize that a Slater determinant, though composed of single-particle functions, is itself an exact solution to the Schroedinger equation for the multiparticle system when electronic interactions are neglected.

Consider the Schroedinger equation for a molecule using the fixed nuclei approximation

$$\hat{H}\Psi(1, 2, \ldots, N) = E\Psi(1, 2, \ldots, N) \tag{4.35}$$

where

$$\hat{H} = \sum_i \hat{h}_i(i) + \sum_{i>j} V_{ij}(i, j)$$

is the molecule's Hamiltonian, $\hat{h}_i(i)$ is a single-electron Hamiltonian (the Hamiltonian for the problem concerning the ith electron moving in the field of nuclei) and $V_{ij}(i, j)$ is the Coulomb electron interaction operator.

We have already mentioned that each Slater determinant is called an electronic configuration. Returning once more to the configuration interaction method, in which the solution was sought as a linear combination of configurations, and given that each configuration is designated ϕ, then the approximate wave function is written as:

$$\tilde{\tilde{\Psi}} = \sum_{k=1}^{n} C_k \phi_k(1, 2, \ldots, N) \tag{4.36}$$

We substitute the function of equation (4.36) into equation (4.35) to obtain

$$\hat{H} \sum_{k=1}^{n} C_k \phi_k - E \sum_{k=1}^{n} C_k \phi_k = f(1, 2, 3, \ldots, N) \tag{4.37}$$

where function $f(1, 2, \ldots, N)$ is the error function of equation (4.35).

The Bubnov–Galerkin method makes the function f orthogonal to the basis elements ϕ_k. By multiplying equation (4.37) by ϕ_l and integrating it over the co-ordinates of all electrons we arrive at the following system of algebraic equations:

$$\sum_{k=1}^{n} C_k[\langle \phi_l | H | \phi_k \rangle - E \langle \phi_l | \phi_k \rangle] = 0 \qquad l = 1, 2, \ldots, n \tag{4.38}$$

where

$$\langle \phi_l | \hat{H} | \phi_k \rangle = \int \phi_l(1, \ldots, N) \hat{H} \phi_k(1, \ldots, N) \, \mathrm{d}V_1 \, \mathrm{d}V_2 \cdots \mathrm{d}V_N$$

$$\langle \phi_l | \phi_k \rangle = \int \phi_l(1, \ldots, N) \phi_k(1, \ldots, N) \, \mathrm{d}V_1 \, \mathrm{d}V_2 \cdots \mathrm{d}V_N$$

When the basis set is orthogonal we also have

$$\langle \phi_l | \phi_k \rangle = \delta_{lk}$$

where δ_{lk} is the Kronecker delta.

It is apparent that equation (4.38) coincides with that obtained by Ritz's variational method. This proves the applicability of the Bubnov–Galerkin method for linear differential and integral equations.

We shall now derive the Hartree–Fock equation for an approximate solution of equation (4.35). We shall consider a single-determinant approximation and find the equations of the respective single-particle functions that comprise the Slater determinant

$$\phi(1, 2, \ldots, N) = \frac{\det}{\sqrt{N!}} \begin{bmatrix} \varphi_1(1) & \varphi_1(2) & \cdots & \varphi_1(N) \\ \varphi_2(1) & \varphi_2(2) & \cdots & \varphi_2(N) \\ \cdots & \cdots & \cdots & \cdots \\ \varphi_N(1) & \varphi_N(2) & \cdots & \varphi_N(N) \end{bmatrix} \tag{4.39}$$

where $\varphi_i(i)$ is the single-particle wave function or spin–orbital.

This problem is, in a sense, the inverse of the Bubnov–Galerkin problem because the coefficients C_k are already known and equal to $+1$ and -1, and the basis functions $u_i = \prod_i \varphi_i(i)$ themselves that must be found so that they satisfy equation (4.38) and be the best for the selected representation.

Here an arbitrary shell is being considered and may either be open or closed. After substituting equation (4.39) into equation (4.35) we obtain the relation

$$\hat{H} \det[\varphi_1(1), \varphi_2(2), \ldots, \varphi_N(N)] - E \det[\varphi_1(1), \varphi_2(2), \ldots, \varphi_N(N)] = f(1, 2, \ldots, N)$$

By applying Kantorovich's method and by orthogonalizing the error functions $f(1, 2, \ldots, N)$ in the co-ordinate space of the second,... and Nth electrons to the product of the spin–orbitals $\varphi_2(2) \cdots \varphi_N(N)$, we arrive at an equation with respect to $\varphi_1(1)$ in the form

$$\begin{aligned} &\langle \varphi_2(2) \cdots \varphi_N(N) | \hat{H} | \det[\varphi_1(1), \ldots, \varphi_N(N)] \\ &\quad - E \langle \varphi_2(2) \cdots \varphi_N(N) | \det[\varphi_1(1), \ldots, \varphi_N(N)] \rangle = 0 \end{aligned} \tag{4.40}$$

where

$$\begin{aligned} &\langle \varphi_2(2) \cdots \varphi_N(N) | \hat{H} | \det[\varphi_1(1), \ldots, \varphi_N(N)] \rangle \\ &\quad = \int \varphi_2(2) \cdots \varphi_N(N) \hat{H} \det[\varphi_1(1), \ldots, \varphi_N(N)] \, \mathrm{d}V_1 \, \mathrm{d}V_2 \cdots \mathrm{d}V_N \end{aligned}$$

The second integral is handled similarly with a change from the $\hat{H}$ operator to I.

Clearly, the error can be orthogonalized to the product of the spin–orbitals with any permutation of indices. The corresponding permutation must be made in the determinant by taking into account that the value of the error function, f, is multiplied by -1 raised to a power equal to the number of permutations. The form of equation (4.40) does not change. We can obtain equations for the remaining spin–orbitals $\varphi_2(2), \ldots, \varphi_N(N)$, in a similar way.

We shall calculate the integrals in equation (4.40) assuming the orthogonality of the set of spin–orbitals $\varphi_1(1)$, $\varphi_2(2), \ldots, \varphi_N(1)$. The second integral on the left-hand side of equation (4.40) is equal to $E\varphi_1(1)$. We shall write matrix elements from the single-electron Hamiltonian $\hat{h}_i(i)$ by taking one element of the sum $\varphi_{p1}(1)$, $\varphi_{p2}(2) \cdots \varphi_{pN}(N)$ in $\det[\varphi_1(1), \varphi_2(2), \ldots, \varphi_N(N)]$, where $p1, \ldots, pN$ refer to the permutation of the indices $1, 2, \ldots, N$. Hence

$$\begin{aligned} &\int \varphi_2(2) \cdots \varphi_N(N) \hat{h}_i \varphi_{p1}(1) \varphi_{p2}(2) \cdots \varphi_{pN}(N) \, \mathrm{d}V_2 \, \mathrm{d}V_3 \cdots \mathrm{d}V_N \\ &\quad = \begin{cases} \text{if} \quad p1 = 1, \ldots, pN = N \\ \text{then} \;\; \hat{h}_1(1) \varphi_1(1), \quad i = 1 \\ \qquad \varphi_1(1) \displaystyle\int \varphi_i(i) \hat{h}_i(i) \varphi_i(i) \, \mathrm{d}V_i, \quad i \neq 1 \\ \text{and } 0 \text{ if } p_k \neq k \text{ for any } k. \end{cases} \end{aligned} \tag{4.41}$$

The integral of the kind

$$\int \varphi_i(i)\hat{h}(i)\varphi_1(i)\,\mathrm{d}V_i \tag{4.42}$$

is omitted in equation (4.41) because it can be cancelled by an orthogonal transformation of the set of spin–orbitals $\varphi_i(1)$, with the Slater determinant being invariant.

We now consider the two-electron integrals, designating them g'_{ij}:

$$g'_{ij} = \int \varphi_2(2)\cdots\varphi_N(N)V(i,j)\det[\varphi_1(1),\ldots,\varphi_N(N)]\,\mathrm{d}V_2\cdots\mathrm{d}V_N \tag{4.43}$$

We obtain the following for individual components of the determinant:

$$\begin{aligned} g''_{ij} &= \int \varphi_2(2)\cdots\varphi_N(N)V(i,j)\varphi_{p1}(1)\cdots\varphi_{pN}(N)\,\mathrm{d}V_2\cdots\mathrm{d}V_N \\ &= \varphi_{p1}(1)\int \varphi_2(2)\cdots\varphi_N(N)V(i,j)\varphi_{p2}(2)\cdots\varphi_{pN}(N)\,\mathrm{d}V_2\cdots\mathrm{d}V_N \end{aligned} \tag{4.44}$$

When $j=1$ we arrive at a matrix element that is reminiscent of the single-electron integral (4.41). Hence the result can be written out immediately:

$$g''_{ij} = \begin{cases} \varphi_{p1}(1)\displaystyle\int \varphi_i(i)V(i,1)\varphi_{pi}(i)\,\mathrm{d}V_i, \quad p_1 = p_i = 1, i \\ \text{all } p_k = k \text{ given } k \neq i, I \\ 0, \text{ if } p_k \neq k \text{ given some } k \neq i, I \end{cases} \tag{4.45}$$

Given $j>1$ we have

$$\begin{aligned} g''_{ij} &= \varphi_{p1}(1)\int \varphi_2(2)\cdots\varphi_N(N)V(i,j)\varphi_{p2}(2)\cdots\varphi_{pN}(N)\,\mathrm{d}V_2\cdots\mathrm{d}V_N \\ &= \varphi_{p1}(1)\int \varphi_i(i)\varphi_j(j)V(i,j)\varphi_{pi}(i)\varphi_{pj}(j)\,\mathrm{d}V_i\,\mathrm{d}V_j \end{aligned} \tag{4.46}$$

for all $p_k = k$ given that $k \neq i, j$.

Because of the orthogonality of the spin–orbitals $\varphi_{pi}(i)$, we only obtain two possibilities for the indices pi and pj, i.e. $pi=i$ and $pj=j$, or $pi=j$ and $pj=i$. The index $p1=1$ because $i>j>I$. In total:

$$\varphi_1(1)\int \varphi_i(i)\varphi_j(j)V(i,j)\varphi_{pi}(i)\varphi_{pj}(j)\,\mathrm{d}V_i\,\mathrm{d}V_j \tag{4.47}$$

p_i, $p_j = i, j$; for all $p_k = k$ for $k \neq i, j$

0, if given that some $k \neq i, j, p_k \neq k$

When calculating matrix element g'_{ij} it is necessary to take into account that it is a linear combination of elements g''_{ij} with coefficients $+1$ and -1 and the v_{ij} depend on the parity (even or odd) of the given permutation of indices $p2, p3, \ldots, pN$.

We have thus calculated all the matrix elements in equation (4.40), and it can be written as

$$\hat{h}(1)\varphi_1(1) + \left(\sum_{i=1} \int \varphi_i(i)\hat{h}\varphi_i(i)\,\mathrm{d}V_i\right)\varphi_1(1) + \sum_{i\neq 1}\left(\left(\int \varphi_i(i)V(i,1)\varphi_i(i)\,\mathrm{d}V_i\right)\varphi_1(1)\right.$$
$$- \sum_{i\neq 1}\left(\left(\int \varphi_i(i)V(i,1)\varphi_1(i)\,\mathrm{d}V_i\right)\varphi_i(1) - \left[\mathrm{E} - \sum_{i\neq 1}\int \varphi_i(i)\hat{h}_i(i)\varphi_i(i)\,\mathrm{d}V_i\right.\right.$$
$$- \sum_{i>j>1}\left(\int \varphi_i^2(i)V(i,j)\varphi_j^2(j)\,\mathrm{d}V_i\,\mathrm{d}V_j\right.$$
$$\left.\left.\left. - \int \varphi_i(i)\varphi_j(j)V(i,j)\varphi_i(j)\varphi_j(i)\,\mathrm{d}V_i\,\mathrm{d}V_j\right)\right)\right]\varphi_1(1) = 0 \tag{4.48}$$

The expression in the square brackets is a constant, which we shall call ε_1. Finally, the desired equation for the spin–orbitals $\varphi_i(i)$ after a substitution of I for i can be written as

$$\hat{h}_i(i)\varphi_i(i) + \left[\sum_{j\neq 1}\int \varphi_j^2(j)V(i,j)\,\mathrm{d}V_j\right]\varphi_i(i)$$
$$- \sum_{j\neq 1}\left[\left[\int \varphi_j(j)\varphi_i(j)V(i,j)\,\mathrm{d}V_j\right]\varphi_j(i) = \varepsilon_i\varphi_i(i) \quad i = 1, 2, \ldots, N \right. \tag{4.49}$$

We obtain the value of ε from equation (4.48) after multiplying it by spin–orbital $\varphi_1(1)$ and integrating with respect to the variables of the first electron:

$$\varepsilon = \sum_{i=1}^{N}\int \varphi_i(i)\hat{h}_i(i)\varphi_i(i)\,\mathrm{d}V_i + \sum_{i>j}\int \varphi_i^2(i)\varphi_j^2(j)V(i,j)\,\mathrm{d}V_i\,\mathrm{d}V_j$$
$$- \sum_{i>j}\int \varphi_i(i)\varphi_j(j)V(i,j)\varphi_i(j)\varphi_j(i)\,\mathrm{d}V_i\,\mathrm{d}V_j \tag{4.50}$$

In order to make equation (4.49) more symmetric we remove the $i \neq j$ condition from the sums because the additional components compensate for each other.

As a result we obtain the Hartree–Fock[4–6] equation:

$$\hat{h}_i(i)\varphi_i(i) + \left[\sum_{j=1}^{N}\int \varphi_j^2(j)V(i,j)\,\mathrm{d}V_j\right]\varphi_i(i)$$
$$- \sum_{j=1}^{N}\left[\int \varphi_j(j)\varphi_i(j)V(i,j)\,\mathrm{d}V_j\right]\varphi_j(i) = \varepsilon_i\varphi_i(i) \quad i = 1, 2, \ldots, N \tag{4.51}$$

As the Hartree–Fock equation is solved the eigenvalue ε and the Slater determinant of equation (4.23) are calculated, and they have the physical sense of approximate eigenvalues and eigenfunctions of the self-consistent field theory. It is for this reason, generally speaking, that it is not particularly useful to assign physical meanings to individual orbitals φ_i and levels ε_i, because it is only when they are all considered together that a physical picture of a given electron state of a system can be obtained.

Further use of orthogonality with respect to spin for the spin–orbitals in a non-relativistic approximation, which only takes into account that the spin and spatial variables of the electrons are divided, allows us to write an equation for the spatial function of the spin–orbitals, i.e. simple orbitals, instead of equation (4.51).

We designate the spatial part of a spin–orbital $\varphi_i(i)$ as $\varphi_i'(i)$, while $\alpha(i)$ and $\beta(i)$ are given as the spin functions, where

$$\alpha = \begin{bmatrix} 1 \\ 0 \end{bmatrix} \quad \text{and} \quad \beta = \begin{bmatrix} 0 \\ 1 \end{bmatrix}$$

In the case of a closed electron shell we have an even number of electrons, and every pair has electrons of opposite spin which have the same spatial parts φ_i'. We can thus write equation (4.51) as

$$\hat{h}(i)\varphi_i'(i) + \left[2\sum_{j=1}^{N/2} \int \varphi_j'^2(j) V(i,j)\,\mathrm{d}V_j \right] \varphi_i'(i) - \sum_{j=1}^{N/2} \left[\int \varphi_j'(j)\varphi_i'(j) V(i,j)\,\mathrm{d}V_j \right] \varphi_j'(i) = \varepsilon_i \varphi_i'(i) \tag{4.52a}$$

where $i = 1, 2, \ldots, N/2$.

In the case of open electron shells we have m spin–orbitals with spin functions α, and $N - m$ spin–orbitals with spin functions β. As a result, equation (4.51) is changed to a set of two systems of Hartree–Fock equations, one system for the m orbitals with α spin and one for the $N - m$ orbitals with β spin:

$$\hat{h}_i(i)\varphi_i'(i) + \left[\sum_{j=1}^{N} \int \varphi_j'^2(j) V(i,j)\,\mathrm{d}V_j \right] \varphi_i'(i) - \sum_{j=1}^{m} \left[\int \varphi_j'(j)\varphi_i'(j) V(i,j)\,\mathrm{d}V_j \right] \varphi_j'(i) = \varepsilon_i \varphi_i'(i)$$

where $i = 1, \ldots, m$, and

$$\hat{h}_i(i)\varphi_i'(i) + \left[\sum_{j=1}^{N} \int \varphi_j'^2(j) V(i,j)\,\mathrm{d}V_j \right] \varphi_i'(i) - \sum_{j=m+1}^{N} \left[\int \varphi_j'(j)\varphi_i'(j) V(i,j)\,\mathrm{d}V_j \right] \varphi_j'(i) = \varepsilon_i \varphi_i'(i) \tag{4.52b}$$

where $i = m + 1, \ldots, N$.

Note that the orbitals of the first system of equation (4.52b) are not, generally speaking, orthogonal to the orbitals of the second system of equations. The system of Hartree–Fock equations may also be written if we go from co-ordinate to momentum representation of orbitals[7].

In order to solve the self-consistent field equations of equation (4.52) numerically, we must make the following approximation[8]. The $\varphi_i'(i)$ orbitals are represented in the form of a linear combination of basis functions which satisfy the quadratic integrability condition, i.e.

$$\varphi_i'(i) = \varphi_i(i) = \sum_{k=1}^{n} C_{ki}\chi_k(i) \tag{4.53}$$

where the C_{ik} are the coefficients that must be determined. Substituting equation (4.53) into equation (4.52a) we obtain

$$\begin{aligned}
&\hat{h}_i(i)\sum_{k=1}^{n} C_{ik}\chi_k(i) + \left[2\sum_{j=1}^{N/2}\int \sum_{k,k'}^{n} C_{jk}C_{jk'}\chi_k(j)\chi_{k'}(j)V(i,j)\,\mathrm{d}V_j\right]\sum_{k=1}^{n} C_{ik}\chi_k(i)\\
&\quad -\sum_{j=1}^{N/2}\left[\int \sum_{k=1}^{n} C_{jk}\sum_{k'=1}^{n} C_{ik'}\chi_k(j)\chi_{k'}(j)V(i,j)\,\mathrm{d}V_j\right]\sum_{k=1}^{n} C_{jk}\chi_k(i)\\
&\quad -\varepsilon_i\sum_{k=1}^{n} C_{ik}\chi_k(i) = f_i(i)
\end{aligned} \tag{4.54}$$

We then apply the Bubnov–Galerkin method to calculate the C_{ik} coefficients. By assuming that the error $f_i(i)$ is orthogonal to the basis functions $\chi_k(i)$, regrouping the terms in k and reorganizing the indices, we obtain the following system of algebraic equations:

$$\begin{aligned}
&\sum_{k=1}^{n} C_{ik}.\int \chi_l(i)\hat{h}_i(i)\chi_k(i)\,\mathrm{d}V_i + \sum_{k=1}^{n} C_{ik}\sum_{p=1}^{n}\sum_{q=1}^{n}\sum_{j=1}^{N/2} C_{jp}C_{jq}\\
&\quad\times\left[2\int \chi_l(i)\chi_k(i)V(i,j)\chi_p(j)\chi_q(j)\,\mathrm{d}V_i\,\mathrm{d}V_j\right.\\
&\quad\left.-\int \chi_l(i)\chi_p(i)V(i,j)\chi_k(j)\chi_q(j)\,\mathrm{d}V_i\,\mathrm{d}V_j\right]\\
&\quad-\varepsilon_i\sum_{k=1}^{n} C_{ik}\int \chi_l(i)\chi_k(i)\,\mathrm{d}V_i = 0
\end{aligned} \tag{4.55}$$

Next we take into account that the integration variables are not real and can formally be replaced by 1 and 2, and we introduce the following notation for the integrals:

$$\begin{aligned}
h_{lk} &= \int \chi_l(1)\hat{h}(1)\chi_k(1)\,\mathrm{d}V_1\\
g_{lkpq} &= \int \chi_l(1)\chi_k(1)V(1,2)\chi_p(2)\chi_q(2)\,\mathrm{d}V_1\,\mathrm{d}V_2\\
S_{lk} &= \int \chi_l(1)\chi_k(1)\,\mathrm{d}V_1
\end{aligned} \tag{4.56}$$

Using equation (4.56) we can rewrite equation (4.55) as a matrix equation:

$$\sum_{k=1}^{n} C_{ik}h_{lk} + \sum_{k=1}^{n} C_{ik} \sum_{p,q=1}^{n} P_{pq}(2g_{lkpq} - g_{lpkq}) = \varepsilon_i \sum_{k=1}^{n} C_{ik}S_{lk} \tag{4.57}$$

where the

$$P_{pq} = \sum_{j=1}^{N/2} C_{jp}C_{jq} \tag{4.58}$$

are elements of the density matrix, $l = 1, \ldots, n$.

At first, the elements of the C matrix needed to construct the density matrix P can be taken from a solution of the problem with the $\hat{h}$ operator. Then using this density matrix ($P^{(0)}$) a solution of the Hartree–Fock equation can be found, leading to a new density matrix which is then used to construct new elements for the Hartree–Fock matrix (F). This process is repeated until $P^{(k)} = P^{(k+1)}$. It is then said that a self-consistent solution has been achieved. Even though we know the process converges, there is no proof that it should do so. Nevertheless, repeated numerical solutions using this approach have demonstrated that, as a rule, self-consistent solutions are actually obtained. Since the final solution obtained by diagonalizing the quadratic form with the Hartree–Fock matrix, all the eigenfunctions, in the LCAO form, that have been found this way will be orthogonal and normalized.

The self-consistent field method for finding the single-electron orbitals has become widespread because of its relative simplicity, reasonable accuracy and convenient computer implementation. The main problem in this respect arises due to the necessity of calculating a large number (of the order n^4 for n atomic orbitals) of integrals of the form

$$\int \chi_k(1)\chi_l(1)\chi_p(2)\chi_q(2)\frac{1}{r_{12}}\,dV_1\,dV_2$$

for which the atomic functions have different centres.

When approximating, these integrals are assumed to be zero if one of the centres is significantly far removed from the others. This is because the χ_k-function is only noticeably non-zero in the region of its atom. If the χ_k and χ_l functions are close together, and the χ_p and χ_q are also close, but the two pairs are far from each other, then the integral may be approximated as:

$$\int \chi_k(1)\chi_l(1)\chi_p(2)\chi_q(2)V(1,2)\,dV_1\,dV_2 \approx \frac{s_{kl}s_{pq}}{r_{kl,pq}} \tag{4.59}$$

where s_{kl} and s_{pq} are the respective overlap integrals, and $r_{kl,pq}$ is the distance between the midpoints of the lines connecting χ_k, χ_l and χ_p, χ_q. We give two more estimates:

$$\int \chi_k^2(1)\chi_l^2(2)V(1,2)\,dV_1\,dV_2 \approx \frac{1}{r_{kl}} \tag{4.60}$$

$$\int \chi_k^2(1)\chi_l(2)\chi_p(2)V(1,2)\,\mathrm{d}V_1\,\mathrm{d}V_2 \approx \frac{S_{lp}}{r_{k,lp}} \tag{4.61}$$

where $r_{k,lp}$ is the distance between the centre of χ_k to the midpoint of the segment connecting the centre of χ_l and χ_p.

If the molecule possesses spatial symmetry, then the matrix F made up of the elements of Ritz's equation may be reduced to block form using a transformation to symmetry orbitals (see Chapter 5).

The single-electron orbitals found by the self-consistent field method are the best, as can be shown, in that they are the easiest for constructing the determinantal functions, which correspond to the electron configurations. The matrix elements between the determinantal functions constructed from self-consistent orbitals have many of the properties we considered in Section 4.1. The only difference are the elements $\int \phi_k \hat{h} \phi_n \mathrm{d}V$, which are not zero when $\phi_k \neq \phi_n$ because self-consistent orbitals are not solutions to the equation $\hat{h}\Psi_0 = E\Psi_0$. However, they are zero whenever the configurations ϕ_n and ϕ_k differ by more than two orbitals.

4.5 Approximate Expressions for the Matrix Elements of the Hartree–Fock Operator

4.5.1 Constructing a semi-empirical theory of electron shells

In the last few sections we considered some general methods of calculating energy levels and the density matrix for the electron shells of a polyatomic molecule. In principle, these approximate calculations yield reasonably accurate results that correlate well with experiment. However, when we carry out these calculations in practice we run into serious numerical problems.

Apart from the necessity of solving high-order systems of equations (moreover, for the self-consistent method iterations are necessary), it is necessary to calculate a very large number of differing integrals which are contained in the expression for the corresponding operator matrix elements. This is especially great for two-electron integrals. We have already stated that this number is proportional to n^4, where n is the number of basis atomic orbitals included in the calculation. If the molecule has some symmetry, then many of these integrals will be similar, just as the single-electron Coulomb and resonance integrals are similar. However, even in this case the overall number of different integrals is exceedingly large for complicated molecules.

Another situation should also be noted. The whole calculation is carried out using a wave function in the form of a linear combination of atomic orbitals, but if the basis set is limited, then the function introduces considerable errors in all the subsequent calculations. The good accuracy we noted earlier requires a significant extension, by an order or more, of the basis with respect to the minimum basis. Even so, it often turns out that for complicated systems an accurate estimation is still in poor agreement with experiment. Unfortunately, the present status of solving multi-electron problems is such that we cannot estimate, *a priori*, the error which occurs in the final result from this and other approximations. Hence whenever these calculations are carried out we must bear in mind that there is a real danger of poor agreement with experiment. The theory of molecular electron shells for integrating and predicting the spectra of complicated molecules will only be useful when they can be done fast enough to try a large number of different variants, as this will give a worthwhile basis for subsequent analysis. The necessity of performing large numbers of calculations therefore requires that the theory itself to be simplified and its exact variants substituted by approximate ones based on introduced semi-empirical parameters[9,10]. We shall dwell on this in some detail.

We shall consider the Coulomb and exchange two-electron integrals in the expression for the average energy H separately. These integrals have the form

$$\int \varphi_i^2(1)\varphi_j^2(2)\frac{1}{r_{12}}\,\mathrm{d}V_1\,\mathrm{d}V_2$$

and

$$\int \varphi_i(1)\varphi_j(1)\varphi_i(2)\varphi_j(2)\frac{1}{r_{12}}\,\mathrm{d}V_1\,\mathrm{d}V_2$$

In the expression for the Coulomb integral there are squares of wave functions, φ_i^2 and φ_j^2, which determine the spatial density of the electrons occupying the i and j orbitals. Since the greatest electronic density is concentrated around the nuclei and since the distribution of it in the neighbourhood of the nuclei is given (if the constant coefficients are neglected) by the squares of the atomic orbitals, we find that we do not apparently make too great an error if we retain in the expression for the distribution of the charge density due to the electron cloud only whose terms which contain squares of the atomic wave functions. Hence the integral can be approximated as

$$\int \varphi_i^2(1)\varphi_j^2(2)\frac{1}{r_{12}}\,\mathrm{d}V_1\,\mathrm{d}V_2 = \sum_{k,l,m,n} C_{ki}C_{li}C_{mj}C_{nj}\int \chi_k(1)\chi_l(1)\chi_m(2)\chi_n(2)\frac{1}{r_{12}}\,\mathrm{d}V_1\,\mathrm{d}V_2$$

$$\approx \sum_{k,l} C_{ki}^2 C_{lj}^2 \int \chi_k^2(1)\chi_l^2(2)\frac{1}{r_{12}}\,\mathrm{d}V_1\,\mathrm{d}V_2$$

In other words, we retain only the single- and double-centred two-electron Coulomb integrals. The three- and four-centred integrals contained in the integrals of the products of more than two different atomic orbitals disappear.

The physical sense of the two-electron integrals is as follows. Each has the form

$$\int \chi_k^2(1)\chi_l^2(2)\frac{1}{r_{12}}\,\mathrm{d}V_1\,\mathrm{d}V_2$$

and give the energy of the electrostatic interaction of two unit charges distributed in space with densities of χ_k^2 and χ_l^2.

It is clear that if standard atomic orbitals are used for a series of molecules which all contain the same structural grouping, then when the geometry of the grouping remains unchanged over the series of molecules the values of the respective integrals of the form $\int \chi_k^2(1)\chi_l^2(2)1/r_{12}\,dV_1\,dV_2$ that correspond to atoms in the group will be constant. If the distance between the centres of the k and l atomic orbitals is large enough, then these integrals can be estimated as

$$\int \chi_k^2(1)\chi_l^2(2)\frac{1}{r_{12}}\,dV_1\,dV_2 \approx \frac{1}{r_{kl}}$$

Consequently, the size of the integrals diminishes as the distance between the centres of the orbitals increases. An approximate theory need therefore retain only the integrals for neighbouring atoms, and at the limit the integrals for the atoms of a bond only.

We now consider the exchange integral, rewriting it as

$$\int \varphi_i(1)\varphi_j(1)\varphi_i(2)\varphi_j(2)\frac{1}{r_{12}}\,dV_1\,dV_2 = \int\left(\sum_k C_{ki}\chi_k(1)\right)\cdots\left(\sum_n C_{nj}\chi_n(2)\right)\frac{1}{r_{12}}\,dV_1\,dV_2$$

By retaining in the integral only the terms containing the squares of the wave functions of one centre, we obtain

$$\int \varphi_i(1)\varphi_j(1)\varphi_i(2)\varphi_j(2)\frac{1}{r_{12}}\,dV_1\,dV_2 \approx \sum_{k,m} C_{ki}C_{kj}C_{mi}C_{mj}\int \chi_k^2(1)\chi_m^2(2)\frac{1}{r_{12}}\,dV_1\,dV_2$$

i.e. we obtain an expression that is similar to that considered earlier. The three- and four-centred integrals disappear.

One again, integral values appear that are conserved for atoms forming an unchanged grouping. These integrals may either be calculated and tabulated once and for all (like the one-electron Coulomb and resonance integrals which also are localized, see Chapter 3), or may be determined from experiment via a solution of the inverse spectral problem. They can then be used to construct the matrix elements and subsequently to estimate the electron spectra of the series of molecules containing the additive grouping. The methods for solving the inverse spectral problem for electron and vibrational spectra are basically quite similar, and we shall consider them in Chapter 13.

The approximate expressions we have introduced for the Coulomb and exchange integrals in a two-electron system correspond to the retention of the two-electron Coulomb terms in only the diagonal elements in the Hartree–Fock matrix. The relevant integrals here have the form

$$\int \chi_i^2\varphi_j^2(2)\frac{1}{r_{12}}\,dV_1\,dV_2 \approx \sum_k C_{kj}^2\int \chi_i^2(1)\chi_k^2(2)\frac{1}{r_{12}}\,dV_1\,dV_2$$

(including $k = 1$), i.e. only the products of the squares of the atomic orbitals are contained. The non-diagonal elements of the F matrix contain terms of the form

$$\int \chi_k(1)\chi_l(1)\varphi_m^2(2)\frac{1}{r_{12}}\,dV_1\,dV_2$$

in the Coulomb integral, and this cannot be transformed into a form containing only products of squares of atomic orbitals.

The exchange integrals in the diagonal elements of the Hartree–Fock matrix will contain expressions of the form

$$\int \chi_k(1)\varphi_j(2)\varphi_j(1)\chi_k(2)\frac{1}{r_{12}}\mathrm{d}V_1\,\mathrm{d}V_2 \approx C_{ki}^2\int \chi_k^2(1)\chi_k^2(2)\frac{1}{r_{12}}\mathrm{d}V_1\,\mathrm{d}V_2$$

if only integrals containing squares of the atomic orbitals are included. Integrals of this sort will be conserved in value over a series of molecules containing standard atomic orbitals. In fact, the exchange integrals need not be included in the diagonal elements of the F matrix because they have the same form as the Coulomb two-electron integrals and can be calculated from them simply by deducting the value of the Coulomb terms.

In the non-diagonal elements of the Hartree–Fock matrix the expression

$$\int \chi_k(1)\varphi_j(2)\varphi_j(1)\chi_m(2)\frac{1}{r_{12}}\mathrm{d}V_1\,\mathrm{d}V_2 \approx C_{kj}C_{mj}\int \chi_k^2(1)\chi_m^2(2)\frac{1}{r_{12}}\mathrm{d}V_1\,\mathrm{d}V_2$$

will correspond to an exchange term if we retain only the terms containing the squares of atomic orbitals.

In this way we see that the expression for the density of the electronic charge of a sum of atomic orbital squares reduces to substantially simpler forms for the matrix elements in the self-consistent case. Obviously, a simpler reduction can be achieved when also calculating the configuration interaction matrix elements. This enables the corresponding estimates for complicated molecules to be calculated in real time. Similar approximations with small variations in detail lie at the basis of all of the most popular calculation methods for complicated systems. As we have shown, the simplification in calculating the remaining integrals may be achieved either by tabulation or by parameterizing the values of the individual integrals included in a matrix element, and then rejecting or estimating the remaining integrals in the matrix element expressions using approximation formulae. The semi-empirical procedures for calculating the electronic shells of a polyatomic molecule are developed in this way[9,10].

The value of semi-empirical theory is that first, quite complicated calculations can be carried out in a relatively short time on medium-sized computers. Moreover, what is especially important is that using a reasonable set of parameters determined by experiment the errors in the theory itself (due to the assumptions that underlie it) can be compensated for. Unfortunately, calculating the different parameters of a molecule such as its energy levels, dipole moments, force constants, etc. requires, as experience has shown, a number of different or numerically valued parameter systems even when the most modern semi-empirical methods are used. Thus, in order to solve each problem a new system of parameters must be chosen, which, it is found, can only be used for a given class of molecules and then only for analysing the molecular property originally studied. This is the major weakness of all the semi-empirical schemes used in quantum mechanical studies of polyatomic molecules (see also Chapter 1).

Recently another scheme, which could be called a hybrid method, has become popular. In this method some of the integrals in the relevant matrix elements, such as the single- and two-electron single-centred ones, are calculated accurately, whilst the remaining integrals are estimated and given in parametric form. There are grounds for believing that this approach might yield better results and could lead to the selection of sets of parameters that will apply for a much larger category of molecules and that will also be suitable for describing a range of different properties.

Many other different approximation methods have been developed but they will not be discussed here.

4.6 Expressions for the Matrix Elements of Dipole Transitions

We know that the probability of an electronic transition is given by the size of the square dipole moment's matrix element. This has the form

$$\vec{\mu}_{mn} = \sum_i \int \Psi''_v(Q)\Psi'_v(Q)\,\mathrm{d}Q \int \Psi''_e(r,Q)\vec{r}_i\Psi'_e(r,Q)\,\mathrm{d}V$$

$$\approx \left(\sum_i \int \Psi''_e(r,0)\vec{r}_i\Psi'_e(r,0)\,\mathrm{d}V\right)\int \Psi''_v(Q)\Psi'_v(Q)\,\mathrm{d}Q$$

The first multiplicand corresponds to a purely electronic transition. If we now take advantage of the representation of an electronic wave function as a linear combination of determinants, then we obtain the following for the electronic part of the matrix element:

$$\sum_i \int \Psi''_e\vec{r}_i\Psi'_e\,\mathrm{d}V = \sum_{i,k,n} C''_k C'_n \int \phi_n\vec{r}_i\phi_n\,\mathrm{d}V \tag{4.62}$$

Moreover, the matrix elements between the determinantal functions ϕ_k and ϕ_n are non-zero only when these functions correspond to the same spin state. It is appropriate here to note that transitions are also not allowed (their integrals vanish) when the eigenfunctions of the combining electron states have the same overall spin projection, but correspond to a different multiplet. Thus, for example, a transition between an excited singlet level (s) and a component of a triplet (t) also with a zero overall spin projection state is not allowed.

We showed in Section 4.2 that in this case

$$\Psi_t = \frac{1}{\sqrt{2}}(\phi_3 + \phi_4) \quad \text{and} \quad \Psi_s = \frac{1}{\sqrt{2}}(\phi_3 - \phi_4)$$

Hence

$$\int \Psi_t\left(\sum_i \vec{r}_i\right)\Psi_s\,\mathrm{d}V = \frac{1}{2}\int (\phi_3^2 - \phi_4^2)\left(\sum_i \vec{r}_i\right)\mathrm{d}V = 0$$

because the average value of the electronic components of the dipole moments, as given by the integrals $\int \phi_3^2(\sum_i \vec{r}_i)\mathrm{d}V$ and $\int \phi_4^2(\sum_i \vec{r}_i)\mathrm{d}V$, are equal (the ϕ_3 and ϕ_4 states have the same spatial structure).

A transition from the non-excited ground singlet state to an excited singlet one is allowed, whilst a transition from the ground (singlet) state to an excited triplet one (or vice versa) is not allowed. This can be seen from a two-level system with the orbitals φ_1 and φ_2.

Since the ground singlet state corresponds to a full shell case, then it has the following corresponding determinantal function:

$$\phi_0 = \frac{1}{\sqrt{2}} \det \begin{bmatrix} \varphi_1(1)\alpha(1) & \varphi_1(2)\alpha(2) \\ \varphi_1(1)\beta(1) & \varphi_1(2)\beta(2) \end{bmatrix}$$

That is, both electrons, which have opposite spins, occupy a state with the spatial orbital φ_1. A function in the form of a difference of determinant (see Section 4.2), i.e.

$$\phi_s = \frac{1}{\sqrt{2}}(\phi_3 - \phi_4)$$

corresponds to the singlet excited state. Above we have

$$\phi_3 = \frac{1}{\sqrt{2}} \det \begin{bmatrix} \varphi_1(1)\alpha(1) & \varphi_1(2)\alpha(2) \\ \varphi_2(1)\beta(1) & \varphi_2(2)\beta(2) \end{bmatrix}$$

(the electron with the α spin part jumps to the φ_2 orbital), and

$$\phi_4 = \frac{1}{\sqrt{2}} \det \begin{bmatrix} \varphi_1(1)\beta(1) & \varphi_1(2)\beta(2) \\ \varphi_2(1)\alpha(1) & \varphi_2(2)\alpha(2) \end{bmatrix}$$

(the electron with the β spin part jumps to the φ_2 orbital).

The matrix element of the dipole transition with a singlet ground state is in this case equal to

$$\vec{\mu}_{mn} = \frac{1}{\sqrt{2}} \int (\phi_3 - \phi_4)(\vec{r}_1 + \vec{r}_2)\phi_0 \, \mathrm{d}V_1 \, \mathrm{d}V_2 = \frac{2}{\sqrt{2}} \int \varphi_2(1)\vec{r}_1\varphi_1(1) \, \mathrm{d}V_1 \tag{4.63}$$

Therefore

$$\int \phi_3(\vec{r}_1 + \vec{r}_2)\phi_0 \, \mathrm{d}V_1 \, \mathrm{d}V_2$$

$$= \frac{1}{2} \int \{\varphi_1(1)\alpha(1)\varphi_2(2)\beta(2) - \varphi_2(1)\beta(1)\varphi_1(2)\alpha(2)\}(\vec{r}_1 + \vec{r}_2)$$

$$\times \{\varphi_1(1)\alpha(1)\varphi_1(2)\beta(2) - \varphi_1(1)\beta(1)\varphi(2)\alpha(2)\} \, \mathrm{d}V_1 \, \mathrm{d}V_2$$

which, after expanding the terms in brackets and summing over the spin variables, gives

$$\int \phi_3(\vec{r}_1 + \vec{r}_2)\phi_0 \, \mathrm{d}V_1 \, \mathrm{d}V_2$$

$$= \frac{1}{2} \int \varphi_1(1)\varphi_2(2)(\vec{r}_1 + \vec{r}_2)\varphi_1(1)\varphi_1(2) \, \mathrm{d}V_1 \, \mathrm{d}V_2$$

$$+\frac{1}{2}\int \varphi_2(1)\varphi_1(2)(\vec{r}_1+\vec{r}_2)\varphi_1(1)\varphi_1(2)\,\mathrm{d}V_1\,\mathrm{d}V_2$$

$$=\frac{1}{2}\int \varphi_1(1)\vec{r}_1\varphi_1(1)\,\mathrm{d}V_1\int \varphi_2(2)\varphi_1(2)\,\mathrm{d}V_2 \tag{4.64}$$

$$+\frac{1}{2}\int \varphi_1^2(1)\,\mathrm{d}V_1\int \varphi_2(2)\vec{r}_2\varphi_1(2)\,\mathrm{d}V_2$$

$$+\frac{1}{2}\int \varphi_2(1)\vec{r}_1\varphi_1(1)\,\mathrm{d}V_1\int \varphi_1^2(2)\,\mathrm{d}V_2$$

$$+\frac{1}{2}\int \varphi_1(2)\vec{r}_2\varphi_1(2)\,\mathrm{d}V_2\int \varphi_2(1)\varphi_1(1)\,\mathrm{d}V_1=\int \varphi_2(1)\vec{r}_1\varphi_1(1)\,\mathrm{d}V_1$$

if we take into account that the integrals $\int\varphi_2(2)\vec{r}_2\varphi_1(2)\,\mathrm{d}V_2$ and $\int\varphi_2(1)\vec{r}_1\varphi_1(1)\,\mathrm{d}V_1$ are equal and the orbitals φ_1 and φ_2 are orthogonal.The integral

$$\int \phi_4(\vec{r}_1+\vec{r}_2)\phi_0\,\mathrm{d}V_1\,\mathrm{d}V_2$$

$$=\frac{1}{2}\int \{\varphi_1(1)\beta(1)\varphi_2(2)\alpha(2)-\varphi_2(1)\alpha(1)\varphi_1(2)\beta(2)\}(\vec{r}_1+\vec{r}_2)$$

$$\times\{\varphi_1(1)\alpha(1)\varphi_1(2)\beta(2)-\varphi_1(1)\beta(1)\varphi_1(2)\alpha(2)\}\,\mathrm{d}V_1\,\mathrm{d}V_2$$

after expanding the term in braces and summing over the spin variables is transformed into the form

$$\int \phi_4(\vec{r}_1+\vec{r}_2)\phi_0\,\mathrm{d}V_1\,\mathrm{d}V_2=-\frac{1}{2}\int \varphi_1(1)\varphi_2(2)(\vec{r}_1+\vec{r}_2)\varphi_1(1)\varphi_1(2)\,\mathrm{d}V_1\,\mathrm{d}V_2$$

$$-\frac{1}{2}\int \varphi_2(1)\varphi_1(2)(\vec{r}_1+\vec{r}_2)\varphi_1(1)\varphi_1(2)\,\mathrm{d}V_1\,\mathrm{d}V_2$$

By comparing this expression with the expression for $\int\phi_3(\vec{r}_1+\vec{r}_2)\phi_0\,\mathrm{d}V_1\,\mathrm{d}V_2$ we obtain

$$\int \phi_3(\vec{r}_1+\vec{r}_2)\phi_0\,\mathrm{d}V_1\,\mathrm{d}V_2=-\int \phi_4(\vec{r}_1+\vec{r}_2)\phi_0\,\mathrm{d}V_1\,\mathrm{d}V_2$$

$$=-\int \varphi_2(1)\vec{r}_1\varphi_1(1)\,\mathrm{d}V_1$$

For the triplet component with $S_z=0$ the eigenfunction is

$$\phi_t=\frac{1}{\sqrt{2}}(\phi_3+\phi_4)$$

and so the matrix element of a singlet–triplet transition will be

$$\frac{1}{\sqrt{2}}\int (\phi_3+\phi_4)(\vec{r}_1+\vec{r}_2)\phi_0\,\mathrm{d}V_1\,\mathrm{d}V_2=0$$

Thus singlet–triplet transitions are prohibited. As was shown in Chapter 3, matrix elements of the type

$$\int \phi_k \left(\sum_i \vec{r}_i \right) \phi_n \, dV$$

for the same complete spin projections will be non-zero only when the ϕ_k and ϕ_n determinantal functions differ only by one spatial orbital.

The probability of singlet–singlet transition is determined (see equation (4.63)) by the value of the matrix element of the dipole transition between two one-electron orbitals of the corresponding symmetry. This question will be discussed in the next chapter.

4.7 Use of Overcomplete Basis Sets

Numerical computations of electronic structural parameters of atoms and molecules by LCAO–MO methods using large bases of AO often leads to the problem of the so-called 'overcomplete' basis. This problem arises from the singularity of matrix of overlap integrals. Such a situation can arise owing to computer round-off errors of overlap matrix elements and because of the peculiarities of the LCAO–MO method, as well as when distances between some nuclei of a molecule become small and AO on different close centres coincide, thus leading to an overcompleteness of the basis set. In calculations of Rydberg states such a situation arises quite often. In this case the overlap matrix will contain many small eigenvalues.

In the literature attention has been paid repeatedly to the difficulties caused by the usual technique of construction of an LCAO–MO, and these have led to recommendations that variational calculations utilizing large bases should not be carried out.

We shall describe the corresponding procedure for the solution of electronic problems by Hartree–Fock–Roothaan LCAO–MO method. In this case the values of energy levels are obtained as a result of simultaneous diagonalization of two matrices: the matrix of overlap integrals S and the Hartree–Fock matrix F. This operation is equivalent to the simultaneous reduction of two quadratic forms of matrices S and F into a sum of squares, with matrix S being transformed to a unit matrix. The matrices S and F are square and symmetric and matrix S is positive defined. Let the order of these matrices be N.

Independently of its singularity, matrix S can be transformed into a diagonal form Λ_s by an orthogonal transformation L_s. For the construction of matrix L_s one of the variants of the rotation method can be used as it is non-sensitive to singularity of the matrix to be diagonalized.

Hence we have:

$$\tilde{L}_s S L_s = \Lambda_s \tag{4.65}$$

If the matrix S is singular then, permutating the columns of the matrix L_s, we can write without loss of generality:

$$\begin{bmatrix} \tilde{L}_{s1} \\ \tilde{L}_{s2} \end{bmatrix} S[L_{s1} \quad L_{s2}] = \begin{bmatrix} \Lambda_{s1} & 0 \\ 0 & 0 \end{bmatrix} \tag{4.66}$$

Here, L_{s1} is a rectangular submatrix of $N \times (N-n)$ size (n is the singularity degree of the matrix S), composed with the eigenvectors of the matrix L_s, corresponding to non-zero eigenvalues of the matrix S; L_{s2} is the rectangular submatrix of the matrix L_s, composed with the eigenvectors, corresponding to the zero eigenvalues of the matrix S;

$$\begin{bmatrix} \Lambda_{s1} & 0 \\ 0 & 0 \end{bmatrix}$$

is the diagonal matrix which contains the diagonal submatrix Λ_{s1} of $(N-n)$ order with all non-zero eigenvalues λ_s of the matrix S on its diagonal.

Evidently, relationship (4.66) is equivalent to

$$\tilde{L}_{s1} S L_{s1} = \Lambda_{s1} \tag{4.67}$$

The transformation of the original basis of AOs with the matrix $L_{s1}\Lambda_{s1}^{-1/2}$ converts the original matrix S to the unit matrix of smaller (for value n) order. This is equivalent to a transition to a new orthonormal set including independent linear combinations of original AOs whose number is less than the number of original AOs.

Simultaneous investigation of eigenvectors composing the submatrix L_{s2} can be used for the separation of the independent original AOs when the co-ordinates of these vectors are the coefficients in the linear combinations of the original AO. Nevertheless, note that there is no such necessity in practical calculations.

The transition to the new reduced basis of orthonormal, independent linear combinations of the original AO needs the corresponding transformation of the matrix F:

$$F' = \Lambda_{s1}^{-1/2} \tilde{L}_{s1} F L_{s1} \Lambda^{-1/2} \tag{4.68}$$

Thus the square symmetric matrix whose order is reduced by n has to be transformed by the terminal orthogonal transformation with the matrix L_F into a diagonal form $E = \tilde{L}_F F' L_F$, and this ends the solution of the problem. The full matrix of transformation from the original AO basis to the new orthonormal MO basis, which diagonalizes the matrix F, has the form:

$$\phi = L_{s1} \Lambda_{s1}^{-1/2} L_F \Phi = C\Phi \tag{4.69}$$

where ϕ is a column-matrix of the original AO and Φ is a column-matrix of the MOs. The matrix C has the size $N \times (N-n)$.

Relationship (4.69) allows us to express the original AO through the MO set. The inverse relationship, by means of which it is possible to find the MOs and corresponding coefficients for AOs has the form:

$$\Phi = \tilde{L}_F \Lambda_{s1}^{1/2} \tilde{L}_{s1} \phi = C'\phi \tag{4.70}$$

It is clear, that $C'C = 1$ (the unit matrix of order $(N-n)$), but $CC' \neq 1$.

Therefore relationship (4.70) is obtained from equation (4.69) simply by multiplying it on the left by the rectangular matrix C' of dimensions $(N-n) \times N$. To this end, we see that the variational problem can be successfully solved in overcomplete AO basis sets with practically no changes in computational procedure.

Note that in some cases the deliberate use of dependent AO orbitals can be caused by a physical peculiarity of a model is convenient. For example, if we choose functions $3d \simeq x^2e^{-r}$, y^2e^{-r}, z^2e^{-r}; $3s \simeq r^2e^{-r}$ instead of hydrogen-like $3d$-functions for an atom, then it follows from the relationship $3d_{x^2} + 3d_{y^2} + 3d_{z^2} = 3s$ that the basis $3s$, $3d_{x^2}$, $3d_{y^2}$, $3d_{z^2}$ is overcomplete.

References

1. R. McWeeny and B. T. Sutcliffe. *Methods of Molecular Quantum Mechanics*, Academic Press, London/New York (1969).
2. B. K. Novosadov and L. A. Gribov. *Theoret. Chim. Acta (Berl.)*, **67**, 449 (1985).
3. S. G. Mikhlin. *Variational Methods in Mathematical Physics* 2nd edn, Nauka, Physico-Mathematical Literature, Moscow (1970) (in Russian).
4. D. R. Hartree. *Proc. Camb. Phil. Soc.*, **24**, 89, 111, 426 (1928).
5. V. A. Fock. *Foundation of Quantum Mechanics*, Nauka, Physico-Mathematical Literature, Moscow (1976) (in Russian).
6. J. C. Slater. *The Self-Consistent Field for Molecules and Solids*, McGraw-Hill, New York (1974).
7. B. K. Novosadov. *J. Molec. Struct.*, **54**, 269 (1979).
8. C. C. J. Roothaan. *Rev. Modern Physics*, **23**, No. 2, 69 (1951).
9. G. A. Segal (ed.). *Semiempirical Methods of Electronic Structure Calculation*, Vols 1 and 2, Plenum Press, New York/London, 1977.
10. M. J. S. Dewar. *The Molecular Orbital Theory of Organic Chemistry*, McGraw-Hill, New York (1969).
11. L. A. Gribov and B. K. Novosadov. *J. Molec. Struct.* (*Theochem.*), **136**, 387 (1986).

Chapter 5

The Symmetries of Molecular Orbitals and Normal Co-ordinates

5.1 The Symmetry Elements of Polyatomic Molecules

A most important concept in the theory of molecular structures is symmetry, and we now introduce the definition of its elements and operations.

(1) Centre of symmetry Consider a domain V in space, and a point A outside the domain. We draw a straight line from an arbitrary point B in V through A to another point B′ such that the length of AB is equal to that of AB′ (see Figure 5.1). If we carry out a similar operation for a point C, and every other point in V, we shall obtain a domain V' similar to V. It is said that V is symmetric to V' with respect to the centre of symmetry at A.

If the centre of symmetry coincides with the origin of the co-ordinate then the co-ordinates of all the points in V' can be obtained from those in V by simply changing their signs. This operation is called an 'inversion with respect to the centre of symmetry'.

(2) Plane of symmetry We now isolate a plane outside the domain V. We draw a straight line from an arbitrary point A within V perpendicular to the plane and end the line at a point A′ such that the distance between A and the plane equals the distance between the plane and A′ (see Figure 5.2). If a similar procedure is used for every point in V, then a domain V' is obtained that is symmetric to V with respect to the plane. Such a plane is called a 'plane of symmetry'.

If the co-ordinate axes are chosen so that one of them, say the x axis, is

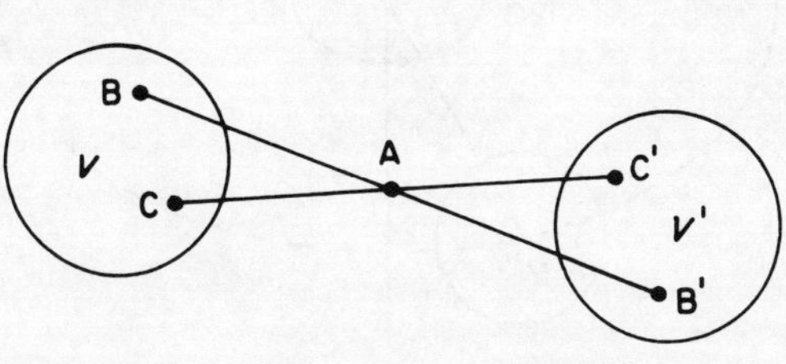

Figure 5.1. Symmetry with respect to a centre

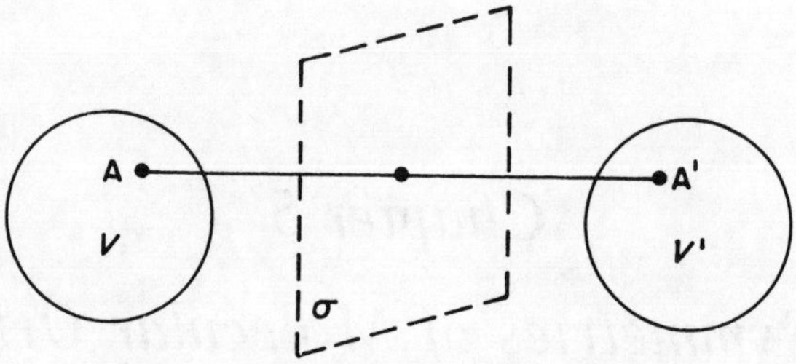

Figure 5.2. Symmetry with respect to a plane

perpendicular to the plane of symmetry so that the origin lies in the plane, then all the co-ordinates of V' can be obtained from those of V simply by changing the signs of their x-co-ordinates. This operation is called a 'reflection in a plane'.

(3) Axis of symmetry If there is a domain V' in space, or several domains (V', V'',...) obtained from the original domain V by a rotation about an axis through an angle that is a multiple of $2\pi/p$, then it is said that the domains are symmetric to each other with respect to the axis of symmetry C_p of order p (see Figure 5.3).

The appropriate co-ordinate transformations for rotation are well known, and we can write the matrix for the symmetry transformation. If the z-axis coincides with the axis of symmetry we obtain:

$$\begin{array}{c} \\ x \\ y \\ z \end{array} \begin{array}{c} \begin{array}{ccc} x' & y' & z' \end{array} \\ \begin{bmatrix} \cos\dfrac{2\pi}{p}; & -\sin\dfrac{2\pi}{p}; & 0 \\ \sin\dfrac{2\pi}{p}; & \cos\dfrac{2\pi}{p}; & 0 \\ 0 & 0 & 1 \end{bmatrix} \end{array} \tag{5.1}$$

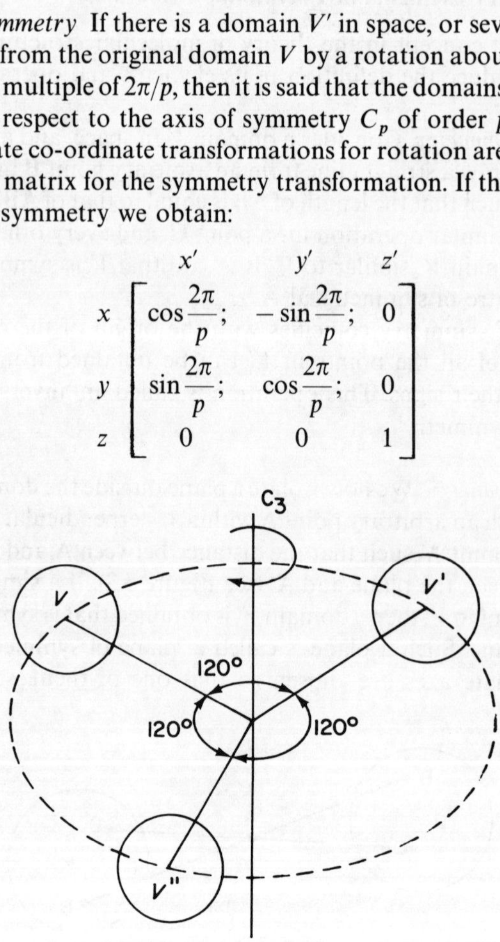

Figure 5.3. Symmetry with respect to a third-order axis

The final symmetry element is called the 'reflection–rotation axis' S_p and combines a rotation about the axis and a reflection in the plane perpendicular to the axis.

If the entire space of a molecule can be divided into separate parts that are distributed according to certain symmetry relations, then it is said that the molecule possesses symmetry. The set of a molecule's symmetry elements (centre, plane and axis of symmetry) is characterized by the molecule's symmetry point group. A group is called a point group if, after all the possible symmetry operations have been completed, one point of space remains unchanged. Some of the possible point groups for a molecule are shown in Table 5.1.

Conventionally, every molecule that has symmetry elements can be categorized as a molecule of low, medium or high symmetry. The set of low-symmetry molecules contains those which have only a centre, a plane or a second-order axis of symmetry; the medium-symmetry molecules are those that have one axis of symmetry that is third order or higher; and high-symmetry molecules may have several axes of symmetry, third order or higher. The presence of symmetry elements within a molecule leads to a number of very important consequences.

Table 5.1. The symmetry elements for several point groups

Designation of group	Number and type of symmetry elements
C_2	C_2
C_3	C_3
C_i	i
S_6	$S_6; C_3(\| S_6)$
C_s	σ
C_{2v}	$C_2; 2\sigma_v$
C_{3v}	$C_3; 3\sigma_v$
C_{4v}	$C_4; C_2(\| C_4); 4\sigma_v$
C_{6v}	$C_6; C_3(\| C_6); C_2(\| C_6); 6\sigma_v$
D_2	$3C_2(C_2^{(1)} \perp C_2^{(2)} \perp C_2^{(3)})$
D_3	$C_3; 3C_2(C_2^{(1)}, C_2^{(2)}, C_2^{(3)} \perp C_3)$
C_{2h}	$C_2; \sigma_h; i$
C_{3h}	$C_3; \sigma_h; S_3(\| C_3)$
D_{2d}	$3C_2(C_2^{(1)} \perp C_2^{(2)} \perp C_2^{(3)}); S_4(\| C_2); 2\sigma_d(\| S_4)$
D_{3d}	$C_3; 3C_2(\perp C_3); S_6(\| C_3); i; 3\sigma_d$
D_{2h}	$C_2(C_2^{(1)} \perp C_2^{(2)} \perp C_2^{(3)}); 3\sigma(\sigma^{(1)} \perp \sigma^{(2)} \perp \sigma^{(3)})$
D_{3h}	$C_3; 3C_2(\perp C_3); 3\sigma_v; \sigma_h$
D_{4h}	$C_4; 4C_2(\perp C_4); 4\sigma_v; \sigma_h; C_2(\| C_4); S_4(\| C_4); i$
D_{6h}	$C_6; 6C_2(\perp C_6); 6\sigma_v; 6_h; C_2(\| C_6); C_3(\| C_6); S_6(\| C_6); i$
T	$3C_2(C_2^{(1)} \perp C_2^{(2)} \perp C_2^{(3)}); 4C_3$
T_d	$3C_2(C_2^{(1)} \perp C_2^{(2)} \perp C_2^{(3)}); 4C_3; 6\sigma; 3S_4(\| C_2)$

Notes: C_p: axis of order P; S_p: reflection rotation axis of order p; i: centre of symmetry; σ: plane of symmetry; $\bar{\sigma}_v$: plane of symmetry parallel to an axis of symmetry; σ_h: plane of symmetry perpendicular to an axis of symmetry; σ_d: plane of symmetry that bisects the angle between two second-order axes of symmetry; $C_p(\perp C_{p'})$: axis of symmetry perpendicular to the axis of symmetry C_p; $C_p\,(\| C_{p'})$: axis of symmetry parallel to the axis of symmetry $C_{p'}$.

First, it is plain that the density of states must be the same for the equilibrium position in different domains of a molecule's space, the domains being transformable into each other via the molecule's symmetry operations. This means that the square of the wave function that defines this density must not be changed from one symmetric domain to another. Hence we must have the following for two points whose co-ordinates are x, y, z and x', y', z' and which are symmetric with respect to a centre or plane of symmetry:

$$\Psi^2(x, y, z) = \Psi^2(x', y', z')$$
$$\Psi(x, y, z) = \pm\Psi(x', y', z')$$

An operation which substitutes the co-ordinates in a function of a point in space with the co-ordinates of another point that is symmetric to the first with respect to a given symmetry element is called a 'symmetry operation'. The operation may be compared to the appropriate symmetry operator. Then we can say that when the symmetry operator that corresponds to a second-order symmetry element acts on a function, the eigenfunction either remains unchanged or changes sign. In the first case, the function acted upon is called symmetric with respect to the given symmetry element, while in the second it is anti-symmetric with respect to the element. The square of the wave function is unchanged by the action of the symmetry operator on it. We can arrive at the same conclusion by considering the invariance of $\bar{H}$ and the integrand in the normalized integral during symmetry operations.

When the symmetry elements of third or higher order exist, then we already have three or more symmetric points for which the squares of the wave function must be invariant. Here, if the action of a second-order symmetry operation causes the co-ordinates of points to be transformed in pairs, i.e. co-ordinate x of one point is transformed to co-ordinate x' of the other and co-ordinate y to y', then a rotation about, say, a third-order axis that coincides with the z axis changes the x, y pair of one point into the x', y' pair of another, which in turn is changed simultaneously into the x'', y'' pair of the third symmetric point. Finally the x'', y'' pair is transformed to the x, y pair of the first point. Thus a consistent simultaneous transformation of co-ordinates in the different points of space takes place.

It can be shown that the presence in a molecule of symmetry elements of third order or higher causes degenerate solutions, i.e. values of the energy of the molecule to which two or more eigenfunctions correspond. In this case the density of the state is defined by the sum of the squares of the degenerate wave functions, and since the value of the state density function must be invariant at symmetry points in a molecule's space, so the sum of the squares of the wave functions of the states degenerate with respect to the appropriate symmetry operation must also remain invariant.

An analogous situation is true for normal co-ordinates, Q, when describing vibrations of polyatomic molecules. In this case the potential energy must be invariant with respect to symmetry operations. In normal co-ordinates and in the harmonic approximation it takes the form

$$u_{\text{pot}} = \tfrac{1}{2}\sum_k \lambda_k Q_k^2 + \tfrac{1}{2}\sum_m \lambda_m (Q_{m_1}^2 + Q_{m_2}^2) + \tfrac{1}{2}\sum_l \lambda_l (Q_{l_1}^2 + Q_{l_2}^2 + Q_{l_3}^2)$$

The second and third terms of this expression correspond to doubly and triply degenerate vibrations. When a symmetry operator acts upon this expression the squares of the normal co-ordinates for the non-degenerate vibrations and the sum of their squares for degenerate vibrations must not be unchanged. Therefore the normal co-ordinates for non-degenerate vibrations, like the wave functions, may only be symmetric or anti-symmetric with respect to second-order symmetry elements, or degenerate in the presence of third-order elements. The normal co-ordinates in the latter case are simultaneously changed in pairs or in triplets, but the sum of their squares remains invariant.

The symmetry of the wave functions and normal co-ordinates has great significance when working out the various kinds of integrals that are encountered in the theory of electronic and vibrational spectra. A basic requirement is that the integral does not change for any symmetry operation that is allowable for the molecule. If this is not true then the integral vanishes.

It follows in particular from this that $S_{nm} = \int \Psi_n \Psi_m \, dV = 0$ and that $h_{nm} = \int \Psi_n \hat{H} \Psi_m \, dV = 0$ if the functions Ψ_n and Ψ_m belong to different symmetry types, i.e. the actions of symmetry operators change them in different ways. That the first integral equals zero is apparent, whilst the equality to zero of the second is because the action of the operator $\hat{H}$ on a function does not change its symmetry.

When the quantum-mechanical problem is being solved by the variational method the use of functions having a certain symmetry as the basis functions Φ_k makes it possible to simplify the task considerably, because a large number of the S_{nm}- and h_{nm}-matrix elements vanish. An analogous simplification is achieved if symmetry co-ordinates, which depend on a symmetry operation, are made use of when calculating vibrations. We will become more familiar with this aspect below.

5.2 Symmetricized Linearly Combined Atomic Orbitals: Hybridization

We shall consider in this section the important topic of the construction of electronic wave functions that possess properties of symmetry. We have shown in the previous section that if a molecule has symmetry elements then its eigenfunctions must definitely be transformed in full during all symmetry operations that are intrinsic to the molecule. In particular, for example, during symmetry operations corresponding to second-order elements the electronic eigenfunctions must either be symmetric or antisymmetric.

In order to explain how symmetricized functions (we will call them molecular orbitals of symmetry (MOS)) can be constructed we will use linearly combined atomic orbitals (LCAO) as the basis. We introduce the concepts of equivalent atoms and an atomic orbital (AO). We call atoms in symmetric molecules equivalent if they are of a single sort and are either transformed into each other during all the symmetry operations of a molecule or else remain unchanged. Thus, for example, the two hydrogen atoms in acetylene are equivalent, as are the two carbons, the four hydrogen atoms in ethylene and methane are equivalent, etc.

In the LCAO method each atom has its own atomic orbitals centred on it. We call the atomic orbitals of equivalent atoms that do not change when transformed from one to another during all of the molecule's symmetry operations 'equivalent' atomic orbitals. Clearly, any *s* orbital is an equivalent atomic orbital because it has spherical symmetry and only includes a radial part. Of course, this implies that the exponential factors in the atomic orbitals of equivalent atoms must be chosen to be the same. The situation is very much more difficult for the *p*, *d*, etc. orbitals. These atomic orbitals include an angular component as well as a radial one and depend on the orientation of their local co-ordinate axes with respect to the Cartesian co-ordinate axes that are common for the whole molecule. It is more the exception than the rule that the orientations of *p*- and *d*-type atomic orbitals can be chosen for equivalent atoms so that the atomic orbitals themselves are equivalent. Fortunately, the problem can, in many cases, be resolved if we go from the usual atomic orbitals (hydrogen-like or Slater orbitals) to linear combinations of the atomic orbitals centred on an atom which will have the equivalence property with respect to the molecule's symmetry elements.

The new atomic orbitals constructed in this way are called hybrid atomic orbitals (HAOs). They have a clearly expressed directional effect which leads to increased electronic density in the space between the two atoms forming a bond, and this is exactly the condition required for a molecule's stable geometric configuration to be formed.

We shall now consider how to construct hybrid orbitals for a few examples that, whilst being special cases, are nonetheless of very wide occurrence. Here and below we suppose that *s* AO corresponds to 2*s* atomic state.

(1) sp hybridization Let us assume that we have an atom with one *s* orbital and one p_x orbital, oriented along the *x* axis (see Figure 5.4). We wish to construct two hybrid orbitals that will be equivalent with respect to a centre of symmetry coincident with the atom or a plane of symmetry perpendicular

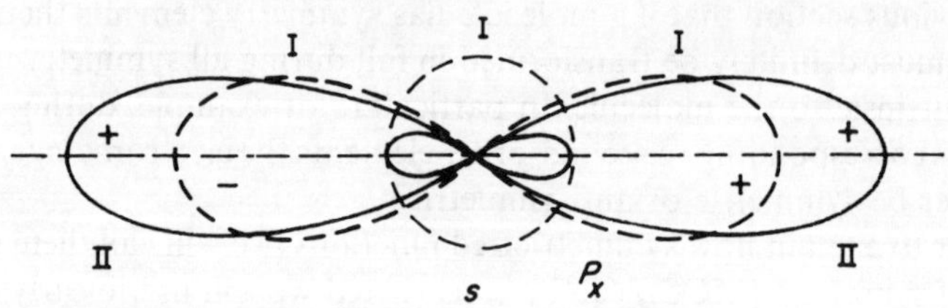

Figure 5.4. *s* and p_x atomic orbitals (I) and hybrid *sp* orbitals symmetric with respect to a second-order element (II)

to the x axis. The p_x function does not have this property because it changes sign when reflected in a plane or rotated about an axis.

We compile two new hybrid functions (HAOs) or linear combinations of the originals:

$$\begin{aligned}\chi_1^{(h)} &= c_{1s}\chi_s + c_{1p}\chi_{px}\\ \chi_2^{(h)} &= c_{2s}\chi_s + c_{2p}\chi_{px}\end{aligned} \tag{5.2}$$

and we require that the hybrid orbitals are normalized, orthogonal and symmetric with respect to both the plane and axes of symmetry. We have

$$\begin{aligned}1 &= \int \chi_1^{(h)2}\,\mathrm{d}v = c_{1s}^2 \int \chi_s^2\,\mathrm{d}v + c_{1p}^2 \int \chi_{px}^2\,\mathrm{d}v + 2c_{1s}c_{1p}\int \chi_s\chi_{px}\,\mathrm{d}v = c_{1s}^2 + c_{1p}^2\\ &= \int \chi_2^{(h)2}\,\mathrm{d}v = c_{2s}^2 + c_{2p}^2;\quad 0 = \int \chi_1^{(h)}\chi_2^{(h)}\,\mathrm{d}v = c_{1s}c_{2s} + c_{1p}c_{2p}\end{aligned} \tag{5.3}$$

We consider here that χ_s and χ_{p_x} each have one centre, are normalized and are orthogonal.

The requirement of symmetry with respect to the plane perpendicular to the x axis means that

$$\chi_1^{(h)}(x) = -\chi_2^{(h)}(-x)$$

or

$$c_{1s}\chi_s(x) + c_{1p}\chi_{p_x}(x) = c_{2s}\chi_s(-x) + c_{2p}\chi_{p_x}(-x) \tag{5.4}$$

Since $\chi_s(x) = \chi_s(-x)$ and $-\chi_{p_x}(x) = -\chi_{p_x}(-x)$, we must conclude that the following must be true:

$$c_{1s} = c_{2s} \quad \text{and} \quad c_{1p} = -c_{2p}$$

By taking advantage of equation (5.3) we find that, for normalization to be achieved,

$$0 = c_{1s}^2 - c_{1p}^2 \quad \text{or} \quad c_{1s} = \pm c_{1p} = \frac{1}{\sqrt{2}} \tag{5.5}$$

The symmetric hybrid atomic orbital we are looking for will have the form

$$\chi_1^{(h)} = \frac{1}{\sqrt{2}}(\chi_s + \chi_{px});\quad \chi_2^{(h)} = \frac{1}{\sqrt{2}}(\chi_s - \chi_{px}) \tag{5.6}$$

The form of these orbitals is shown in Figure 5.4. It is clear that they have directional properties and, moreover, the maxima of $\chi_1^{(h)2}$ and $\chi_2^{(h)2}$ are distributed to the right and left, respectively, with respect to the atom. If we take a linear symmetric grouping of atoms ABA with the hybrid orbitals we are considering on atom B, then the orbitals will provide a larger electron density in an AB bond than the usual s or p orbitals.

(2) sp^2 hybridization We shall consider how to construct three equivalent hybrid atomic orbitals on one centre with C_{3h} symmetry, i.e. the orbitals will

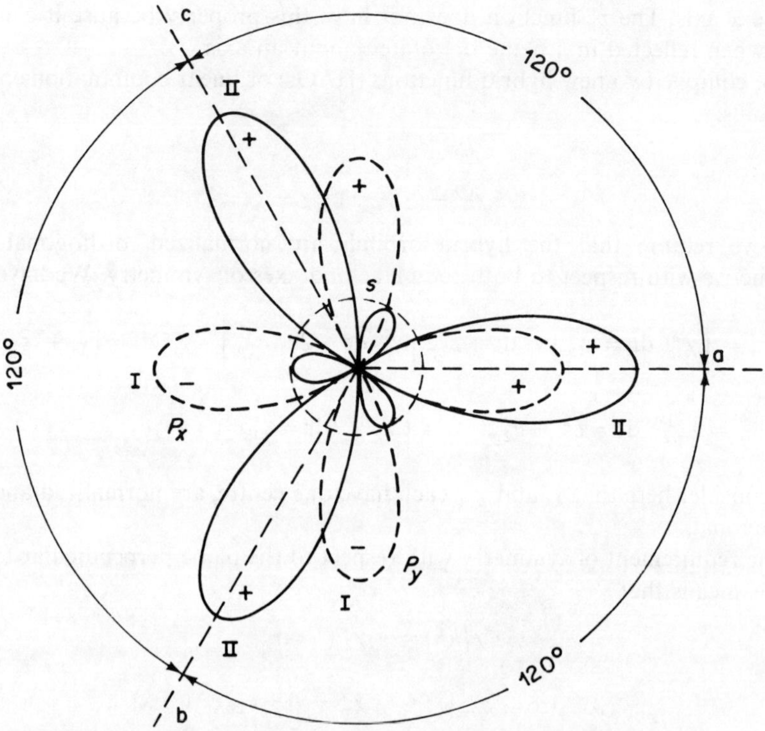

Figure 5.5. s, p_x and p_y orbitals (I) and symmetricized hybrid orbitals (II). The third-order axis is perpendicular to the plane of the page through the atom

be symmetric with respect to one third-order axis and a plane perpendicular to it. These orbitals may be constructed from one s orbital and two p orbitals (p_x and p_y). These hybrids are encountered in A_3B-type planar symmetric molecules (see Figure 5.5), or in atomic groupings that have this type of local symmetry.

We choose the distribution of the p_x- and p_y-orbitals as shown in Figure 5.5. Our task is to construct three equivalent hybrid orbitals that correspond to C_{3h} symmetry, i.e. they lie in one plane (the xy plane) and can be transformed into each other by a rotation about a third-order axis that is perpendicular to the plane, or can be transformed by reflections in any of three planes each of which passes through this third-order axis and one of the straight lines a, b or c. The lines a, b and c intersect at the axis and make angles of 120° ($2\pi/3$) with each other.

We are looking for a HAO in the form of a LCAO function of the following kind:

$$\chi_i^{(h)} = c_{is}\chi_s + (c_{ip_x}x + c_{ip_y}y)f_p(r), \quad i = 1, 2, 3 \tag{5.7}$$

We have assumed here that the radial parts of the p orbitals are the same and the orbitals themselves can be represented as the products

$$\chi_{p_x} = x f_p(r)$$
$$\chi_{p_y} = y f_p(r) \tag{5.8}$$

where $f_p(r)$ is the radial part of the atomic orbital.

The normalization and orthogonality conditions for the HAOs can be written in the following way (it is assumed that the atomic orbitals are orthonormalized):

$$1 = c_{is}^2 + c_{ip_x}^2 + c_{ip_y}^2 \quad (i, j = 1, 2, 3; i \neq j)$$
$$0 = c_{is}c_{js} + c_{ip_x}c_{jp_x} + c_{ip_y}c_{jp_y} \tag{5.9}$$

We now take advantage of the equivalence condition for the HAOs. In order to do this we shall consider three equivalent points, the intersections of the three lines a, b and c (see Figure 5.5), and the unit circle. We obtain (1.0), (1/2, $\sqrt{3}/2$), and ($-1/2$, $-\sqrt{3}/2$), respectively, for the co-ordinates of these points. The HAOs must be equal to each other at these points. By comparing them with each other one by one, and by taking into account that their *s* components and $f_p(r)$ do not change from one of the points to another, we find that by equating the coefficients for the radial parts of the atomic orbitals in the HAOs we obtain:

$$c_{1s} = c_{2s} = c_{3s} = c_s$$
$$c_{1p_x} = -\tfrac{1}{2}c_{2p_x} + \frac{\sqrt{3}}{2}c_{2p_y} = -\tfrac{1}{2}c_{3p_x} - \frac{\sqrt{3}}{2}c_{3p_y} \tag{5.10}$$

Hence

$$c_{2p_x} = c_{3p_x}; \quad c_{2p_y} = -c_{3p_y}$$

If we then use the normalization condition of the second HAO and its orthogonality to the third HAO, we obtain

$$c_s^2 + c_{2p_x}^2 + c_{2p_y}^2 = 1$$
$$c_s^2 + c_{2p_x}^2 - c_{2p_y}^2 = 0$$

that is,

$$c_{2p_y} = \pm\frac{1}{\sqrt{2}}, \quad c_s^2 + c_{2p_x}^2 = \tfrac{1}{2} \tag{5.11}$$

We note further that the p_y orbital is anti-symmetric to the x axis and its contribution to the HAO along the x axis (the straight line a) must carry anti-symmetry with it into the HAO. Since we wish to construct a HAO that is symmetric with respect to the σ_x plane, we must set $c_{1py} = 0$.

We obtain, allowing for equation (5.11), the following from the normalization and orthogonality conditions for the first and second HAOs:

$$c_s^2 + c_{1p_x}^2 = 1; \quad c_s^2 + c_{2p_x}^2 = \tfrac{1}{2}; \quad c_s^2 + c_{1p_x}c_{2p_x} = 0 \tag{5.12}$$

On the other hand, by using equation (5.10) we have

$$2c_{1p_x} + c_{2p_x} = \sqrt{3}c_{2p_y}$$

or, by squaring,

$$4c_{1p_x}^2 + c_{2p_x}^2 + 4c_{1p_x}c_{2p_x} = \tfrac{3}{2}$$

i.e.

$$4c_{1p_x}^2 + c_{2p_x}^2 - 4c_s^2 = \tfrac{3}{2}$$

This, together with equation (5.12), yields a new system of equations in terms of the squares of the coefficients we are looking for:

$$\begin{aligned} c_s^2 + c_{1p_x}^2 &= 1 \\ c_s^2 + c_{2p_x}^2 &= \tfrac{1}{2} \\ 4c_{1p_x}^2 + c_{2p_x}^2 - 4c_s^2 &= \tfrac{3}{2} \end{aligned} \tag{5.13}$$

Solving these equations we find that

$$c_s^2 = \tfrac{1}{3}, \quad c_{1p_x}^2 = \tfrac{2}{3}, \quad c_{2p_x}^2 = \tfrac{1}{6} \tag{5.14}$$

It is not difficult to be convinced that all the requirements that we set for the HAOs are satisfied by the following solution:

$$c_s = \frac{1}{\sqrt{3}}, \quad c_{1p_x} = \sqrt{\frac{2}{3}}, \quad c_{2p_x} = -\frac{1}{\sqrt{6}}, \quad c_{1p_y} = 0,$$

$$c_{2p_y} = \frac{1}{\sqrt{2}}, \quad c_{3p_y} = -\frac{1}{\sqrt{6}}, \quad c_{3p_y} = -\frac{1}{\sqrt{2}}$$

Finally, we obtain the following forms for the HAOs:

$$\begin{aligned} \chi_1^{(h)} &= \frac{1}{\sqrt{3}}\chi_s + \frac{2}{\sqrt{6}}\chi_{p_x} \\ \chi_2^{(h)} &= \frac{1}{\sqrt{3}}\chi_s - \frac{1}{\sqrt{6}}\chi_{p_x} + \frac{1}{\sqrt{2}}\chi_{p_y} \\ \chi_3^{(h)} &= \frac{1}{\sqrt{3}}\chi_s - \frac{1}{\sqrt{6}}\chi_{p_x} - \frac{1}{\sqrt{2}}\chi_{p_z} \end{aligned} \tag{5.15}$$

A representation of the HAOs is shown in Figure 5.5.

(3) sp^3 hybridization We shall now consider the last case of how to construct four equivalent hybrid atomic orbitals which have T_d symmetry. These orbitals are encountered, for example, in the carbon atom of the methane (CH_4) molecule.

The co-ordinate axes and the distribution of the *s*- and *p*-orbitals are shown in Figure 5.6(a). The hybrid orbitals point to the vertices of a cube (Figure 5.6(b)). The co-ordinates of these vertices to which the HAOs 'point' are equivalent points, and their co-ordinates are: (a, a, a); $(-a, a, -a)$; and $(a, -a, -a)$. The sides of the cube are $2a$ long.

The hybrid orbitals are constructed from one *s*-orbital and the three *p*-orbitals (p_x, p_y and p_z), their coefficients being c_{is}, c_{ip_x}, c_{ip_y} and c_{ip_z}, respectively. The

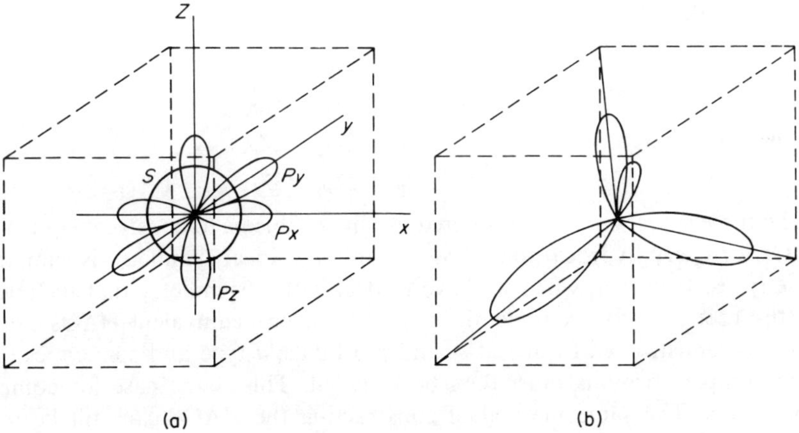

Figure 5.6. s, p_x, p_y and p_z orbitals (a), and symmetricized HAOs (b). In order to simplify the figure the 'small lobes' of the hybrid orbitals are not shown

orthogonality and normalization conditions yield:

$$\begin{aligned} &c_{is}^2 + c_{ip_x}^2 + c_{ip_y}^2 + c_{ip_z}^2 = 1, \quad i = 1, 2, 3, 4; \\ &c_{is}c_{js} + c_{ip_x}c_{jp_x} + c_{ip_y}c_{jp_y} + c_{ip_z}c_{jp_z} = 0 \ (i \neq j) \end{aligned} \tag{5.16}$$

By comparing the values of the hybrid orbitals at the cube vertices, by analogy with the sp^2 hybridization case, we obtain

$$\begin{aligned} &c_{1s} = c_{2s} = c_{3s} = c_{4s} = c_s \\ &ac_{1p_x} + ac_{1p_y} + ac_{1p_z} = -ac_{2p_x} - ac_{2p_y} - ac_{2p_z} \\ &\quad = -ac_{3p_x} + ac_{2p_y} - ac_{3p_z} = ac_{4p_x} - ac_{4p_y} - ac_{4p_z} \end{aligned} \tag{5.17}$$

These identities are satisfied if we set

$$\begin{aligned} c_{1p_x} &= -c_{2p_x} = -c_{3p_x} = c_{4p_x} = c_{p_x} \\ c_{1p_y} &= -c_{2p_y} = c_{3p_y} = -c_{4p_y} = c_{p_y} \\ c_{1p_z} &= c_{2p_z} = -c_{3p_z} = -c_{4p_z} = c_{p_z} \end{aligned}$$

By taking equation (5.16) into account, we have

$$\begin{aligned} c_s^2 + 3c_{p_x}^2 &= 1 \\ c_s^2 - c_{p_x}^2 - c_{p_y}^2 + c_{p_z}^2 &= 0 \\ c_s^2 - c_{p_x}^2 + c_{p_y}^2 - c_{p_z}^2 &= 0 \\ c_s^2 + c_{p_x}^2 - c_{p_y}^2 - c_{p_z}^2 &= 0 \end{aligned} \tag{5.18}$$

Solving equation (5.18) gives us

$$c_s^2 = \tfrac{1}{4}; \quad c_{p_x}^2 = c_{p_y}^2 = c_{p_z}^2 = \tfrac{1}{4}$$

Consequently, the desired HAOs must be

$$\begin{aligned}
\chi_1^{(h)} &= \tfrac{1}{2}\chi_s + \tfrac{1}{2}\chi_{p_x} + \tfrac{1}{2}\chi_{p_y} + \tfrac{1}{2}\chi_{p_z} \\
\chi_2^{(h)} &= \tfrac{1}{2}\chi_s - \tfrac{1}{2}\chi_{p_x} - \tfrac{1}{2}\chi_{p_y} + \tfrac{1}{2}\chi_{p_z} \\
\chi_3^{(h)} &= \tfrac{1}{2}\chi_s - \tfrac{1}{2}\chi_{p_x} + \tfrac{1}{2}\chi_{p_y} - \tfrac{1}{2}\chi_{p_z} \\
\chi_4^{(h)} &= \tfrac{1}{2}\chi_s + \tfrac{1}{2}\chi_{p_x} - \tfrac{1}{2}\chi_{p_y} - \tfrac{1}{2}\chi_{p_z}
\end{aligned} \tag{5.19}$$

The three cases we have considered usually enable us to construct equivalent HAOs for every hydrocarbon whose overall and local symmetry is similar to these types. However, when we have to deal with a symmetry that includes a fourth-order axis (for example, the C_{4h} point group) equivalent HAOs can no longer be constructed from just *s*- and *p*-orbitals; *d*-type and sometimes even more complex orbitals must then be included. This is the case for complex compounds. The same method of constructing the HAOs can still be used, although the mathematical operations are more complicated.

The formulae we have just obtained for hybrid atomic orbitals, or for equivalent atomic orbitals (EAOs) which are the same thing, enable us to formulate some general rules for finding the equivalent orbitals in any arbitrary case.

In order to do this we rewrite the relationships we obtained above for the atomic and hybrid orbitals in symbolic matrix form, thus:

$$\chi^{(h)} = C\chi \tag{5.20}$$

where C is the matrix made up of the coefficients for the atomic orbitals. Hence

$$\chi = C^{-1}\chi^{(h)} = \tilde{C}\chi^{(h)} = \tilde{C}\chi_{\text{EAO}} \tag{5.21}$$

since C is orthogonal. It lays down condition of orthogonality of HAO of one atom.

Let us now assume that the equivalent orbitals χ_{EAO} for a given bond node are given. If the node has a local symmetry, then a symmetricized atomic orbital (SAO) can be constructed. This aspect will be considered in detail below. We have to find a transformation $\chi_{\text{SAO}} = C'\chi_{\text{EAO}}$. The symmetricized orbitals obtained in this way must be mutually orthogonal by construction and must be transformed in the appropriate way during the symmetry operations allowable for the node (local symmetry). We note further that the usual atomic orbitals of one node are also mutually orthogonal, and, given that they have the appropriate arrangement with respect to the node's symmetry elements, they will be transformed according to certain symmetry rules. These may coincide with those of the transformation of the symmetricized atomic orbitals that were constructed from the equivalent orbitals. Hence these symmetric atomic orbitals can be simply regarded as coincident with the ordinary atomic orbitals. This yields the rule for choosing from all of the orbitals those that will coincide with the symmetricized orbitals in their symmetry properties. We only have to include those atomic orbitals in the problem. As a result, we obtain

$$\chi_{\text{SAO}} = \chi_{\text{AO}} = C'\chi_{\text{EAO}} = C\chi^{(h)}$$

Here C' is identical to the $\tilde{C}$ from equation (5.21). It is important that initial AO must be of the same set of degenerated AO and have the same general quantum number.

In real cases strict local symmetry of the bond connection is not always realized. Then, in order to construct the symmetricized orbitals, we can obtain in the equivalent orbitals starting from the nearest symmetry configuration. To construct the elements of the C matrix we can use the simplest and most obvious method. This is based on the fact that the symmetry of the equivalent orbitals is reflected in the specific form of the H, S, F, T and U matrices which are encountered in spectra. (For T and U matrices see Chapter 7.)

In this respect, if we only have two equivalent orbitals (local symmetry leading to sp hybridization), then the H and S matrices will clearly have the structure

$$H, S = \begin{bmatrix} a & b \\ b & a \end{bmatrix}$$

For four equivalent orbitals, given T_d symmetry for the node (sp^3 hybridization), the H and S matrices will be:

$$H, S = \begin{bmatrix} a & b & b & b \\ b & a & b & b \\ b & b & a & b \\ b & b & b & a \end{bmatrix}$$

Transferring to a symmetricized or an ordinary atomic orbital (which requires the construction of an orthogonal combination of equivalent orbitals) means that these matrices, which have this special structure, need to be diagonalized. This can be achieved as follows.

We now return to the question as to how molecular orbitals of symmetry can be constructed if we use HAOs. In order to make the procedure more understandable, we shall consider ethylene as an example.

The atomic orbital basis consists of 1_s-orbitals centred on the hydrogen atoms, sp^2 orbitals centred on the carbon atoms and oriented in the plane of the molecule so that one of the hybrid orbitals is along the $C{=}C$ axis, and two p_z orbitals, also centred on the carbon atoms. The distribution of the atomic orbitals is indicated conventionally in Figure 5.7.

All the $1s$ orbitals on the hydrogen orbitals are equivalent because they can be interchanged via the symmetry operations allowed within the molecule. We recall that the ethylene molecule belongs to the V_h symmetry group and has a centre of symmetry, three mutually perpendicular second-order axes of symmetry that intersect at the centre of symmetry and three mutually perpendicular planes of symmetry. All these symmetry elements are shown in Figure 5.7.

The four HAOs on the carbons (marked 1–4) are also equivalent, as are the two HAOs on carbons 5 and 6. The p_z atomic orbitals of the carbon atoms are not equivalent. This deficiency can be eliminated by introducing an sp_z hybrid orbital that mixes the p_z state with the $3s$ atomic orbital. It is, however, possible to act in another way, as shown below.

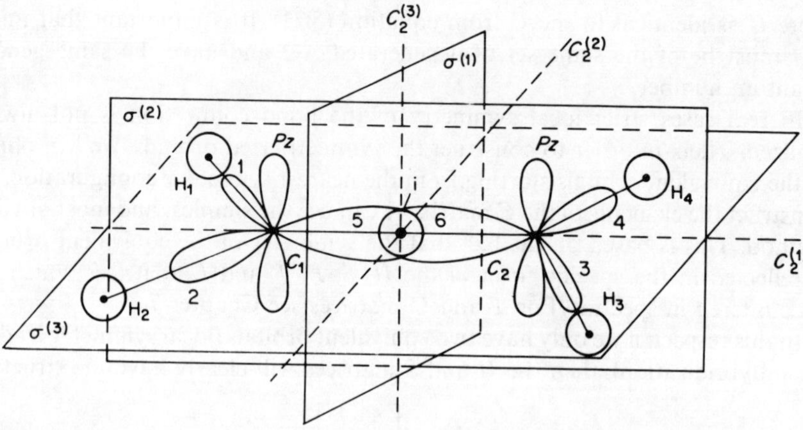

Figure 5.7. The HAOs in ethylene. The groups of equivalent orbitals are the four s orbitals, the sp^2 orbitals (1–4) on the carbon atoms, and the sp^2 orbitals (5, 6) between the carbon atoms

We return to the groups of equivalent orbitals and see what properties the s_{kl}- and h_{kl} matrix elements will have in the presence of equivalent orbitals within the molecule. It follows from a simple argument that all the Coulomb integrals, $h_{kk} = h_H$, will be the same for equivalent orbitals. Moreover, it must be true that $h_{H_1,H_2} = h_{H_3,H_4} = h'_H$, $h_{H_1,H_3} = h_{H_2,H_4} = h''_H$ and $h_{H_1,H_4} = h_{H_2,H_3} = h'''_H$. Note that although the resonance integrals for distant atomic orbitals are very small and are usually neglected, they are retained here.

The part of matrix H which corresponds to the s orbitals of the hydrogen atoms will consequently have the form:

$$\begin{bmatrix} h_H & h'_H & h''_H & h'''_H \\ h'_H & h_H & h'''_H & h''_H \\ h''_H & h'''_H & h_H & h'_H \\ h'''_H & h''_H & h'_H & h_H \end{bmatrix} \tag{5.22}$$

It is worth noting that the molecule's symmetry is reflected in the appearance of similar elements in the H matrix. The S matrix will have an analogous structure for groups of the same atomic orbitals.

All that we have said also applies in equal measure to the HAOs 1, 2, 3 and 4 of the equivalent carbon atoms. Furthermore, formulae that are completely analogous to equations (3.10), (3.11) and (3.15) can be obtained for a HAO only changing E for the average energy of the hybrid orbital or a linear combination of E for every atomic orbital in the hybrid orbital. In general, it is not difficult to express the h_{kk}- and h_{kl} integrals of a hybrid orbital in terms of linear combinations of the integrals of the atomic orbitals included in it.

The H matrix for HAOs 5 and 6 will have the form

$$\begin{bmatrix} h_c & h'_c \\ h'_c & h_c \end{bmatrix} \tag{5.23}$$

where h_c is the Coulomb integral for HAO 5 and h'_c is the resonance integral for orbitals 5 and 6. The S matrix has a similar structure. A very important consequence of this situation is that these matrices can be reduced to the diagonal form using standard transformations of the original basis without needing the numerical values of the matrix elements. Later (Section 5.3) we will study this topic in greater detail, but for now we will indicate, without any proof, that if we go over to a new basis of the form

$$\begin{aligned}\varphi_{1H}^{(s)} &= \tfrac{1}{2}\chi_{1s,H_1} + \tfrac{1}{2}\chi_{1s,H_2} + \tfrac{1}{2}\chi_{1s,H_3} + \tfrac{1}{2}\chi_{1s,H_4}\\ \varphi_{2H}^{(s)} &= \tfrac{1}{2}\chi_{1s,H_1} - \tfrac{1}{2}\chi_{1s,H_2} - \tfrac{1}{2}\chi_{1s,H_3} + \tfrac{1}{2}\chi_{1s,H_4}\\ \varphi_{3H}^{(s)} &= \tfrac{1}{2}\chi_{1s,H_1} - \tfrac{1}{2}\chi_{1s,H_2} + \tfrac{1}{2}\chi_{1s,H_3} - \tfrac{1}{2}\chi_{1s,H_4}\\ \varphi_{4H}^{(s)} &= \tfrac{1}{2}\chi_{1s,H_1} + \tfrac{1}{2}\chi_{1s,H_2} - \tfrac{1}{2}\chi_{1s,H_3} - \tfrac{1}{2}\chi_{1s,H_4}\end{aligned} \tag{5.24}$$

(the analogous for the four equivalent hybrid orbitals, 1–4, on the carbon atoms) and

$$\varphi_{1c}^{(s)} = \frac{1}{\sqrt{2}}(\chi_{c_5}^{(h)} + \chi_{c_6}^{(h)}); \quad \varphi_{2c}^{(s)} = \frac{1}{\sqrt{2}}(\chi_{c_5}^{(h)} - \chi_{c_6}^{(h)}) \tag{5.25}$$

where χ_{1s,H_1} is the atomic orbital for the hydrogen atoms and $\chi_{c_5}^{(h)}$ and $\chi_{c_6}^{(h)}$ are the hybrid orbitals on the carbon atoms, then all the non-diagonal elements in matrix (5.22), their analogues for the hybrid orbitals Nos 1–4 and the non-diagonal elements in matrix (5.23) will disappear. Many of the matrix elements in those parts of the H and S matrices that contain the s_{kl}- and h_{kl} integrals that embrace both the ordinary and HAOs from different equivalent groups will also disappear.

In order to explain this we shall consider how the functions $\varphi^{(s)}$ behave with respect to the symmetry operations permitted for the molecule. We start with the simplest, $\varphi_{1c}^{(s)}$ and $\varphi_{2c}^{(s)}$. We can conclude immediately from the form of the hybrid orbitals $\chi_{c_5}^{(h)}$ and $\chi_{c_6}^{(h)}$ that the value of the $\varphi_{1c}^{(s)}$ function will not change for any symmetry operation that is permissible for the ethylene molecule. Thus the function is fully symmetric. The $\varphi_{2c}^{(s)}$ function changes sign when reflected in the $\sigma^{(1)}$ plane, when rotated about the second-order axes $C_2^{(2)}$ and $C_2^{(3)}$, and when inverted with respect to the centre of symmetry, i.e. under the action of any of the symmetry operations which put the $\chi_{c_5}^{(h)}$ function in place of the $\chi_{c_6}^{(h)}$ function and vice versa. Consequently, the $\varphi_{2c}^{(s)}$ orbital is anti-symmetric with respect to these symmetry elements.

We know consider the orbitals of the hydrogen atoms. The symmetry operations will make equivalent atomic orbitals change places. Clearly, the $\varphi_{1H}^{(s)}$ functions will not be changed in the process and so are fully symmetric.

The $\varphi_{2H}^{(s)}$ function changes sign when reflected in the $\sigma^{(2)}$ plane, or when rotated about the $C_2^{(1)}$ and $C_2^{(3)}$ axes, i.e. is anti-symmetric with respect to these elements (see Figure 5.7). During the other symmetry operations it does not change sign and so is symmetric with respect to them.

The function $\varphi_{3H}^{(s)}$ is anti-symmetric with respect to the $\sigma^{(2)}$ plane and the $C_2^{(1)}$ axis, and is symmetric with respect to the others. The $\varphi_{4H}^{(s)}$ function is anti-symmetric with respect to the $\sigma^{(1)}$ plane, the centre of symmetry and the

$C_2^{(2)}$ and $C_2^{(3)}$ axes, whilst it is symmetric with respect to the other symmetry elements.

An similar situation is true for the functions constructed according to rule (5.24) from the four equivalent hybrid orbitals 1–4 of the carbon atoms.

The functions with (*s*) index thus prove to behave in a fully defined way with respect to the molecule's symmetry operations. We therefore call them symmetrized molecular orbitals (SMO). SMOs can be constructed whenever the molecule has at least one symmetry element.

It is clear from this example that the various orbitals of symmetry for a group of equivalent ordinary or HAOs are differentiated by such a property that, if any pair of them is taken, then it will always be true that at least one symmetry element can be found with respect to which of the pair's SMOs will be symmetric and the other anti-symmetric. This symmetry element will divide the space of definition of the given functions into two symmetric parts in which the products of the functions will be only different by their signs. It is for this reason that the integrals $\int \varphi_1^{(s)} \varphi_2^{(s)} \, dv$, taken over all space, must be zero for such functions. Integrals of the kind $\int \varphi_1^{(s)} \hat{H} \varphi_2^{(s)} \, dv$ will also be zero.

In this respect the energy operator $\hat{H}$ is always fully symmetric and its action on any function never changes the function's symmetry properties. The operator $\hat{H} = \hat{T}_e + V$ contains a potential part which remains invariant during any symmetry operation and multiplying the function $\varphi_2^{(s)}$ by it cannot change the function's symmetry properties. The $\hat{H}$ operator also contains a kinetic part in which the $\hat{T}$-function is doubly differentiated. Let us clarify what happens in double differentiation. If the $\varphi_2^{(s)}$ function is symmetric with respect to a symmetry element, then this means that if the function grows in some domain 'on the left' of the symmetry element, then it must diminish in the symmetric domain 'on the right'. The reverse is true for anti-symmetric functions. Hence the first derivative of the symmetry element will have different signs in the 'left' and 'right' symmetric domains, while the second derivatives will have the same sign in both domains. Thus, repeated differentiation leads to a change in the parity of the function. Consequently, double differentiation restores the parity.

We can arrive at the same conclusion simply by using the definition of a partial derivative as the limit of the ratio of the difference in the functions, given two close arguments to the difference between the arguments. Then the rule for transforming the function during symmetry operations is applied to the resulting expression. Thus the integral $\int \varphi_1^{(s)} \hat{H} \varphi_2^{(s)} \, dv$ must behave in exactly the same way as the product $\varphi_1^{(s)} \varphi_2^{(s)}$ at least with respect to the molecule's symmetry operations. Consequently, when the integral $\int \varphi_1^{(s)} \varphi_2^{(s)} \, dv$ is zero due to the symmetry of the SMOs, $\int \varphi_1^{(s)} \hat{H} \varphi_2^{(s)} \, dv$ must also be zero. What has been said is clearly true when the molecular orbitals of symmetry $\varphi_1^{(s)}$ and $\varphi_2^{(s)}$ belong to different groups of equivalent ordinary or hybrid atomic orbitals. If, however, $\varphi_1^{(s)}$ and $\varphi_2^{(s)}$ have the same symmetry type then these integrals will be non-zero.

One very important conclusion follows from this, i.e. if a SMO is used as a basis in a test function instead of an ordinary or HAO, then the expression for $\bar{H}$ will take on the appearance of a sum of quadratic forms, each of which will only contain the C_k coefficients of one type of SMO. The quadratic form of the orthogonality condition will also have an analogous appearance. The task of reducing $\bar{H}$ to a sum of squares then boils down to a sum of squares, either a series of quadratic forms for functions of each symmetry type or diagonalizing a number of low-order H or S-matrices. We have thus the ability to simplify the solution of the whole task quite considerably.

For ethylene (considered above) instead of having to diagonalize tenth-order H and S matrices (excluding the p_z-function), we need only diagonalize two second-order and two third-order matrices.

The introduction of SMOs, as we have shown, allows us to divide the wave functions into groups according to the criterion of general behavior during the molecule's symmetry operations. The quantum-mechanical problem can then be solved for each group separately.

Linear and planar molecules represent important classes, since all their functions divide along symmetric and antisymmetric lines with respect to the molecule's planar axis. The first type are σ functions, the second π functions. We have to choose the atomic orbitals and SMOs with this in mind.

The overall problem then divides into two separate parts. The first is concerned with σ MOs and σ electrons whilst the second involves π MOs and π electrons.

5.3 The Construction of Molecular Orbitals of Symmetry and Symmetry Co-ordinates

5.3.1 General principals for calculating the eigenvectors of circulant matrices

The material in this chapter is applicable both to the theory of electron shells and to the that of vibrational spectra of polyatomic molecules. Thus we will anticipate somewhat, and use some of the terminology and concepts from vibrational theory to be presented in Chapters 7–9. The strength of this unified approach is that when we construct the SMOs in electronic theory and symmetry co-ordinates in vibrational theory, we encounter matrices in both cases that can be diagonalized in the same way, independent of their physical nature.

We saw in Section 5.2 that when, for example, we have three equivalent orbitals, the matrices H and S must have the following form:

$$\begin{bmatrix} a & b & b \\ b & a & b \\ b & b & a \end{bmatrix}$$

The matrices of the kinematic or force constants have an entirely analogous structure for three equivalent natural vibrational co-ordinates (see Chapter 8 and 9). When, for example, there are two groups of three equivalent atomic orbitals, or two groups of three equivalent natural vibrational co-ordinates, we obtain a matrix of the following form:

$$\begin{bmatrix} a & b & b & e & f & f \\ b & a & b & f & e & f \\ b & b & a & f & f & e \\ e & f & f & c & d & d \\ f & e & f & d & c & d \\ f & f & e & d & d & c \end{bmatrix} \tag{5.26}$$

It is important here that particular cells have the same elements, but it does not matter what their numerical values or physical natures are. This enables us, when considering the diagonalization methods of these matrices, to move away from atomic orbitals or natural vibrational co-ordinates and simply pay attention to groups of equivalent objects that can be changed into each other during the permitted symmetry operations. The simplest objects of this sort are simple unit vectors, which can be transformed into each other by, for example, reflections in a plane, rotations about an axis of symmetry, etc. Thus, for instance, if we were to consider two unit vectors that are transformed into each other by the symmetry operation corresponding to a reflection in a plane, and were to construct a scalar product from them which also includes the products of the vectors with themselves, we would obtain a matrix with the following

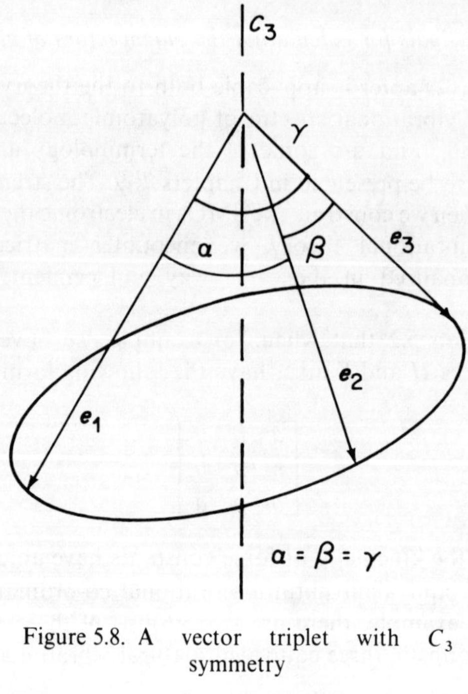

Figure 5.8. A vector triplet with C_3 symmetry

structure:

$$\begin{bmatrix} 1 & \vec{e}_1\vec{e}_2 \\ \vec{e}_1\vec{e}_2 & 1 \end{bmatrix} \text{ or } \begin{bmatrix} a & b \\ b & a \end{bmatrix}$$

If we have three vectors (see Figure 5.8) that are transformed into each other by rotations about a third-order axis, the matrix of all the possible scalar products of the vectors will have the form:

$$\begin{bmatrix} 1 & \vec{e}_1\vec{e}_2 & \vec{e}_1\vec{e}_3 \\ \vec{e}_1\vec{e}_2 & 1 & \vec{e}_2\vec{e}_3 \\ \vec{e}_1\vec{e}_3 & \vec{e}_2\vec{e}_3 & 1 \end{bmatrix} \text{ or } \begin{bmatrix} a & b & b \\ b & a & b \\ b & b & a \end{bmatrix} \tag{5.27}$$

since $\vec{e}_1\vec{e}_2 = \vec{e}_2\vec{e}_3 = \vec{e}_1\vec{e}_3$.

If we consider two groups of unit vectors such as those shown in Figure 5.9, each of which can be transformed into another by a symmetry operation corresponding to a third-order axis of symmetry and planes of symmetry, then the following matrix is encountered:

$$\left[\begin{array}{ccc|ccc} 1 & \vec{e}_1\vec{e}_2 & \vec{e}_1\vec{e}_3 & \vec{e}_1\vec{e}_4 & \vec{e}_1\vec{e}_5 & \vec{e}_1\vec{e}_6 \\ \vec{e}_1\vec{e}_2 & 1 & \vec{e}_2\vec{e}_3 & \vec{e}_2\vec{e}_4 & \vec{e}_2\vec{e}_5 & \vec{e}_2\vec{e}_6 \\ \vec{e}_1\vec{e}_3 & \vec{e}_2\vec{e}_3 & 1 & \vec{e}_3\vec{e}_4 & \vec{e}_3\vec{e}_5 & \vec{e}_3\vec{e}_6 \\ \hline \vec{e}_1\vec{e}_4 & \vec{e}_1\vec{e}_5 & \vec{e}_1\vec{e}_6 & 1 & \vec{e}_4\vec{e}_5 & \vec{e}_4\vec{e}_6 \\ \vec{e}_2\vec{e}_4 & \vec{e}_2\vec{e}_5 & \vec{e}_2\vec{e}_6 & \vec{e}_4\vec{e}_5 & 1 & \vec{e}_5\vec{e}_6 \\ \vec{e}_3\vec{e}_4 & \vec{e}_3\vec{e}_5 & \vec{e}_3\vec{e}_6 & \vec{e}_4\vec{e}_6 & \vec{e}_5\vec{e}_6 & 1 \end{array}\right] \tag{5.28}$$

This is similar to matrix (5.26).

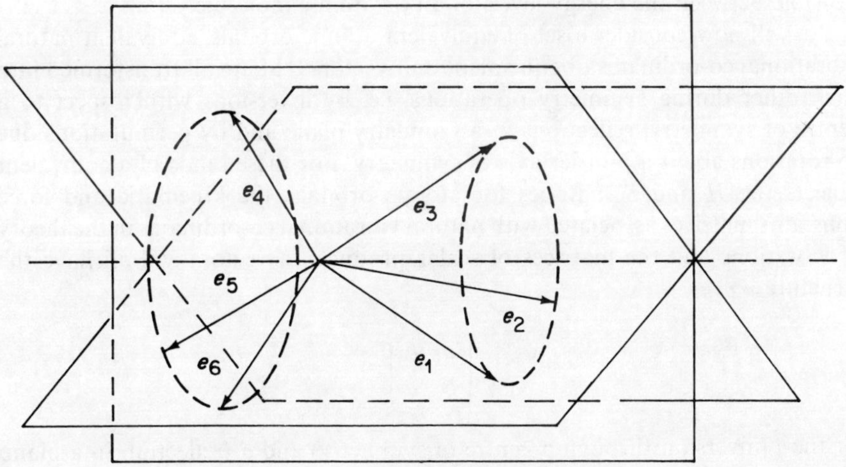

Figure 5.9. A triplet of equivalent vectors

In this way the appearance of matrices with special structures is not related to the physical character of the object being studied, but is due to the particular properties of the symmetry of the space. This is why the diagonalization procedure for this type of matrix is completely independent of the choice of object. Usually these matrices are diagonalized using the methods of group theory. However, it can be shown that there is a simpler way of solving the problem that is no less general and does not require the introduction of new concepts[1].

If the H and S (or the T and U) matrices both have the same structure as discussed above, then using the approaches described in this chapter they can be reduced to quasi-diagonal form by an orthogonal transformation simultaneously with the symmetry coefficients matric C. Since these matrices are symmetric, C will be orthogonal, a fact which guarantees that both T and U, in particular, will simultaneously be reduced to quasi-diagonal form. This is true even though T corresponds to the quadratic form of momenta and U to the quadratic form of co-ordinates.

The reduction of the H, S, T and U matrices is independent of the value of their elements and, because of this trait of this special form, it is possible to solve both the electron and vibrational problems for each quasi-diagonal block separately. Thus we can break up one complicated problem into a series of easier ones. The more symmetric a molecule (i.e. the more symmetry elements it possesses), the greater the simplification possible. As a result we can find the SMOs, and the vibrational symmetry co-ordinates, as LCAOs, or natural vibration co-ordinates. This, in turn, enables us to classify molecular orbitals, the wave functions of the electronic problem, the normal vibrations and the wave functions of the vibrational problem according to their symmetry attributes. This allows us to introduce 'selection rules according to symmetry' into the discussion about whether optical dipole (or other) transitions are possible between the energy levels of a polyatomic molecule.

We will now consider a set of equivalent atomic orbitals, equivalent natural vibrational co-ordinates or equivalent unit vectors that are all transformed into each other during symmetry operations, i.e. by inversions with respect to a centre of symmetry, reflections in a symmetry plane and by permutations due to rotations about a p-order axis of symmetry. For these kinds of a equivalent objects the H and S matrices for atomic orbitals, the kinematic and force constants matrices associated with natural vibrational co-ordinates in the theory of vibrations and the matrices of scalar products of vectors will all have the structure

$$\begin{bmatrix} a & b \\ b & a \end{bmatrix}$$

for the i (inversion through a centre of symmetry) and σ (reflection in a plane of symmetry) operations. These matrices will have the form:

$$\begin{bmatrix} a & b_1 & b_2 & \cdots & b_{p/2} & \cdots & b_2 b_1 \\ b_1 & a & b_1 & \cdots & b_{p/2} & \cdots & b_2 \\ \cdots & \cdots & \cdots & \cdots & \cdots & \cdots & \cdots \\ b_1 & b_2 & & \cdots & b_{p/2} & \cdots & b_1 a \end{bmatrix} \tag{5.29}$$

for rotations about an even p-order axis and

$$\begin{bmatrix} a & b_1 & b_2 & \cdots & b_{p-1/2} & b_{p-1/2} & \cdots & b_1 \\ b_1 & a & b_1 & \cdots & b_{p-1/2} & b_{p-1/2} & \cdots & b_2 \\ \cdots & \cdots & \cdots & \cdots & \cdots & \cdots & \cdots & \cdots \\ b_1 & & & \cdots & b_{p-1/2} & b_{p-1/2} & \cdots & a \end{bmatrix} \tag{5.30}$$

for rotations about an odd p-order axis. A characteristic of these matrices is that each row is obtained by a cyclic permutation of the previous one. We will call these matrices 'circulants'. In general, their eigennumbers and eigenvectors can easily be calculated. It is found that the form of the elements of the eigenvectors only depends on the order of the matrices, but not on the numerical values of their variables. This allows us to construct the SMOs or symmetry co-ordinates even when the values of their coefficients are only determined by the symmetries of the system and not by concrete masses of the atoms, bond lengths, vibrational force constants or only by the form of the atomic orbitals when constructing SMOs.

Since every row of a circulant matrix contains the same elements, although in a different order, it is plain that one of the eigenvectors of this type of matrix must consist entirely of ones, and the corresponding eigenvalue will simply be the sum of all the elements in a row. We choose cyclic matrices that are real and symmetric, and must therefore be diagonalizable by orthogonal transformations. If one of a matrix's eigenvalues consists entirely of ones, then it follows from this and the orthogonality condition that the sum of all the elements in the remaining eigenvectors must be zero. In order to determine the numerical values of the elements of the remaining eigenvectors we will search for them in the form $\{1, x_1^{(k)}, x_2^{(k)}, \ldots, x_{p-1}^{(k)}\}$, where k is the index of the eigenvector.

By substituting this vector into the equation that defines the matrix's eigennumbers and eigenvectors, we obtain the following in an explicit form:

$$\begin{aligned} a + b_1 x_1^{(k)} + b_2 x_2^{(k)} + \cdots + b_1 x_{p-1}^{(k)} &= \lambda_k \cdot 1 \\ b_1 + a x_1^{(k)} + b_1 x_2^{(k)} + \cdots + b_2 x_{p-1}^{(k)} &= \lambda_k \cdot x_1^{(k)} \\ b_1 + b_2 x_1^{(k)} + b_3 x_2^{(k)} + \cdots + a x_{p-1}^{(k)} &= \lambda_k x_{p-1}^{(k)} \end{aligned} \tag{5.31}$$

We see that the left-hand side of the second equation must equal the left-hand side of the first multiplied by $x_1^{(k)}$, i.e.

$$x_1^{(k)}(a + b_1 x_1^{(k)} + b_2 x_2^{(k)} + \cdots + b_1 x_{p-1}^{(k)}) = b_1 + a x_1^{(k)} + b_1 x_2^{(k)} + \cdots + b_2 x_{p-1}^{(k)}$$

or

$$a x_1^{(k)} + b(x_1^{(k)})^2 + b_2 x_2^{(k)} x_2^{(k)} + \cdots + b_1 x_1^{(k)} x_{p-1}^{(k)} = b_1 + a x_1^{(k)} + b_1 x_2^{(k)} + \cdots + b_2 x_{p-1}^{(k)}$$

Now because of the cylic permutation of the elements of each row of a matrix we note that the following must be satisfied:

$$(x_1^{(k)})^2 = x_2^{(k)}, \quad x_1^{(k)} x_{p-1}^{(k)} = 1 \tag{5.32}$$

The left-hand side of the third equation of formula (5.31) differs from the left-hand side of the first by the multiplication of the first by $x_2^{(k)}$. Since, on the other hand, $x_2^{(k)} = (x_1^{(k)})^2$, then by bearing in mind the cyclic distribution of the a and b_i elements in each row and by comparing the coefficients of the b_1 element immediately following the a term in the first line (the line having been multiplied by $(x_1^{(k)})^2$) and the b_1 element in the third line, we obtain $x_3^{(k)} = (x_1^{(k)})^3$. If we continue this argument we find that, in general, $x_n^{(k)} = (x_1^{(k)})^n$, and since, aditionally, $x_1^{(k)} \cdot x_{p-1}^{(k)} = x_1^{(k)} \cdot (x_1^{(k)})^{p-1} = (x_1^{(k)})^p = 1$, we finally obtain

$$x_1^{(k)} = x^{(k)} = p\sqrt{1} = \cos\frac{2\pi k}{p} + i\sin\frac{2\pi k}{p}, \quad k = 0, 1, 2, \ldots, (p-1) \tag{5.33}$$

In this way, the values of the elements of the eigenvectors of the circulant matrix are

$$1, (p\sqrt{1}), (p\sqrt{1})^2, \ldots, (p\sqrt{1})^{p-1} \tag{5.34}$$

The matrix we considered was symmetric and real, and so all its eigennumbers are real. Consequently, we must obtain degenerate eigenvectors and eigennumbers for the complex conjugates of $P\sqrt{1}$.

Since we encountered pairs of the same elements in the rows of the matrix, we will obtain the following for eigennumbers λ_k for odd p:

$$\lambda_k = a + b_1[x^{(k)} + (x^{(k)})^{p-1}] + b_2[x^{(k)^2} + x^{(k)^{p-2}}] + \cdots$$
$$+ b_m^{(k)}[(x^{(k)})^m + (x^{(k)})^{p-m}] + \cdots \tag{5.35}$$

However,

$$(x^{(k)})^m + (x^{(k)})^{p-m} = \cos\frac{2\pi k}{p}m + i\sin\frac{2\pi k}{p}m + \cos\frac{2\pi k}{p}(p-m)$$
$$+ i\sin\frac{2\pi k}{p}(p-m) = 2\cos\frac{2\pi k}{p}m$$

Consequently we have, for odd p,

$$\lambda_k = a + 2\sum_{m=1}^{(p-1)/2} b_m \cos\frac{2\pi k}{p}m, \quad k = 0,1,2,\ldots,(p-1) \tag{5.36}$$

An additional term enters the expression for the λ_k of even p, and takes the form:

$$b_{p/2}(x^{(k)})^{p/2} = b_{p/2}\cos k\pi = (-1)^k b_{b/2}$$

Thus we obtain in this case

$$\lambda_k = a + (-1)^k b_{p/2} + 2 \sum_{m=1}^{p/2-1} b_m \cos \frac{2\pi k}{p} m \tag{5.37}$$

The eigenvectors obtained from degenerate λ_k will be complex conjugates. We can, however, eliminate this situation if we take into account that when the eigenvalues are real, the real and imaginary parts of the complex conjugate eigenvectors will themselves be eigenvectors if taken separately. This means that we can take the following as the eigenvectors for degenerate eigennumbers:

$$\begin{bmatrix} 1 \\ \cos \dfrac{2\pi k}{p} \\ \cos \dfrac{2\pi k}{p} \cdot 2 \\ \cdots \\ \cos \dfrac{2\pi k}{p}(p-1) \end{bmatrix}, \quad \begin{bmatrix} 0 \\ \sin \dfrac{2\pi k}{p} \\ \sin \dfrac{2n k}{p} \cdot 2 \\ \cdots \\ \sin \dfrac{2\pi k}{p}(p-1) \end{bmatrix} \tag{5.38}$$

Here we will make an important comment. Suppose we consider a real symmetric matrix A of order p that is being diagonalized using an orthogonal transformation C and that it has degenerate eigenvalues. Let there be, for example, two of them (this will not lessen the generality of the argument). Then an additional orthogonal transformation will leave the diagonal matrix unchanged, the transformation's matrix being

$$\begin{bmatrix} I & 0 \\ 0 & C' \end{bmatrix}$$

where I is the identity submatrix of order $(p-2)$ and C' is an orthogonal submatrix of an order equal to the degree of the degeneracy. It may therefore be said that a matrix can only be diagonalized with respect to orthogonal transformations when there are degenerate eigenvalues, and in this sense diagonalization is uncertain. The canonical orthogonal matrix C' has, for the special case of twofold degeneracy, the form

$$C' = \begin{bmatrix} \cos\varphi & -\sin\varphi \\ \sin\varphi & \cos\varphi \end{bmatrix} \tag{5.39}$$

By post-multiplying the pair of eigenvectors in equation (5.38) for the degenerate eigennumbers by the matrix C', then normalizing all the eigenvectors to unity, and taking into account that

$$\sum_{m=0}^{p-1} \cos^2 \frac{2\pi k}{p} m = \sum_{m=0}^{p-1} \sin^2 \frac{2\pi k}{p} m = \frac{p}{2}$$

we finally obtain the orthogonal matrix containing the eigenvectors of the circulant matrices we are considering, i.e.

$$C=\sqrt{\frac{2}{p}}\times\begin{bmatrix} 1/\sqrt{2} & -1/\sqrt{2} & \cdots & \cos\varphi & -\sin\varphi & \cdots \\ 1/\sqrt{2} & 1\sqrt{2} & \cdots & \cos\left(\frac{2\pi k}{p}-\varphi\right) & \sin\left(\frac{2\pi k}{p}-\varphi\right) & \cdots \\ \cdots & \cdots & \cdots & \cdots & \cdots & \cdots \\ 1\sqrt{2} & (-1)^m 1\sqrt{2} & \cdots & \cos\left(\frac{2\pi k}{p}m-\varphi\right) & \sin\left(\frac{2\pi k}{p}m-\varphi\right) & \cdots \\ \cdots & \cdots & \cdots & \cdots & \cdots & \cdots \\ 1\sqrt{2} & (-1)^p 1\sqrt{2} & \cdots & \cos\left[\frac{2\pi k}{p}(p-1)-\varphi\right] & \sin\left[\frac{2\pi k}{p}(p-1)-\varphi\right] & \cdots \end{bmatrix} \tag{5.40}$$

$k=1,2,\ldots(p-1)$, though not including $p/2$ for even p.

The second column only appears in the matrix for even p and corresponds to the symmetry orbitals and co-ordinates in the molecule that are antisymmetric with respect to the element of symmetry. We must choose one k for the columns of the matrix in equation (5.40) which corresponds to a degenerate eigenvalue. A pair of the column in the matrix corresponds to each degenerate vibration or molecular orbital of symmetry.

Different choices of the φ parameter (the rotation of the orthogonal co-ordinate axes) will yield different numerical values for the elements of the vectors. Moreover, both vectors change simultaneously and in close agreement with a simultaneous transformation of the degenerate SMOs or symmetry co-ordinates. Being able to change the linear combination of atomic orbitals in a SMO or the combination of natural co-ordinates in a symmetry co-ordinate by changing the parameter φ allows us, as we shall see later, to carry out an orientation operation to obtain a consistent choice of SMOs for equivalent atomic orbitals of different kinds for example, s-orbitals or HAOs, or to obtain a consistent choice of symmetry co-ordinates for the equivalent natural co-ordinates of angles and bonds.

The general formula we have just obtained for the C matrix in the case of circulants enables us to find the values of the elements of this matrix easily for particular cases. Thus when $p=2$ for an original matrix whose structure is

$$\begin{bmatrix} a & b \\ b & a \end{bmatrix}$$

we have the following diagonalizing C matrix:

$$C = \frac{1}{\sqrt{2}}\begin{bmatrix} 1 & -1 \\ 1 & 1 \end{bmatrix}$$

its eigennumbers being

$$\lambda_1 = a + b, \quad \lambda_2 = a - b$$

when $p = 3$ and the matrix

$$\begin{bmatrix} a & b & b \\ b & a & b \\ b & b & a \end{bmatrix}$$

we obtain

$$\lambda_1 = a + 2b, \quad \lambda_2 = \lambda_3 = a - b$$

and

$$C = \sqrt{\frac{2}{3}}\begin{bmatrix} 1/\sqrt{2} & 1 & 0 \\ 1/\sqrt{2} & -1/2 & \sqrt{3}/2 \\ 1/\sqrt{2} & -1/2 & -\sqrt{3}/2 \end{bmatrix}$$

For $p = 4$ and the matrix

$$\begin{bmatrix} a & b_1 & b_2 & b_1 \\ b_1 & a & b_1 & b_2 \\ b_2 & b_1 & a & b_1 \\ b_1 & b_2 & b_1 & a \end{bmatrix}$$

we obtain the following:

$$\lambda_1 = a + 2b_1 + b_2, \quad \lambda_2 = a - 2b_1 + b_2, \quad \lambda_3 = \lambda_4 = a - b_2$$

$$C = \frac{1}{2}\begin{bmatrix} 1 & -1 & 2/\sqrt{2} & 0 \\ 1 & 1 & 0 & -2/\sqrt{2} \\ 1 & -1 & -2/\sqrt{2} & 0 \\ 1 & 1 & 0 & 2/\sqrt{2} \end{bmatrix}$$

For the two matrices ($p = 3, 4$) the columns corresponding to the degenerate eigenvalues depend on the value of φ and can be given in another way. For example, φ was chosen above as equal to zero, but if it were $\pi/4$ then for $p = 4$ the relevant pair of columns would be, respectively,

$$\frac{1}{2}\begin{bmatrix} 1 & -1 \\ 1 & 1 \\ -1 & 1 \\ -1 & -1 \end{bmatrix}$$

If we denote columns of equivalent objects by q and the columns of symmetry objects by q^s, then we obtain the relationship $q = Cq^s$ or $q^s = \tilde{C}q$. This is why the elements of the columns of C are called the symmetry coefficients.

5.3.2 *Calculating the symmetry coefficients when there are several symmetry elements*

Several symmetry elements may be present in real molecules. In such cases equivalent orbitals or natural co-ordinates are transformed into each other by transformations with respect to all the symmetry elements in the molecule.

We shall now consider how the matrix of symmetry coefficients can be found in this context. We shall stay initially with the example of the ethylene molecule (Figure 5.10). It has four equivalent CH bonds which are transformed into each other by rotations about a second-order axis of symmetry coinciding with the C=C bond. They can also be reflected in a plane that is perpendicular to the axis. The general submatrix for the Coulomb and resonance integrals or the overlap integrals for the equivalent HAOs that are directed along the CH bond has the structure

$$\left[\begin{array}{cc|cc} a & b_1 & b_2 & b_3 \\ b_1 & a & b_3 & b_2 \\ \hline b_2 & b_3 & a & b_1 \\ b_3 & b_2 & b_1 & a \end{array}\right] \tag{5.41}$$

The submatrix containing the kinematic coefficients for the equivalent co-ordinates for the extensions of the CH bonds has the same structure, as does the submatrix of the force constant matrix for the same co-ordinates. Exactly the same structure is obtained for the scalar products of the unit vectors that are directed along the CH bonds and pointing from the carbon atoms towards the hydrogen atoms. Matrix (5.41) is not a circulant matrix, but its eigenvalues and eigenvectors can be determined easily if it is broken up into second-order blocks (as shown by the dotted lines). This means that all the equivalent objects (atomic orbitals, natural co-ordinates or unit vectors) are divided into two subjects in which the equivalent objects are transformed into each other by rotational operations about the second-order axis of symmetry, which coincides

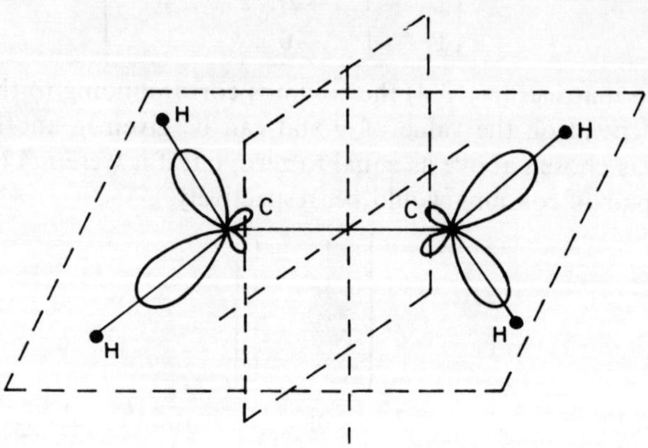

Figure 5.10. Equivalent HAO for C atoms of ethylene

with the C=C bond. Thus the atomic orbitals or natural co-ordinates of the left-hand pair of CH bonds fall into one subgroup of equivalent objects, whilst those of the right-hand pair fall into the second subgroup (see Figure 5.10).

Each of these second-order blocks for the subjects of the equivalent objects will possess the structure of the simplest circulants, and can be diagonalized by an orthogonal transformation with a matrix C having the form

$$C = \frac{1}{\sqrt{2}}\begin{bmatrix} 1 & -1 \\ 1 & 1 \end{bmatrix}$$

Consequently, an orthogonal transformation whose matrix is

$$C_1 = \frac{1}{\sqrt{2}}\begin{bmatrix} 1 & -1 & 0 & 0 \\ 1 & 1 & 0 & 0 \\ 0 & 0 & 1 & -1 \\ 0 & 0 & 1 & 1 \end{bmatrix}$$

will change the original matrix to the form

$$\left[\begin{array}{cc|cc} \underline{a+b_1} & 0 & \underline{b_2 b_3} & 0 \\ 0 & \underline{\underline{a-b_1}} & 0 & \underline{\underline{b_2-b_3}} \\ \underline{b_2+b_3} & 0 & \underline{a+b_1} & 0 \\ 0 & \underline{\underline{b_2-b_3}} & 0 & \underline{\underline{a-b_1}} \end{array}\right]$$

It is clear that this matrix can in turn be divided into two second-order submatrices (the elements of each of these submatrices above have been underlined either once or twice). First, the submatrices do not interact with each other, and second, once again they are circulants. Hence a second orthogonal transformation, with a matrix

$$C_2 = \frac{1}{\sqrt{2}}\begin{bmatrix} 1 & 0 & -1 & 0 \\ 0 & 1 & 0 & -1 \\ 1 & 0 & 1 & 0 \\ 0 & 1 & 0 & 1 \end{bmatrix}$$

will diagonalize matrix (5.41), thus:

$$\begin{bmatrix} a_1+b_1+b_2+b_3 & 0 & 0 & 0 \\ 0 & a-b_1+b_2-b_3 & 0 & 0 \\ 0 & 0 & a+b_1-b_2-b_3 & 0 \\ 0 & 0 & 0 & a-b_1-b_2+b_3 \end{bmatrix}$$

The overall matrix C of the eigenvectors of the original matrix is equal to the product of the two matrices C_1 and C_2, i.e.

$$C = C_1 C_2 = \frac{1}{2}\begin{bmatrix} 1 & -1 & -1 & 1 \\ 1 & 1 & -1 & -1 \\ 1 & -1 & 1 & -1 \\ 1 & 1 & 1 & 1 \end{bmatrix} \tag{5.42}$$

We can construct the SMOs or symmetry co-ordinates using this matrix, and they will have the forms

$$q_1^{(s)} = \tfrac{1}{2}(q_1 + q_2 + q_3 + q_4)$$
$$q_2^{(s)} = \tfrac{1}{2}(-q_1 + q_2 - q_3 + q_4)$$
$$q_3^{(s)} = \tfrac{1}{2}(-q_1 - q_2 + q_3 + q_4)$$
$$q_4^{(s)} = \tfrac{1}{2}(q_1 - q_2 - q_3 + q_4)$$

where q_k may refer either to an atomic orbital or to the natural co-ordinate of a CH-bond extension.

As another example we now consider the construction of the SMOs or symmetry co-ordinates from the equivalent HAOs or natural CH-bond extension co-ordinates in the ethane molecules (see Figure 5.11). Ethane has D_{3h} symmetry and equivalent objects can be transformed into each other by symmetry operations that include rotations about a third-order axis of symmetry, which coincides with the C—C bond, and by reflections in a plane perpendicular to the symmetry axis. The matrices containing the Coulomb and resonance integrals, the overlap integrals, the kinematic coefficients or force constants for natural CH-bond-stretching co-ordinates and the scalar products of the unit vectors are directed along the CH bond from the carbon to the hydrogen atoms, and all have the following structure:

$$A = \left[\begin{array}{ccc|ccc} a & b_1 & b_1 & b_2 & b_3 & b_3 \\ b_1 & a & b_1 & b_3 & b_2 & b_3 \\ b_1 & b_1 & a & b_3 & b_3 & b_2 \\ \hline b_2 & b_3 & b_3 & a & b_1 & b_1 \\ b_3 & b_2 & b_3 & b_1 & a & b_1 \\ b_3 & b_3 & b_2 & b_1 & b_1 & a \end{array}\right] \tag{5.43}$$

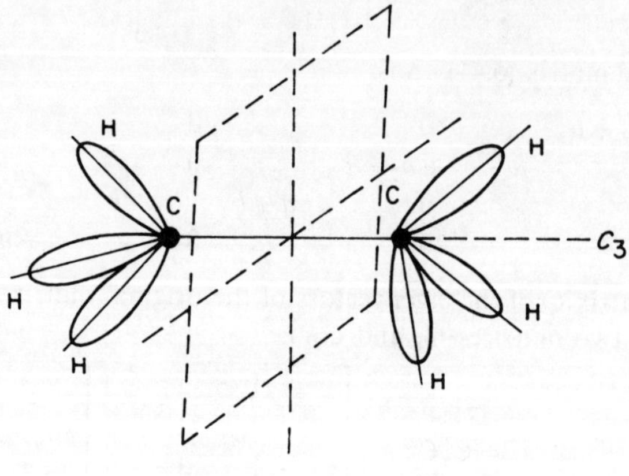

Figure 5.11. Equivalent HAO to C atoms of ethane

This matrix is also not a circulant, but it can be broken down into third-order blocks that are circulants. Such a matrix can be diagonalized by two successive orthogonal transformations with matrices C_1 and C_2. The C_1 matrix has the form:

$$C_1 = \frac{1}{\sqrt{6}} \left[\begin{array}{ccc|ccc} \sqrt{2} & 2 & 0 & 0 & 0 & 0 \\ \sqrt{2} & -1 & \sqrt{3} & 0 & 0 & 0 \\ \sqrt{2} & -1 & -\sqrt{3} & 0 & 0 & 0 \\ \hline 0 & 0 & 0 & \sqrt{2} & 2 & 0 \\ 0 & 0 & 0 & \sqrt{2} & -1 & \sqrt{3} \\ 0 & 0 & 0 & \sqrt{2} & -1 & -\sqrt{3} \end{array}\right]$$

and takes the original matrix A to the form

$$\left[\begin{array}{ccc|ccc} \underline{a+2b_1} & 0 & 0 & \underline{b_2+2b_3} & 0 & 0 \\ 0 & \underline{\underline{a-b_1}} & 0 & 0 & \underline{\underline{b_2-b_3}} & 0 \\ 0 & 0 & \underline{\underline{\underline{a-b_1}}} & 0 & 0 & \underline{\underline{\underline{b_2-b_3}}} \\ \hline \underline{b_2+2b_3} & 0 & 0 & \underline{a_1+2b_1} & 0 & 0 \\ 0 & \underline{\underline{b_2-b_3}} & 0 & 0 & \underline{\underline{-b}} & 0 \\ 0 & 0 & \underline{\underline{\underline{b_2-b_3}}} & 0 & 0 & \underline{\underline{\underline{a-b_1}}} \end{array}\right]$$

This can then be broken down into three second-order blocks (their elements are, respectively, underlined once, twice or three times). The blocks have the standard structure.

Matrix C_2 is

$$C_2 = \frac{1}{\sqrt{2}} \begin{bmatrix} 1 & 0 & 0 & -1 & 0 & 0 \\ 0 & 1 & 0 & 0 & -1 & 0 \\ 0 & 0 & 1 & 0 & 0 & -1 \\ 1 & 0 & 0 & 1 & 0 & 0 \\ 0 & 1 & 0 & 0 & 1 & 0 \\ 0 & 0 & 1 & 0 & 0 & 1 \end{bmatrix}$$

and diagonalizes A, giving it the elements:

$$\begin{aligned} \lambda_1 &= a + 2b_1 + b_2 + 2b_3; \quad & \lambda_2 &= a - b_1 + b_2 - b_3 \\ \lambda_3 &= a - b_1 + b_2 - b_3; \quad & \lambda_4 &= a + 2b_1 - b_2 - 2b_3 \\ \lambda_5 &= a - b_1 - b_2 + b_3; \quad & \lambda_6 &= a - b_1 - b_2 + b_3 \end{aligned}$$

We thus have two pairs of degenerate eigenvalues, i.e. $\lambda_2 = \lambda_3$ and $\lambda_5 = \lambda_6$.

The overall matrix for the complete transformation is

$$C = C_1 C_2 = \frac{1}{\sqrt{12}} \begin{bmatrix} \sqrt{2} & 2 & 0 & -\sqrt{2} & -2 & 0 \\ \sqrt{2} & -1 & \sqrt{3} & -\sqrt{2} & 1 & -\sqrt{3} \\ \sqrt{2} & -1 & -\sqrt{3} & -\sqrt{2} & 1 & \sqrt{3} \\ \sqrt{2} & 2 & 0 & \sqrt{2} & 2 & 0 \\ \sqrt{2} & -1 & \sqrt{3} & \sqrt{2} & -1 & \sqrt{3} \\ \sqrt{2} & -1 & -\sqrt{3} & \sqrt{2} & -1 & -\sqrt{3} \end{bmatrix}$$

We can therefore obtain the following for the symmetry co-ordinates of ethane's CH bonds or the SMO's for the CH HAOs:

$$q_1^{(s)} = \frac{1}{\sqrt{6}}(q_1 + q_2 + q_3 + q_4 + q_5 + q_6)$$

$$q_2^{(s)} = \frac{1}{\sqrt{6}}(2q_1 - q_2 - q_3 + 2q_4 - q_5 - q_6)$$

etc. A SMO or symmetry co-ordinate $q_1^{(s)}$ corresponds to a molecular orbital or atomic displacement that is symmetric with respect to all the elements of symmetry. A SMO or symmetry co-ordinate $q_2^{(s)}$ describes a function (or co-ordinate) that is degenerate with respect to the C_3 axis, symmetric to one of the σ_v planes (see Figure 5.11) and symmetric with respect to the σ_h plane, which is perpendicular to the third-order axis of symmetry.

The procedures we have described above for constructing the coefficients of symmetry for the ethylene and ethane molecules can be represented in another way if we take advantage of what is called direct (or Kronecker) matrix multiplication.

As an example we obtain a C matrix as a result of direct product of matrices

$$A = \begin{bmatrix} a_{11} & a_{12} \\ a_{21} & a_{22} \end{bmatrix} \quad \text{and} \quad B = \begin{bmatrix} b_{11} & b_{12} \\ b_{21} & b_{22} \end{bmatrix}$$

of the following form:

$$C = A \times B = \begin{bmatrix} a_{11}b_{11} & a_{11}b_{12} & a_{12}b_{11} & a_{12}b_{12} \\ a_{11}b_{21} & a_{11}b_{22} & a_{12}b_{21} & a_{12}b_{22} \\ a_{21}b_{11} & a_{21}b_{12} & a_{22}b_{11} & a_{22}b_{12} \\ a_{21}b_{21} & a_{21}b_{22} & a_{22}b_{21} & a_{22}b_{22} \end{bmatrix}$$

In this way the Kronecker product, which we will denote by a multiplication sign, will yield a matrix whose rank is larger than that of either of the original matrices.

It is convenient to write the matrices (5.41), obtained when there are equivalent objects that can be transformed into each other when there are several symmetry elements in the molecule, in the form of Kronecker products, thus:

$$\begin{bmatrix} 1 & 0 \\ 0 & 1 \end{bmatrix} \times \begin{bmatrix} a & b_1 \\ b_1 & a \end{bmatrix} + \begin{bmatrix} 0 & 1 \\ 1 & 0 \end{bmatrix} \times \begin{bmatrix} b_2 & b_3 \\ b_3 & b_2 \end{bmatrix} \tag{5.44}$$

Note that after the Kronecker multiplication the resultant matrices are added together normally. The matrix for the ethane molecule analogous to matrix (5.43) can be written as:

$$\begin{bmatrix} 1 & 0 \\ 0 & 1 \end{bmatrix} \times \begin{bmatrix} a_1 & b_1 & b_1 \\ b_1 & a & b_1 \\ b_1 & b_1 & a \end{bmatrix} + \begin{bmatrix} 0 & 1 \\ 1 & 0 \end{bmatrix} \times \begin{bmatrix} b_2 & b_3 & b_3 \\ b_3 & b_2 & b_3 \\ b_3 & b_3 & b_2 \end{bmatrix} \tag{5.45}$$

An important property of Kronecker products is that if two matrices C_1 and C_2 can diagonalize two other matrices A and B, respectively, then the Kronecker product of A and B can be diagonalized by the Kronecker product of C_1 and C_2. We can take advantage of this in our analysis. The matrices in equation (5.44) that are involved in the Kronecker products both have the simple structure of

$$\begin{bmatrix} a & b \\ b & a \end{bmatrix}$$

Consequently, each of the multiplicands in the product must be diagonalizable by the transformation

$$\frac{1}{\sqrt{2}} \begin{bmatrix} 1 & -1 \\ 1 & 1 \end{bmatrix}$$

The complete transformation will thus have the form

$$\frac{1}{\sqrt{2}} \begin{bmatrix} 1 & -1 \\ 1 & 1 \end{bmatrix} \times \begin{bmatrix} 1 & -1 \\ 1 & 1 \end{bmatrix}$$

Having completed the Kronecker multiplication, we obtain

$$\frac{1}{2} \begin{bmatrix} 1 & -1 & -1 & 1 \\ 1 & 1 & -1 & -1 \\ 1 & -1 & 1 & -1 \\ 1 & 1 & 1 & 1 \end{bmatrix}$$

This matrix is exactly the same as that found, i.e. matrix (5.42), containing the symmetry coefficients for the ethylene's CH bonds.

The matrix of symmetry objects (CH bonds) for the ethane molecule will have the form of matrix (5.45), i.e.

$$\begin{bmatrix} 1 & 0 \\ 0 & 1 \end{bmatrix} \times \begin{bmatrix} a & b_1 & b_1 \\ b_1 & a & b_1 \\ b_1 & b_1 & a \end{bmatrix} + \begin{bmatrix} 0 & 1 \\ 1 & 0 \end{bmatrix} \times \begin{bmatrix} b_2 & b_3 & b_2 \\ b_3 & b_2 & b_3 \\ b_3 & b_3 & b_2 \end{bmatrix}$$

The Kronecker multiplicands are circulants in structure, i.e.

$$\begin{bmatrix} a & b \\ b & a \end{bmatrix} \text{ and } \begin{bmatrix} a & b & b \\ b & a & b \\ b & b & a \end{bmatrix}$$

and must therefore be diagonalized, respectively by:

$$C_1 = \frac{1}{\sqrt{2}}\begin{bmatrix} 1 & -1 \\ 1 & 1 \end{bmatrix} \quad \text{and} \quad C_2 = \frac{\sqrt{2}}{\sqrt{3}}\begin{bmatrix} 1/\sqrt{2} & 1 & 0 \\ 1/\sqrt{2} & -1/2 & \sqrt{3}/2 \\ 1/\sqrt{2} & -1/2 & -\sqrt{3}/2 \end{bmatrix} \tag{5.46}$$

The complete matrix of the transformation has the form

$$C = C_1 \times C_2 = \frac{1}{\sqrt{2}}\begin{bmatrix} 1 & -1 \\ 1 & 1 \end{bmatrix} \times \frac{\sqrt{2}}{\sqrt{3}}\begin{bmatrix} 1/\sqrt{2} & 1 & 0 \\ 1/\sqrt{2} & -1/2 & \sqrt{3}/2 \\ 1/\sqrt{2} & -1/2 & -\sqrt{3}/2 \end{bmatrix}$$

which is exactly what we found for the matrix containing the symmetry coefficients for ethane.

In general, whenever there are several symmetry elements in the molecule and if the complete matrices H, S, etc. can be written as a sum of Kronecker products for the equivalent objects, then there is a diagonalization matrix C which is common for all the components if a diagonalization transformation can be shown for each of the Kronecker multiplicands individually. Moreover, these individual transformations must be the same for all the terms. The complete transformation, or the full matrix of the symmetry coefficients, will be obtained as the Kronecker product of all the partial transformations. This method of constructing the symmetry coefficients for equivalent objects is very convenient, and allows us, in principle, to write a formula for the symmetry coefficients from the general matrix (5.40) in any specific case, and enables us to perform the corresponding calculations on a computer. In general, several multiplicands may appear in the Kronecker product.

We have considered the ethane molecule, which has only one third-order axis. We can conclude from the properties of circulants that every time we encounter an axis that is of order three or more, pairs of degenerate eigenvalues will arise. However, whenever axes that are of order three or more cross, we may obtain threefold or possibly an even greater degree of degeneracy. We will show this using the methane molecule as an example.

Methane has four third-order axes that intersect, and each of these axes passes through only one of the CH bonds. All the atomic orbitals associated with the CH bonds or their natural stretching vibrational co-ordinates are equivalent. The matrices containing the Coulomb and resonance integrals, the kinematic and force constants the scalar products of the bonds' directional vectors all have the structure

$$\begin{bmatrix} a & b & b & b \\ b & a & b & b \\ b & b & a & b \\ b & b & b & a \end{bmatrix}$$

These matrices are the analogues of the submatrices for the CH bonds in the ethylene molecule if we set $b_1 = b_2 = b_3 = b$ in the latter. Since the eigenvectors of all the matrices we are considering and for the whole set of equivalent objects must not depend on the values of their elements, it is clear that the matrices of the eigenvectors in the methane molecule for the CH bonds will coincide with the C matrices in the ethylene molecule. However, the physical sense of its columns is changed.

Since the b coefficients in the matrix for the methane molecule are all the same its eigenvalues will be, respectively, $a + 3b$, $a - b$, $a - b$, $a - b$, i.e. the last three columns of the C matrix for methane must belong to a triply degenerate eigenvalue. As was the case for double degeneracy, the choice of the elements of the eigenvectors of the C matrices can only be determined without respect for orthogonal transformations when there is triple degeneracy. The canonical orthogonal third-order matrix can be represented as a product of three orthogonal matrices, each of which depends on the parameters φ, ψ and θ, respectively (the angles of rotation of the co-ordinate axes or orthogonal transformations of the variables). The canonical matrix has the form:

$$\begin{bmatrix} \cos\varphi & -\sin\varphi & 0 \\ \sin\varphi & \cos\varphi & 0 \\ 0 & 0 & 1 \end{bmatrix} \begin{bmatrix} \cos\psi & 0 & -\sin\psi \\ 0 & 1 & 0 \\ \sin\psi & 0 & \cos\psi \end{bmatrix} \begin{bmatrix} 1 & 0 & 0 \\ 0 & \cos\theta & -\sin\theta \\ 0 & \sin\theta & \cos\theta \end{bmatrix} \tag{5.47}$$

Consequently, each of the elements of the eigenvectors, given triple degeneracy, is a function of all three parameters φ, ψ and θ. If any of these are changed, the elements of these eigenvectors are altered simultaneously. This enables us to effect orientation operations, i.e. to organize the choice of co-ordinates of degenerate types when there are equivalent co-ordinates or equivalent orbitals of different kinds in a molecule.

5.3.3 *Classifying SMOs or symmetry co-ordinates by symmetry type*

If there are several groups of equivalent atomic orbitals (for example, s orbitals and hybrid atomic orbitals) in a molecule or several groups of equivalent natural vibration co-ordinates (angles and bonds), then we can construct SMOs or symmetry co-ordinates for each such group of equivalent objects according to the rules given above.

Having done this, we must co-ordinate the symmetry co-ordinates, or SMOs, together when the symmetries are of different types. In order to solve this problem we need to estimate the changes in the SMOs, or in the symmetry co-ordinates, during all the symmetry operations allowed by the molecule, after they have been constructed. All the SMOs, or symmetry co-ordinates, that change in the same way during a symmetry operation are gathered together into one group. It is convenient, for simplicity, to proceed in the following way: each equivalent atomic orbital is made to coincide with a unit directional vector. The direction of the unit vector is obvious for hybrid orbitals which clearly have directional properties. The vector can be chosen arbitrarily for s orbitals if all the vectors for equivalent orbitals can be transformed into each other during all the symmetry operations allowed by the molecule. However, it is convenient to direct the unit vectors for s orbitals along the equivalent bonds. We have already noted that when choosing symmetry coefficients the nature of the objects with which we are dealing is irrelevant. They may be AOs, HAOs, natural vibration co-ordinates or simply unit directional vectors. We will call the linear combinations of the unit directional vectors for equivalent objects that are constructed from symmetry coefficients, directional vectors of symmetry. These vectors are transformed during the symmetry operations of a molecule in a well-defined way analogous to that of the transformation of the SMOs or symmetry co-ordinates. For this reason, it is convenient from now on to make the SMOs or symmetry co-ordinates agree for different kinds of atomic orbitals or natural vibrational co-ordinates using the more visual case, in which these objects are substituted for by equivalent vectors and their corresponding vectors of symmetry.

We shall consider an appropriate operation in the special case of the ethylene molecule and divide the molecule into two groups of equivalent atomic orbitals. All the s-orbitals of the hydrogen atoms are gathered in the first group. In the second we place the equivalent hybrid orbitals of the carbon atoms that are directed along the CH bonds. The corresponding directional vectors for these orbitals are distributed as shown in Figure 5.12.

It follows from what has been said that the formulae for the SMOs are:

$$\begin{aligned}
\psi_1^{(s)} &= \tfrac{1}{2}(s_1 + s_2 + s_3 + s_4) & \psi_5^{(s)} &= \tfrac{1}{2}(P_{CH_1} + P_{CH_2} + P_{CH_3} + P_{CH_4}) \\
\psi_2^{(s)} &= \tfrac{1}{2}(s_1 + s_2 - s_3 - s_4) & \psi_6^{(s)} &= \tfrac{1}{2}(P_{CH_1} + P_{CH_2} - P_{CH_3} - P_{CH_4}) \\
\psi_3^{(s)} &= \tfrac{1}{2}(s_1 - s_2 + s_3 - s_4) & \psi_7^{(s)} &= \tfrac{1}{2}(P_{CH_1} - P_{CH_2} + P_{CH_3} - P_{CH_4}) \\
\psi_4^{(s)} &= \tfrac{1}{2}(s_1 - s_2 - s_3 + s_4) & \psi_8^{(s)} &= \tfrac{1}{2}(P_{CH_1} - P_{CH_2} - P_{CH_3} + P_{CH_4})
\end{aligned} \tag{5.48}$$

For simplicity, we shall denote an s function of the hydrogen atoms by the symbol s_k and a hybrid orbital of a carbon atom, so far as it depends on its direction, by the symbol P_{CH_k}. The distribution of the vectors $\vec{e}_k$ according to function is clear from Figure 5.12.

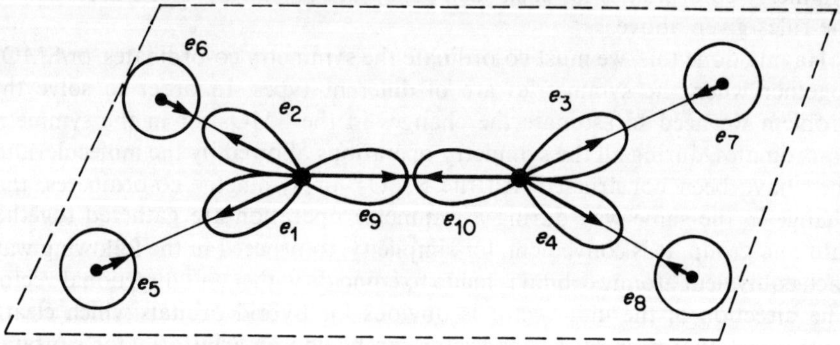

Figure 5.12. The system of unit directional vectors for the ethylene molecule. The groups of equivalent vectors are $(\vec{e}_1, \vec{e}_2, \vec{e}_3, \vec{e}_4)$, $(\vec{e}_5, \vec{e}_6, \vec{e}_7, \vec{e}_8)$ and $(\vec{e}_9, \vec{e}_{10})$

The vectors of symmetry for these objects, in conformity with equation (5.48), will have the form:

$$\vec{e}_1^{(s)} = \tfrac{1}{2}(\vec{e}_1 + \vec{e}_2 + \vec{e}_3 + \vec{e}_4)$$
$$\vec{e}_2 = \tfrac{1}{2}(\vec{e}_1 + \vec{e}_2 - \vec{e}_3 - \vec{e}_4), \ldots$$

It is plain that the vectors of symmetry for the atomic orbitals of the hydrogen and carbon atoms can be united into pairs by their parallelism thus: $\vec{e}_1^{(s)}$ and $\vec{e}_5^{(s)}$, $\vec{e}_2^{(s)}$ and $\vec{e}_6^{(s)}$, $\vec{e}_3^{(s)}$ and $\vec{e}_7^{(s)}$, and $\vec{e}_4^{(s)}$ and $\vec{e}_8^{(s)}$.

The vectors of symmetry of each pair are transformed into each other in the same way during symmetry operations, i.e. they have the same type of symmetry. We obtain the following for the orbitals of symmetry that are constructed from the carbon atom orbitals in the pair of hybrid orbitals that are directed along the C=C bond according to the directional vectors shown in Figure 5.12:

$$\psi_9^{(s)} = \frac{1}{\sqrt{2}}(P_{\mathrm{cc}_1} + P_{\mathrm{cc}_2}) \quad \text{and} \quad \psi_{10}^{(s)} = \frac{1}{\sqrt{2}}(P_{\mathrm{cc}_1} - P_{\mathrm{cc}_2})$$

and respectively,

$$\vec{e}_9^{(s)} = \frac{1}{\sqrt{2}}(\vec{e}_9 + \vec{e}_{10}), \quad \vec{e}_{10}^{(s)} = \frac{1}{\sqrt{2}}(\vec{e}_9 - \vec{e}_{10})$$

It is clear that the first vector of symmetry is transformed during symmetry operations in exactly the same way as the vectors $\vec{e}_{10}^{(s)}$ and $\vec{e}_5^{(s)}$, whilst the $\vec{e}_{10}^{(s)}$ vector is transformed like $\vec{e}_2^{(s)}$ and $\vec{e}_6^{(s)}$. These vectors are, moreover, collinear (i.e. they lie along one straight line). This is a necessary and sufficient condition for the corresponding SMOs to belong to the same symmetry type. It may happen here that a vector of symmetry is zero, and then it is necessary to consider each of its constructions separately to estimate its orientation. This method of adjusting different kinds of co-ordinates using symmetry vectors helps us in many cases to choose quickly the requisite set of different SMOs or symmetry co-ordinates.

The symmetry of appropriate co-ordinates can be estimated for molecule vibrations by studying the displacements of the atoms when the given symmetry co-ordinates are changed. The displacements of the equivalent atoms of a given collection when one of the symmetry co-ordinates is changed is given by the vectors $\vec{e}$, $\vec{f}$, etc. These vectors, taken with coefficients equal to those in front of the natural co-ordinates in the symmetry co-ordinate being studied, comprise th matrix $\vec{B}$ (see Section 8.3). Of course, the orientation of the equilibrium unit vectors of equivalent co-ordinates must be chosen so that they are transformed into each other during symmetry operations. By substituting the symmetry of the displacements of equivalent atoms, which are obtained in the way just described, for the symmetry co-ordinates (for example, the angles or bond co-ordinates), we can group together the co-ordinates according to their symmetry.

We will consider this procedure for the example of the planar vibrations of

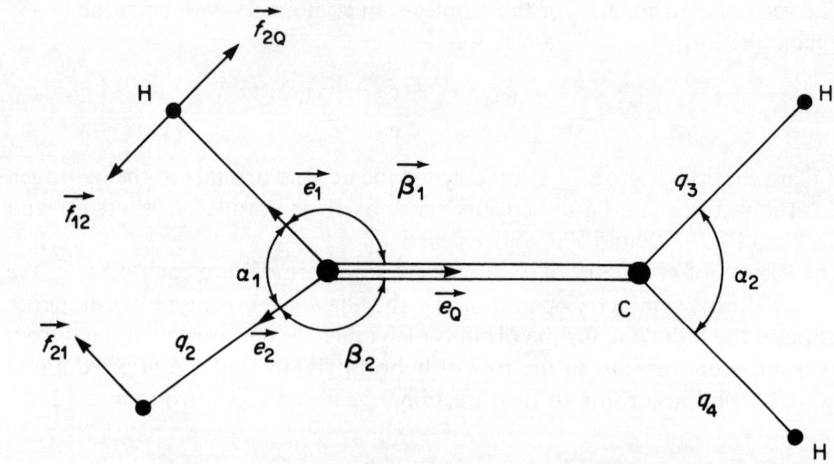

Figure 5.13. Co-ordinates and vectors of ethylene

ethylene, and introduce the co-ordinates and co-ordinate vectors for the molecule as shown in Figure 5.13.

The submatrix of the interactions of the co-ordinates of equivalent CH bonds (q_i) and that of the H—C—C angles (β_i) will have the same structure, i.e. that in equation (5.28), and the submatrix of the co-ordinates α_1 and α_2 will have the same structure as in equation (5.23). The symmetry co-ordinates must now be constructed according to the rule given above (see page 138), i.e.

$$q_1^{(s)} = \tfrac{1}{2}(q_1 + q_2 + q_3 + q_4)$$
$$q_2^{(s)} = \tfrac{1}{2}(q_1 - q_2 + q_3 - q_4)$$

We obtain similar linear combinations of natural angular $B_i^{(s)}$ co-ordinates for the B_i co-ordinates, while all the H—C—H angle co-ordinates will be

$$\alpha_1^{(s)} = \frac{1}{\sqrt{2}}(\alpha_1 + \alpha_2), \quad \alpha_2^{(s)} = \frac{1}{\sqrt{2}}(\alpha_1 - \alpha_2)$$

We can form an opinion on the symmetry of the vibrations from the displacements of the hydrogen atoms which form a set of equivalent atoms. Note that overall, in the great majority of practical cases, it is enough to consider only one or two sets of equivalent atoms in the molecule in order to classify the symmetry co-ordinates according to symmetry. Hence for the co-ordinate $q_1^{(s)} = 1/2(q_1 + q_2 + q_3 + q_4)$ we obtain $\delta\vec{r}_1 = \vec{e}_1$, $\delta\vec{r}_2 = \vec{e}_2$, $\delta\vec{r}_3 = \vec{e}_3$ and $\delta\vec{r}_4 = \vec{e}_4$ (Figure 5.14(a)) according to the rule we mentioned.*

The displacements are therefore symmetric with respect to all symmetry elements, and so the $q_1^{(s)}$-co-ordinate must belong to the fully symmetric A_g-symmetry type.

*We need not consider the normalizing factor in the C matrix.

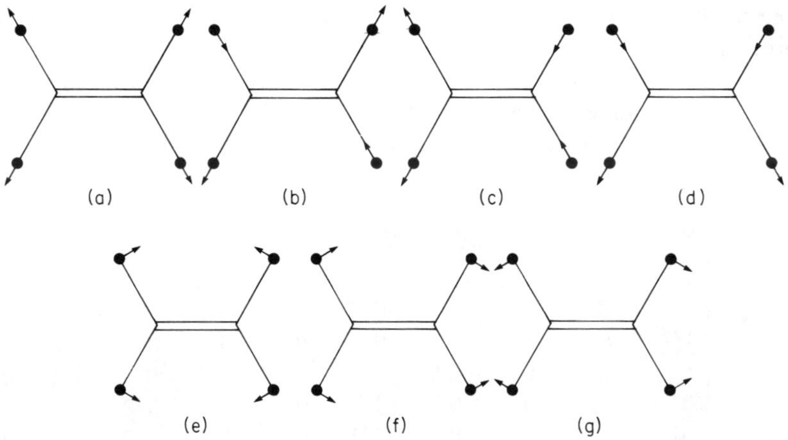

Figure 5.14. Symmetry for displacements of H atoms of an ethylene molecule

We obtain $\delta\vec{r}_1 = \vec{e}_1, \delta\vec{r}_2 = -\vec{e}_2, \delta\vec{r}_3 = \vec{e}_3$ and $\delta\vec{r}_4 = -\vec{e}_4$ (see Figure 5.14(b)) for the co-ordinate $q_2^{(s)} = 1/2\ (q_1 - q_2 + q_3 - q_4)$. That is, the displacements are symmetric with respect to the centre of symmetry, the second-order axis of symmetry and to all the planes of symmetry (except for the plane of the molecule). Consequently, the $q_2^{(s)}$-co-ordinate must belong to the B_{1g} type.

Using similar arguments we obtain the displacements shown in Figures 5.14(c) and (d) for the $q_3^{(s)}$ and $q_4^{(s)}$-co-ordinates. Consequently, they must belong to the B_{3u} and B_{2u} symmetry types, respectively. For the co-ordinate $\alpha_1^{(s)} = (1/\sqrt{2})(\alpha_1 + \alpha_2)$ we find $\delta\vec{r}_1 = -\vec{f}_{12}$, $\delta\vec{r}_2 = -\vec{f}_{21}$, $\delta\vec{r}_3 = -\vec{f}_{34}$ and $\delta\vec{r}_4 = -\vec{f}_{41}$ (Figure 5.14(e)). That is, the displacements are symmetric with respect to all the symmetry elements. This means that the $\alpha_1^{(s)}$ co-ordinate must also belong to the fully symmetric A_g type.

The atom displacement pattern for the co-ordinate $\alpha_2^{(s)} = (1/\sqrt{2})(\alpha_1 - \alpha_2)$ will have the form shown in Figure 5.14(f), and by its symmetry it coincides with the displacement pattern for the $q_3^{(s)}$ co-ordinate. Hence the $\alpha_2^{(s)}$ co-ordinate is of the B_{3u} type.

The $\delta\vec{r}_i$ displacements for the $B_1^{(s)}$ co-ordinate will have the form shown in Figure 5.14(g), and they are similar in symmetry to the $q_1^{(s)}$ co-ordinate. Consequently, the $B_1^{(s)}$ co-ordinate must belong to the A_g type.

By similar arguments we find that the $B_2^{(s)}$ co-ordinate belongs to the B_{1g} symmetry type, the $B_3^{(s)}$ co-ordinate to the B_{3u} type and $B_4^{(s)}$ co-ordinate to the B_{2u} type.

The Q-co-ordinate for the C=C bond remains. It is an unrepeated (singular) bond and therefore coincides with a symmetry co-ordinate. A cursory consideration makes it clear that when the co-ordinate is changed, fully symmetric displacements of all the atoms are obtained. Hence it belongs to the A_g symmetry type.

Note that, in general, the changes in unrepeated (singular) bonds and angles belong to the fully symmetric type. This, however, is not true for non-planar co-ordinates.

If we construct the matrix containing, say, the symmetry coefficients for all the orbitals of symmetry that lie in the plane of the ethylene molecule using the rule we have described, and we group them according to their symmetry types, we obtain:

$$
\begin{array}{ll}
 & \begin{array}{cccccccccc} \psi_1^{(s)} & \psi_5^{(s)} & \psi_9^{(s)} & \psi_2^{(s)} & \psi_6^{(s)} & \psi_{10}^{(s)} & \psi_3^{(s)} & \psi_7^{(s)} & \psi_4^{(s)} & \psi_8^{(s)} \end{array} \\
C = \begin{array}{c} s_1 \\ s_2 \\ s_3 \\ s_4 \\ P_{CH_1} \\ P_{CH_2} \\ P_{CH_3} \\ P_{CH_4} \\ P_{CC_1} \\ P_{CC_2} \end{array} &
\left[\begin{array}{cccccccccc}
1/2 & 0 & & 1/2 & 0 & 0 & 1/2 & 0 & 1/2 & 0 \\
1/2 & 0 & 0 & 1/2 & 0 & 0 & -1/2 & 0 & -1/2 & 0 \\
1/2 & 0 & 0 & -1/2 & 0 & 0 & 1/2 & 0 & -1/2 & 0 \\
1/2 & 0 & 0 & -1/2 & 0 & 0 & -1/2 & 0 & 1/2 & 0 \\
0 & 1/2 & 0 & 0 & 1/2 & 0 & 0 & 1/2 & 0 & 1/2 \\
0 & 1/2 & 0 & 0 & 1/2 & 0 & 0 & -1/2 & 0 & -1/2 \\
0 & 1/2 & 0 & 0 & -1/2 & 0 & 0 & 1/2 & 0 & -1/2 \\
0 & 1/2 & 0 & 0 & -1/2 & 0 & 0 & -1/2 & 0 & 1/2 \\
0 & 0 & 1/\sqrt{2} & 0 & 0 & 1/\sqrt{2} & 0 & 0 & 0 & 0 \\
0 & 0 & 1/\sqrt{2} & 0 & 0 & -1/\sqrt{2} & 0 & 0 & 0 & 0
\end{array}\right]
\end{array}
\tag{5.49}
$$

We thus have for the symmetry co-ordinates of the planar vibrations of the ethylene molecule:

$$
\begin{array}{ll}
 & \begin{array}{ccccccccccc} q_1^{(s)} & \beta_1^{(s)} & \alpha_1^{(s)} & Q^{(s)} & q_2^{(s)} & \beta_2^{(s)} & q_3^{(s)} & \beta_3^{(s)} & \alpha_2^{(s)} & q_4^{(s)} & \beta_4^{(s)} \end{array} \\
C = &
\left[\begin{array}{c|ccccccccccc}
q_1 & 1/2 & 0 & 0 & 0 & 1/2 & 0 & 1/2 & 0 & 0 & 1/2 & 0 \\
q_2 & 1/2 & 0 & 0 & 0 & -1/2 & 0 & 1/2 & 0 & 0 & -1/2 & 0 \\
q_3 & 1/2 & 0 & 0 & 0 & 1/2 & 0 & -1/2 & 0 & 0 & -1/2 & 0 \\
q_4 & 1/2 & 0 & 0 & 0 & -1/2 & 0 & -1/2 & 0 & 0 & 1/2 & 0 \\
Q & 0 & 0 & 0 & 1 & 0 & 0 & 0 & 0 & 0 & 0 & 0 \\
\beta_1 & 0 & 1/2 & 0 & 0 & 0 & 1/2 & 0 & 1/2 & 0 & 0 & 1/2 \\
\beta_2 & 0 & 1/2 & 0 & 0 & 0 & -1/2 & 0 & 1/2 & 0 & 0 & -1/2 \\
\beta_3 & 0 & 1/2 & 0 & 0 & 0 & 1/2 & 0 & -1/2 & 0 & 0 & -1/2 \\
\beta_4 & 0 & 1/2 & 0 & 0 & 0 & -1/2 & 0 & -1/2 & 0 & 0 & 1/2 \\
\alpha_1 & 0 & 0 & 1/\sqrt{2} & 0 & 0 & 0 & 0 & 0 & 1/\sqrt{2} & 0 & 0 \\
\alpha_2 & 0 & 0 & 1/\sqrt{2} & 0 & 0 & 0 & 0 & 0 & -1/\sqrt{2} & 0 & 0
\end{array}\right]
\end{array}
\tag{5.50}
$$

The symbol q_k refers to a CH bond-stretching co-ordinate, Q to the C=C bond-stretching co-ordinate and α and β to the co-ordinates of the changes in the HCH and HCC angles.

As a result of the multiplications $\tilde{C}HC$, $\tilde{C}SC$, $\tilde{C}TC$, $\tilde{C}UC$, the original matrices H, S, T and U are quasi-diagonalized. Here T and U are the matrices containing the force and kinematic coefficients, respectively, in natural vibration co-ordinates. The $H^{(s)}$ and $S^{(s)}$ matrices contain two second-order and two third-order blocks. For the vibration problem, we obtain one fourth-order, one third-order and two second-order blocks. Hence instead of one tenth- or eleventh-order equation we have only to solve several lower-order ones.

In conclusion, we note once more that if we have an unrepeated singular single equivalent orbital, or bond or angle co-ordinate, then it always belongs to the fully symmetric type.

5.3.4 Adjustment of degenerate SMOs or symmetry co-ordinates for different types of symmetry: orientation operations

In molecules of high symmetry the presence of degenerate eigenvalues in the H, S, T and U submatrices gives rise to degenerate eigenvectors, whose elements act as symmetry coefficients. These eigenvectors can only be uniquely defind by ignoring the effects of orthogonal transformations when constructing the co-ordinates or orbitals of symmetry, and they depend on one parameter (φ) when there is double degeneracy and on three (φ, ψ and θ) when there is triple degeneracy. Different choices of all the parameters yield different numerical values for the elements of the eigenvectors, and consequently different coefficients in the linear combinations of atomic orbitals or natural co-ordinates that make up the symmetry orbitals or symmetry co-ordinates.

If each natural equivalent co-ordinate or atomic orbital is assigned a vector then we obtain different symmetry vectors. Moreover, if they are non-zero then every choice of the parameters φ, ψ and θ will yield different directions in space for these symmetry vectors. This does not lead to any ambiguity for one set of equivalent natural co-ordinates or atomic orbitals because the diagonalization of the H, S, T and U submatrices is wholly without dependence on the choice of φ, ψ or θ. However, if a molecule has groups of equivalent atomic orbitals or natural co-ordinates that are of different kinds (for example, angle and bond co-ordinates or s and p orbitals) and this is always the case in practice, then an arbitrary choice of φ, ψ and θ may be made that are maladjusted (not in agreement) for different molecular orbitals or co-ordinates of symmetry. This may yield formally constructed orbitals of symmetry co-ordinates* that can describe the transformations of the wave functions or atom displacements when the molecule is vibrating, but that cannot be divided into the groups corresponding to their different symmetry types. Such linear combinations are not SMOs or symmetry co-ordinates and they cannot be reached by quasi-diagonalizing (dividing into blocks) the original H, S, T or U matrices. This means that the terms corresponding, say, to the resonance integrals for the orbitals of symmetry of different kinds, or the interactions between new vibration co-ordinates (the linear combination of angles and bonds), do not vanish.

Thus in order to construct SMOs or symmetry co-ordinates we must do the following. One of the sets of equivalent objects must be selected and the SMOs,

*These are the linear combinations of either the atomic orbitals or the natural co-ordinates whose coefficients are the elements of the eigenvectors of the submatrices of either the resonance integrals or the interaction between the equivalent co-ordinates.

symmetry co-ordinates or symmetry vectors constructed for it using the rule described. When the corresponding degeneracy appears we must endeavour to make the symmetries of the MOs (or those of the displacements of the atoms from their equilibrium positions during vibrations) of the other sets of equivalent objects analogous to the symmetries of the orbitals (or those of the displacements) of the degenerate molecular orbitals (or symmetry co-ordinates) of the first group by varying φ, ψ and θ. The flexibility in the choice of φ, ψ and θ is retained, but it is preconditioned for all the sets of equivalent objects by the flexibility of the choice of the first set. All the remaining objects of symmetry must be adjusted to the co-ordinates of the first set of equivalent objects. This adjustment operation for every degenerate SMO, or vibrational co-ordinate of symmetry, when they are of different kinds has become known as the 'orientation operation.'

Obviously, only one order of degenerate objects are adjusted to each other. We have to deal with orientation operations very frequently when studying vibrations in molecules, and so we will now discuss the introduction of symmetry co-ordinates using, by way of example, the planar vibrations of a molecule XY_4 which has D_{4h} symmetry.

We take the bond-stretching co-ordinates q_1, q_2, q_3, q_4 and the bond-angle co-ordinates (see Figure 5.15) as the natural vibrational co-ordinates. The symmetry matrix for the four symmetric objects has already been obtained (see page 135). Accordingly, the symmetry co-ordinates must be written in the form:

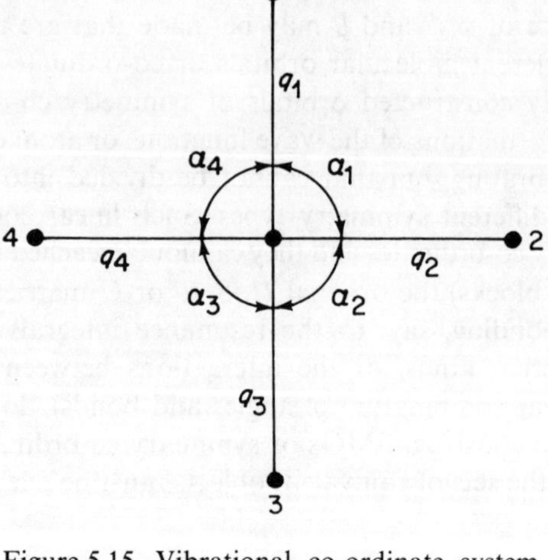

Figure 5.15. Vibrational co-ordinate system for an XY_4 molecule

$$
\begin{aligned}
q_1^{(s)} &= \tfrac{1}{2}(q_1 + q_2 + q_3 + q_4) \\
q_2^{(s)} &= \tfrac{1}{2}(q_1 - q_2 + q_3 - q_4) \\
q_3^{(s)} &= \frac{1}{\sqrt{2}}(q_1 - q_3) \\
q_4^{(s)} &= \frac{1}{\sqrt{2}}(q_2 - q_4)
\end{aligned}
\Bigg\} \text{Doubly degenerate co-ordinates}
$$

$$
\begin{aligned}
\alpha_1^{(s)} &= \tfrac{1}{2}(\alpha_1 + \alpha_2 + \alpha_3 + \alpha_4) \\
\alpha_2^{(s)} &= \tfrac{1}{2}(\alpha_1 - \alpha_2 + \alpha_3 - \alpha_4)
\end{aligned}
$$

To begin with, we will use the same symmetry coefficients for the angle co-

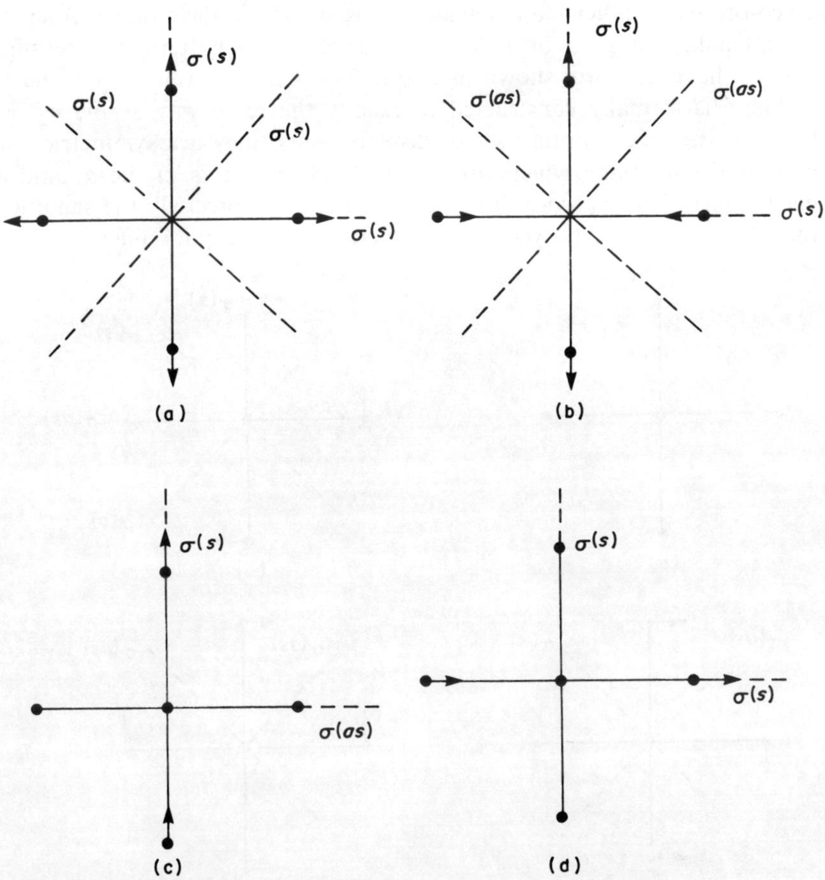

Figure 5.16. The displacement of the equivalent atoms (indicated by arrows) correspond to the following symmetry co-ordinates: (a) $q_1^{(s)}$, (b) $q_2^{(s)}$, (c) $q_3^{(s)}$ (d) $q_4^{(s)}$. The planes shown are designated $\sigma^{(s)}$ and $\sigma^{(as)}$, and the displacements of the equivalent atoms are, respectively, symmetric and anti-symmetric with respect to them

ordinates of the degenerate pair as we used for the bonds. We have, therefore,

$$\alpha_3^{(s)} = \frac{1}{\sqrt{2}}(\alpha_1 - \alpha_3); \quad \alpha_4^{(s)} = \frac{1}{\sqrt{2}}(\alpha_2 - \alpha_4)$$

For the first symmetry co-ordinate all the bonds must stretch to the same extent, while for the second the first bond will stretch and the second will contract by the same amount, and the third will stretch and the fourth will contact to the same extent. The contraction (or extension) of a bond accounts for the plus (or minus) in front of the natural co-ordinate for the bond extension in the expression for the symmetry co-ordinate. The displacement pattern of the terminal atoms is shown in Figure 5.16 for all the symmetry co-ordinates.

In the symmetry co-ordinates for the angles the pluses or minuses before the angle co-ordinates indicate an increase or a reduction in their values. A simple argument makes it easy to conclude that the displacement pattern of the terminal atoms will have the form shown in Figure 5.17. Clearly, even though the $\alpha_3^{(s)}$ co-ordinate is formally constructed in exactly the same way as the $q_3^{(s)}$ co-ordinate, it describes another set of displacements (they are symmetric with respect to the σ_v plane which runs through the bisectors of the α_1 and α_3 valency angles). The $\alpha_4^{(s)}$ co-ordinate is constructed in precisely the same way as the $q_4^{(s)}$ co-ordinate, and yet their symmetry types are quite different.

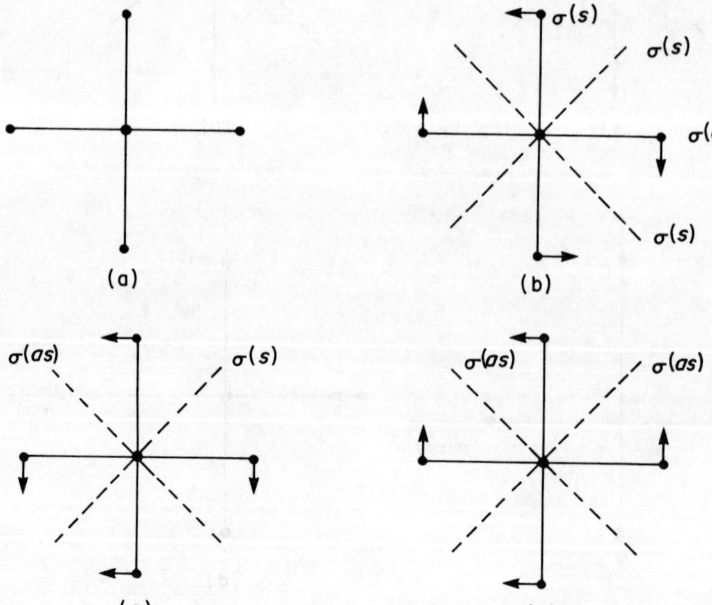

Figure 5.17. The displacements of the equivalent atoms (indicated by arrows) for the symmetry co-ordinates: (a) $\alpha_1^{(s)}$, (b) $\alpha_2^{(s)}$, (c) $\alpha_3^{(s)}$ and (d) $\alpha_4^{(s)}$. The displacements are, respectively, symmetric to the two planes $\sigma^{(s)}$ and $\sigma^{(as)}$ (shown)

If, however, we use these symmetry co-ordinates the problem will not simplify to the maximum possible extent: we will obtain a fourth-order block instead of the two second-order ones we should get from a correct choice of co-ordinates.

We now turn our attention to the directions of the vectors of symmetry. By studying the directions we can change the co-ordinates so as to simplify the task as much as possible. In order to do this, we must strive to make the vectors of symmetry of the co-ordinates being adjusted collinear. This is achieved by rotating the unit vector of the degenerate co-ordinate α by $\pi/4$ or making an additional transformation of the $\alpha_3^{(s)}$- and $\alpha_4^{(s)}$ co-ordinates from $\varphi = \pi/4$. Hence we obtain the following matrix for the columns of the degenerate pair of eigen-vectors of the interaction between the angle co-ordinates:

$$1/2\begin{bmatrix} 1 & -1 \\ 1 & 1 \\ -1 & 1 \\ -1 & -1 \end{bmatrix}$$

The symmetry co-ordinates have the form

$$\alpha_3^{(s)} = \tfrac{1}{2}(\alpha_1 + \alpha_2 - \alpha_3 - \alpha_4); \quad \alpha_4^{(s)} = \tfrac{1}{2}(\alpha_1 - \alpha_2 - \alpha_3 + \alpha_4)$$

The displacements corresponding to these co-ordinates are shown in Figure 5.18. The $\alpha_4^{(s)}$ co-ordinate coincides in symmetry with the $q_3^{(s)}$ one and the $\alpha_3^{(s)}$ co-ordinate with that of $q_4^{(s)}$. We have completed the orientation operation for the degenerate angle co-ordinates and obtained correct symmetry ones.

When there is twofold degeneracy, sorting out the orientations of the different kinds of quantities is quite easy as a rule, because we need only select the value of one parameter, φ. Two rules can be borne in mind for convenience as the help in the choice of symmetry co-ordinates or molecular orbitals of symmetry:

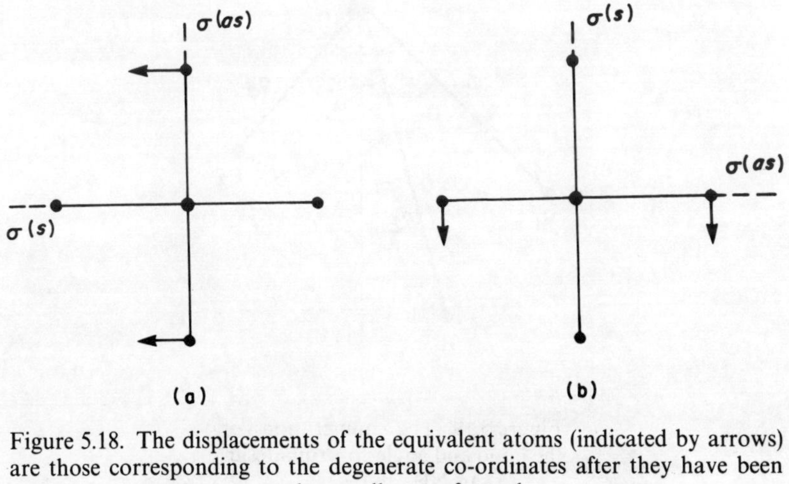

Figure 5.18. The displacements of the equivalent atoms (indicated by arrows) are those corresponding to the degenerate co-ordinates after they have been orthogonally transformed

(1) If a molecule has an odd-order axis of symmetry then when the symmetry co-ordinates (or molecular orbitals or symmetry) are constructed for the different groups of the angles and bond natural co-ordinates (or atomic orbitals) that are equivalent with respect to the axis, a numbering of the natural co-ordinates should be chosen so that bond number 1 of one set lies opposite a bond (or atomic orbital) of another, while the angles are numbered according to the bonds opposite them. For example, the numbering of the angles and bonds for the ammonia molecule should be selected as shown in Figure 5.19.

(2) If a molecule has an even-order axis of symmetry then the numbering of the angles should, as a rule, follow the numbering of the bonds. Moreover, when constructing the symmetry co-ordinates for the angles we should take $\varphi = \pi/p$, where p is the order of axis.

We have already shown that the symmetry vectors are very convenient. It can be shown that in the case of vibrations the vectors characterize the movement of the centre of mass of the set of equivalent atoms during a vibration of the given symmetry type. The symmetry vectors must be orthogonal for vibrations of different symmetry. In fact, a pair of degenerate quantities gives rise to a pair orthogonal symmetry vectors that swivel together as a unit when the φ parameter is changed. Triply degenerate symmetry co-ordinates or molecular orbitals of symmetry gives rise to a triplet of rigidly interconnected orthogonal vectors of symmetry. Since there cannot be more than three non-zero orthogonal vectors in ordinary space, zero-valued vectors of symmetry will appear when a molecule has more than three symmetry elements. Hence we can, in order to retain simplicity, represent these vectors as sums of oppositely orientated

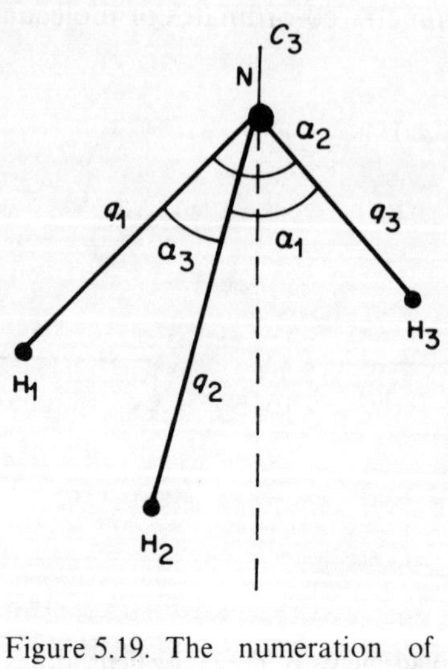

Figure 5.19. The numeration of the bond and angle co-ordinates in an NH_3 molecule

directional vectors, each of which describes the motion of the centre of mass of some of the atoms in a set of equivalent atoms or of some of the equivalent atomic orbitals. The distribution of a vector of symmetry, or that of its constituents, with respect to the molecule's symmetry elements obviously describe the symmetry of the vibrations or that of the wave functions, and so it enables us to classify the co-ordinate and wave functions according to their symmetries. It also helps us complete an orientation operation if we select the φ, ψ and θ parameters according to the angles that are formed between the vectors of symmetry of degenerate co-ordinates, or molecular orbitals, that are different in kind.

For example, the symmetry vectors in a methane molecule for the triply degenerate bond co-ordinates are pointed along the three mutually perpendicular second-order axes of the molecule. In order to obtain an analogous set of directions of the symmetry vectors for the triply degenerate angle co-ordinates we must take the following as these symmetry co-ordinates:

$$\alpha_1^F = \frac{1}{\sqrt{2}}(\alpha_{12} - \alpha_{34}), \quad \alpha_2^F = \frac{1}{\sqrt{2}}(\alpha_{13} - \alpha_{24}), \quad \alpha_3^F = \frac{1}{\sqrt{2}}(\alpha_{14} - \alpha_{24})$$

This can be seen from fairly obvious considerations. The subscripts to the natural angle co-ordinates refer to the numbers of the bonds that form the relevant valency angles.

In molecular vibrational theory both non-planar vibration co-ordinates (see Section 8.4) and in some cases linear combinations of co-ordinates in addition to the bond extension and valency angle change co-ordinates are used. The symmetries of these complex co-ordinates can be obtained formally from those of the constituent co-ordinate or by studying the displacement of the atoms from their equilibrium positions when the co-ordinates are changed.

Note that, in contrast to angle and bond co-ordinates, non-planar co-ordinates, even singular ones, may not belong to a fully symmetric type. For example, the singular co-ordinate of the rotations of the planes of CH bonds in the ethylene molecule towards each other and about the C=C axis is not fully symmetric. When constructing these co-ordinates it is convenient to break them up into their constituents which have the equivalence property. Hence all the rules we have described can be formally applied to them.

For example, in order to construct correct non-planar symmetry co-ordinates in the ethylene molecule it is useful to introduce eight equivalent co-ordinates for the shifts of a CH bond out of the HCC plane (respectively, above and below the plane of the molecule). The correct symmetry co-ordinates for a CCH group in an acetonitryl molecule are obtained as linear combinations of the three equivalent co-ordinates related to the bending of the collinear CCN grouping, the unit vectors of which lie in the molecule's appropriate σ_v-planes.

An XY_4 molecule with D_{4h} symmetry has two independent co-ordinates for bond shifts out of the plane. In order to introduce the symmetry co-ordinates correctly, we must add six 'excess' natural co-ordinates that have this symmetry type, directing the co-ordinates' unit vectors to above and below the plane.

Then, according to our rules, because all these co-ordinates are equivalent and comprise the complete set, we can construct eight symmetry co-ordinates, six of which will in fact be zero. This does not, however, adversely affect our ability to diagonalize the corresponding H, S, T and U matrices.

5.3.5 *Special cases of the C_n and C_{nh} group*

The C_n, C_{nh}, S_{2n}, T and I_h groups have a special place among those of molecular symmetry. The non-diagonal submatrices in these groups, i.e. $H(S)$ and $T(U)$, which correspond to the interactions between the sets of different equivalent atomic orbitals or natural co-ordinates may also be circulant in structure. However, in contrast to the cases we considered above, they are not symmetric. For example, the overall matrix T for the bond-stretching co-ordinates for an X_3Y_3 molecule that, in structure, is like two equilateral pyramids joined at a vertex and that has an overall third-order axis with respect to which both pyramids are turned through an angle not equal to 120° (C_3 symmetry), has the form

$$\left[\begin{array}{ccc|ccc} a & b & b & c & d & e \\ b & a & b & e & c & d \\ b & b & a & d & e & c \\ \hline c & e & d & f & g & g \\ d & c & e & g & f & g \\ e & d & c & g & g & f \end{array}\right]$$

As can be seen, the three third-order submatrices (marked out as shown) are circulant in structure, but the non-diagonal submatrix is not symmetric. Analogous situations are true for any molecule in a C_n group.

As we have shown, all submatrices of this form can be diagonalized simultaneously by the same unitary transformation whose matrix has the following column elements:

$$1; \left(\cos\frac{2\pi k}{p} m + i \sin\frac{2\pi k}{p} m \right), \quad \begin{array}{l} k = 0, 1, 2, \ldots, (p-1) \\ m = 0, 1, 2, \ldots, (p-1) \end{array}$$

The columns' eigennumbers are:

$$\lambda_k = a + \sum_m b_m \left(\cos\frac{2\pi k}{p} m + i \sin\frac{2\pi k}{p} m \right)$$

The difference between symmetric and non-symmetric circulant matrices is that all the eigenvalues in the symmetric type will be real, and complex-conjugate eigenvectors will correspond to pairs of the equal eigenvalues. Complex-conjugate eigenvectors of non-symmetric circulant matrices correspond to complex-conjugate pairs of eigenvalues.

As a result of such a transformation the original matrix is broken down into a series of matrices whose order equals the number of groups of equivalent objects. For example, we obtain for the X_3Y_3 the following:

$$\begin{bmatrix} \alpha & \gamma \\ \gamma & \varphi \end{bmatrix}, \begin{bmatrix} \beta & d^+ \\ d^+ & \varepsilon \end{bmatrix}, \begin{bmatrix} \beta & d^- \\ d^- & \varepsilon \end{bmatrix}$$

where α, β, φ, ε, γ, d^+ and d^- are the eigenvalues of the two diagonal and one non-diagonal third-order matrices, respectively, the eigenvalues d^+ and d^- being complex conjugates. Even so, the greatest simplification of the original task has been achieved by the greatest use of its symmetry.

The inconvenience of this approach comes when complex numbers appear. These bring additional complications into the theory, and although it is not an insurmountable problem it is very inconvenient when using a computer.

In order to avoid the problem we introduce an additional unitary transformation whose matrix is:

$$k = \frac{1}{\sqrt{2}} \begin{bmatrix} i & 1 \\ -i & 1 \end{bmatrix}$$

for every pair of vectors that corresponds to a complex-conjugate eigenvalue. It is plain that this is equivalent to introducing two new columns, one containing the real parts of the original elements and the other the coefficients of the imaginary parts. Thus pairs of real columns are obtained in the overall matrices of the transformations. This additional transformation clearly does not affect the diagonal form of the diagonal symmetric submatrices, since the columns obtained in this way for symmetric circulant matrices are again the eigenvectors.

The non-diagonal submatrices (non-symmetric circulants) are not diagonalized but will contain second-order blocks arranged along their diagonal, each block corresponding to a pair of complex-conjugate eigenvalues. If we call the real and imaginary parts of these eigenvalues d_1 and d_2, respectively, then the transformation gives us:

$$\frac{1}{2}\begin{bmatrix} -i & i \\ 1 & 1 \end{bmatrix}\begin{bmatrix} d^+ & 0 \\ 0 & d^- \end{bmatrix}\begin{bmatrix} i & 1 \\ -i & 1 \end{bmatrix} = \frac{1}{2}\begin{bmatrix} -i & i \\ 1 & 1 \end{bmatrix}\begin{bmatrix} d_1 + id_2 & 0 \\ 0 & d_1 - id_2 \end{bmatrix}\begin{bmatrix} i & 1 \\ -i & 1 \end{bmatrix}$$
$$= \begin{bmatrix} d_1 & d_2 \\ -d_2 & d_1 \end{bmatrix}$$

The appearance of second-order blocks along the diagonals of non-symmetric circulant matrices (they correspond to interactions of the co-ordinates or atomic orbitals between different equivalent groups) means that the problem will not be simplified to the maximum extent, and we will have to take blocks of twice the order we would had to for the degenerate pairs of eigenvalues. For example, for the same X_3Y_3 molecule we have one second-order block

$$\begin{bmatrix} \alpha & \gamma \\ \gamma & \varphi \end{bmatrix}$$

and one fourth-order block:

$$\left[\begin{array}{cc|cc} \beta & d_1 & 0 & d_2 \\ d_1 & \varepsilon & -d_2 & 0 \\ \hline 0 & -d_2 & \beta & d_1 \\ d_2 & 0 & d_1 & \varepsilon \end{array}\right]$$

The rule for constructing such a block has been explained, and no difficulty will be encountered. Once completed, the resultant problem can be easily solved.

For more details and a description of algorithms for computers see reference 2.

References

1. M. V. Volkenshtein, L. A. Gribov, M. A. Elyashevich and B. I. Stepanov. *Vibrations of Molecules*, 2nd edn, Nauka, Physico.-Mathematical Literature, Moscow (1972) (in Russian).
2. A. B. Kovrikov *et al. Spatial Symmetry and Optimization of Molecular Spectra Calculations*, Belorussian State University, Minsk (1983).

Chapter 6

Separation of Motions of Nuclei of Polyatomic Molecules: Equation for Vibrations

6.1 Separation of Types of Nuclear Motions in Polyatomics

In Chapter 2 we touched upon the question concerning the separation of motions in polyatomics. We now shall discuss this more thoroughly. First, we must remember that in the adiabatic approximation it is possible to obtain different equations for electron and nuclei motions. The next task is to separate internal (vibrational and internal rotational) nuclear motions from those involving external translation and rotation. The traditional methods of solution assume relatively small vibrations when the molecule can still be considered as a quasi-solid, and rotations can be described on the basis of the mechanics of solids.[1–9] Obviously, in the general case such an approach is not suitable. Moreover, the restrictions on the vibrational amplitudes are imposed on the potential and not the kinetic part of the Hamilton function. Meanwhile, the molecular deformation co-ordinates and the notion of rotation must also be valid for the description of the purely kinetic operator, and the form of the corresponding term at a given choice of the curvilinear co-ordinates should be the same, no matter how large or small are the vibrational amplitudes, or even whether they are present at all (molecule dissociation). Note that the traditional approaches do not satisfy this latter condition.

We suggest, here, a new approach to the separation of motions in a system of interacting particles, which is both simple and sufficiently general, i.e. it does not require knowledge of the vibrations and, which is important from the methodical standpoint, is analogous to the procedure for the construction of the vibrational part of the Hamilton function in the theory of molecular vibrational spectra[10]. We limit ourselves to the construction of the classical Hamiltonian function (of its kinetic term), since the transition to the quantum operator can be performed by the conventional rules.

We obtain the kinetic term of the Hamilton function in curvilinear co-ordinates in the usual way. Let the Cartesian co-ordinates of atoms (particles) comprising the molecule be related to certain arbitrary generalized co-ordinates by the relationships $\vec{r}_k = f_k(q)$. Differentiating these with respect to time and using matrix symbolism, we obtain $\dot{\vec{r}} = (\partial f/\partial q)\dot{q} = A\dot{q}$. Here $\dot{\vec{r}}$ is the column-matrix of the Cartesian velocities of atoms, $\dot{q}$ is the similar column-matrix of the velocities of the generalized co-ordinates and A is the square matrix of the elements $\partial f_k/\partial q_i$. Since the co-ordinates $\vec{r}_k$ and q_i and their velocities must be interrelated in a unique way, an inverse matrix should exist, A^{-1}, such that $\dot{q} = A^{-1}\dot{\vec{r}}$. The kinetic energy T_{kin} in curvilinear co-ordinates has the form $T_{\text{kin}} = 1/2\tilde{\dot{q}}(\tilde{A}MA)\dot{q}$, where M is the diagonal matrix of atomic masses. Using the definition of the generalized momentum conjugated to the given curvilinear co-ordinate, we obtain the expression for the column of generalized momenta $p = \partial L/\partial \dot{q} = \partial T/\partial \dot{q} = \tilde{A}MA\dot{q}$. Thence $\dot{q} = (\tilde{A}MA)^{-1}p$. Thus $T_{\text{kin}} = 1/2p(\tilde{A}MA)^{-1}(\tilde{A}MA)(\tilde{A}MA)^{-1}p = 1/2\tilde{p}A^{-1}M^{-1}\tilde{A}^{-1}p$. Hence, we can conclude that for Cartesian and generalized momenta the relationship $\vec{p} = \tilde{A}^{-1}p$ is valid, where A^{-1} is the matrix describing the velocities of curvilinear co-ordinates via the velocities of the Cartesian co-ordinates of atoms. As usual, we denote this matrix by the symbol B. Thus the following procedure is suggested for the determination of the square symmetric matrix whose elements are the coefficients of the quadratic form corresponding to the kinetic energy of the system of particles in the momentum representation (canonical form): it is necessary to find the matrix B and to perform the multiplication, $(BM^{-1}\tilde{B})$, multiplying the elements of matrices B as scalars. It is essential that there is no need to determine the matrix A and convert it. Instead, we can determine matrix B, which is much simpler. In this sense, the matrices A and B are considered as independent since, in fact, it is the transformation of velocities and not of co-ordinates which is essential for finding the kinetic energy in the canonical form. The same result can be obtained via differentiation of the Lagrange function which depends, in the general case, on the arbitrary co-ordinates and velocities with respect to the Cartesian co-ordinates. Then, in matrix symbolism: $\vec{p} = \partial L/\partial \dot{\vec{r}} = (\partial \dot{q}/\partial \dot{\vec{r}})(\partial L/\partial \dot{q}) = (\partial \dot{q}/\partial \dot{\vec{r}})p$. The matrix of derivatives $(\partial \dot{q}_i/\partial \dot{\vec{r}}_k)$ is just the matrix B.

Thus our problem is reduced to the choice of the generalized co-ordinates and the construction of the matrix B. To characterize vibrations (relative nucleus positions), we use the conventional vibrational co-ordinates (the changes in bond lengths, valency and dihedral angles, etc. relative to their initial values or to those in the equilibrium state). The properties of the conventional vibrational co-ordinates and the methods of expressing their velocities via the corresponding Cartesian velocities of atoms are well known from the theory of vibrations for polyatomic molecules (see Chapters 7 and 8). It is important that the corresponding rows in the matrix B possess the following properties: first, the sum of the row elements is equal to zero and, second, the sum of the pairwise vector products of the instantaneous values of the matrix elements of any row in vibrational co-ordinates by the instant radius vectors of the atoms relative to any origin is also equal to zero (see Section 8.3). The first condition provides the zero velocities of the generalized co-ordinates in molecular translations and the second is responsible for zero velocities of the rotation of a molecule, as a whole, around any axis.

Thus we must assume that for the rows of matrix B in vibrational co-ordinates at any moment the following expressions are fulfilled:

$$\sum_k \vec{b}_{ik} = 0 \tag{6.1}$$

$$\sum \vec{b}_{ik} \times \vec{r}_k = 0 \tag{6.2}$$

Note that the vectors $\vec{b}_{ik}$ are given in the external (laboratory) system of Cartesian co-ordinates.

The molecular translation may be characterized by the velocity of the centre of mass. Then the corresponding row of the matrix B contains the elements m_k/M_t, where m_k are the masses of atoms and M_t is the total mass of the molecule.

Now proceed to rotation. First, define the rotation of the system as a whole. Usually this notion is associated with the motion of the major axes of the point system inertia. The axes can vary relative to the laboratory system of co-ordinates, both due to the rotation of the whole system and to its deformation, which results in the well-known complications and, at the usual methods of separation of vibrations and rotations, also in the appearance of non-diagonal terms corresponding to the momentum products of vibrational and rotational co-ordinates. Moreover, the method cannot be used for the motions leading to the system's decomposition. Therefore, we suggest another method. We asked ourselves about the characteristic of a system rotating as a whole. Obviously, such a characteristic is the non-zero term of the total moment of momentum relative to the centre of mass. Such a moment can appear only under the action of external forces, and it is its non-zero value which should be taken as an indicator of the rotation of the particle system as a whole. This moment can be related to an effective angular velocity defined by the expression

$$\vec{W} = \frac{\vec{N}}{I_p} = \tfrac{3}{2}\sum_k m_k(\vec{r}_{ck} \times \dot{\vec{r}}_k)/I_p$$

the summation being carried out over all atoms of the molecule. The values $\vec{r}_{ck}$ are the instantaneous radius vectors of the atoms relative to the centre of mass, I_p is the polar instantaneous moment of the molecular inertia relative to the centre of mass and $\dot{\vec{r}}_k$ are the Cartesian velocities of the molecular atoms in the laboratory system of co-ordinates. The factor 3/2 is introduced to make valid the relationship $\vec{W} = \vec{w} = \vec{N}/I$ for the rigid spherical gyroscope, where $\vec{w}$ is the instantaneous velocity and I is the moment of the gyroscope inertia, relative to an instantaneous rotational axis passing through its centre.

The dimensionality of the vector $\vec{w}$ is that of the angular velocity, and becomes zero (see below) in cases where all the atoms are displaced in the same way or if their motions correspond to purely vibrational motions resulting in deformations. Thus, this vector is non-zero only for purely rotational motions, i.e. when the total moment of momentum is non-zero, and therefore can be accepted as a generalized rotational velocity.

Then, in matrix form,

$$\vec{\omega} = \tfrac{3}{2}[I_p^{-1} m((r_c))]\dot{\vec{r}} \tag{6.3}$$

Here, $\vec{\omega}$ is the column-matrix of order three with the elements $\omega_x, \omega_y, \omega_z$; $\dot{\vec{r}}$ is the column-matrix of the co-ordinates of the atomic Cartesian velocities and $[I_p^{-1} m((r_c))]$ is the row-matrix of the elements

$$b_{\omega k} = I_p^{-1} m_k \begin{bmatrix} 0 & -z_k & y_k \\ z_k & 0 & -x_k \\ -y_k & x_k & 0 \end{bmatrix}$$

where x_k, etc. are the instantaneous co-ordinates of atomic vectors relative to the centre of mass of the molecule. Note that $I_p = (I_1 + I_2 + I_3)/2$ is invariant relative to translations and rotations, whereas I_1, etc. are the instantaneous major moments of inertia of the molecule.

The row-matrix $[I_p^{-1} m((r_c))]$ is a row of velocity co-ordinates determining the rotation in the total matrix B of the transition from Cartesian co-ordinates to the generalized ones. The sum of its elements must be equal to zero. Indeed, the moment of gravitation forces relative to the suspension point brought into coincidence with the centre of mass is

$$\vec{F} = \sum_k m_k (\vec{r}_{ck} \times \vec{g})$$

Vector $\vec{F}$ can be expressed in the matrix form as

$$\vec{F} = \sum_k m_k((r_{ck}))\vec{g} = \left\{ \sum_k m_k((r_{ck})) \right\} \vec{g}$$

where $\vec{F}$ is the column-matrix with the co-ordinates F_x, F_y, F_z and $\vec{g}$ is the column-matrix with the co-ordinates g_x, g_y, g_z. By definition, the moment of $\vec{F}$ equals zero, then

$$\sum_k m_k((r_{ck})) = 0$$

Thus the invariance of $\vec{\omega}$ relative to the translation motions has been proved.

To prove the invariance of the velocity w relative to vibrations remember that $\vec{p} = \tilde{B}p$, or $\dot{\vec{r}} = M^{-1}\tilde{B}p$. The velocities of the purely vibrational displacements of atoms must be the functions of the generalized momenta alone which correspond to the vibrational co-ordinates. In other words, such motions of atoms may be described by the relation $\vec{r}_{\text{vibr}} = M^{-1}\tilde{B}_{\text{vibr}} P_{\text{vibr}}$, where B_{vibr} is the rectangular submatrix corresponding to the velocities of vibrational co-ordinates. Its dimensionality is $(3n-6)3n$, where n is the number of atoms in the molecule. The elements of each row-matrix obey relationships (6.1) and (6.2). For pure vibrations we obtain

$$\vec{W} = \tfrac{3}{2}[I_p^{-1} m((r_c))] M^{-1} \tilde{\vec{B}}_{\text{vibr}} P_{\text{vibr}}$$

Applying equation (6.2) to the element of the row-matrix, which is the product

$[\frac{3}{2}I_p^{-1}m((r_c))]M^{-1}\tilde{\vec{B}}_{\text{vibr}}$ we obtain

$$\tfrac{3}{2}I_p^{-1}[m((r_c))]M^{-1}\vec{B}_{i\text{vibr}}=\tfrac{3}{2}I_p^{-1}\sum_k \vec{r}_{ck}\times\vec{b}_{ik}=0$$

Here, $\vec{B}_{i\text{vibr}}$ is the ith column of the matrix $\tilde{\vec{B}}_{\text{vibr}}$ (i is the row of the matrix B_{vibr}). Thus, vector $\vec{w}$ is independent of vibrational movements.

From the above we can write the total matrix of transition from Cartesian to the generalized co-ordinates in the form

$$B=\begin{bmatrix}\vec{B}_{\text{vibr}}\\ \tfrac{3}{2}[I_p^{-1}m((r_c))]\\ [m/M_t]\end{bmatrix} \tag{6.4}$$

Hence the matrix of the kinematic coefficients is

$$T=BM^{-1}\tilde{B}=\begin{bmatrix}\vec{B}_{\text{vibr}}M^{-1}\tilde{\vec{B}}_{\text{vibr}} & 0 & 0\\ 0 & \tfrac{9}{4}I_p^{-2}((I)) & 0\\ 0 & 0 & M_t^{-1}\end{bmatrix}=\begin{bmatrix}T_{\text{vibr}} & 0 & 0\\ 0 & T_{\text{rot}} & 0\\ 0 & 0 & T_{\text{trans}}\end{bmatrix}$$

Taking into account the properties (6.1) and (6.2) in multiplication, we obtain zero values of $\vec{B}_{\text{vibr}}M^{-1}[\widetilde{m/M_t}]$;

$$\tfrac{3}{2}[I_p^{-1}m((r_c))]M^{-1}\tilde{\vec{B}}_{\text{vibr}} \quad \text{and} \quad \tfrac{3}{2}[I_p^{-1}m((r_c))]M^{-1}[\widetilde{m/M_t}]$$

Now the matrix T becomes quasi-diagonal, and it is in this sense that the movements can be separated. The submatrix of order $3n-6$, $T_{\text{vibr}}=\vec{B}_{\text{vibr}}M^{-1}\tilde{\vec{B}}_{\text{vibr}}$, is the conventional matrix of kinematic coefficients for the vibrational co-ordinates used in the theory of molecular vibrationals. This matrix is invariant to rotations and translational motions, because its elements are combinations of the scalar products of vectors $\vec{b}_{ik}$. In the case of vibrations of an arbitrary amplitude it is necessary to take into consideration the dependence of the matrix elements on the vibrational co-ordinates. It is essential that rotations do not affect vibrations. The reverse effect is revealed in the dependence of the elements of the matrix T_{rot} on the vibrational co-ordinates. The submatrix of order three is $T_{\text{rot}}=9/4I_p^{-2}((I))$, where $((I))$ is the tensor of the moment of inertia relative to the centre of mass of the molecule and corresponds to the rotational motion. The diagonal matrix of order three, T_{trans}, describes the translational motion. Its elements are equal to the inverse mass of the whole molecule. The transition to the normal vibrational co-ordinates and major axes of inertia provides the diagonal form of the matrix T and, respectively, the complete separation of the generalized momenta. Using the normal co-ordinates of the molecular vibrations, we obtain the kinetic energy for the separated variables as

$$T_{\text{kin}}=\frac{1}{2}\sum_{i=1}^{3n-6}P_i^2+\frac{1}{2}\frac{P_c^2}{M_c}+\frac{1}{2}\sum_{\alpha=1}^{3}\tfrac{9}{4}I_p^{-2}I_\alpha P_{w\alpha}^2$$

$$I_p=I_p(Q);\quad I_\alpha=I_\alpha(Q) \tag{6.5}$$

Here, P_i^2 and other squares of the generalized momenta and Q are the normal vibrational co-ordinates. Note that in our case such normal co-ordinates are non-linear functions of the initial vibrational co-ordinates.

Expression (6.5) corresponds to the complete separation of the nuclear motions in a polyatomic molecule. Its structure coincides with that of the well-known equation for the kinetic part of the Hamiltonian function in vibrational, translational and rotational co-ordinates and is adequate to it in the case of small vibrations of the spherical gyroscope. Yet in this case there is no explicit vibrational angular momentum whose value depends on the method of introduction of movable co-ordinate axes associated with the molecule. The corresponding term is absent in our equation, since we have not introduced such axes in the explicit form. Note that the usual approach does not assume the expression for the differences between the angular and vibrational momenta in the Hamiltonian function. The physical meaning of these differences corresponds to the kinetic moment which we have introduced into our consideration and which, in turn, corresponds to the generalized velocity, $\vec{w}$.

In conclusion, note that in the derivation of equation (6.5) no restrictions have been imposed on the value of co-ordinate changes due to molecular deformation. Therefore this expression is valid for both small and large vibrational amplitudes. Moreover, the transition to the quasi-diagonal form of the matrix of kinematic coefficients can be performed by the suggested methods even for the dissociating atomic system, which experience no vibrations at all.

In this section we have shown that the choice of suitable velocities of the generalized co-ordinates and the corresponding transformations allows us to separate the term in the total kinetic energy of the system of nuclei which is related only to internal deformations from translational and rotational movements. As soon as a potential function depends only upon internal co-ordinates in the curvilinear co-ordinates system then the classical expression for the corresponding Hamiltonian function in the case of internal nuclei motion must have the following form:

$$H = \tfrac{1}{2}\sum_{i,j} P_i \tau_{ij}(q) P_j + V(q) \tag{6.6}$$

where $\tau_{ij}(q)$ are the elements of symmetrical matrix, $T = \vec{B}M^{-1}\vec{\tilde{B}}$, or in the relative units of inverse mass, $T = \vec{B}E\vec{\tilde{B}}$, where $E = \text{const.}M^{-1}$ (T is the kinematic coefficients matrix). Our problem now is to obtain the Schroedinger equation, corresponding to the function in the form of equation (6.6). We shall discuss this problem in the next section.

6.2 Derivation of an Expression for the Kinetic Energy Operator in the Theory of Vibration of Polyatomic Molecules

Vibrational motions of the nuclei of polyatomic molecules (or in fact all relative motions, which include internal rotations) can be studied using a set of special curvilinear 'natural' co-ordinates. If it becomes necessary to write the relevant operator in the same notation, difficulties are encountered with its kinetic energy component.

What form the kinetic operator should take in curvilinear co-ordinates has been discussed repeatedly in the literature[11]. It has become a classic problem, and traditionally it has been approached using tensor analysis with the aid of Laplace transforms written in the ordinary Cartesian co-ordinate system. A number of other approaches have been suggested over the last few years[12–19], based on introducing the momentum operator explicitly in curvilinear co-ordinates and on a close analogy with the method for obtaining the expression for the kinetic energy in the impulse representation in classic theoretical mechanics. Besides the accessibility of methodical simplifications, this approach makes it possible to construct a quantum-mechanical operator for the kinetic energy part of the overall Hamiltonian, even when transforming the co-ordinates also results in a simultaneous change in the dimensions of the space[20]. In this situation tensor analysis is simply inapplicable. This situation arises when the Schroedinger equation is being written for purely vibrational (or, in general, relative) motions of the nuclei in a polyatomic molecule.

In this section we shall consider this new approach in detail. We shall start from the classic expression for the kinetic energy of the nuclei in a generalized system of co-ordinates. This has the quadratic form

$$T_{\text{kin}} = \tfrac{1}{2}\sum_{i,j} p_i \tau_{ij}(q) p_j = \tfrac{1}{2}\tilde{p}\,T(q)p \tag{6.7}$$

where p_i is a generalized momenta conjugate to the selected curvilinear co-ordinates and $\tau_{ij}(q)$ is an element of the symmetric matrix $T(q) = \vec{B}M^{-1}\tilde{\vec{B}}$. Here, $\vec{B} = \vec{B}_{\text{vib}}$ is a matrix which expresses the relationship between the rate of change of the generalized co-ordinate $\dot{q}_i$ and the Cartesian velocity of the nucleus $\dot{\vec{r}}_m$ (or a physical point) of the molecule in terms of the laboratory frame of reference. It is worth mentioning that the $\vec{B}$ matrix may be either square or rectangular, i.e. the expression for the kinetic energy may be formally written in exactly the same way when the dimensionality of the space is conserved and when it is changed by the transition to the curvilinear co-ordinates. This must occur if we are to write the expression for the energy of purely relative (for example, vibrational) motions of a molecule's nuclei while using the natural vibrational co-ordinates. (This equation will be dealt with in detail in Chapter 8.)

We then carry out the new non-linear transformation of the co-ordinates, obtaining normal co-ordinates such that the kinetic energy takes the form:

$$T_{\text{kin}} = \tfrac{1}{2}\sum_{k} P_k^2 \tag{6.8}$$

This is equivalent to a substitution of the generalized momenta without changing the space's dimensionality, during which the $T(q)$ matrix is transformed into unit form. This can clearly always be done, because the $T(q)$ matrix is symmetric and real for every value of the q co-ordinate.

In order to find the necessary transformation we note that, according to the general rules (see Chapter 7), a transformation from Cartesian generalized co-ordinates to the initial curvilinear ones may be made if the $L(Q)$ matrix is known. This matrix occurs in the equation $\dot{q} = L(Q)\dot{Q}$, and the relevant momentum transformation thus takes the form $P = \tilde{L}(Q)p$. By substituting the momenta into the expression for the kinetic energy in Cartesian generalized co-ordinates and returning, subsequently, to the initial curvilinear co-ordinates, we find that $T(q) = L(Q)\tilde{L}(Q)$. When it is necessary to calculate the $L(Q)$ matrix, which must be related to the $L'(Q)$ matrix that transforms the $T(q)$ matrix into unit form, a way is thus indicated via the relation $L'(Q) = [\tilde{L}(Q)]^{-1}$.

Since the $\dot{q}_i$ and Q_k functions are full derivatives, it follows from the equation $\dot{q} = L(Q)\dot{Q}$ that $\mathrm{d}q = L(Q)\,\mathrm{d}Q$ for the exact differentials of the co-ordinates q_i and Q_k. We can thus conclude that the $L(Q)$ matrix is none other than the matrix of partial derivatives $L(Q) = [\partial q/\partial Q]$.

In the system of the Q_k normal co-ordinates the variables in the kinetic part of the Hamiltonian will be separated, and the Q_k thus form a generalized orthonormalized Cartesian system. If we now proceed to a quantum consideration, then independent of the method we use (Poisson quantum brackets, an invariant requirement with respect to translation or a self-conjugation requirement), we obtain the very simple expression $P_k = -i\hbar\partial/\partial Q_k$ for this operator in Cartesian co-ordinates. This follows in particular from the mathematical equivalence of all the Cartesian co-ordinates. Thus we shall in future rely on the assumption that in any N-dimensional Cartesian co-ordinate system the momentum operator will have the form $P_k = -i\hbar\partial/\partial Q_k$, while the normalization integral is

$$\int \Psi^*(Q)\Psi(Q)\prod_k \mathrm{d}Q_k = 1$$

From here[21], we can obtain an important relationship for the eigenfunctions $\Psi(q)$ and $\Psi(Q)$ in the q_i and Q_i co-ordinate systems, where the generalized curvilinear co-ordinates q_i are not orthogonal, and can find a relationship between the Jacobian of the transformation and the matrix $T(q)$ that enters the expression for the kinetic energy in terms of an initially selected unorthonormalized curvilinear q_i co-ordinate system (the natural co-ordinate system in particular). What is important is that the dimensionality does not change during the transformation from the q_i co-ordinates to the Q_k ones. Above all, note that in an unorthonormalized co-ordinate system the normalization integral must, according to the rule for a change of variables in a multidimensional integral, be transformed into the form:

$$1 = \int \Psi^*(Q)\Psi(Q)\prod_k \mathrm{d}Q_k = \int \Psi^*(q)\Psi(q)I\prod_i \mathrm{d}q_i \tag{6.9}$$

where I is the Jacobian $I = D(\partial Q_k/\partial q_i)$ of the relevant transformation.

The quantity I is a 'weight function' which must be taken into account both when calculating the matrix elements and when formulating the Hermiticity condition for operators expressed in terms of the q_i co-ordinates and p_i momenta.

In order to determine the Jacobian I we take into account that $I = D(\partial Q_k/\partial q_i) = [D(\partial q_i/\partial Q_i)]^{-1}$ because the Jacobian of the inverse transformation is the inverse (reciprocal) in value. We showed above that $[\partial q/\partial Q] = L(Q)$, and so

$$D\left(\frac{\partial q_i}{\partial Q_k}\right) = D[L(Q)] \tag{6.10}$$

Moreover, $L(Q)\tilde{L}(Q) = T(q)$, and so we can write $D(T) = \{D[L(Q)]\}^2$ and hence the Jacobian $I = \{D(T)\}^{-1/2} = t^{-1/2}$, where t is the determinant of the T matrix.

In this way we have expressed the Jacobian I in terms of the known parameters of the quadratic form which represents the kinetic energy with respect to the motions of the molecule's nuclei in the canonic (momentum) form.

We now set about deriving the kinetic energy operator in curvilinear co-ordinates by using some of the ideas in reference[19]. Since we shall only consider a transformation of the kinetic part of the Hamiltonian, we can omit the potential component and limit ourselves to the case of independent particles. This is entirely possible because the kinetic part of the Hamiltonian does not depend on whether the molecule is a stable system. The potential part in quantum and classical mechanics have identical forms, and so the equation for the eigenvalues of the momentum operator in the multidimensional Cartesian co-ordinate system can be written as:

$$\nabla\Psi = -i\hbar\sum_k \vec{e}_k \frac{\partial}{\partial Q_k}\Psi = E_p\Psi \tag{6.11}$$

This equation in separated variables is satisfied if the whole set of equations

$$-i\hbar\frac{\partial}{\partial Q_k}\Psi_k = P_k\Psi_k \tag{6.12}$$

are satisfied for all k. Moreover, we have

$$\Psi = \prod_k \Psi_k \quad \text{and} \quad E_p = \sum_k P_k$$

where Ψ_k and P_k are an eigenfunction and an eigenvalue of one of the equations for a Cartesian component of the total momentum.

The equation system (6.12) can be written in matrix notation as

$$\left[-i\hbar\frac{\partial}{\partial Q} - P\right]\Psi(Q) = 0 \tag{6.13}$$

Here, $\partial/\partial Q$ and P are column-matrices, respectively, for the $\partial/\partial Q_k$ component of the momentum, operator and its eigenvalue P_k. Strictly speaking, the eigenfunction Ψ for separated variables must be represented as a product $\Psi = \prod_k \Psi_k$. This must be borne in mind when using matrix notation, and means that each equation, which corresponds to the column-matrix element

$[-i\hbar\partial/\partial Q - P]$, must contain its own function Ψ_k. However, it is clear that if one of the Ψ_k functions is an eigenfunction of the operator $\hat{P}_k = -i\hbar\partial/\partial Q_k$ then the product Ψ of functions of the form $\Psi = \prod_k \Psi_k$ will also be an eigenfunction for each of the operator's components, because the operator only acts on one of the functions of the products.

We now change the variables in matrix equation (6.13), bearing in mind that the eigenvalue P_k must, according to the convergence principle, behave as a component of classic momentum and hence be transformed via the relationship $P = \tilde{L}(Q)p$, where $L(Q)$ is the matrix relating the speed component in curvilinear and Cartesian co-ordinates.

By then using the chain differentiation formula, i.e.

$$\frac{\partial \phi}{\partial Q_k} = \sum_i \frac{\partial \phi}{\partial q_i} \cdot \frac{\partial q_i}{\partial Q_k}$$

we find the relationship of the operators $\partial/\partial Q_k$ and $\partial/\partial q_i$. In matrix notation this has the form

$$\frac{\partial}{\partial Q} = \tilde{L}(Q)\frac{\partial}{\partial q}$$

or

$$\frac{\tilde{\partial}}{\partial Q} = \frac{\tilde{\partial}}{\partial q} L(Q)$$

By substituting these matrices into equation (6.13) and taking into account that the variables must be changed in the eigenfunctions, we obtain

$$\tilde{L}(Q)\left[-i\hbar\frac{\partial}{\partial q} - P\right]\Psi(q) = 0$$

Since $L(Q) \neq 0$, then this equation will be satisfied if the matrix equation

$$\left[-i\hbar\frac{\partial}{\partial q} - P\right]\Psi(q) = 0 \tag{6.14}$$

is satisfied.

As shown above, when calculating the matrix elements with the functions $\Psi(q)$ it is essential to allow for the 'weight factor'. It is more convenient, however, to introduce the eigenfunctions normalized in another way, that is, so that the normalization integral does not contain the 'weight factor' as occurs in the case of Cartesian co-ordinates. In order to do this the $\Psi(q)$ function is substituted by the functions $\Psi'(q) = I^{1/2}\Psi(q)$. The latter will satisfy a new equation with the operator $\hat{p}'$ which, however, must contain the previous eigenvalues. The $\hat{p}'$ operator may then be obtained from the original operator $\hat{p}$ using the substitution

$$\hat{p}J^{-1/2}J^{1/2}\Psi(q) = pJ^{-1/2}J^{1/2}\Psi(q)$$

It follows that by left-multiplying this identity by $I^{1/2}$ we find

$$J^{1/2}\hat{p}J^{-1/2}\Psi'(q) = p\Psi'(q)$$

or

$$\hat{p}'\Psi'(q) = p\Psi'(q)$$

where

$$\hat{p}' = J^{1/2}\hat{p}J^{-1/2}$$

By taking this into account, we can rewrite equation (6.15) as

$$\left[-i\hbar J^{1/2}\frac{\partial}{\partial q}J^{-1/2} - p\right]\Psi'(q) = 0$$

Since the p column must be a column of the eigenvalues for the momentum components in curvilinear co-ordinates, and since it can be obtained simply by a formal change of variables, as in the $\partial/\partial q$ column, it is natural to take the operators

$$\hat{p}_i = -i\hbar J^{1/2}\frac{\partial}{\partial q_i}J^{-1/2} \tag{6.15}$$

as components of the vector (column-matrix) of the momentum operator in curvilinear co-ordinates. Hence it is possible to write the equation for these components in a way analogous to the equation in Cartesian co-ordinate space, i.e.

$$[\hat{p} - p]\Psi'(q) = 0$$

where

$$\hat{p} = -i\hbar J^{1/2}\frac{\partial}{\partial q}J^{-1/2}$$

It is easy to see that the operators we have obtained in curvilinear co-ordinates correspond to the basic rules for comutating the co-ordinate and momenta operators. We have the following for the effect of the $\{q_i, p_i\}$ commutator on a function $\phi(q)$:

$$\left(-i\hbar J^{1/2}\frac{\partial}{\partial q}J^{-1/2}q + i\hbar qJ^{1/2}\frac{\partial}{\partial q}J^{-1/2}\right)\phi(q) = -i\hbar\phi(q)$$

or

$$\{q, \hat{p}\} = -i\hbar$$

Further, it is not difficult to verify that $\{\hat{p}_i, \hat{p}_j\} = 0$.

Hence the vector operator with the components

$$\hat{p}_i = -i\hbar J^{1/2}\frac{\partial}{\partial q_i}J^{-1/2}$$

is in fact the vector operator of momentum in curvilinear co-ordinates.

We now proceed to the investigation of the kinetic energy operator. We have already mentioned that in any Cartesian co-ordinates the components of the momentum for any kinetic energy operator will have the form $\hat{P}_k = -i\hbar\partial/\partial Q_k$, while the corresponding operator of the kinetic energy has the form

$$\hat{T}_{\text{kin}} = -\frac{\hbar^2}{2}\sum_k\frac{\partial^2}{\partial Q_k^2} = -\frac{\hbar^2}{2}\sum_k\hat{P}_k^2$$

We assume that all the masses are units. This expression may be obtained using the formal matrix product of the row-matrix which have the components $\hat{P}_k = -i\hbar\partial/\partial Q_k$ and the relevant column-matrix. In curvilinear co-ordinates normalized operator it is necessary to take advantage of the row-matrices which contain operators conjugate to the selected ones.

Thus we find the operator p^+ which is conjugate to the opeator

$$\hat{p} = -i\hbar I^{1/2}\frac{\partial}{\partial q}J^{-1/2}$$

By definition, we call the operator $\hat{p}^+ = (p^T)^*$; where $\hat{p}^T$ is the transposed operator corresponding to the relationship

$$\int f_2\hat{p}^T f_1\,\mathrm{d}q = \int f_1\hat{p}f_2\,\mathrm{d}q$$

is called the conjugate operator to the one given.

We shall consider the following integral (the constant $-i$ is insignificant):

$$\int f_1\hat{p}f_2\,\mathrm{d}q = \int f_1 I^{1/2}\frac{\mathrm{d}}{\mathrm{d}q}I^{-1/2}f_2\,\mathrm{d}q$$

$$= \int f_1\frac{\mathrm{d}}{\mathrm{d}q}f_2\,\mathrm{d}q - \frac{1}{2}\int f_1 f_2 I^{-1}\frac{\mathrm{d}I}{\mathrm{d}q}\mathrm{d}q$$

$$= f_1 f_2 - \int f_2\frac{\mathrm{d}f_1}{\mathrm{d}q}\mathrm{d}q - \frac{1}{2}\int f_1 f_2\frac{\mathrm{d}\ln I}{\mathrm{d}q}\mathrm{d}q$$

Since the integration is over one variable, the partial derivative is substituted by the total one. If, as is usually taken, the functions f_1 and f_2 die away to zero at the boundaries in which they are defined, we then find that

$$\int f_1\frac{\mathrm{d}f_2}{\mathrm{d}q}\mathrm{d}q = -\int f_2\frac{\mathrm{d}f_1}{\mathrm{d}q}\mathrm{d}q$$

That is, the integral changes sign when the order of the functions f_1 and f_2 is reversed. However, the integral

$$\int f_1 f_2\frac{\mathrm{d}\ln I}{\mathrm{d}q}\mathrm{d}q$$

does not depend on the order of the f_1 and f_2 functions. However, we can write

$$\int f_1\hat{p}f_2\,\mathrm{d}q = \int f_1\left(\frac{\mathrm{d}}{\mathrm{d}q} - \frac{1}{2}\frac{\mathrm{d}\ln J}{\mathrm{d}q}\right)f_2\,\mathrm{d}q$$

and see immediately that in order that the relationship

$$\int f_1\hat{p}f_2\,\mathrm{d}q = \int f_2\hat{p}^T f_1\,\mathrm{d}q$$

be fulfilled, the operator $\hat{p}^T$ must be taken as

$$\hat{p}^T = -\left(\frac{d}{dq} + \frac{1}{2}\frac{d\ln J}{dq}\right) = -J^{-1/2}\frac{d}{dq}J^{1/2}$$

which leads to the $\hat{p}_i^+$ operator in the form

$$\hat{p}_i^+ = -i\hbar J^{-1/2}\frac{\partial}{\partial q_i}J^{1/2} \tag{6.16}$$

In this way the elements of the row-matrix of the components of the operators conjugate to the components of the momentum operator in curvilinear co-ordinates must be in the form of equation (6.16). This yields the following matrix relationships for the rows and columns from the operator components in Cartesian and curvilinear co-ordinates, i.e.

$$\begin{aligned} P &= \frac{\partial}{\partial Q} = \tilde{L}(Q)J^{1/2}\frac{\partial}{\partial q}J^{-1/2} \\ \tilde{P} &= \frac{\tilde{\partial}}{\partial Q} = J^{-1/2}\frac{\tilde{\partial}}{\partial q}J^{1/2}L(Q) \end{aligned} \tag{6.17}$$

The change of variables in the kinetic energy operator in Cartesian co-ordinates for the operator in curvilinear co-ordinates may now be carried out in a way completely analogous to the classical expression using matrix substitution and multiplications. We have

$$\begin{aligned} \hat{T}_{\text{kin}} &= -\frac{\hbar^2}{2}\sum_k \frac{\partial^2}{\partial Q_k^2} = -\frac{\hbar^2}{2}\tilde{P}P = -\frac{\hbar^2}{2}J^{-1/2}\frac{\tilde{\partial}}{\partial q}J^{1/2}L(Q) \\ &\times \tilde{L}(Q)J^{1/2}\frac{\partial}{\partial q}J^{-1/2} = -\frac{\hbar^2}{2}J^{-1/2}\frac{\tilde{\partial}}{\partial q}JT(q)\frac{\partial}{\partial q}J^{-1/2} \end{aligned} \tag{6.18}$$

We can obtain a direct verification that this operator is self-conjugate. When constructing it there is a complete analogy with the corresponding construction in classical mechanics. Moreover, since obtaining the $T(q)$ matrix (kinematic coefficients) is basic, we may change the dimensionality of the space. In this respect, when going from the natural co-ordinates to the laboratory Cartesian system we shall use the transformation $T(q) = \vec{B}M^{-1}\tilde{\vec{B}}$ to construct the $T(q)$ matrix. The $\vec{B}$ matrix is rectangular and $3N.(3N-6)$ in size, and reflects the relationship between the rates of change in Cartesian and natural co-ordinates, while M^{-1} is the diagonal matrix containing the reciprocals of the masses of the atoms. This is a peculiarity of this transformation, and differs from the traditional one.

In order to obtain the final expression we must take into account that $I = t^{-1/2}$. Hence we obtain the following for the $\hat{T}_{\text{kin}}$ operator in a curvilinear co-ordinate:

$$T_{\text{kin}} = -\frac{\hbar^2}{2}\sum_{i,j} t^{1/4}\frac{\partial}{\partial q_i}\tau_{ij}(q)t^{-1/2}\frac{\partial}{\partial q_j}t^{1/4} \tag{6.19}$$

For subsequent investigations the expression for the kinetic energy operator should be rewritten in another form. Using matrix notation we write the column and row components of the momentum operator (and its conjugate) in the form:

$$\hat{p} = -i\hbar t^{-1/4}\frac{\partial}{\partial q}t^{1/4} = -i\hbar\left(\frac{\partial}{\partial q} + \frac{1}{4}\frac{\partial \ln t}{\partial q}\right) = -i\hbar\left(\frac{\partial}{\partial q} + F\right);$$

$$\hat{p}^{+} = -i\hbar t^{1/4}\frac{\partial}{\partial q}t^{-1/4} = -i\hbar\left(\frac{\tilde{\partial}}{\partial q} - \hat{F}\right) \tag{6.20}$$

Here, all the notations $\partial/\partial q$, F, $\widetilde{\partial/\partial q}$, $\tilde{F}$, etc. are the relevant row and column matrices. Hence the kinetic energy operator is written in matrix notation as:

$$\hat{T}_{\text{kin}} = -\frac{\hbar^2}{2}\left(\frac{\tilde{\partial}}{\partial q} - \tilde{F}\right)T\left(\frac{\partial}{\partial q} + F\right)$$

$$= -\frac{\hbar^2}{2}\left(\frac{\tilde{\partial}}{\partial q}T\frac{\partial}{\partial q} - \tilde{F}TF + \frac{\tilde{\partial}}{\partial q}TF - \tilde{F}T\frac{\partial}{\partial q}\right) \tag{6.21}$$

The last two terms comprise a commutator. If it is expanded, then we obtain the following expression:

$$\hat{T}_{\text{kin}} = -\frac{\hbar^2}{2}\sum_{i,j}\left\{\frac{\partial}{\partial q_i}\tau_{ij}(q)\frac{\partial}{\partial q_j} - \frac{1}{16}\tau_{ij}\frac{\partial \ln t}{\partial q_i}\cdot\frac{\partial \ln t}{\partial q_j}\right.$$
$$\left. + \frac{1}{4}\tau_{ij}\frac{\partial^2 \ln t}{\partial q_i \partial q_j} + \left(\frac{1}{4}\frac{\partial \tau_{ij}}{\partial q_i}\cdot\frac{\partial \ln t}{\partial q_j} + \frac{1}{4}\frac{\partial \tau_{ij}}{\partial q_j}\frac{\partial \ln t}{\partial q_i}\right)\right\}$$

This may be rewritten in yet another form, in which we obtain

$$\frac{\partial \ln t}{\partial q_i} = Sp\left(T^{-1}\frac{\partial T}{\partial q_i}\right)$$

where T is the matrix of kinematic coefficients and its elements are $\tau_{ij}(q)$.

This relationship can be obtained if we take into account that $t = \prod_k \lambda_{\tau k}$, where $\lambda_{\tau k}$ is an eigenvalue of the T matrix. Hence we obtain:

$$\frac{\partial \ln t}{\partial q_i} = \sum_k \frac{\partial \ln \lambda_{\tau k}}{\partial q_i} = \sum_k \lambda_{\tau k}^{-1}\frac{\partial \lambda_{\tau k}}{\partial q_i} = Sp\left(T^{-1}\frac{\partial T}{\partial q_i}\right)$$

Furthermore,

$$\frac{\partial^2 \ln t}{\partial q_i \partial q_j} = Sp\left(T^{-1}\frac{\partial^2 T}{\partial q_i \partial q_j} - T^{-1}\frac{\partial T}{\partial q_i}T^{-1}\frac{\partial T}{\partial q_j}\right) \tag{6.23}$$

Hence equation (6.22) may be rewritten yet again in the form

$$\hat{T}_{\text{kin}} = -\frac{\hbar^2}{2}\sum_{i,j}\left\{\frac{\partial}{\partial q_i}\tau_{ij}(q)\frac{\partial}{\partial q_j} - \frac{1}{16}\tau_{ij}Sp\left(T^{-1}\frac{\partial T}{\partial q_i}\right)\right.$$

$$\times Sp\left(T^{-1}\frac{\partial T}{\partial q_j}\right)+\frac{1}{4}Sp\left(T^{-1}\frac{\partial^2 T}{\partial q_i \partial q_j}-T^{-1}\frac{\partial T}{\partial q_i}T^{-1}\frac{\partial T}{\partial q_j}\right)$$
$$+\frac{1}{4}\left[\frac{\partial \tau_{ij}}{\partial q_i}Sp\left(T^{-1}\frac{\partial T}{\partial q_i}\right)+\frac{\partial \tau_{ij}}{\partial q_j}\left(T^{-1}\frac{\partial T}{\partial q_j}\right)\right]\Bigg\} \tag{6.24}$$

It can be seen that the differential part is only in the first term: the last three terms are simply functions, and can be taken into the potential part. However, note that the last three terms are themselves basically functions of the co-ordinates which were selected for describing the relative motions of the molecule's nuclei. If we consider motion about the equilibrium state, then the values of the τ_{ij} and $\ln t$ coefficients may be given as in the form of a series in terms of the curvilinear co-ordinates. We therefore take the τ_{ij} and $\ln t$ quantities to be equal to

$$\tau_{ij}=\tau_{ij}(0)+\sum_k a_{ijk}q_k+\tfrac{1}{2}\sum_{k,n}a_{ijkn}q_k q_n+\cdots$$

$$\ln t=\ln t_0+\sum_k b_k q_k+\frac{1}{2}\sum_{k,n}b_{kn}q_k q_n+\cdots$$

Hence we obtain the following for the correction $\Delta\hat{H}$ by differentiating, substituting and grouping similar terms:

$$\Delta\hat{H}=\text{const}+\sum_k A_k q_k+\tfrac{1}{2}\sum_{k,n}B_{kn}q_k q_n+\cdots \tag{6.25}$$

By using perturbation theory it can be shown that the first two terms may not give general corrections to differences of energy levels of the Hamiltonian with the operator

$$\hat{T}_{\text{kin}}=-\frac{\hbar^2}{2}\sum_{i,j}\frac{\partial}{\partial q_i}\tau_{ij}(q)\frac{\partial}{\partial q_j}$$

Hence these need not be taken into account. The third and higher terms in equation (6.25) may give such an error, but first, they do so only for the individual matrix elements that are non-zero, and second, only products of the $a_{ij}k$ and b_k coefficients or the values $a_{ij}k_n$ and b_{kn} are included in the B_{kn} coefficients, i.e. the terms are too small to consider. Consequently all the corrections to the energy levels due to terms with the operator $\Delta\hat{H}$ will be insignificant. This has been confirmed by specific calculations.

Hence when dealing with atomic vibrations in a molecule whose amplitude is not large we can take advantage of the simplified Hamiltonian with a kinetic operator in the form:

$$\hat{T}=-\frac{\hbar^2}{2}\sum_{i,j}\frac{\partial}{\partial q_i}\tau_{ij}(q)\frac{\partial}{\partial q_j} \tag{6.26}$$

In order to obtain the total operator for the Schroedinger equation in the case of relative nuclear motions it is necessary to add the potential function to the operators written in the form of equations (6.19) or (6.24).

For small vibrations we can set $\tau_{ij} = \text{const}$, and

$$V(q) = \tfrac{1}{2}\sum_{i,j} u_{ij} q_i q_j$$

and obtain the simple equation, corresponding to harmonic vibrations

$$\hat{H} = \tfrac{1}{2}\sum_{i,j} (\tau_{ij}\hat{p}_i\hat{p}_j + u_{ij}q_iq_j) \tag{6.27}$$

where

$$\hat{p}_i = -i\hbar\frac{\partial}{\partial q_i}$$

References

1. C. Eckart. *Phys. Rev.*, **47**, 552 (1935).
2. A. Sayvetz. *J. Chem. Phys.*, **6,** 383 (1939).
3. G. A. Natanzon and M. N. Adamov. *Bull. Leningrad State Univ.*, **4**, 28 (1974); **10**, 24 (1974) (in Russian).
4. M. N. Adamov and G. A. Natanzon. *Bull. Leningrad State Univ.*, **22**, 30 (1970); **4** 50 (1970) (in Russian).
5. A. N. Petelin and A. A. Kiselev. *Bull. Leningrad State Univ.*, **16**, 24 (1970) (in Russian).
6. A. Ye. Bolonkin and Yu. S. Makuskin. *Opt. and Spectr.*, **32**, 264 (1972) (in Russian).
7. J. D. Louck. *J. Molec. Spectrosc.*, **6**, 107 (1976).
8. J. D. Louck and H. W. Galbraith. *Rev. Modern Phys.*, **48**, 69 (1976).
9. A. A. Kiselev. *Canad. J. Phys.*, **56**, No.6, 615 (1978).
10. L. A. Gribov. *J. Molec. Struct.* (*Theochem*), **87**, 133 (1982).
11. B. Podolsky. *Phys. Rev.*, **32**, 812 (1938).
12. G. Gruber. *Foundations of Phys.*, **1**, 227 (1971).
13. G. R. Gruber. *Intern. J. Theor. Phys.*, **31**, 31 (1972); **7**, 253 (1974).
14. G. R. Gruber. *Foundation of Phys.*, **6**, 111 (1976).
15. F. I. Bloore and L. Routh. *Nuovo Cimento*, **25B**, 78 (1975).
16. K. R. Brownstein. *Amer. J. Phys.*, **44**, 677 (1976).
17. P. Vollasenor-Gonzalez and Cisneros-Parra. *Amer. J. Phys.*, **49**, 754 (1981).
18. V. D. Krivchenkov. *Russian Phys. Usp.*, **135**, 337 (1981) (in Russian).
19. Yu. I. Ponomarev. *Opt. and Spectr.*, **45**, 611 (1978) (in Russian).
20. L. A. Gribov and N. I. Prokofieva. *Rev. of TACA*, **2**, 174 (1983) (in Russian).
21. L. A. Gribov and N. I. Prokofieva. *Opt. and Spectr.*, **44**, 1032 (1978) (in Russian).

Chapter 7

Solving the Vibrational Problem in the Harmonic Approximation

7.1 The Harmonic Approximation

In Chapter 6 we considered a general equation for relative nucleus motions of atoms and the properties of the terms they include. In this chapter we will set about solving this equation. Initially we shall consider the simplest, but most important, case, the harmonic approximation. This approximation is made if we restrict our consideration to small-amplitude vibrations. The dependence of the kinematic coefficient matrix elements on the nuclear co-ordinates can be neglected by calculating them using equilibrium bond lengths, valency angles, etc. We can also assume that the matrix elements are constant and take the potential function to be a quadratic polynomial.

The Schroedinger equation for the nuclear vibrations of a molecule at the harmonic approximation is therefore

$$[\hat{T}_{\text{vib}}(\hat{p}) + U_{\text{pot}}]\Psi(q) = E_{\text{vib}}\Psi(q) \tag{7.1}$$

where

$$\hat{T}_{\text{vib}}(\hat{p}) = \tfrac{1}{2}\sum_{i,j}\tau_{ij}\hat{p}_i p_j \tag{7.2}$$

is the kinetic energy operator with constant coefficients $\tau_{ij} = \tau_{ji}$ (kinematic coefficients) and

$$U_{\text{pot}} = \tfrac{1}{2}\sum_{i} u_{ij}q_i q_j \tag{7.3}$$

is the potential energy operator with constant coefficients $u_{ij} = u_{ji}$ (force constants), with the $\hat{p}_i$ operators being conjugates to the vibrational co-ordinates and equal to $\hat{p}_i = -i\hbar(\partial/\partial q_i)$.

In matrix notation the expression for the operators for kinetic energy and potential energy of the vibrations take the forms, respectively:

$$\hat{T}_{\text{vib}} = \tfrac{1}{2}\tilde{\hat{p}}T\hat{p} \quad \text{and} \quad U_{\text{pot}} = \tfrac{1}{2}\tilde{q}Uq$$

Here T and U are square symmetric matrices having elements τ_{ij} and u_{ij}:

$\tilde{\hat{p}} = [\hat{p}_1, \hat{p}_2, \ldots, \hat{p}_n]$ and $\tilde{q} = [q_1, q_2, \ldots, q_n]$ are row matrices and

$$\hat{p} = \begin{bmatrix} p_1 \\ \vdots \\ p_n \end{bmatrix} \quad \text{and} \quad q = \begin{bmatrix} q_1 \\ \vdots \\ q_n \end{bmatrix}$$

are column matrices of the momentum operator and the operator for the co-ordinates conjugate to them with respect to nuclear positions. The T matrix is called the kinematic coefficient matrix, and the U matrix the force constant matrix. The Schroedinger equation for the vibrational motion of the nuclei is therefore, in matrix notation:

$$\hat{H}\Psi_{\text{vib}} = \tfrac{1}{2}(\tilde{\hat{p}}T\hat{p} + \tilde{q}Uq)\Psi_{\text{vib}} = E_{\text{vib}}\Psi_{\text{vib}} \tag{7.4}$$

In order to solve equation (7.4) we must first transform the equation, via a change of variables, to a form in which the $\hat{H}$ operator is a sum of $\hat{H}_k$ operators each of which acts on only one variable. In this case we know that the eigenfunction Ψ_{vib} has the form $\Psi_{\text{vib}} = \prod_k \psi_k$ and the energy E has the form $E = \sum_k E_k$ where the ψ_k and E_k are the eigenfunctions and eigenvalues of the $\hat{H}_k$ operators and satisfy the equation

$$\hat{H}_k\psi_k = E_k\psi_k \tag{7.5}$$

Before considering the solution for a particular case of equation (7.4) we wish to make one important point. Let us assume we changed the co-ordinates $q_1, \ldots, q_n$ to new co-ordinates $q'_1, \ldots, q'_n$ so that there are linear relationships between them with constant coefficients, l_{ik}, i.e.

$$\begin{cases} q_1 = l_{11}q'_1 + l_{12}q'_2 + \cdots + l_{1n}q'_n \\ \cdots\cdots\cdots\cdots\cdots\cdots\cdots\cdots\cdots\cdots \\ q_n = l_{n1}q'_1 + l_{n2}q'_2 + \cdots + l_{nn}q'_n \end{cases} \tag{7.6}$$

In matrix notation this would be $q = Lq'$. Let us assume also that we consider that the number of q co-ordinates is the same as the number of q' co-ordinates and that both q_i and q'_i are independent. Then the matrix L will be square and non-singular, i.e. it may have an inverse. Consequently, we will have $q' = L^{-1}q$.

The momentum operator conjugate to the q_i co-ordinate, i.e. $\hat{p}_i = -i\hbar(\partial/\partial q_i)$ must be changed during the linear change of variables $q = Lq'$ in agreement with the formula $\hat{p} = \tilde{L}^{-1}p'$, where $\hat{p}'$ is the column of the momentum operators $\hat{p}'_k = -i\hbar(\partial/\partial q'_k)$ which are conjugate to the q'_k-co-ordinates. This relationship can easily be obtained from the rule for differentiating a complicated function, i.e.

$$\partial/\partial q_i = \sum_i \partial q'_j/\partial q_i \cdot \partial/\partial q'_i$$

or

$$\partial/\partial q_i = l'_{1i}\partial/\partial q'_1 + l'_{2i}\partial/\partial q_2 + \cdots + l'_{ni}\partial/\partial q'_n$$

where l'_{ji} is an element of the ith column of the matrix L^{-1}. From here, again using matrix notation, we obtain the relation needed, i.e.

$$\hat{p} = \tilde{L}^{-1}\hat{p}' \tag{7.7}$$

Using this, we can express the relationship between the sets of new and old momenta if the appropriate linear co-ordinate transformation is given, and vice versa.

In order to reduce the operator

$$\hat{H}_{\text{vib}} = \tfrac{1}{2}\tilde{\hat{p}}Tp \pm \tfrac{1}{2}\tilde{q}Uq$$

to the form $\hat{H}_{\text{vib}} = \sum_k \hat{H}_k$ we take advantage of the fact that both of the operator's terms are quadratic forms and can be simultaneously reduced to a sum of squares. We transform initially the operator's first term using the linear transformation of the momenta that has a matrix $L^{(T)}$ such that $\hat{p} = L^{(T)}\hat{p}'$ and $\tilde{L}^{(T)}TL^{(T)} = \Lambda^{(T)}$, where $\Lambda^{(T)}$ is the diagonal matrix of the eigenvalues $\lambda_{\tau i}$ of the T matrix. Since T is symmetric and real its transformation must be orthogonal, and consequently the $L^{(T)}$ matrix has the property $L^{(T)^{-1}} = \widetilde{L^{(T)}}$.

Since kinetic energy can not have a negative value it follows directly that all the elements of $\Lambda^{(T)}$ will be positive. We now generate the diagonal matrix $\Lambda^{(T)^{-1/2}}$, whose elements are $\lambda_{\tau i}^{-1/2}$, and left and right multiply the expression $\tilde{L}^{(T)}TL^{(T)} = \Lambda^{(T)}$ by $\Lambda^{(T)^{-1/2}}$. As a result we get $\Lambda^{(T)^{-1/2}}\tilde{L}^{(T)}TL^{(T)}\Lambda^{(T)^{-1/2}} = 1$. The matrix $L' = L^{(T)}\Lambda^{(T)^{-1/2}}$ can be taken as the transformation matrix for the momenta operators $\hat{p} = L'\hat{p}''$. Moreover, this transformation will not be orthogonal.

In the q_i'' co-ordinates, which are conjugate to the momenta $\hat{p}_i''$, the kinetic energy matrix T will be reduced to the identity matrix, i.e. the expression for $\hat{T}_{\text{vib}}$ will take the form

$$\hat{T}_{\text{vib}} = \tfrac{1}{2}\tilde{\hat{p}}''\hat{p}'' = \tfrac{1}{2}\sum_k \hat{p}_k''^2 \tag{7.8}$$

On the basis of equations (7.6) and (7.7) the co-ordinate transformation corresponding to the momentum transformation must have the form

$$q = \tilde{L}'^{-1}q'' = L^{(T)}\Lambda^{(T)1/2}q''$$

Consequently, the quadratic form $\tfrac{1}{2}\tilde{q}Uq$ becomes $\tfrac{1}{2}\tilde{q}''U'q'' = \tfrac{1}{2}\tilde{q}''\Lambda^{(T)^{1/2}}\tilde{L}^{(T)}UL^{(T)}\Lambda^{(T)^{1/2}}q''$. The matrix $U' = \Lambda^{(T)^{1/2}}\tilde{L}^{(T)}UL^{(T)}\Lambda^{(T)^{1/2}}$ once again becomes symmetric and may be reduced to the diagonal form Λ by the orthogonal transformation $q'' = L^{(u)}Q$ with the matrix $L^{(u)}$. The momenta transformation corresponding to the co-ordinate transformation is written as $\hat{p}'' = \tilde{L}^{(u)-1}\hat{P}$.*

Since the $L^{(u)}$ matrix is orthogonal, it follows that $L^{(u)^{-1}} = \tilde{L}^{(u)}$ and consequently equation (7.8) becomes

$$\hat{T}_{\text{vib}} = \tfrac{1}{2}\tilde{\hat{P}}L^{(u)^{-1}}L^{(u)}\hat{P} = \tfrac{1}{2}\tilde{\hat{P}}\hat{P} = \tfrac{1}{2}\sum_k \hat{P}_k^2$$

*In future, the momenta conjugate to normal co-ordinates (see below) will be designated by P.

The whole sequence of co-ordinate transformations can be written as a single matrix:

$$L_q = L^{(T)}\Lambda^{(T)^{1/2}}L^{(u)} \tag{7.9}$$

The momentum transformation will thus be accomplished by the matrix

$$L_p = L^{(T)}\Lambda^{(T)^{-1/2}}L^{(u)} = \tilde{L}_q^{-1} \tag{7.10}$$

These matrices transform the quadratic form $\hat{T}_{\text{vib}} = \frac{1}{2}\tilde{\hat{p}}T p$ to

$$\hat{T}_{\text{vib}} = \tfrac{1}{2}\tilde{\hat{P}}\hat{P} = \tfrac{1}{2}\sum_k \hat{P}_k^2 \tag{7.11}$$

and the form $U_{\text{vib}} = \frac{1}{2}\tilde{q}Uq$ to

$$U_{\text{vib}} = \tfrac{1}{2}\tilde{Q}\Lambda Q = \tfrac{1}{2}\sum_k \lambda_k Q_k^2 \tag{7.12}$$

Hence as a result of the linear co-ordinate transformation $q = L_q Q$ the operator $\hat{H}_{\text{vib}} = \frac{1}{2}\tilde{\hat{p}}T\hat{p} + \frac{1}{2}\tilde{q}Uq$ is transformed into

$$H_{\text{vib}} = \frac{1}{2}\left(\sum_k \hat{P}_k^2 + \lambda_k Q_k^2\right)$$

i.e. it is a sum of the operators

$$\hat{H}_k = \tfrac{1}{2}\hat{P}_k^2 + \tfrac{1}{2}\lambda_k Q_k^2 \tag{7.13}$$

Co-ordinates Q_k, which have the property of making the $\hat{H}_{\text{vib}}$ operator a sum of $\hat{H}_k$ operators when it is expressed in terms of them, are called normal co-ordinates.

The $\hat{H}_k\psi_k = E_k\psi_k$ equation with $\hat{H}_k$ operator as in equation (7.13) is the equation of state of a harmonic wave oscillator. The eigenfunctions E_k (energy levels) of such oscillators can be expressed thus:

$$E_k = \hbar\nu_k(v_k + \tfrac{1}{2}) \tag{7.14}$$

where $\nu_k = \sqrt{\lambda_k}$,* as usual, corresponds to the frequency of the oscillator's vibrations, and v_k is a vibrational quantum number taking the values $0, 1, 2, \ldots$.

The eigenfunctions of a harmonic oscillator have the form

$$\Psi_k(Q_k) = N_{v_k}\mathrm{e}^{-1/2\beta_k Q_k^2}H_{v_k}(\beta^{1/2}Q_k) \tag{7.15}$$

where N_{v_k} is the normalizing multiplier equal to

$$\left[\left(\frac{\beta_k}{\pi}\right)^{1/2}\frac{1}{2^{v_k}(v_k)!}\right]^{1/2}$$

and H_{v_k} is a Chebyshev–Hermitian polynomial of order v_k with respect to Q_k;

$$\beta_k = \frac{4\pi^2 M\nu_k}{h}$$

*Omitting the dimensionality multiplier.

(M is a factor which has the dimensions of mass). The first three terms of the polynomial are

$$H_0(x) = 1, \quad H_1(x) = 2x, \quad H_2(x) = 4x - 2 \tag{7.16}$$

where $x = \beta_k^{1/2} Q_k$. Higher-order polynomials can be obtained using recurrence formulae, i.e.

$$H_{v+1}(x) - 2xH_v(x) + 2vH_{v-1}(x) = 0 \tag{7.17}$$

The eigenfunctions of a harmonic oscillator are orthogonal and normalized, i.e.

$$\int \Psi_v^* \Psi_{v'} \, \mathrm{d}Q = \delta_{vv'} \tag{7.18}$$

where the asterisk indicates conjugation and $\delta_{vv'}$ is the Kronecker symbol, so we have $\delta_{vv'} = 1$ if the vibrational quantum numbers v and v' of the functions ψ_v^* and $\psi_{v'}$ are identical, while $\delta_{vv'} = 0$ if $v \neq v'$.

The disposition of the vibrational energy levels of a polyatomic molecule are defined by

$$E_{\text{vib}} = \sum_k E_k = h \sum_k \nu_k (v_k + \tfrac{1}{2}) \tag{7.19}$$

Each energy level corresponds to each set of vibrational quantum number v_k. The lowest energy level, for which all the $v_k = 0$, has the energy

$$E_0 = \frac{h}{2} \sum_k \nu_k$$

This is the zero-level energy of the molecule and can reach high values in polyatomic molecules with a large number of degrees of freedom. Energy levels for which all but one of the quantum numbers are zero (the exception equalling unity) are called 'fundamental' levels; energy levels with one quantum number greater than unity are called 'overtones'; energy levels with several non-zero quantum numbers represent combination levels. Figure 7.1 represents the energy levels of a polyatomic molecule.

The emission or absorption of light is accompanied by different types of transitions between the levels of a molecule. The frequency of the light emitted and absorbed by the molecule is given by

$$h\nu_{\text{vib}} = E'_{\text{vib}} - E''_{\text{vib}} = h \sum_k \nu_k (v'_k - v''_k) \tag{7.20}$$

where E'_{vib} is the energy of the upper state and E''_{vib} is that of the lower one. For example, one quantum of light with a frequency, say, ν_1, is absorbed with a transition from the ground state (all the $v''_k = 0$) into one of the fundamental levels with $v'_1 = 1$. When a collection of identical molecules is irradiated by electromagnetic radiation with a continuous range of frequencies, quanta with frequency ν_1 are absorbed, and when the light propagating through the substance is viewed we observe an absorption band whose frequency is ν_1. These bands

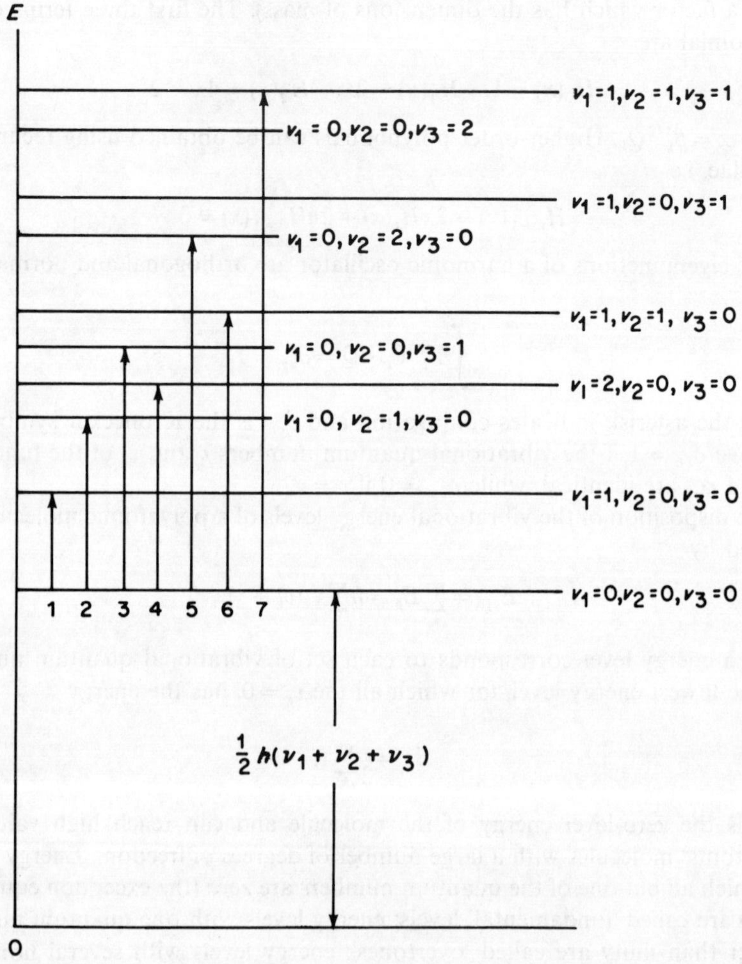

Figure 7.1. A diagram of the vibrational energy levels of a tri-atomic, non-linear molecule

are called 'fundamental absorption' bands. When there are transitions from a lower level to overtone or combination levels the corresponding bands are observed and named accordingly.

The frequencies of these bands can also be easily determined from equation (7.20). For example, given a transition from the ground state to an overtone level with $v_1'' = 2$, an energy quantum of frequency $2\nu_i$ is absorbed. Given a transition from the ground state to a compound level with $v_1' = 1$ and $v_2' = 1$ the frequency of the absorbed quantum is equal to $\nu_1 + \nu_2$. Transitions may take place from a low-energy state other than the ground state, i.e. higher than the ground state. Difference absorption frequencies thus arise. An example is a transition from the fundamental level $v_1'' = 1$ (the remaining levels $v_k'' = 0$) to the level $v_1' = 1$ (the remaining $v_k^1 = 0$) which may be higher if $\nu_2 > \nu_1$. According to equation (7.20) the frequency of the absorbed quantum is $\nu_2 - \nu_1$.

If the molecule has sufficient symmetry it may happen that the frequencies of two or more of the normal oscillators are the same. The expression for the molecule's energy levels then gains terms of the form $[v_k + (g_k/2)]$, where $v_k = \sum_i v_{ik}$, v_{ik} is a quantum number of a normal oscillator and g_k is the number of oscillators that have the same v_k. The position of the molecule's energy level is then defined by the quantum number $v_k = \sum_i v_{ik}$ and may have the same value for different permissible sets of v_{ik}. Thus if one frequency has two oscillators, the state $v_k = 1$ is accomplished when either $v_{1k} = 1$ and $v_{2k} = 0$, or $v_{1k} = 0$ and $v_{2k} = 1$. Such an energy level is called 'doubly degenerate'. The state with $v_k = 2$ is achieved when $v_{1k} = 2$ and $v_{2k} = 0$, or $v_{1k} = 1$ and $v_{2k} = 1$, or $v_{1k} = 0$ and $v_{2k} = 2$. Such a level is called triply degenerate. In the general case a level with $v_k = v_{1k} + v_{2k}$ will be $(v_k + 1)$ fold degenerate. A similar line of argument can be followed, and for $g_k = 3$ we get

$$v_k = \sum_{i=1}^{3} v_{ik}$$

The degree of degeneracy an energy level has will then be expressed by the formula $\frac{1}{2}(v_k + 1)\,(v_k + 2)$.

It is evident that the ground state will never be degenerate, and the degeneracy of the fundamental levels will coincide with g_k (the degree of the degeneracy of normal oscillators).

Before turning to a description of a molecule's state at different permissible vibrational levels, we shall demonstrate one more method of calculating the L_q matrix for the transformation to normal co-ordinates and the matrix's eigenvalues λ_i.

This method is based on the use of a matrix that is the product of the kinematic coefficient matrix and the force constant matrix. We denote this matrix $D = TU$, and will in future call it the generalized dynamic coefficient matrix, or simply the molecule's dynamic matrix. This matrix has an important role in the theory of molecular vibrational spectra, and can be used to obtain a whole series of important conclusions. Hence we shall consider the matrix $D = TU$ and its application to the similarity transformation $L_q^{-1} D L_q$, using the matrix we obtained above for the transformation to normal co-ordinates as the matrix L_q. By rewriting the product $L_q^{-1} D L_q$ in the form $L_q^{-1} T \tilde{L}_q^{-1} \tilde{L}_q U L_q$ and allowing for the fact that according to equation (7.12) $L_q^{-1} = \tilde{L}_p$, we find that $L_q^{-1} D L_q = \tilde{L}_p T L_p \tilde{L}_p U L_q$. However, the $\tilde{L}_p T L_p$ and $\tilde{L}_q U L_q$ products must be, respectively, $\tilde{L}_p T L_p = I$ and $\tilde{L}_q U L_q = \Lambda$ (see above, page 177). We can conclude from this that the similarity transformation $L_q^{-1} D L_q$ reduces D to the diagonal matrix Λ. By left-multiplying the equation $L_q^{-1} D L_q = \Lambda$ by L_q we obtain $D L_q = L_q \Lambda$, which is equivalent to a set of equations of the form

$$D l^{(k)} = \lambda_k l^{(k)} \tag{7.21}$$

where $l^{(k)}$ is the kth column of the L_q matrix and λ_k is the eigenvalue corresponding to the kth column.

Clearly, the L_q matrix's columns and the λ_k satisfy the matrix equation

$$Dl = \lambda l \quad \text{or} \quad (D - \lambda I)l = 0 \tag{7.22}$$

This is nothing but the characteristic secular equation defining the eigenvalues and eigenvectors of $D = TU$. It can only be solved for those values of λ that satisfy the vector equation

$$\det(D - \lambda I) = 0 \tag{7.23}$$

Each of the λ_k of equation (7.21) has its own set of l'_{ik} ($i = 1, \ldots, n$). However, the definition of this set has an arbitrary factor and the set is indeterminate in this sense. In order to eliminate this indeterminacy we set one of the l'_{ik} numbers to some value, for example unity. The other set of values l''_{ik} can clearly be then found by multiplying the column $l'^{(k)}$ by an appropriate factor N_k. One of the $l^{(k)''}$ columns obtained in this way must be the same as the corresponding $l^{(k)}$ column of the L_q transformation matrix of $q = L_q.Q$. Consequently, we have

$$l^{(k)''} = N_k l^{(k)'} = l^{(k)}$$

Since L_q can reduce the matrix U to the Λ form, i.e. $\tilde{L}_q U L_q = \Lambda$, we find that $\tilde{l}^{(k)} U l^{(k)} = \lambda_k$. By substituting the $\tilde{l}^{(k)}$ row and the $l^{(k)}$ column with the $\tilde{l}^{(k)'}$ row and $l^{(k)'}$ column, which were multiplied by N_k, we obtain $N_k \tilde{l}^{(k)'} U l^{(k)'} N_k = \lambda_k$, whence

$$N_k^2 = \lambda_k / \tilde{l}^{(k)'} U l^{(k)'} \tag{7.24}$$

7.2 Description of a Molecule's Configuration at Different Vibrational Levels

We shall now consider the state of a molecule at different vibrational levels. As is known, the probability of various co-ordinate values is defined by the squares of the appropriate wave function. The co-ordinate probability distribution for a one-dimensional harmonic oscillator is shown in Figure 7.2.

We see that we have a single probability maximum at $q = 0$ for the ground state, while excited states produce a series of maxima, the greatest probability occurring around the turning points of the classic motion. The larger the vibrational quantum number, the closer is the overall character of the quantum-mechanical probability distribution, as a function of a vibrational co-ordinate, to the distribution obtained using classic theory. The disposition of the maxima of the squares of the wave functions can easily be found using the usual methods for calculating the extreme values of functions. For example, the wave function of the kth normal oscillator for the state $v_k = 1$ has, according to equation (7.15), the form

$$\psi_k = 2N_1 \beta_k^{1/2} Q_k \mathrm{e}^{-1/2\beta k Q_k^2} \tag{7.25}$$

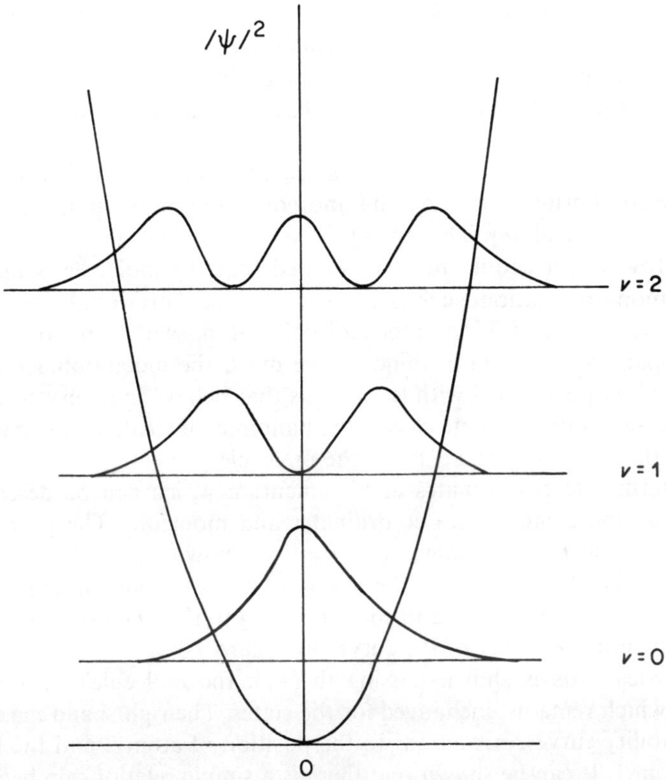

Figure 7.2. Co-ordinate probability distribution for harmonic oscillator

The extreme values of ψ_k^2 can be found from

$$2\psi_k \frac{d\psi_k}{dQ_k} = 0$$

We obtain $Q_{k1} = 0$ and $Q_{k2,3} = \pm \beta_k^{-1/2}$. It can easily be verified that the value $Q_{kl} = 0$ corresponds to a minimum of ψ_k^2, while the $Q_{k2,3} = \pm \beta_k^{1/2}$ values are maxima. The $\beta_k^{-1/2}$ corresponds to the classical vibration amplitude.

For the state $v_k = 2$ the maxima of ψ_k^2 for a harmonic oscillator are found for values of Q_k equal to zero and $\pm(\frac{5}{2}\beta_k)^{-1/2}$. The probabilities of the co-ordinate values at the point $(\frac{5}{2}\beta_k)^{-1/2}$ and at zero are very close together, a fact not difficult to verify. In fact, the ratio between them is 1.32:1.00, and this emphasizes the real difference between the quantum and classical oscillators.

In view of the importance of clearly understanding the difference between quantum and classical oscillators we shall study the topic in more detail. In order to achieve a more visual description we shall initially consider a diatomic molecule whose state can be completely described by one co-ordinate, i.e. the change in the bond length. Since in this very simple case this co-ordinate is itself a normal one, omitting the coefficient, we will call it Q. We obtain the curve shown in Figure 7.2 for the square of the wave function, which defines the probability of the co-ordinate having a particular value, for the state when $v = 0$.

We can say that the value of Q is 'smeared' over a region around its zero value (the equilibrium position), and, moreover, the molecule in this state has a non-zero energy of $\frac{1}{2}h\nu$. This energy is conventionally called the 'zero-point' energy. However, it should not be assumed that the molecule is undergoing small harmonic oscillations such as classic vibrations. This would mean ascribing defined trajectories of motion to the molecule's atoms which are not permissible for microparticles, bearing in mind, as we must, the indeterminacy principle. In the microworld we deal with formations that behave in a very specific way. All we can say in the concrete case of the biatomic molecule we are considering is that in the ground state, i.e. $v = 0$, the molecule possesses a non-zero energy and indeterminate co-ordinates and momentum which can be described, for example, by the mean square co-ordinates and momenta. The probability of the value of the Q co-ordinate with $v = 0$ is shown in the lowest curve in Figure 7.2. The distribution of the x co-ordinates of both atoms along the straight line connecting them and corresponding to this Q co-ordinate distribution is shown in the lower of the curves in Figure 7.3.

The vertical axis is shown passing through the molecule's centre-of-mass position, which remains unchanged for the states. The right-hand maximum of the probability curve corresponds to the position of atom 2 and the left-hand one to atom 1. It can be shown that there is a simple relationship between the normal co-ordinate Q and the x_1 and x_2 co-ordinates of the atoms which are measured along the molecular axis and to the left and right, respectively, of the co-ordinates' origin which passes through the molecule's centre of mass. The

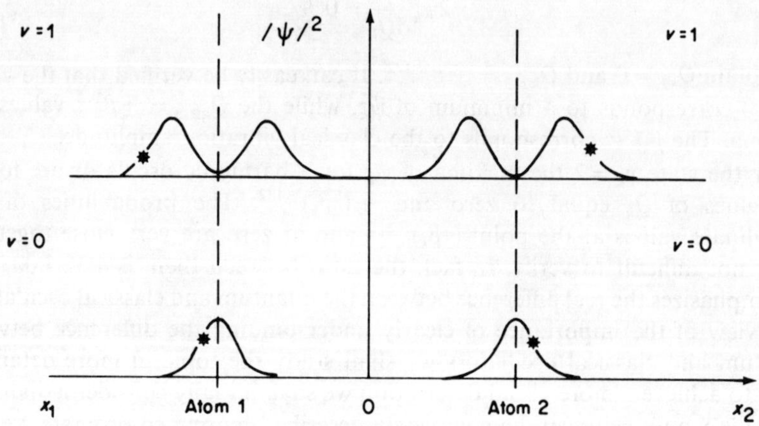

Figure 7.3. Conjugated points for vibrations of two atoms

relationship is:

$$x_1 = -(M_1M)^{-1}Q, \quad x_2 = (M_2M)^{-1}Q \tag{7.26}$$

where M_1 and M_2 are the masses of the atoms and M is the reduced mass of the molecule, i.e. $M = M_1M_2/(M_1 + M_2)$. The width of the probability distribution curve for the position of each atom can be given in terms of mean square co-ordinates which will be proportional to M_1^{-2} for the first atom and to M_2^{-2} for the second (see Section 7.3).

Thus the locality of each atom depends substantially on its mass, and decreases as its mass grows. It is important to bear in mind the following. There must be a link between the probability distribution of the 1-atom's co-ordinates in its locality and the respective distribution of the 2-atom's co-ordinates in its locality, and this link is satisfied by the requirement, also retained in quantum mechanics, that the centre of mass of the molecule remains unchanged during vibrations. This means that for each point on the first atom's probability distribution curve there must be a corresponding point conjugate to it on the same curve of the second atom. Two such points on each curve are shown in Figure 7.3 by asterisks. Thus the 'smearing' of the co-ordinates of one atom automatically entails that the co-ordinates of the other are also 'smeared'. It is the imposition of this requirement that makes it possible for us to retain the assumption that the position of the molecule's centre of mass is unchanged, whilst at the same time being able to make the quantum-mechanical assumption of the uncertainty of the positions of each of the atoms within their localities.

We shall now consider the molecule's state at the level whose vibrational quantum number is. $v = 1$. The probability function for the Q co-ordinate (Figure 7.2) has two maxima, and consequently the probability distribution of the co-ordinates of both atoms will take the form shown in Figure 7.3 (upper curve). The distance between the maxima of the atom's co-ordinates probability distributions depends on the masses of the atoms and the frequency of the vibration. Moreover, there is a relationship between the atoms' distribution functions which involves the fact that the left-hand maximum of the left-hand atom corresponds to the right-hand maximum of the right-hand one, and vice versa. Thus we find that, just as was the case for the state with $v = 0$, the co-ordinate probability distribution of one atom ('smeared' co-ordinates) is associated in a definite way with the distribution ('smearing') of the co-ordinates of the other.

As already mentioned, the separation between the probability functions' maxima for a harmonic oscillator approximates to the amplitude of the oscillations of a classical oscillator. This situation allows us to use the term 'vibrational form' even in quantum mechanics, but investing it with a somewhat different physical meaning.

In order to explain this better, we shall consider the processes that take place in a molecule when affected by light. According to the classical view, the molecule's electrons begin to oscillate under the influence of the changing electric field of the light waves dragging their nucleus after them. As a result, given a perturbation (which in quantum mechanics corresponds to a transition from the ground state, $v'' = 0$, to the first perturbed state, $v' = 1$) the atomic vibrations take place as harmonic vibrations with the frequency of the incident light. Moreover, whilst the wave is having its effect they are periodically in the position corresponding to the maximum extension (amplitude) of the bond and periodically in that corresponding to the maximum contraction of the bond. This situation can be visualized as in Figure 7.4.

Each of the atoms oscillates about its own equilibrium position. Moreover, the amplitude of this oscillation decreases as the atoms' masses and the oscillation frequency increase. The vibrational amplitude of each atom and their trajectories during motion can be called the form of the atoms' vibrations for the normal vibration of the diatomic molecule.

According to quantum mechanics, when the molecule absorbs a quantum of light it makes a transition from one stationary state to a new one (obviously, its life in the new state may not be long). The situation here is not that one state is distinguished from another by the fact that atoms are initially immobile and then vibrate harmonically in the second state, but rather that in the first state all the molecule's atoms are 'smeared out' in co-ordinate space according to one rule and in the second higher state they are 'smeared out' according to a second rule. The usual classic view of the harmonic vibration of atoms does not coincide with this, and so it is impossible to imagine the situation as being one in which the molecule 'leaps' from one 'peak' of the probability function to another with the frequency of the incident light. The analogy with the classic case only holds in that both require an electromagnetic wave of the same frequency in order for the atoms to oscillate and the molecule to go from one state to another. Moreover, the distance between the peaks of the distribution function of the atoms in quantum mechanics approximates to the amplitudes of the deviations of the atoms from their equilibrium positions during classical vibrations. This enables us to use analogies with classically vibrating systems

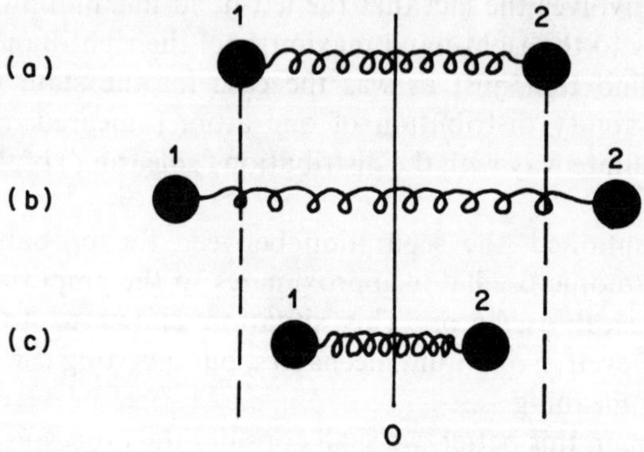

Figure 7.4. (a) Equilibrium position; (b) maximum extension; (c) maximum contraction. Zero is a position of centre of mass

in order to obtain a clear picture of the situation when describing molecules. Quite commonly, the classic properties of the mathematical characteristics of the vibrations are even retained using Lagrange equations. From now on we will constantly turn to classic analogies, and speak of a molecule's vibrations, even though, strictly, the usual conception of a vibration does not hold true in quantum-mechanical arguments. The effect of an electromagnetic wave on a quantized system is to make it change the nucleus 'smearing out' picture abruptly and retain the new situation throughout the life of the new state.

Bearing in mind the comments we have made concerning the quantum-mechanical treatment of vibrations in diatomic molecules, we will now continue with the description of a polyatomic molecule's states. In principle, there is nothing new here, but a number of additional features do appear.

The original co-ordinates in a polyatomic molecule are related to the normal co-ordinates thus:

$$\begin{aligned} q_1 &= l_{11}Q_1 + l_{12}Q_2 + \cdots + l_{1n}Q_n \\ q_2 &= l_{21}Q_1 + l_{22}Q_2 + \cdots + l_{2n}Q_n \\ &\cdots\cdots\cdots\cdots\cdots\cdots \\ q_n &= l_{n1}Q_1 + l_{n2}Q_2 + \cdots + l_{nn}Q_n \end{aligned} \tag{7.27}$$

The value of a l_{ij} coefficient can be found in one of the ways we have described above. As demonstrated, the relative signs of the columns of the L_q matrix remain indeterminate. However, because the square of a probability distribution's wave function is symmetric with respect to zero, this situation does not have any importance, and we can also consider all the l_{1j} elements in the first equation to be positive.

We have already noted that the relationship between the original relative nuclear co-ordinates and the normal co-ordinates is the same for both classic and quantum-mechanical theory. Given the right initial conditions, it is possible to stimulate any normal vibration in a classic system. Let us assume that there is a vibration in which, for example, the Q_1 co-ordinates change harmonically with frequency $\nu_1 = \sqrt{\lambda_1}$, the remaining normal co-ordinates being zero. The q_i coordinate is then given by

$$\begin{aligned} q^1 &= l_{11}Q_1 \\ q_2 &= l_{21}Q_1 \\ &\cdots\cdots \\ q_n &= l_{n1}Q_1 \end{aligned} \tag{7.28}$$

i.e. the relative nuclear co-ordinates change in proportion to Q with amplitudes given by the l_{i1} coefficients.

Thus, for example, we obtain for a linear symmetric triatomic molecule the following relationships as in equation (7.27) for vibrations along the molecule's axis, if the extension of the first bond (q_1) and that of the second (q_2) are taken as the relative co-ordinates of the nuclei:

$$q_1 = Q_1 + Q_2, \quad q_2 = Q_1 - Q_2 \tag{7.29}$$

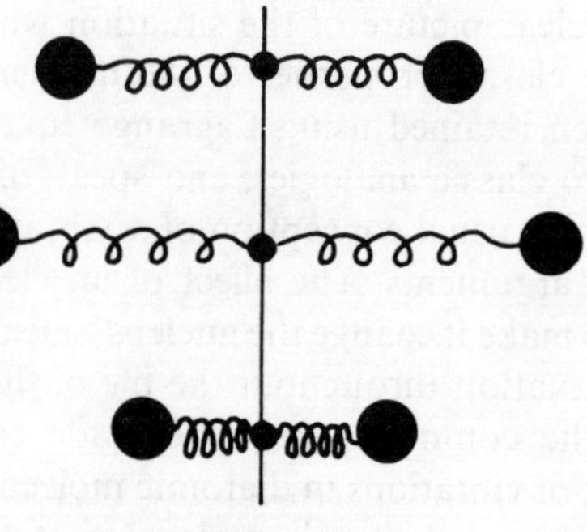

Figure 7.5. Position of the centre of mass

For the first normal vibration we have $q_1 = q_2 = Q_1$, i.e. in phase modes of bonds with ν_1 frequency (Figure 7.5). During vibrations of the second normal vibration Q_2 we obtain $q_1 = Q_2$ and $q_2 = -Q_2$, i.e. only one bond stretches during the vibration while the other contracts, and the bonds' vibrations are thus out of phase with each other. This kind of vibration has the form shown in Figure 7.6.

Consequently the l_{ik} coefficient of one of the L_q matrix's columns characterize the amplitudes of the changes in the internuclear separation during perturbations of one of the normal co-ordinates. The coefficients allow us to give a clear picture of the vibrations of the separate parts of the molecule, and this is why the set of elements of a column of the L_q matrix is conventionally called the form of the normal vibration. Often diagrams are drawn with arrows near the atoms, indicating the direction of the atoms' displacements during a perturbation of one normal vibration for a quarter of its period and the amplitude of these displacements. For example, the vibrations ν_1 and ν_2 of the molecule XY_2 are shown thus in Figure 7.7. The forms of the normal vibrations of a molecule such as that of water will be as shown in Figure 7.8.

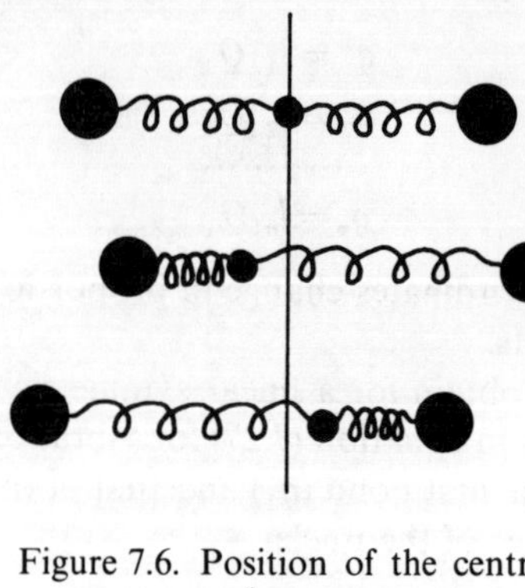

Figure 7.6. Position of the centre of mass

Figure 7.7. Symmetric and anti-symmetric normal vibrations of the linear molecule XY_2

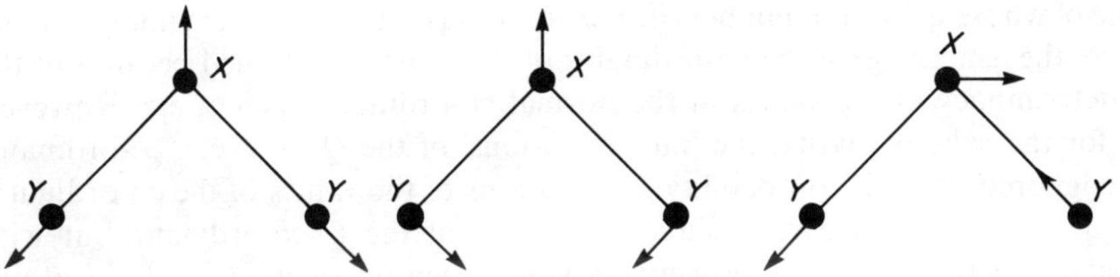

Figure 7.8. Schematic descriptions of the normal vibrations of a non-linear XY_2 molecule

The first and third normal vibrations of the H_2O molecule predominantly change the bond lengths (such vibrations are called 'valence' vibrations), while the second mainly changes the bond angle ('deformation' vibrations). If the changes in the bond lengths, q_1 and q_3, and those in the valence angle γ are taken as the co-ordinates, then the relationships between these and the normal co-ordinates, in a dimensionless scale, can be written approximately as

$$\begin{aligned} q_1 &= Q_1 + 0.1Q_2 + Q_3 \\ q_2 \equiv \gamma &= 0.1Q_1 + Q_2 + 0.Q_3 \\ q_3 &= Q_1 + 0.1Q_2 - Q_3 \end{aligned} \tag{7.30}$$

We call the 1, 0.1, 1 set of coefficients in front of the Q_1 co-ordinate the form of the first normal vibration. This will mean that both bonds will change in the same way in this vibration, while the angle will be insignificantly changed (an order of magnitude less). In the second vibration the form is given by the coefficients 0.1, 1, 0.1, i.e. mainly the angle changes, while the form of the third vibration is 1, 0, -1, i.e. the angle is not changed but the bonds vibrate out of phase to each other.

Thus the classic idea of the forms of vibrations has a very clear sense, and a normal vibration can be defined as one in which all the atoms oscillate harmonically with the same frequency and pass through their equilibrium positions simultaneously.

In a quantum-mechanical argument the situation is rather more complex. We have already demonstrated that the concept of vibrations, as periodic changes in the relative positions of the atoms, loses its sense. A transition from one state to another abruptly changes the 'smearing out' picture of the nuclei. Even when the quantum numbers are zero-valued, each of the Q_k co-ordinates are characterized by a definite distribution function, and consequently, as was shown for a diatomic molecule, each of the molecule's nuclei is localized within a 'smear' region and not to a point. However, the probability maxima of the atomic localizations correspond to the equilibrium positions of the atoms, and in this sense we can speak of the analogy with the dispositions of the atoms in the classic model without vibrations. When the molecule transfers to a state one of whose quantum numbers (for example, v_1) is unity, the remaining staying zero, the 'smearing' of the co-ordinates of the atoms is retained because of the indeterminacy of the values of the normal co-ordinates Q_2, Q_3, etc. However, if, for the sake of clarity, the 'null vibrations' of the Q_2, Q_3, etc. co-ordinates are ignored, then the probability distributions of the values of the co-ordinates $q_1, q_2, \ldots, q_n$ will clearly be similar to those of the Q co-ordinate, but with coefficients $l_{11}, l_{21}, \ldots, l_{n1}$, respectively. Consequently, we find that all the atoms will be 'smeared out' in areas with two maxima, as was the case for the diatomic molecule, the positions of the maxima approximating to the turning points of the classic vibrations. We can in this sense carry over the concept of vibrational forms (described above) from the classic case to the quantum-mechanical system. Having to account for the 'null vibrations' distorts the simple picture. A non-zero probability does appear for the location of an atom even in its non-classic trajectory during the first normal vibration. However, this probability is reduced in any direction perpendicular to the trajectory following a law reminiscent of the Gauss curve.

Obviously, in states, several of whose quantum numbers are non-zero, the 'smearing' of the nuclei will be considerably more complicated. The full picture can only be obtained by analysing the multidimensional function

$$\Psi = \prod_k \psi_k(Q_k)$$

in which the Q_k are replaced by the q co-ordinates. However, this problem will not encroach upon our task. We shall dwell only on the calculation of the mean square values of the relative co-ordinates for a given state, which can, to some degree, be used as a measure of the 'smearing' of the nuclear co-ordinates.

Our only comment here is that it is the presence of the 'null vibrations' that enables the requirement of positional indeterminacy to be fulfilled for a molecule's atoms in all its states. This can be demonstrated with clarity for the example of the linear symmetric molecule XYX, to which we have already referred. The linear vibrations can be completely described by two normal co-ordinates Q_1 and Q_2, which are related to the atoms' Cartesian co-ordinates on the molecule's axis thus:

$$x_1 = \alpha Q_1 + \beta Q_2, \quad x_2 = \gamma Q_2, \quad x_3 = -\alpha Q_1 + \beta Q_2 \tag{7.31}$$

where α, β and γ are coefficients dependent on the atomic masses. When the first co-ordinate Q_1 is changed, the outer atoms must be displaced to the right and left, respectively, and by the same amount, the central atom remaining immobile. The 'smearing' of the Q co-ordinate clearly leads to a proportionate 'smearing' of those of the outer atoms, but it has no effect on the central atom.

Superficially there seems to be a contradiction with the uncertainty principle, but this can be removed if we take into account that, for a full description of the position of any of the atoms, all the normal co-ordinates must be introduced. Then the 'smearing' of the central atom in any state is proportionate to that of the Q_2 co-ordinate, and will be associated with the uncertainty of this co-ordinate.

7.3 Calculating the Mean Square Amplitudes of the Vibrational Co-ordinates

By definition, the mean square average value of a relative co-ordinate in a molecule's given state is given by the integral

$$\int \Psi^* q_j^2(Q_1, \ldots, Q_n)\Psi \, dQ = \bar{q}_j^2 \tag{7.32}$$

Then because

$$\Psi = \prod_k \psi_k(Q_k)$$

and

$$q_j = \sum_k l_{jk} Q_k$$

we obtain

$$q_j^2 = \sum_k l_{jk}^2 Q_k^2 + 2 \sum_{\substack{m,m' \\ m>m'}} l_{jm} l_{jm'} Q_m Q_{m'}$$

and

$$\begin{aligned}
\bar{q}_j^2 &= \int \prod_k \psi_k^*(Q_k)\left(\sum_k l_{jk}^2 Q_k^2 + 2\sum_{m,m'} l_{jm} l_{jm'} Q_m Q_{m'} \right) \prod_k \psi_k(Q_k) \prod_k dQ_k \\
&= \sum_k l_{jk}^2 \int \psi_k^*(Q_k) Q_k^2 \psi_k(Q_k) \, dQ_k \prod_{r \neq k} \int \psi_r^*(Q_r)\psi_r(Q_r) \, dQ_r \\
&\quad + 2\sum_{m,m'} l_{jm} l_{jm'} \int \psi_m^*(Q_m) Q_m \psi_m(Q_m) \, dQ_m \\
&\quad \times \int \psi_{m'}^*(Q_{m'}) Q_{m'} \psi_{m'}(Q_{m'}) \, dQ_{m'} \prod_{p \neq m \neq m'} \int \psi_p^* \psi_p \, dQ_p
\end{aligned} \tag{7.33}$$

Integrals such as $\int \psi_k^* Q_k^2 \psi_k \, dQ_k$ are matrix elements of the square of a co-ordinate for a harmonic oscillator, and are equal to $Q_{0k}^2(2v_k + 1)$ for any energy level, where $Q_{0k} = \sqrt{(h/8\pi^2 M v_k)}$ is the so-called zero amplitude. Integrals such as $\int \psi_m^*(Q_m) Q_m \psi_m(Q) \, dQ_m$ are zero for a harmonic oscillator, whilst $\int \psi_k^* \psi_k \, dQ_k = 1$ by normalization condition. Finally, as a result we obtain

$$\bar{q}_j^2 = \sum_k l_{jk}^2 Q_{0k}^2 (2v_k + 1) \quad \text{and} \quad \sqrt{(\bar{q}_j^2)} = \sqrt{\left[\sum_k l_{jk}^2 Q_{0k}^2 (2v_k + 1) \right]} \tag{7.34}$$

This formula makes it possible for us to define the mean of the standard deviations of the relative co-ordinates from the equilibrium position for any of the states of the molecule if only the vibrational frequency of the normal oscillators ν_k and the L_q matrix's elements of the L_q matrix are known. For the ground state, when all the quantum numbers v_k are zero, we obtain

$$\sqrt{(\bar{q}_j^2)} = \sqrt{\left(\sum_k l_{jk}^2 Q_{0k}^2\right)} \tag{7.35}$$

The deviations from the equilibrium position (zero vibrations) will in this case be given by the quantity $Q_{0k} = \sqrt{(h/8\pi^2 M\nu_k)}$, which is known as the zero amplitude of the vibration of a normal oscillator.

In addition to the average standard deviations of the relative co-ordinates, it is useful to know the size of the average potential energy of the vibrations for a polyatomic molecule in any vibrational state. This energy, $\bar{U}(q)$, is defined as $\bar{U}(q) = \int \Psi^* \bar{U}(q) \Psi \, \mathrm{d}Q$, and the next task is to solve this integral. We take the potential energy of the harmonic oscillations to be $U(q) = \frac{1}{2}\tilde{q}Uq$, or $U(Q) = \frac{1}{2}\sum_{k'} \lambda_k Q_k^2$ in normal co-ordinates, from which we obtain

$$\bar{U}Q) = \frac{1}{2}\sum_k \int \left(\prod_m \psi_m^*\right) \lambda_k Q_k^2 \left(\prod_m \psi_m\right) \prod_m \mathrm{d}Q_m = \frac{1}{2}\sum_k \lambda_k Q_{0k}^2 (2v_k + 1) \tag{7.36}$$

In order to calculate the vibrational energy distribution with respect to the relative co-ordinate we must know the value of $\bar{U}(q)$ expressed in terms of changes in these co-ordinates. Hence, since $\lambda_k = \tilde{l}^{(k)} U l^{(k)}$, we obtain, according to equation (7.36), the following:

$$\bar{U}(q) = \frac{1}{2}\sum_k \tilde{l}^{(k)} U l^{(k)} Q_{0k}^2 (2v_k + 1) \tag{7.37}$$

Thus the average potential energy in any vibrational state will equal the sum of the potential energies of all the normal oscillators. Each term in equation (7.37) equals $\tilde{l}^{(k)} U l^{(k)} Q_{0k}^2 (2v_k + 1)$, and is the potential energy of a classic normal vibration of the molecule. All the natural relative co-ordinates will change with the amplitudes, which equal $l_{jk} Q_{0k} \sqrt{(2v_k + 1)}$, where l_{jk} is an element of the kth column of the matrix L_q, and with phases given by the relative signs of these elements.

In conclusion, we note that unique relationships between the co-ordinate amplitudes, as given by the elements of a particular column of L_q, will only be obtained for non-degenerate vibrations when one normal oscillator is perturbed. In the case of a degenerate vibrations, whose corresponding normal co-ordinates are formed during symmetry operations in pairs, triplets, etc. depending on symmetry, the relationship between the elements in the column of L_q that corresponds to one of the normal co-ordinates in the set of degenerate co-ordinates will be changed according to the choice of the unit vectors for the degenerate co-ordinates (the choice of the matrix for the transformation from the symmetry co-ordinates to the atoms' relative co-ordinates). Hence when there is degeneracy corresponding to degenerate normal co-ordinates, the columns of L_q may be called forms of the normal vibrations only tentatively. A similar analysis for the motions of the atoms for the clear classic description demonstrates that the atoms in this case either move in circles (double degeneracy) or on spheres (triple degeneracy) (see also Chapter 5). This situation does not have any import when calculating the mean square amplitudes of the relative co-ordinates or the average potential energy. For example, in equation (7.34) there are terms such as $(\sum_i (l_j)^2_{ki}) Q^2_{0k}(2v_k + 1)$, where the summation is over all degenerate vibrations. Such sums are invariant with respect to the choice of the unit vectors for the degenerate normal co-ordinates (see Chapter 5).

7.4 Calculating Partial Derivatives of Vibration Frequencies with Respect to a Molecule's Parameters

An investigation of how a vibration's form and frequency depend on the various parameters of the force constant and kinematic matrices is very important when analysing the vibrations of a polyatomic molecule. It is evident that this dependency is determined by the partial derivatives of the vibration's form and frequency with respect to a selected parameter. Calculations of the partial derivatives of a vibration's frequency occurs regularly. The task arises particularly in the following cases:

(1) When evaluating the degree of localization of a group in a molecule and the sensitivity of its frequency on small disturbances within the group;
(2) When solving the inverse spectral problem (see Chapter 13);
(3) When evaluating the influence of small external disturbances (such as intermolecular interactions) on a molecule's spectrum.

Since the derivative $(\partial \nu_k/\partial x)$, where x is the parameter of differentiation, is related to the derivative $(\partial \lambda_k/\partial x)$ very simply by the equation $2\nu_k(\partial \nu_k/\partial x) = (\partial \lambda_k/\partial x)$ and the $(\partial \lambda_k/\partial x)$ derivative can be calculated much more easily, we shall only concern ourselves with the latter below.

In order to calculate the partial derivative $(\partial \lambda_k/\partial x)$ we shall use the relation $DL_q = L_q\Lambda$, which is true both for dependent and independent co-ordinates. Moreover, since in both cases $\tilde{L}_p \cdot L_q = I$ is true, we have

$$\tilde{L}_p D L_q = \Lambda \tag{7.38}$$

By differentiating, we obtain

$$\frac{\partial \tilde{L}_p}{\partial x} D L_q + \tilde{L}_p \frac{\partial D}{\partial x} L_q + \tilde{L}_p D \frac{\partial L_q}{\partial x} = \frac{\partial \Lambda}{\partial x} \tag{7.39}$$

For a given eigenvalue λ_k we have

$$\frac{\partial \tilde{l}_p^{(k)}}{\partial x} D l_q^{(k)} + \tilde{l}_p^{(k)} \frac{\partial D}{\partial x} l_q^{(k)} + \tilde{l}_p^{(k)} D \frac{\partial l_q^{(k)}}{\partial x} = \frac{\partial \lambda_k}{\partial x} \tag{7.40}$$

where $l_p^{(k)}$ and $l_q^{(k)}$ are the kth columns of the L_p and L_q matrices, respectively. We then consider together the characteristic equations $(D-\lambda I)l=0$ and $(\tilde{D}-\lambda I)l'=0$ and demonstrate that the eigenvalues and eigenvectors of the D and $\tilde{D}$ matrices are related. Since $\mathrm{Det}\,(D-\lambda I)=\mathrm{Det}\,(\tilde{D}-\lambda I)$ the eigenvalues of the D and $\tilde{D}$ matrices coincide. Hence we have for the kth eigenvalue of $\tilde{D}$

$$\tilde{D}l'^{(k)}=\lambda_k l'^{(k)} \quad \text{or} \quad \tilde{l}'^{(k)}D=\lambda_k \tilde{l}'^{(k)} \tag{7.41}$$

where $l'^{(k)}$ is the eigenvector corresponding to the kth eigenvalue. We now right-multiply the latter equation by $l_q^{(m)}$, i.e. by the eigenvector with the m superscript from the D matrix. We thus obtain

$$\tilde{l}'^{(k)}Dl_q^{(m)}=\lambda_k \tilde{l}'^{(k)}l_q^{(m)} \quad \text{or} \quad \tilde{l}'^{(k)}\lambda_m l_q^{(m)}=\lambda_k \tilde{l}'^{(k)}l_q^{(m)}$$

i.e.

$$(\lambda_m-\lambda_k)\tilde{l}'^{(k)}l_q^{(m)}=0 \tag{7.42}$$

from which we find that for $m \neq k$ we get $\tilde{l}'^{(k)}l_q^{(m)}=0$, and for $m=k$ $\tilde{l}'^{(k)}l_q^{(k)}=\alpha_k$. If the eigenvectors of the $\tilde{D}$ matrix are normalized so that $\alpha_k=1$, then we obtain the following for the matrices containing the eigenvectors of the $\tilde{D}$ and D matrix:

$$\tilde{L}'L_q=1$$

In this way the L' matrix has the same properties as the L_p matrix. Consequently, it is a solution of the equation

$$\tilde{D}L_p=L_p\Lambda \tag{7.43}$$

If this is true then we must have the relationship $\tilde{D}l_p^{(k)}=\lambda_k l_p^{(k)}$ as well as $Dl_q^{(k)}=\lambda_k l_q^{(k)}$. Using these we can rewrite equation (7.40) as

$$\frac{\partial \tilde{l}_p^{(k)}}{\partial x}\lambda_k l_q^{(k)}+\lambda_k \tilde{l}_p^{(k)}\frac{\partial l_q^{(k)}}{\partial x}+\tilde{l}_p^{(k)}\frac{\partial D}{\partial x}l_q^{(k)}=\frac{\partial \lambda_k}{\partial x} \tag{7.44}$$

or

$$\lambda_k\left(\frac{\partial \tilde{l}_p^{(k)}}{\partial x}l_q^{(k)}+\tilde{l}_p^{(k)}\frac{\partial l_q^{(k)}}{\partial x}\right)+\tilde{l}_p^{(k)}\frac{\partial D}{\partial x}l_q^{(k)}=\frac{\partial \lambda_k}{\partial x} \tag{7.45}$$

We now differentiate $\tilde{L}_pL_q=I$ to obtain

$$\frac{\partial \tilde{L}_p}{\partial x}L_q+\tilde{L}_p\frac{\partial L_q}{\partial x}=0$$

from which

$$\partial \tilde{l}_p^{(k)}/\partial x \cdot l_q^{(k)}+\tilde{l}_p^{(k)}\cdot \partial l_q^{(k)}/\partial x=0$$

and finally

$$\frac{\partial \lambda_k}{\partial x}=\tilde{l}_p^{(k)}\frac{\partial D}{\partial x}l_q^{(k)} \tag{7.46}$$

Since $D=T\cdot U$ this formula can be written in another form, i.e.

$$\frac{\partial \lambda_k}{\partial x}=\tilde{l}_p^{(k)}\frac{\partial T}{\partial x}Ul_q^{(k)}+\tilde{l}_p^{(k)}T\frac{\partial U}{\partial x}l_q^{(k)} \tag{7.47}$$

If only the T matrix depends on the parameter x then we will obtain

$$\frac{\partial \lambda_k}{\partial x} = \tilde{l}_p^{(k)} \frac{\partial T}{\partial x} U l_q^{(k)} \tag{7.48}$$

and if only the U matrix depends on x we have

$$\frac{\partial \lambda_k}{\partial x} = \tilde{l}_p^{(k)} T \frac{\partial U}{\partial x} l_q^{(k)} \tag{7.49}$$

Both equations (7.48) and (7.49) are true when using both dependent and independent co-ordinates. This fact follows directly as corollaries of both formulae.

When making numerical estimates, $l_p^{(k)}$, T, U and $l_q^{(k)}$ are replaced by their values for a specific choice of T and U. The $(\partial\lambda_k/\partial x)$ derivative then has the sense of a derivative taken from a neighbourhood of a point defined by the numerical choice of T and U. The T matrix must always be chosen in the form $T = \vec{B}E\vec{\tilde{B}}$ in order to make a definite choice even for the dependent co-ordinates (see Section 7.6).

By taking $\tilde{L}_p L_q = L_q \tilde{L}_p = L_p \tilde{L}_q = I$ into account for independent co-ordinates we can simplify the formula for $(\partial\lambda_k/\partial x)$. To do this, we left-multiply $\tilde{L}_p T L_p = I$ by L_q and $\tilde{L}_q U L_q = \Lambda$ by L_p to find that $T L_p = L_q$ and $U L_q = L_q \Lambda$. Hence we find that $\tilde{l}_p^{(k)} T = \tilde{l}_q^{(k)}$ and $U l_q^{(k)} = \lambda_k l_p^{(k)}$. By substituting these expressions into equation (7.47) we finally obtain

$$\frac{\partial \lambda_k}{\partial x} = \lambda_k \tilde{l}_p^{(k)} \frac{\partial T}{\partial x} l_p^{(k)} + \tilde{l}_q \frac{\partial U}{\partial x} l_q^{(k)} \tag{7.50}$$

If we use symmetry co-ordinates when differentiating the T and U matrices (and their symmetries are not distorted in doing so) then the formula for $(\partial\lambda_k/\partial x)$ will retain its form, with the T and U being replaced by $T^{(s)}$ and $U^{(s)}$, and the $l_p^{(k)}$ and $l_q^{(k)}$ by $l_p^{(s)(k)}$ and $l_q^{(s)(k)}$, where s is the index of the $T^{(s)}$ or $U^{(s)}$ submatrices in symmetry co-ordinates for the given symmetry type. When a $(\partial\lambda_k/\partial x)$ derivative is calculated for symmetrically degenerate frequencies any of the degenerate components can be taken, since the value of the derivative will not depend on the choice of component. When there is random degeneracy, the $(\partial\lambda_k/\partial x)$ derivative will depend on the choice of the $l_p^{(k)}$ and $l_q^{(k)}$ vectors for the degenerate components and will, in general, be indeterminate. We shall return to this topic below.

If we choose a diagonal element of the force constant matrix, u_{jj}, as the differentiation parameter and all other elements are, moreover, assumed to be independent of u_{jj}, then the derivative of the force constant matrix with respect to the jth diagonal element will have the form:

$$\frac{\partial U}{\partial u_{jj}} = \begin{bmatrix} 0 & \cdots & 0 & \cdots & 0 \\ \cdots & \cdots & \cdots & \cdots & \cdots \\ 0 & \cdots & 1 & \cdots & 0 \\ \cdots & \cdots & \cdots & \cdots & \cdots \\ 0 & \cdots & 0 & \cdots & 0 \end{bmatrix} \begin{matrix} \\ \\ j\text{th row} \\ \\ \\ \end{matrix} \tag{7.51}$$

jth column

That is, all the elements of this matrix are zero except for one equal to unity, situated in the jj'th position; i.e. similar to the position of the element with respect to which the differentiation was carried out.

Consequently

$$\frac{\partial \lambda_k}{\partial u_{jj}} = l_{jk}^2 \tag{7.52}$$

i.e. a derivative of the square of the vibrational frequency with respect to a diagonal element of the force constant matrix is simply equal to the square of the corresponding element in the k column of matrix L_q.

The derivative matrix $(\partial U/\partial u_{ij})$ with respect to a non-diagonal element has the following form if all the remaining elements do not depend on the differentiating element:

$$\frac{\partial U}{\partial u_{ij}} = \begin{bmatrix} \cdots\cdots\cdots\cdots \\ \cdots 0 \cdots 1 \cdots 0 \\ \cdots\cdots\cdots\cdots \\ \cdots 1 \cdots 0 \cdots 0 \\ \cdots\cdots\cdots\cdots \\ \cdots 0 \cdots 0 \cdots 0 \end{bmatrix} \begin{matrix} \\ i\text{th row} \\ \\ j\text{th row} \\ \\ \\ \end{matrix} \tag{7.53}$$

ith column $\quad$ jth column

i.e. all the elements are zero except for the two unit elements situated symmetrically with respect to the leading diagonal in cells corresponding to the interaction force constants of the co-ordinate pair. Hence we obtain

$$\frac{\partial \lambda_k}{\partial u_{ij}} = 2 l_{ik} l_{jk} \tag{7.54}$$

i.e. the derivative with respect to the non-diagonal element u_{ij} is equal to twice the product of the elements of the corresponding co-ordinate pair in the column of L_q. These formulae can conveniently be written as

$$\partial \lambda_k / \partial x = (2 - \delta_{ij}) l_{ik} l_{jk} \tag{7.55}$$

where δ_{ij} is the Kronecker symbol.

This formula is very simple and is widely used when solving the explicit problems that often occur in chemical spectroscopy. In particular it is easy to show using the formula that in the presence of random degeneracy the sum of $(\partial \lambda_k/\partial x)$ derivatives calculated for all degenerate components is invariant with respect to the choice of the actual numerical values of these components.

In order to prove this, consider, for example, the case of double degeneracy. For simplicity, we will set the i index to 1 and the j index to 2. Thus we will have the following for the submatrices of the elements of the eigenvectors of the pair of degenerate eigenvalues whose indices are k' and k'':

$$\begin{bmatrix} l_{1k'} & l_{1k''} \\ l_{2k'} & l_{2k''} \end{bmatrix} \tag{7.56}$$

When this matrix is right-multiplied by the orthogonal matrix

$$\begin{bmatrix} \alpha_{11} & \alpha_{12} \\ \alpha_{21} & \alpha_{22} \end{bmatrix} \tag{7.57}$$

we obtain a submatrix of the elements of eigenvectors having the form

$$\begin{bmatrix} l_{1k'}\alpha_{11} + l_{1k''}\alpha_{21}; & l_{1k'}\alpha_{12} + l_{1k''}\alpha_{22} \\ l_{2k'}\alpha_{11} + l_{2k''}\alpha_{21}; & l_{2k'}\alpha_{12} + l_{2k''}\alpha_{22} \end{bmatrix} \tag{7.58}$$

We obtain the following for the sum of derivatives $(\partial\lambda_{k'}/\partial u_{12}) + (\partial\lambda_k''/\partial u_{12})$ in the first case:

$$(2 - \delta_{12})(l_{1k'}l_{2k'} + l_{1k''}l_{2k''}) \tag{7.59}$$

whilst for the derivative with respect to the diagonal element we get $(l_{1k'})^2 + (l_{2k''})^2$. In the second case we obtain the following for the same sum:

$$\begin{aligned} &(2 - \delta_{12})(l_{1k'}\alpha_{11} + l_{1k''}\alpha_{21})(l_{2k'}\alpha_{11} + l_{2k''}\alpha_{21}) \\ &\quad + (l_{1k'}\alpha_{12} + l_{1k''}\alpha_{22})(l_{2k'}\alpha_{12} + l_{2k''}\alpha_{22}) \\ &= (2 - \delta_{12})l_{1k'}l_{2k'}(\alpha_{11}^2 + \alpha_{12}^2) + l_{1k''}l_{2k'}(\alpha_{11}\alpha_{21} + \alpha_{12}\alpha_{22}) \\ &\quad + l_{1k'}l_{2k''}(\alpha_{11}\alpha_{21} + \alpha_{12}\alpha_{22}) + l_{1k''}l_{2k''}(\alpha_{21}^2 + \alpha_{22}^2) \end{aligned} \tag{7.60}$$

Since the matrix in equation (7.57) is orthogonal, and

$$\alpha_{11}^2 + \alpha_{12}^2 = \alpha_{21}^2 + \alpha_{22}^2 = 1, \quad \alpha_{11}\alpha_{21} + \alpha_{12}\alpha_{22} = 0 \tag{7.61}$$

Finally, the derivative sum $(\partial\lambda_{k'}/\partial x) + (\partial\lambda_{k''}/\partial x)$ also equals $(2 - \delta_{12}) \times (l_{1k'}l_{2k'} + l_{1k''}l_{2k''})$ in the second case, thus proving the assertion. It is not too difficult to widen this proof to the general case of an arbitrary multiplicity of degeneracy.

In order to verify the calculation of a derivative the following clear relationship is useful:

$$\sum_k \frac{\partial\lambda_k}{\partial x} = \frac{\partial}{\partial x}(SpD) = Sp\left(\frac{\partial D}{\partial x}\right) \tag{7.62}$$

which follows directly from $\sum_k \lambda_k = SpD$. Since a $(\partial\lambda_k/\partial u_{jj})$ derivative, which is with respect to a diagonal element of the U matrix, is simple $(l_{jk})^2$ in independent co-ordinates, it must therefore always be positive. This means that if the diagonal elements of U increase, all the frequencies may only grow or, in any case, not decrease (if $l_{jk} = 0$). However, the derivatives $(\partial\lambda_k/\partial u_{ij}) = 2l_{ik}l_{jk}$ may be either positive or negative. Thus an increase in U_{ij} may increase or decrease the frequencies.

When calculating partial derivatives of the squares of vibrational frequencies with respect to parameters on which only the elements of the T matrix depend, the following should be borne in mind. If the mass of an atom is a differentiation parameter (or, more simply, ε_i is chosen as the parameter), then

$$\frac{\partial T}{\partial\varepsilon_i} = \vec{B}\frac{\partial E}{\partial\varepsilon_i}\tilde{\vec{B}} = \vec{B}\frac{\partial}{\partial\varepsilon_i}[\varepsilon_i\delta_{ik}]\tilde{\vec{B}} \tag{7.63}$$

Since the matrix $(\partial E/\partial \varepsilon_i)$ is a diagonal one each of whose elements are zero except for the ith element, which is unity, the product $\vec{B}(\partial E/\partial \varepsilon_i)/\tilde{\vec{B}}$ will be a matrix whose elements are formed as if they were the scalar products of the ith column of the $\vec{B}$ matrix and the ith row of the $\tilde{\vec{B}}$ matrix. The diagonal elements of this matrix will be the squares of the corresponding elements of the ith column of the $\vec{B}$ matrix. In the case of independent co-ordinates we have

$$\frac{\partial \lambda_k}{\partial \varepsilon_i} = \lambda_k \tilde{l}_p^{(k)} \frac{\partial T}{\partial \varepsilon_i} l_p^{(k)} = \lambda_k \tilde{l}_p^{(k)} \vec{B}_k \frac{\partial E}{\partial \varepsilon_i} \tilde{\vec{B}}_k l_p^{(k)} \tag{7.64}$$

However, on the other hand, $\vec{p} = \tilde{\vec{B}}p$, where p is the column-matrix of the momenta conjugate to the natural co-ordinates and $\vec{p}$ is the column-matrix of the atom momenta in Cartesian ones. From here we find that $\tilde{\vec{B}} l_p^{(k)} = l_{\vec{p}}^{(k)}$, where $l_{\vec{p}}^{(k)}$ is an eigenvector of the L_p matrix that transforms normal momenta $P\,(\vec{p} = L_{\vec{p}}P)$ to Cartesian atom momenta.

Consequently

$$\frac{\partial \lambda_k}{\partial \varepsilon_i} = \lambda_k \tilde{l}_{\vec{p}}^{(k)} \frac{\partial E}{\partial \varepsilon_i} l_{\vec{p}}^{(k)} = \lambda_k (l_{\vec{p}ki})^2 \tag{7.65}$$

Hence it would seem that $(\partial \lambda_k/\partial \varepsilon_i) > 0$ always, i.e. given an increase in ε_i or a decrease in m_i, the frequency of a vibration may only grow, or at least not decrease (if $(l_{\vec{p}})_{ki} = 0$, i.e. the ith atom is not displaced in the given normal vibration).

In order to calculate a partial derivative $(\partial \lambda_k/\partial x)$ with respect to a geometrical parameter it is useful to take advantage of the formula

$$\frac{\partial T}{\partial x} = \frac{\partial \vec{B}}{\partial x} E \tilde{\vec{B}} + \vec{B} E \frac{\partial \tilde{\vec{B}}}{\partial x} \tag{7.66}$$

7.5 Relationships between the Squares of Vibrational Frequencies in Isotopic Molecules

When studying polyatomic molecules it is now usual for investigators to consider the spectra of isotopically substituted molecules as well as those of the basic compounds. The use of isotopically substituted species makes it much easier to assign reliable characteristic vibrational frequencies to vibrations associated with particular atomic groupings, and, more important, to obtain additional equations for calculating force constants on the basis of experimental data.

In order to assign vibrational frequencies it is more advantageous to have isotopic substitution for which the change of mass is not too large (for example, O^{16} for O^{18}, C^{12} for C^{14}, etc.). This is because, on the one hand, the frequency shifts thus obtained are easily registered (of the order of 10–20 cm^{-1}) and, on the other, they do not substantially distort the form of the spectrum, which makes it easier to match bands from two isotopic modifications. However, in order to obtain additional equations with which to determine force constants it is better to substitute atoms with isotopes that are significantly different in mass (for example, deuterium for hydrogen). In this case the vibrations are substantially changed and extra independent equations can be obtained.

We have a number of comments to make here. First, it is clear that if the masses of all the molecule's atoms are increased proportionately then, because the matrix $T = \vec{B}E\tilde{\vec{B}}$ is divided by a constant factor, this mass change will result in a proportionate reduction of the vibrational frequencies. In Section 7.4 it was shown by a study of the $(\partial\lambda_k/\partial\varepsilon_i)$ derivative that all these derivatives are positive or zero (the latter being the special case when the ith atom, in the appropriate normal co-ordinate, is not displaced from its equilibrium position). Consequently when the mass of one atom is increased all the vibration frequencies are reduced, or at least remain unchanged.

In order to estimate frequency shifts due to an isotopic substitution with a small change in the masses of the atoms we can calculate the $(\partial\lambda_k/\partial\varepsilon_i)$ derivative when the T and U matrices of the molecule are known. However, there are a number of other relations between the squares of the vibration frequencies $v_k^2 = \lambda_k$ that do not depend on the U matrix and so can be used when the U matrix is unknown. These relations also show that not every absorption band detected for isotopically substituted molecules can be used to calculate force constants, i.e. are useful in solving the inverse spectral problem.

The relationships are corollaries of those that exist between different combinations of the eigenvalues λ_k and the elements of the matrix $D = TU$. Thus according to a theorem concerning the leading minors of a matrix and its eigenvalues we have

$$\sum_k \prod_m \lambda_m = \sum_k M_k^{(m)} \tag{7.67}$$

Here, $M_k^{(m)}$ is a leading minor of order m of the D matrix and is the product of m of the matrix's eigenvalues (the number of multiplicands in the products equalling the order of the minor, and the number k of different products corresponding to all the possible sets of n from m elements).

In the general case the first and last relations of equation (7.67) between $\sum_k M_k^{(m)}$ and $\prod_k \lambda_k$ have the simple form:

$$SpD = \sum_k \lambda_k \quad \text{and} \quad \det D = \prod_k \lambda_k \tag{7.68}$$

Here $\det D$ is the determinant of D. Since D is equal to the product of the force constants and kinematic matrices, its determinant will be the product of those of these two matrices, i.e.

$$\det D = \det T \cdot \det U \tag{7.69}$$

Consequently

$$\prod_k \lambda_k = \det T \cdot \det U \tag{7.70}$$

By analogy, we obtain for isotopically substituted molecules:

$$\det D^* = \det T^* \det U = \prod_k \lambda_k^*$$

from which

$$\left(\prod_k \lambda_k\right) \det T^* = \left(\prod_k \overset{*}{\lambda}_k\right) \det T \tag{7.71}$$

This equation is the product rule for the frequencies of isotopic molecules, and it is usually used to verify the assignment to fundamental modes of observed absorption frequency bands in normal and isotopic molecules. In addition, it enables us to calculate a missing frequency of a normal vibration of a pair of isotopically substituted molecules if the remaining frequencies are known. It can also be used to verify calculations when the vibration frequencies are estimated.

The product rule can be written in a somewhat different form containing the molecule's principal moment of inertia. However, this form is less convenient for practical calculations, especially when using a computer.

We shall now consider the matrix E, which contains the reciprocals of the atomic masses for a molecule. Each atom has a certain number corresponding to the number of the ith element in E. Obviously, this number remains unchanged after an isotopic substitution. When an atom is substituted for by one of its isotopes we shall obtain a different E matrix, and, moreover, the ε_i element corresponding to the atom's number must also be changed. If the molecule has a number of equivalent atoms, then when each of the atoms in this set is substituted for successively by an isotope we shall obtain a series of E matrices. These differ from to each other only by the successive changes in the ε_i values situated in the positions corresponding to the number of this group of substituted atoms.

For example, if the H atoms in H_2O are numbered 1 and 2, then when hydrogen atom (1) is substituted for by a deuterium atom, we obtain an E matrix of the form

$$\begin{bmatrix} \varepsilon_D & 0 & 0 \\ 0 & \varepsilon_H & 0 \\ 0 & 0 & \varepsilon_O \end{bmatrix} \tag{7.72}$$

When hydrogen atom (2) is substituted we obtain

$$\begin{bmatrix} \varepsilon_H & 0 & 0 \\ 0 & \varepsilon_D & 0 \\ 0 & 0 & \varepsilon_O \end{bmatrix} \tag{7.73}$$

We shall further assume that we can choose the isotopically substituted molecules so that a linear combination of their matrices $E^{(m)}$ with coefficients α_m vanish, i.e. $\sum_m \alpha_m E^{(m)} = 0$. Obviously, when we construct each of these matrices we must take care that the order of their elements follows the same atom numeration for each of the isotopically substituted species. We can thus confirm that

$$\sum_m \alpha_m \left(\sum_k \lambda_k^{(m)}\right) = 0 \tag{7.74}$$

i.e. the sum of the squares of the frequencies of all the isotopically substituted molecules is zero with the α_m coefficients. To prove this, we left-multiply the equation $\sum_m \alpha_m E^{(m)} = 0$ by the matrix $\vec{B}$, the transformation matrix from Cartesian to natural or symmetry co-ordinates, and right-multiply it by $\tilde{\vec{B}}U$. Since these matrices are unchanged by isotopic substitution, we find

$$\vec{B}\left(\sum_m \alpha_m E^{(m)}\right)\cdot\tilde{\vec{B}}U = \sum_m \alpha_m(\vec{B}E^{(m)}\tilde{\vec{B}})U = \sum_m \alpha_m D^{(m)} = 0 \tag{7.75}$$

where $D^{(m)} = \vec{B}E^{(m)}\tilde{\vec{B}}U$ is the dynamic coefficient matrix for the mth isotopic modification. Then by allowing for the relation $S_p D^{(m)} = \sum_k \lambda_k^{(m)}$ and that the trace of a matrix sum is equal to the sum of the traces, we obtain

$$Sp\sum_m \alpha_m D^{(m)} = \sum_m \alpha_m SpD^{(m)} = \sum_m \alpha_m\left(\sum_k \lambda_k^{(m)}\right) \tag{7.76}$$

The rule we have just obtained is called the sum rule for the squares of frequencies, and is used in the same way as the product rule. The advantage of this sum rule is that it can be used without any calculation associated with the elements of the kinematic coefficient matrix. Its disadvantage is that for the general case of non-symmetric matrices data on a minimum of four isotopic molecules is necessary before it can be used. However, given symmetric matrices, it is possible to establish relationships between the sum of the squares of vibrational frequencies for light, heavy and intermediate modifications.

For example, we can write the following equation for the molecules H_2O, D_2O and HDO:

$$E^{H_2O} + E^{D_2O} - E^{HDO} - E^{DHO} = 0 \tag{7.77}$$

and hence, because $\sum_k \lambda_k$ must be the same for HDO and DHO, we obtain

$$\sum_k \lambda_k^{H_2O} + \sum_k \lambda_k^{D_2O} - 2\sum_k \lambda_k^{HDO} = 0$$

or

$$\sum_k \lambda_k^{H_2O} + \sum_k \lambda_k^{D_2O} = 2\sum_k \lambda_k^{HDO} \tag{7.78}$$

For methane and its deuterated modifications CD_4 and CHD_3 we have

$$E^{CH_4} + 3E^{CD_4} - E^{CHDDD} - E^{CDHDD} - E^{CDDHD} - E^{CDDDH} = 0 \tag{7.79}$$

Since the $\sum_k \lambda_k$ in the last four modifications must be the same due to symmetry, we obtain

$$\sum_k \lambda_k^{CH_4} + 3\sum_k \lambda_k^{CD_4} = 4\sum_k \lambda_k^{CHD_3} \tag{7.80}$$

Similarly, the following relationships between the squares of frequencies for other isotopic species of methane can be obtained:

$$\sum_k \lambda_k^{CH_4} + \sum_k \lambda_k^{CH_4} = 2\sum_k \lambda_k^{CH_2D_2} \tag{7.81}$$

$$3\sum_k \lambda_k^{CH_4} + \sum_k \lambda_k^{CD_4} = 4\sum_k \lambda_k^{CH_3D} \tag{7.82}$$

The existence of other relationships between combinations of the squares of frequencies of the normal vibrations of isotopically substituted molecules (relationships that are independent of the force constants) leads to a reduction in the number of independent equations that are necessary to solve the inverse spectral problem explicitly. It may, in particular, be proved that the sum of the squares of the vibrational frequencies in a symmetric molecule in which a group of equivalent atoms has been partially substituted for can be expressed as a linear combination of the squares of the frequencies of a light and a fully substituted modification.

Let us assume that the first m atoms of some molecule belonged to a group of equivalent atoms and we substitute for them isotopes of the same mass. We number all the atoms in the set in the same order in which they are moved during the molecule's symmetry operations. We compile the diagonal matrix E^{H} for the light symmetric modification, the matrix E^{D} for the modification fully substituted by isotopes and the matrices E^{HD}, which are for the partially substituted modifications and which differ from each other by the number of isotopically substituted equivalent atoms in the group. For example, for singly substituted ammonia these matrices are

$$\begin{bmatrix} \mathrm{H} & 0 & 0 & 0 \\ 0 & \mathrm{H} & 0 & 0 \\ 0 & 0 & \mathrm{H} & 0 \\ 0 & 0 & 0 & \mathrm{N} \end{bmatrix} \begin{bmatrix} \mathrm{D} & 0 & 0 & 0 \\ 0 & \mathrm{D} & 0 & 0 \\ 0 & 0 & \mathrm{D} & 0 \\ 0 & 0 & 0 & \mathrm{N} \end{bmatrix} \begin{bmatrix} \mathrm{D} & 0 & 0 & 0 \\ 0 & \mathrm{H} & 0 & 0 \\ 0 & 0 & \mathrm{H} & 0 \\ 0 & 0 & 0 & \mathrm{N} \end{bmatrix}$$

$$\begin{bmatrix} \mathrm{H} & 0 & 0 & 0 \\ 0 & \mathrm{D} & 0 & 0 \\ 0 & 0 & \mathrm{H} & 0 \\ 0 & 0 & 0 & \mathrm{H} \end{bmatrix} \begin{bmatrix} \mathrm{H} & 0 & 0 & 0 \\ 0 & \mathrm{H} & 0 & 0 \\ 0 & 0 & \mathrm{D} & 0 \\ 0 & 0 & 0 & \mathrm{N} \end{bmatrix} \tag{7.83}$$

For trans-substituted ethylene we have

$$\begin{bmatrix} \mathrm{H} & 0 & 0 & 0 & 0 & 0 \\ 0 & \mathrm{H} & 0 & 0 & 0 & 0 \\ 0 & 0 & \mathrm{H} & 0 & 0 & 0 \\ 0 & 0 & 0 & \mathrm{H} & 0 & 0 \\ 0 & 0 & 0 & 0 & \mathrm{C} & 0 \\ 0 & 0 & 0 & 0 & 0 & \mathrm{C} \end{bmatrix} \begin{bmatrix} \mathrm{D} & 0 & 0 & 0 & 0 & 0 \\ 0 & \mathrm{D} & 0 & 0 & 0 & 0 \\ 0 & 0 & \mathrm{D} & 0 & 0 & 0 \\ 0 & 0 & 0 & \mathrm{D} & 0 & 0 \\ 0 & 0 & 0 & 0 & \mathrm{C} & 0 \\ 0 & 0 & 0 & 0 & 0 & \mathrm{C} \end{bmatrix}$$

$$\begin{bmatrix} \mathrm{D} & 0 & 0 & 0 & 0 & 0 \\ 0 & \mathrm{D} & 0 & 0 & 0 & 0 \\ 0 & 0 & \mathrm{H} & 0 & 0 & 0 \\ 0 & 0 & 0 & \mathrm{H} & 0 & 0 \\ 0 & 0 & 0 & 0 & \mathrm{C} & 0 \\ 0 & 0 & 0 & 0 & 0 & \mathrm{C} \end{bmatrix} \begin{bmatrix} \mathrm{H} & 0 & 0 & 0 & 0 & 0 \\ 0 & \mathrm{H} & 0 & 0 & 0 & 0 \\ 0 & 0 & \mathrm{D} & 0 & 0 & 0 \\ 0 & 0 & 0 & \mathrm{D} & 0 & 0 \\ 0 & 0 & 0 & 0 & \mathrm{C} & 0 \\ 0 & 0 & 0 & 0 & 0 & \mathrm{C} \end{bmatrix} \tag{7.84}$$

Hence we obtain

$$2E^{NH_3} + E^{ND_3} = E^{NDH_2} + E^{NHDH} + E^{NH_2D} \tag{7.85}$$

and

$$E^{C_2H_4} + E^{C_2D_4} = E^{C_2D_2H_2} + E^{C_2H_2D_2} \tag{7.86}$$

from which

$$2\sum_k \lambda_k^{NH_3} + \sum_k \lambda_k^{ND_3} = 3\sum_k \lambda_k^{NH_2D} \tag{7.87}$$

and

$$\sum_k \lambda_k^{C_2H_4} + \sum_k \lambda_k^{C_2D_4} = 2\sum_k \lambda_k^{C_2H_2D_2} \tag{7.88}$$

If the symmetry of the set of isotopically substituted molecules is the same, then clearly the product and sum rules must be obeyed for each symmetry type of vibration individually. If the symmetries are not the same then all the vibrational frequencies of the isotopic modifications must participate in the rules.

We obtain the following from the product rule for a diatomic molecule due to a substitution of a H atom for a D atom in an XH group with $\varepsilon_H \gg \varepsilon_x$:

$$\nu_H/\nu_D = \sqrt{2} \tag{7.89}$$

and for deformational vibrations of the H–X–H group given a substitution of H by D (D–X–D) and $\varepsilon_H \gg \varepsilon_x$:

$$\nu_H/\nu_D = \sqrt{2} \tag{7.90}$$

This kind of change in a frequency, by $\sqrt{2}$, due to deuteration will only occur for a diatomic molecule or a triatomic molecule for deformation vibrations with $m_x \gg m_H$. In every other case these rules are only satisfied approximately, and are broken if the vibrations are not characteristic in form or frequency for the group. Therefore even when only one H atom is substituted for by a D atom in a polyatomic molecule, all the absorption bands might be displaced. This displacement happens most when the vibration is directly associated with that particular H atom, and commonly several frequencies are displaced as well by 20–30 cm^{-1}, and sometimes by as much as 100 cm^{-1}. In addition, deuteration can lead to a distortion of the characteristic property of the vibrations and to changes in their forms, and consequently to changes in the intensities of the corresponding absorption bands. All these phenomena substantially complicate the search for the analogues of the non-substituted modification's bands in the spectra of the deuterated compounds. It must also be taken into account that the rules we are considering are only true for small harmonic vibrations of polyatomic molecules. Hence they will only be approximately satisfied by the observed vibrational frequencies. The accuracy to which they satisfy the rules will, however, be sufficiently good for most practical purposes.

These rules are the ones most frequently needed in practice. However, other relationships exist that can be used to increase the number of independent equations available for calculating force constants for polyatomic molecules from experimental vibrational frequencies. This is called the 'inverse spectral problem', and is very important for all theories of molecular vibrational spectra.

It is considered in detail in Chapter 13. We shall limit ourselves here to a few comments concerning the use of isotopically substituted molecules.

Relations (7.67) for a natural system of equations for calculating force constants, i.e.

$$\sum_k M_k^{(m)} = \sum_k \prod_k \lambda_k^{\exp}$$

in which the experimental frequencies of the normal vibrations are substituted into the right-hand sides of the equations. The unknown force constants enter the left-hand sides of the equations as coefficients in the expressions for $M^{(m)}$. Similar equations can be compiled for all isotopic modifications, and the problem is to find out whether these equations are independent, and consequently whether they contain any extra information that can be used to calculate the force constants.

In order to solve this problem we use a Cartesian system of co-ordinates. In this case the quantity for the ith atom will only enter three diagonal elements of the matrix E. Since the matrix D_r equals the matrix product $D_{\bar{r}} = EU_r$, it follows directly that the ε_i of the atom can only be contained in the elements of three of the rows of $D_{\bar{r}}$. This property of $D_{\bar{r}}$ for a special choice of co-ordinates leads to the leading minors $M_k^{(m)}$ of the D matrix of any order being homogeneous functions of degree m in terms of ε_i. The minors must therefore be expressible as linear combinations of products of ε_i, the degree of each ε_i, moreover, being no more than three.

This means that the sum of all the leading minors $\sum_k M_k^{(m)}$ of order m of the D matrix must be written in the form

$$\sum_k M_k^{(m)} = \sum b_{\alpha_1,\alpha_2\ldots\alpha_m} \varepsilon_{\alpha_1} \varepsilon_{\alpha_2} \cdots \varepsilon_{\alpha_m} \tag{7.91}$$

where the α index cannot be repeated more than three times. Note that for a Cartesian co-ordinate system the presence of null frequencies means that the sum of leading minors whose order is greater than $3N$-6 will be zero.

Choosing one of the atoms (say, the first one) equation (7.91) can be written as

$$\sum M_k^{(m)} = \varepsilon_1^3 \phi_1^{(m)} + \varepsilon_1^2 \phi_2^{(m)} + \varepsilon_1 \phi_3^{(m)} + \phi_4^{(m)} \tag{7.92}$$

if $m \geqslant 3$. For $m = 1$ and 2 we obtain, respectively,

$$\varepsilon_1 \phi_1^{(1)} + \phi_2^{(1)} = SpD_{\bar{r}} \tag{7.93}$$

and

$$\varepsilon_1^2 \phi_1^{(2)} + \varepsilon_1 \phi_2^{(2)} + \phi_3^{(2)} = \sum_k M_k^2 \tag{7.94}$$

Here, the $\phi_k^{(m)}$ coefficients depend on the masses of the remaining atoms, the molecule's geometry, the interatomic forces and the co-ordinate system adopted. We further assume that the first atom is subsequently substituted for by an isotope. Then, for three isotopes, for example, we obtain:

$$\varepsilon_1\phi_1^{(1)} + \phi_2^{(1)} = SpD_{\bar{r}} = \sum_k \lambda_k$$
$$\varepsilon_1^*\phi_1^{(1)} + \phi_2^{(1)} = SpD_{\bar{r}}^* = \sum_k \lambda_k^* \tag{7.95}$$
$$\varepsilon_1^{**}\phi_1^{(1)} + \phi_2^{(1)} = SpD_{\bar{r}}^{**} = \sum \lambda_k^{**}$$

It is obvious that by considering the first two relations to be a system of non-homogeneous equations with respect to the unknown $\phi_1^{(1)}$ and $\phi_2^{(1)}$ we can find these unknowns. By substituting then into the third equations we can express $\sum_k \lambda_k^{**}$ in terms of $\sum_k \lambda_k$ and $\sum_k \lambda_k^*$. It is not difficult to see that we then obtain

$$(\varepsilon_1^* - \varepsilon_1^{**})\sum_k \lambda_k + (\varepsilon_1^{**} - \varepsilon_1)\sum_k \lambda_k^* = (\varepsilon_1^* - \varepsilon_1)\sum_k \lambda_k^{**} \tag{7.96}$$

Therefore the $\sum_k \lambda_k^{**}$ sum for the third isotopic modification is not an independent parameter, and the corresponding equation cannot be used as extra information to calculate the force constants.

In an analogous way it is found that any fourth equation of the form

$$\varepsilon_1^2\phi_1^{(2)} + \varepsilon_1\phi_2^{(2)} + \phi_3^{(2)} = \sum_{i>j} \lambda_i\lambda_j \tag{7.97}$$

is not independent, and hence has no additional information about the force constants. There must be a linear relationship between the $\sum_{i,j}\lambda_i\lambda_j$ sums for four isotopic modifications. The independent equations constructed from

$$\sum \underbrace{\lambda \cdots \lambda}_{m} = \varepsilon_1^3\phi_1^{(m)} + \varepsilon_1^2\phi_2^{(m)} + \varepsilon_1\phi_3^{(m)} + \phi_4^{(m)} \tag{7.98}$$

for any m higher than three, where m is the rank of the $D_{\bar{r}}$ matrix, are exhausted by the time the fourth equation is reached. Further substitutions of the selected atom by isotopes will therefore not enable us to obtain any new information about the force constants. There must be linear relationships between all these sums of the products of the squares of the frequencies for the vibrations of five isotopic modifications.

Obviously, in practice the substitution of one atom by a large number of different isotopes is only possible for relatively large atoms, and this in itself puts certain limits on the solutions of inverse spectral problems.

However, the argument above does become important for the successive substitution of hydrogen by deuterium and tritium. For example, when solving the problem concerning the stretching vibrations of HCN using the isotopic modifications DCN and TCN, the frequencies of the latter carry no extra information on the force constants. The equation det $D_{\bar{r}} = \lambda_1\lambda_2$ for both DCN and TCN only differ from the corresponding equation for HCN by a constant, whilst the equation $SpD_{\bar{r}} = \lambda_1 + \lambda_2$ for TCN is a corollary of the six frequency-force constant-relations for the stretching vibrations of HCN, DCN and TCH are independent. We will show in Chapter 13 that we can only obtain two systems of force constants for HCN from these equations, and then only if all the elements of the molecule's U matrix can be considered to be independent.

When a molecule that has symmetry elements it is possible to use symmetry co-ordinates. These are linear combinations of the Cartesian co-ordinates which describe the motion of an atom with respect to one of three non-coplanar axes. A symmetry co-ordinate constructed from Cartesian ones will describe the motion of a set of equivalent atoms or the motion of a centre of gravity. The matrix equation of the vibrations then reduces to a block form, and the expression $\sum b_{\alpha_1 \ldots \alpha_m} \varepsilon_{\alpha_1} \cdots \varepsilon_{\alpha_n}$ reduces to a form that does not contain ε_{α_i} products and powers of them for equivalent atoms of the same set.

Since the force field of a molecule and its geometry does not change during isotopic substitutions (or, at least, any change is so small that it is irrelevant for vibrational spectra), and since the coefficients $b_{\alpha_1 \ldots \alpha_m}$ only depend on force and geometric parameters, then the form of the representation of $\sum b_{\alpha_1 \ldots \alpha_m} \varepsilon_{\alpha_1} \cdots \varepsilon_{\alpha_m}$ must also be retained for molecules in which some of the equivalent atoms have been substituted for by isotopes if the symmetry co-ordinates of the unsubstituted molecule can be used as co-ordinates for the substituted ones. All the molecules will then contain terms that include first, second and third powers of ε_i for each group of equivalent atoms (each atom is allocated its own number that is independent of the isotopic substitution). However, there will be no terms with mixed products of ε_i, the index referring to the different atoms within a set of equivalent atoms.

As a result, if the E matrices of partially substituted molecules are expressed in terms of those of the unsubstituted and fully substituted molecules (and this is always possible), then even the matrices containing the second and third powers of ε_i instead of ε_i itself can be expressed in terms of the corresponding matrices of the light and heavy molecules. Consequently, all the sums of the squared vibrational frequencies of the partially substituted molecules can be expressed in terms of the corresponding sums belonging to the heavy and light modifications. In consequence, they are not independent, and cannot be used to obtain extra information on the force constants for the molecule.

The relations between the sums of products of the squares of frequencies of isotopic molecules are conventionally called the sum rules for products of squared frequencies. They can be deduced in a somewhat different form to that given above. Readers wishing to consult the generalized sum rules should refer to the original literature[1–8].

To conclude, we have a number of additional comments to make. All the vibrations of linear and planar molecules can be categorized as stretching or deformation (planar and non-planar) for any isotopic substitutions. All expressions such as equation (7.92) for stretching and deformation vibrations will contain only first powers of ε. Hence relationships like equation (7.96) will not only arise between the sums of the squares of the vibrational frequencies of the three isotopic substitutes but also between the sums of the products of the frequencies squared of the same molecules.

This remains true for both the sum of the squares and the sum of the products of the squares of the frequencies of the non-planar vibrations of three isotopic modifications of a planar molecule. Again all the equation (7.94)-type expressions for the non-planar vibrations of planar molecules will only contain the first powers of ε_i. The highest power of ε_i in such expressions for planar vibrations of planar molecules will be two. Consequently there must be linear relationships between the sums of products containing two or more squares of the frequencies of the planar vibrations of isotopic modifications.

When H_nX internal groups vibrate (where n is the number of hydrogen of atoms bonded to the X atom) and the mass of the X atom is significantly greater than the summed masses of all the n hydrogen atoms, the X atom remains practically stationary. Clearly, a substitution of a D atom for an H atom will then result in a proportionate shift of all the vibrational frequencies. The equations thus obtained for calculating the force constant matrix elements will all be dependent, and not provide any new force constants not already found from the light modification. These equations will have no use other than for constructing redundant equation systems to average the solutions of inverse spectral problems.

Similar situations happen so frequently in practice that using isotopically substituted molecules for determining force constant systems is not really as powerful a method as it would seem at first sight.

7.6 The Use of Dependent Co-ordinates

In many cases dependent co-ordinates are used when solving problems concerning the spectra of polyatomic molecules. This is done either for the convenience of constructing symmetry co-ordinates or by choosing a special model of the potential function of a polyatomic molecule. For example, too many co-ordinates are used in a potential function such as the Urey–Bradley potential. In the usual quadratic potential function, based on the idea of each bond and valency angle having their own potentials, excessive co-ordinates may also appear if the number of bonds and valency angles in the molecule exceeds $3N-6$ N being the number of atoms in the molecule. This phenomenon is frequently encountered in practice, especially when dealing with cyclic molecules.

We shall first consider how to express the kinetic and potential energies when a system is described by a number of co-ordinates that is larger than the number of degrees of freedom. In such cases there must be certain relationships between the equations given by a correspondent equations.

The form of the expression for kinetic energy when there are excess co-ordinates is fairly obvious. In fact, because the expression for the kinetic energy of a collection of particles is in every case initially in terms of Cartesian co-ordinates associated with the molecule, then it must be a quadratic form in terms of the other co-ordinates, which may include dependent ones, so long as the transition from Cartesian to the selected co-ordinates is accomplished via a linear transformation.

The situation is somewhat more complex for potential energy. When we were describing potential functions we mentioned that in the area around the equilibrium position (at the bottom of the potential well, of which in this case there is only one) representing the function as a quadratic form with constant coefficients is a natural approximation. This representation finds its justification in the expansion of an arbitrary type of potential function into a series in terms of vibrational co-ordinates. The disappearance of the linear terms in the expansion when independent co-ordinates are being used is related to equilibrium conditions, and, as is known, these can be formulated in different ways for dependent and independent co-ordinates. In the case of independent co-ordinates the equilibrium position is defined by the system of equations $(\partial U/\partial q_i)=0$ $(i=1,2,\ldots,3N-6)$. It follows directly from this that should vibrational co-ordinates be selected so that they vanish at the equilibrium position, the coefficients in the linear terms in the expansion of the electronic energy into a series with respect to vibration co-ordinates will also vanish.

We now assume that the definite relationships between the co-ordinates are expressible as the equations $\varphi_k(q_1,\ldots,q_k)=0$ $(k=1,\ldots,n)$. Hence equilibrium will occur at the point which satisfies the system of equations

$$\frac{\partial U}{\partial q_i}+\sum_k \eta_k \frac{\partial \varphi_k}{\partial q_i}=0 \quad \text{(for all } i) \tag{7.99}$$

$$\varphi_k=0 \quad \text{(for all } k)$$

where η_k is a Lagrange multiplier.

If again the vibrational co-ordinates q_i are selected so that they vanish at the equilibrium position, then we find from these equations that the coefficients a_i belonging to the linear terms in the expansion of the potential energy into a series with respect to vibration co-ordinates, i.e.

$$U=U_0+\sum_i a_i q_i+\tfrac{1}{2}\sum_{i,j} u_{ij} q_i q_j$$

must obey the equations

$$a_i+\sum_k \eta_k \frac{\partial \varphi_k}{\partial q_i}=0 \tag{7.100}$$

and consequently may be non-zero.

By multiplying equations (7.100) by the relevant q_i and collecting like terms together, we find that

$$\sum_i a_i q_i+\sum_k \eta_k \sum_i \frac{\partial \varphi_k}{\partial q_i} q_i=0 \tag{7.101}$$

The sum $\sum a_i q_i$ consequently vanishes only when

$$\left(\sum_i \frac{\partial \varphi_k}{\partial q_i}\right) q_i=0$$

for all k, i.e. when the φ_k functions are homogeneous (and, in particular, linear with respect to the vibrational co-ordinates).

We now consider what happens when we use small increments in geometrical parameters or total differentials as the vibrational co-ordinates when calculating a molecule's vibrations. In this case any relationship between these co-ordinates can, in principle, only be linear, since each independent co-ordinate can be linearly expressed in terms of the Cartesian ones that are associated with the molecule. That this is possible follows directly from constructing the co-ordinates as linear combinations with the vector coefficients of the atom displacements. The existence of a relationship between the co-ordinates means that there is a relationship between the unit vectors of these co-ordinates which are vectors in normal three-dimensional space, and this relationship must also be linear.

We come therefore to the next important conclusion, which is that when solving a vibrational problem in the harmonic approximation using a system of natural co-ordinates, the kinetic and potential energies can be given as quadratic forms with constant coefficients, and there are no linear terms in the expression for the potential energy.

We shall now consider the important problem of determining the kinetic and potential energy matrices when using dependent co-ordinates. The kinetic and potential energies of a system do not depend on whether dependent or independent co-ordinate systems were used to obtain them. Also, the elimination of dependent co-ordinates is not unique when there are several relationships (even linear ones) between them, in the sense that the 'excess' co-ordinates can be expressed in terms of the independent co-ordinates by a series of relationships. Then it is clear that there may be an infinite number of possible matrices for the potential and kinetic energies, i.e. matrices giving the same expression after a transition to normal co-ordinates. In this sense we can speak of an indeterminacy in the definition of the kinetic and potential energy matrices in a system of dependent co-ordinates. However, it is possible to introduce a definite, physically justified limitation on the choice of these matrices as used in the theory of the vibration of polyatomic molecules.

We naturally choose the matrix $T = \vec{B}E\tilde{\vec{B}}$ as the kinematic one. This is constructed in the same way as that of the kinematic matrix in independent co-ordinates, except that the $\vec{B}$ matrix contains, in addition to rows corresponding to independent co-ordinates, rows expressing 'excess' co-ordinates in terms of the displacements of atoms from their equilibrium positions. At the same time, the rows corresponding to independent co-ordinates will not be changed by the introduction of the 'extra' co-ordinates. This is because the change of a geometric parameter of the molecule (or a combination of them) that has been taken as the vibrational co-ordinate is expressed, as we will see in Chapter 8, in terms of atomic displacements from their equilibrium positions in Cartesian co-ordinates, and in a way that is natural and completely independent of the remaining co-ordinates. As a result, the kinematic matrix that we construct in this way will differ from that for all independent systems of co-ordinates by the addition of rows and columns corresponding to the introduction of the 'extra' co-ordinates. Since there is the same relationship between the rows and columns of the $\vec{B}$ matrix as there is between the co-ordinates, this relationship must also exist between the rows and columns of the T matrix that is equal to the product $\vec{B}E\tilde{\vec{B}}$.

Hence when the 'extra' co-ordinates are introduced, the additions to the kinematic matrix reduce to the addition of rows and columns to the submatrix T that remains unchanged for the set of independent co-ordinates, the additional rows and columns being combinations of the independent ones. Elimination can be accomplished simply by striking out those rows and columns of the kinematic matrix that correspond to the 'extra' co-ordinates.

During practical calculations the kinematic matrix is always compiled initially for a system of normal co-ordinates, and then later it is transformed to a new co-ordinate system, depending on the demands of the problem. The property we mentioned above (i.e. that the linear relationship between the rows and columns in the original kinematic matrix is the same as that between the natural co-ordinates themselves) enables us to verify the compilation of the matrix if the relationship is known or can be found by the eigenvectors corresponding to the matrices' zero eigenvalues. The elements of these eigenvectors are exactly those coefficients which would be contained in a linear combination of co-ordinates that equalled zero.

The situation is more complicated when choosing a definite form for the force constant matrix, since a specific model of the interatomic forces must be given in this case. The potential energy of this system will be the sum of those of its separate elements, given appropriate deformations. In the general case of a polyatomic molecule, the transition from the force constant matrix for dependent co-ordinates, the form of which is wholly determined by the model of the interatomic forces, to the force constant matrix for independent co-ordinates is accomplished according to the rules for transforming matrices during a change of variables in a quadratic form.

Let us assume that dependent co-ordinates had been used for the original matrix. We can thus divide the set of co-ordinates into two groups q^{I} and q^{II}. We shall, moreover, assume that the q^{II} co-ordinates can be expressed in terms of the q^{I} co-ordinates via linear relationships, the elements of which form the matrix A, i.e. $q^{\mathrm{II}} = Aq^{\mathrm{I}}$. Hence the column of co-ordinates

$$\begin{bmatrix} q^{\mathrm{I}} \\ q^{\mathrm{II}} \end{bmatrix}$$

can be expressed in terms of columns of the q^{I} co-ordinates (the independent co-ordinates) via a matrix relationship that contains a rectangular matrix:

$$\begin{bmatrix} q^{\mathrm{I}} \\ q^{\mathrm{II}} \end{bmatrix} = \begin{bmatrix} I \\ A \end{bmatrix} q^{\mathrm{I}}$$

By writing the force constant matrix for the q^{I} and q^{II} co-ordinates as:

$$U = \begin{bmatrix} U_{11} & U_{12} \\ U_{21} & U_{22} \end{bmatrix} \tag{7.102}$$

and transforming to the q^1 variables using the rule

$$[I, \tilde{A}] \begin{bmatrix} U_{11} & U_{12} \\ U_{21} & U_{22} \end{bmatrix} \begin{bmatrix} I \\ A \end{bmatrix} \tag{7.103}$$

we find that the force constant matrix for independent co-ordinates U_0 will be expressed in terms of submatrices U_{ij} thus:

$$U_0 = U_{11} + U_{12}A + \tilde{A}U_{21} + \tilde{A}U_{22}A \tag{7.104}$$

If only one co-ordinate is excluded then the A matrix is changed by a row with elements a_i. This transformation reduces to the addition of the excluded column of the original dependent-co-ordinate matrix to all the remaining columns with a elements, and then the analogous operation on the rows of the matrix obtained. In the resultant matrix the row and column of the excluded co-ordinate must then be deleted.

This method of excluding dependent co-ordinates is not unique. We can clearly take advantage of new ones that are formed as independent linear combinations of all the molecule's original natural co-ordinates. Such combinations are automatically constructed using a method that is based on the property mentioned above concerning the linear relationship between the rows and columns of a kinematic matrix of the form $\vec{B}E\tilde{\vec{B}}$, where the $\vec{B}$ matrix is constructed bearing in mind all the natural co-ordinates of the molecule that are being used.

Note that since only the elements of the U matrix from the solution of the inverse spectral problem can be defined in independent co-ordinates, the force constants in a system of dependent co-ordinates may only be found if the number of constants to be determined in the selected model is no larger than the number of U matrix elements that are, in principle, calculable for the given conditions in the system of independent co-ordinates, and if equation (7.106) yields a system of equations sufficient to do so.

As mentioned earlier, 'extra' co-ordinates are most often introduced in order to simplify the account of a molecule's symmetry, since this is more simply solved when using symmetry co-ordinates as the initial system of natural vibrational ones. For example, in order to solve a vibrational problem fully for the ethane molecule it is sufficient to introduce only five angles out of the whole set of angles which have hydrogen atoms as vertices. However, the introduction of symmetry in this case becomes quite difficult. When, indeed, all the co-ordinates are introduced, they form two groups of equivalent co-ordinates, i.e. α_1, α_2 and α_3 (co-ordinates of the changes in the H–C–H angles) and $\beta_1, \beta_2, \beta_3$ (co-ordinates of the changes in the H–C–C angles). Symmetry co-ordinates can be compiled for each of these groups in the usual way:

$$\frac{1}{\sqrt{3}}(\alpha_1 + \alpha_2 + \alpha_3) = \alpha_1^{(s)} \quad \frac{1}{\sqrt{2}}(\alpha_1 - \alpha_2) = \alpha_2^{(s)} \tag{7.105}$$

etc.

After transforming to these co-ordinates the T and U matrices are changed into a quasi-diagonal form. At the same time there is the relationship between the co-ordinates, i.e. $\alpha_1^{(s)} + \beta_1^{(s)} = 0$ (see below). This relationship can be used to eliminate the co-ordinate in the appropriate blocks of the T and U matrices using the rule described above.

In order to transfer to independent co-ordinates we need to know the A

matrix, by which the 'extra' co-ordinates are expressed in terms of the independent ones. The matrix can easily be found if the linear combinations relating the selected vibrational co-ordinates together have been determined. Since the relationship between the co-ordinates must be the same as that between the rows and columns of the kinematic matrix (if it has been constructed as described above), then, as we have already demonstrated, finding these relationships can be reduced to a search for the system of the fundamental solutions of the corresponding homogeneous equation system or to a calculation of the eigenvalues and eigenvectors of the matrix $T = \vec{B}E\tilde{\vec{B}}$. Since the relationship between the co-ordinates must not be related to the masses of the atoms, the T matrix can, in these calculations, be replaced by the matrix $(\vec{B}\tilde{\vec{B}})$.

This method makes it possible to look for the relationships between the co-ordinates both before and after the transition to the symmetry co-ordinates, and can easily be automated. It is worth noting in this respect that if there are several linear relationships between the co-ordinates, then any linear combination of these relationships is also a linear relationship. Hence in such a case the A matrix will not have an unique form.

In a few very simple cases the relationships between the co-ordinates can be found using some simple approaches. These reduce to a standard form, knowledge of which is, as a rule, enough to solve some concrete problems (for example, associated with the estimation of the spectra of organic molecules). These standard relationships are most simply found for natural co-ordinates. The link between the symmetry co-ordinates can then be easily be established if the C matrix is given (the matrix that expresses the natural co-ordinates in terms of symmetry ones).

We shall now consider an example in which there is more than one relationship between the natural co-ordinates—the ethane molecule. This has the following relationships between the angle co-ordinates of the changes in the H–C–H-bond angles (α_i) and the H–C–C-bond angles (β_i), the vertices being the hydrogen atoms:

$$\sum_i^3 \alpha_i + \sum_i^3 \beta_i = 0$$

$$\sum_i^3 \alpha'_i + \sum_i^3 \beta'_i = 0 \tag{7.106}$$

It is very clear that the overall number of different relationships between the symmetry co-ordinates must be equal to that of linearly independent relationships between the natural co-ordinates. It can be shown that the coefficients in one of the rows of the A matrix are the same as those of symmetry for the same co-ordinates in the expressions for the symmetry ones. Hence these symmetry co-ordinates will, by definition, be zero and can thus generally be disregarded.

There are three standard relationships between natural co-ordinates that are most commonly satisfied. Suppose we have a planar element that consists of three bonds meeting at a single vertex. Since the sum of the angles between these bonds is always 2π during a planar deformation of the molecule, the following relationship between the angle co-ordinates (changes in the angles) must always exist:

$$\sum_{i}^{3} \alpha_i = 0 \tag{7.107}$$

Now let us consider the case for four bonds meeting at a single vertex and forming various tetrahedral angles. We introduce six angle co-ordinates: α_1, α_2, α_3, α_4, α_5 and α_6. By compiling the $\vec{B}$ matrix for them and adding their rows together, it can be shown that they sum to zero. Since the relationship that exists between the rows of this matrix must also exist between the co-ordinates, we can conclude that

$$\sum_{i}^{6} \alpha_i = 0 \tag{7.108}$$

Thus we obtain sums of the vectors $\vec{f}_{ij}$, which are perpendicular to a single bond (and so lie in the same plane) and which are at 120° to each other, this being a consequence of the angles between all the bonds being equal.

In general, whenever there are four or more non-coplanar bonds meeting at a single atom there are definite relationships between the valence-angle co-ordinates which describe the mutual displacement of the bonds. This is because the orientating vectors of each bond $\vec{e}_i, i \geqslant 4$, can be expressed as a linear combination of the first three non-coplanar vectors.

We can act in an analogous way for the more complicated case when the equilibrium angles have other values and are not equal.

For cyclic molecules we have a clear relationship for the internal angles, $\sum_i \gamma_i = 0$. In order to deduce additional relationships between changes in bond lengths and those in angles we must adapt the cyclical condition $\sum_i (s_i \vec{e}_i) = 0$, where s_i is a bond length and $\vec{e}_i$ is its orientating vector. By differentiating this equation and expressing the differentials thus obtained in terms of natural co-ordinates we can find the desired relationship.

Although the relationships between the co-ordinates can always be found by applying various artificial approaches, these operations are frequently very cumbersome, and, more importantly, they are peculiar to each group of co-ordinates. For this reason, when the calculations are done by computer none of these approaches are used because it is too difficult to cast them into algorithms. The search for dependent co-ordinates is therefore accomplished by diagonalizing the T matrix. This operation can easily be carried out, and enables the problem to be solved reliably even when there are a large number of dependent co-ordinates. Thus, for example, if the changes in all the bonds and angles are introduced as co-ordinates for the adamantane molecule, 48 dependent coordinates are obtained. Obviously, it is not possible to find them all by any other method.

We shall consider the calculation algorithm in more detail. As has been said, when linearly dependent co-ordinates are used, any linear relationship that may exist between the set of rows and columns of the T matrix must also exist between the co-ordinates.

Therefore matrix T (kinematic coefficients) is singular and the number of T-matrix eigenvalues with zero values corresponds to that of dependent co-ordinates in the original set. By transposing rows and columns it is always possible to make zero eigenvalues occur last.

A non-singular orthogonal matrix $L^{(T)}$ which makes the T matrix diagonal,

$$\begin{bmatrix} \Lambda^{(T)} & 0 \\ 0 & 0 \end{bmatrix}$$

can be found as usual by the Jacobi method. This matrix can be divided into two parts: the first will include those columns of Λ that correspond to non-zero eigenvalues of matrix T and the second part of $L^{(T)}$ has columns which correspond to zero eigenvalues. Denote the first block (rectangular matrix) by symbol L^{indep}. The remaining columns of the $L^{(T)}$ matrix, denoted by L^{dep}, form a rectangular matrix whose columns include coefficients of the co-ordinates q_i in linear relationship between them. If linear combinations of all initial co-ordinates q_i, constructed in accordance with matrix equality $q' = \tilde{L}^{\text{indep}} q$, are taken as new co-ordinates q_i', then the matrix T is evidently transformed into the form $\Lambda^{(T)}$. It will include only diagonal non-zero elements. All new co-ordinates q_i' obtained in this way will be independent.

Let us write matrix $L^{(T)}$ in the form $L^{(T)} = [L^{\text{indep}}, L^{\text{dep}}]$. After diagonalization of matrix T we will have for the kinetic energy

$$T_{\text{kin}} = 1/2[\tilde{p}^{\text{indep}}, \tilde{p}^{\text{dep}}] \begin{bmatrix} \Lambda^{(T)} & 0 \\ 0 & 0 \end{bmatrix} \begin{bmatrix} p^{\text{indep}} \\ p^{\text{dep}} \end{bmatrix} = 1/2 \tilde{p}^{\text{indep}} \Lambda^{(T)} p^{\text{indep}}.$$

As matrix $L^{(T)}$ is orthogonal and non-singular the corresponding transformation of co-ordinates will be

$$q = [L^{\text{indep}}, L^{\text{dep}}] q'$$

or

$$q' = \begin{bmatrix} L^{\text{indep}} \\ L^{\text{dep}} \end{bmatrix} q$$

However, the elements of columns L^{dep} of matrix $L^{(T)}$ which correspond to zero eigenvalues represent coefficients in linear combinations between the initial co-ordinates. This means that the product $L^{\text{dep}} q$ will be zero. Thus there are zero co-ordinates among the q' set. When transforming to co-ordinates $q_i'^{\text{indep}} \neq 0$, in accordance with the transformation $\tilde{L}^{(T)} U L^{(T)} = U'$ blocks of the matrix U' which correspond to zero co-ordinates q' will be multiplied by zero, and there will not be any input into the potential energy expression. So we have $V_{\text{pot}} = 1/2 \tilde{q}'^{\text{ind}} (\tilde{L}^{\text{ind}} U L^{\text{ind}}) q'^{\text{ind}}$.

The further transformation of matrices T' and U' to normal co-ordinates will be made as in the case of independent co-ordinates. (We have discussed this problem above.) As a result, matrices L_q and L_p for the total co-ordinates and momenta transformations $q = L_q Q$ and $p = L_p P$ will be found. Matrices L_q and L_p will be rectangular, in which the number of rows will be equal that of the initial co-ordinates, and the number of columns is equal to that of the degrees of freedom of vibrations for the given type of symmetry. This method of solution can be easy done by computers, and it enables us to determine not only the frequencies or normal modes of vibration whilst using dependent initial co-ordinates but also to lower the order of matrices for diagonalization by transforming to independent (linear combinations of initial co-ordinates) co-ordinates at the second stage.

In dependent co-ordinates matrix L_q will be

$$L_q = L^{(T)\text{indep}} \Lambda^{(T)1/2}(\neq 0) L^{(U)} \tag{7.109}$$

where $L^{(T)\text{ind}}$ is the rectangular matrix described above which contains only columns of the total matrix $L^{(T)}$ corresponding to non-zero eigenvalues of the T matrix, $\Lambda^{(T)1/2}$ is a diagonal matrix of square roots of non-zero eigenvalues of the T matrix and $L^{(U)}$ is a square orthogonal matrix which diagonalizes matrix $U'' = \Lambda^{(T)1/2}_{(\neq 0)} \tilde{L}^{(T)\text{ind}} U L^{(T)\text{indep}} \Lambda^{(T)1/2}_{(\neq 0)}$.

The corresponding rectangular matrix for momenta transformation is

$$L_p = L^{(T)\text{indep}} \Lambda^{(T)-1/2}(\neq 0) L^{(U)} \tag{7.110}$$

It is not difficult to prove that formal use of the algorithm of one-to-one diagonalization of T and U matrices without any changes in the case of dependent co-ordinates reveals vibrations of zero frequency and the corresponding normal modes.

There is the correlation $\tilde{L}_p L_q = I$ for the case of dependent co-ordinates, but already $\tilde{L}_q L_p \neq I$. Also the equality $L_q \tilde{L}_q = T$ does not exist.

When solving the problem in terms of dependent co-ordinates we can also use the equation in the form $(D - \lambda I)q = 0$, where the matrix $D = TU$ is obtained by multiplying the T and U matrices written out in the system of initial co-ordinates, including the dependent ones.

To prove this, let us return to page 182. For the independent co-ordinates q^{I} we have

$$(T^{\text{I}} U^{\text{I}} - \lambda I) q^{\text{I}} = 0 \tag{7.111}$$

We have also

$$\begin{bmatrix} q^{\text{I}} \\ q^{\text{II}} \end{bmatrix} = \begin{bmatrix} I \\ A \end{bmatrix} q^{\text{I}} = A' q^{\text{I}}$$

and $U^{\text{I}} = \tilde{A}' U^{\text{I,II}} A'$, where $U^{\text{I,II}}$ is a force constant matrix for the initial set of co-ordinates q^{I} and q^{II}. Thus we can write $(T^{\text{I}} \tilde{A}' U^{\text{I,II}} A' - \lambda I) q^{\text{I}} = 0$.

If this expression is multiplied by A' to the left and after taking into account that $A' T^{\text{I}} \tilde{A}' = T^{\text{I,II}}$, the following is obtained:

$$(A' T^{\text{I}} \tilde{A}' U^{\text{I,II}} A' - \lambda A' I) q^{\text{I}} = (T^{\text{I,II}} U^{\text{I,II}} - \lambda I) \begin{bmatrix} q^{\text{I}} \\ q^{\text{II}} \end{bmatrix} = 0$$

Since this equality is obtained from equation (7.111) for independent co-ordinates by means of simple substitutions it will have the same non-zero eigenvalues. The corresponding eigenvectors must have some additional elements which are obtained as linear combinations of the elements for independent co-ordinates. We must have also some zero solutions and the number of these would be equal to the number of dependent co-ordinates

For additional information about general problems of harmonic vibrations investigations the reader showed consult references 9–16.

References

1. L. M. Sverdlov. *Doklady USSR Acad. Sci.*, **86**, 513 (1952) (in Russian).
2. L. M. Sverdlov. *Opt. and Spectr.*, **8**, 36 (1960) (in Russian).
3. S. Brodersen, A. Langseth. *J. Molec. Spectr.*, **3**, 114 (1959).
4. S. Brodersen. *J. Molec. Struct.*, **3**, 450 (1959).
5. W. J. Lehman. *J. Molec. Spectr.*, **7**, 261 (1961).
6. J. Heicklen. *J. Chem. Phys.*, **36**, 721 (1962).
7. K. Machida. *J. Chem. Phys.*, **38**, 1360 (1963).
8. A. J. Barbes and W. J. Orville-Thomas (eds). *Vibrational Spectroscopy—Modern Trends*, Elsevier, Amsterdam/Oxford/New-York (1977), Ch. 12.
9. M. V. Volkenshtein, L. A. Gribov, M. A. Elyashevich and B. I. Stepanov. *Vibrations of Molecules*, 2nd edn, Nauka, Physico-Mathematical Literature, Moscow (1972) (in Russian).
10. L. M. Sverdlov, M. A. Kovner and E. P. Krainov. *Vibrational Spectra of Polyatomic Molecules*, John Wiley, New York (1974).
11. E. B. Wilson, J. C. Decius and P. C. Cross. *Molecular Vibrations*, McGraw-Hill, New York (1955).
12. S. Califano. *Vibrational States*, John Wiley, London/New York (1976).
13. S. J. Cyvin. *Molecular Vibrations and Mean Square Amplitudes*, Elsevier, Amsterdam (1968).
14. J. M. Mills. *Infrared Spectroscopy and Molecular Structure*, Elsevier, Amsterdam (1963).
15. D. Steele. *Theory of Vibrational Spectroscopy*, W. B. Saunders Philadelphia/London (1971).
16. L. A. Gribov and V. A. Dementiev. *Methods and Algorithms for Calculations in the Theory of Vibrational Spectra of Molecules*, Nauka, Moscow (1981).

Chapter 8

Computation of Matrices of Transition to Curvilinear Co-ordinates and of Kinematic Coefficients

8.1 Natural Vibrational Co-ordinates and Classification of Vibrations

It was shown in Chapter 6 that computation of various characteristics of the internal motions of atoms in polyatomic molecules can be reduced, as an adiabatic approximation, to the solution of the Schroedinger equation using curvilinear co-ordinates specifically selected for the purpose. The equation can be then presented as

$$\left\{-\frac{\hbar^2}{2}\sum_{i,j} t^{1/4}\frac{\partial}{\partial q_i}\tau_{ij}(q)t^{-1/2}\frac{\partial}{\partial q_j}t^{1/4} + V(q)\right\}\Psi_v(q) = E_v\Psi_v(q)$$

where q_i are the curvilinear co-ordinates chosen to solve the problem; $\tau_{ij}(q)$ are the kinematical coefficients dependent upon the vibrational co-ordinates selected and being the elements of the symmetrical matrix $T = \vec{B}E\tilde{\vec{B}}$; t is the determinant of matrix T; and $V(q)$ is a potential function.

Matrix $\vec{B}$ provides the relationship between the Cartesian velocities of motion of the nuclei and the rates of change of the curvilinear vibrational co-ordinates (generalized velocities). Here and below the symbol E is used to denote the matrix M^{-1}. This nomenclature simplifies a number of formulae, and it is convenient to use, provided that we employ dimensionless values to represent atomic masses (for example, $\varepsilon_k = m_0/m_k$, where m_0 is some standard mass which can be taken equal to the hydrogen mass or the atomic mass unit).

It was further demonstrated in Chapter 6 that the equation for the internal motion of atoms within a molecule can also be represented as

$$\left\{-\frac{\hbar^2}{2}\sum_{i,j}\frac{\partial}{\partial q_i}\tau_{ij}(q)\frac{\partial}{\partial q_j} + \Delta T(q) + V(q)\right\}\Psi_v(q) = E_v\Psi_v(q)$$

where ΔT is an additional term containing no differential operators, being thus capable of being brought into the potential term. In some cases it can be entirely neglected. To summarize, the priority is to produce the matrices $\vec{B}$ and T and to study their properties[1].

When seeking practical solutions to the problem of molecular vibrations we are faced, first, with the question of how to choose a reasonable frame of reference which transforms the equations of motion to a form comprising these co-ordinates. Certain constraints are imposed upon the selection of the reference frame. First, in order to ensure the most clear way of describing the form of vibration, enabling us more easily to classify the normal vibration patterns, we must introduce co-ordinates which can be closely associated with individual structural elements of a molecule and with the characteristics of the equilibrium state of the molecule. It is shown below that another reason for this condition is that, because of the cumbersome nature of the numerical solution to the Schroedinger equation for molecular vibrations within the frame of a consistent quantum-mechanical approach, we are forced to employ semi-empirical theories. In these a molecule, for practical purposes, is considered to be a system consisting of separate relatively closed groups. Using semi-empirical theories it is necessary to find parameters by solving so-called 'inverse spectral problems', or we can transfer parameters of known values from other molecules possessing the same structural elements. This means that we are forced from the beginning to use such structural elements, and also to utilize the properties of the electron shells related to these structural groups in order to derive simple potential functions.

Further, it is worth noting that in solving anharmonic problems and those involving internal rotation we have to take into account the dependence of the elements of the matrices of kinematic coefficients upon vibrational co-ordinates. To simplify solutions of all the subsequent problems, it is necessary, first, to express these coefficients via the generalized co-ordinates in the simplest and most clear way. Because of this, it becomes inconvenient, in particular, to use as the generalized co-ordinates those which do not offer a simple physical meaning.

It has become common practice to specify the equilibrium state of a polyatomic by using equilibrium bond lengths, valence angles and angles between different groups. This convention leads to the usual structural formulae and spatial models of molecules. Furthermore, general approaches and ideas elaborated in structural chemistry prove most useful in studying internal motions within molecules. That is, such motions are accounted for only by variations in the equilibrium configuration of the system. It seems that the minimum number of parameters describing the equilibrium configuration of a molecule must be equal to the number of internal degrees of freedom, i.e. $3N-6$ for a non-linear molecule and $3N-5$ for a linear one. A certain degree of flexibility is available in choosing a set of independent variables. For example, we can choose as independent parameters either bond lengths and valence angles or bond lengths and distances between unbound atoms. In the case of many chemical compounds (especially organic) it is more convenient to describe a molecule through a set of bond lengths and angles. Such an approach is in accord with the up-to-date concepts of quantum chemistry.

The equilibrium configuration of a molecule is disturbed by vibration. It appears here that, in exact conformity with the description of the equilibrium state, any instantaneous configurations of the system can be characterized by a set of instantaneous bond lengths and valence angles, dihedral angles, etc. The generalized velocities will then represent the rate of change of the bond lengths, valence angles and dihedral angles between the planes, etc. Clearly, these have an exact and understandable physical meaning.

An important note should be made here. If the potential surface of a molecule is such that an appreciable minimum is found along all or part of the co-ordinates, then the motion of atoms, with rather low energies, will take place in the proximity of these minima. In this case we can speak of an equilibrium geometry within the molecule (at least for some of the variables taken as co-ordinates). Then it is convenient to take as vibrational co-ordinates not the instantaneous values of geometrical parameters but the increments in these as compared with the equilibrium values. In the case of the equilibrium configuration they are, of course, reduced to zero. This results, among other things, in a simpler form of the potential functions. To mark out such co-ordinates we shall call them absolute term natural co-ordinates. It is clear that generalized velocities of variations in the absolute values of the geometrical parameters, measured from some initial value, will be the same, and, moreover, they will be expressed in the same way through the Cartesian velocities of atomic motion within a molecule, as presented in the laboratory reference frame. For this reason, this aspect will not be mentioned again, implying that we will use natural vibrational co-ordinates in those cases where, under the variation of some of the geometrical characteristics of a molecule, the potential surface is passing through a minimum, which is fairly deep for the vibration occurring in its proximity. For those parameters for which the cross-section of the potential surface has no minima (or if they are not deep enough) we will just use the instantaneous absolute values of the angles, calling such co-ordinates the internal co-ordinates. In particular, such a situation occurs if internal rotations are present within the system.

Further on, bond tension co-ordinates are denoted by q_i, Q_i(Q for tension of bonds C—C, C=C and some others). These notations should not be confused with those for the arbitrary natural co-ordinates used in solving problems of a general character, or with the notations for the normal co-ordinates which will be given in the following chapters using the same symbols.

Another group of co-ordinates are the changes in the valence angles, which are denoted by γ_{ik}, α_{ik}, β_{ik}, etc.

To characterize vibration in the case of the simplest non-planar molecules it suffices to know variations, q, of bond lengths, and changes γ in the angles of type Y–X–Z, where X, Y and Z are neighbouring atoms. The same applies equally to vibrations of planar molecules, where vibrations occur within the plane of the molecule. In the case of more complex non-planar molecules, and for non-planar vibration of planar molecules, it becomes insufficient to have co-ordinates q and γ which define variations in the location of the adjacent atoms. In these cases we have also to introduce natural co-ordinates of other types. Sometimes, separate parts of complex molecules can turn with respect to one another. For example, the two CH_3 groups of ethane can rotate around the C—C bond. Such rotations can be specified by establishing the changes in angles χ, which define the mutual position of definite groups within the molecule. As applied to the case illustrated in Figure 8.1, where bond $X—Y$ defines the chosen axis, bonds $X—Z$ and $Y—U$ define two planes passing through this axis. We therefore introduce a co-ordinate χ_{ZU} denoting the variation in the equilibrium angle between these planes. In the case of planar molecules comprising at least three bonds lying in a single plane (Figure 8.2), to characterize a non-planar vibration we introduce the bending angle, ρ, of one bond with respect to the planes defined by other bonds. A certain degree of arbitrariness is involved in choosing independent co-ordinates χ and ρ, as is also the case for co-ordinates q and γ. For example, in the case of the ethane molecule (Figure 8.3) having a single degree of freedom accounted for by a relative twist of the CH_3 groups around the C—C axis, the co-ordinate can be chosen to be the variation in any of the angles between the planes passing through C—C bond and the CH bonds, χ, or a linear combination of these co-ordinates. It is only natural that this choice will define the form of the potential function.

With the above-mentioned natural vibrational co-ordinates the theory of vibration of polyatomics also makes use of linear combination of these co-ordinates. This is done both in the case where we have to take into account the symmetrical pattern of a molecule and in describing the concurrent motion of a great number of atoms with respect to one another. In particular, such a situation can occur in describing the rotations of one group of atoms relative to another within a molecule, i.e. in the case of vibrations corresponding to the transition of one rotational isomer into another.

It has already been mentioned that the number of vibrational degrees of freedom for the molecules consisting of N atoms is equal to $3N-6$, so it is clear that to provide a comprehensive description of all the possible deformations or vibrations of polyatomics we have to introduce no less than $3N-6$ natural vibrational co-ordinates. True, in some cases we can neglect some of the vibrational motions, so that the number of natural vibrational co-ordinates can be reduced. This reduction is often used when dealing only with high-frequency oscillations in molecules. For example, when dealing with ethane, we do not take into account the counter-rotation of CH_3 groups about C—C bonds. A similar approach is adopted when dealing with propylene, where rotation of the vinyl and methyl groups around the single C—C bond is usually discarded.

It is often the case that, when introducing in a formal way in a molecule variations of all of the bond lengths, valence angles and angles between the bonds and dihedral angles as the natural vibrational co-ordinates, the number of such co-ordinates exceeds that of the vibrational degrees of freedom of the molecule, i.e. it is greater than $3N-6$. The co-ordinates are therefore dependent and some are redundant (see Chapter 7). This fact is not some kind of a constraint, since in the course of the subsequent computations the reduntant co-ordinates can be easily taken into account or discarded. Such an operation can be made automatically with the aid of a computer. Hence in practice, in

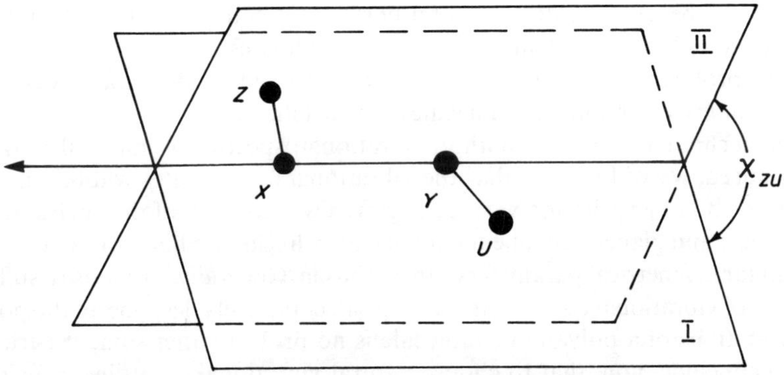

Figure 8.1. The first type of rotational co-ordinate

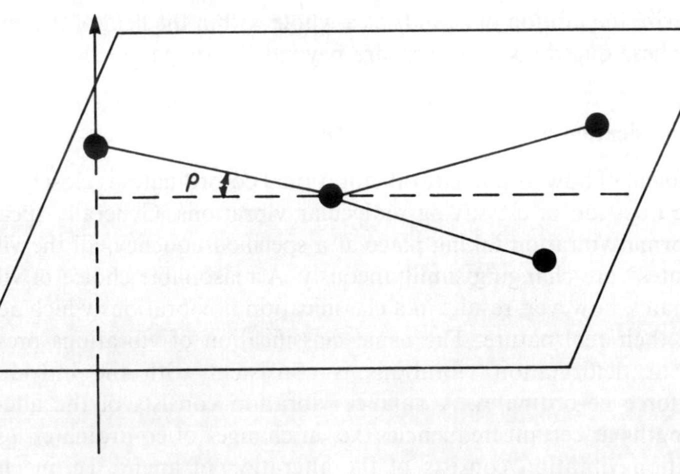

Figure 8.2. The out-of-plane co-ordinate for one bond

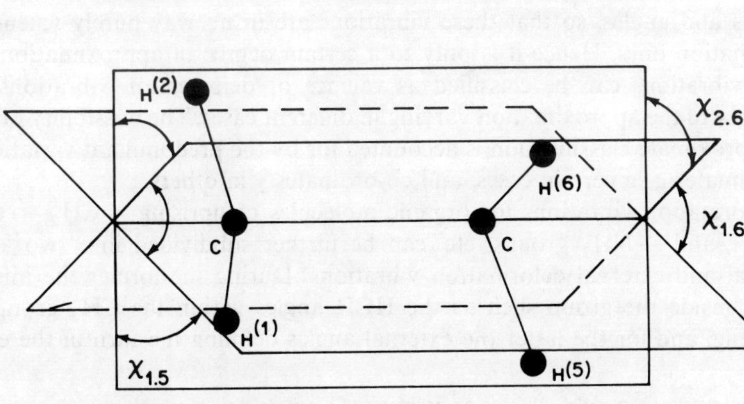

Figure 8.3. The second type of rotational co-ordinate

order not be complicate computations, no attempts are initially made to introduce only independent vibrational co-ordinates.

The problem is that to determine *a priori* only independent vibrational co-ordinates in complicated molecules can be difficult. That is why, while various other techniques for computing vibrational spectra of molecules will be introduced, we will assume that the vibrational co-ordinates will be variations in all the bond lengths and valence angles as well as most of the angles between the bonds and planes and dihedral angles; also, linear combinations of variations of similar geometrical parameters can be chosen. Generally, to choose a sufficient system of vibrational co-ordinates which can completely describe all the possible deformations of a polyatomic molecule is no problem after some experience.

To conclude, note that to choose natural vibrational co-ordinates using the above procedure whilst studying certain classes of molecules (for example, complexes accommodating heavy central atoms) seems to be unreasonable in many cases. For such systems it is advisable to introduce co-ordinates which characterize the motion of ligands as a whole within the field of the complexing atoms. These questions, however, are beyond the scope of this book.

8.2 Classification of Molecular Vibrations

The problem of how to use natural vibrational co-ordinates is closely associated with the question of classifying molecular vibrations. Generally speaking, for every normal vibration taking place at a specific frequency, all the vibrational coordinates* are changing simultaneously. A reasonable choice of vibrational co-ordinates, however, results in a classification of vibrations which adequately displays their real nature. The usual classification of vibrations presented as valence or deformation vibrations is consistent with the introduction of valence-force co-ordinates. A valance vibration consists of the alteration of bond lengths at certain frequencies, i.e. in changes of co-ordinates, q; whilst a deformation vibration consists of the alteration of angles, i.e. in changes of co-ordinates, γ. In fact, real vibrations generally involve alteration of both bond lengths and angles, so that these vibrations are in no way purely valence or deformation ones. Hence it is only to a certain degree of approximation that these vibrations can be classified as valence or deformation vibrations, the accuracy of the approximation varying in different cases. The possibility of such an approximate classification is accounted for by the predominant variation of co-ordinates q in certain cases, and co-ordinates γ in others.

Deformation vibrations for organic molecules comprising $=CH_2$, $-CH_3$, $-CH_2$, and $-NH_2$ groups, etc. can be further subdivided into two kinds: internal and external deformation vibrations. During the former the internal angles inside the group such as the HCH angles within the CH_3 group are changing, and for the latter the external angles defining the turn of the entire

*For symmetrical molecules—all the co-ordinates of a certain type of symmetry.

group as a whole are altering; for example, the C=C—H$^{(1)}$ and C=C—H$^{(2)}$ angles for the =CH_2 group which is bonded through a double link to another atom of carbon (as is the case in ethylene) are altering.

In working out a more detailed classification of vibrations of complex molecules we should discriminate between valence vibrations of different kinds of bonds as well as deformation vibrations of different kinds of angles. In the case of organic molecules we can encounter valence vibrations of single, double and triple CC bonds as well as CX bonds, where X=H, halogen, oxygen, nitrogen, etc. We can also identify $\widehat{\mathrm{CCC}}$, $\widehat{\mathrm{HCH}}$ and $\widehat{\mathrm{ClCCl}}$ bond angle deformation vibrations. In the simplest case the vibration of an individual bond (or angle) is associated with a characteristic frequency. It is worth noting that in several cases a certain frequency can prove to be fairly characteristic, i.e. it will always appear once a given bond occurs in a molecule and maintain its value whilst in different compounds. At the same time, variations in a given major bond co-ordinate, q_i, or angle co-ordinate, γ_{ij}, will be accompanied by substantial variations in some other co-ordinates. Hence to consider a frequency to be characteristic it can be sufficient to have a predominant change in the co-ordinate under consideration, q_i or γ_{ij}, accompanied by simultaneous variations in other co-ordinates as well. For example, valence vibrations of certain bonds are often accompanied by substantial variations in the bond angles embracing these bonds.

When a molecule has a number of bonds of one kind (for example, single CC bonds, forming a carbon chain) we are often able to identify vibrational frequencies which correspond to simultaneous alteration of these co-ordinates as the major co-ordinates. That is, we can identify those vibrational frequencies for which the co-ordinates of one type are being primarily changed. These are the characteristics vibrations associated with a combination of bonds or angles of one type. The number of such vibrations equals the number of co-ordinates of this kind.*

In the general case of N-atomic branched molecules, which have no closed rings, the number of bonds is equal to $N-1$. So we have $N-1$ co-ordinates q_i associated with $N-1$ valence vibrations, and $(3N-6)-(N-1)=2N-5$ angular co-ordinates associated with $2N-5$ deformation vibrations.†

Generally speaking, while computing vibrations, from the beginning it is not worth taking into account additional relationships and discarding unnecessary co-ordinates. It is more convenient to start with the introduction of such 'unnecessary' co-ordinates and taking the additional terms into account at a later stage of the computation.

*Where a symmetrical pattern is present, frequencies of some vibrations can coincide with one another, and then the number of different characteristic frequencies will be less.

†In the case of more complicated molecules the overall number of angular co-ordinates comprises co-ordinates χ, which characterize rotation around the bonds interlinking portions of a molecule, and for simpler molecules ρ co-ordinates, which characterize non-planar vibrations.

8.3 Derivation of Formulae for Elements of Matrix $\vec{B}$

It has been shown in Section 6.1 that to write the operator of the kinetic energy in curvilinear co-ordinates, which co-ordinates have been chosen to be represented by variations in the structural geometric parameters of molecules, we have to form a matrix $\vec{B}$ so that, by multiplication, $T = \vec{B}E\tilde{\vec{B}}$. To do this, we express the generalized velocities through the velocities of atoms within a molecule given in a laboratory reference framework. It appears that in the general case the generalized velocities q_i (rate of change in the natural co-ordinates) will be expressed through the velocities of atoms $\dot{\vec{r}}$, written in the external Cartesian system of co-ordinates, according to the following relationship:

$$\dot{q}_j = \varphi_j(\dot{\vec{r}}_1, \dot{\vec{r}}_2, \ldots, \dot{\vec{r}}_n) \tag{8.1}$$

Moreover, as will be demonstrated below, the functions φ_j are linear with respect to $\dot{\vec{r}}_k$. Taking this into account, and bearing in mind that the generalized velocities q_j are scalar quantities, we will write equation (8.1) in the following form[1]:

$$\dot{q}_j = \sum_k \vec{b}_{jk} \dot{\vec{r}}_k \tag{8.2}$$

Here, $\vec{b}_{jk}$ are vector coefficients which have not yet been determined. Products $\vec{b}_{jk} \cdot \dot{\vec{r}}_k$ are scalars. In the latter formula all the values are related to a certain fixed moment in time.

Relations (8.2) can be written using the compact matrix notation as follows:

$$\dot{q} = \vec{B}\dot{\vec{r}} \tag{8.3}$$

The elements of the matrix $\vec{B}$ will be represented by coefficients $\vec{b}_{jk}$ which are in fact vectors, as shown previously, being defined in the same Cartesian co-ordinate system as that for $\dot{\vec{r}}_k$.

Note that, generally speaking, matrix $\vec{B}$ can be selected only to within the accuracy of a certain matrix $\vec{A}$ such that $\vec{A} \cdot \dot{\vec{r}} = 0$. Then $(\vec{B} + \vec{A})\dot{\vec{r}} = B\dot{\vec{r}}$. That is why, for the sake of definiteness, we will assume below that the elements of matrix $\vec{B}$ are calculated with the formulae given.

From equations (8.1)–(8.3) we can easily conclude that

$$\vec{p} = \tilde{\vec{B}}p$$

(see also Sections 6.1 and 7.1). Since we choose the vibrational co-ordinates so as to characterize the state of the structural elements of a molecule, the rate of change of these co-ordinates must be a function of the velocities of the atoms forming the given structural element. Moreover, in order to enable the co-ordinates being introduced to really describe internal motions alone, it must be ensured that the values of all $\dot{q}_j$ be equal to zero for the translational motions of the structural element or for its rotation as a whole around any instantaneous axis.

The first condition is satisfied if we choose coefficients $\vec{b}_{jk}$ in such a way as to ensure that they differ from zero for every co-ordinate q_j only for those atoms comprising the structural element being considered. To meet the second and third conditions, the relations

$$\sum_k \vec{b}_{jk} = 0 \quad \text{and} \quad \sum_k (\vec{b}_{jk} \times \vec{r}_k) = 0 \tag{8.4}$$

have to be obeyed for sets of coefficients, $\vec{b}_{jk}$, for each of the co-ordinates q_j (i.e. for each line of matrix $\vec{B}$). Here, $\vec{b}_{jk} \times \vec{r}_k$ is a vector product of the respective element $\vec{b}_{jk}$ and instantaneous radius-vector $\vec{r}_k$ of the atom.

Indeed, in this case for the translatory motion of an element the velocities of its atoms must be equal ($\dot{\vec{r}}_k = \dot{\vec{r}}$) and hence

$$\dot{q}_j = \sum_k \vec{b}_{jk} \cdot \dot{\vec{r}}_k = \dot{\vec{r}} \sum_k \vec{b}_{jk} = 0$$

because of the condition that $\sum_k b_{jk} = 0$.

When a structural element is rotating without disturbing the geometrical configuration $\dot{\vec{r}}_k = \vec{r}_k \times \vec{\omega}$, where $\vec{\omega}$ is an instantaneous angular velocity, therefore

$$\dot{q}_j = \sum_k \vec{b}_{jk}(\vec{r}_i \times \vec{\omega}) = \vec{\omega} \sum_k (\vec{b}_{jk} \times \vec{r}_k) = 0$$

if the condition is true that $\sum_k \vec{b}_{jk} \times \vec{r}_k = 0$.

Thus the problem to be solved is that of expressing the rates of variation of the generalized co-ordinates via velocities of atomic motions in the Cartesian system of co-ordinates[1].

For the present consider the simplest case when the co-ordinates consist of variations in bond lengths and valence angles. Let us distinguish in a molecule a certain valence angle formed by atoms 0, 1 and 2, where atom 0 is located at the vertex of the angle (Figure 8.4). Define the instantaneous positions of atoms 0, 1 and 2 through respective radius-vectors $\vec{r}_0$, $\vec{r}_1$ and $\vec{r}_2$.*

Then it is true that at any instant

$$\vec{r}_1 - \vec{r}_0 = s_1 \vec{e}_1 \tag{8.5a}$$

and

$$\vec{r}_2 - \vec{r}_0 = s_2 \vec{e}_2 \tag{8.5b}$$

where s_1 and s_2 are the instantaneous lengths of the first and second bonds of the valence angle chosen and $\vec{e}_1$ and $\vec{e}_2$ are the instantaneous unit direction vectors of these bonds, determined within the same Cartesian co-ordinate system. Vectors $\vec{e}_1$ and $\vec{e}_2$ are drawn from atom 0 towards atoms 1 and 2.

Differentiating equations (8.5a) and (8.5b) with respect to time, we find

$$\dot{\vec{r}}_1 - \dot{\vec{r}}_0 = \dot{s}_1 \vec{e}_1 + s_1 \dot{\vec{e}}_1 \tag{8.6a}$$

$$\dot{\vec{r}}_2 - \dot{\vec{r}}_0 = \dot{s}_2 \vec{e}_2 + s_2 \dot{\vec{e}}_2 \tag{8.6b}$$

where $\dot{\vec{e}}_1 \perp \vec{e}_1$ and $\dot{\vec{e}}_2 \perp \vec{e}_2$ for any instant of time.

*It is assumed here that the values comprised by the formulae are functions of time, i.e. $\vec{r}_0 = \vec{r}_0(t)$, etc.

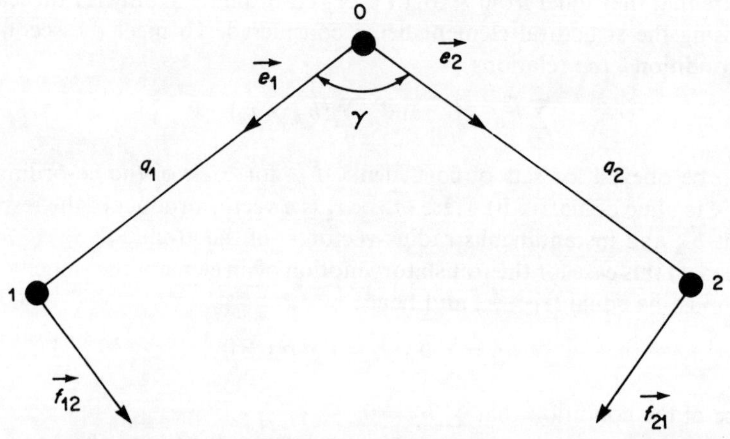

Figure 8.4. Unit vectors (orts) for co-ordinates of bond and valency angle distortions

By multiplying equations (8.6a) and (8.6b) in a scalar way by $\vec{e}_1$ and $\vec{e}_2$, we obtain

$$\dot{s}_1 = (\dot{\vec{r}}_1 - \dot{\vec{r}}_0)\vec{e}_1 = \dot{q}_1 \tag{8.7a}$$

$$\dot{s}_2 = (\dot{\vec{r}}_2 - \dot{\vec{r}}_0)\vec{e}_2 = \dot{q}_2 \tag{8.7b}$$

(since $s_1 = s_1(0) + \Delta s_1$, where $s_1(0)$ is the bond length at the initial moment of time, and $\Delta s_1 = q_1$ is the increment of the bond length in the course of vibration, then $\dot{s}_1 = \dot{q}_1$).

Equations (8.7a) and (8.7b) enable us to find the elements of matrix $\vec{B}$ in the case of co-ordinates consisting of variations in the bond lengths, which are thus equal to $\vec{e}_n$ and $-\vec{e}_n$, if the bond is formed by the nth and kth atoms, and vector $\vec{e}_n$ is drawn from the kth to nth atom.

The rate of change of the valence angle, $\dot{\gamma}$, is defined by the sum of rates $\dot{\gamma}_1 + \dot{\gamma}_2$ of changes of the direction vectors of the first and second bonds towards each other in the instantaneous plane of the valence angle. To obtain this rate, multiply equation (8.6a) in a scalar way by the unit vector $\vec{f}_{12}$ drawn within the instantaneous plane of the valence angle normally to the instantaneous direction vector $\vec{e}_1$ inside the angle. We obtain

$$\dot{\gamma}_1 = \dot{\vec{e}}_1\vec{f}_{12} = \frac{1}{s_1}\vec{f}_{12}(\dot{\vec{r}}_1 - \dot{\vec{r}}_0) \tag{8.8a}$$

In a similar way from equation (8.6b) we obtain

$$\dot{\gamma}_2 = \dot{\vec{e}}_2\vec{f}_{21} = \frac{1}{s_2}\vec{f}_{21}(\dot{\vec{r}}_2 - \dot{\vec{r}}_0) \tag{8.8b}$$

A positive value of $\dot{\gamma}_1$ corresponds to a movement of the first bond towards vector $\vec{f}_{12}$, whilst a positive value of $\dot{\gamma}_2$ is associated with the movement of the second bond towards $\vec{f}_{21}$. Since vectors $\vec{f}_{12}$ and $\vec{f}_{21}$ are directed towards one another, rate of change of the valence angle, which decreases, will be equal to

$$\dot{\gamma} = \dot{\gamma}_1 + \dot{\gamma}_2 = \frac{1}{s_1}\vec{f}_{12}\dot{\vec{r}}_1 + \frac{1}{s_2}\vec{f}_{21}\dot{\vec{r}}_2 - \left(\frac{1}{s_2}\vec{f}_{12} + \frac{1}{s_2}\vec{f}_{21}\right)\dot{\vec{r}}_0 \tag{8.9}$$

Later, in order to include the case $\dot{\gamma} > 0$, where an increase in the valence angle occurs, we assume that

$$\dot{\gamma} = -(\dot{\gamma}_1 + \dot{\gamma}_2) = -\frac{1}{s_1}\vec{f}_{12}\dot{\vec{r}}_1 - \frac{1}{s_2}\vec{f}_{21}\dot{\vec{r}}_2 + \left(\frac{1}{s_1}\vec{f}_{12} + \frac{1}{s_2}\vec{f}_{21}\right)\dot{\vec{r}}_0 \tag{8.10}$$

Thus the elements of matrix $\vec{B}$ for the valence angle variation co-ordinate will finally acquire the following form (if the angle $\widehat{iks}$ is formed by atoms i, j, k at the vertex of the angle):

$$-\frac{1}{s_i}\vec{f}_{ij}, \quad -\frac{1}{s_j}\vec{f}_{ji}, \quad \vec{F}_{ij} = \frac{1}{s_i}\vec{f}_{ij} + \frac{1}{s_j}\vec{f}_{ji}$$

The elements of matrix $\vec{B}$ thus obtained for the co-ordinates of bond-stretching and valence angle variation obey the above-mentioned conditions for vibrational co-ordinates. These are that the sums of elements for each co-ordinate at any instant of time are equal to zero; also equal to zero are the sums of pair-by-pair vector products of elements $\vec{b}_{jk}$ and instantaneous radius-vectors of the respective atoms. The first condition can be inferred easily from equations (8.7a), (8.7b) and (8.10). The second condition can be verified as follows.

For the case of bond stretching between atoms 0 and 1, condition $\sum_k(\vec{b}_{jk} \times \vec{r}_k)$ amounts to

$$\vec{e}_1 \times \vec{r}_1 + (-\vec{e}_1) \times \vec{r}_0 = \vec{e}_1 \times (\vec{r}_1 - \vec{r}_0) = \vec{e}_1 \times s_1\vec{e}_1 = 0$$

For the angle variation co-ordinate between atoms 0, 1 and 2 a similar condition in open form can be written as follows:

$$\begin{aligned}
&\left(-\frac{\vec{f}_{12}}{s_1}\right) \times \vec{r}_1 + \left(-\frac{\vec{f}_{21}}{s_2}\right) \times \vec{r}_2 + \vec{F}_{12} \times \vec{r}_0 \\
&= \left(-\frac{\vec{f}_{12}}{s_1}\right) \times \vec{r}_1 + \left(-\frac{\vec{f}_{21}}{s_2}\right) \times \vec{r}_2 + \left(\frac{\vec{f}_{12}}{s_1} + \frac{\vec{f}_{21}}{s_2}\right) \times \vec{r}_0 \\
&= \frac{1}{s_2}\vec{f}_{12} \times (\vec{r}_0 - \vec{r}_1) + \frac{1}{s_2}\vec{f}_{21} \times (\vec{r}_0 - \vec{r}_2) = \vec{f}_{12} \times (-\vec{e}_1) + \vec{f}_{21} \times (-\vec{e}_2) \\
&= 0
\end{aligned}$$

since vectors $\vec{f}_{12} \times \vec{e}_1$ and $\vec{f}_{21} \times \vec{e}_2$ are equal in magnitude but opposite in sign.

Using equations (8.7) and (8.10) we obtain matrix $\vec{B}$ for all bond-stretching and valence angles deformation co-ordinates (see below), arriving at the kinetic energy matrix in the momentum representation:

$$T = \vec{B}E\tilde{\vec{B}} \tag{8.11}$$

Having multiplied these matrices in a scalar way, we can see that the elements of the kinetic energy matrix in the momentum representation (i.e. the matrix of kinematical coefficients) consist of linear combinations of scalar products of the direction vectors $\vec{e}_i, \vec{f}_{ij}$, being thus dependent only upon the instantaneous values of the angles between these vectors, rather than upon the position of these vectors in the space of the stationary (external) co-ordinates.

Indeed, for a triatomic molecule with atoms 0, 1 and 2, writing equations (8.7a), (8.7b) and (8.10) in matrix form, we obtain

$$\begin{bmatrix} \dot{q}_1 \\ \dot{q}_2 \\ \dot{\gamma} \end{bmatrix} = \begin{bmatrix} -\vec{e}_1 & \vec{e}_1 & 0 \\ -\vec{e}_2 & 0 & \vec{e}_2 \\ \vec{F}_{12} & -\vec{f}_{12}/s_1 & -\vec{f}_{21}/s_2 \end{bmatrix} \begin{bmatrix} \dot{\vec{r}}_0 \\ \dot{\vec{r}}_1 \\ \dot{\vec{r}}_2 \end{bmatrix} = \vec{B}\dot{\vec{r}} \tag{8.12}$$

According to equation (8.11), it can then be inferred from the above that

$$T = \vec{B}E\tilde{\vec{B}}$$

$$= \begin{bmatrix} \varepsilon_0 + \varepsilon_1 & \varepsilon_0 \cos\theta & -\dfrac{1}{s_2}\varepsilon_0 \sin\theta \\ \varepsilon_0 \cos\theta & \varepsilon_0 + \varepsilon_2 & -\dfrac{1}{s_1}\varepsilon_0 \sin\theta \\ -\dfrac{\varepsilon_0}{s_2}\sin\theta & -\dfrac{\varepsilon_0}{s_1}\sin\theta & \dfrac{\varepsilon_1}{s_1^2} + \dfrac{\varepsilon_2}{s_2^2} + \varepsilon_0\left(\dfrac{1}{s_1^2} + \dfrac{1}{s_2^2} + 2\dfrac{1}{s_1 s_2}\cos\theta\right) \end{bmatrix} \tag{8.13}$$

where θ is the instantaneous angle between vectors $\vec{e}_1$ and $\vec{e}_2$.

With the vibrational motion of a molecule the elements of the matrix of kinematic coefficients will appear to be periodic functions of time, and their values will oscillate around some magnitude corresponding to the equilibrium state of the molecule and to the equilibrium values of the bond lengths and valence angles. Later, it will be seen that we can deal with the vibrational problem of a polyatomic molecule within the system of equilibrium unit vectors $\vec{e}_i(0), \vec{f}_{ij}$, etc. without any loss of generality. Then the instantaneous direction vectors of bonds and valence angles will oscillate about their equilibrium values. In this case instantaneous radius-vectors $\vec{r}_k$ of the atoms can be defined in the Cartesian co-ordinate system, which is stationary with respect to the equilibrium unit vectors of the molecule.

In the most general form when the equilibrium geometry of a molecule is taken as a reference point, then the matrix T can be represented as $T(q) = T(0) + \Delta T(q)$. As the amplitude of vibration decreases, the correction $\Delta T(q)$ becomes smaller, tending to the zero in the limit. That is why, for infinitesimal vibrations, kinetic energy can be put equal to $T = 1/2\tilde{p}T(0)p$, where $T = \vec{B}(0)E\tilde{\vec{B}}(0)$ is the matrix of kinematical coefficients written as for the equilibrium state. Accordingly, transformations of instantaneous velocities within the region close to the equilibrium position (cf. equations (8.7) and (8.10)) will be as follows:

$$\dot{q}_1 = (\dot{\vec{r}}_1 - \dot{\vec{r}}_2)\vec{e}_1(0) \tag{8.14}$$

$$\dot{\gamma} = -\frac{1}{s_1(0)}\vec{f}_{12}(0)\dot{\vec{r}}_1 - \frac{1}{s_2(0)}\vec{f}_{21}(0)\dot{\vec{r}}_2 + \vec{F}_{12}(0)\dot{\vec{r}}_0 \tag{8.15}$$

Condition $\sum_k \vec{b}_{jk} = 0$ will remain valid and $\sum_k \vec{b}_{jk} \times \vec{r}_k$ will be transformed to $\sum \vec{b}_{jk}(0) \times \vec{r}_k(0)$.

The product $\vec{e}_1(0)(\dot{\vec{r}}_1 - \dot{\vec{r}}_0)$ is a projection of the difference of atomic velocities taken in the equilibrium bond direction, or, in other words, the rate of variation of the projection of the instantaneous bond vector $s_1\vec{e}_1$ onto $\vec{e}_1(0)$. The magnitude of the projection of vector $s_1\vec{e}_1$ onto $\vec{e}_1(0)$ is equal to $s_1 \cos\varphi$, where φ is the angle between the equilibrium and instantaneous bond directions. Bearing in mind that $\cos\varphi \simeq 1$ up to an angle of $\varphi = 3°$, it becomes clear that the bond-length variation rate must be equal, within a good degree of accuracy, to the rate of variation of its projection onto the equilibrium bond direction. Therefore to arrive at the expression for the rate of variation of the vibrational co-ordinate q_i we can use equation (8.7).

Similar reasoning can be used in the case of angular co-ordinates: for small angles between $\vec{f}_{ij}$ and $\vec{f}_{ij}(0)$ and for the small changes in bond length occurring during vibration, the rate of change in the valence angle can be represented as a sum of rates for the projections of the difference between the atomic velocities taken onto the equilibrium unit vectors; the atoms involved form the arms of an angle, related to the unit bond, and this rate can be found from equation (8.14).

To obtain expressions for the vibrational co-ordinates q_i and γ_{ij} we have to integrate equations (8.14) and (8.15) with respect to time, assuming that the vectors $\vec{e}_1(0), \vec{f}_{12}(0)$ are invariant. Being fully consistent with what has been said above on the transformation of instantaneous velocities, we can therefore take as vibrational co-ordinates variations of the projections of the instantaneous bond vectors $s_i\vec{e}_i$ and valence angles $1/s_i \cdot \vec{f}_{ij}$ and $\vec{f}_{ji}/s_j$ onto the equilibrium unit vectors of a molecule. Absolute values of these projections will differ from the increments of the bond lengths and valence angles by the values of the respective cosines, so that this difference can be neglected for small vibrations.

Thus finally,

$$q_1 = \vec{e}_1(0)(\delta\vec{r}_1 - \delta\vec{r}_0) \tag{8.16}$$

$$\gamma = -\frac{\vec{f}_{12}}{s_1(0)}\delta\vec{r}_1 - \frac{\vec{f}_{21}(0)}{s_2(0)}\delta\vec{r}_2 + \vec{F}_{12}(0)\delta\vec{r}_0 \tag{8.17}$$

In equations (8.16) and (8.17) all the vectors and bond lengths are related to the equilibrium state of a molecule. Increments $\delta\vec{r}_k$ are in fact the variations of the radius-vectors of the vibrating atoms in a moving frame of reference.

These co-ordinate projections can also be employed in computations. However, they will no longer be exact vibrational co-ordinates, which factor is clearly manifested as the amplitude increases, since in this case the condition $\sum_k \vec{b}_{jk} \times \vec{r}_k = 0$ is no longer satisfied. It is only with small vibrational amplitudes of the order of magnitude not exceeding several hundred Å that these co-ordinates can be utilized. In so doing, an essential fact should be borne in mind that, though for low vibration levels, changes in the bond lengths and valence angles coincide, in terms of the absolute value, with the variations in their projections onto the equilibrium unit vectors of a molecule, the systems of the unit vectors are different for the accurate and approximate vibrational co-ordinates.

That is, a movable system of unit vectors corresponds to the accurate co-ordinates, since vectors $\vec{e}_i$ and $\vec{f}_{ij}$ are functions of time, whilst a stationary system is associated with approximate co-ordinates. It has been mentioned already that for small-amplitude vibrations this difference can be discarded from the expression for kinetic energy. Nonetheless, we will always employ accurate co-ordinates, and not resort to co-ordinate projections.

To conclude, it will be noted that the derivation of the expression for the matrix of kinematical coefficients given above clearly demonstrates that the elements of this matrix are independent of the choice of co-ordinates which characterize the rotation of a molecule as a whole.

8.4 Derivation of Expressions for Non-planar Co-ordinates

For a comprehensive description of the vibrations of polyatomic molecules it is necessary to use more than the two types of co-ordinates described above. In addition, we have to introduce co-ordinates of variations of dihedral angles between planes, co-ordinates of variations of the angles between a bond and a plane (a particular case of such co-ordinates are those of the points where bonds are out of the plane passing through some atoms in the molecule), etc. A general term—non-planar— is used to denote such co-ordinates. Expressions for the elements of the $\vec{B}$ matrix for these co-ordinates can be obtained in a way similar to that described in Section 8.3.

Let us take three adjacent non-collinear atoms within a molecule by the radius-vectors $\vec{r}_0$, $\vec{r}_1$ and $\vec{r}_2$ (Figure 8.5). As was the case previously, we assume that these vectors are determined in a stationary rectangular co-ordinate system, being functions of time. Now introduce two direction unit vectors $\vec{e}_1$ and $\vec{e}_2$, drawing these from an atom with a zero index to atoms with indices 1 and 2, respectively, along the lines connecting these atoms. Since atoms 0, 1 and 2 define a plane, its instantaneous position will be determined by the normal, equal to

$$\vec{n} = \frac{1}{\sin\theta}(\vec{e}_1 \times \vec{e}_2)$$

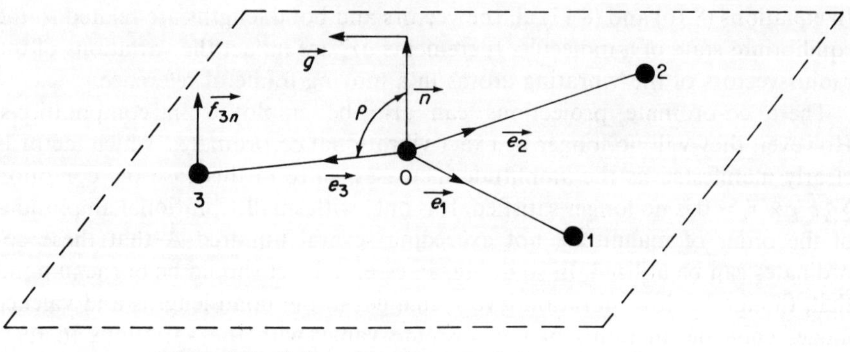

Figure 8.5. Unit vectors for out-of-plane co-ordinate

where $\theta = \widehat{\vec{e}_1, \vec{e}_2}$. Here all values are related to the same given instant of time and are time-dependent.

When arbitrary displacements of the atoms occur, the normal will be changed, and its rate of variation is expressed by

$$\dot{\vec{n}} = -\frac{\dot{\theta}\cos\theta}{\sin^2\theta}(\vec{e}_1 \times \vec{e}_2) + \frac{1}{\sin\theta}(\dot{\vec{e}}_1 \times \vec{e}_2) + \frac{1}{\sin\theta}(\vec{e}_1 \times \dot{\vec{e}}_2)$$

$$= \frac{1}{\sin\theta}(\dot{\vec{e}}_1 \times \vec{e}_2) - \frac{1}{\sin\theta}(\dot{\vec{e}}_2 \times \vec{e}_1) - \frac{\dot{\theta}\cos\theta}{\sin^2\theta}(\vec{e}_1 \times \vec{e}_2) \tag{8.18}$$

The speed of rotation $\dot{\kappa}$ of a plane around a certain axis seems to be determined by the magnitude of projection $\dot{\vec{n}}$ onto a unit vector $\vec{g}$ drawn normally to the instantaneous position of the normal and the instantaneous axis of rotation of the plane. It is advisable to choose the direction of the vector towards the fall of the normal under the rotation of the plane in order to ensure that the respective projection be positive. Thus

$$\dot{\kappa} = \vec{g}\,\dot{\vec{n}} = \frac{\vec{g}}{\sin\theta}(\dot{\vec{e}}_1 \times \vec{e}_2) - \frac{\vec{g}}{\sin\theta}(\dot{\vec{e}}_2 \times \vec{e}_1) - \left(\frac{\dot{\theta}\cos\theta}{\sin^2\theta}\right)\vec{g}(\vec{e}_1 \times \vec{e}_2)$$

$$= \frac{1}{\sin}\dot{\vec{e}}_1(\vec{e}_2 \times \vec{g}) - \frac{1}{\sin\theta}\dot{\vec{e}}_2(\vec{e}_1 \times \vec{g}) \tag{8.19}$$

The product $\vec{g}(\vec{e}_1 \times \vec{e}_2) = 0$, since $\vec{g} \perp \vec{n}$, and $\vec{n} \parallel (\vec{e}_1 \times \vec{e}_2)$. According to equations (8.6a) and (8.6b):

$$\dot{\vec{e}}_1 = \frac{1}{s_1}(\dot{\vec{r}}_1 - \dot{\vec{r}}_0 - \dot{q}_1\vec{e}_1) \quad \text{and} \quad \vec{e}_2 = \frac{1}{s_2}(\vec{e}_2 - \vec{e}_0 - \dot{q}_2\vec{e}_1)$$

Hence the mixed product

$$\dot{\vec{e}}\,(\vec{e}_2 \times \vec{g}) = \frac{\dot{\vec{r}}_1 - \dot{\vec{r}}_0}{s_1}(\vec{e}_2 \times \vec{g}) - \frac{\dot{q}_1}{s_1}\vec{e}_1(\vec{e}_2 \times \vec{g})$$

$$= \frac{\dot{\vec{r}}_1 - \dot{\vec{r}}_0}{s_1}(\vec{e}_2 \times \vec{g}) = \frac{\vec{n}(\dot{\vec{r}}_1 - \dot{\vec{r}}_0)}{s_1}\sin(\widehat{\vec{e}_2, \vec{g}}) \tag{8.20}$$

Mixed product $\vec{e}_1(\vec{e}_2 \times \vec{g})$ is reduced to zero since vectors $\vec{e}_1 \perp \vec{n}$, and $(\vec{e}_2 \times \vec{g}) \parallel \vec{n}$. A similar expression is obtained for the term $\dot{\vec{e}}_2(\vec{e}_1 \times \vec{g})$:

$$\dot{\vec{e}}_2(\vec{e}_1 \times \vec{g}) = \frac{\vec{n}(\dot{\vec{r}}_2 - \dot{\vec{r}}_0)}{s_2}\sin(\widehat{\vec{e}_1, \vec{g}}) \tag{8.21}$$

Finally,

$$\dot{\kappa} = \frac{\vec{n}}{\sin\theta}\left\{\frac{\dot{\vec{r}}_1 - \dot{\vec{r}}_0}{s_1}\sin(\widehat{\vec{e}_2, \vec{g}}) - \frac{\dot{\vec{r}}_2 - \dot{\vec{r}}_0}{s_2}\sin(\widehat{\vec{e}_1, \vec{g}})\right\} \tag{8.22}$$

Thus we have expressed the speed of rotation of the plane around a given axis through the values of instantaneous unit vectors ($\vec{n}$ and $\vec{g}$) and the velocities of displacement of the atoms which form the plane. Further, it should be taken into account that, in general, two types of co-ordinates can be encountered:

(1) Variation of an angle between a bond and a plane measured from the variation of an angle between a bond unit vector and the normal to the surface; and
(2) Variation of a dihedral angle between the planes, measured from the angle between the normals to both of the planes.

Then, by analogy with the case of valence angles, the instantaneous rate of change of the angle between a bond and a plane must be equal to the sum of the projection of the rates of change of the bond unit vector onto the instantaneous plane drawn through the bond and the normal to the plane formed by the three atoms, and the projection, on that same plane, of the velocity of counter-rotation of the chosen plane.

Incidentally, the role of the unit vectors of the angle between the bond and plane will be played, respectively, by vector $\vec{f}_{in}$, drawn normally to bond i inside the angle between the bond and the normal to the plane, and by the vector $\vec{g}$, drawn in that same plane perpendicularly to the normal toward vector $\vec{f}_{in}$ (Figure 8.5). Then the rate of decrease in the angle will be equal to

$$\dot{\rho} = \dot{\gamma}_3 + \dot{\kappa} = \frac{1}{s_3}\vec{f}_{3n}(\dot{\vec{r}}_3 - \dot{\vec{r}}_0)$$
$$+ \frac{\vec{n}}{\sin\theta}\left[\frac{\sin(\vec{e}_2, \vec{g})}{s_1}(\dot{\vec{r}}_1 - \dot{\vec{r}}_0) - \frac{\sin(\widehat{\vec{e}_1, \vec{g}})}{s_2}(\dot{\vec{r}}_2 - \dot{\vec{r}}_0)\right] \tag{8.23}$$

where the bond is formed by atoms 0 and 3.

The rate of change of the dihedral angle between the planes will be equal to that of the angle between the normals to the planes. By analogy with the valence angle, we have to introduce instantaneous unit vectors $\vec{g}_{\mathrm{I}}$ and $\vec{g}_{\mathrm{II}}$, which are perpendicular to the instantaneous positions of the normals, lying in the instantaneous plane passing through these normals and being directed opposite to one another (Figure 8.6).

Then the rate of change (decrease) of the dihedral angle between the planes formed by atoms 0, 1, 2 and 0′, 1′, 2′ with normals $\vec{n}_{\mathrm{I}}$ and $\vec{n}_{\mathrm{II}}$ will be equal to

$$\dot{\kappa} = \dot{\kappa}_1 + \dot{\kappa}_2 = \frac{\vec{n}_{\mathrm{I}}}{\sin\theta_1}\left[\frac{\sin(\widehat{\vec{e}_2, \vec{g}_{\mathrm{I}}})}{s_1}(\dot{\vec{r}}_1 - \dot{\vec{r}}_0) - \frac{\sin(\widehat{\vec{e}_1, \vec{g}})}{s_2}(\dot{\vec{r}}_2 - \dot{\vec{r}}_1)\right]$$
$$+ \frac{\vec{n}_{\mathrm{II}}}{\sin\theta_2}\left[\frac{\sin(\widehat{\vec{e}_{2'}, \vec{g}_{\mathrm{II}}})}{s_{1'}}(\dot{\vec{r}}_{1'} - \dot{\vec{r}}_{0'}) - \frac{\sin(\widehat{\vec{e}_{1'}, \vec{g}_{\mathrm{II}}})}{s_{2'}}(\dot{\vec{r}}_{2'} - \dot{\vec{r}}_{0'})\right] \tag{8.24}$$

These formulae show the relationship between the rates of change of the non-planar natural co-ordinates through the velocities of displacements of atoms which form these co-ordinates. As is the case for the co-ordinates of variations of a bond and angle, these enable us to produce the matrix $\vec{B}$ and the matrix of kinematical coefficients.

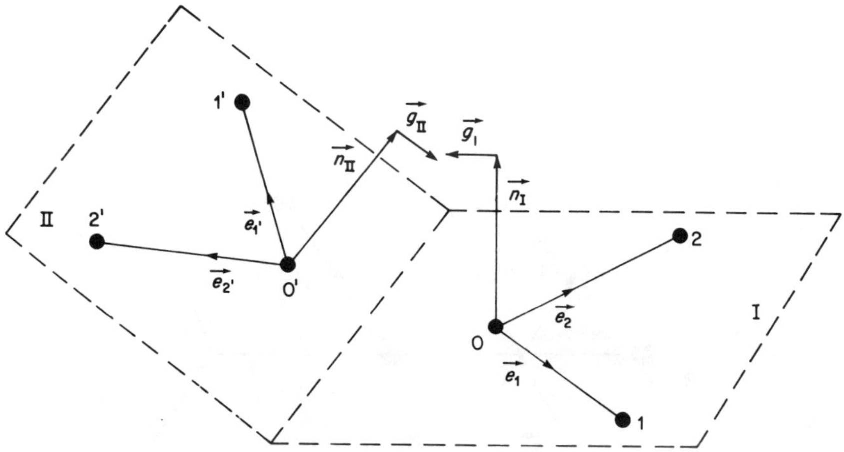

Figure 8.6. Unit vectors for dihedral angle

When using equation (8.22) we should measure the angles between vectors $\vec{e}_1$ and $\vec{g}$ ($\vec{e}_2$ and $\vec{g}$) in one direction (usually counterclockwise), where the vector $\vec{e}_1$ ($\vec{e}_2$) always serves as the origin. That is why $\sin(\widehat{\vec{e}_1, \vec{g}})$ ($\sin(\widehat{\vec{e}_2, \vec{g}})$) can be either positive (for angles less than 180°) or negative (for angles greater than 180°). This can be attributed to the necessity of having the same direction (coinciding with $\vec{n}$) for the vector products $\vec{e}_1 \times \vec{g}$ and $\vec{e}_2 \times \vec{g}$. The value of $\sin\theta$ is always taken to be positive. The vectors $\vec{e}_1$ and $\vec{e}_2$ can always be chosen so that the angle between them is less than 180°. Further, when establishing the rotation of each plane associated with a pair of bonds forming a valence angle it is always convenient to assume that both of the direction bond vectors point away from the vertex of the angle.

We shall now consider some examples of setting up non-planar co-ordinates and obtaining expressions for these through the velocities of atoms in a laboratory framework.

(1) Variation of a dihedral angle between the planes, where the axis of rotation forming the edge of the dihedral angle coincides with one of the bonds In this case (Figure 8.7) the axis of rotation coincides with bonds 2–3. A change in the dihedral angle formed by planes I and II, occurring in the course of a displacement of the respective atoms (co-ordinate χ), will be equal to the sum of rotations of both planes toward each other. For this reason, we choose the directions of the normals $\vec{n}_I$ and $\vec{n}_{II}$ to planes I and II towards the inside of the dihedral angle, taking bonds 1–2 and 3–4 as the origin. Consistently with the chosen directions of rotation of the planes, we introduce vectors $\vec{g}_I$ and $\vec{g}_{II}$ normal to bonds 2–3 (Figure 8.7).

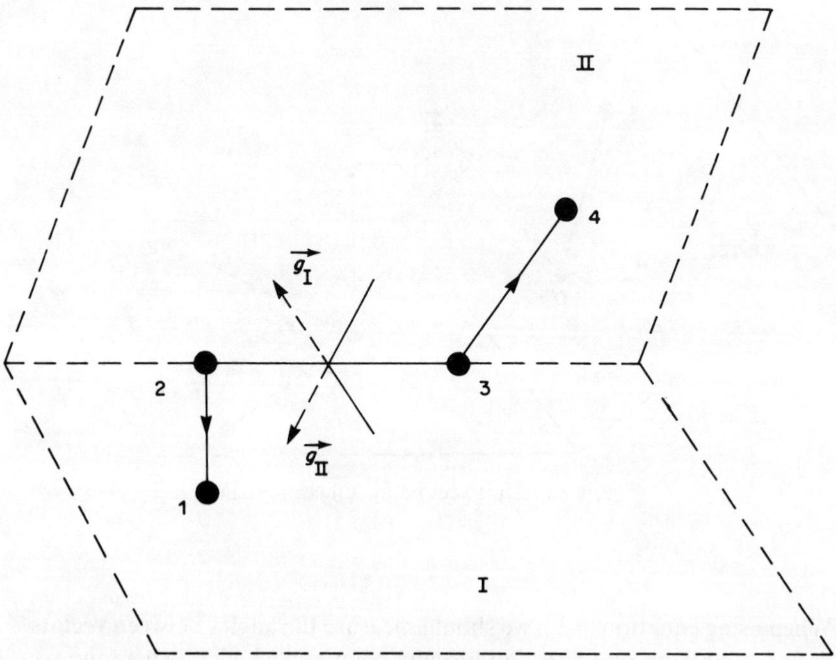

Figure 8.7. Unit vectors ($\vec{g}$) for the first type of rotational co-ordinate

Applying equation (8.24) we obtain*

$$\dot{\chi} = \frac{(\dot{\vec{r}}_1 - \dot{\vec{r}}_2)}{s_1}\frac{n_1}{\sin\theta_{13}} - \frac{\operatorname{ctg}\theta_{13}}{s_3}\vec{n}_1(\dot{\vec{r}}_3 - \dot{\vec{r}}_2)$$

$$+ \frac{\vec{n}_{\mathrm{II}}}{s_4 \sin\theta_{24}}(\dot{\vec{r}}_4 - \dot{\vec{r}}_3) - \frac{\operatorname{ctg}\theta_{24}}{s_2}\vec{n}_{\mathrm{II}}(\dot{\vec{r}}_2 - \dot{\vec{r}}_3) \tag{8.25}$$

Depending upon the angles between bonds 1–2 and 2–3 (3–4 and 2–3) ctg θ_{13}, can be either positive or negative. In a particular case all the bonds can lie in a single plane. Then $\vec{n}_{\mathrm{I}} = \vec{n}_{\mathrm{II}}$ and $\vec{g}_{\mathrm{I}} = -\vec{g}_{\mathrm{II}}$

(2) Co-ordinate of rotation of two planes around a common axis, which coincides with neither of the bonds A particular case of such co-ordinates is represented by a co-ordinate of rotation of CH_2 planes around a C=C bond in ethylene (Figure 8.8). This co-ordinate describes a torsional vibration. Because of the two facts that the axis of rotation passes through the C=C bond, and that the CH_2 groups must rotate in opposite directions, $\vec{g}_{\mathrm{I}}$ and $\vec{g}_{\mathrm{II}}$ must lie within the plane of the molecule, being directed in opposite directions. In this case we chose the same directions for the normals to the planes of both CH_2 groups. That is why in the left-hand CH_2 group the first bond is that of bond 2–1, whilst in the right-hand

*s_i is used to denote a bond terminating at atom i, so that $s_2 = s_3$.

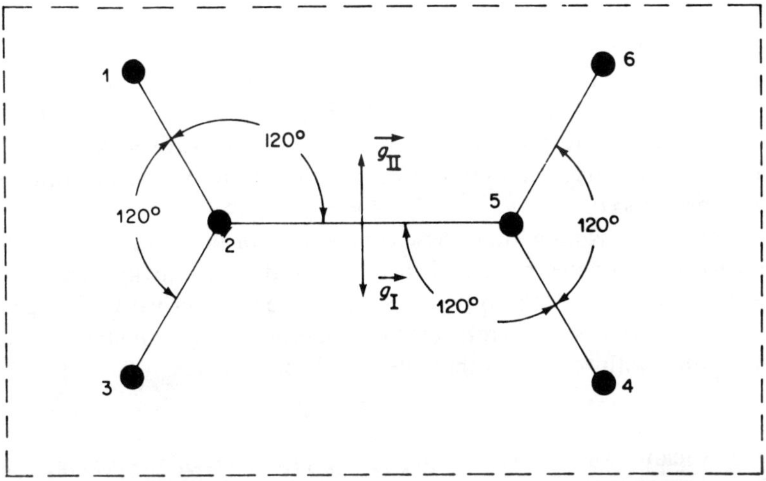

Figure 8.8. Unit vectors ($\vec{g}$) for the second type of rotational co-ordinate (ethylene)

group it is bond 5–4. From equation (8.24) we obtain

$$\dot{\chi} = \frac{\Delta\dot{\vec{r}}_{12}}{S_{CH}}\cdot\vec{n}\frac{\sin 30^\circ}{\sin 120^\circ} - \frac{\Delta\dot{\vec{r}}_{32}}{S_{CH}}\cdot\vec{n}\frac{\sin 30^\circ}{\sin 120^\circ} + \frac{\Delta\dot{\vec{r}}_{45}}{S_{CH}}\cdot\vec{n}\frac{\sin 30^\circ}{\sin 120^\circ}$$

$$-\frac{\Delta\dot{\vec{r}}_{65}}{S_{CH}}\cdot\vec{n}\frac{\sin 30^\circ}{\sin 120^\circ} = \frac{\vec{n}}{S_{CH}\sqrt{3}}\cdot(\Delta\dot{\vec{r}}_{13} + \Delta\dot{\vec{r}}_{46}) \tag{8.26}$$

where $\Delta\dot{\vec{r}}_{ij} = \dot{\vec{r}}_i - \dot{\vec{r}}_j$.

(3) Out-of-plane bond variations It has been shown that the angle variation co-ordinate between the plane of atoms *i, j, k* and bond *k, l* when all atoms are displaced is the sum of the rotation of the plane about the axis perpendicular to bond *kl* lying in plane *ijk* and the counter-rotation of bond *kl* about the same axis toward the normal to the plane. For the case of ethylene (Figure 8.8) the velocity of the defined out-of-plane motion of the C=C bond from the plane of the left-hand CH_2 group can be immediately obtained with equation (8.23), i.e.

$$\dot{\rho} = \frac{\Delta\dot{\vec{r}}_{12}}{S_{CH}}\vec{n}\frac{\sin 120^\circ}{\sin 120^\circ} - \frac{\Delta\dot{\vec{r}}_{32}}{S_{CH}}\vec{n}\frac{\sin 240^\circ}{\sin 120^\circ} + \frac{\Delta\dot{\vec{r}}_{52}}{S_{CC}}\vec{n}$$

$$= \frac{\vec{n}}{S_{CH}}(\Delta\dot{\vec{r}}_{12} + \Delta\dot{\vec{r}}_{32}) + \frac{\Delta\dot{\vec{r}}_{52}}{S_{CC}}\vec{n} \tag{8.27}$$

Thus we can express rates of change with line of various natural co-ordinates through the differences of vectors of atomic displacements from the equilibrium position due to the vibration.

It is worth noting that it is normal to introduce planar co-ordinates in such a way that their positive value corresponds to an increase in bond lengths or to an increase in the valence angle. Conversely, for nonplanar co-ordinates, a positive sign stands for a reduction of the dihedral angle. This can be attributed to the fact that, usually, nonplanar co-ordinates define distortions of the planar portions of a molecule, and in this case, for example, respective dihedral angles can only be reduced.

In conclusion, it is noted that these expressions for the rates of change of the generalized co-ordinates presented in terms of atomic velocities expressed in Cartesian co-ordinates of the form $\dot{q} = \vec{B}\dot{\vec{r}}$ comprise total derivatives of the values of q and radius-vectors $\vec{r}$ with respect to time. Consequently, similar relationships can be written with respect to the differentials $dq = \vec{B}\,d\vec{r}$.

8.5 Co-ordinates Suitable for Specifying the Vibrations of Ligands in the Proximity of Heavy Complexing Atoms

As mentioned in Section (8.1), the co-ordinates introduced in subsequent sections are not always convenient to use in practice. Thus, for example, in describing the motion of ligands in the proximity of heavy complexing atoms the valence-force system of co-ordinates fails to reflect the nature of the metal–ligand chemical bond. Such bonds are often of a delocalized nature, and the motion of ligands in the proximity of the central atoms occurs at low frequencies. To a fairly good approximation we can assume that ligands move as an integral unit with respect to the central atom. In this case, we describe the motion of a ligand toward a complexing atom by introducing a co-ordinate which represents the change in the distance between this atom and the centre of mass of the ligand. We can easily find the respective relationship between the rates of change of such generalized co-ordinates and those of the Cartesian co-ordinates of the atom. Indeed, since the centre of mass of a ligand is determined as $\vec{R} = (1/M)\sum_k m_k\vec{r}_k$, where M_k are the masses of the atoms of a ligand and M its total mass, we have

$$S_{cm}\vec{e}_{cm} = \vec{R} - \vec{r}_0 = \frac{1}{M}\sum_k m_k\vec{r}_k - \vec{r}_0$$

for the vector connecting the central atom ($\vec{r}_0$) and the centre of mass of the ligand, where S_{cm} is the length of this vector.

By differentiating this expression with respect to time and multiplying the resultant relationship by $\vec{e}_{cm}$ in a scalar way we obtain

$$\dot{S}_{cm} = \dot{q}_{cm} = \frac{1}{M}\sum_k m_k\vec{e}_{cm}\dot{\vec{r}}_k - \vec{e}_{cm}\dot{\vec{r}}_0$$

Thus we obtain the relationship for calculating the elements of matrix $\vec{B}$ for the co-ordinate selected.

8.6 Properties Element of the Matrix of Kinematic Coefficients

We derived above a set of formulae which express the rates of change of natural vibrational co-ordinates via the velocities of atoms. These enable us to make up the matrix $\vec{B}$ for the transition from Cartesian co-ordinates of atoms to natural ones. For example, matrix $\vec{B}$ for ethylene is given in Table 8.1 and the system of co-ordinates and vectors is shown in Figure 8.9.

In the table $\vec{F}_{jk}$ denotes the sum $\vec{f}_{jk}/s_j + \vec{f}_{kj}/s_k$ and σ and σ_Q represents s_{CH}^{-1} and s_{CC}^{-1}, respectively. To calculate matrix T we have to make up matrix $\vec{B}$ followed by the multiplication $(\vec{B}E\tilde{\vec{B}})$.

All the elements of matrix $\vec{B}$ are vectors. In order to use a computer for making such multiplications a rectangular co-ordinate system has to be introduced by finding the projections of all of the direction vectors $\vec{e}_i, \vec{f}_{ik}$, etc. onto the axes of this system and substituting the projections of these vectors in matrix $\vec{B}$.

As a result, we obtain three matrices B_x, B_y and B_z (or two matrices, if the molecule is planar and in-plane vibrations alone are being considered). Taking into account that matrix T is formed as a dot product of vectors of matrices $\vec{B}$ and $\tilde{\vec{B}}$, we obtain

$$T = B_x E\tilde{B}_x + B_y E\tilde{B}_y + B_z E\tilde{B}_z \tag{8.28}$$

It will be helpful to use

$$\vec{f}_{12} = \frac{1}{\sin\theta}(\vec{e}_1 \times \vec{e}_2) \times \vec{e}_1 \tag{8.29}$$

Obviously, this equation is not applicable when $\theta = 180°$, i.e. if the bonds are collinear.

Vector $\vec{n}$ in non-planar co-ordinates is also capable of being expressed through

Table 8.1. Natural vibrational co-ordinates for the planar vibrations of ethylene as expressed through atomic displacements with the aid of matrix $\vec{B}$

$$\begin{bmatrix} Q \\ q_1 \\ q_2 \\ q_3 \\ q_4 \\ \alpha \\ \beta_1 \\ \beta_2 \\ \alpha_2 \\ \beta_3 \\ \beta_4 \end{bmatrix} = \begin{bmatrix} 0 & 0 & 0 & 0 & \vec{e}_1 & -\vec{e}_5 \\ \vec{e}_1 & 0 & 0 & 0 & -\vec{e}_1 & 0 \\ 0 & \vec{e}_2 & 0 & 0 & -\vec{e}_2 & 0 \\ 0 & 0 & -\vec{e}_3 & 0 & 0 & -\vec{e}_3 \\ 0 & 0 & 0 & \vec{e}_4 & 0 & -\vec{e}_4 \\ -\sigma\vec{f}_{12} & -\sigma\vec{f}_{21} & 0 & 0 & \vec{F}_{12} & 0 \\ -\sigma\vec{f}_{1Q} & 0 & 0 & 0 & \vec{F}_{1Q} & -\sigma_Q\vec{f}_{Q1} \\ 0 & -\sigma\vec{f}_{2Q} & 0 & 0 & \vec{F}_{2Q} & -\sigma_Q\vec{f}_{Q2} \\ 0 & 0 & -\sigma\vec{f}_{34} & -\sigma\vec{f}_{43} & 0 & \vec{F}_{34} \\ 0 & 0 & -\sigma_Q\vec{f}_{Q3} & 0 & -\sigma_Q\vec{f}_{Q3} & \vec{f}_{3Q} \\ 0 & 0 & 0 & -\sigma\vec{f}_{4Q} & -\sigma_Q\vec{f}_{Q4} & \vec{F}_{4Q} \end{bmatrix} \begin{bmatrix} \vec{r}_1 \\ \vec{r}_2 \\ \vec{r}_3 \\ \vec{r}_4 \\ \vec{r}_5 \\ \vec{r}_6 \end{bmatrix}$$

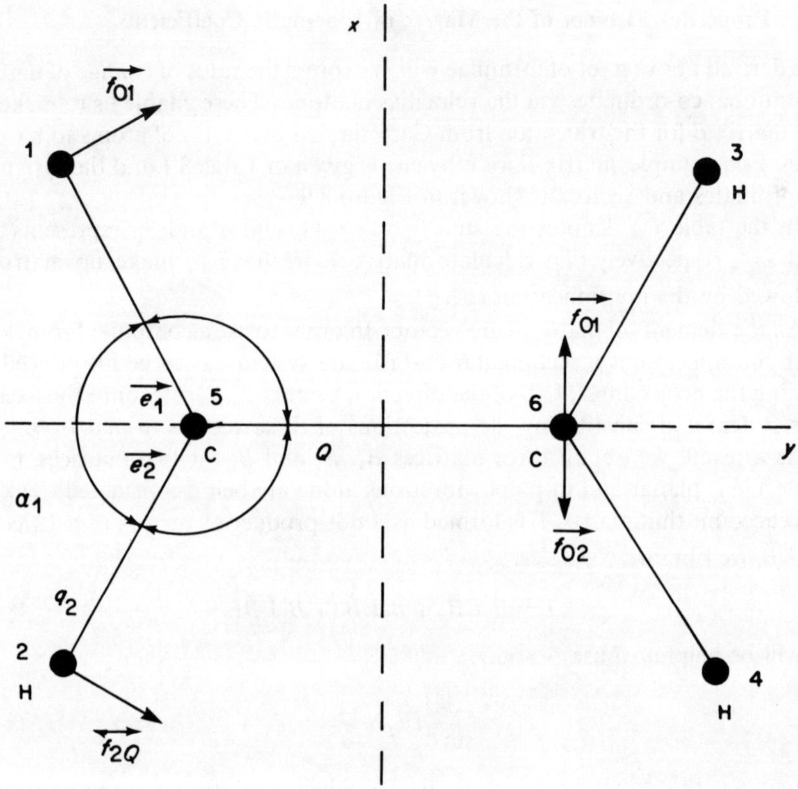

Figure 8.9. The co-ordinates and vectors system of ethylene molecule (plane vibrations)

the co-ordinates of the direction vectors $\vec{e}_1$ and $\vec{e}_2$ according to

$$\vec{n} = \frac{1}{\sin\theta}(\vec{e}_1 \times \vec{e}_2)$$

Let us now consider some properties of the matrices $\vec{B}$ and T. It has already been mentioned that the sum of elements in each row of matrix $\vec{B}$ and that of the vector products of the elements in each row $\vec{B}$ and the equilibrium radius-vectors of respective atoms must equal zero. Moreover, if there is a linear relationship between certain natural co-ordinates, then a similar relationship must also hold true for the rows of matrix $\vec{B}$ related to the co-ordinates.

When an element of matrix T lying at the crossing of a line corresponding to the ith co-ordinate and a column corresponding to the jth co-ordinate is identified then it can easily be seen that the element of interaction between the ith and jth co-ordinates will be equal to

$$\tau_{ij} = \sum_{k}^{n} \varepsilon_k \vec{b}_{ik} \vec{b}_{jk} \tag{8.30}$$

Subscript k in $\vec{b}_{ik}$ corresponds to the kth atom. From this it follows that if the given structural elements do not possess atoms in common, and in which changes in the geometry of the elements are taken as natural vibrational co-ordinates, then the element of kinematical interaction of such co-ordinates will equal zero. In fact, in this case no non-zero elements will be found in the rows of co-ordinates with indices i and j in matrix $\vec{B}$, which might be positioned in one column, i.e. there will be non-zero elements $\vec{b}_{ik}$ and $\vec{b}_{jk}$ with the same indices k of the atoms.

Now let us consider a certain group of atoms (designated by index 1) coupled by elastic forces. In a Cartesian system of co-ordinates the kinetic energy of this group is characterized by a matrix $T_1 = E_1$ and the positions of the atoms are characterized by radius-vectors, the rate of change of which form a column matrix $\dot{\vec{r}}^{\mathrm{I}}$.

Let us introduce natural co-ordinates designated by a column matrix, q^{I}, so that they characterize the vibrational motion of group I as a free molecule and, call these co-ordinates the internal co-ordinates with respect to group 1. Then, within the system of the natural co-ordinates, velocities of co-ordinates of totality $\dot{q}^{\mathrm{I}}$ must be associated with the velocities of co-ordinates of totality $\dot{\vec{r}}^{\mathrm{I}}$ through the relationship $\dot{q}^{\mathrm{I}} = \vec{B}^{\mathrm{I}}\dot{\vec{r}}^{\mathrm{I}}$ since they are dependent only upon the selection of the co-ordinates and the geometry of group I. Next, we introduce group I without any changes into another molecule, i.e. we add a totality II of atoms. A new molecule produced in this way can be described through the set of previous natural co-ordinates, plus a new set of additional ones designated as q^{II}. Since the natural co-ordinates q^{I} are expressed only through the displacements of atoms of group I, we obtain

$$\begin{bmatrix} \dot{q}^{\mathrm{I}} \\ \dot{q}^{\mathrm{II}} \end{bmatrix} = \begin{bmatrix} \vec{B}^{\mathrm{I}} & 0 \\ \vec{B}^{\mathrm{II,I}} & \vec{B}^{\mathrm{II}} \end{bmatrix} \begin{bmatrix} \dot{\vec{r}}^{\mathrm{I}} \\ \dot{\vec{r}}^{\mathrm{II}} \end{bmatrix}$$

where $\dot{\vec{r}}^{\mathrm{II}}$ is a column matrix of the changes in the atomic radius-vectors of the atoms added to group I. Then the complete matrix of kinematical coefficients acquires the form

$$T = \begin{bmatrix} \vec{B}^{\mathrm{I}}E^{\mathrm{I}}\tilde{\vec{B}}^{\mathrm{I}} & \vec{B}^{\mathrm{I}}E^{\mathrm{I}}\tilde{\vec{B}}^{\mathrm{II,I}} \\ \vec{B}^{\mathrm{II,I}}E^{\mathrm{I}}\tilde{\vec{B}}^{\mathrm{I}} & \vec{B}^{\mathrm{II,I}}E^{\mathrm{I}}\tilde{\vec{B}}^{\mathrm{II,I}} + \vec{B}^{\mathrm{II}}E^{\mathrm{II}}\tilde{\vec{B}}^{\mathrm{II}} \end{bmatrix}$$

This can be demonstrated by carrying out the multiplication $\vec{B}E\tilde{\vec{B}}$, where

$$\vec{B} = \begin{bmatrix} \vec{B}^{\mathrm{I}} & 0 \\ \vec{B}^{\mathrm{II,I}} & \vec{B}^{\mathrm{II}} \end{bmatrix} \quad \text{and} \quad E = \begin{bmatrix} E^{\mathrm{I}} & 0 \\ 0 & E^{\mathrm{II}} \end{bmatrix}$$

Thus the elements of matrix T related to all the internal co-ordinates of group I are not changed when a group of atoms is transferred from one molecule to another, provided that the geometrical parameters and masses of the atoms in this group do not vary. This property of the elements of matrix T is most useful for the successive exploration of homologous series of molecules with increasingly complicated structure. In this case, the unit arrays or blocks in matrix T, relating to the internal co-ordinates of the groups common to a row of molecules, are not changed provided that the groups exhibit the property of additivity. As a result, all the computations are considerably reduced.

8.7 Standard Formulae for Kinematic Coefficients

In several cases of general investigation it has prove helpful to use standard formulae for the elements of the matrix of kinematic coefficients. To derive these formulae it is sufficient to choose rows in matrix $\vec{B}$ relating to the co-ordinates whose kinematic interaction are to be determined. We then multiply by one another the elements in these rows located in the columns whose numbers coincide with those of atoms common to the pair of co-ordinates chosen. These products are then multiplied by the number of common atoms and the results added together. For example, for a group comprising atoms 0, 1, 2 (see Figure 8.4 and matrix (8.13)) the kinematic interaction element for co-ordinates q_1 and q_1 is

$$\tau_{11} = \varepsilon_0 \vec{e}_1^{\,2} + \varepsilon_1 \vec{e}_1^{\,2} = \varepsilon_0 + \varepsilon_1 \tag{8.31}$$

for co-ordinates q_1 and q_2 it is

$$\tau_{12} = \varepsilon_0 \vec{e}_1 \vec{e}_2 = \varepsilon_0 \cos\theta \tag{8.32}$$

and for co-ordinates q_1 and γ we finds

$$-\varepsilon_0 \frac{1}{s_2} \sin\theta = -\varepsilon_0 \frac{1}{s_2} \vec{e}_1 \vec{f}_{21} \tag{8.33}$$

It was shown in Section 8.6 that the elements of the T matrix are dependent only upon the properties of a given structural element. That is why the form of the equations derived above for the kinematic interactions of the co-ordinates remains invariable in all of those molecules which are described in such a way that changes in the bond lengths and valence angles are taken as co-ordinates, and hence these can be tabulated as standard formulae.

In addition, other sets of general formulae can be derived for polyatomic groups, since there are three possible types of interaction of angles with bonds:

(1) The interaction discussed above (equation (8.33)) of an angle with a bond forming one of the arms of the angle;
(2) Interaction between an angle and a bond possessing an atom lying at the vertex of the angle; and,
(3) Interaction between an angle and a bond having an atom lying at the end of one arm of the angle.

The corresponding formulae are

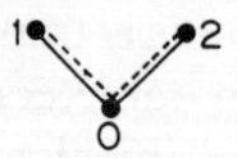

$$-\varepsilon_0 \frac{1}{s_2} \vec{e}_1 \vec{f}_{21} \tag{8.34}$$

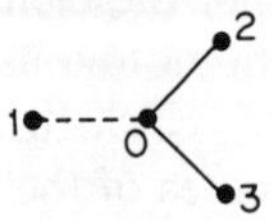

$$-\varepsilon_0 \vec{e}_1 \vec{F}_{23} \tag{8.35}$$

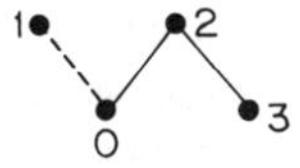

$$-\varepsilon_0 \vec{e}_1 \frac{\vec{f}_{23}}{s_2} \tag{8.36}$$

There are six types of angle/angle interaction, and these consist of interaction of an angle with itself and two types of interaction of angles having one common link:

$$\varepsilon_0 \vec{F}_{12}^2 + \frac{\varepsilon_1 \vec{f}_{12}^2}{s_1^2} + \frac{\varepsilon_2 \vec{f}_{12}^2}{s_2^2} \tag{8.37}$$

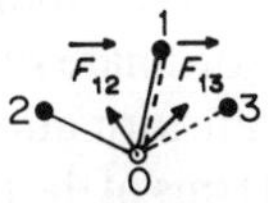

$$\varepsilon_0 \vec{F}_{12} \vec{F}_{13} + \varepsilon_1 \frac{1}{s_1^2} \vec{f}_{12} \vec{f}_{13} \tag{8.38}$$

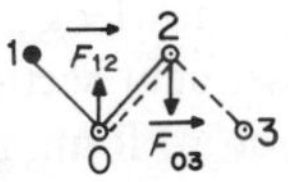

$$-\varepsilon_0 \frac{1}{s_2} \vec{f}_{03} \vec{F}_{12} - \varepsilon_2 \frac{1}{s_2} \vec{f}_{21} \vec{F}_{03} \tag{8.39}$$

Additionally, there are three types of interaction involving angles having one common atom:

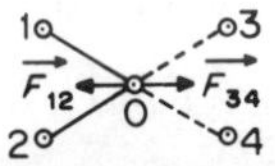

$$\varepsilon_0 \vec{F}_{12} \vec{F}_{34} \tag{8.40}$$

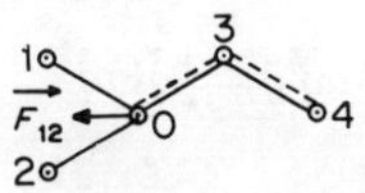

$$\frac{-\varepsilon_0 \vec{F}_{12} \vec{f}_{34}}{s_3} \tag{8.41}$$

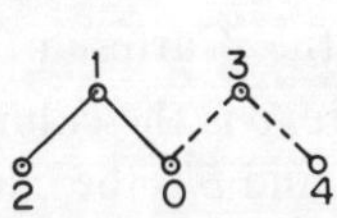

$$\frac{\varepsilon_0 \vec{f}_{12} \vec{f}_{34}}{s_1 s_3} \tag{8.42}$$

The types of interaction are schematically illustrated alongside the formulae. Thus we have nine types of interaction of angles with bonds and of interaction of angles with each other. Including bond/bond interactions we have, in all, eleven major types of kinematic interactions. Naturally, all these interactions should be taken into account in setting up secular equations.

Similarly, formulae for the kinematic coefficients involving non-planar co-ordinates (and for the interaction of these with planar co-ordinates) can be derived. The overall number of various types of interaction is rather high, numbering 43 in addition to those already mentioned.

When making manual calculations it is helpful to use the tables of kinematic coefficients when vibrations are described in terms of co-ordinates involving changes in valence angles and bond lengths. It is not advisable to use the corresponding standard formulae for the kinematic coefficients when there are different types of co-ordinates since the number and complexity of these would increase considerably. In this case it is much easier to compile the $\vec{B}$ matrix, since, in essence, knowledge of only three expressions of the type (8.7), (8.10) and (8.22) is needed, and the matrix itself contains a large number of zeros. We then obtain the kinematic coefficient matrix by simply multiplying together matrices $\vec{B}$, E and $\tilde{\vec{B}}$.

8.8 Expression for the Displacements of Atoms from the Equilibrium Positions through the Natural Co-ordinates

Matrix $\vec{B}$ expresses the rates of change of the natural vibrational co-ordinates through the velocities of atoms in a Cartesian co-ordinate system. In addition to knowing $\vec{B}$ we also need to know the inverse transformation matrix as well as certain properties of the kinetic energy matrix in terms of the natural co-ordinates in the co-ordinate representation chosen. This knowledge will prove indispensable in coping with problems raised in subsequent sections such as infrared band intensities and the degree of polarization of Raman lines.

In practice, we seek a matrix $\vec{B}$ which effects the transformation*

$$\dot{\vec{r}} = \vec{B}'\dot{q}$$

Substitutions of this expression in that for the kinetic energy

$$T_{\text{kin}} = \tfrac{1}{2}\sum_i m_i \dot{\vec{r}}_i^{\,2} = \tfrac{1}{2}\dot{\tilde{\vec{r}}} M \dot{\vec{r}},$$

where $M = [m_i \delta_{ik}]$, then gives,

$$T_{\text{kin}} = \tfrac{1}{2}\dot{\tilde{q}}\tilde{\vec{B}}' M \vec{B}'\dot{q} = \tfrac{1}{2}\dot{\tilde{q}}\Theta\dot{q} \tag{8.43}$$

To determine $\vec{B}'$ we recall that the column-matrix $\vec{p}$ of momenta $\vec{p}_i$ for the atoms of a molecule conjugate to the Cartesian co-ordinates, which are stationary in space, equals $\vec{p} = \tilde{\vec{B}}p$. Here, p is the column-matrix of momenta p_j conjugate to the natural co-ordinates q_i and $\vec{B}$ is the matrix of transition from the velocities of atoms in the Cartesian co-ordinates to the rates of change in the natural co-ordinates dealt with above. Note that in the formula $\dot{q} = \vec{B}\dot{\vec{r}}$ column $\dot{\vec{r}}$ can comprise arbitrary $\dot{\vec{r}}_i$, which correspond to both purely vibrational motion and to the motions of a molecule or of its separate parts as a whole. In the latter case the values of $\dot{q}_j$ will equal zero. On the other hand, in the formula $\vec{p} = \tilde{\vec{B}}p$ the momenta $\vec{p}_i$ will be associated with purely vibrational motion, which is in precise agreement with the fact that momenta p_j are related to the purely vibrational co-ordinates alone. Indeed, for any set p_j, the momenta $\vec{p}_i$ obtained via the transformation ($\vec{p} = \tilde{\vec{B}}p$) obey the conditions of a purely vibrational motion, namely, $\sum_i \vec{p}_i = 0$ and $\sum_i(\vec{p}_i \times \vec{r}_i) = 0$ (equality to zero of momenta and angular momenta of a system of particles), which can be directly inferred from the

*Again we assume that the elements of this matrix and vectors in the subsequent formulae are functions of time.

properties of the elements of a row in matrix $\vec{B}$ or a column in matrix $\tilde{\vec{B}}$ (see Sections 8.3 and 8.4). Incidentally, the transformation $\vec{p} = \tilde{\vec{B}}p$ is determined to within the accuracy of the translatory motions and rotations of a molecule as a whole.

Since $\dot{\vec{r}}_i = \varepsilon_i \vec{p}_i$ for the atoms of the system, expressions for the column of velocities of atoms under vibration will be obtained by multiplying $\vec{p} = \tilde{\vec{B}}p$ from the left by a diagonal matrix $E = [\varepsilon_i \delta_{ik}]$. Therefore we obtain

$$\dot{\vec{r}} = E\tilde{\vec{B}}p = E\tilde{\vec{B}}\Theta\dot{q} \tag{8.44}$$

since $p_j = \partial T_{\text{kin}}/\partial \dot{q}_j$ by definition, and hence, taking into account equation (8.43), $p = \Theta\dot{q}$. Hence the sought-after form for matrix $\vec{B}$ is as follows:

$$\vec{B}' = E\tilde{\vec{B}}\Theta \tag{8.45}$$

We should further bear in mind that the transformation $\dot{q} = \vec{B}\dot{\vec{r}}$ holds true both in the case where $\dot{\vec{r}}_i$ obeys the conditions for vibrational motions and where the $\dot{\vec{r}}_i$ are of an arbitrary nature. That is why if a set $\dot{q}_j$ is given, then using the transformation $\vec{B}'\dot{q} = \dot{\vec{r}}$ we can obtain $\dot{\vec{r}}_i$ for the vibrational motions, and then, by applying a transformation $\vec{B}\dot{\vec{r}} = \dot{q}$, again shift to the same $\dot{q}_j$. From this it can be concluded that matrices $\vec{B}$ and $\vec{B}'$ must be related to one another by the correlation

$$\vec{B}\vec{B}' = I \tag{8.46}$$

However, $\vec{B}'\vec{B} \neq I$. This can be attributed to the irreversibility of the transformation of arbitrary Cartesian velocities of atoms into the velocities of the natural co-ordinates, and vice versa.

Employing the above property of matrices $\vec{B}$ and $\vec{B}'$, we obtain

$$I = \vec{B}\vec{B}' = \vec{B}E\tilde{\vec{B}}\Theta = T\Theta \tag{8.47}$$

from which it follows that $T\Theta = I$. If the vibrational co-ordinates are chosen to be independent, then matrix T is a non-singular matrix, possessing a reciprocal matrix ($T^{-1} = \Theta$).

Finally

$$\vec{B}' = E\tilde{\vec{B}}T^{-1} \tag{8.48}$$

and hence

$$\dot{\vec{r}} = E\tilde{\vec{B}}T^{-1}\dot{q} \tag{8.49}$$

It therefore follows that for exact differentials, or for small displacements of atoms $\mathrm{d}\vec{r}_i$ from the equilibrium positions in the system of co-ordinates associated with the instantaneous unit vectors of a molecule,

$$\mathrm{d}\vec{r} = E\tilde{\vec{B}}T^{-1}q \quad (\mathrm{d}q = q) \tag{8.50}$$

When shifting to normal co-ordinates, and taking into account that $q = L_q Q$, we obtain the following relationship:

$$\mathrm{d}\vec{r} = E\tilde{\vec{B}}T^{-1}L_q Q \quad (\mathrm{d}Q = Q) \tag{8.51}$$

Therefore, for a particular normal vibration described by a normal co-ordinate with index k, we obtain

$$\mathrm{d}\vec{r}^{(k)} = E\tilde{\vec{B}}T^{-1}l_q^{(k)}Q_k \tag{8.52}$$

This is the so-called Crawford formula[2].

An expression for the displacements of atoms from the equilibrium positions, using normal co-ordinates, can be obtained in a somewhat different form. To this end we gain use relationship $\vec{p} = \tilde{\vec{B}}p$ or $\dot{\vec{r}} = E\tilde{\vec{B}}p$ where p is a column of momenta conjugate to the natural co-ordinates. These momenta are related to those conjugate to the normal co-ordinates through matrix L_p so that $p = L_p \cdot P$. From this it follows that $\dot{\vec{r}} = E\tilde{\vec{B}}L_p \cdot P$. However, because of the normalization of the normal co-ordinates $P = \dot{Q}$. Therefore

$$\dot{\vec{r}} = E\tilde{\vec{B}}L_p \cdot \dot{Q} \tag{8.53}$$

and

$$\mathrm{d}\vec{r} = E\tilde{\vec{B}}L_p \cdot Q \quad (\mathrm{d}Q = Q) \tag{8.54}$$

For the kth vibration

$$\mathrm{d}\vec{r}^{(k)} = E\tilde{\vec{B}}l_p^{(k)}Q_k \tag{8.55}$$

Note that while equations (8.50), (8.51) and (8.52) are true only for the independent co-ordinates, equations (8.53) and (8.54) are also true for the dependent co-ordinates.

It is worth emphasizing one important property of the kinetic energy matrix in the co-ordinate representation $\tilde{\vec{B}}'M\vec{B}'$. This is that, as distinct from the kinetic energy matrix in the momentum representation, the elements of the former matrix related to the internal co-ordinates of a certain group of atoms will not be invariant within the series of molecules all containing this group. This is accounted for by the requirement for the conservation of momentum and angular momentum, which requirement is imposed upon the entire molecule rather than on its separate parts.

This property of non-invariance can be further inferred from the very form of matix $\Theta = \tilde{\vec{B}}' M \vec{B}'$, wherein all the elements are generally non-zero (in contrast to matrix $T = \vec{B}E\tilde{\vec{B}}$, where some elements are always zero in the case of a complex molecule).

8.9 Expression for Derivatives of Bond Vectors with Respect to the Vibrational Co-ordinates

For anharmonic, infra-red and Raman intensity calculations (Chapters 11 and 15) it is necessary to know the expressions for the first- and the second-order unit vectors of bond derivatives with respect to vibrational co-ordinates. To construct the corresponding formulae let us return to expressions (8.6), rearranged to give

$$\dot{\vec{e}}_1 = 1/s_1(\Delta\dot{\vec{r}}_1 - \dot{q}_1\vec{e}_1)$$
$$\dot{\vec{e}}_2 = 1/s_2(\Delta\dot{\vec{r}}_2 - \dot{\vec{q}}_2\vec{e}_2) \text{ etc.}$$

Here, $\Delta\dot{\vec{r}}_1 = \dot{r}_1 - \dot{\vec{r}}_0$ is the difference between the Cartesian velocities of the first and the end atoms for the corresponding bond.

Using matrix notation for all bonds we have

$$\dot{\vec{e}} = S^{-1}(\Delta\dot{\vec{r}} - \vec{E}'\dot{q})$$

where S^{-1} is the diagonal matrix of the reciprocal bond lengths; Δ is the matrix in which the number of rows is equal to the number of bonds and the number of columns is equal to the number of atoms in the molecule (in each row of this matrix there is $+1$ to represent the end atom of the bond and a -1 for the beginning atom of the bond and the remaining elements are equal to zero); and $\vec{E}' = [\vec{E}, 0]$ is the matrix in which the number of rows is also equal to the number of bonds, and the number of columns is equal to those of vibrational co-ordinates. Every row of this matrix contains only one non-zero element equal to the unit vector of the corresponding bond and located in the mesh corresponding to the number of the co-ordinate of the given bond.

After taking into account the results of Section 8.8 we can write

$$\dot{\vec{e}} = S^{-1}(\Delta E\tilde{\vec{B}}\Theta - \vec{E}')\dot{q} \tag{8.56}$$

or

$$\dot{\vec{e}} = S^{-1}(\Delta E\tilde{\vec{B}}L_p - \vec{E}'L_q)\dot{Q} \tag{8.57}$$

The total differential of the bond unit vectors from equation (8.57) gives

$$\mathrm{d}\vec{e} = S^{-1}(\Delta E\tilde{\vec{B}}L_p - \vec{E}'L_q)\,\mathrm{d}Q \tag{8.58}$$

Hence the matrix of the partial derivatives of the unit bond vectors is given by

$$\left(\frac{\partial\vec{e}}{\partial Q}\right) = S^{-1}(\Delta E\tilde{\vec{B}}L_p - \vec{E}'L_q) \tag{8.59}$$

and therefore

$$\left(\frac{\partial\vec{e}}{\partial q}\right) = S^{-1}(\Delta E\tilde{\vec{B}}\Theta - \vec{E}') \tag{8.60}$$

where $(\partial\vec{e}/\partial q)$ is the matrix of partial derivatives of unit bond vectors with respect to the natural co-ordinates and Θ is the matrix of the kinetic energy in the co-ordinate representation (inverse to the kinematic one). Note that Θ must also be calculated for the equilibrium state.

It follows from equation (8.60) that the first derivatives of the unit bond vectors are expressed in terms of the co-ordinate unit vectors $\vec{e}_k, \vec{f}_{kk}$, etc. which appear in the formulae for the elements of the matrices $\vec{B}$ and $\vec{E}'$.

Differentiation of the matrix of the second derivatives $(\partial^2\vec{e}_k/\partial q_i\partial_q q_j)$ with respect to a particular co-ordinate with subscript j yields

$$\left(\frac{\partial^2 \vec{e}_k}{\partial q_i \partial q_j}\right) = S^{-1}\left(\Delta E \frac{\partial \tilde{\tilde{B}}}{\partial q_j}\Theta + \Delta E \tilde{\tilde{B}} \frac{\partial \Theta}{\partial q_j} - \frac{\partial \vec{E}'}{\partial q_j}\right) + \left(\frac{\partial}{\partial q_j} S^{-1}\right)(\Delta E \tilde{\tilde{B}} \Theta - \vec{E}')$$

Under differentiation of matrix $\vec{B}$ as well as obtaining the first derivatives of the vectors $\vec{e}_k$ (this can be done according to the known rules) the vectors $\vec{f}_{kk'}$ and others must also be differentiated (with respect to angular co-ordinates). For non-linear regions of a molecule the derivatives of the vectors $\vec{f}_{kk'}$ can be calculated in the simplest way by differentiating the expression

$$\vec{f}_{kk'} = \frac{1}{\sin(\vec{e}_k, \vec{e}_{k'})}(\vec{e}_k \times \vec{e}_{k'}) \times \vec{e}_k$$

In calculating the derivatives of vector $\vec{f}_{kk'}$ for the collinear regions the method used in valence-optical theory for Raman spectral intensities is applied. Thus (see Chapter 15)

$$\partial \vec{f}_{kk'} / \partial q_i = -(\vec{f}_{kk'} \cdot \partial \vec{e}_k / \partial q_i) \vec{e}_k$$

With non-planar co-ordinates the unit vectors $\vec{n}$ and $\vec{g}$ appear, which in dihedral angles play the same role as the unit vector $f_{kk'}$ in the angular co-ordinates. Therefore the calculations of the above derivatives with respect to the internal co-ordinates can be performed in exactly the same way.

Having outlined the methods for calculating the derivatives of the elements of the matrix $\vec{B}$ with respect to the internal co-ordinates it is significant that all first derivatives are calculated through the values of the co-ordinate unit vectors. This fact permits us to obtain the second derivatives $(\partial^2 \vec{e}_k / \partial q_i \partial q_j)_0$, which are also expressed in terms of the unit vectors of the internal co-ordinates. Since the corresponding expressions are cumbersome, they are not given in detail. Only a description of the calculation procedure is included. Based on the same considerations, the derivatives $(\partial^3 \vec{e}_k / \partial q_i \partial q_j \partial q_e)_0$ may also be calculated.

References

1. M. V. Volkenshtein, L. A. Gribov, M. A. Elyashevich and B. I. Stepanov. *Vibrations of Molecules*, Nauka, Physico-Mathematical Literature, Moscow (1972) (in Russian).
2. B. L. Crawford and E. W. H. Fletcher. *J. Chem. Phys.*, **19**, 141 (1951).

Chapter 9

The Potential Function of a Polyatomic Molecule and its Properties

9.1 An Analytical Form of the Potential Function

The movement of atoms in a molecule takes place in a potential field which is given by the appropriate potential function that describes an adequately complex potential surface. We showed in Chapter 2 that in the adiabatic approximation this surface is a geometric locus corresponding to the complete energy (the purely electronic and nucleus–nucleus repulsions) of the molecule for different nuclear dispositions. When the complete energy is calculated for each point on the locus (which is, in principle, possible using methods already described) all nuclei are assumed to be immobile.

Clearly, since a molecule's complete energy calculated in this way will not depend on the translational and rotational motions of the system as a whole, it is obvious that is must only be a function of internal co-ordinates, i.e. in most cases bond lengths, valency angles, angles between bonds and planes, dihedral angles, etc.

Unfortunately, our knowledge of the form of a potential function during an arbitrary change in internal co-ordinates is very meagre, particularly so at the dissociation threshold. This is due to the very cumbersome nature of the quantum-mechanical calculations of potential surfaces, which has not enabled us so far estimate sufficient points on the surface of a complex system to make it possible to come to any generalized conclusions about the analytical form of the corresponding function. It has also not been possible to study the potential surface in adequate detail experimentally by processing spectral data and solving what is known as the inverse spectral problem (see below). Thus the literature does not contain any procedures for establishing an analytical description of a potential surface, even though many different such representations have been attempted. Since none of these can in any way be thought satisfactory, we shall not dwell on the subject.

What we have just said does not, of course, mean that we do not know anything of the typical forms of potential surfaces for what may be termed 'normal' molecules when they have a state that has a characteristic form. We know that the potential surface for the majority of sections has a single minimum, though a few may have to (for example, molecules that can be inverted or have *cis* and *trans* isomers or in the presence of chair- and bath-type conformations) or more. The energy grows in a monotonous way for most cases, starting from a minimum which corresponds to what is called the molecule's 'geometrical equilibrium' (and only in this case can we speak of the existence of such a geometry). It grows to a limit that corresponds to the system's dissociation energy along some internal co-ordinate, or, alternatively, it tends to infinity if the change in the co-ordinates make atoms approach each other closely. For some co-ordinates (for example, ammonia inversion ones) the potential cross-section has two minima and two wings leading to dissociation. These correspond to the positive and negative values of the inversion co-ordinate, the origin being at the point where the nitrogen nucleus passes through the centre of the plane in which the hydrogen atom nuclei lie.

A section of a potential surface need not have any dissociation limits at all, having instead several minima that occur, say, in the presence of internal rotations, and the co-ordinates correspond to this motion. The section then has the form of a periodic function.

Proceeding from these general considerations (at present there are unfortunately no others) we can suggest the following analytical expression in order to describe the main features of a potential surface:

$$\begin{aligned} E(q) = V(q,\chi,\rho) &= \tfrac{1}{2}\sum_{i,j} D_{ij}\alpha_i\alpha_j + \tfrac{1}{6}\sum_{i,j,k} D_{ijk}\alpha_i\alpha_j\alpha_k \\ &\quad + \tfrac{1}{24}\sum_{i,j,k,l} D_{ijkl}\alpha_i\alpha_j\alpha_k\alpha_l + \sum_{n,m}(V_{no} + a_{mn}\cdot\cos m\chi_n + b_{mn}\sin m\chi_n) \qquad (9.1) \\ &\quad + \sum_t \{b_t(\rho_t - \rho_{et})^2 + \exp[-c_t(\rho_t - \rho_{et})]^2\} \\ \alpha_i &= \{1 - \exp[-a_i(q_i - q_{ie})]\} \end{aligned}$$

Here the first three expressions are terms in a Dirichlet series (an expansion with respect to Morse braces $\alpha_i = \{1 - \exp[-a_i(q_i - q_{ei})]\}$) having coefficient D_{ij} which describe the dissociation energy; q_{ei} is the equilibrium value (corresponding to the energy minimum) of a valency angle or bond; and a_i is a parameter of a Morse function for individual constituents. These terms only contain angle and bond co-ordinates. The fourth term refers to co-ordinates which describe internal rotations and is therefore a periodic function (Fourier series) with parameters a_{mn}, b_{mn}, V_{on}. The fifth term characterizes inversion motions.

This potential function is quite general. The only terms that are absent are those that correspond to interactions between inversion motions and internal rotations and other forms of motions relative to nuclear motions in the molecule. A whole series of results from practical estimates have shown that this approximation can in fact be made, even though it is obviously not too strict.

The potential function is very general, because any smooth surface (for example, one with a single minimum) can always be represented by a Dirichlet series. It is, in principle, possible to conceive that all this function's parameters could be determined if the complete adiabatic electronic energy could be calculated quantum-chemically, taking nuclear–nuclear repulsions into account, for a large number of independent sets of nuclear dispositions and sets of the co-ordinates we use to describe this disposition. As a result, a system of equation is obtained in terms of the parameters being sought and a solution of this set of equations must then be found.

We have one comment to make here. The function $E(q)$ can be written both in terms of a system of independent co-ordinates or of a system of dependent ones. We mean by the latter that the deformation of each of the molecule's structural elements requires it own energy independent of the others. In terms of mechanical models this means that each bond and valency angle is assigned its own spring. Each of these springs is introduced as being independent by the definition of the problem itself. By contrast, in real deformations, changes, for example, in two angles inevitably lead to a deformation in the third if they are dependent. Obviously this must be borne in mind when calculating, point by point, the parameters of a potential surface. Hence such a relationship will not hinder the determination of the parameters separately (see Section 9.5).

We now consider a simple example. Let us assume we have a mechanical system consisting of three balls connected together by springs, as shown in Figure 9.1. The deformation energy will be measured from the state in which all the springs are stationary. We take the lengths of the springs as the co-ordinates, i.e. X_1, X_2 and $X_3 = X_1 + X_2$. The system's deformation energy will then be calculated by

$$E(X) = k_1X_1^2 + k_2X_2^2 + k_3X_3^2 \quad \text{or} \quad E(X) = k_1X_1^2 + k_2X_2^2 + k_3(X_1 + X_2)^2$$

Let us assume that we have in some way measured the deformation energy for the cases $X_1 = 1$, $X_2 = 0$; $X_1 = 0$, $X_2 = 1$; and $X_1 = 1$, $X_2 = 1$. Hence we obtain the equation system:

$$E_1 = k_1 + k_3$$
$$E_2 = k_2 + k_3$$
$$E_3 = k_1 + k_2 + 4k_3$$

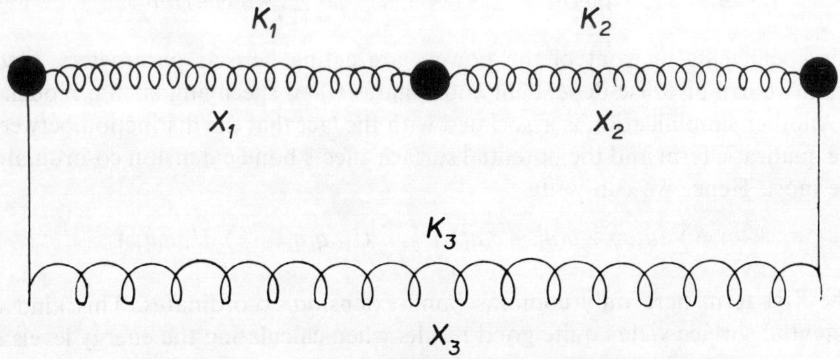

Figure 9.1. Mechanical model of linear three-atomic molecule

and by solving for the parameters k_1, k_2 and k_3 we find

$$k_1 = 3/2E_1 + 1/2E_2 - 1/2E_3, \quad k_2 = 3/2E_2 + 1/2E_1 - 1/2E_3$$

and

$$k_3 = 1/2(E_3 - E_1 - E_2)$$

Hence we have solved the problem even though there were dependent co-ordinates. We must note, though, that this would be impossible by observing the vibration frequencies only because we would only obtain two equations, and this is insufficient for determining three coefficients. We shall return to this later.

We can simplify equation (9.1) if we transfer the co-ordinate origin to the minimum and measure those co-ordinates from it which attain the minimum. This corresponds to a transformation from absolute internal co-ordinates to natural ones, i.e. increments in bond lengths and valency angles, etc. with respect to their equilibrium values. This transformation, which was mentioned in Chapter 6, cannot be put into the form of a kinetic energy operator, but it is useful when describing the potential energy.

It can be simplified further if it is remembered that, at low energies, atoms move about their equilibrium positions and the dissociation threshold is never approached. We will obtain only finite motions which correspond to discrete energy levels. Obviously, this limits our computational ability, but unfortunately we do not know any way of finding a full solution. Moreover, in most cases experimentally this is precisely the kind of motions with which we have to deal.

Remember at this juncture that, in general, when solving the Schroedinger equation of a polyatomic molecule many different isomers may have the same gross formula. Hence in order to choose one for analysis we must restrict the possible atom motions, and this will automatically yield a potential surface that will have a good, deep, unique minimum for most sections.

Different simplifications can be made here. It is usual to substitute a function having a dissociation threshold for one not having a threshold (a well with infinite walls). A power function truncated at some point (for example, a fourth-degree polynomial) is often chosen. In this case we obtain

$$V(q) = \tfrac{1}{2}\sum_{i,j} U_{ij}q_iq_j + \tfrac{1}{6}\sum_{i,j,k} U_{ijk}q_iq_jq_k + \tfrac{1}{24}\sum_{i,j,k,l} U_{ijkl}q_iq_jq_kq_l + \cdots \tag{9.2}$$

The coefficients in front of the powers are acting here as parameters. Once again we can also use dependent coordinates when specifying such a model.

Another simplification is associated with the fact that the distinction between the quadratic term and the potential surface affects bond extension co-ordinates the most. Hence we can write

$$V(q) = \sum_k (a_kq_k^2 + b_kq_k^3 + c_kq_k^4) + \tfrac{1}{2}\sum_{k,k'} U_{kk'}q_kq_{k'} + \tfrac{1}{2}\sum_{i,j} U_{ij}q_iq_j + \cdots$$

The first term here only contains bond-extension co-ordinates. This kind of potential surface yields quite good results when calculating the energy levels of a polyatomic molecule. When calculating very high energy levels it is better to take Dirichlet series as a first term.

Finally, the most simple functions only contain quadratic terms for bond and angle co-ordinates and the potential function takes the form:

$$V(q) = \tfrac{1}{2}\sum_{i,j} U_{ij}q_iq_j + V_{\text{inverse}} + V_{\text{int.rot}}$$

It is possible to take one last step. When the minima in the sections for internal rotation and inversion co-ordinates are deep enough, we need only consider vibrations in the region of the minimum. Thus, for instance, we can calculate the energy levels of a system with many minima separately for each minimum and the potential function for all the degrees of freedom can be selected as a quadratic form (a multidimensional paraboloid):

$$V(q) = \tfrac{1}{2}\sum_{i,j} U_{ij}q_iq_j$$

In the great majority of cases this approximation can be used to estimate the vibrations of quite complex systems, although, of course, sometimes deviations from the experimental data are significant. Nonetheless, solving problems using these very simple potential functions is the basis of the theory of molecular vibrational spectra, because it permits the most correct approach to be used.

The coefficients U_{ij} in the quadratic form representing the potential function are known as force constants, and we will be using them in the future. We can go over to the quadratic potential by simply expanding the function into a series in terms of vibration co-ordinates about a selected minimum of the potential surface, i.e. the equilibrium position. This is usually done in the theory of polyatomic molecular vibrational spectra. It is worth noting that when differentiating a potential function in the region of the equilibrium position, all the first derivatives vanish only when the differentiation is with respect to independent co-ordinates. When dependent co-ordinates are used, the $(\partial E/\partial q)_0$ derivatives do not individually vanish. However, the whole of the energy's first differential, i.e. $\sum_i (\partial E/\partial q_i)_0 q_i$, does vanish. We discussed this topic in Section 7.1.

Another comment is that in practical calculations we do not, as a rule, actually compute the $(\partial^2 E/\partial q_i \partial q_j)_0$ derivatives which form the force constant matrix. Instead we set up a model by relying on certain *a priori* representations with respect to the molecule's potential function. The function can be expressed in a system of either independent or dependent co-ordinates, as the investigator wishes. In either case the coefficients that are included will have a clear physical sense and are independent (even though the co-ordinates themselves may be dependent—see the example given above of the system of three balls connected by springs). This approach should not be confused with the superficially very similar differentiation of the potential function with respect to dependent vibration co-ordinates which can lead to definite relationships between the second derivatives $(\partial^2 E/\partial q_i \partial q_j)_0$. This is one of the reasons why the parameters of the potential surface function can be found in any co-ordinate system by obtaining a grid of energy values at different points on the potential surface but not by knowing vibration frequencies.

Since potential functions in the form of a quadratic polynomial are very important for the theory of internal motions in a molecule, we shall describe some of their properties in more detail. We shall only consider one more topic. Let us assume that we have selected a potential function for some system of co-ordinates and we wish to find equivalent parameters for another potential function in another co-ordinate system. The problem is very simple for a quadratic potential function and, given linear relationships between the co-ordinates, we need only to change the variables according to the usual rules for a quadratic form. The situation is quite different for more complex functions, such as the general form of equation (9.1). Even if the relationships between the two co-ordinate systems are known (and in most cases they are non-linear relationships), then the change of variables operation which seems most reasonable (i.e. substituting the new co-ordinates into the old function) may be practically impossible for a complex case. We could then select a collection of vibrational co-ordinates from the first system and calculate the potential energy for the collection using the function with the molecule's atoms given in terms of Cartesian co-ordinates. This can clearly be done uniquely. Then the values of the internal co-ordinates of the second system can be found using the knowledge of the atoms' Cartesian co-ordinates. Again this will yield a unique answer. The collection of field values that have been obtained for the co-ordinates of the new system is then substituted into the new potential function. If the new function can be used for every point on the potential surface then the energy calculated using the second function must coincide with that from the first calculation. If this procedure is repeated for many points it will then yield a system of equations for defining some parameters in terms of others. By solving the system—and this may prove difficult, though it is possible in principle—the relationships between the parameters can be found comparing the potential functions even when there is non-linearity.

In the simplest case of polynomial, the parameters being the coefficients for the powers of the co-ordinates, we will obtain the following by truncating at the quadratic terms:

$$\tfrac{1}{2}\sum_{i,j} U'_{ij}q'_iq'_j + \tfrac{1}{6}\sum_{i,j,k} U'_{ijk}q'_iq'_jq'_k + \tfrac{1}{24}\sum_{i,j,k,l} U'_{ijkl}q'_iq'_jq'_kq'_l = \tfrac{1}{2}\sum_{i,j} U''_{ij}q''_i q''_j + \cdots$$

The whole set of such equations for different deformations can be written as a matrix equation:

$$A'U' = A''U''$$

where A' is a matrix with the elements $q_i'^{(1)}q_j'^{(1)}$, $q_i'^{(2)}q_j'^{(2)}$, etc., whose values are known and correspond to the same deformation, while the column matrices containing the U parameters correspond to the left and right functions. Clearly,

$$U' = A'^{-1}A''U''$$

9.2 General Properties of the Force Constants of a Polyatomic Molecule

We showed above that the simplest case of a molecule's potential energy function can be given as

$$U(q) = \tfrac{1}{2}\sum_{j}^{n} U_{ij} q_j^2 + \sum_{j,k}^{n} U_{jk} q_j q_k \tag{9.3}$$

We can separate out of equation (9.3) those terms which depend on only one co-ordinate and those that depend on a product of two and which thus describes their interaction.

The force changing a co-ordinate q_j is

$$f_j = -\frac{\partial U}{\partial q_j} = -U_{ij} q_j - \sum_{k}^{n} U_{jk} q_k \tag{9.4}$$

i.e. it is made up of the usual quasi-elastic forces that depend on all the other co-ordinates. In principle, all the $U_{jk} \neq 0$. However, those corresponding to q co-ordinates that refer to widely separated bonds or angles can, in practice, be set equal to zero for complicated molecules.

Proceeding from expression (9.1) for the potential energy we can decide how to find a rational choice of vibration co-ordinates. Since vibration problems are severely complicated by the presence of non-diagonal terms it is best to choose co-ordinates which yield such terms as small as possible. The usual changes in bond lengths and angles are such co-ordinates, as can be seen from a simple visualization of the molecule's structure.

We will look at the general properties of force constants, and in doing so will always use by implication the natural vibrational co-ordinates, q_j, when we refer to vibrational co-ordinates. By starting from the general expression (9.3) for the potential energy of the vibrations in natural co-ordinates, which has the following form in matrix notation:

$$U(q) = \tfrac{1}{2}\tilde{q} U q \tag{9.5}$$

where U is a square symmetric matrix with elements U_{ij}, we obtain the diagonal element U_{jj} as double the system's energy* for the state defined by the co-ordinates $q_j = 1$, $q_k = 0$ for all $k \neq j$.

It follows directly that all the U_{jj} are positive because the molecule's potential energy in any displaced state is always larger than the equilibrium configuration energy, which we set equal to zero.

The non-diagonal element U_{jk} is naturally defined as the molecule's energy for the state given by the co-ordinates $q_j = 1$, $q_k = 1$ and $q_{l \neq j \neq k} = 0$ minus the molecule's summed potential energy for the states given by $q_j = 1$, $q_k = 0$ and

*Omitting the constant coefficient.

$q_l = 0$; and $q_j = 0$, $q_k = 1$ and $q_l = 0$. For example.

$$U_{12} = U_{21} = U(1,1,0,\ldots,0) - U(1,0,\ldots 0) - U(0,1,0,0,\ldots,0) \qquad (9.6)$$

In general, U_{ij} may be positive or negative.

Obviously, these definitions are only suitable if all the q_j are independent. If some of them are dependent then the state $q_j = 1$, $q_{k \neq j} = 0$ (for the remaining co-ordinates) will not occur. However, elements of the force-constant matrix can formally be included using the above definition if fragments of the molecule are being considered by selecting them so that the states of the fragment are described solely by independent sets of vibrational co-ordinates. When the dependent co-ordinates are eliminated, linear combinations of elements corresponding to the independent and excluded co-ordinates appear in the force-constant matrix.

We shall divide all the natural vibrational co-ordinates into two groups q^{I} and q^{II}. We assume that we can change the variables in such a way that some of the co-ordinates (q^{I}) remain as before, while the others (q^{II}) can be substituted by linear co-ordinates of the former. We will call the co-ordinate we thus obtain q^{III}. The overall co-ordinate transformation can consequently be written as

$$\begin{bmatrix} q^{\mathrm{I}} \\ q^{\mathrm{III}} \end{bmatrix} = \begin{bmatrix} I & 0 \\ A_{21} & A_{22} \end{bmatrix} \begin{bmatrix} q^{\mathrm{I}} \\ q^{\mathrm{II}} \end{bmatrix} \qquad (9.7)$$

from which

$$\begin{bmatrix} q^{\mathrm{I}} \\ q^{\mathrm{II}} \end{bmatrix} = \begin{bmatrix} I & 0 \\ -A_{22}^{-1}A_{21}; & A_{22}^{-1} \end{bmatrix} \begin{bmatrix} q^{\mathrm{I}} \\ q^{\mathrm{III}} \end{bmatrix} \qquad (9.8)$$

According to the rules governing changes of variables in the quadratic form (9.5) the linear transformation with the matrices in equation (9.8) we will obtain the following expression for the submatrix U'_{11} of the U matrix in terms of the new co-ordinates q^{I} and q^{III}:

$$U'_{11} = U_{11} - \tilde{A}_{21}\tilde{A}_{22}^{-1}U_{21} - U_{12}A_{22}^{-1}A_{21} + \tilde{A}_{21}\tilde{A}_{22}^{-1}U_{22}A_{22}^{-1}A_{21} \neq U_{11} \qquad (9.9)$$

where U_{11} is the analogous submatrix of the U matrix in terms of the co-ordinate q^{I} and q^{II}. Thus the partial change of variables results in a change in those elements of U which belong to the q^{I} co-ordinates, which remain as they were. Obviously if $A_{21} = 0$ then $U'_{11} = U_{11}$.

The dependency that we noted for the force constants on the system of co-ordinates can easily be understood if we turn to the definition of these quantities given above. Suppose, for example, that the q^{I} set contains only one co-ordinate q_1. The elements U_{11} and U'_{11} will be equal to twice the potential energy in the states, respectively, given by $q_1 = 1$, $q^{\mathrm{II}} = 0$ and $q_1 = 1$, $q^{\mathrm{III}} = 0$. However, these co-ordinate values will clearly only describe different displaced configurations of the molecule if $A_{21} \neq 0$. This is why the corresponding values of the potential energy must also be different.

For example, the deformations of a triatomic non-linear molecule may be completely described by two different sets of co-ordinates: (1) bond extensions (q_1 and q_2) and changes in the bond angle (γ); and (2) bond extensions (q_1 and q_2) and changes in the separations between unconnected atoms X–X (q_3). Obviously, the deformation that is defined by the co-ordinate set $q_1 = 1$, $q_2 = 0$ and $\gamma = 0$ (one bond is extended while the other and the angle remain unchanged) and $q_1 = 1$, $q_2 = 0$, $q_3 = 0$ (one bond is extended, the other and separation being unchanged) are not identical. This is because it is impossible for the X–X separation to remain unchanged ($q_3 = 0$) when one of the bonds is extended without simultaneously reducing the value of the XYX valency angle.

If the force-constant matrix U is non-diagonal for a given co-ordinate system, then during the deformation described by the co-ordinates $q_j = 1$ and $q_{k \neq j} = 0$ forces will arise within the molecule that are conjugate to these co-ordinates and, according to equation (9.4), will be

$$f_j = -U_{jj}, \quad f_{k \neq j} = -U_{kj} \tag{9.10}$$

If all the $U_{kj} = U_{jk} \neq 0$ then the deformation may only occur when external forces equal and opposite to the forces $f_k = -U_{kj}$ are applied to all co-ordinates $q_{k \neq j}$. We call such a deformation constructed. It is clear that the potential energy of a system deformed in this way will, in general, differ from the deformation energy that occurs when one co-ordinate is changed by unity ($q_j = 1$) while all the others remain free and no external forces are applied to the relevant bonds, angles, etc. We shall determine the energy of this unconstrained deformation.

The value of the function U for $q_j = 1$ and arbitrary co-ordinates $q_{k \neq j}$ is defined by

$$U'' = \tfrac{1}{2}U_{jj} + \frac{1}{2}\sum_{k \neq j}^{n} U_{kk}q_k^2 + \sum_{k \neq j}^{n} U_{jk}q_k + \sum_{k \neq l \neq j}^{n} U_{kl}q_kq_l \tag{9.11}$$

To make the following manipulations more convenient we shall once again use matrix notation. We shall use q'' for the column matrix of all the vibration co-ordinates except q_j and $U^{(j')}$ for the jth column of the U matrix but without the diagonal element U_{jj}. We designate the square symmetric matrix obtained from U by deleting its jth row and jth column by $U^{(j'j')}$. Expression (9.11) thus becomes

$$U'' = \tfrac{1}{2}U_{jj} + \tilde{U}^{(j')}q' + \tfrac{1}{2}\tilde{q}'U^{(j'j')}q' \tag{9.12}$$

During unconstrained deformations the co-ordinates $q_{k \neq j}$ of the q' set must take on values corresponding to an energy minimum, i.e. they must satisfy the equation system

$$\partial U''/\partial q_k = 0 \tag{9.13}$$

for all $k \neq j$. By differentiating equation (9.12) with respect to all the co-ordinates $q_{k \neq j}$ we obtain the following for the column matrix derivative $[\partial U''/\partial q_k]$:

$$\left[\frac{\partial U''}{\partial q_k}\right] = U^{(j')} + U^{(j'j')}q' \tag{9.14}$$

From this by using equation (9.13) we obtain the following expression for the co-ordinates $q_{k\neq j}$ corresponding to the energy minimum during an unconstrained deformation, given $q_j = 1$:

$$U^{(j')} + U^{(j'j')} q'_{\min} = 0 \tag{9.15}$$

i.e.

$$q'_{\min} = -\, {U^{(j'j')}}^{-1} U^{(j')} \tag{9.16}$$

By substituting this expression into equation (9.12) we find that the molecule's potential energy corresponding to the minimum for the selected unconstrained deformation is equal to

$$U''_{\min} = \tfrac{1}{2} U_{ij} - \tfrac{1}{2} \tilde{U}^{(j')} {U^{(j'j')}}^{-1} U^{(j')} \tag{9.17}$$

The quantity $Y_{jj} = U''_{\min} = 1/2\, U_{jj} - 1/2 \tilde{u}^{(j')} {U^{(j'j')}}^{-1} u^{(j')}$ could be called the force constant of the structural element which is described by the co-ordinate q_j during the unconstrained deformation. Since the energy of an unconstrained deformation for the given first co-ordinate clearly does not depend on the choice of all the other co-ordinates, the constant Y_{jj} is invariant with respect to a partial change of variables.

It is not difficult to show that $Y_{jj} < u_{jj}$. This is because the $q_{k\neq j}$ co-ordinates can be selected so that $U^{(j'j')} = I$. Such a selection has no effect on the quantity $U_{jj} > 0$. Hence we obtain the expression $U''_{\min} = 1/2\, U_{jj} - \tilde{u}^{(j')} u^{(j')}$. However, $\tilde{u}^{(j')} \cdot u^{(j')} > 0$, hence it follows that

$$Y_{jj} < U_{jj} \tag{9.18}$$

This result also follows from simple physical arguments. In order to create a constrained deformation of the molecule an unconstrained deformation must first be induced and then those $q_{k\neq j}$ co-ordinates that are not zero in the unconstrained deformation must be changed so that they disappear.

If the elements of the column $u^{(j')}$, and hence of the row $\tilde{u}^{(j')}$, are zero in some co-ordinate system, then $Y_{jj} = U_{jj}$, i.e. the difference between the constrained and unconstrained deformations disappears. It is not difficult to conclude from this very simple argument that a U_{jj} coefficient will then describe the potential energy of the deformation of an appropriate structure element as if it were unconstrained molecule. The constant U_{jj} may then be introduced as a characteristic that is inherent in the structural elements and depends solely on its properties.

Thus the force-constant matrix will have a diagonal form in the co-ordinate system that describes changes in a molecule's structural elements if and only if the force constants reflect the properties of the structural elements. The force-constant matrix will have a quasi-diagonal form in a co-ordinate system containing changes in the geometrical parameters of one of a molecule's atomic groupings as co-ordinates if and only if the set of force constants (the diagonal submatrix of the force-constant matrix) depend solely on the properties of the interatomic forces of the atomic grouping and reflect only its properties. In the general case the matrix's non-diagonal elements will be non-zero and the force constants (diagonal and non-diagonal) will thus correspond to the whole molecule and depend on all of its parts.

Since the form of the initial U matrix is associated with the selected set of vibrational co-ordinates and since the same model of the interatomic forces is expressed in different ways in different systems of natural co-ordinates (as we saw above), we can say that the potential energy's matrix elements are reflections of the elastic properties of the interatomic forces in the particular co-ordinate system that is selected.

The choice of the working system of natural vibration co-ordinates must be defined by the model of the interatomic forces that is accepted. Moreover, we must try to make the force-constant matrix that is obtained as close to the diagonal form as possible. Thus the molecule's vibrations should first, be described with as few parameters (force constants) as possible, and second, a force constant should, as we have already said, be considered as a characteristic of the structural element, a change in which was taken as the vibration co-ordinate. As a result, if a series of molecules all contain given structural elements, the structures of the elements remaining unchanged and the non-diagonal terms of the force-constant matrices being either very small or zero, then we can expect the force constants to remain the same for the whole series of compounds, i.e. the force constants will be transferable. This important property of force constants, which they possess for a very large number of organic molecules, enables us to use the force constants of individual bonds, angles and the other structural elements that are found in some molecules in order to estimate the spectra of related compounds. From now on we shall call atomic groupings that possess transferable force constants 'additive atomic groupings'.

A variety of interatomic force models are used when carrying out practical calculations. We shall briefly discuss the general properties of a few of the most important.

The central force-field approximation One of the simplest approximations (and historically the first to be used to reduce the number of force constants) is based on the concept of a central force. The idea is that the forces confining the atoms to their equilibrium positions are solely dependent on the separations between the atoms and act along the straight lines connecting atom pairs. The concept results in a diagonal force-constant matrix in the relevant system of natural co-ordinates. Calculations for many specific molecules have shown that the approximation is not always satisfied. The method's deficiency is especially apparent for the deformational vibrations of linear molecules, and the frequencies for these vibrations that have been obtained using the approximation contradict experimental evidence. It should be stated, however, that the model is useful when investigating inorganic and co-ordinated compounds.

The valence force-field approximation This approximation does not suffer from the difficulties encountered with the central force-field approximation when dealing with linear molecules. It is closer to the chemical idea of an interatomic force in an organic molecule with covalent bonds.

It is, therefore, no accident that the valence force-field approximation yields significantly better results in the great majority of molecules that can in some way be fitted into the valence scheme. This approximation considers the forces which are associated with changes in bond lengths, valency angles and the other elements forming natural vibration co-ordinates. Calculations for many molecules have shown that the valence force-field approximation with the addition of a series of non-diagonal elements to the force-constant matrix in a system of natural vibration co-ordinates is very satisfactory. The model's main advantage is its ability to include considerations of separate structural elements of a molecule. It can thus be expected that there will be sufficiently closed formations to reduce the problem of constructing the molecule's potential field to the task of investigating the properties of each of its structural elements separately. The valence force-field approximation is thus at the heart of the construction of anharmonic potential functions.

A molecule's potential function in the valence force-field approximation is a sum of the potentials of the bonds and valency angles. However, the approximation has severe limitations. All the elements in the potential energy matrix must be considered as independents, whereas an analysis of the force constants obtained experimentally has shown that the appearance of non-diagonal elements very frequently results in changes of appropriate diagonal elements, i.e. there is a definite functional link between diagonal and non-diagonal elements. This link does not follow from the valence force-field model and can only be accounted for by empirical means.

A number of difficulties arise when explaining the physical significance of non-diagonal force constants. This is particularly true when dealing with the unexpectedly large number of force constants for the interactions between pairs of valency angles that have one bond in common. These deficiencies, although substantial when constructing anharmonic functions and models of the various interaction potentials, are not a hindrance when using the system practically. The matrix elements in this situation are introduced as empirical constants or are calculated quantum mechanically.

9.3 The Quantum Theory of Force Constants

Let us assume that q is a vibrational co-ordinate—for example, a co-ordinate in Cartesian space associated with a molecule or a natural vibrational co-ordinate. We shall now assume that the energy of nuclear repulsions are included in the energy of an electronic state. Since

$$E_n = \int \Psi_n \hat{H} \Psi_n \mathrm{d}V \tag{9.19}$$

we can, by differentiating the integrand, write

$$\frac{\partial E_n}{\partial q} = \int \Psi_n \frac{\partial \hat{H}}{\partial q} \Psi_n \mathrm{d}V + \int \frac{\partial \Psi_n}{\partial q} \hat{H} \Psi_n \mathrm{d}V + \int \Psi_n \hat{H} \frac{\partial \Psi_n}{\partial q} \mathrm{d}V \tag{9.20}$$

We will go on to assume that we know accurately any eigenfunction Ψ_n of a given electronic state in this formula.

As a result of the normalizing condition for eigenfunctions we have

$$\frac{\partial}{\partial q}\int \Psi_n^2 \mathrm{d}V = 2\int \Psi_n \frac{\partial \Psi_n}{\partial q}\mathrm{d}V = 0 \tag{9.21}$$

If we then allow the Schroedinger operator $\hat{H}$ to be self-conjugate, we then obtain in equation (9.21) the following:

$$\int \frac{\partial \Psi_n}{\partial q}\hat{H}\Psi_n \mathrm{d}V = \int \Psi_n \hat{H}\frac{\partial \Psi_n}{\partial q}\mathrm{d}V = E_n \int \frac{\partial \Psi_n}{\partial q}\Psi_n \mathrm{d}V = 0$$

(since $\hat{H}\Psi_n = E_n\Psi_n$). Hence we find that

$$\frac{\partial E_n}{\partial q} = \int \Psi_n \frac{\partial \hat{H}}{\partial q}\Psi_n \mathrm{d}V \tag{9.22}$$

This result is known as the Hellman–Feynman theorem[1].

In the special case in which we are interested the Hamiltonian operator is

$$\hat{H} = \hat{T}_e + V_{pot}$$

where $\hat{T}_e$ is the kinetic energy operator and V_{pot} is the potential operator. If the q co-ordinate is a parameter corresponding solely to nuclear movement we find

$$\frac{\partial \hat{H}}{\partial q} = \frac{\partial V_{pot}}{\partial q}$$

The potential part V_{pot} of the $\hat{H}$ operator can be divided into three: the part corresponding to electron–electron interactions, that corresponding to nucleus–nucleus interactions and that corresponding to electron–nucleus interactions.

We obtain the purely classic expression of the potential of Coulomb repulsions of all the nuclei for the term containing the nucleus–nucleus interactions. The electron–electron interaction operator does not depend explicitly on the nuclear co-ordinates, and hence the corresponding term disappears when differentiated. The electron–nucleus interaction term has the following form in detailed notation:

$$V(r_{i\alpha}) = -\sum_{i,\alpha}\frac{z_\alpha}{r_{i\alpha}}$$

(the summations being over all electrons, i, and all nuclei, α).

It follows from the derivation of equation (9.22) that it is also true for the Hartree–Fock operator $\hat{F}$ and

$$\frac{\partial E_n^{(F)}}{\partial q} = \int \Psi_n^{(F)}\frac{\partial \hat{F}}{\partial q}\Psi_n^{(F)}\mathrm{d}V = \int \Psi_n^{(F)}\frac{\partial \hat{H}}{\partial q}\Psi_n^{(F)}\mathrm{d}V$$

The square of the eigenfunction $(\Psi_n)^2$ determines the electron charge density in the molecule, and hence equation (9.22) can finally be written as

$$-\frac{\partial E_n}{\partial q} = \int \Psi_n \frac{\partial \hat{H}}{\partial q} \Psi_n \, \mathrm{d}V = \sum_{i,\alpha} \int \Psi_n^2 z_\alpha \frac{\partial}{\partial q}\left(\frac{1}{r_{i\alpha}}\right) \mathrm{d}V - \tfrac{1}{2} \sum_{\alpha,\beta} \frac{\partial}{\partial q}\left(\frac{z_\alpha z_\beta}{r_{\alpha\beta}}\right)$$

$$= \sum_{i,\alpha} \int \rho_{in} z_\alpha \frac{\partial}{\partial q}\left(\frac{1}{r_{i\alpha}}\right) \mathrm{d}V - \tfrac{1}{2} \sum_{\alpha,\beta} \frac{\partial}{\partial q}\left(\frac{z_\alpha z_\beta}{z_{\alpha\beta}}\right) \tag{9.23}$$

This is the same as the classic expression of a force acting on an atomic or nuclear co-ordinate due to all the other nuclei and the negative charge that is distributed in space with density ρ_n. The quantity $1/r_{i\alpha}$, where $r_{i\alpha}$ is the distance between the nucleus indexed α and the electron indexed i, equals

$$1/r_{i\alpha} = [(x_\alpha - x_i)^2 + (y_2 - y_i)^2 + (z_\alpha - z_i)^2]^{-1/2}$$

Thus we must have

$$\frac{\partial}{\partial q}\left(\frac{1}{r_{i\alpha}}\right) = \frac{r_{i\alpha x} \dfrac{\partial x_\alpha}{\partial q} + r_{i\alpha y} \dfrac{\partial y_2}{\partial q} + r_{i\alpha r} \dfrac{\partial z_\alpha}{\partial q}}{r_{i\alpha}^3} \tag{9.24}$$

Here, $r_{i\alpha x} = (x_i - x_\alpha)$ is the projection of $\vec{r}_{i\alpha}$ on the x axis and $r_{i\alpha y}$ and $r_{i\alpha z}$ are its analogous.

If q is a Cartesian co-ordinate of the α atom (the radius vector of the α atom) then

$$\frac{\partial x_\alpha}{\partial q} = \vec{i}, \quad \frac{\partial y_\alpha}{\partial q} = \vec{j} \quad \text{and} \quad \frac{\partial z_\alpha}{\partial q} = \vec{k}$$

where $\vec{i}$, $\vec{j}$ and $\vec{k}$ are the co-ordinate system's unit vectors. Hence equation (9.24) can be written as

$$\frac{\partial}{\partial \vec{r}_\alpha}\left(\frac{1}{r_{i\alpha}}\right) = \frac{\vec{r}_{i\alpha}}{r_{i\alpha}^3} = \vec{E}_\alpha^{(i)} \tag{9.25}$$

This corresponds to the expression for the intensity of the electric field at the site occupied by the α atom due to a unit charge at the point defined by the $\vec{r}_{i\alpha}$ vector. By substituting equation (9.25) into equation (9.23) we can then write the following for the Cartesian force acting on the α atom due to the molecule's electronic cloud:

$$-\frac{\partial E_n}{\partial \vec{r}_\alpha} = \vec{f}_\alpha = \sum_i \int \rho_{ni} z_\alpha \vec{E}_\alpha^{(i)} \, \mathrm{d}V = \sum_i \int \rho_{ni} \vec{F}_\alpha^{(i)} \mathrm{d}V$$

This is exactly the same as the classic expression for the electrostatic force acting on a point charge z_α due to a charge distribution ρ_n which is created by all the electrons.

The expression

$$\sum_{i,\alpha} \frac{\partial}{\partial q}(z_\alpha / r_{i\alpha})$$

for arbitrary nuclear co-ordinates is a projection on the appropriate unit vectors of the electrostatic force acting on a nucleus. Thus, for example, a force conjugate to a bond-extension co-ordinate will be equal to the projection on the bond-orientating vector of the difference between the Cartesian forces acting on the nuclei of the atoms forming the bond.

We will pursue our quantum investigation only for the negative charge distribution. Since the stationary state of no interactions other than Coulomb ones are either allowed for or considered in the original Schroedinger equation, the result could be predicted. All the forces acting on the atoms or co-ordinates are zero in the equilibrium position. When the molecule is deformed, forces that increase must appear and they will depend first, on changes in the Coulomb interactions of all the molecule's atoms, and second, on changes in the electron density. The forces can then be imagined to be of the form

$$f_{q_k} = \left(\frac{\partial f_{q_k}}{\partial q_k}\right)_0 q_k + \sum_{l \neq k} \left(\frac{\partial f_{q_k}}{\partial q_l}\right)_0 q_l = -\left[\frac{\partial}{\partial q_k}\left(\frac{\partial V}{\partial q_k}\right)\right]_0 q_k - \sum_{l \neq k}\left[\frac{\partial}{\partial q_l}\left(\frac{\partial V}{\partial q_k}\right)_0 q_l\right] = -u_{kk} q_k - \sum_{l \neq k} u_{kl} q_l \tag{9.26}$$

A u_{kk} coefficient is a force constant for the appropriate co-ordinate and a u_{kl} coefficient is that for the interaction between the two co-ordinates. In the harmonic approximation we limit ourselves to only the linear terms in the expansion of the forces in terms of vibrational co-ordinates.

The result of differentiating equation (9.23) in detailed notation is:

$$u_{kl} = \frac{\partial f_{q_k}}{\partial q_l} = \int \rho_n \frac{\partial}{\partial q_l}(F_{q_k})\mathrm{d}V + \int F_{q_k} \frac{\partial \rho_n}{\partial q_l}\mathrm{d}V - \tfrac{1}{2}\sum_{\alpha,\beta} \frac{\partial^2}{\partial q_l \partial q_k}\left(\frac{z_\alpha z_\alpha}{r_{\alpha\beta}}\right) \tag{9.27}$$

Here

$$F_{q_k} = \sum_{i,\alpha} \frac{\partial}{\partial q_k}\left(\frac{z_\alpha}{r_{i\alpha}}\right) = \sum_i F_{q_k}^{(i)}$$

is the electrostatic force acting on the q_kth co-ordinate due to a unit of negative charge.

Thus a force constant is defined by three terms. The physical significance of the first term is that it is the contribution to the growing force as a nucleus is displaced in space through a homogeneously distributed electronic cloud; the second is the contribution due to the force that arises when the electronic distribution is changed for an unchanged disposition of nuclei; and the third is the fraction of the force associated with changes in the Coulomb repulsions of the nuclei.

Since the operators in the integrals in equations (9.23) and (9.27) are expressed as sums of single-electron operators, when the electronic functions are cast in the form of unideterminant functions we obtain

$$f_q = \sum_i \int \rho_{n_i} F_q^{(i)} \, \mathrm{d}V_i - \frac{1}{2}\frac{\partial}{\partial q}\left(\sum_{\alpha,\beta} \frac{Z_\alpha Z_\beta}{r_{\alpha\beta}}\right)$$

$$u_{kl} = \sum_i \left\{ \int \rho_{n_i} \frac{\partial}{\partial q_l}(F_{q_k}^{(i)}) \mathrm{d}V_i + \int F_{q_k}^{(i)} \frac{\partial}{\partial q_l} \rho_{n_i} \, \mathrm{d}V \right\} - \frac{1}{2}\frac{\partial}{\partial q_l}\left(\frac{\partial}{\partial q_k} \sum_{\alpha,\beta} \frac{Z_\alpha Z_\beta}{r_{\alpha\beta}}\right) \tag{9.28}$$

Here $F_{q_k}^{(i)}$ is the part of the electrostatic force that is conjugate to q and acts due to the unit charge of the ith electron.

The unideterminant functions most frequently used to calculate a polyatomic molecule's force constants are those that are obtained for a fundamental state using the self-consistent field method. Calculations have shown that it is not a bad approximation. If we use a linear combination of atomic orbitals for each of the single-electron functions in the determinant then we obtain

$$u_{kl} = \sum_m \sigma_m \left\{ \int \left(\sum_\alpha c_{\alpha m} \chi_\alpha\right)^2 \frac{\partial}{\partial q_l}(F_{q_k}) \mathrm{d}V + \int F_{q_k} \frac{\partial}{\partial q_l}\left(\sum_\alpha c_{\alpha m}\chi_\alpha\right)^2 \mathrm{d}V \right\} - \frac{1}{2}\frac{\partial^2}{\partial q_k \partial q_l}\left(\sum \frac{Z_\alpha Z_\beta}{r_{\alpha\beta}}\right) \tag{9.29}$$

Here

$$(\partial/\partial q_l)(F_{q_k}) = (\partial^2/\partial q_l \partial q_k)\left(\sum_\alpha \frac{\vec{r}_\alpha Z_\alpha}{|\vec{r}_\alpha|^3}\right)$$

where $\vec{r}_\alpha$ is the radius vector from the α nucleus to the selected point of space. The summation of the expression in the braces is carried out over all occupied electron states allowing for the shell number.

In matrix notation we obtain

$$\begin{aligned} u_{kl} &= \sum_m \sigma_m \left\{ \tilde{c}^{(m)} \left[\int \chi_\alpha \chi_\beta \frac{\partial}{\partial q_l}(F_{q_k}) \mathrm{d}V \right] c^{(m)} + \int F_{q_k} \frac{\partial}{\partial q_l}(\tilde{c}^{(m)} \chi \tilde{\chi} c^{(m)}) \mathrm{d}V \right\} - \frac{1}{2}\frac{\partial^2}{\partial q_k \partial q_l}\left(\sum_{\alpha,\beta} \frac{Z_\alpha Z_\beta}{r_{\alpha\beta}}\right) \\ &= Sp\left(C \sum \tilde{C} \cdot \frac{\partial}{\partial q_l} F_{q_k}\right) + Sp\left(\frac{\partial}{\partial q_l}(C\sum\tilde{C}) \cdot F_{q_k}\right) - \frac{1}{2}\frac{\partial^2}{\partial q_k \partial q_l}\left(\sum_{\alpha,\beta} \frac{Z_\alpha Z_\beta}{r_{\alpha\beta}}\right) \\ &= Sp\left(P \frac{\partial}{\partial q_l} F_{q_k}\right) + Sp\left(\frac{\partial p}{\partial q_l} \cdot F_{q_k}\right) - \frac{1}{2}\frac{\partial^2}{\partial q_k \partial q_l}\left(\sum_{\alpha,\beta} \frac{Z_\alpha Z_\beta}{r_{\alpha\beta}}\right) \end{aligned} \tag{9.30}$$

Here F_{q_k} is a matrix with elements $\int \chi_\alpha \chi_\beta (F_{q_k}) \mathrm{d}V$. We assume in addition that while differentiating, the atomic orbitals remain unchanged and their centres do not follow the nuclei.

By using this expression we arrive at a series of conclusions about the physical significance of the diagonal and non-diagonal force constants.

First, we note that in the linearly combined atomic orbital (LCAO) approximation the F_{q_k} matrix is none other than the matrix derivative of the H matrix of Coulomb and resonance single-electron integrals. This follows directly from equations (9.23), (9.28) and (9.29). When these integrals are differentiated, the kinetic parts and the atomic orbitals are considered to be constants. The appropriate expressions for the derivatives can easily be found from the general formulae for these integrals given in Section 3.1. This is the reason why equation (9.30) can be rewritten as

$$\begin{aligned} u_{kl} = Sp\left(P\frac{\partial^2}{\partial q_k \partial q_l} H \right) + \tfrac{1}{2} Sp\left(\frac{\partial}{\partial q_l} P \frac{\partial}{\partial q_k} H \right) \\ \tfrac{1}{2} Sp\left(\frac{\partial}{\partial q_k} P \frac{\partial}{\partial q_l} H \right) + \frac{1}{2}\frac{\partial^2}{\partial q_k \partial q_l}\left(\sum_{\alpha,\beta} \frac{z_\alpha z_\beta}{r_{\alpha\beta}} \right) \end{aligned} \qquad (9.31)$$

We shall see that in such cases the force constants for the interactions between remote atomic groupings must be zero. We must bear in mind that all the matrix's elements must die out fairly rapidly as the distance between two groupings (co-ordinates) increases. We can therefore speak of the 'localization' of these elements within the atomic grouping, i.e. a short-range feature. Hence if two co-ordinates q_k and q_l refer to two different atomic groupings, then all the elements of the second derivatives $(\partial^2 H/\partial q_k \partial q_l)$ of the Coulomb and resonance integrals will be zero, and thus the first term in equation (9.31) will also vanish. The situation is also relevant for the third term in equation (9.31), the elements of which are short range in nature.

We now turn to the term $Sp(\partial/\partial q_l)P(\partial/\partial q_k)H$. The matrix element derivatives of the density matrix P and the integral matrix H do not vanish. However, when q_k and q_l refer to different atomic groupings and the electron density in the neighbourhood of the q_k co-ordinate does not change during a deformation of the molecule as described by the q_l co-ordinate, then, by allowing for an additional 'short-rangedness' in the H matrix, we obtain the following structures for $(\partial/\partial q_l)P$ and $(\partial/\partial q_k)H$, respectively:

$$\left[\begin{array}{c|c} 0 & 0 \\ \hline 0 & \neq 0 \end{array}\right] \begin{array}{l} q_k \text{ group} \\ q_l \text{ group} \end{array} \quad \text{and} \quad \left[\begin{array}{c|c} \neq 0 & 0 \\ \hline 0 & 0 \end{array}\right] \begin{array}{l} q_k \text{ group} \\ q_l \text{ group} \end{array}$$

The trace (Sp) of the product of two such matrices will be zero. Since the properties of the H matrix do not depend on the electronic cloud distribution and the matrix derivative $(\partial^2/\partial q_k \partial q_l)H$ is always zero for remote co-ordinates, significantly sized non-diagonal force constants can only appear for co-ordinates which do not share atoms if the electronic cloud is very mobile and the electron density is significantly sensitive to changes in geometrical parameters in any part of the molecule.

Finally, we can conclude from equation (9.31) that the force constants must depend linearly on the charges on both the bonds and the atoms (i.e. on the density matrix). This result agrees with the experimentally obtained curves showing how the diagonal force constants of bonds depend on the bond charges in the molecule. These plots have a small degree of curvature which is due to the second and third terms in equation (9.31).

If a molecule's electrons can be divided into groups (for example, delocalized π electrons and localized σ electrons) then an additional summation over the electronic groups must be added to equation (9.31). It is not difficult to conclude that when all the valence electrons can be divided into localized pairs in individual bonds (a crude model of saturated hydrocarbons), the diagonal force constants of the bonds will only depend on the distribution of the electrons of the given bond (if the nuclear repulsions are ignored), whilst the non-diagonal force constants will also be zero if the internuclear repulsions are not considered.

If the molecule contains delocalized electrons (for example, in aromatic compounds), then the diagonal force constant will no longer depend solely on the electronic distribution in a given bond. This conclusion is also in agreement with experiment.

To conclude this section we will look at an example of the influence of polar substituents on the electron shell in order to demonstrate the weak influence of remote interactions in ordinary molecules.

This type of interaction is much greater than the interaction that any deformation in one of a molecule's groupings may have on the electron density distribution in another grouping (this is the effect that we showed above was reflected as non-zero non-diagonal force constants). Hence if the effect of a polar substituent is not large then the effect described by non-diagonal force constants will be even less. This will show whether we can set these non-diagonal elements to zero for co-ordinates describing the deformations of unconnected groupings within a molecule. Moreover, since, as we have demonstrated above, the forces acting on the nuclei of a molecule's atoms due to the other atoms and the electron cloud have an electrostatic character, it can be shown at once that the force constants within an atomic grouping are also conserved, thus ensuring their transferability.

If the electron density of a molecule is divided into separate portions, the contribution of any portion to the formation of the total force acting on a given nucleus in the molecule will decrease in inverse proportion to the square of the distance between this portion and the nucleus (i.e. very rapidly). This point of view is confirmed by direct calculations of the partial forces which act on the molecular nuclei from various parts of the electron cloud (pertinent to polyatomic molecules). Moreover, the attraction produced by the electronic cloud is, to a great extend, compensated by the nuclear repulsion. All this suggests that the chemical bond is a local formation. Changing a chemical bond requires a change in the electronic density around this bond.

If the variations of the given electron density are small, the forces acting on the nuclei in a given part of the molecule, or its associate, must be proportional to these variations. To estimate the effect of a distant substituent from the viewpoint of the rapidity of its decrease along a chain of chemical bonds (including conjugated ones), the reader should refer to Table 9.1, which shows the effect of electron (valency) density redistribution in a benzene molecule after introducing a hydroxyl group. Density redistribution is indicated by the differences of the Mulliken charges between two molecules with and without substitution. Tables 9.1–9.4 were obtained by calculating the electronic state of the molecule by the CNDO method. An absolute electron (valency) density in some cross-sections in molecular space has also been calculated. The coefficients in LCAO for the electron density calculations were also obtained by the CNDO procedure. As is seen, the difference of the electron densities in

Table 9.1. Difference between the Mulliken charges for the phenol and benzene molecules

Numbering of atoms	Change of charge	
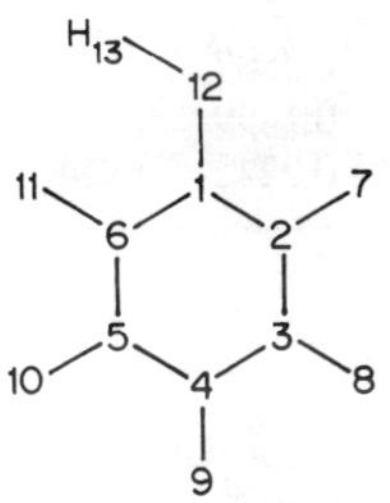	C1	0.1790
	C2	−0.0667
	C3	0.0236
	C4	−0.0293
	C5	0.0223
	C6	−0.0667
	H7	0.0147
	H8	0.0004
	H9	0.0004
	H10	0.0016
	H11	0.0147
	O12	−0.245
	H13	0.1411

Table 9.2. Difference between the Mulliken charges for the nitrobenzene and benzene molecules

Numbering of atoms	Change of charge	
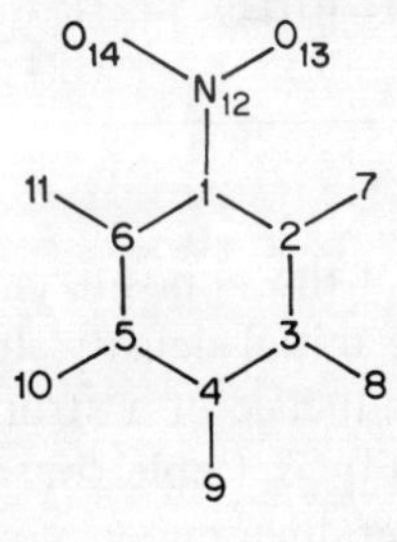	C1	0.0240
	C2	0.0163
	C3	−0.0011
	C4	0.0228
	C5	−0.0011
	C6	0.0163
	H7	0.0266
	H8	0.0038
	H9	0.0008
	H10	0.0038
	H11	0.0266
	N12	0.4761
	O13	−0.3357
	O14	−0.3357

Table 9.3. Difference between the Mulliken charges for the *o*-hydroxyazobenzene and azobenzene molecules

Numbering of atoms

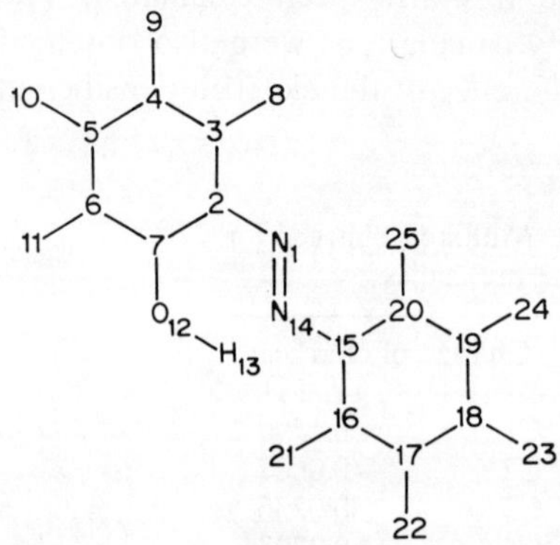

Change of charge

N1	0.0292	C15	0.0029
C2	−0.0791	C16	−0.0032
C3	0.0274	C17	0.0014
C4	−0.0328	C18	0.0013
C5	0.0280	C19	0.0013
C6	−0.0622	C20	−0.0020
C7	0.1822	H21	−0.0001
H8	−0.0005	H22	0.0012
H9	0.0009	H23	0.0002
H10	0.0008	H24	0.0009
H11	0.0152	H25	0.0000
O12	−0.2887		
H13	0.1899		
N14	−0.0242		

Table 9.4. Difference between the Mulliken charges for the *o*-hydroxyazobenzene ions and the *o*-hydroxyazobenzene molecule

Numbering of atoms

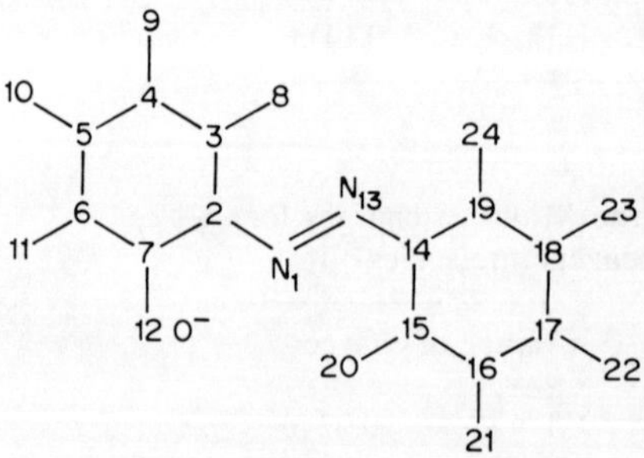

Change of charge

N1	−0.0097	N13	−0.0041
C2	−0.0556	C14	0.0081
C3	0.0008	C15	−0.0155
C4	−0.0701	C16	−0.0044
C5	−0.0023	C17	−0.0283
C6	−0.1640	C18	−0.0039
C7	0.0021	C19	−0.0213
H8	−0.0496	H20	−0.0002
H9	−0.0471	H21	−0.0237
H10	−0.0531	H22	−0.0257
H11	−0.0449	H23	−0.0249
O12	−0.5573	H24	−0.0100

phenol or benzene, measured in the vicinity of the opposite carbon atoms, is very small and does not exceed 3–5% of the initial density. In other words, even in the case of classic conjugation, the influence of a strongly disturbing group on the distribution of electronic density is feeble for relatively close portions of the molecule. In any case, the general character of chemical bonds in these portions is not dependent on the type of substituent. In Table 9.2 are given the corresponding values for benzene and nitrobenzene. The result appears to be the same. Table 9.3 presents the electron density redistribution for a hydroxy group which attacks the orthoposition relative to the azo-group in azobenzene. It is clearly seen that variations in the electron density of the second ring are negligible (the variation of absolute electron density is less than 0.001 amu^{-3}). Table 9.4 shows the redistribution of the electron density in o-hydroxyazobenzene after ionization. It is seen that the electron density is changed in the region of the hydroxy group, while in distant parts of the molecule the electron density undergoes minor change only.

In general, in all the cases investigated the maximum variations of the electron density at a distance equivalent to more than 4-5-bond distances from the changed portion of the molecule do not exceed 3–5%.

On the basis of these calculations we can assume that the influence of distant substituents attenuates very rapidly even along a chain of conjugated bonds. Hence the non-diagonal force constants for the co-ordinates describing remote grouping can effectively be set equal to zero. This docs not, however, mean that we can do this for neighbouring bonds. Thus significantly sized non-diagonal force constants appear in conjugated systems for the next bond but one, and sometimes for the next bond but two. Additional information about the physical meaning of force constants can be found in references 2–8.

9.4 Calculating the Force Constants

We have considered above the general properties of the force constants that can be explained using quantum concepts and considerations of the behaviour of the electronic cloud. We shall now present one of the effective techniques of calculating force constants which uses quantum chemistry methods. We shall not, as is usual, solve an operator problem but shall reduce it to the algebraic problem of simultaneously diagonalizing the H and S (or F and S) matrices, where H is the matrix of single-electron integrals (if, for example, the common Hückel method is used) or the matrix with elements $h_{kn} = \int \phi_k \hat{H} \phi_n \mathrm{d}V$ (if the configuration superposition method is used), where F is the Hartree–Fock matrix (we limit ourselves to closed shells) and S is a matrix containing overlap integrals. If the basis for the variational problem is orthonormal then $S = I$. When diagonalizing the pair of matrices H and S (or F and S) simultaneously, eigenvalues are found whose sum takes account of occupation numbers (or simply one eigenvalue if the configuration superposition method is being used).

The calculation of force constants of a polyatomic molecule by quantum-chemical methods has now been fairly well worked out[8], and there are both numerical and analytical approaches. However, virtually all the analytical work has been based on the use of operators and explicit expressions for the problem's eigenfunctions and their derivatives. Then approximations of the eigenfunctions that have been obtained by the usual MO, LCAO and CI methods are substituted into these expressions.

This substitution, however, is not always correct, and therefore amendments become necessary which complicate the problem and lead to the appearance of terms that do not always have a clear physical significance. This is particularly true of the widely applied 'force method'[9].

Another way of solving this problem is available which, from the start, is based on the fact that all the current practical methods of solving the electronic problem reduce the operator problem to an algebraic one.

Depending on the solution method (Hartree–Fock–Roothaan or configuration superposition) we will obtain matrices H and S, or F and S or simply H (for $S = I$), and when these are diagonalized simultaneously we will have found eigenvalues whose sum allows for occupation numbers (or simply one eigenvalue if the CI method is used). Thus we can find values for the electronic energy, which is what we want. Determining the force constants consequently reduced to calculating the second derivatives of the eigenvalues of matrix pair H and S (or F and S).

We know from Section 4.6 that the matrix of eigenvectors of matrices F (or H) and S is $C = L_s\Lambda_s^{-1/2}L_F$ and we also have $\tilde{C}SC = I$ and $\tilde{C}FC$ (or CHC) $= E$, where matrix I is the unit and E is the diagonal matrix of corresponding eigenvalues (orbitals' energy). We assume now that matrix S is not singular and there is the inverse matrix S^{-1}.

We turn now to the calculation of the derivatives of the eigenvalues of the matrix pair with respect to the nuclear co-ordinates q_1. To do this, we form the matrix $S^{-1}H = A$. Using the equations $\tilde{C}SC = I$ and $\tilde{C}HC = E$ we can write $C^{-1}AC = E$, or $C_t^{-1}\tilde{A}C_t = E$, where $C_t = \tilde{C}^{-1} = L_s\Lambda_s^{1/2}L_H$. We can also obtain $AC = CE$ and $\tilde{C}_tA = E\tilde{C}_t$ or $\tilde{A}C_t = C_tE$. Hence we obtain the following for one eigenvector:

$$AC^{(n)} = E_nC^{(n)} \tag{9.32}$$

or

$$\tilde{C}_t^{(n)}A = E_n\tilde{C}_t^{(n)} \tag{9.33}$$

Since $\tilde{C}_tAC = E$, then, by differentiating, we obtain the following for some eigenvalue E_n (an orbital energy or configuration energy):

$$\frac{\partial E}{\partial q} = \frac{\partial \tilde{C}_t^{(n)}}{\partial q}AC^{(n)} + \tilde{C}_t^{(n)}A\left(\frac{\partial}{\partial q}C^{(n)}\right) + \tilde{C}_t^{(n)}\frac{\partial A}{\partial q}C^{(n)} \tag{9.34}$$

Using equation (9.33) we obtain

$$\frac{\partial E}{\partial q} = E_n\left[\left(\frac{\partial}{\partial q}\tilde{C}_t^{(n)}\right)C^{(n)} + \tilde{C}_t^{(n)}\left(\frac{\partial}{\partial q}C^{(n)}\right)\right] + \tilde{C}_t^{(n)}\frac{\partial A}{\partial q}C^{(n)} \tag{9.35}$$

Then, by differentiating $\tilde{C}_t \cdot C = I$, or $\tilde{C}_t^{(n)}C_t^{(n)} = I$ for one diagonal element, we find that the expression in the square brackets is zero and hence we obtain:

$$\frac{\partial E}{\partial q} = \tilde{C}_t^{(n)}\frac{\partial A}{\partial q}C^{(n)} \tag{9.36}$$

By differentiating this expression repeatedly we find that for the derivative $(\partial^2 E_n/\partial q_i \partial q_j)$. This also enables us to calculate the force constants U_{ij} by summing, when necessary, these derivatives over all occupied molecular orbitals taking into account the occupation numbers (see below). We thus have

$$\frac{\partial^2 E_n}{\partial q_i \partial q_j} = \left(\frac{\partial}{\partial q_j}\tilde{C}_t^{(n)}\right)\frac{\partial A}{\partial q}C^{(n)} + \tilde{C}_t^{(n)}\frac{\partial^2 A}{\partial q_i \partial q_j}C^{(n)} + \tilde{C}_t^{(n)}\frac{\partial A}{\partial q_i}\left(\frac{\partial}{\partial q_j}C^{(n)}\right) \quad (9.37)$$

The next task is to calculate the derivatives of the eigenvectors of the C and C_t matrices. We start by differentiating the equation $AC^{(n)} = E_n C^{(n)}$. This yields

$$\frac{\partial A}{\partial q_j}C^{(n)} + A\left(\frac{\partial}{\partial q_j}C^{(n)}\right) = E_n\left(\frac{\partial}{\partial q_j}C^{(n)}\right) + \frac{\partial E_n}{\partial q_j}C^{(n)}$$

Multiplying this equation on the left by the row $\tilde{C}_t^{(k)}$ $(K \neq n)$ and rearranging, we obtain

$$\tilde{C}_t^{(k)}(A - E_n)\left(\frac{\partial}{\partial q_j}C^{(n)}\right) = \tilde{C}_t^{(k)}\frac{\partial A}{\partial q_j}C^{(n)} + \frac{\partial E_n}{\partial q_j}\tilde{C}_t^{(k)}C^{(r)} \quad (9.38)$$

Let us consider the particular case, i.e. $E_k \neq E_n (k \neq n)$, i.e. when the orbital energies are not degenerate, which is the condition required for the adiabatic approximation to be fulfilled. It will become clear lower down that the column derivative can be represented in the form

$$\frac{\partial}{\partial q_j}C^{(n)} = \sum_{m \neq n} a_{mn}C^{(m)}$$

The a_{mn} are not yet defined. We substitute this expression into equation (9.38) and obtain

$$\sum_{m \neq n} a_{mn}\{\tilde{C}_t^{(k)}AC^{(m)} - E_n\tilde{C}_t^{(k)}C^{(m)}\} = \tilde{C}_t^{(k)}\frac{\partial A}{\partial q_j}C^{(n)} + \frac{\partial E_n}{\partial q_j}\tilde{C}_t^{(k)}C^{(n)} \quad (9.39)$$

Noting that, since $\tilde{C}_t C = I$, then $\tilde{C}_t^{(k)} \cdot C^{(m)} = \delta_{km}$ and that $A \cdot C^{(m)} = E_m C^{(m)}$, we can simplify the last equation to the form

$$a_{kn}(E_k - E_n) = \tilde{C}_t^{(k)}\frac{\partial A}{\partial q_j}C^{(n)}$$

or

$$a_{kn} = \tilde{C}_t^{(k)}\frac{\partial A}{\partial q_j}C^{(n)}/E_k - E_n \quad (9.40)$$

Since the vector $C_t^{(n)}$ satisfies the equation $\tilde{A}C_t^{(n)} = E_n C_t^{(n)}$, if we do exactly the same as above and substitute A by $\tilde{A}$, $C^{(n)}$ by $C_t^{(n)}$ and $C_t^{(n)}$ by $\tilde{C}^{(n)}$ we can write

$$\frac{\partial}{\partial q_i}C_t^{(n)} = \sum_{k \neq n} a'_{kn}C_t^{(k)}$$

where

$$a'_{kn} = \tilde{C}^{(k)}\frac{\partial \tilde{A}}{\partial q_i}C_t^{(n)}/(E_k - E_n) = \tilde{C}_t^{(n)}\frac{\partial A}{\partial q_i}C^{(k)}/(E_k - E_n) \quad (9.41)$$

This enables us to write a formula for $(\partial^2 E_n/\partial q_i \partial q_j)$, i.e.

$$\frac{\partial^2 E_n}{\partial q_i \partial q_j} = \tilde{C}_t^{(n)} \frac{\partial^2 A}{\partial q_i \partial q_j} C^{(n)} + \sum_{k \neq n} \left\{ \frac{\tilde{C}_t^{(n)} \frac{\partial A}{\partial q_j} C^{(k)} \cdot \tilde{C}_t^{(k)} \frac{\partial A}{\partial q_i} C^{(n)}}{E_k - E_n} + \frac{\tilde{C}_t^{(k)} \frac{\partial A}{\partial q_j} C^{(n)} \cdot \tilde{C}_t^{(n)} \frac{\partial A}{\partial q_i} C^{(k)}}{E_k - E_n} \right\} \tag{9.42}$$

This formula can be subjected to a further transformation taking into account first, that $A = S^{-1}H$, and hence we must have

$$\frac{\partial A}{\partial q_i} = \frac{\partial S^{-1}}{\partial q_i} H + S^{-1} \frac{\partial H}{\partial q_i}$$

and

$$\frac{\partial^2 A}{\partial q_i \partial q_j} = \frac{\partial^2 S^{-1}}{\partial q_i \partial q_j} H + \frac{\partial S^{-1}}{\partial q_i} \cdot \frac{\partial H}{\partial q_j} + \frac{\partial S^{-1}}{\partial q_j} \frac{\partial H}{\partial q_i} + S^{-1} \frac{\partial^2 H}{\partial q_i \partial q_j}$$

and second, that since $\tilde{C}HC = E$ while $\tilde{C}_t S^{-1} C_t = I$, we must have the following relations for these eigenvectors:

$$HC^{(n)} = E_n C_t^{(n)}$$

and

$$S^{-1} C_t^{(n)} = C^{(n)}$$

Hence the expression for the matrix elements in equation (9.42) can be written as

$$\begin{aligned}
\tilde{C}_t^{(n)} \frac{\partial^2 A}{\partial q_i \partial q_j} C^{(n)} &= \tilde{C}_t^{(n)} \left[\frac{\partial^2 S^{-1}}{\partial q_i \partial q_j} H + \frac{\partial S^{-1}}{\partial q_i} \cdot \frac{\partial H}{\partial q_j} + \frac{\partial S^{-1}}{\partial q_j} \cdot \frac{\partial H}{\partial q_i} \right. \\
&\quad \left. + S^{-1} \frac{\partial^2 H}{\partial q_i \partial q_j} \right] C^{(n)} = E_n \tilde{C}_t^{(n)} \frac{\partial^2 S^{-1}}{\partial q_i \partial q_j} C^{(n)} + \tilde{C}^{(n)} \frac{\partial^2 H}{\partial q_i \partial q_j} C^{(n)} \\
&\quad - \tilde{C}^{(n)} \frac{\partial S}{\partial q_i} S^{-1} \frac{\partial H}{\partial q_j} C^{(n)} - \tilde{C}^{(n)} \frac{\partial S}{\partial q_j} S^{-1} \frac{\partial H}{\partial q_i} C^{(n)} \\
&= \tilde{C}^{(n)} \left\{ \frac{\partial^2 H}{\partial q_i \partial q_j} - E_n \frac{\partial^2 S}{\partial q_i \partial q_j} - \frac{\partial S}{\partial q_i} S^{-1} \left(\frac{\partial H}{\partial q_j} - E_n \frac{\partial S}{\partial q_j} \right) \right. \\
&\quad \left. - \frac{\partial S}{\partial q_j} S^{-1} \left(\frac{\partial H}{\partial q_i} - E_n \frac{\partial S}{\partial q_i} \right) \right\} C^{(n)}
\end{aligned} \tag{9.43}$$

Taking into account also that

$$\frac{\partial S^{-1}}{\partial q_i} = - S^{-1} \frac{\partial S}{\partial q_i} S^{-1}$$

and

$$\frac{\partial^2 S^{-1}}{\partial q_i \partial q_j} = S^{-1}\frac{\partial S}{\partial q_j}S^{-1}\frac{\partial S}{\partial q_i}S^{-1} + S^{-1}\frac{\partial S}{\partial q_i}S^{-1}\frac{\partial S}{\partial q_j}S^{-1} - S^{-1}\frac{\partial^2 S}{\partial q_i \partial q_j}S^{-1}$$

we similarly obtain

$$\begin{aligned}\tilde{C}_t^{(k)}\frac{\partial A}{\partial q_i}C^{(n)} &= \tilde{C}_t^{(n)}\left\{-S^{-1}\frac{\partial S}{\partial q_i}S^{-1}H + S^{-1}\frac{\partial H}{\partial q_i}\right\}C^{(n)} \\ &= \tilde{C}^{(k)}\left(\frac{\partial H}{\partial q_i} - E_n\frac{\partial S}{\partial q_i}\right)C^{(n)} \qquad (9.44)\end{aligned}$$

In this way we have managed to get rid of the C_t matrix, and what is particularly important, we have obtained explicit derivatives of the H and S matrices with respect to nuclear co-ordinates in the final expressions for the matrix elements. The formulae we have obtained are themselves quite simple, and practical calculations using them are not difficult. The corresponding derivatives of the H and S matrix elements are also found without too much trouble from expressions for these matrix elements taken from the approximation.

A final comment connected with the calculation of the derivatives of a Hartree–Fock matrix must be made. These elements, together with their single-electron parts which are explicitly dependent on nuclear co-ordinates, also contain two-electron terms, and these do not depend explicitly on nuclear co-ordinates. Thus when calculating the derivatives of the elements of the F matrix with respect to the vibrational co-ordinates, we can limit ourselves only to differentiating the single-electron constituents. This conclusion also follows both from the fact that at each iteration stage a self-consistent solution is being sought which is not related to the previous one, and that the two-electron terms in the matrix elements are constructed so that the corresponding density matrix is induced 'from the outside' and remains unchanged, and can be assumed to be constant during new variations of the energy. This is even more true in the region where self-consistency is attained. These considerations show that when both the Hartree–Fock–Roothaan method and a single determinant approximation is used, a force constant may be defined using a summation taking into account the occupation numbers of the problem's orbital energies, whilst the derivatives $\partial^2 \bar{H}/\partial q_i \partial q_j$ may not be defined.

When using equations (9.42)–(9.44) to make practical estimates, we should remember that the $(\partial^2 E_n/\partial q_i \partial q_j)$ derivatives are taken at a minimum of the potential surface. This makes the numerical values of the $C^{(n)}$ vectors, the eigenvalues E_n and the matrix S^{-1} fully determinate, and ensures that these calculations can be done by computer. Since we have demonstrated some of the properties of the derivatives of E_n and $C^{(n)}$ it is not too difficult to find expressions for derivatives such as $(\partial^3 E_n/\partial q_i \partial q_j \partial q_l)$ or higher, even though the expressions may be quite cumbersome.

9.5 Effective Algorithm of a Quantum-mechanical Search of Parameter Values of the Potential Function of Polyatomic Molecules in an Arbitrary System of Internal Co-ordinates

It was noted above that the question of the quantum-chemical determination of the equilibrium geometry of polyatomic molecules at a position of minimum energy of the potential surface, and of the parameters of the potential function, has been discussed elsewhere. However, despite the lengthy history of this problem and achievements in the field, the technique of the respective calculations has not yet been standardized and, what is more, is insufficiently developed, especially in relation to large polyatomic (for example, polycyclic) compounds. In a number of cases the available calculation methods happen to be simply inapplicable.

Thus, for example, the analytic methods allow us to find only the second derivatives of the potential function according to the internal co-ordinates (independent only). This makes it impossible to use them in operation with anharmonic, especially model, potential functions and for polycyclic compounds, where indication of the system of independent co-ordinates is itself a problem. Employment of the Cartesian co-ordinates in these cases is also rather inconvenient, for certain reasons. Employment of the numerical methods also creates difficulties that grow considerably in operation in the system of dependent natural vibration co-ordinates. This, however, must inevitably be introduced in operation with a number of molecules, when it is necessary, (1) to obtain those parameters of the potential functions that would be closely connected with separate structural elements of molecules, (2) that would have an obvious physical meaning and (3) that could be found at the presence values of the others.

All this led to the development of a new, sufficiently universal algorithm[10] in the search for the parameters of the potential surface, assigned in the simplest case in the form of a fourth-degree polynomial in the system of dependent natural vibration co-ordinates.

The algorithm allows us to find the equilibrium geometry of molecules of an arbitrary structure and the parameters of the model polynomial function. It gives the possibility of determining the structural and dynamic characteristics in parts and of utilizing the special computer library of molecular fragments.

The sequence of operation of the whole algorithm is as follows. At the first stage a search of the equilibrium geometry of the molecule is carried out (if desirable). To this end, a certain zero approximation of the geometry and the simplest diagonal quadratic potential function are selected. The bond length and valence angles, angles between the planes and bonds, as well as the dihedral angles between the planes laid through the triplets of the molecule atoms are introduced as co-ordinates. In this case, no limitations as to the type of the molecule is made. The quadratic potential function is introduced with arbitrary parameters. Thus if a problem is put to allow subsequent strain only on separate parts of the molecule, it is effected by assignment of very high values of elasticities for all the rest of the introduced natural co-ordinates.

On the basis of these data, methods described in Chapters 7 and 8 are used to solve the vibration problem and to transfer to independent normal co-ordinates with an automatic registration of symmetry. Then, with the aid of a random numbers transmitter, a set of values of completely symmetric normal co-ordinates is assigned, and the shifts of atoms from the selected initial position and the new Cartesian co-ordinates of atoms are calculated by equation (8.55) for each point in the space of the co-ordinates thus obtained. Separation of only assigned strains is also effected automatically by determining the high and low frequencies. A quantum-chemical energy problem is solved for all independent molecule strains thus obtained. In this case, solution of the problem for each subsequent point starts with the assignment of the density matrix of the preceding one, so as to accelerate the process of self-consistency.

The next procedure is the construction of an energy functional (quadratic polynomial) in completely symmetric normal co-ordinates that would approximate the obtained points and the slope to be minimum with the aid of Newton–Ruffson's iteration procedure[11]. The possibility of automatic separation of only

Table 9.5. Results of optimization of molecule geometry of cyclopropane, cyclobutane, cubane and propellane 1.1.1, 2.1.1, 2.2.1

Molecule	Number of co-ordinates[a]	Geometric characteristic[b]
	27, 21, 3	CC 1.503; CH 1.104; HCH 108.0
	36, 30, 3	CC 1.533; CH 1.113; HCH 106.6
	68, 42, 2	CC 1.566; CH 1.106
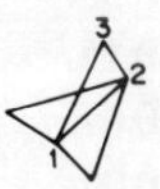	43, 27, 5	C^1C^2 1.661; C^1C^3 1.534; CH 1.087; CC^3C 65.6; HCH 105.5
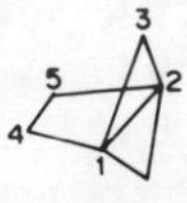	54, 36, 16	C^1C^2 1.658; C^1C^2 1.523; C^1C^4 1.548; C^4C^5 1.533; CH 1.106; CC^3C 66.2, CC^4C 92.9; CC^3H 121.7; C^1C^4H 113.4; C^5C^4H 115.5; HC^3H 108.8; HC^4H 105.9; Dihedral angle 127.9
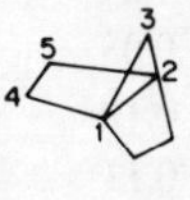	61, 45, 13	C^1C^2 1.654; C^1C^3 1.523; C^1C^4 1.550; C^4C^5 1.535; CH 1.115; CC^3C 65.6; CC^4C 92.2; CC^3H 119.9; C^1C^4H 114.8; C^5C^4H 116.8; HC^3H 107.1; HC^4H 103.0; Dihedral angle 123.9

[a]Given are the full numbers of natural dependent and independent and fully symmetric co-ordinates.
[b]Values of bond lengths in Angströms (angles in degrees).

independent fully symmetric co-ordinates provides for a quick run of the entire solution, and sharply reduces the number of variables for symmetric molecules.

Thus, for example, the number of original natural co-ordinates in a cubane molecule equals 68, while that of fully symmetric co-ordinates of geometry optimization equals only two. In one of the propellanes these numbers equal 61 and 13, respectively (see Table 9.1).

On achievement of the minimum of the potential surface, a search of the parameters of the potential function is begun which in the given variant takes the form of a fourth-degree polynomial assigned in the system of dependent natural co-ordinates:

$$E = \tfrac{1}{2}\sum_i U_{ij} q_i q_i + \tfrac{1}{6}\sum_i U_{iii} q_i^3 + \tfrac{1}{24}\sum_i U_{iiii} q_i^4 \qquad (9.45)$$

Introduction or exclusion of coefficients at the respective degrees of co-ordinates makes it possible to vary the form of the potential function, depending on the peculiarities of the specific problem. This gives us an opportunity for carrying out calculations both for the whole molecule and for any of its structural fragments. This procedure is especially convenient in operation with large molecules, incorporating the atom grouping already studied.

Table 9.6. Force constants of molecules of propellanes calculated in approximation MINDO/3

Force constants	Types of propellanes 1.1.1	2.1.1	2.2.1
K_1	1.915 ± 0.51	3.281 ± 0.73	3.777 ± 0.79
K_2		5.631 ± 0.36	5.130 ± 0.31
K_3		5.474 ± 0.42	5.077 ± 0.51
K_5	4.739 ± 0.16	5.150 ± 0.25	5.563 ± 0.51
$K_9 = K_{13}$	6.745 ± 0.40	5.913 ± 0.57	6.474 ± 0.98
K_{23}		0.65 ± 0.26	0.570 ± 0.38
$K_{25} = K_{57}$	1.212 ± 0.04	1.469 ± 0.09	1.261 ± 0.15
$K_{29} = K_{5,13}$	$0.584 + 0.02$	$0.546 + 0.05$	$0.523 + 0.07$
$K_{9,10} = K_{13,14}$	0.324 ± 0.05	0.316 ± 0.08	0.536 ± 0.08
A_2^{25}		0.598 ± 0.13	0.503 ± 0.15
A_2^{29}		0.216 ± 0.08	0.210 ± 0.10
$A_5^{5\ 13}$	0.202 ± 0.04	$0.166 + 0.07$	0.316 ± 0.11
A_5^{57}	0.497 ± 0.04	0.512 ± 0.12	
A_5^{25}		0.465 ± 0.11	0.653 ± 0.21
A_3^{39}		0.210 ± 0.07	0.316 ± 0.08
$A_9^{29} = A_{13}^{5\ 13}$	0.245 ± 0.05	0.304 ± 0.11	0.218 ± 0.10
$A_9^{9\ 10} = A_{13}^{13\ 14}$	0.440 ± 0.08	0.265 ± 0.12	0.372 ± 0.11

Note: Designations of force constants conform to 11, numeration of bonds for propellane molecule (2.1.1) is given in Figure 9.2, dimensions of quantities in mdyne/Å, mdyne, mdyne Å for bond–bond, bond–angle and angle–angle co-ordinates, respectively.

The unknown parameters are found with the aid of a solution by the method of least squares of the redundant system of linear equations obtained at substitution into equation (9.45) of the energy values calculated for different independent strains of the molecule about the minimum of the potential surface. These strains are obtained by the method described above, i.e. as a result of the increments of the normal co-ordinates assigned by chance. Since, for each strain of this kind, the values of all the bond lengths and valence angles, as well

Table 9.7. Values of force constants of a propylene molecule obtained by three methods of calculation: (1) computation of all parameters simultaneously; (2) computation of parameters by fragments; (3) computation of parameters at angular co-ordinates only

Parameter	(1)	(2)	(3)
K_I	11.242 ± 0.30	11.615 ± 0.32	
H_1^2	0.343 ± 0.10	0.457 ± 0.07	
A_1^{12}	0.174 ± 0.08	0.274 ± 0.05	
K_2	7.059 ± 0.43	6.754 ± 0.20	
A_2^{12}	0.209 ± 0.04	0.249 ± 0.03	
A_2^{23}	0.231 ± 0.05	0.208 ± 0.03	
K_{12}	0.389 ± 0.03	0.382 ± 0.02	0.387 ± 0.03
K_{12}^{14}	0.041 ± 0.04		0.015 ± 0.03
l_{13}^{14}	0.084 ± 0.04		0.048 ± 0.03
K_{23}	0.396 ± 0.06	0.362 ± 0.03	0.395 ± 0.05
K_4	6.890 ± 0.48	7.184 ± 0.49	
A_4^{14}	0.193 ± 0.07	0.190 ± 0.06	
A_4^{45}	0.255 ± 0.07	0.217 ± 0.05	
A_1^{15}	0.349 ± 0.09	0.334 ± 0.10	
A_5^{15}	0.392 ± 0.09	0.448 ± 0.09	
K_{14}	0.432 ± 0.04	0.440 ± 0.04	0.450 ± 0.04
K_{45}	0.418 ± 0.05	0.419 ± 0.04	0.462 ± 0.04
K_{15}	0.499 ± 0.07	0.500 ± 0.08	0.571 ± 0.04
l_{15}^{57}	0.061 ± 0.02		0.022 ± 0.02
l_{15}^{56}	0.016 ± 0.03		0.043 ± 0.02
K_5	7.248 ± 0.27	7.323 ± 0.37	
A_5^{56}	0.141 ± 0.07	0.188 ± 0.09	
K_6	7.058 ± 0.38	6.571 ± 0.30	
A_6^{56}	0.175 ± 0.04	0.210 ± 0.03	
A_6^{67}	0.181 ± 0.03	0.192 ± 0.02	
K_{56}	0.524 ± 0.03	0.487 ± 0.03	0.510 ± 0.05
l_{56}^{67}	0.063 ± 0.02		0.056 ± 0.01
K_{67}	0.505 ± 0.03	0.567 ± 0.03	0.567 ± 0.04

Note: Designations are the same as in Table 9.6; for numeration of bonds, refer to Figure 9.3.

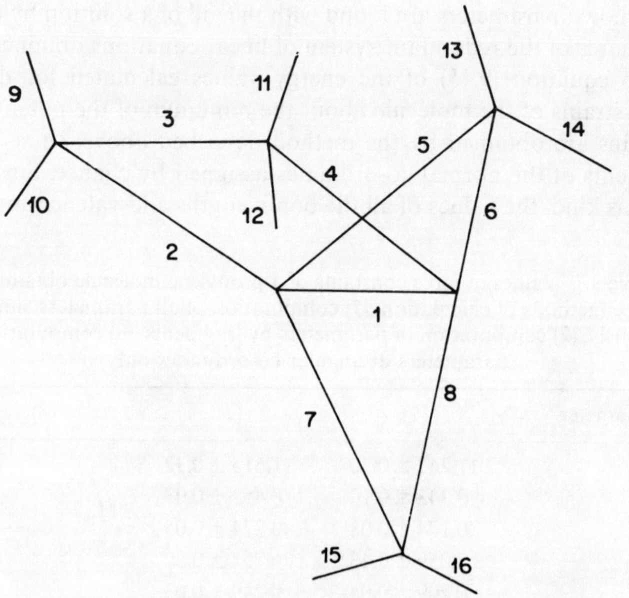

Figure 9.2. Numeration of bonds of propellane molecule 2.1.1

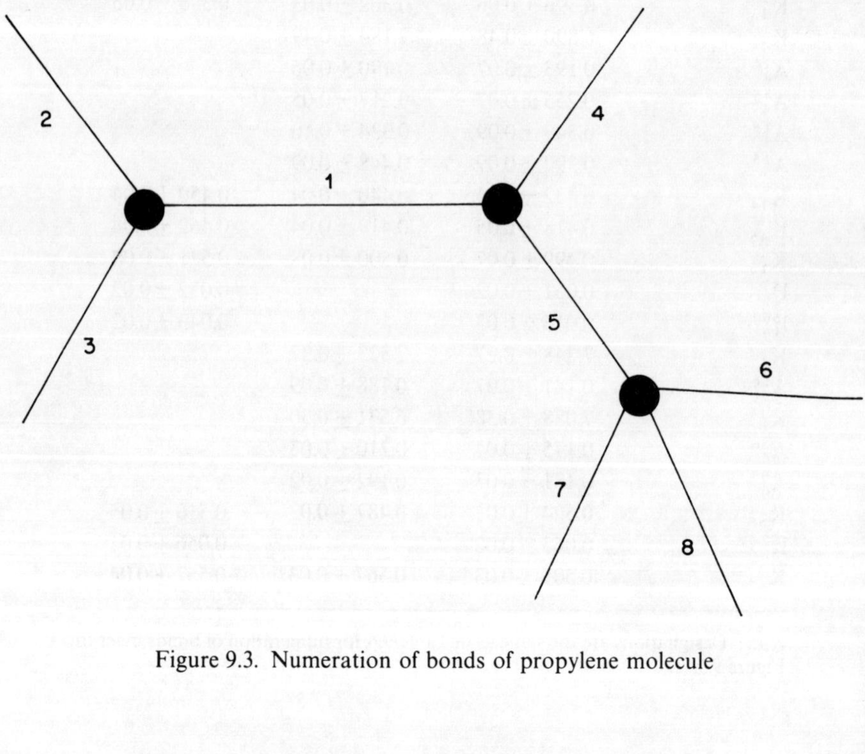

Figure 9.3. Numeration of bonds of propylene molecule

of the other geometric characteristics, are determined quite precisely and unambiguously (even if dependent co-ordinates are available between the respective natural ones), this allows us to form a system of equations linear with respect to the desired coefficients U_{ij}, U_{iii}, etc. by substituting all these values into the right side of equation (9.45).

It should now be noted that in those cases when there are linear dependences between the natural co-ordinates, similar dependences appear between their squares and paired products. This results in a singularity of the matrix of normal equations, and determination of coefficients in a potential function becomes impossible. However, there is an easy way out of this situation: from the very beginning, some of the undiagonal quadratic force constants should not be introduced into the potential function (assumed to be equal to zero).

Such an approach always proves to be accurate if the dependence involves a large number of natural co-ordinates arranged at a great distance from each other. This case is typical for cyclic and polycyclic molecules.

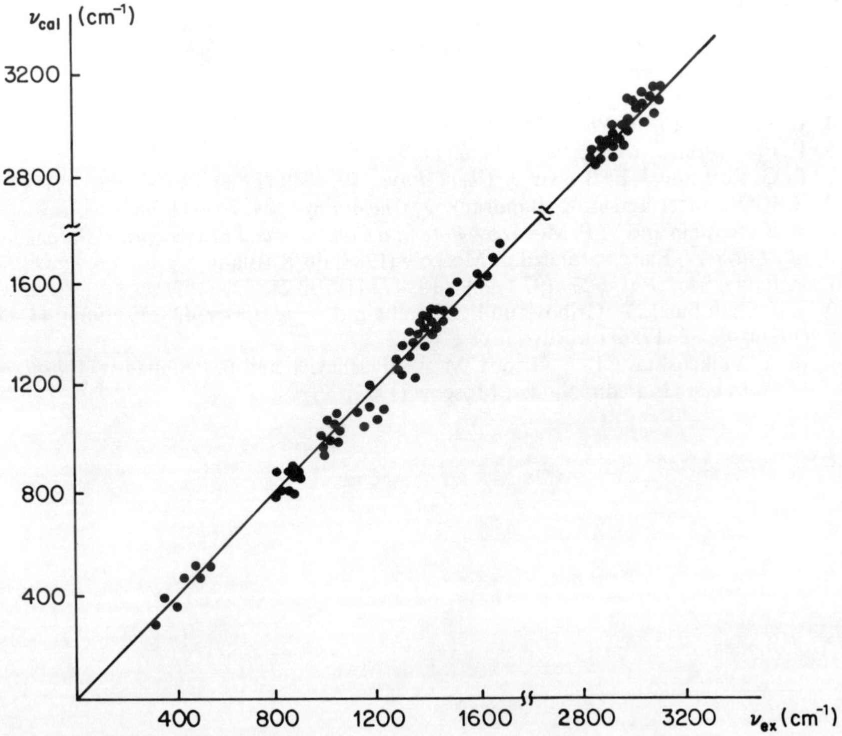

Figure 9.4. Relation between experimental frequency (ν_{ex}) and frequency calculated in approximation *MINDO*/3 (ν_{calc}) of normal vibrations of *n*-paraffins, allenes, acetylenes and some aromatic hydrocarbons. In calculations use was made of the following sealing factors: 0.7; 1.2; 1.0 for force constants of types K_q, K_α and H_{qx}, respectively. Constants K_{qq} and $K_{\alpha\alpha}$ were assumed to be equal to 0

In the simplest situations (when, for example, dependences between angular co-ordinates arise in plane or tetrahedral structures) the problem becomes solvable, if even one of the undiagonal force constants of interaction of angular co-ordinates is nullified. Also it does not disturb the approach community, since in fact the strain energy is determined not by all parameters separately but by their linear combination.

As a result, we obtain a universal procedure suitable for the analysis of molecules of any structure under the conditions of a monotonous universal assignment of the source information. This makes the use of the respective programs rather convenient. It is important that, in the construction of a potential function, use may be also made of the methods of its fragmentary design using a library of standard fragments. The problem is then greatly simplified, and involves the search of only those parameters of the function that approximate the joints of these fragments. Operation of the algorithm is illustrated in Tables 9.5–9.7 and in Figure 9.4.

References

1. R. P. Feynman. *Phys. Rev.*, **56**, 340 (1939).
2. W. B. Brown. *Proc. Camb. Phys. Soc.*, **54** (1958).
3. W. L. Clinton. *J. Chem. Phys.*, **33**, 1603 (1960).
4. L. Salem. *J. Chem. Phys.*, **38**, 1227 (1963).
5. P. Phillipson. *J. Chem. Phys.*, **44**, 633 (1966).
6. R. G. Parr and J. E. Brown. *J. Chem. Phys.*, **49**, 4849 (1968).
7. R. F. W. Bader and A. D. Bandrauk. *J. Chem. Phys.*, **49**, 1666 (1968).
8. V. V. Rossihin and V. P. Morozov. *Potential Constants and Electrooptical Parameters of Molecules*, Energoatomizdat, Moscow (1983) (in Russian).
9. P. Pylay. *Mol. Phys.*, **17**, 197 (1969); **18**, 473 (1970); **21**, 329 (1971).
10. S. A. Shatohin. L. A. Gribov and I. S. Pereligin. *J. Appl. Spectr.* **43**, 259 (1985); **44**, 849 (1986); **45**, 88 (1986) (in Russian).
11. M. V. Volkenshtain, L. A. Gribov, M. A. Elyashevich and B. I. Stepanov. *Vibrations of Molecules*, 2nd edn, Nauka, Moscow (1972).

Chapter 10

Solution of the Problem for Anharmonic Vibrations and Vibrations with Large Amplitudes

10.1 General Comments

We have considered the vibrational problem in the harmonic approximation when the full equation for the vibrations has the form:

$$\left\{-\frac{\hbar^2}{2}\sum_{i,j} t^{1/4}\frac{\partial}{\partial q_i}\tau_{ij}(q)t^{-1/2}\frac{\partial}{\partial q_j}t^{1/4}+v(q)\right\}\Psi_v(q)=E_v\Psi_v(q) \quad (10.1)$$

which can be replaced by the simpler

$$\left\{-\frac{\hbar^2}{2}\sum_{i,j}\tau_{ij}(0)\frac{\partial^2}{\partial q_i\partial q_j}+\tfrac{1}{2}\sum_{i,j}U_{ij}q_iq_j\right\}\Psi_v(q)=E_v\psi_v(q)$$

This approximation turns out to be reasonably good for small vibrations around the equilibrium position when the potential surface has an adequately deep minimum and good curvature. In reality, however, molecules are observed to have considerable deviations (sometimes leading to modifications of the order of $100\,\text{cm}^{-1}$) from the harmonic approximation when the vibrations of bonds with light terminal atoms are being considered. These deviations become very noticeable in excited electronic states when shallow and fairly wide potential wells are encountered.

The harmonic approximation cannot be applied at all when the potential surface has two minima that are separated by a low threshold (for example, inversion in ammonia or several minima as for internal rotations). It is therefore necessary to find a solution for these situations in the general case.

Before describing the methods of solution, we recall that the general equation (10.1) can be rewritten in the form

$$\left\{-\frac{\hbar^2}{2}\sum_{i,j}\frac{\partial}{\partial q_i}\tau_{ij}(q)\frac{\partial}{\partial q_j}+V(q)+\Delta T(q)\right\}\psi_v(q)=E_v\psi_v(q) \quad (10.2)$$

where $\Delta T(q)$ is an increment associated with the kinetic energy operator. It does not, however, contain a differential operator, and hence can be included in the potential term. Furthermore, this increment can generally be neglected for small amplitudes. We will return to this point later.

When desired, however, the increment can be taken into account, calculating it point by point as the potential surface is being ascertained. In other words, having constructed a new effective surface as a locus that corresponds, for different sets of natural co-ordinates, to

$$V'(q) = E'(q) = E_e(q) + V_{nn}(q) + \Delta T(q) = V(q) + \Delta T(q)$$

where $E_e(q)$ is the complete adiabatic electron energy, $V_{nn}(q)$ the energy of nuclear–nuclear repulsions and $\Delta T(q)$ the modification discussed in Chapter 6 (Section 6.2), calculation of the modification for each set of natural co-ordinates is not difficult.

To summarize, we will in all cases proceed from the expression for the Hamiltonian of the relative nuclear movements that has the simple but very general form:

$$\left\{ -\frac{\hbar^2}{2} \sum_{i,j} \frac{\partial}{\partial q_i} \tau_{ij}(q) \frac{\partial}{\partial q_j} + V'(q) \right\} \psi_v(q) = E_v \psi_v(q)$$

We cannot, however, solve the problem in a general form for this Hamiltonian because of the complexity of the expressions for the kinematic coefficients $\tau_{ij}(q)$. We must therefore find approximate solutions. There are two possible approaches, both discussed below, but first we must make one very important point.

The kinetic energy operator is made up of a symmetry matrix $T(q)$ containing the kinetic coefficients. Clearly, it can always be transformed into the unit matrix by the transformation matrix $L(Q)$ which enters the equation $\dot{q} = L(Q)\dot{Q}$. The matrix $L(Q)$ will itself be a function of vibrational co-ordinates. The potential function must also be transformed in a corresponding way. If this function is written as a power series, then a substitution of variables will not change it, and again we can separate the quadratic part. The harmonic approximation is then constructed. Subsequent orthogonal transformations of the co-ordinates will ensure that the potential function's quadratic part will be diagonalized. We call the whole transformation the 'transition to non-linear normal co-ordinates'.

We can actually call the resultant co-ordinates 'normal' ones, because co-ordinates in which the kinetic operator can be made into a unit matrix and the potential part acquires the form of a sum of squares are termed 'normal' by definition.

At first sight, this co-ordinate transformation is quite attractive. However, we must bear the following important points in mind. First, the process of finding $L(Q)$ and ascertaining the explicit dependence of the original co-ordinates q_i on the resulting Q_k is not at all easy. Furthermore, the transformation of the potential term to these co-ordinates is also not very conveniently obtained by a computer even when the potential function is in the form of a polynominal. Even so, we have not yet got rid of the anharmonicity in the kinetic term, but have merely relegated it to the potential part. For example, when substituting for the original co-ordinates in the expressions in terms of the normal co-ordinates, which are usually chosen in the form of a series

$$q = L(0)Q + \sum_k \left[\left(\frac{\partial L}{\partial Q_k} \right)_0 Q_k \right] Q + \tfrac{1}{2} \sum_{k,n} \left[\left(\frac{\partial^2 L}{\partial Q_k \partial Q_n} \right)_0 Q_k Q_n \right] Q$$

if the original potential has been chosen even as a quadratic form, then quadratic, cubic, etc. terms will appear.

However, even this aspect is not the worst feature. The most critical aspect of this procedure is that the non-linear substitution of variables makes it practically impossible to use any expression for the potential functions other than polynomials (multidimensional power series) when solving the problem.

The non-linear transformations are also not very convenient when estimating the energy levels of molecules with internal rotations (i.e. when estimating the interactions of the vibrations and rotations).

All these considerations force us to conclude that a transition to non-linear normal co-ordinates in the general case is not particularly promising when solving problems concerned with large-amplitudes movements of a molecule's atoms.

The second approach is more rewarding. Without limiting generality, we can represent the kinetic energy operator as

$$\begin{aligned} \hat{T}_{\text{kin}} &= -\frac{\hbar^2}{2} \sum_{i,j} \tau_{ij}(0) \frac{\partial^2}{\partial q_i \partial q_j} - \frac{\hbar^2}{2} \sum \frac{\partial}{\partial q_i} \Delta\tau_{ij}(q) \frac{\partial}{\partial q_j} + \Delta T(q) \\ &= -\frac{\hbar^2}{2} \frac{\tilde{\partial}}{\partial q} T(0) \frac{\partial}{\partial q} + \Delta T' \left(q, \frac{\partial}{\partial q} \right) \end{aligned}$$

We can do this, for example, by expanding the matrix of the kinetic coefficients $T(q)$ into a series around some chosen orgin. For example, we can take the equilibrium values of the geometric parameters as such an origin. We can also act in an analogous way for the case of the internal rotations (see below). We will expand on this question later, and for the time being limit ourselves to studying the vibrations corresponding to potential surfaces with one minimum.

Therefore we can accept as harmonic (see Chapter 7) the approximation in which the kinetic operator is written in the simple form $\hat{T} = 1/2\tilde{\hat{p}}T(0)\hat{P}$ and the potential term is a quadratic form in the original co-ordinates. Both terms can later be turned into sums of squares by a linear transformation of the co-ordinates $q = L_q(0)Q$. We call the new co-ordinates obtained in this way 'normal linear' ones, and will use them exclusively in future. These co-ordinates possess the main advantage that they do not drastically change the form of the potential function and can be operated upon easily by using computers.

Clearly, when we are solving an anharmonic problem, an additional term whose operator is $\Delta T'$ will arise, i.e. what is conventionally called 'kinematic anharmonicity' appears. As we will see below, this does not lead to any noticeable difficulty. Furthermore, using only linear co-ordinate transformations and partitioning some of the constant parts of the T matrix enables us to construct a relatively simple theory and methodology for estimating energy levels, even when there are internal rotations.

This is why a transition to these normal co-ordinates is preferable, promising and universal. Note also that by partitioning only the operator

$$\hat{T} = -\frac{\hbar^2}{2}\sum_{i,j}\tau_{ij}(0)\partial^2/\partial q_i\partial q_j$$

in the kinetic term (whilst conserving the initial sense of the internal co-ordinates used) means that we have gone from the true kinetic energy operator to some effective one. It can be shown that the application of non-linear normal co-ordinates, the transfer of kinetic anharmonicity to the potential part and the application of linear normal co-ordinates all lead to the same results[1]. However, the second approach is preferable.

We shall now discuss how to solve anharmonic problems, in which the traditional method has been to apply perturbation theory in the form of a series[2–7]. Clearly, by using the matrix elements for the power operators for momentum and co-ordinate we can, here and elsewhere, write, by considering only terms to the second order,

$$E = \sum_k \nu_k(v_k + 1/2) + \sum_{k,n} \varkappa_{kn}(v_k + 1/2)(v_n + 1/2) \tag{10.3}$$

Since every transformation of the co-ordinates and form of the original potential functions in terms of the natural co-ordinates are known, then the $\varkappa_{kn}$ coefficients can be found as functions of the coefficients in the original potential terms. Obviously, they are equivalent for both linear and non-linear normal co-ordinates. The corresponding expressions, even for the simplest triatomic molecules, are quite cumbersome, and we will not provide them here because, as we shall see below, this approach is not the best possible.

The chief deficiency of the perturbation method in the form of series is that it cannot be applied when there is degeneracy, but degeneracy is typical for polyatomic molecules. Hence in real cases where degeneracy does occur, (and this leads to the appearance of what is called 'resonance') we must complicate the logic of the solution considerably by including secular equations for these resonance levels. Strictly speaking, even equation (10.3) (which is so attractive because if its simplicity and is hence widely used) has already become incorrect. This is why there has not been enough progress in solving an harmonic problems in practice, despite the many years that have been spent using the techniques of perturbation theory, and solving these problems remains exceedingly difficult in the case of molecules that have more than five or six atoms. We should also note that the application of perturbation techniques greatly complicates either the expressions obtained or the logic of the estimates when going to orders higher than two. It has also been shown that writing universal algorithms to solve inverse spectral problems is very tedious, requiring several stages of calculation and the use of individual algorithms for each type of molecule.

These considerations indicate that the use of perturbation techniques in the theory of anharmonic vibrations is not a rational one when making large-scale estimates of energy levels and wave functions.

The alternative approach is one based on the variational solution of the problem[7-20]. This has become especially widespread recently, and because of its simplicity, universality and adaptability to computer solution it can be claimed that it has become a truly standard method for solving anharmonic problems, including analysis of inversions and internal rotations. True, it has made it necessary for us to operate with matrices that sometimes have orders of several thousands, but this is not impossible for modern computers. The situation is even better, because the matrices that are obtained have a special structure that lends itself to diagonalization and to a whole series of simplifying techniques.

In particular, we can consider the eigenvalues of a matrix being obtained to be all positive and can thus be expanded into a series (in terms of increasing values). Since experimental observations are only done for low energy levels and their interactions with high levels are small, good results can be obtained when a primary diagonal submatrix of order 300–400 is isolated in the H matrix. This is entirely adequate for the majority of applied problems. The submatrix can be diagonalized in the usual way and then the modifications to the eigenvalues of this approximation thus obtained are considered to be solely due to the effect of the diagonal elements of high energies. This frees us from having to operate with matrices of excessively high orders.

10.2 The Variational Method for Solving Anharmonic Problems

We proceed from the equation

$$\left\{-\frac{\hbar^2}{2}\sum_{i,j}\frac{\partial}{\partial q_i}\tau_{ij}(q)\frac{\partial}{\partial q_j}+V(q)\right\}\psi_v=E_v\psi_v$$

after assuming that the additional kinematic term, which does not contain differential operators, is zero. Trial calculations have shown that, in fact, corrections to allow for it are less than 5–10 cm^{-1} to energy levels with quantum numbers up to 7–8.

By partitioning the matrix for the kinetic energy to obtain a part that is calculated using the geometric parameters corresponding to the equilibrium position, and partitioning the matrix for the potential energy to obtain a purely quadratic part, we complete and solve an equation for the polyatomic molecule's vibrations in the harmonic approximation having transformed to linear normal co-ordinates. As a result, the harmonic energy levels

$$E=\sum_k \nu_k(v_k+\tfrac{1}{2})$$

the eigenfunctions

$$\psi_v=\prod_k \psi_{v_k}$$

the matrices L_q and L_p and the transformations for the co-ordinates and momenta are found.

The matrix containing the kinematic coefficients in the new normal co-ordinates will have the form $T = \tilde{L}_p \vec{B} E \tilde{\vec{B}} L_p$, and it is important to note that the matrix L_p could be either square or rectangular (the latter in the system of natural co-ordinates is redundant).

Since the original matrix $T = \vec{B} E \tilde{\vec{B}}$ is dependent on the natural co-ordinates, so will be the matrix $T = \tilde{L}_p \vec{B} E \tilde{\vec{B}} L_p$. Furthermore, L_p is a matrix with constant coefficients because it is calculated from the equilibrium values of all the geometric parameters.

The dependence of the elements of the T matrix on the nuclear co-ordinates (bond lengths, valency angles, etc.) is weak enough (hyperbolic or sinusoidal) for it to be given as a power series with a small number of terms for quite large deviations of the geometric parameters from their equilibrium values. Practical calculations have shown that it is entirely possible to limit the expansion to only two terms for bond extensions of the order of 0.1 Å and changes of the angles of the order of 5°–10°. Hence by taking into account the choice of the L_p matrix, we will obtain the following for the matrix containing kinematic coefficients:

$$
\begin{aligned}
T = I + \sum_k \tilde{L}_p \left[\left(\frac{\partial \vec{B}}{\partial Q_k} \right)_0 E \tilde{\vec{B}}_0 + \vec{B}_0 E \left(\frac{\partial \tilde{\vec{B}}}{\partial Q_k} \right)_0 \right] L_p Q_k \\
+ \tfrac{1}{2} \sum_{k,n} \tilde{L}_p \left[\left(\frac{\partial^2 \vec{B}}{\partial Q_k \partial Q_n} \right)_0 E \tilde{\vec{B}}_0 + \left(\frac{\partial \vec{B}}{\partial Q_k} \right)_0 E \left(\frac{\partial \tilde{\vec{B}}}{\partial Q_n} \right)_0 \right. \\
\left. + \left(\frac{\partial \vec{B}}{\partial Q_n} \right)_0 E \left(\frac{\partial \tilde{\vec{B}}}{\partial Q_k} \right)_0 + \vec{B}_0 E \left(\frac{\partial^2 \tilde{\vec{B}}}{\partial Q_k \partial Q_n} \right)_0 \right] L_p Q_k Q_n
\end{aligned}
\tag{10.4}
$$

Scalar products such as $(\partial \vec{B}/\partial Q_k)_0 E \tilde{\vec{B}}$ are invariant with respect to rotations of the molecule as a whole. Hence we can fix in space some disposition of all the unit vectors of the natural coefficients (which are included in the $\vec{B}$ matrix) that correspond to the molecule's equilibrium. The derivatives $(\partial \vec{B}/\partial Q_k)_0$ and $(\partial^2 \tilde{\vec{B}}/\partial Q_k \partial Q_n)_0$ around the equilibrium position can be calculated. The method of calculating these derivatives was indicated in the last section of Chapter 8.

Further, we will use the notation

$$
T' = \tilde{L}_p \left[\left(\frac{\partial \vec{B}}{\partial Q_k} \right)_0 E \tilde{\vec{B}}_0 + \vec{B}_0 E \left(\frac{\partial \tilde{\vec{B}}}{\partial Q_k} \right)_0 \right]
$$

and

$$
T'' = \tilde{L}_p \left[\left(\frac{\partial^2 \vec{B}}{\partial Q_k \partial Q_n} \right)_0 E \tilde{\vec{B}} + \cdots \right]
$$

The kinetic energy operator is then rewritten as

$$\hat{T}_{\text{kin}} = -\frac{\hbar^2}{2}\left(\sum_k \frac{\partial^2}{\partial Q_k^2} + \sum_{k,n,m} \tau'_{knm}\frac{\partial}{\partial Q_k}Q_n\frac{\partial}{\partial Q_m} + \tfrac{1}{2}\sum_{k,n,m,l}\tau''_{knml}\frac{\partial}{\partial Q_k}Q_nQ_m\frac{\partial}{\partial Q_l}\right)$$

$$= -\frac{h^2}{2}\sum_k \partial^2/\partial Q_k^2 + \Delta\hat{T}\left(Q, \frac{\partial}{\partial Q}\right)$$

If we use a series up to the fourth power for the potential function, then we obtain

$$V(Q) = \tfrac{1}{2}\sum_k \lambda_k Q_k^2 + \tfrac{1}{6}\sum_{k,n,m} u'_{knm}Q_kQ_nQ_m + \tfrac{1}{24}\sum_{k,n,m,l} u''_{knml}Q_kQ_nQ_mQ_l$$

$$= \tfrac{1}{2}\sum_k \lambda_k^2 Q_k^2 + \Delta V(Q)$$

where the u' and u'' coefficients are clearly linear combinations of the coefficients in the natural vibration co-ordinates.

In this way we obtain the following Schroedinger equation:

$$\left\{-\frac{\hbar^2}{2}\sum_k \partial^2/\partial Q_k + \tfrac{1}{2}\sum_k \lambda_k Q_k^2 + \Delta\hat{T}\left(Q, \frac{\partial}{\partial Q}\right) + \Delta V(Q)\right\}\psi_v(Q) = E_v\psi_v(Q) \tag{10.5}$$

We will look for a solution using Ritz's variational method using a basis of harmonic functions

$$\psi_v = \prod_i \psi_i(Q_i)$$

This basis makes the matrix elements very simple, because all variables are separable. As an example, we have

$$\int \prod_i \psi'_{v_i} Q_kQ_nQ_m \prod_i \psi''_{v_i}\,\mathrm{d}Q$$

$$= \langle k'|Q|_k k''\rangle\langle n'|Q_n|n''\rangle\langle m'|Q_m|m''\rangle \prod_S \langle S'|S''\rangle \quad (S \neq k, n, m)$$

where

$$\prod_S \langle S'|S''\rangle = 0 \quad \text{if } S' \neq S'' \quad \text{and} \quad \prod_S \langle S'|S''\rangle = 1 \quad \text{if } S' = S''$$

Since the basis functions are orthogonal the task will be reduced to the diagonalization of one symmetric matrix H, whose elements are

$$h_{ij} = \delta_{ij}E_i^{(0)} + \left\langle i\left|\Delta\hat{T}\left(Q, \frac{\partial}{\partial Q}\right) + \Delta V(Q)\right|j\right\rangle$$

As we will see, this requires the calculation of appropriate linear combinations of matrix elements from various collections of momenta operators and co-ordinates of the form

$$\langle i|\hat{P}_kQ_n\hat{P}_m|j\rangle, \quad \langle i|\hat{P}_kQ_nQ_m\hat{P}_l|j\rangle \quad i = v'_k, v'_n, v'_m$$

$$\langle i|Q_kQ_nQ_m|j\rangle\cdots \qquad j = v''_k, v''_n, v''_m$$

All the h_{ij} matrix elements must be explicit linear functions of the parameters that characterize the potential function in the original natural co-ordinates.

This makes it possible for us to solve the inverse spectral problem easily for anharmonic vibrations, completing them in a single stage and programming them easily for a computer, in contrast to the situation when using the perturbation method.

We can use recurrence relations to calculate the matrix elements of the co-ordinate and momentum powers, i.e.

$$Q\Psi_v = \left(\frac{v+1}{2\gamma}\right)^{1/2}\Psi_{v+1} + \left(\frac{v}{2\gamma}\right)^{1/2}\Psi_{v-1}$$

$$\hat{P}\psi_v = i\hbar\left(\gamma\frac{v+1}{2}\right)^{1/2}\psi_{v+1} - i\hbar\left(\gamma\frac{v}{2}\right)^{1/2}\Psi_{v-1} \quad \gamma = 2\pi\nu/\hbar \tag{10.6}$$

In turn, these are corrollaries of the recurrence relations of Hermite's polynomial. The recurrence relations together with the well-known matrix elements for a normal oscillator's co-ordinate and momentum, which have the form

$$Q_v^{v+1} = \left(\frac{v+1}{2\gamma}\right)^{1/2}, \quad Q_v^{v-1} = \left(\frac{v}{2\gamma}\right)^{1/2}, \quad \hat{P}_v^{v+1} = -i\hbar\frac{[\gamma(v+1)]^{1/2}}{\sqrt{2}}$$

$$\hat{P}_v^{v-1} = \frac{i\hbar}{\sqrt{2}}(\gamma v)^{1/2}$$

enable us to calculate any of the matrix elements we may encounter when solving an anharmonic problem using this method.

We could also take advantage of another approach. The set of matrix elements $\langle k|Q|n\rangle$ and $\langle k|\hat{P}|n\rangle$ can be represented as matrices with the following elements (written here omitting the constant coefficients):

$$Q_k^n = \begin{cases}(v_n+1)^{1/2} & \text{if} \quad v_n = v_k + 1 \\ (v_n-1)^{1/2} & \text{if} \quad v_n = v_k - 1\end{cases}; \quad P_k^n = \begin{cases}-(v_k+1)^{1/2} & \text{if} \quad v_k = v_n + 1 \\ (v_k-1)^{1/2} & \text{if} \quad v_k = v_n - 1\end{cases}$$

In all the remaining cases $Q_k^n = 0$ and $P_k^n = 0$. Hence the Q matrix, for example, will have the form:

	$n=1$, $v_n=0$	$n=2$, $v_n=1$	$n=3$, $v_n=2$	$n=4$, $v_n=3$	$n=5$, $v_n=4$	⋯
$k=1$, $v_k=0$	0	$\sqrt{1}$	0	0	0	⋯
$k=2$, $v_k=1$	$\sqrt{1}$	0	$\sqrt{2}$	0	0	⋯
$k=3$, $v_k=2$	0	$\sqrt{2}$	0	$\sqrt{3}$	0	⋯
$k=4$, $v_k=3$	0	0	$\sqrt{3}$	0	$\sqrt{4}$	⋯
⋯	⋯	⋯	⋯	⋯	⋯	

(10.7)

The P matrix will look exactly the same, the only difference being that the upper off-diagonal terms will be negative.

As a result of the recurrence formula being obtained from Hermite's polynomial, and consequently from harmonic functions, the theorem that states that the matrix elements from any set of co-ordinate and momentum operators are equal to the products of the corresponding set of matrices P and Q is true. Thus, for example, the matrix elements

$$\langle k|\hat{P}QQ\hat{P}|n\rangle$$

from the $\hat{P}QQ\hat{P}$ operator can be found as elements of the matrix

$$(PQQP) = PQQP$$

whose elements are

$$(PQQP)_k^n = \sum_{m,l,g} P_k^m Q_m^l Q_l^g P_g^n \tag{10.8}$$

This method of ascertaining the matrix elements from different sets of co-ordinate and momentum operators is particularly convenient when implementing the variational procedure on digital computers, since only the usual matrix multiplications are needed.

When the potential function is used in natural co-ordinates as a fourth-order Taylor series, the Hamiltonian matrix that is obtained using the variational procedure has a directly expressed band structure. It enables us in the general case to evaluate reasonably accurately the minimum basis dimensions needed for some vibrational energy level to achieve its variational limit.

Thus when the Hamiltonian matrix has a band structure and its non-diagonal elements are at least an order of magnitude lower than the diagonal elements, all the elements from the band of non-zero non-diagonal elements must be included in the row containing the diagonal element in order for that element of the matrix to achieve its variational limit. When the potential function in the form of a fourth-order Taylor series is being used, only those elements of the Hamiltonian matrix H for which the summed change in the quantum numbers does not exceed four during a transition from the k state to the n state will be non-zero.

We can conclude from this that in order for a vibrational energy level that is described by the quantum numbers $v_1, v_2, \ldots, v_{3N-6}$ to achieve its variational limit in a calculation, we must use the sufficient number of basis harmonic wave functions, i.e.

$$N_{\text{var}} = \frac{(N_1 + 3N - 6 + 4)!}{(N_1 + 4)!(3N - 6)!} \tag{10.9}$$

where

$$N_1 = \sum_{k}^{3N-6} v_k$$

and N is the number of the molecule's normal vibrations.

Note that in most cases, practical estimates corroborate equation (10.9) very well. Only in a few unfavourable cases (for instance, in the presence of strong Fermi resonance or when calculating higher vibrational energy levels) is it necessary to substitute the four in equation (10.9) for a five or a six.

In order to diagonalize the Hamiltonian matrices we can use the wide variety of methods that have been developed in mathematics to diagonalize symmetric matrices. Thus, depending on the particular problem being dealt with, the following methods have proved to be effective: (1) the Jacobian method, (2) reduction to a tridiagonal form, (3) the *QLR* algorithm, (4) the algorithm for diagonalizing band matrices and (5) the algorithm for calculating the individual eigenvalues and eigenvectors.

To conclude this section we will make a few comments on the solution of the inverse spectral problems for anharmonic potential functions. In many cases these problems coincide with the inverse problems for harmonic vibrations, since they reduce to the determination of the coefficients in the matrix elements, Moreoever, the elements depend linearly on the required coefficients according to the given eigenvalues of the matrix. The least squares method is also very useful for solving reverse anharmonic problems and is considered in detail in Chapter 13. Thus we will not dwell on this aspect here, limiting ourselves to a few points.

First, we note that the abundance of spectral lines and bands, obtained if the overtones and combination frequencies as well as the fundamentals are experimentally measured, greatly increases the possibility of finding solutions. Even if we consider that in many cases it is permissible for energy levels in the anharmonic approximation to obey the simple formula (10.3), then when determining the coefficients on the potential surface and given sufficient experimental data, $(3N-6).(3N-3)/2$ pairs of values are obtained. This is considerably more than can be observed in the harmonic approximation. Since this formula is not accurate, even more experimental data can be obtained. If judicious physical arguments are used to decrease the number of coefficients in the anharmonic potential function, say by only obtaining the anharmonic terms for the bond-stretching co-ordinates, then the inverse spectral problems become soluble for one molecule, as shown by experience.

A last important comment. The application of the variational procedure does not require accurate harmonic functions to be used. This makes it possible in particular to avoid repeated changes in the basis and consequently repeated recalculations of the matrix elements in the *H* matrix. The basis calculated during the first iteration can thus be retained as the harmonic one throughout the solution of the inverse problem. A transformation to the true harmonic basis can be carried out at the very end of the calculation after the desired accuracy of solution has been reached.

10.3 Examples of the Solutions of Anharmonic Vibrational Problems

In this section we are going to provide a number of examples of the solution of anharmonic vibration problems and judge the results obtained. In every case a general form of the Hamiltonian equation (10.1) was used. The $\tau_{ij}(q)$ coefficients are expanded into a series of normal co-ordinates up to the quadratic terms only.

This Hamiltonian does not take account of rotations of the molecules, but this, however, is not significant in our case because if a molecule's vibrations have any marked effect on its rotational energy levels, the reverse effect is practically absent.

Hence if we only need to find the vibrational energy levels for a given potential surface, or, conversely, find the parameters of the surface from the experimentally determined vibrational energy levels calculated from the spectra of molecules in the gas phase or those of molecules in a solid matrix, then we can completely ignore rotations. This view has been corroborated by calculations we carried out specially to test this point. These demonstrated that the solutions of the direct spectral problem that took rotations into account practically coincided with solutions calculated with no allowance for rotations, both solutions being obtained using variational methods. At the same time, the parameters of some molecules' potential surfaces determined by processing their gas phase spectra and those determined by processing their matrix phase spectra were not found to differ significantly. For instance, the variational estimates we did for a water molecule without allowing for rotations and similar estimates[9] allowing for rotations and the corresponding interactions—shown in parentheses—are as follows (to save space, we give only a few of the purely vibrational frequencies, all in cm^{-1}): 1599(1598), 3164(3159), 3716(3718), 3818(3821), 5303(5301), 5404(5404), 7416(7424), 7518(7522) and 7615(7620). We used the force field presented in reference 22 and the same set of harmonic basis eigenfunctions. In both our estimates and those of reference 9 the application of the potential parameters given in reference 22 gave poor agreement with experiment, where the respective anharmonic frequencies are 1594, 3152, 3657, 3756, 5235, 5331, 7201, 7250 and 7445 cm^{-1}. Note that the inverse anharmonic problem in reference 22 was solved within the bounds of the usual method from the theory of second-order perturbations, and by its very formulation is generally unjustifiable when there are multiple resonances. Both we and the authors of reference 9 consider that this accounts for the discrepancies between the solution of the direct problem using the usual techniques and that of using the variational method even when the same potential function is used. Obviously, the basis had to be large enough to obtain a stable variational limit, and in fact the solution then obtained for the given potential was practically exact. There is, therefore, some doubt in a whole range of published papers concerning the accuracy of the force fields found using the more approximate technique.

The kinematic contribution to the anharmonicity of the vibrations is about 15% for all molecules. Expanding the coefficient of the Hamiltonian's kinematic part in terms of vibrational co-ordinates up to the quadratic terms provides estimates of the kinematic contributions to the general energy levels being calculated that are accurate for all practical purposes. The contribution of the terms containing Sp or $Tr\,[T^{-1}(\partial T/\partial q_i)]$ in the Hamiltonian used in this section is found to be negligibly small (of the order of 1 cm^{-1} and less). This indicates that the anharmonic problem can be completely solved (through obviously not for large amplitudes) by the very simple Hamiltonian with

$$\hat{T} = -\frac{\hbar^2}{2}\sum_{i,d}\frac{\partial}{\partial q_i}\tau_{ij}(q)\frac{\partial}{\partial q_j}; \quad \text{and} \quad \Delta T = 0$$

It is useful to compare the effects of the anharmonic terms on the position of the vibrational energy levels in molecules with light nuclei and those with heavy ones. Even though, for example, the order of magnitude of the significant coefficients in the potential functions of the NO_2, H_2O (D_2O), and H_2Se molecules is the same, the influence of the anharmonic terms in the potentials of the molecules with heavy atoms on the vibration frequencies is significant, though less than is the case for the molecules with light atoms. Thus the corrections to the anharmonicity for the energy levels with quantum numbers 100, 010, 001, 200, 020 and 002 are 175.07, 54.81, 186.73, 462.84, 146.87 and 440.40 for the H_2O molecule and 33.24, 10.00, 54.07, 79.20, 20.70 and 140.40 for NO_2. The higher terms in the potential functions of heavy molecules are abstracted from experimental frequencies due to low transitions which have much poorer accuracies than have the analogous values for the molecules containing light atoms. The f_{rrr} and f_{rrrr} coefficients and their determination errors are, respectively, -51.77 ± 3.1 and 185.42 ± 41.10 for H_2O and -53.54 ± 10.43 and 328.10 ± 196.81 for O_3.

Tables 10.1 and 10.2 indicate the values of coefficients in the linear combinations of basis functions for H_2O and NO_2 molecules. We can see that in many cases there are sufficiently great contributions from non-diagonal elements of the corresponding C matrix. These are the so-called 'resonance effects', which are, as usual, greater in H_2O then in NO_2.

The next examples demonstrating the technique of solving anharmonic problems are the results for CO_2 and HCN. These problems have already been considered[23–8]; however, functions describing the rotations of bent molecules were included in the basis functions together with the purely vibrational ones. This made it necessary to take explicit account of the vibrations and rotations, and consequently the computational procedure was unnecessarily complicated. On the other hand, the approach in which two vibrational co-ordinates of a bent molecule are introduced from the very start in mutually perpendicular directions is completely equivalent. This makes it possible to retain the whole computational scheme in its entirety to describe polyatomic linear molecules and not just triatomic linear ones.

When solving the anharmonic problem for $^{12}CO_2$ ($^{13}CO_2$) and HCN(DCN) we limited ourselves by using a few simplified exponential potential functions. Being able to do this holds out promise that the anharmonic vibrations of polyatomic molecules will be estimated, since it is clear that their complete potential functions cannot be found due to the complexity of the task and the deficiencies of the experimental data. Inverse spectral problems can only be solved for basic isotopic modifications because the task is only well defined in this case (the numbers of parameters being varied are 7 and 9 for CO_2 and HCN, respectively, while those of vibrational frequencies available to solve the

Table 10.1. The elements of the C matrix for H_2O

Diagonal element	Magnitude at non-diagonal elements for the corresponding Energy level				
000	100	102	200	300	030
0.9880	−0.1455	−0.0322	0.0225	−0.0214	0.0129
010	110	112	020	210	040
0.9828	−0.1698	−0.0327	0.0308	0.0298	0.0276
100	200	000	002	020	300
−0.9301	0.2878	−0.1486	0.0875	0.0786	−0.0696
020	120	030	100	050	220
0.9620	−0.1905	0.0916	0.0814	0.0470	0.0376
200	300	100	002	102	400
−0.7462	0.3920	−0.3107	−0.2668	0.2245	−0.1302
002	102	200	300	012	202
−0.8528	0.3208	0.3191	−0.1589	0.1005	−0.0991
110	210	010	030	012	310
−0.9082	0.3167	−0.1750	0.1086	0.0871	−0.0838
011	111	211	001	113	021
0.9431	−0.2957	0.0800	0.0706	0.0607	0.0435
101	201	001	301	003	203
−0.8067	0.4261	−0.2801	−0.1525	0.1364	0.0962
030	130	040	110	020	060
−0.9424	0.2052	−0.1660	−0.1175	0.0952	−0.0716
300	200	400	202	102	002
−0.4568	−0.4187	0.3854	0.3700	0.3416	0.2314
003	103	201	301	203	013
0.8083	−0.3862	−0.2741	0.2014	0.1353	−0.1346
201	301	101	103	401	003
−0.5493	0.4746	−0.4706	0.2626	−0.2120	−0.1750
102	300	400	202	002	200
−0.5572	0.4432	−0.3418	0.3406	−0.3171	0.1817

Note: Sets of quantum numbers are indicated.

inverse problem are 29 and 26, respectively). In solving the inverse problems, statistical weights equal to one were used for the fundamental frequencies, primary overtones and combination frequencies. The levels corresponding to transitions with summed changes in quantum numbers of 3 and 4 were allocated the statistical weights − 0.3 and − 0.1, respectively. A stable variational limit was observed for each of the energy levels being calculated. The number of basis functions taken when solving the problem for the CO_2 molecule was 503, and 338 were taken for the HCN molecule. The presence of symmetry enables the variational matrix for CO_2 to be quasi-diagonalized to eight blocks and to four blocks for HCN. The greatest block order in the CO_2 problem was 110 and the least 30. The respective values were 126 and 50 for HCN. The quasi-diagonalization of the matrix of the overall variational problem allowed the dimensions of the basis to be substantially increased without increasing the

Table 10.2. The elements of C the matrix for NO_2

Diagonal elements	Magnitude of non-diagonal elements for the corresponding energy level				
000	100	010	102	300	200
0.9890	−0.1350	0.4196	−0.0327	−0.1727	0.0155
010	110	020	000	112	040
0.9842	−0.1479	0.0612	−0.0462	−0.0329	0.0209
100	200	000	020	110	002
0.9428	−0.2412	0.1311	0.1038	0.0949	−0.0785
001	101	201	103	011	111
0.9490	−0.2913	0.0673	−0.0562	0.0628	−0.0281
020	120	100	030	010	210
0.9729	−0.1609	0.0920	0.0765	−0.0728	0.0359
200	300	100	210	120	102
−0.8619	0.3421	−0.2431	−0.1334	−0.1389	0.1169
002	102	202	200	012	104
−0.8833	0.4068	−0.1346	0.0919	−0.0823	0.0776
110	210	030	010	120	100
0.9125	−0.2532	0.1777	0.1407	0.1301	−0.1151
011	111	001	211	021	113
−0.9364	0.3036	0.0832	−0.0766	−0.0898	0.0563
101	201	001	301	003	011
0.8147	−0.4210	0.2884	0.1303	−0.1220	0.0916
030	130	110	020	040	120
0.9554	−0.1735	−0.1528	−0.0991	0.0885	−0.0562
300	400	200	220	012	310
0.7380	−0.4176	0.3406	0.1631	−0.1452	0.1412
003	103	203	201	013	101
0.8137	−0.4829	0.1912	−0.1269	−0.0956	0.0955
201	301	101	401	103	111
−0.6506	0.4965	0.4269	−0.1860	0.1632	−0.1388
102	202	002	302	004	012
−0.6456	0.5084	−0.4088	−0.2125	0.1478	−0.1278

Note: Sets of quantum numbers are indicated.

computer calculation time. The force field found by solving the inverse anharmonic problem was later used without change to solve the direct problems for isotopic modifications of the same molecules. Very favourable agreement was obtained between the calculated and experimental frequencies, which demonstrated the validity of the coefficients found for the potential functions.

The coefficients calculated for the potential functions are given in Table 10.3. Data from other authors are also given, as are the coefficients of the potential surface of HCN, estimated using quantum calculations. As can be seen, the agreement between our results and those of other authors is very satisfactory. The quantum calculations corroborate the reasonableness of the approximation selected for the potential function. Estimates of the energy levels (counting from the ground state) are given in Tables 10.4 and 10.5 together with the necessary

Table 10.3. The coefficients of the potential functions of the CO_2 and HCN molecules

Coefficient and its dimensions	Our Results	Data from ref. 28	Data from ref. 29
$f_{rr}/md\,(\text{Å}^{-1})$	$15{,}96 \pm 0.10$	$16{,}022 \pm 0.004$	16.017 ± 0.008
$f_{rr'}/md\,(\text{Å}^{-1})$	$1{,}25 \pm 0.07$	$1{,}261 \pm 0.004$	1.249 ± 0.07
$f_{\alpha\alpha}/md\,(\text{Å})$	0.80 ± 0.01	0.785 ± 0.004	0.784 ± 0.001
$f_{rrr}/md\,(\text{Å}^{-2})$	-131.07 ± 5.06	$-113.94 + 0.16$	-114.331 ± 0.018
$f_{r\alpha\alpha}/md$	-3.77 ± 0.53	-1.218 ± 0.006	-1.209 ± 0.016
$f_{rrrr}/md\,(\text{Å}^{-3})$	1121.0 ± 56.0	630.0 ± 4.0	640.068 ± 12.0
$f_{\alpha\alpha\alpha\alpha}/md\,(\text{Å})$	9.51 ± 0.65	1.106 ± 0.048	0.828 ± 0.048

HCN

$f_{RR}/md\,(\text{A}^{-1})$	6.25 ± 0.07	6.216 ± 0.006[a]	6.894[b]	6.66[c]
$f_{Rr}/md\,(\text{Å}^{-1})$	-0.16 ± 0.18	-0.242 ± 0.016	-0.184	-0.498
$f_{rr}/md\,(\text{Å}^{-1})$	18.40 ± 0.31	18.644 ± 0.031	24.318	23.36
$f_{\alpha\alpha}/md\,(\text{Å})$	0.26 ± 0.004	0.261 ± 0.003	—	—
$f_{RRR}/md\,(\text{Å}^{-2})$	-33.34 ± 1.39	-32.34 ± 0.06	-38.00	-38.10
$f_{rrr}/md\,(\text{Å}^{-2})$	-130.9 ± 4.2	-125.88 ± 0.15	-151.96	-151.5
$f_{RRRR}/md\,(\text{Å}^{-3})$	121.8 ± 3.7	133.68 ± 0.30	197.2	192.1
$f_{rrrr}/md\,(\text{Å}^{-3})$	810.0 ± 138.9	960.0 ± 0.4	807.8	808.3
$f_{\alpha\alpha\alpha\alpha}/md\,(\text{Å}^{-3})$	0.49 ± 0.09	0.2424 ± 0.0012	—	—

[a] Data from reference 30.
[b] Data from reference 31.
[c] Data from reference 32.

figures for interpretation. These are taken as

$$\alpha_{ki} = \frac{C_{ki}^2 E_k^{(0)}}{E_k}$$

where C_{ki} is a coefficient in the linear combination of basis functions, $E_k^{(0)}$ is the energy of the corresponding level in the harmonic approximation and E_k is the energy of the given anharmonic level.

Table 10.5 contains only the set of figures for interpretation for a few of the vibrational states of CO_2, but it shows the contributions of harmonic vibrations into the vibration levels in CO_2 when there is strong Fermi resonance. It is worth noting that the interpretation of the vibrational states of $^{12}CO_2$, $^{13}CO_2$, HCN and DCN all conform to the traditional interpretation.

The curves in Figure 10.1 demonstrate the behaviour of calculated energy levels as the basis is expanded. It can clearly be seen that as the basis increases, the values of all the levels tend monotonically to the variational limit. This

Table 10.4. The experimental and calculated vibrational frequencies (in cm^{-1}) of $^{12}CO_2$, $^{13}CO_2$, HCN and DCN

$^{12}CO_2$		$^{13}CO_2$		HCN		DCN	
Experimental	Calculated $\Delta\nu = \nu_c - \nu_{ex}$	Experimental	Calculated $\Delta\nu = \nu_c - \nu_{ex}$	Experimental	Calculated $\Delta\nu = \nu_c - \nu_{ex}$	Experimental	Calculated $\Delta\nu = \nu_c - \nu_{ex}$
667.1	−1.0	648.5	−1.0	712.0	0.2	569.1	−1.0
1285.4	−2.1	1265.2	−3.7	1411.4	6.7	1130.0	1.2
1335.1	7.9	1292.2	7.5	1426.5	8.2	1137.9	−0.1
1388.2	4.3	1370.1	5.5	2096.9	−8.8	1690.2	7.6
1932.5	−7.3	1896.5	−9.8	2106.1	13.1	1706.6	0.5
2003.3	−1.5	2037.1	6.6	2143.8	17.4	1925.3	2.5
2076.9	6.4	2283.5	−8.5	2802.7	−18.0	2497.1	−2.7
2349.1	−8.5	2920.3	−20.4	2803.0	29.5	2630.3	6.4
2548.3	−10.6	3527.7	−27.6	2818.2	−12.3	3060.7	−5.2
2585.0	10.7	3632.9	3.5	3315.5	19.0	3068.6	−3.2
2671.2	−14.7	4147.2	−41.7	3502.1	−14.9	3183.7	−1.8
2672.8	19.8	4287.7	−7.1	3516.9	−13.6	3623.9	−27.0
2760.7	24.6	6780.2	15.7	4004.2	7.5	3729.1	−5.5
2797.2	13.2			4173.1	−22.1	3736.9	7.1
3004.0	21.3			4201.3	−9.4	3836.4	3.6
3181.5	−17.0			4684.3	2.0	4411.0	−5.9
3240.6	−0.9			4699.2	3.3	4523.3	5.7
3339.3	−10.9			4878.3	39.3	5079.5	−8.2
3391.8	−29.0			5366.9	−40.0	5220.2	9.4
3442.1	7.4			5393.7	7.9	5757.9	−8.3
3550.6	−25.1			5571.9	−45.9		
3612.8	−22.5			6083.4	−11.7		
3659.3	−23.9			6519.6	−27.9		
3714.8	−2.1			6767.3	−42.6		
4247.7	−36.9			7192.8	−10.4		
4314.8	−41.5			8585.6	−5.2		
4390.6	−11.6						
4673.3	−4.6						
6972.6	18.2						

Table 10.5. Figures for interpretation for several vibrational states of CO_2

Experimental	(1000)*	(0200)	(0020)	(1200)	(1020)	(2000)		
1388.2	0.4604	0.2285	0.2280	0.0253	0.0252	0.0489		
1285.4	(1000)	(0200)	(0020)	(1200)	(1020)	(2000)		
	0.4478	0.2448	0.2448	0.0241	0.0241	0.0549		
1335.1		(0200)	(0020)	(1200)	(1020)			
		0.4724	0.4736	0.0489	0.0491			
2797.2	(2000)	(1020)	(1200)	(0400)	(0040)	(2200)	(0220)	(2020)
	0.1162	0.1742	0.1440	0.1532	0.1530	0.0442	0.0350	0.0442
2548.3	(2000)	(1020)	(1200)	(0400)	(0040)	(2200)	(0220)	(2020)
	0.1253	0.1976	0.1980	0.0815	0.0817	0.0536	0.1447	0.0535
2672.8	(2000)	(1400)	(1040)	(0400)	(0040)	(0220)	(1000)	(3000)
	0.3070	0.0365	0.0365	0.2100	0.2099	0.1062	0.0224	0.0701
2760.7	(1200)	(1020)	(0400)	(0040)	(1400)	(1040)	(2200)	(2020)
	0.1574	0.1573	0.2556	0.2558	0.0539	0.0539	0.0376	0.0376
2585.0	(1200)	(1020)	(0400)	(0040)	(1400)	(1040)	(2200)	(2020)
	0.2298	0.2295	0.1907	0.1906	0.0274	0.0274	0.0640	0.0639
2671.2	(2000)	(0220)	(1220)	(3000)				
	0.2366	0.6061	0.1124	0.0563				

Note: The harmonic eigenfunctions' quantum number sets are indicated in parentheses. The first digit refers to the in-phase normal co-ordinate, the fourth to the out-of-phase co-ordinate and the second and third to bending vibrations.

confirms that the solutions obtained are not in fact approximate, but are accurate wave functions and eigenvalues of the Schroedinger equation. The inaccuracy of the solutions of the anharmonic problem is associated solely with the simplifications embedded in the equations themselves, and are determined by the form of the potential functions and by the limitation of the general functional form of the kinematic coefficients on the selected vibrational co-ordinates to second-order polynomials. This shows up the sharp distinction between the variational solution of an anharmonic problem and the traditional one by the perturbation method. Not only do assumptions have to be made when tackling the problem by the latter method, but obtaining the solution itself can only be done approximately. Moreover, the influence of these approximations have on the final result is not controlled. To obtain this control would require a transition to higher approximations in perturbation theory, which would make the whole scheme very cumbersome. Verifying that the variational limit has been achieved and consequently ensuring the accuracy of the selected equations can, in our case, be done very easily without any change to the computational scheme.

The energy levels of CO_2 and HCN are shown schematically in Figure 10.2, and the quasi-continuum shows up very clearly. It is worth noting that because the basis is limited, the upper energy levels are allowed to a greater extent than is true in reality. Tentative values for the energy levels, which cannot be calculated when solving the variational problem on the basis of selected functions and which we estimated, are indicated by dashed lines.

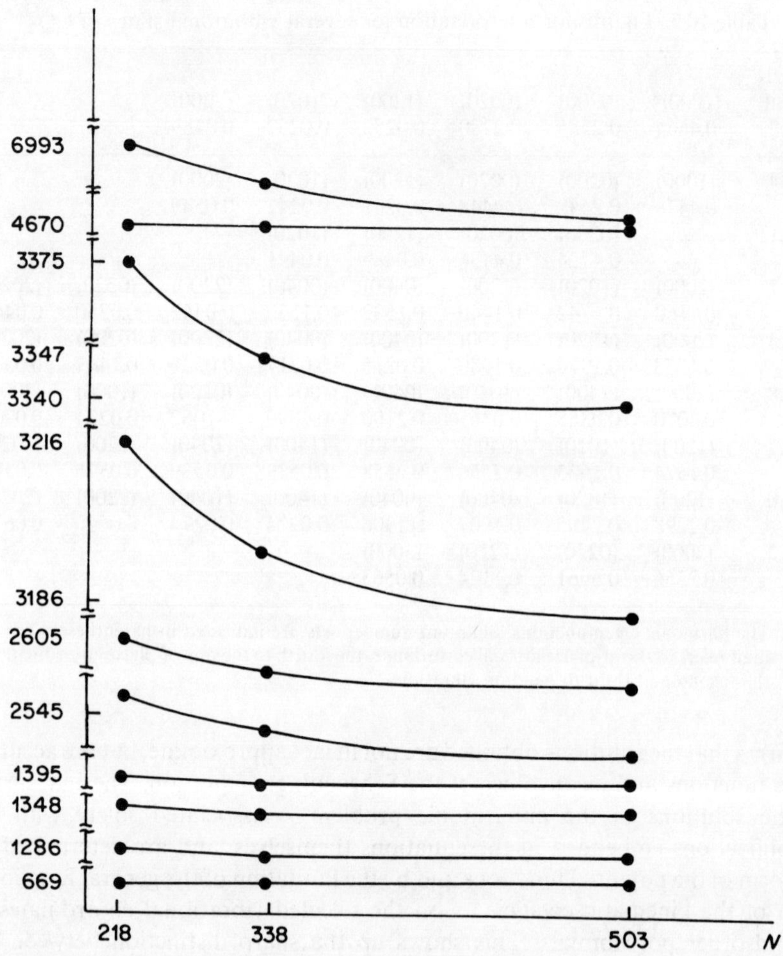

Figure 10.1. Diagram of the behaviour of the calculated energy levels of a CO_2 molecule in the vicinity of a variational limit that has been reached. N is the number of basis functions

In conclusion, note that when there is twofold degeneracy for bending vibrations in linear molecules there is no sense, in practical calculations, in including an explicit pseudo-rotation co-ordinate or in using the concept of a vibrational moment of momentum, because this requires a special mathematical apparatus and solution programmes for the anharmonic problems of linear molecules. The much simpler approach is to use the universal purely vibrational approach, having included bending vibration co-ordinates in two perpendicular directions.

We shall discuss the variational estimates of the anharmonic vibration levels for acetylene as the next example. The direct and inverse problems were solved

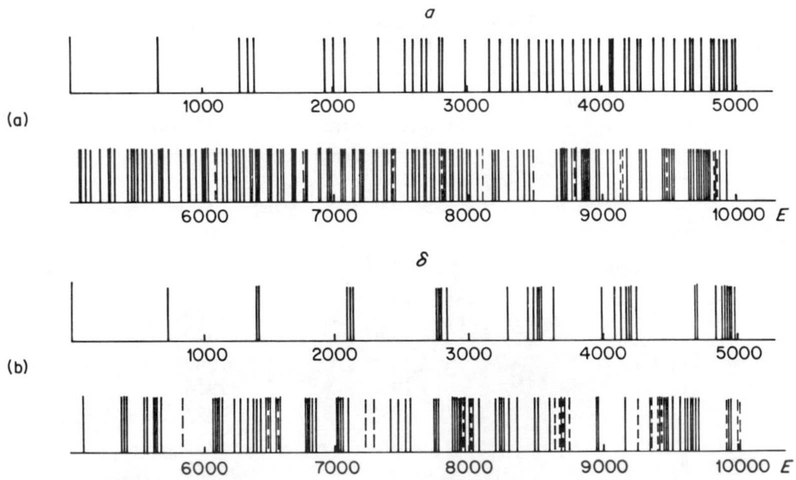

Figure 10.2. Calculated vibration energy levels for (a) CO_2 and (b) HCN

using a basis of 813 harmonic functions, and because of the symmetry and infinitesimal nature of the separate elements, the Hamiltonian matrix for this molecule can be broken down into eight blocks (457, 112, 100, 100, 100, 100, 72, 72). We must note immediately that the variational limit was only achieved for the fundamental transitions, the primary overtones and the combination frequencies. Frequencies for which the summed quantum number values was greater than two did not, as a rule, achieve their variational limits. When solving the inverse problems we used statistical weights equal to one for the fundamental frequencies, primary overtones and combination frequencies. The levels corresponding to summed quantum number changes of 3 and 4 were allocated statistical weights of 0.3 and 0.1, respectively. The inverse problem can be solved only for basic isotopic modifications of the molecule. Only the direct problem was solved for the deutero-modification to verify the field that was obtained. When solving both the direct and inverse anharmonic mechanical spectral problems we chose models of the potential function that contained the least number of anharmonic coefficients. Such coefficients were not included in the potential function for the C≡C bond in the molecule because there are no experimental data about the overtones of this bond. In addition, it has been made clear that the deformational vibrations in C_2H_2 have a small enough degree of harmonicity for the anharmonic coefficients to be omitted from the molecule's potential function for these vibrations.

The results are given in Table 10.6. We considered above that for both HCN and CO_2 the anharmonic vibration problem for linear molecules could be formulated within the bounds of the purely vibrational approach, taking degenerate vibrations into account. It is not therefore necessary to introduce explicit pseudo-rotational co-ordinates or use the concept for the vibrational moment of momentum. Furthermore, the problem of having to account for

Table 10.6. Calculated and experimental frequencies for C_2H_2 and C_2D_2 vibrations

Interpretation of the levels	Experimental	Calculated
	C_2H_2	
ν_4	613	603
ν_5	730	730
		1334
$\nu_4 + \nu_5$	1328	1333
	1940	1950
$2\nu_4 + \nu_5$		1949
ν_3	1974	1974
ν_1	3282	3282
	3295	3322
$\nu_3 + \nu_4 + \nu_5$		3321
ν_2	3374	3365
	3420	3349
$\nu_3 + 2\nu_5$		3349
$\nu_2 + \nu_5$	4092	4103
	4674	4735
$\nu_2 + \nu_4 + \nu_5$		4735
	4723	4776
$\nu_1 + 2\nu_5$		4776
	4798	4862
$\nu_2 + 2\nu_5$		4863
$2\nu_1$	6502	6511
$2\nu_1 + \nu_5$	6557	6563
$2\nu_2$	6709	6704
$2\nu_2 + \nu_5$	7219	7305
$\nu_2 + \nu_3 + \nu_1$	8512	8612
	C_2D_2	
ν_4	511	504
ν_5	539	537
$2\nu_4$	1028	1009
		1009
$2\nu_5$	1086	1075
		1075
$\nu_3 + \nu_5$	2298	2292
ν_1	2439	2437
ν_2	2705	2709
$\nu_1 + \nu_4$	2945	2943
$\nu_1 + \nu_5$	2973	2976
$\nu_2 + \nu_5$	3236	3248
$\nu_3 + \nu_1$	4191	4180
$\nu_1 + \nu_3 + \nu_4$	4693	4706
$\nu_2 + \nu_1$	5097	5104
	5198	5257
$\nu_1 + \nu_2 + \nu_4 + \nu_5$		5257
$\nu_2 + \nu_1 + \nu_4$	5595	5640
$2\nu_2 + \nu_5$	5909	5968

vibration and rotation interactions does not arise in this approach. In the case of C_2H_2 the traditional two-rotation and two-deformation co-ordinates were also replaced by four bending vibrational co-ordinates in the C_2H_2 molecule in two mutually perpendicular directions. This enables us to retain the whole calculation scheme and to use a single software package, even when describing linear molecules, for both the direct and inverse anharmonic mechanical spectral problems. The relative abundance of experimental material for C_2H_2 allowed us to refine several of the harmonic coefficients in its potential function at the same time. It became clear that the anharmonic coefficients satisfy the Morse approximation for bonds very well and can be determined with a high degree of accuracy.

When solving anharmonic vibrational problems using the method described it is found that the main difficulty encountered is having to diagonalize very large matrices. The calculation of the matrix elements themselves, especially when the potential function used is in the form of a power series, is very fast, even when the basis is of the order of several thousand. To overcome this limitation we tried a special approach of stagewise diagonalization of the large-order matrices encountered solving an anharmonic problem.

The large matrix diagonalization algorithm can use the fact that the matrices which arise using the variational solution method have very clearly expressed band structures. Thus if two large enough contiguous diagonal blocks can be isolated, then the corresponding non-diagonal submatrices of their interactions will only contain non-zero elements at the corners very close to the 'join' of the quasi-diagonal matrices. All the remaining elements will be zero. Moreover, if the non-diagonal elements are not large, then the eigenvalues of the diagonal submatrices will, to a very good degree of accuracy, be equal to the eigenvalues of the whole matrix. This enables us to solve the problem in two stages. In the first, by automatically isolating a series of large diagonal blocks, they can be diagonalized accurately. At the second stage, corrections are made to the eigenvalues that have been found using the perturbation method. The perturbations can be induced as a few additional small 'turns' of the co-ordinate system in the appropriate orthogonal Jacobian transformations. Even though, strictly speaking, the non-diagonal submatrices, which correspond to the interactions between neighbouring diagonal blocks, may be non-zero during these transformations, it can be shown that, due to the normalizing properties of the eigenvectors of the diagonal blocks, the values of these elements are reduced in size inversely proportionally to the sizes of the diagonal blocks. Furthermore, because the Jacobian method has quadratic convergence even after the first 'turn' of the system, there will be elements in the places of the non-diagonal interaction submatrices that will not exceed the squares of the initial elements of the submatrix. Hence when solving the variational problem, and in the absence of symmetry, there will be as much simplification as can be achieved when solving harmonic problems for large non-symmetric molecules using the perturbational method. Infact it is more convenient to do this the opposite way round. First, the most significant non-diagonal interaction elements between the diagonal submatrices are elimited by one or two Jacobian interactions. Then the diagonal submatrices are diagonalized individually. Special computer experiments have shown that with an appropriate choice of cutoff for levels with respect to the triviality of neighbouring submatrix interactions, eigenvalues can be obtained for the full anharmonic vibration problem that are exact for all practical purpose. In these cases there is no need to resort to a direct diagonalization algorithm for every large matrix. The computer time spent evaluating the eigenvalues of the matrices does not become excessively large for any problem, even when the basis has several thousand functions, so long as the sizes of a matrix's diagonal blocks do not exceed 200–300. Note that trivial non-diagonal interaction elements mean that the order of the diagonal blocks is small. Concrete calculations have shown that this algorithm gives good results, not only with band Hamiltonian matrices but also with 'planar' Hamiltonians which have two or more weakly interacting sub-blocks. In the latter case the energy levels may interact with arbitrary strength within any of the sub-blocks. Note that when there is weak anharmonicity in a molecule due to weak interactions between the vibration energy levels within the whole Hamiltonian, the algorithm treats the matrix's diagonal elements as weakly interacting submatrices.

A criterion for isolating weakly interacting blocks in the general matrix of the anharmonic problem that can be used is the smallness of the quantity $S = \sin\varphi$ after each Jacobian iteration. Two or more sub-blocks of the Hamiltonian matrix are considered not to be interacting if, for any of their interaction elements, the quantity $S = \sin\varphi$ arising from the first Jacobian turn is less than some level of discrimination given as an operator.

In order to check this procedure we solved the direct anharmonic problem for ethane and propane. Since no data about the anharmonic energy levels are available for either molecule or their anharmonic potential functions, and since our task was to verify the efficiency of a computer algorithm, we simply carried out a computer experiment to see if it was possible to solve anharmonic problems for these reasonably complicated molecules with bases large enough to attain stable variational limits. The constants of the potential function's quadratic part that were needed for the calculation were taken from the solution of the purely harmonic problems. The anharmonic constants were assumed to be equal in size to those for the C—H and C—C-bonds, while the corresponding angles were found by us earlier when solving anharmonic problems for molecules with similar structural elements. The problem was solved using dependent co-ordinates and rotation around the C—C bond was ignored. The number of co-ordinates were 19 and 28 for C_2H_6 and C_3H_8, and the numbers of degrees of freedom were 17 and 25, respectively.

During the machine experiment the discrimination level S_m ($S = \sin\varphi$) was found from our method so that the eigenvalues of the large matrices could be ascertained with accuracies not worse than 1 cm^{-1}. It was found that to achieve this, $S_m \leqslant 0.1$. With $S_m = 0.15$ the calculation error in the eigenvalues grew to 5 cm^{-1}. The whole Hamiltonian then broke up into a large number of weakly interacting blocks both for C_2H_6 and C_3H_8. This part of the experiment showed the algorithm's reliability, and that in future it would allow direct anharmonic problems to be solved with bases up to 1900 functions in size for both C_2H_6 and C_3H_8. Sufficiently exact variational limits were then obtained for all first- and second-order levels, which are the only ones that are experimentally observed for large molecules.

The second stage of the computer experiment was to ascertain whether anharmonic problems could be solved taking into account only partial anharmonicity. In fact, it is well known that the vibrations of peripheral X—H bonds are very characteristic, and are only weakly mixed with the skeletal vibrations. At the same time, they are quite anharmonic, since they are associated with the movements of light atoms, whilst the vibrations of the molecule's skeleton are almost harmonic. This suggested that it may be possible to estimate approximate anharmonic corrections to frequencies only for those vibrations which need appreciable corrections. This might be possible using the variational calculation principle by including only those excited eigenfunctions in the general basis which are intimately associated with the vibrations of the peripheral atoms.

In order to demonstrate the usefulness of this approach for ethane and propane the anharmonic energy levels were estimated with the same potential function but with different bases. Both complete bases (1000 or more functions) and bases consisting of subsets of functions were applied. The subsets included excited basis functions only for CH bond vibrations, or only for deformational

Table 10.7. Calculated energies of the transitions between the levels corresponding to the valency vibrations of the CH bond in C_2H_6

Transition interpretation	Transition energies in the harmonic approximation	First estimate for $N = 930$	Second estimate for $N = 1875$	Transition interpretation	Transition energies in the harmonic approximation	First estimate for $N = 930$	Second estimate for $N = 1875$
ν_1	3006	2973	2982	$\nu_1 + \nu_2$	6017	5980	6015
ν_2	3010	2983	2992	$\nu_1 + \nu_3$	6081	6014	6026
ν_3	3074	3039	3038	$\nu_1 + \nu_4$	6081	6016	6026
ν_4	3074	3040	3038	$\nu_1 + \nu_6$	6094	6015	6026
ν_5	3088	3053	3051	$\nu_2 + \nu_3$	6085	6081	6082
ν_6	3088	3053	3051	$\nu_2 + \nu_4$	6085	6081	6084
$2\nu_1$	6012	5977	6006	$\nu_2 + \nu_5$	6099	6083	6083
$2\nu_2$	6012	5990	6012	$\nu_2 + \nu_6$	6099	6084	6087
$2\nu_3$	6149	6100	6105	$\nu_3 + \nu_4$	6149	6121	6122
$2\nu_4$	6149	6119	6110	$\nu_3 + \nu_5$	6163	6111	6118
$2\nu_5$	6177	6136	6134	$\nu_4 + \nu_6$	6163	6147	6135
$2\nu_6$	6177	6147	6147	$\nu_5 + \nu_6$	6177	6150	6137

Notes:(1) The first estimates used a basis containing basis wave functions that only correspond to the valency vibrations of the CH bonds.

(2) The second estimates used a basis containing basis were functions that only correspond to all the normal vibrations of C_2H_6.

HCH or C—C bond vibrations. The estimates also showed that in the anharmonic approximation the valency and deformational vibrations of terminal groups with light atoms at their ends are poorly mixed together. Moreover, if we limit the accuracy to a few cm^{-1}, then it is quite possible to reproduce the main features of the distribution of the anharmonic energy levels by using a model in which only anharmonic corrections are considered for the vibrations of terminal bonds with light atoms. This enables us to reduce substantially the initial basis of a variational treatment, since the number of basis functions N_{var} (see equation (10.9)) needed to ascertain the vibration energy levels as their variational limits grows very quickly as the number of normal vibrations whose harmonic eigenfunctions are included in the basis increases. The weak dependency we have noted of the anharmonic energy levels on the selected approximation (only taking account of the excited wave functions for normal co-ordinates most associated with the terminal atoms and given the inclusion of an adequately full set of functions) for the vibrations of the terminal atoms is demonstrated in Table 10.7.

Table 10.8. Corrections for anharmonicity for the CH valency vibrations of C_2H_6 and C_3H_8[a]

Ethane		Propane			
Transition	$\Delta\nu$	Transition	$\Delta\nu$	Transition	$\Delta\nu$
ν_1	−39	ν_1	−40	$\nu_5+\nu_6$	−78
ν_2	−35	ν_2	−37	$2\nu_6$	−76
ν_3	−42	ν_3	−36	$\nu_1+\nu_7$	−78
ν_4	−42	ν_4	−43	$\nu_2+\nu_7$	−81
ν_5	−42	ν_5	−42	$\nu_3+\nu_7$	−59
ν_6	−42	ν_6	−34	$\nu_4+\nu_7$	−94
$2\nu_1$	−84	ν_7	−35	$\nu_5+\nu_7$	−67
$2\nu_2$	−70	ν_8	−38	$\nu_6+\nu_7$	−83
$2\nu_3$	−99	$2\nu_1$	−83	$2\nu_7$	−76
$2\nu_4$	−77	$\nu_1+\nu_2$	−72	$\nu_1+\nu_8$	−86
$2\nu_5$	−90	$2\nu_2$	−70	$\nu_2+\nu_8$	−64
$2\nu_6$	−77	$\nu_1+\nu_3$	−59	$\nu_3+\nu_8$	−57
$\nu_1+\nu_2$	−97	$\nu_2+\nu_3$	−68	$\nu_4+\nu_8$	−76
$\nu_1+\nu_3$	−134	$2\nu_3$	−64	$\nu_5+\nu_8$	−75
$\nu_1+\nu_4$	−131	$\nu_1+\nu_4$	−68	$\nu_6+\nu_8$	−75
$\nu_1+\nu_5$	−146	$\nu_2+\nu_4$	−80	$\nu_7+\nu_8$	−76
$\nu_1+\nu_6$	−145	$\nu_3+\nu_4$	−76	$2\nu_8$	−78
$\nu_2+\nu_3$	−50	$2\nu_4$	−65		
$\nu_2+\nu_5$	−63	$\nu_2+\nu_5$	−79		
$\nu_2+\nu_6$	−62	$\nu_3+\nu_5$	−74		
$\nu_3+\nu_4$	−75	$\nu_4+\nu_5$	−81		
$\nu_3+\nu_5$	−99	$2\nu_5$	−52		

[a]Calculated with a basis containing basis wave functions corresponding only to the valency vibrations of the CH bonds. The basis size for ethane was 930 and that for propane was 1303 functions.

It is interesting to compare the magnitudes of the anharmonic corrections to the harmonic energy levels found by a reasonably strict solution of the anharmonic problems for ethane and propane, assuming that the anharmonic coefficients for corresponding bonds and other elements in both molecules are the same. It is evident from Table 10.8 that the anharmonic corrections for the CH valence vibrations, for which they are the greatest, are found to be quite close under these conditions, being 35–45 cm^{-1} for the fundamental frequencies and 70–100 cm^{-1} for the primary overtones. Thus the anharmonic calculations account for the well-known fact that within the harmonic approximation a correction for anharmonicity can be included using spectroscopic masses, a set of force constants or other similar approaches.

The results discussed clearly demonstrate that, under certain conditions, characteristic behaviour must be observable not only for the fundamental frequencies but also for the overtones and compound frequencies. This reveals a method for estimating the anharmonic energy levels in complicated molecules based on the idea of fragmented estimates similar to the method already successfully constructed for the analysis of harmonic vibrations. It is possible, therefore, to say with some conviction that calculations of the anharmonic energy levels for very large molecules seems not only realistic but to be possible on average-sized computers.

10.4 Use of Variational Method for Calculation of Highly Excited Vibration Levels of Polyatomic Molecule Energy

In solving the vibration problems of large amplitudes we encounter a feature that is usually neglected but nevertheless is of very great importance. The question is, that for polyatomic molecules the presence of the so-called 'topological isomers' that have an identical empirical formula but quite a different geometry is typical. The number of such isomers may be very high.

Therefore, even such a simple system such as system C_6H_6 has 217 isomers, the usual benzene being among them. Its own minimum and potential correspond well to each isomer of this kind on the common polymetric potential surface. Therefore, when we say that, for example, the vibration problem will be solved for a molecule of benzene and not of prismane (which also has the formula C_6H_6), we really mean that the solution will be searched for only in the vicinity of a quite definite minimum. Hence it follows that when vibration problem themselves are put, it is necessary to limit the probable motions of atoms so that they would occur in the vicinity of the preset minimum.

As a rule, the topological isomers feature an essentially differing geometry, therefore to transform one isomer to another it is first necessary to rearrange its skeleton. If we change, for example, only the end bond lengths, we shall not manage to transform one isomer to another. This means that it is important to limit the motion only along those degrees of freedom which can cause an essential change in the form of the geometrical figure of the given isomer, while the tensions of the end bonds may be not limited and their rupture may be allowed (outflow of the potential to the dissociation limit).

Mathematically, putting limitations on the probable motions of atoms means a selection of a quite definite form of the potential surface.

For those degrees of freedom, for which a danger of 'transfer' to another minimum arises at abrupt changes of the respective co-ordinates, such sections of the potential surface must be selected in the solution of vibration problems that would lead to a well with infinitely high walls. First it concerns the strains of the molecule skeletons. For the end-bond valence co-ordinates and, in certain cases, for the internal bonds (in linear molecules and for linear section of non-linear molecules) it is good practice to use potentials with a dissociation limit (for example, of the Morse type).

This is especially rational as, as specific calculations have shown, the anharmonicity of skeleton vibrations of polyatomic molecules is not high, and the quasi-continuum is attained mainly due to the vibrations of the end bonds with light atoms at the ends.

On the basis of the above and available experimental calculations we assume, in the study of vibrations of large amplitudes for polyatomic molecules, that a sufficiently common potential function may be taken in the form

$$V(q) = \tfrac{1}{2}\sum_{i,j} D_{ij}d_id_j + \tfrac{1}{6}\sum_{i,j,k} D_{ijk}d_id_jd_k + \tfrac{1}{24}\sum_{i,j,k,l} D_{ijkl}d_id_jd_kd_l + V(r,d) \qquad (10.10)$$

where $d_i = (1 - e^{-a_i \cdot \Delta r_i})$ (see Section 9.1). Here the first three terms concern the tension co-ordinates of the end (and some internal) bonds, the fourth term presented by a usual fourth-degree polynomial with a full or truncated set of coefficients characterizes the potential part that is a bond with angular co-ordinates and the group of valence co-ordinates for which no rupture is allowed, as well as with interaction between these and other co-ordinates.

In work with such functions a problem arises of the calculation of matrix elements of type $X_{n,m} = \langle F_n(Q)/e^{-2bq/j_nQ}\rangle$ in the basis of harmonic functions $F_n(Q)$. In practical calculations it is convenient to present these in the form

$$X_{n,m} = \sum_{i=0}^{n+m} Z_{n,m,i} b_i$$

where $Z_{n,m,i}$ = constant coefficients

$$Z_{n,m,i} = \frac{e^{b^2/4}}{\sqrt{(\pi 2^{n+m} n!m!)}} \sum_{p=0}^{n/2} \frac{(-1)^p n!}{2(n-2p)2^p p!} \times \sum_{q=0}^{m/2} \frac{(-1)^q m!}{2(m-2q)2^q q!} \cdot \frac{(n+m-2p-2q)!}{(n+m-2p-2q-i)!r!} W_i \qquad (10.11)$$

and

$$W_i = \int_{-\infty}^{+\infty} a^i e^{-\alpha^2} d\alpha = \begin{cases} (i-1)!! \sqrt{\dfrac{\pi}{e^i}} & \text{if} \quad i \text{ is even} \\ 0 & \text{if} \quad i \text{ is odd} \end{cases} \tag{10.12}$$

As specific calculations have shown, the computer time required to fill in the matrix of the variational problem at calculation of the matrix elements with a function incorporating exponential terms does exceed that spent in construction of a similar matrix, but with a potential in the form of a fourth-degree polynomial.

Let us dwell on one question. In solving the problem of motions of atoms, with account taken of the probable high tension of separate bonds, the expansion of kinematic coefficients in terms of natural co-ordinates accurate up to the quadratic terms accepted above becomes insufficient. First, it concerns the reciprocal bond lengths and the reciprocal squares of them entering the expression for these coefficients. In order not to pass over to expansions of higher orders (especially, as for the other magnitudes, it is not necessary) we took advantage of the fact that the reciprocal bond lengths and the reciprocal squares of them can be approximated within a very wide range with a good precision by linear combinations of simple exponents in the form of $e^{-\Delta r/r_0}/r_0$, where r_0 = equilibrium lengths and Δr = link increments.

The kinematic coefficients are then presented in the form of a sum of products of the exponential components by the polynomial ones bound with the strain co-ordinates. Calculation of the matrix elements with such expressions for kinematic coefficients creates no problems.

We used such a technique in calculating the anharmonic vibration spectrum for the HCN and H_2C_4 molecules. The coefficients of the potential function in the form of equation (10.10) were so selected that in its lowest part it would be close to the eighth-degree polynomial for these molecules obtained in references 33 and 34 on the basis of quantum-mechanical calculations and giving in these operations good agreement with the calculated and experimental energy levels.

For results of our calculations of energy vibration levels appear in Table 10.9, which gives, for comparison, some results[33,34] of calculations of the limitation of the potential function for these molecules by the fourth-degree polynomial obtained at differentiation of the potential function in the form of equation (10.10). The energies obtained of the CH and CN bond dissociation for the HCN molecule, equal to 43108 and 92025 cm^{-1}, agree with the experimental values equal to 42.800 and 77.600 cm^{-1}.

The results of these calculations indicate the higher value of influence of the terms, beginning from the fifth up to the eighth order at a polynomial presentation of the potential function on the values of the calculated energy levels. In the solution of anharmonic problems at a considerable tension of the bonds the usual variational basis constructed on the functions of harmonic approximation

Table 10.9. Calculated and experimental frequencies of vibrations (cm^{-1}) of HCN and H_2C_4 molecules

Level interpretation	Experiment	Data[33,34]	Calculation with potential function, type (10.10)	Calculation with fourth-degree polynominal
		HCN		
ν_2	2096.9	2096.9	2100.5	2103.2
ν_1	3311.5	3311.5	3301.0	3334.2
$2\nu_3$	4173.1	4170.8	4177.9	4186.2
$\nu_1+\nu_3$	5993.7	5392.8	5385.9	5429.4
$2\nu_1$	6519.6	6508.4	6485.4	6627.9
$2\nu_1+\nu_3$	8585.6	8572.0	8552.6	8719.7
$3\nu_1$	9627.0	9597.4	9555.8	9925.0
$3\nu_1+\nu_3$	11674.4	11643.7	11603.5	12016.1
$4\nu_1$	12635.9	12594.6	12521.5	13257.8
$2\nu_1+3\nu_3$	12657.9	12633.9	12623.0	12851.8
$3\nu_1+2\nu_3$	13107.2	13652.3	13636.8	14090.9
$2\nu_1+4\nu_3$	14654.0	14624.0	14632.8	14901.7
$4\nu_1+\nu_3$	14670.0	14637.0	14568.6	15349.0
$5\nu_1$	15552.0	15532.8	15434.8	16646.2
		H_2C_4		
ν_3	874.0	874.0	889.2	
$2\nu_3$	—	1742.0	1770.0	
ν_5	2022.3	2022.0	2028.3	
ν_2	2184.0	2182.0	2182.2	
$3\nu_3$	—	2610.0	2658.6	
$\nu_3+\nu_5$	—	2901.0	2917.3	
$\nu_2+\nu_3$	—	3057.0	3070.8	
ν_1	3329.0	3328.0	3320.1	
ν_4	3329.0	3329.0	3329.0	

does not appear to be successful. This is associated with the fact that the 'harmonic' potential well is considerably narrower than the 'anharmonic' one, and the construction of the 'right' sides of the anharmonic functions in the variational procedure occurs only due to the 'tails' of the harmonic ones, which requires a large number of basis functions.

This can be eliminated by constructing the basis from the harmonic functions which correspond to considerably 'wider' parabolic potential wells than the 'true' ones, and are effective well into the larger part of the region of atom motions at vibrations of large amplitudes.

The results of the use of such a technique are illustrated in Figure 10.3 for the first, second and third harmonics of the CH bond vibrations in molecule HCN. It is seen that in the use of 'wider' basis harmonic oscillation functions the solutions approach their variational limit more quickly.

In conclusion, we can say that the variational technique of solution of anharmonic vibration problems with minor additions and preservation of

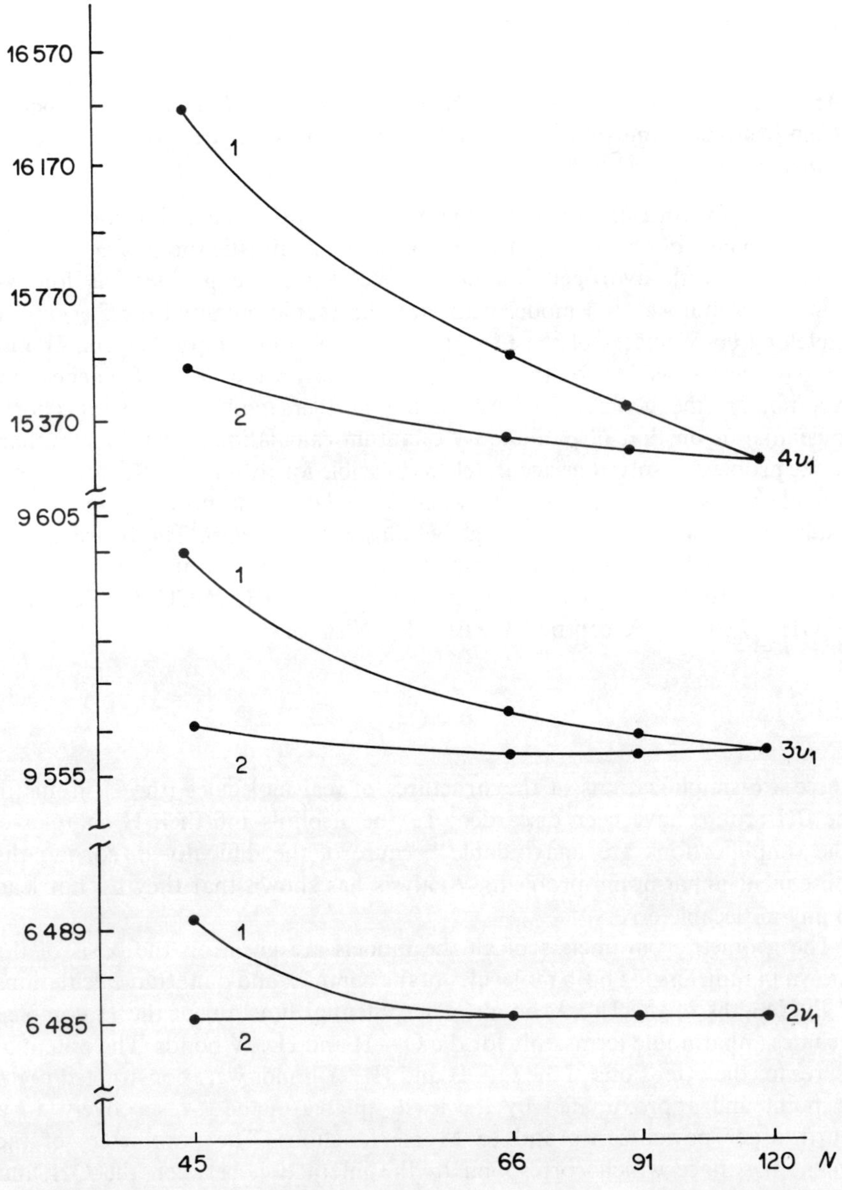

Figure 10.3. Values of calculated energies of the first, second and third harmonics of CH link of molecule HCN versus the number and type of used basis harmonic functions (use was made of 'true' (1) and 'widened' (2) harmonic vibration functions)

computer software almost unchanged can also be used for the calculation of highly excited energy levels up to the dissociation limit, which makes it universal and convenient for practical use.

10.5 An Anharmonic Analysis of the Vibrations of A—H and A...H bonds in Ion-Molecular and Intermolecular H Complexes of Hydroxyl-containing Compounds

As an example of estimates for large molecules with a partial account of the anharmonicity of the vibrations we consider an investigation of a series of compounds with hydrogen bonding[35]. We define the problem as follows. Consider a multicentred model with an anharmonic potential for the natural stretching co-ordinates of the O—H and H...*X* bonds only. We can choose the parameters needed for the calculation either on the basis of independent experiments (the geometry of interacting partners and the skeleton's force constants) or on that of preliminary quantum calculations. The direct anharmonic problem is solved as accurately as possible for the model. No adjustment is made to the experimental data when formulating the problem and so the results of the whole estimation can be considered *a priori*. The Schroedinger equation solved has the general form of equation (10.1), and it is solved for the following structures: COH, COH...*X*, CCOH, CCOH...NCC, CCCOH and CCOH...*X*, where *X* can be Cl^-, Br^-, I^-, NCC or

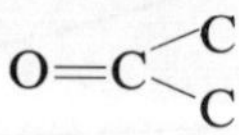

These are simplifications of the structures of real molecules (the H atoms in the CH groups have been discarded), i.e. the alcohols and their H complexes. The simplifications are unavoidable because of the difficulty in solving the subsequent anharmonic problems. Analysis has shown that they do not lead to any noticeable errors.

The geometric parameters of all the models are given on the basis of the known parameters for both molecules of the complex and quantum calculations of the lengths of the H...*X* bonds. The potential function of the H complex includes anharmonic terms only for the O—H and H...*X* bonds. The potential curves for the extensions of the O—H and H...*X* bonds were constructed point by point and approximated by the least squares method, respectively, to a fourth-order power series and to Morse functions. The parameters of the potential surface which correspond to the interaction between the OH and *X*...H bond-extension co-ordinates are assumed to be zero. The remaining parameters of the potential surface are taken to be equal to the harmonic terms, and we have refined these using solutions of the inverse spectral problems taking the literature's data into account.

The problem was solved using the usual system of valence-force natural co-ordinates, and the basis we used included between 85 and 110 harmonic functions. Investigation showed that several of the lower vibration levels (i.e. those corresponding to the valency vibrations of the O—H and H...X bonds that we are interested in) attained stable variational limits.

The estimates of the OH bond valency vibrational frequencies are given in Table 10.10 for all the structures together with the experimental values of the fundamental transition frequencies and primary overtones. It is evident from the data in the table that the technique described is capable of achieving very satisfactory agreement between the calculated and experimental vibrational frequencies not only for the fundamental transitions but also for the overtones. The distinction between the ion-molecular and intermolecular H complexes is clearly seen. It is interesting to note that despite the close proximity of the potential curves for OH bond extensions in the Br^- and methanol complex to that in the I^- and methanol one, the relationship between the vibrational frequencies found experimentally was reproduced correctly in the calculated values. In accordance with the experimental results, the estimated frequencies

Table 10.10. Experimental and calculated frequencies of the valence vibrations of the O—H bonds in methanol, ethanol, and isopropyl alcohol and in their H bond complexes with the halide anions Cl^-, Br^- and I^- and with acetonitrile and acetone (in cm^{-1})

	Experiment	Theory	
Complex considered	ν_{OH}, $2\nu_{OH}$	ν_{OH}^{anh}, $2\nu_{OH}^{anh}$	ν_{OH}^{harm}, $2\nu_{OH}^{harm}$
CO—H	3640, 7121	3649, 7190	3754, 7509
CO—H...Cl^-	3272 —	3268, 6439	3385, 6770
CO—H...Br^-	3330 —	3339, 6602	3422, 6844
CO—H...I^-	3385 —	3382, 6684	3468, 6936
CO—H...N≡C—C	3553, 7000	3547, 7007	3669, 7338
C CO—H...O=C C	3534 —	3539, 7006	3626, 7252
CCO—H	3636, 7096	3628, 7134	3750, 7500
CCO—H...N≡C—C	3545, 6992	3554, 7034	3653, 7306
C CO—H C	3626, 7085	3614, 7078	3764, 7528
C CO—H...Cl^- C	3282 —	3291, 6503	3399, 6798
C CO—H...Br^- C	3338 —	3341, 6556	3508, 7016
C CO—H...I^- C	3390 —	3402, 6723	3499, 6998
C CO—H...N≡C—C C	3555, 6982	3533, 6920	3709, 7418

of the ν_{OH} vibrations in $CH_3OH...Br^-$ are lower than those in $CH_3OH...I^-$. The same relationship between the wells of the potential curves of the OH bond extension in this context was found to be the reverse. As the estimates we have provided shown, the anharmonic corrections are significant in all cases and range from 80 to 170 cm^1.

The frequencies calculated for the low-frequency vibrations of H...*X* bonds are given in Table 10.11. An important result is that, despite the significant deviations of the H...*X* bonds' potential functions from parabolic, even in the region of the potential well, the anharmonic corrections are nevertheless very small. We can therefore claim that when investigating low-frequency H bond vibrations the use of an anharmonic approximation is an unnecessary complication of the task. The physical aspects of this are obvious. Since the strength of all intramolecular bonds is much greater than that of intermolecular bonds, and since the frequencies of intromolecular vibrations are considerably higher than those of the intermolecular frequencies, a good approximation must distinguish between movements within molecules and those between them. In reality, $\nu_{H...X}$ vibrations are the vibratory movements of whole molecules or molecules and ions with respect to each other. The vibrational forms considered in the harmonic approximation (changes in natural co-ordinates) for the low-frequency vibrations of the methanol acetonitrile complex were $q_{C-O} = 0.002$, $q_{OH} = -0.007$, $q_{N...H} = -0.275$, $q_{N\equiv C} = -0.002$, $q_{C-C} = -0.003$ and

Table 10.11. Calculated frequencies of valence vibrations $\nu_{H(D)...X}$

Complex considered	$\nu^{harm}_{H(D)...X}$	$\nu^{anharm}_{H(D)...X}$
CO—H...Cl$^-$	183,92	183,89
CO—D...Cl$^-$	179,56	177,05
CO—H...Br$^-$	127,85	128,61
CO—H...I$^-$	90,25	91,49
CO—H...N≡C—C	160,54	160,44
CO—H...O=C(C)C	117,69	116,05
C—CO—H...N≡C—C	140,07	138,83
(C)(C)CO—H...Cl$^-$	175,81	177,72
(C)(C)CO—H...Br$^-$	141,75	145,05
(C)(C)CO – H...I$^-$	111,36	113,93
(C)(C)CO—H...N≡C—C	124,99	124,76

Table 10.12. The coefficients of the expansion of the vibration function Ψ into a series of harmonic vibrational functions φ (a) for $\nu_{O-H(D)}$ levels and (b) for $\nu_{H(D)\ldots X}$ levels

Complex considered	$\varphi(\nu_{OH})$	$\varphi(\nu_{OH}+\nu_{H\ldots x})$	$\varphi(2\nu_{OH})$	$\varphi(\nu_{OH}+2\nu_{H\ldots x})$	$\varphi(2\nu_{OH}+\nu_{H\ldots x})$	φ_0
			a			
CO—H...Cl^-	0.8865	−0.3663	0.2036	0.1055	−0.0851	−0.0796
CO—D...Cl^-	0.9286	−0.2835	0.1813	0.0598	−0.0559	−0.0699
CO—H...Br^-	0.9183	−0.3138	0.1849	0.0777	−0.0635	−0.0727
CO—H...I^-	0.9369	−0.2531	0.1959	0.0534	−0.0533	−0.0772
CO—H...N≡C—C	0.8987	−0.3427	0.2108	0.0816	−0.0787	−0.0814
CO—H...O=C(C)₂	0.9168	−0.3049	−0.1926	0.0590	0.0633	0.0742
C—CO—H...N≡C—C	0.8853	0.3796	0.1884	0.0951	0.0795	−0.0722
(C)₂CO—H...Cl^-	0.9135	0.3069	−0.2067	0.0739	−0.0696	0.0811
(C)₂CO—H...Br^-	0.9012	0.2923	−0.2520	0.0678	−0.0813	0.0911
(C)₂CO—H...I^-	0.9375	0.2400	−0.2065	0.0482	−0.0527	0.0811
(C)₂CO—H...NC—C	0.8979	0.2939	0.2542	0.0518	0.0816	−0.0982

Table 10.12. (*Contd.*)

Complex considered	$\varphi(\nu_{H...X})$	$\varphi(2\nu_{H...X})$	$\varphi(3\nu_{H...X})$	$\varphi(4\nu_{H...X})$	$\varphi(\nu_{H...X}+\nu_{OH})$	φ_0
			b			
CO—H...Cl^{-1}	0.9073	−0.3471	0.0895	−0.0536	0.0717	0.1924
CO—D...Cl^-	0.9337	−0.3035	0.0638	−0.0464	0.0612	0.1543
CO—H...Br^-	0.9243	−0.3183	0.0793	−0.0518	0.0669	0.1693
CO—H...I^-	0.9490	−0.2615	0.0593	−0.0414	0.0737	0.1358
CO—H...N≡C—C	0.8659	−0.4115	0.1194	−0.0735	0.0670	0.2283
CO—H...O=C(C)C	0.8778	−0.3994	0.1016	−0.0690	−0.0632	0.2151
C—CO—H...N≡C—C	0.8659	0.4163	0.1094	0.0633	0.0618	−0.2292
(C)(C)CO—H...Cl^-	0.9256	0.3101	0.0744	0.0459	−0.0731	−0.1771
(C)(C)CO—H...Br^-	0.9218	0.3107	0.0762	0.0446	−0.0900	−0.1847
(C)(C)CO—H...I^-	0.9490	0.2564	0.0579	0.0372	−0.0761	−0.1418
(C)(C)CO—H...NC—C	0.8680	0.4083	0.1029	0.0665	0.0822	0.2276

$q_{COH} = 0.001$. Since with this kind of motion heavy particles with masses approximating the sums of those of the molecules are displaced, they must also be at the very bottom of the potential well. Thus dynamic anharmonicity also does not appear, but it must become noticeable in the overtone regions.

The investigation of isotopic shifts in the valence vibration frequencies of O—H and H...X bonds when the hydrogen atoms are substituted by deuterium has particular interest. We have made some calculations for the deuterium-substituted structure CO—D...Cl^-, and it was found that the ν_{OH} frequency is moved from its original value of 3268 cm^{-1} to the region of 2395 cm^{-1}, whilst the $\nu_{H...Cl}$ vibration's frequency is shifted from 184 cm^{-1} to 177 cm^{-1}. Hence deuterization has a small effect on low-frequency vibrations but is very significant for the higher-frequency vibrations. If we use the primary harmonic approximation, the frequency shift for low-frequency vibrations must be 2 cm^{-1}, which corroborates our earlier conclusion that anharmonic corrections have little effect on the frequencies of proper H bond vibrations.

The appropriate wave function, compiled as a linear combination of harmonic basis eigenfunctions for the fundamental level which correspond to the low-frequency vibrations of the CO—H...Cl^{-1} complex, can be represented as

$$\Psi \simeq 0.91\varphi(\nu_{H...X}) - 0.35\varphi(2\nu_{H...X}) + 0.09(3\nu_{H...X}) + 0.19\varphi_0 \quad (10.13a)$$

Here, $\varphi(\nu_{H...X})$ to $\varphi(3\nu_{H...X})$ are harmonic eigenfunctions in which the dominant roles are played by the vibrations $\nu_{H...X}$ to $3\nu_{H...X}$, and φ_0 is the eigenfunction of the ground state. Clearly, it follows that even though the eigenfunction of a low-frequency vibration is essentially anharmonic, it nevertheless does not include contributions from other vibrational degrees of freedom associated with the H bond complex.

The situation for O—H bond vibrations is somewhat different. The corresponding eigenfunction here has the form

$$\begin{aligned}\Psi \simeq\ & 0.89\varphi(\nu_{OH}) - 0.37\varphi(\nu_{OH} + \nu_{H...X}) \\ & + 0.20\varphi(2\nu_{OH}) + 0.11\varphi(\nu_{OH} + 2\nu_{H...X}) - 0.09\varphi(2\nu_{OH} + \nu_{H...X}) - 0.08\varphi_0\end{aligned} \quad (10.13b)$$

In this way the eigenfunction of O—H bond vibrations must not only be anharmonic but must also depend substantially on the $\nu_{H...X}$ vibrations. This is a natural result, because an O—H vibration is not characteristic of the form of this co-ordinate since an O—H bond extends the H...X bond contracts to almost the same extent. This must be borne in mind when estimating the intensities of the corresponding absorption bonds. Similar situations were observed for all the complexes studied (see Table 10.12).

10.6 Investigation of the Excited Vibrational State of Small Polyatomic Molecules

The appearance of lasers with the ability to create high densities of quanta with a given frequency has given rise to hopes that laser chemical methods could be developed that would be based on the reactions between photo-excited molecules or between free radicals of a specific type created by irradiation. A particularly attractive idea is to create molecular fragments of known composition by the controlled fission of polyatomic molecules excited to high vibrational states lying close to their dissociation thresholds. What is crucial is the ability to initiate dissociation along a particular vibrational degree of freedom in a given molecule as described by appropriate natural vibration co-ordinates, i.e. bond-extension and valency angle-change co-ordinates. In order for the dissociation to be effective the energy of the absorbed quanta must be mainly localized within a given structural element, and not 'smeared out' over the whole molecule, as this might either make the dissociation unlikely or cause dissociation to occur in an undesirable location.

This raises the general question as to how energy is distributed amongst the various structural fragments when a molecule is vibrationally excited and how the geometrical characteristics of these elements are changed. Unfortunately, the state of large-amplitude vibrational theory for molecules does not at present allow us to obtain the information required for high vibrational states. However, by studying the behaviour of molecules as their lower vibrational states are successively excited, starting from the ground and fundamental levels, we can discover important indications about the processes that must also take place under high excitations. The work described in this section was undertaken to do this. Several geometrical parameters in a series of small polyatomic molecules were studied in excited anharmonic vibrational states.

Before continuing, we must make some preliminary comments. A molecule's geometrical configuration in the ground vibrational state can be described by giving definite values to bond lengths and angles if the potential surface has only one sufficiently deep minimum. This is so because the 'spread' of nuclear co-ordinates, arising from zero-point vibrations, is not large. In excited states this 'spread' (which can be given as the root-mean-square amplitudes of the vibrations of the nuclei or changes in the natural vibrational co-ordinates) grows rapidly, and it is then impossible to speak of the existence of a definite geometrical configuration. Nonetheless, we can use an analogue of the concept and describe it by the average values of the vibrational co-ordinates. If the vibrations are harmonic then the average of the increments to the geometric, taken to be the natural vibrational co-ordinates, is zero for all possible vibrational states. This reflects the fact that the 'spread' of the vibrational co-ordinates 'to the right' of their equilibrium values is equal to that 'to the left'. The situation is different for anharmonic vibrations because the 'spreads' to the 'left' and 'right' are different; the 'centre of gravity' of the vibrational distribution thus shifts and the vibrations take place, as it were, about other equilibrium positions. In this sense, it is possible to speak of changes during vibrational excitations by comparison with an equilibrium geometry of the molecule, and to describe the change by the increments to the average values of the natural vibrational co-ordinates that are taken for different structural elements of the system. The average bond length for diatomic molecules will grow regularly and monotonically during successive vibrational excitations, and particularly quickly in the region close to the dissociation threshold. Thus this quantity can be considered to be an indicator of the development of bond rupture during photo-excitation. This simple behaviour of a bond makes it comparatively easy to obtain a prognosis of appropriate photochemical reactions and to choose real conditions in which to carry them out.

The situation, of course, may be much more complicated in polyatomic molecules. If the anharmonic vibrational functions can be given as a linear combination of harmonic functions (this is clearly true for a very general case) then we have the following for the mean and mean-square values, respectively, of the natural vibrational co-ordinates for an arbitrary vibrational energy level:

$$\bar{q}_{im} = \sum_{k,k',n} l_{in} C_k^{(m)} C_{k'}^{(m)} Q_{on}(v+1)^{1/2}$$
$$(k = v_1, v_2, \ldots, v_n, \ldots, v_N;\ k' = v_1, v_2, \ldots, v_n + 1, \ldots, v_N) \tag{10.14}$$

$$\bar{q}^2_{im} = \Bigg\{ \sum_{n,k} l^2_{in} C_k^{(m)^2} Q^2_{on}(2v_n + 1) + \sum_{n,k,k'} l^2_{in} C_k^{(m)} C_{k'}^{(m)} Q^2_{on}(v_n + 1)^{1/2}$$
$$\times (v_n + 2)^{1/2} + 2 \sum_{n,n',k,k''} l_{in} l_{in'} C_k^{(m)} C_{k''}^{(m)} Q_{on} Q_{on'}(v_n + 1)^{1/2} \tag{10.15}$$
$$\times (v_{n'} + 1)^{1/2} \Bigg\} \begin{matrix} (k = v_1, v_2, \ldots, v_n, \ldots, v_N;\ k' = v_1, v_2, \ldots, v + 2, \ldots, v_N) \\ (k'' = v_1, v_2, \ldots, v_n + 1, \ldots, v_{n'} + 1, \ldots, v_N) \end{matrix}$$

Here, l_{in} is an element of the matrix transforming natural to normal co-ordinates, $C_k^{(m)}$ is a coefficient in a linear combination of harmonic vibration functions which can be found by solving the variational vibrational problem, v_n is a vibrational quantum number, Q_{on} is the null amplitude of a normal co-ordinate and N is the number of the degrees of freedom of the molecule.

Both $\bar{q}_i^{(m)}$ and

$$(\overline{q_i^{(n)^2}})^{1/2}$$

are invariant with respect to the choice of relative signs of the L matrix's column signs. This follows directly from the following argument. Reversing the sign of a column of the L matrix means changing the positive and negative directions of the normal co-ordinate axes. This must clearly not be reflected in the values of the anharmonic energy levels $E_n = \int \Psi_n \hat{H} \Psi_n \mathrm{d}Q = \tilde{C}^{(n)}[\langle k|\hat{H}|k'\rangle]C_n^{(n)}$. In turn, this means that reversing the sign of an l_{ik} coefficient must reverse simultaneously the sign of the corresponding product of coefficients $C_p^{(n)}C_r^{(n)}$ $(p \neq r)$. It is only in this case that there will be no effect on the quantity E_n by any arbitrariness in the choice of the signs of the L matrix's column, which will be reflected in those of the non-diagonal elements of the H matrix in those parts that depend on the cubic terms of the potential function. Thus we have in particular for these terms components of the form $u_{ijs} l_{ik} l^2_{jk'} Q_k Q^2_m$ (and others including the first powers of the l_{ik} coefficients), which, when calculating E_n, give $u_{ijs} l_{ik} C_p^{(n)} C_r^{(n)} \langle k|Q_k|k'\rangle \langle k|Q^2_m|k'\rangle$. Specific calculations confirm this argument. Analogous arguments are also justified for the root-mean-square amplitudes of the vibrations, i.e. $(\overline{q_i^{(n)^2}})^{1/2}$.

It is clear from the formulae that $\bar{q}_i^{(m)}$ and $\sqrt{(q_i^{(m)^2})}$ depend on many factors, and their behaviour with increasing values of the quantum numbers must certainly be very complicated. Analogies with diatomic molecules can only be made when some normal vibration basically reduces to a change in a structural element (a bond, for example) that is characteristic of a molecule and to a change, during an excitation of quantum numbers up to various states, during an excitation of quantum numbers up to various states, that is not mixed by other vibrations.

The conditions for the existence of characteristic vibrations of different kinds have been explained in the general theory of vibrations of polyatomic molecules. We can say that a number of limitations make such characteristic vibrations appear quite often, particularly in the spectra of molecules. Mixtures of vibrations are typical for significant excitations. Consequently, simple patterns in the changes of the states of bond length and the other geometrical characteristics of polyatomic molecules during their vibratory excitations are more often the exception rather than the rule. This, then, is the general argument, and it will be interesting to verify it with concrete examples.

To do this, solutions were sought for the direct and inverse problems using a potential function in the form of a power series, and values calculated for $\bar{q}_i^{(m)}$ and $(\overline{q_i^{(m)2}})^{1/2}$ for CO_2, HCN, H_2O, H_2Se, NO_2, O_3, H_2CO, C_2H_2 and a number of their isotopic modifications. Some of these estimates are given in the tables and Figures 10.4–10.7.

An analysis of these data shows that the first aspect that is striking is the lack of monotonicity in the changes in the geometrical parameters for successive vibrational excitations. The picture is markedly more complicated when we go from triatomic to tetra-atomic molecules. At the same time, when we can separate the successive levels corresponding to excitations of mainly one normal vibration

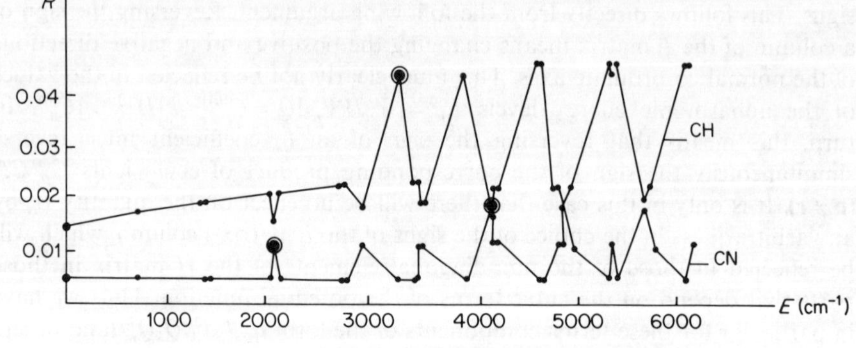

Figure 10.4. The dependence of $\bar{R}_{CH}$ and $\bar{R}_{CN}$ on the energy of HCN (in Å). The calculation was from the ground state

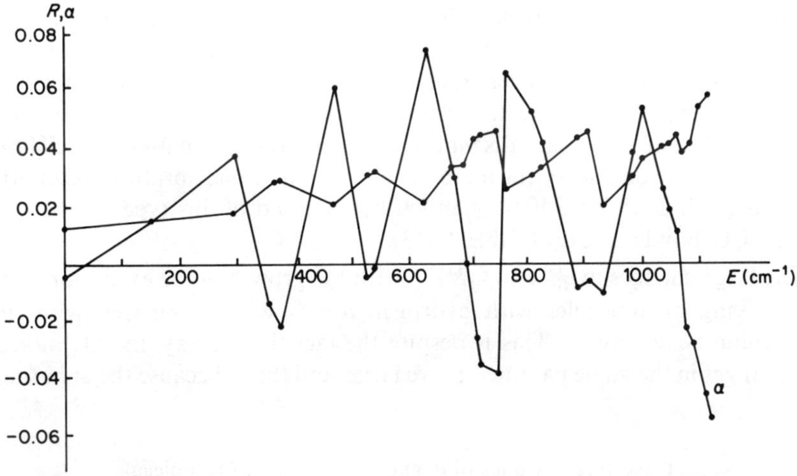

Figure 10.5. The dependence of $\bar{R}$ and $\bar{\alpha}$ on energy for H_2O, taken from the ground state

that is characteristic of a structural element, the behaviours of $\bar{q}_i$ and $\sqrt{\bar{q}_i^2}$ are reminiscent of the situation we have in diatomic molecules.

As an example, for the H_2O molecule the mean values of the bond-change co-ordinates for the $v_q(H_2O)$ valency vibration and those of the valency angle changes in H_2O for the states 000, 001, 002 and 003 for the bond vibrations and for 000, 010, 020, 030 and 040 for the angle vibrations are all separated out. It is clear that $\bar{q}_i^{(m)}$ grows regularly in these cases. Due to changes in the scale of the energy levels corresponding to the valency and deformational vibrations, characteristic 'antiphase' serrated curves are attained. An analogous

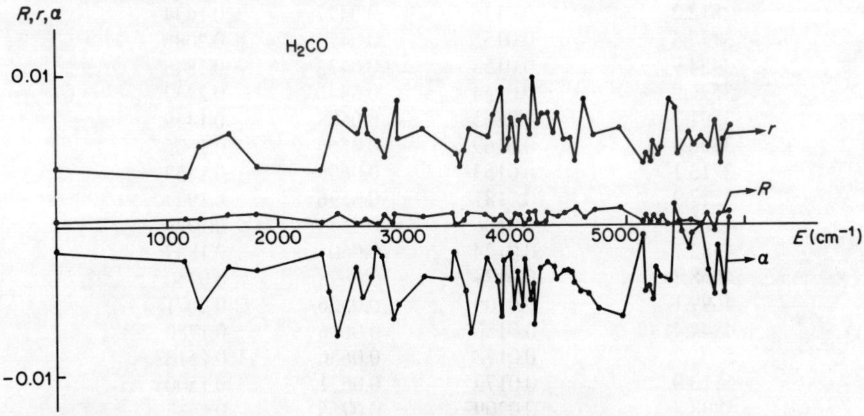

Figure 10.6. The dependence of $\bar{R}_{CO}$, $\bar{r}_{CH}$ and $\bar{\alpha}_{HCO}$ on energy for H_2CO, taken from the ground state

picture is observed for HCN, where the CH and CN vibrations are quite characteristic. Sequences of levels such as these cannot be distinguished in polyatomic molecules, and even in H_2O during transitions to higher vibrational states the simple picture we had above is distorted. All the same, we should note that, despite significant discontinuities, the average length of the CH bond in formaldehyde and acetylene increases somewhat as the vibrational excitations grow as a whole (Figure 10.6). A much better trend of this type can be traced for the CO bonds in CO_2 (Table 10.13).

The large changes in $\bar{q}_i^{(n)}$ and $(\overline{q_i^{(n)}})^{1/2}$ with the growth in the excitation energy are striking in molecules with hydrogen atoms when compared with those containing heavy atoms. This is despite the fact that in, say, the O_3 molecule the changes in the angle parameters are large, and this is because the anharmoni-

Table 10.13. Values of $\bar{q}_i$ and $\sqrt{\bar{q}_i^2}$ for the CO_2 molecule

$\nu_{\text{call.}}(\text{cm}^{-1})$	$\bar{R}_{\text{CO}}(\text{Å})$	$\bar{R}^2_{\text{CO}}(\text{Å})$	$\sqrt{\bar{\alpha}^2}(\text{rad})$
0.0	0.0067	0.0355	0.0922
666.4	0.0089	0.0361	0.0926
1283.2	0.0106	0.0419	0.1341
1342.9	0.0112	0.0369	0.1584
1392.4	0.0100	0.0415	0.1255
1925.6	0.0130	0.0428	0.1900
2002.1	0.0134	0.0382	0.1844
2083.4	0.0122	0.0418	0.1965
2340.7	0.0125	0.0551	0.0936
2540.2	0.0146	0.0465	0.1730
2597.9	0.0147	0.0443	0.1813
2659.5	0.0144	0.0441	0.1777
2694.4	0.0140	0.0456	0.1668
2787.4	0.0143	0.0420	0.1809
2812.2	0.0136	0.0477	0.1634
3177.4	0.0155	0.0473	0.2086
3454.9	0.0158	0.0433	0.1846
3529.2	0.0154	0.0446	0.2432
3591.3	0.0165	0.0588	0.1459
3636.3	0.0169	0.0566	0.1605
3713.1	0.0153	0.0606	0.1153
4217.9	0.0181	0.0596	0.1927
4280.3	0.0186	0.0573	0.1929
4382.7	0.0174	0.0605	0.1887
4668.8	0.0179	0.0698	0.0950
4699.1	0.0109	0.0526	0.2071
4844.4	0.0180	0.0606	0.1780
5001.2	0.0178	0.0636	0.1513
5113.0	0.0179	0.0631	0.1500
5898.8	0.0209	0.0716	0.1513
5933.3	0.0214	0.0704	0.1614
6038.1	0.0198	0.0745	0.1088

city of the vibrations in the molecules of the first group substantially exceeds that found for the second (Figure 10.7).

Fluctuations in the average values for angles are markedly larger than those for bonds. This, too, is understandable, because the angles are much more easily deformed. What is most interesting is that the geometry of the molecule may be changed considerably during a transition between two closely spaced excited energy levels. Moreover, this becomes typical in highly energetic states. Thus, for example, in order for its average bond length to be changed from 0.9721 Å to 0.9881 Å and its angle from 104°34″ to 104°25″ in the region of the fundamental frequencies, an H_2O molecule must absorb or emit an energy quantum of 3657 cm^{-1}. An analogous change in its geometry in the transition from state 021 to state 200 only requires an energy quantum of 330 cm^{-1}. Since this is clearly associated with the convergence of the vibrational energy levels up the potential well and with the complexity of the expressions of their corresponding eigenfunctions due to the 'entanglements' of the basis harmonic functions, there is no reason to think that this is a special result and is only applicable to XY_2 molecules. On the contrary, it does not seem premature to conclude that the ability of a molecule to have its geometry changed substantially by a weak energetic effect, if it is already in an excited state, must be typical. This ability must be more clearly expressed, the more complicated the molecule. The jump-like nature of the changes in the geometrical parameters of a molecule when it is excited vibrationally is also retained at high closely spaced energy levels. This clearly indicates that the molecules must be semi-rigid and easily deformed by weak effects. Since this must, in turn, be reflected in its reactive

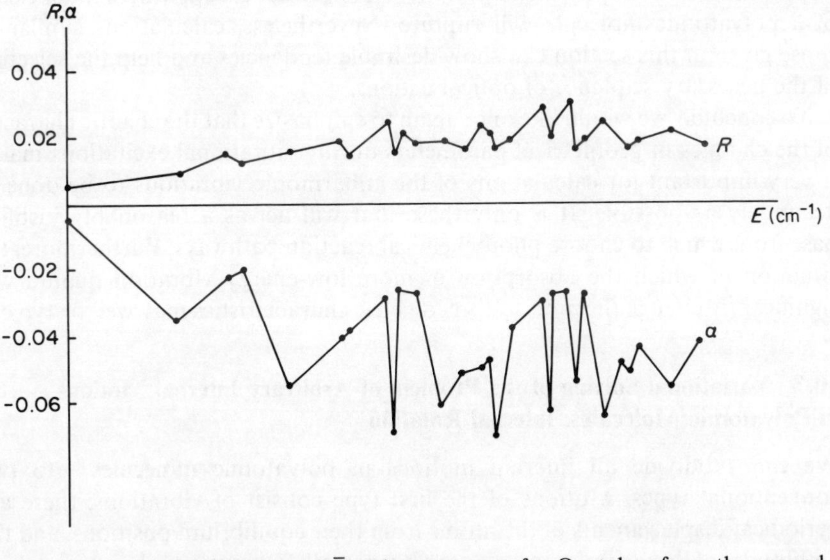

Figure 10.7. The dependence of $\bar{R}$ and $\bar{\alpha}$ on energy for O_3, taken from the ground state

ability, we should expect that the pathway of chemical reactions in systems with 'preheated' molecules can be changed by small energetic effects. This must be of great importance when developing laser chemistry methods.

Although we have only investigated relatively low-energy vibrational states, there is no reason to assume that if we go to higher states the picture will be greatly simplified. On the contrary, the appearance of multiple resonances between harmonic energy levels must lead to an even more complicated situation. We must note that because the vibrational energy is 'smeared' over many degrees of freedom, a transition to another level, differing from the initial one by one or two quantum numbers, may only slightly change the molecule's geometric parameters. However, it may also be that several levels differing by a significant number of quantum numbers may lie close together. Then abrupt jumps in the changes of $\bar{q}_i^{(n)}$ and $(\overline{q_i^{(n)2}})^{1/2}$ will have a large probability. All this leads to the conclusion that a prognosis of the conditions of the photochemical dissociations of certain bonds can be completed using simple considerations and extrapolations, and be correct only when symmetry or considerable differences in atomic masses mean that very characteristic conditions can be fulfilled for the vibrations of individual bonds or atomic groups—and all this when there are large vibrational excitations. We can say from the general theory of vibrations that this situation will probably be observed for terminal bonds XY when $m_x \gg M_y$ and for fully symmetric valency vibrations of highly symmetric (tetrahedrals, octahedrals, etc.) molecules, but will be improbable for skeletal vibrations of complex systems.

The yield from a photochemical reaction can change abruptly upon subsequent vibrational excitations. Moreover, in the general case it is practically impossible, given the present state of theory, to predict exactly where the skeleton of a polyatomic molecule will rupture. Nevertheless, calculations similar to those given in this section can show desirable tendencies and help the selection of the necessary sequence of optical actions.

To conclude, we would like once again to emphasize that the chaotic character of the changes in geometrical parameters during vibrational excitations makes it very important for calculations of the anharmonic vibrations to be done as accurately as possible. It is only these that will act as a reasonably justified base from which to choose photochemical reaction pathways. Furthermore, the situation in which the absorption of more low-energy vibration quanta will significantly change the molecule's geometric characteristics may well be typical.

10.7 Variational Solving of the Problem of Arbitrary Internal Motions in Polyatomic Molecules: Internal Rotation

We can subdivide all internal motions in polyatomic molecules into two conventional types. Motions of the first type consist of vibrations; these are periodical displacements of the atoms from their equilibrium positions, and the amplitudes of these displacements are rather small. Examples are bond stretching (far from a dissociation threshold), changing of valence angles, etc. Another type of internal motion is the rotation of an atomic group around a single valence bond in some polyatomic molecules. Such motions, called 'internal rotations', often occur with large amplitudes.

The available theories consider these two types of molecular motion separately. The theory of molecular vibrations is based usually upon the assumption that internal rotation degrees of freedom are 'frozen'.

The internal rotation of one molecular fragment with reference to another is seen in molecular spectra in two ways. First, for molecules that are symmetrical a splitting (into two components, as a rule) of the corresponding energy levels takes place, and a transition between these sublevels may become possible associated with a microwave frequency. On the other hand, the transitions between torsional levels themselves lead to infra-red absorption bands, and the corresponding frequencies range from some tenths to some hundredths of cm^{-1}, i.e. they are the same order of magnitude as vibrational frequencies. Hence it is clear that the problem of identification and correct interpretation of internal rotation spectral bands is very important[36].

It is well known (see, for example, Lister *et al.*[37]) that the internal rotation problem for symmetrical molecules reduces to a Mathieu equation which permits a precise calculation of the torsional energy levels. However, in cases where the molecule is non-symmetric no precise solution can be obtained.

Recently a variational method (which is also called the method of right numerical diagonalization of a Hamiltonian) has been used for the solution of different problems involving intramolecular motions[38]. The solution of problems involving large-amplitude vibrations of polyatomic molecules with this method has been described above. The same method was used in the work by Lewis *et al.*[39] for the torsional problem; in this work free rotor functions were taken as a variational basis.

Sundius[40] solves the internal rotation problem with the variational method using two different sets of basis functions: a set of Hermitian oscillator functions and one of free rotor functions. He assumes correctly that when a potential pit is deep, the best model for the studying of torsional motions is a harmonic oscillator, but in the opposite case it is a rotor. According to this assumption the author divides the torsional problem into two parts, each of them relating to one type of basis. His calculations confirm numerically that the oscillator basis fits excellently in the case of a high torsional barrier, whereas the free-rotor basis leads to the best numerical results for the case of almost free rotation. Nevertheless, the free-rotor basis, is shown[40] to be quite suitable in cases of an average barrier height.

It is clear, therefore, that methods for separately solving vibrational and one-dimensional internal rotation problems exist. The problem, therefore, is to develop a method suitable for the strict calculation of several internal rotations and their interactions with each other and with vibrational motions.

Such a method can only be numerically based for complicated molecules. The successful application of the variational method for the calculation of vibrational energy levels in the anharmonic approximation (in spite of large bases, this can be accomplished even for rather large molecules also comples the use of the variational procedure for solving the general problem concerning arbitrary atomic motions in a molecule. General principles of such solutions are considered in the next section.

10.7.1 A general scheme for the solving of a problem for vibrations and internal rotations in polyatomic molecules

It is natural to consider internal motions of atoms in polyatomic molecules, irrespective of their characters, by means of an internal co-ordinate system; the internal co-ordinates used are a particular case of generalized curvilinear ones, and the following requirement is essential. The internal co-ordinates must be of such a type that their introduction is possible for arbitrary (including very large) relative displacements of the atoms in a molecule, and that the physical interpretation of the co-ordinates does not change during the atomic motions. In other words, the co-ordinates must be connected with those geometrical concepts which are invariant with respect to values of the displacements. Such co-ordinates are known (see Chapter 8). Normally, they involve distances between atoms with given numbers; angles in triangles which are formed by either three atoms so that one of the atoms is considered as a main apex; (dihedral) angles between two planes which are drawn through triads of atoms so that two of the atoms may be common for every triad; angles between straight lines, which are drawn through pairs of atoms; and planes which are drawn through triads of atoms. The requirement of a correlation between the chosen co-ordinate system and chemical concepts of molecular structure leads to the choice of bond lengths being taken as interatomic distances, the angles are valence ones and the planes of the dihedral angles and the angles between a bond and a plane are drawn through pairs of bonds which form adjacent valence angles. Such a choice of co-ordinates allows one more important condition to be satisfied: we can describe the relative displacements of adjacent atoms, according to the concept of the short-distance action of chemical forces. These important properties concern only the co-ordinates described above and some of their linear combinations, and so they distinguish such co-ordinates and make them very convenient for the study of internal motions.

Let us formulate a general rule for constructing internal rotation co-ordinates. Given two groupings of atoms ABCD... and XYZU... interconnected by a unitary bond through atoms A and X, such groups can revolve about this bond with respect to one another. Atoms B, C, D,... or Y, Z, U,... are located at the ends of the bond associated with atoms A and X.

It is this type of element which is separated out in a molecule due to the division of the comprehensive Schroedinger equation for a molecule over independent degrees of freedom (however, independence here is not mandatory).

By definition, internal rotation is such a motion, unambiguously predetermined in a physical way, where the grouping BCD... is revolving as an integral antity about the axis AX relative to the grouping YZU, whose atoms can be further considered as held at rest, thus placing no restrictions on the generality of the approach. Therefore there must exist the only one generalized co-ordinate, designated here as φ, consistent with such a type of motion.

With the change of the co-ordinate all the atoms BCD... and YZU... must come into relative rotation (since the motion is of a relative nature, we can consider the atoms in one of the groupings as stationary). Bearing in mind that one of the equations $\dot{\vec{r}} = E\vec{\vec{B}}L_p\dot{Q}$ or $\dot{\vec{r}} = E\vec{\vec{B}}\Theta\cdot\dot{q}$ must always hold, we can see that in order for a whole group of atoms to be brought into concerted motion in response to changes of a single co-ordinate, it is necessary for the co-ordinate associated with a structural element to be expressed through Cartesian velocities of displacements of all of the atoms comprising a given structural element. Accordingly, a respective row in the $\vec{B}$ matrix must include non-zero elements in all the cells which are in conformance with the atoms of the structural element under examination.

We cannot find any such element among ordinary natural co-ordinates; there are two non-zero elements in the bonding co-ordinates, three non-zero elements in the angular co-ordinates and a minimum six of these in the dihedron co-ordinates. Clearly, the required natural co-ordinate of internal rotation must be constructed as a linear combination of some of the previously known co-ordinates. Bearing in mind that it is the internal rotation we are dealing with, where only dihedrons of the BAXY or a similar type can change, it becomes only natural that the desired co-ordinate φ must be a linear combination of respective non-planar co-ordinates.

Consequently, it is clear that at least such a collection of co-ordinates of variation in dihedrons between the planes of type BAXY, CAXZ, etc. are involved, would ensure the inclusion of all the end atoms BCD..., YZU.... To make life easier, we can just set up all the possible co-ordinates of variations of the BAXY and other similar angles such that each bond in the left-hand grouping is sequentially combined with all the bonds in the right-hand one. In this case, of course, redundant co-ordinates become involved, but this is not very important.

Now, following the ordinary rules, let us construct the matrix $\vec{B}$ for all the introduced dihedron co-ordinates, such that $\rho = \vec{B}\delta\vec{r}$ (here ρ is a column matrix of non-planar co-ordinates, and $\delta\vec{r}$ displacements of the atoms from their initial positions; in the case of minute displacements the formula is always true).

Now, turn all the bonds in the left-hand grouping AB, AC, AD,... about the axis AX by one and the same small angle. By definition, this really the action of the co-ordinate φ. Let the atoms and bonds in the right-hand grouping remain unchanged. Given the initial unit vector $\vec{e}_k$ of the BA and other bonds, and the bond lengths, we will immediately find increments $\delta\vec{r}_k$ of the Cartesian co-ordinates of atoms B, C, D,... which must evolve from the internal rotation. If the atoms Y, Z, U remain at rest, their $\delta\vec{r}_k = 0$.

Proceeding from the relationship $\rho = \vec{B}\delta\vec{r}$, let us find the values of the elements in column $\rho^{(0)}$ for all introduced dihedron co-ordinates, consistent with small turns of one atomic grouping with respect to the other grouping. Let these values be equal to $\rho^{(0)}$.

On the other hand, equation $\rho = l_\varphi \varphi$ should hold for arbitrary values of elements ρ, where l_φ is a column in the matrix L_q, corresponding to the rotational co-ordinate.

Clearly, the relative values of elements in columns $\rho^{(0)}$ and l_φ must be the same, otherwise φ cannot be an internal rotation co-ordinate, hence, $l_\varphi N\rho^{(0)}$, where N is a constant factor (normalization) to be defined. Therefore we have $\rho = N\rho^{(0)}\varphi$, and hence $(N\tilde{\rho}^{(0)})\rho = [N^2\sum_i(\rho_i^{(0)})^2]\varphi$. To separate co-ordinate φ in a pure form, we impose a requirement that $N^2[\sum_i(\rho_i^{(0)})^2] = 1$. Thus N^2 and l_φ will be defined. Finally, $\varphi = \tilde{l}_\varphi \cdot \rho$.

Thus the required co-ordinate has been found as a linear combination of the simplest natural ones. It is this co-ordinate that must be further worked with, introducing a potential for it rather than for each of the intermediate dihedrons. Again, this co-ordinate should be used in calculating changes of bond unit vectors onto Cartesian axes when computing transition intensities. Once the primary dihedron co-ordinates have accomplished their function they must 'disappear', being entered as a constituent part into the co-ordinate φ.

Case study Given a symmetrical collection of three HCCH dihedrons, and a small turn of one of the CH_3 groups in a molecule of ethane, let

$$\rho^{(0)} = 10^{-1}\begin{bmatrix}1\\1\\1\end{bmatrix}$$

Then $N^2 \cdot 3.10^{-2} = 1$ or $N = (1/\sqrt{3})10$. Accordingly,

$$l_\varphi = N\rho^{(0)} = \frac{1}{\sqrt{3}}\begin{bmatrix}1\\1\\1\end{bmatrix}$$

and $\varphi = (1/\sqrt{3})(\rho_1 + \rho_2 + \rho_3)$, which should be the case according to the rules of symmetry.

The entire procedure can be automated by programming a computer to enter all the dihedrons. Only in some particular cases (for example, in dimethylacetylene molecule) is it necessary to introduce dihedral angles which are formed by planes without common atoms, or to accept the distance between carbon atoms of the end-groups CH_3 as an edge of the dihedral angle.

The Schroedinger equation for relative motions of nuclei in the intrinsic co-ordinate system q_i (which are a particular case of general curvilinear co-ordinates) can be written as equation (10.1) or, with the help of commutation relations, in a simpler form equation (10.2)).

Now let us divide all molecular co-ordinates into two groups: first, the changes of valence bond lengths, valence angles and angles between a bond and a plane;

and, second, the changes of linear combinations of dihedral angles which characterize (as was shown above) the rotations around single bonds. The co-ordinates of the first group will be designated with the symbol q and those of the second group with the symbol χ.

Then, without any limitation of generality, we can write the following expression for the nuclear motion Hamiltonian as follows:

$$\hat{H} = -\frac{\hbar^2}{2}\left\{\sum_{k,n}\frac{\partial}{\partial q_k}\tau_{kn}(q,\chi)\frac{\partial}{\partial q_n} + \sum_{k,i}\left[\frac{\partial}{\partial q_k}\tau_{ki}(q,\chi)\frac{\partial}{\partial \chi_i} + \frac{\partial}{\partial \chi_i}\tau_{ik}(q,\chi)\frac{\partial}{\partial q_k}\right]\right.$$
$$\left. + \sum_{i,j}\frac{\partial}{\partial \chi_i}\tau_{ij}(q,\chi)\frac{\partial}{\partial \chi_j}\right\} + V'(q,\chi) = \hat{T}_v + \hat{T}_{vr} + \hat{T}_r + V'(q,\chi) \qquad (10.16)$$

Here, $\hat{T}_v$ is an operator of the vibrational motion $\hat{T}_{vr}$ is one of the interaction of the vibrations and the internal rotations and $\hat{T}_r$ is an operator of a kinetic energy of the internal rotations.

We must note that although the operator $\hat{T}_v$ contains differentiation only with respect to the vibrational co-ordinates, the values of τ_{kn} can also depend on internal rotation co-ordinates—for example, for non-diagonal kinematic interactions of the angle co-ordinates, the relative disposition of which can change with internal rotation.

If we distinguish the coefficients τ_{kn} which depend on torsional co-ordinates (such coefficients for non-torsional vibrational co-ordinates can only be non-diagonal and correspond to the vibrational ones, which are physically situated near a torsional co-ordinate), equation (10.16) can be written as follows:

$$\hat{H} = -\frac{\hbar^2}{2}\left\{\sum_{k,n}\frac{\partial}{\partial q_k}\tau_{kn}(q)\frac{\partial}{\partial q_n} + \sum_{k',n'}\frac{\partial}{\partial q_{k'}}\tau_{k'n'}(q,\chi)\frac{\partial}{\partial q_{n'}}\right.$$
$$+ \sum_{k,i}\left[\frac{\partial}{\partial q_k}\tau_{ki}(q,\chi)\frac{\partial}{\partial \chi_i} + \frac{\partial}{\partial \chi_i}\tau_{ik}(q,\chi)\frac{\partial}{\partial q_k}\right]$$
$$\left. + \sum_{i,j}\frac{\partial}{\partial \chi_i}\tau_{ij}(q,\chi)\frac{\partial}{\partial \chi_j}\right\} + V'(q,\chi) \qquad (10.17)$$

We can see that it is possible to distinguish in the general Hamiltonian a part of the kinetic operator which depends only on vibrational co-ordinates. Moreover, if we wish to distinguish the purely torsional part it is necessary to make an additional assumption. This consists of the harmonic approximation for the vibrations, when dependence of kinematic coefficients upon the vibrational co-ordinates vanishes, and it is possible to write the Hamiltonian as follows:

$$\hat{H} = -\frac{\hbar^2}{2}\left\{\sum_{k,n}\tau_{kn}\frac{\partial}{\partial q_k}\frac{\partial}{\partial q_n} + \sum_{\substack{k',n'\\k'\neq n'}}\tau_{k'n'}(\chi)\frac{\partial}{\partial q_{k'}}\frac{\partial}{\partial q_{n'}}\right.$$
$$\left. + \sum_{k,i}\left[\frac{\partial}{\partial q_k}\tau_{ki}(\chi)\frac{\partial}{\partial \chi_i} + \frac{\partial}{\partial \chi_i}\tau_{ik}(\chi)\frac{\partial}{\partial q_k}\right] + \sum_{i,j}\frac{\partial}{\partial \chi_i}\tau_{ij}(\chi)\frac{\partial}{\partial \chi_j}\right\} + V'(q,\chi) \quad (10.18)$$

We can see that as a result of the approximations introduced it becomes possible to distinguish two terms in the kinetic part of the Hamiltonian which are connected only with vibrational or torsional co-ordinates. The remaining terms characterize kinematic interactions of these motions. When distinguishing that part of the fourth term connected only with diagonal elements of the matrix of kinematic coefficients for torsional co-ordinates it is possible to write, without any limitation of generality,

$$\sum_{i,j} \frac{\partial}{\partial \chi_i} \tau_{ij}(\chi) \frac{\partial}{\partial \chi_j} = \sum_i \frac{\partial}{\partial \chi_i} \tau_{ii}(\chi) \frac{\partial}{\partial \chi_i} + \sum_{\substack{i,j \\ i \neq j}} \frac{\partial}{\partial \chi_i} \tau_{ij}(\chi) \frac{\partial}{\partial \chi_j} \tag{10.19}$$

The second term here reduces to zero if the co-ordinates for any pair of torsions have no common atoms—a situation that occurs rather frequently.

Now let us turn to the potential function. We shall only consider the case when any internal rotation does not lead to a connection of one part of a complex molecule with another. This amounts to the absence of steric prohibitions, and so the function $V(q, \chi)$ never increases to infinity.

Let us now take into account the following well-known facts:

(1) The distortions in polyatomic molecules which are connected with internal rotations are associated with small energy expenditures;
(2) There are no very large or remote influences even in molecules with conjugated bonds; in particular, it finds a reflection in the stability of the electron density in atomic groups, even when polar substitutors are situated near such groups, and also in the fact that in the set of molecules with the fixed group, the characteristic bond lengths, valence angles, force constants, electro-optical parameters, etc. can be assumed to be invariant to a good approximation;
(3) In molecules with typical chemical bonds non-diagonal force constants for the co-ordinates which have no common atoms are, as a rule, relatively small;
(4) Pure vibrational spectra of the molecules with internal rotations (which are registered, for example, in a gas, liquid crystal or low-temperature matrix) do not differ greatly.

All these facts are possible only in the case when, first, potential interactions between vibrations and torsions are small and, second, when interactions between torsions themselves are also small. This means that the potential function, in general, must have the form:

$$V'(q, \chi) = V_v + V_r + V_{vr}$$

where V_v, V_r and V_{vr} are the terms that correspond to vibrations, internal rotations and their interactions, respectively. These terms can be represented very approximately by the form $V' = V_v + V_r$, where $V_r = \sum_i V_{ri}(\chi_i)$ and $V_{ri}(\chi_i)$ is the potential for one torsional co-ordinate, which is calculated assuming that we take into consideration only the atoms nearest to the given atomic group (the cluster approximation). The potential operator mentioned above will then resemble the kinetic operator in its properties.

In equation (10.18) with the potential

$$V' = V_v(q) + V_r(\chi) = \tfrac{1}{2}\sum_{k,n} U_{kn}q_k q_n + \sum_i V_{ri}(\chi_i)$$

we can distinguish a term with a Hamiltonian in separated variables provided that, in addition, we carry out a transformation to normal vibrational co-ordinates Q:

$$\hat{H}_0 = \left\{-\frac{\hbar^2}{2}\sum_k \frac{\partial^2}{\partial Q_k^2} + \tfrac{1}{2}\sum_k \lambda_k Q_k^2\right\} + \left\{-\frac{\hbar^2}{2}\sum_i \frac{\partial}{\partial \chi_i}\tau_{ii}(\chi_i)\frac{\partial}{\partial \chi_i} + V_{ri}(\chi_i)\right\} \tag{10.20}$$

Its eigenfunctions will be as follows:

$$\Psi = \Psi_v \cdot \Psi_r = \prod_k \Psi_v(Q_k) \cdot \prod_i \Psi_i(\chi_i) \tag{10.21}$$

It is necessary to make one more point. The vibrational part of the Hamiltonian equation (10.20) can be obtained as a result of a reduction of the expression

$$\hat{H}_{v0} = -\frac{\hbar^2}{2}\sum_{k,n} \tau_{kn}\frac{\partial}{\partial q_k}\frac{\partial}{\partial q_n} + \tfrac{1}{2}\sum_{k,n} U_{kn}q_k q_n$$

to the diagonal form, and non-diagonal coefficients must be excluded from this both in the kinetic and the potential parts (as was mentioned above) for those vibrational co-ordinates which have common atoms and the respective unitary vectors of which change their dispositions at internal rotations (for example, the elements that correspond to co-ordinates of the angles ABC and BCD in the dihydral angle, formed by the planes ABC and BCD, where BC is an edge of the dihedral angle). Such a Hamiltonian $\hat{H}_{v0}$ is an 'effective' one. The elimination of some non-diagonal terms of kinematic matrix is a 'violent' operation, which is quite possible from a mathematical viewpoint in order to obtain the general variational solution of the differential equation (10.18).

As mentioned above, in the case of low barrier heights to internal rotation it is worth searching for torsional functions on the basis of one-dimentional free-rotor functions. This means ignoring terms $V_{ri}(\chi_i)$ in the Hamiltonian equation (10.20).

It is natural to chose the functions $V_{ri}(\chi_i)$ as Fourier expansions. The solution of the complete equation can then be found with the variational Ritz method on the basis composed of functions (10.21).

In order to obtain the variational solutions it is necessary to calculate the matrix elements, which all reduce to products of one-dimensional integrals in separated variables. For brevity, we shall write only those terms of equation (10.18) which contain the addenda that are complementary to H_0:

$$
\begin{aligned}
&\text{(a)} \quad \int \Psi'_v \frac{\partial}{\partial Q_{k'}} \frac{\partial}{\partial Q_{n'}} \Psi''_v \mathrm{d}Q \int \Psi'_r \tau_{k'n'} \Psi'' \mathrm{d}\chi \\
&\text{(b)} \quad \int \Psi'_v \frac{\partial}{\partial Q_k} \Psi''_v \mathrm{d}Q \cdot \int \Psi'_r \left[\tau_{ki}(\chi) \frac{\partial}{\partial \chi_i} + \frac{\partial}{\partial \chi_i} \tau_{ki}(\chi) \right] \Psi''_r \mathrm{d}\chi \qquad (10.22) \\
&\text{(c)} \quad \int \Psi'_r \frac{\partial}{\partial \chi_i} \tau_{ij}(\chi) \frac{\partial}{\partial \chi_j} \Psi''_r \mathrm{d}\chi
\end{aligned}
$$

Then let us note that all coefficients $\tau(\chi)$ which depend on torsional co-ordinates will be as follows:

$$
\begin{aligned}
&\tau_{ii}(\chi) = A + B_n \cos n\chi + C_n \sin \chi \quad \text{or} \\
&\tau_{ij}(\chi) = A + \cdots \text{ (double-dimensional Fourier expansion).}
\end{aligned}
$$

Some of the constants A, B_n, C_n may reduce to zero for different coefficients. Terms of the potential part of the complete Hamiltonian can be written in an analogous way.

Therefore a further simplification of the problem can be obtained by assuming that some of the diagonal coefficients $\tau_{ii}(\chi)$ are constant (the calculations show that the dependency of co-ordinates χ_i is not large on these coefficients). It is then possible to take as basis functions one-dimensional Fourier free-rotor functions ψ_r when the problem is solved using the variational method:

$$
\begin{cases}
\phi_0^{(1)} = \dfrac{1}{\sqrt{2\pi}} \\
\phi_l^{(1)} = \dfrac{1}{\sqrt{\pi}} \cos l\chi
\end{cases}
\qquad (10.23)
$$

$$
\phi_l^{(2)}(\chi) = \frac{1}{\sqrt{\pi}} \cdot \sin l\chi, \quad l = 1, 2, 3, \ldots \qquad (10.24)
$$

The perturbing terms for the Hamiltonian H_0 will be some non-diagonal elements in the kinetic part of the Hamiltonian and the terms $V_{ri}(\chi_i)$, $V_{rij}(\chi_i, \chi_j)$ and $V_{vr}(\chi)$ in the potential one. Since all these are periodic functions in terms of the co-ordinates χ, and for the terms which characterize interactions between internal rotations we obtain a polydimensional Fourier series with separable variables, it is clear that the problem is transformed to a calculation of matrix elements of $\cos n\chi$ and $\sin n\chi$ in the basis $\phi_l^{(1)}$ and $\phi_l^{(2)}$. This can be carried out very easily. It is evident that submatrices of the variational matrix for the rotors in bases $\phi_l^{(1)}$ and $\phi_l^{(2)}$ will be separate and differ from each other. This leads to a characteristic doublet structure of torsional levels. To confirm this, consider as an example the following type of Hamiltonian:

$$
\hat{H} = -\frac{\hbar^2}{2I} \cdot \frac{\mathrm{d}^2}{\mathrm{d}\chi^2} + \sum_k \frac{V_k}{2} (1 - \cos k\chi) \qquad (10.25)
$$

where I is a reduced momentum of inertia. We obtain immediately

$$\int_0^{2\pi} \phi_m^{(1)} \hat{H} \phi_n^{(2)} \, d\chi = 0$$

and the variational Hamiltonian will be of the type:

$$H = \begin{bmatrix} H^{(1)} & 0 \\ 0 & H^{(2)} \end{bmatrix}$$

and all torsional energy levels can be found by the separate diagonalization of the submatrices $H^{(1)}$ an $H^{(2)}$, which correspond to the basic functions $\phi_l^{(1)}$ and $\phi_l^{(2)}$, respectively.

Matrix elements of the Hamiltonian equation (10.25) in the basis $\phi_l^{(1)}$ are:

$$\left\langle m \left| \frac{d^2}{d\chi^2} \right| n \right\rangle = -m^2 \delta_{mn}$$

$$\left\langle m \left| \frac{V_k}{2} \right| n \right\rangle = \frac{V_k}{2} \delta_{mn}$$

$$\langle 0 | \cos k\chi | n \rangle = \frac{1}{\sqrt{2}} \delta_{kn} \tag{10.26}$$

$$\langle m | \cos k\chi | n \rangle = \begin{cases} 1/2, & k = \pm(m \pm n) \\ 0 & k \neq \pm(m \pm n) \end{cases}$$

The analogous matrix elements in the basis $\phi_l^{(2)}$ are:

$$\left\langle m \left| \frac{d^2}{d\chi^2} \right| n \right\rangle = -m^2 \delta_{mn}$$

$$\left\langle m \left| \frac{V_k}{2} \right| n \right\rangle = \frac{V_k}{2} \delta_{mn} \tag{10.27}$$

$$\langle m | \cos k\chi | n \rangle = \begin{cases} 1/2, & k = \pm(m - n) \\ -1/2, & k = \pm(m + n) \\ 0, & k \neq \pm(m \pm n) \end{cases}$$

Let us write the Hamiltonian matrix for a concrete torsional problem. We assume that the torsional potential function of the molecule can be written as a Fourier expansion with three coefficients:

$$V(\chi) = \sum_{i=1}^{3} \frac{V_i}{2}(1 - \cos i\chi)$$

The submatrix $H^{(1)}$ of the matrix H is

$$\begin{bmatrix} V_{00}; & -\dfrac{V_1}{2\sqrt{2}}; & -\dfrac{V_2}{2\sqrt{2}}; & -\dfrac{V_3}{2\sqrt{2}}; & 0 & \cdots \\ & V_{00}-\dfrac{V_2}{4}+\tau; & -\tfrac{1}{4}(V_1+V_3); & -\dfrac{V_2}{4}; & -\dfrac{V_3}{4} & \cdots \\ & & V_{00}+4\tau; & -\dfrac{V_1}{4}; & -\dfrac{V_2}{4} & \cdots \\ & & & V_{00}+9\tau-\dfrac{V_1}{4} & & \cdots \end{bmatrix} \tag{10.28}$$

and the submatrix $H^{(2)}$ is

$$\begin{bmatrix} V_{00}+\dfrac{V_2}{4}+\tau; & -\tfrac{1}{4}(V_1-V_3); & -\dfrac{V_2}{4}; & -\dfrac{V_3}{4} & \cdots \\ & V_{00}+4\tau; & -\dfrac{V_1}{4}; & -\dfrac{V_2}{4} & \cdots \\ & & V_{00}+9\tau; & -\dfrac{V_1}{4} & \cdots \end{bmatrix} \tag{10.29}$$

In equations (10.27) and (10.28):

$$\tau = \hbar^2/2I, \quad V_{00} = \tfrac{1}{2}(V_1+V_2+V_3)$$

and both kinematic and force coefficients are measured in cm^{-1}. Only the upper triangles of the symmetrical matrices $H^{(1)}$ and $H^{(2)}$ are written, and we limit ourselves to the case of non-negative values of the rotor quantum number 1.

It can be seen from the latter formulae that the differenes between $H^{(1)}$ and $H^{(2)}$ lead to the splitting of the torsional energy levels which are obtained as a result of the diagonalization. The reason for the splitting is, first, the difference between diagonal elements V_{11} of matrices $H^{(1)}$ and $H^{(2)}$ and, second, the differences in some non-diagonal elements near the main diagonal. In addition, the 'zeroth' line and column, which occurs only in the matrix $H^{(1)}$ but not $H^{(2)}$, also make a contribution to the splitting of the levels. We see that the greater the values of the Fourier coefficients in the torsional potential function, the larger will be the splitting of the torsional energy levels.

10.7.2 Numerical values calculated for torsional frequencies for some polyatomic molecules

We shall discuss here the results of a variational numerical solution of the torsional problem with the help of basic functions (10.21) for molecules of hydrogen peroxide and persulphide, as well as of hydrozine and methyl amine. An *ab initio* torsional potential function as a Fourier expansion has been used.

Sets of torsional energy levels found as solutions of the variational problem for the molecules are shown in Table 10.14. Let us analyse the results obtained with regard to experimental values of torsional frequencies of these molecules.

It can be seen from Table 10.14 that numerical values of the torsional levels for the molecules are grouped in pairs: for low levels the states $l_+ = 0$ and $l_+ = 1$; $l_- = 1$ and $l_+ = 2$, etc. are found to be of a similar energy, but for higher ones the states which differ only in sign of 'l' have almost the same energy. This tendency can be explained by the fact (mentioned above) that in the case when torsional levels are much lower than the torsional potential barrier, the behaviour of the molecule can be described with the harmonic oscillator model, whereas in the contrary case, a rotor model is more suitable. In both instances, however, a free-rotor variational basis can be used successfully. For the molecule of hydrogen persulphide in particular (Table 10.14), the advantages of the rotor model appear later than for thc rest of the molecules being considered. This can be explained by the fact that the torsional potential barrier for this molecule is higher than that for the others. For the molecule with a low barrier (methyl amine) the opposite applies.

It is interesting to compare the calculated with the corresponding experimental values. In some texts 41 the values of the first two pure torsional transitions in the molecule H_2O_2 are adduced; the frequencies are, respectively, 183 and 540 cm^{-1}. As we can see from Table 10.14, the frequencies of the transitions ($l = 0 \rightarrow l_- = 1$) and ($l_- = 1 \rightarrow l_- = 2$) are 155.3 and 515.7 cm^{-1}, respectively. (Both in the text and in Table 10.14 the index + near the torsional quantum number l means that the torsional state is a superposition of rotor states of

Table 10.14. Torsional energy levels for the set of molecules with an internal rotation (cm^{-1})

1	Type of function	H_2O_2	N_2H_4	H_2S_2	CH_3NH_2
0	+	130.7	353.0	0.0	216.4
1	+	149.4	332.2	0.0	219.7
1	−	286.0	1072.9	459.3	219.7
2	+	531.2	1040.3	413.3	568.9
2	−	801.7	1507.2	712.1	568.9
3	+	908.3	1542.9	800.5	801.5
3	−	950.9	1804.9	1334.8	801.5
4	+	1281.6	2465.7	1169.8	976.8
4	−	1794.2	2492.3	1813.9	976.8
5	+	1998.4	3160.2	1939.7	1264.9
5	−	2103.9	3893.7	2237.6	1264.9
6	+	2530.5	3941.0	2504.8	1661.1
6	−	2735.0	4060.7	2495.0	1655.0
7	+	2970.3	4868.2	2519.7	2103.1
7	−	3071.6	4895.2	2895.8	2103.1
8	+	4042.0	5812.6	3418.8	2618.9
8	−	4043.3	5842.1	3497.5	2618.9

the type $\phi^{(1)}$, and similarly for the index $-$.) The comparison can be considered as satisfactory since the precision of the experimental results is not very high[41].

According to Table 10.14, a fundamental torsional frequency for the molecule of hydrazine is near 690 cm^{-1}, whereas the experimental value is approximately half this value[42,43]. Taking into account the reasons given in reference 44, we must, in all probability, regard the assignment of torsional frequencies to be incorrect[42,43]. For accuracy, it seems necessary to take into account all energy levels present, including both vibrational and torsional ones. It is then possible to obtain a real physical picture of transitions in the molecule, and we must allow for the 'hybrid' vibration–torsional transitions to take place, when vibrational and rotation quantum numbers vary their values simultaneously. It is all the more reasonable that, as we can see (for example, for the hydrazine molecule), the value of a torsional quantum is comparable with that of a vibrational one. Hence, we must assign the experimental band around 360–380 cm^{-1} to the 'hybrid' vibration–rotational transition[42,43] $(v=0,\ l_-=1)\rightarrow(v_3=l,\ l=0)$ with a calculated frequency of 378 cm^{-1}.

A calculated value of a fundamental torsional transition for the molecule of H_2S_2 is 413 cm^{-1} (Table 10.14). This coincides closely[45] with the respective observed value of 416 cm^{-1}. An observed value 46 for the torsional frequency for the molecule of methyl amine is 265 cm^{-1}. This is close to the calculated value of 233 cm^{-1} found for the torsional transition $(l_+=1\rightarrow l_+=2)$. We see from this analysis that good agreement is obtained between the calculated and experimental values of torsional frequencies.

10.7.3 *The case of several torsional degrees of freedom*

It is easy to generalize this theory to the case of several torsional degrees of freedom. In particular, we can see, from the most general considerations, that in the situation when there are several internal rotational axes in the molecule the dependence of the kinematic coefficients τ on torsional co-ordinates x_i are expressed as a k-dimensional Fourier expansion in these co-ordinates (if k torsional axes intersect in one point). For example, in the case of two internal rotations a general form for the $\tau_{\chi_i\chi_k}$ kinematic coefficients will be as follows:

$$\begin{aligned}
\tau_{\chi_1\chi_2} &= A\cos\chi_1 + B\\
\tau_{\chi_1\chi_2} &= C\cos\chi_2 + D \qquad\qquad (10.30)\\
\tau_{\chi_1\chi_2} &= A\cos\chi_1\cos\chi_2 + B\sin\chi_1\sin\chi_2 + C\cos\chi_1 + D\cos\chi_2 + F
\end{aligned}$$

The coefficients a, b, etc. can be calculated numerically for example, with the help of a least-squares method from the values of $\tau_i\tau_k$ when we have enough fixed values of the angles χ_i.

The general vibration–torsional problem for such molecules can also be solved using the variational method. A basic function will be

$$\Psi(Q,\chi) = \Psi_v(Q_1, Q_2, \ldots)\cdot\phi_r(\chi_1, \chi_2, \ldots, \chi_N)$$

where ϕ is an orthonormalized internal rotational function

$$\phi_r(\chi_1, \ldots, \chi_N) = \prod_{k=1}^{N} \phi_r^{(i)}(\chi_k) \quad i = 1, 2$$

Each of the torsional levels will be double-degenerate. If we use an *ab initio* torsional potential function for a two-dimensional internal rotor in the form[47]

$$V_r(\chi) = \frac{V_3}{2}(1 - \cos 3\chi_1) + \frac{V_3}{2}(1 - \cos 3\chi_2)$$
$$+ \frac{V'_3}{4}(1 - \cos 3\chi_1)(1 - \cos 3\chi_2) - \frac{V''_3}{4} \sin 3\chi_1 \sin 3\chi_2 \qquad (10.31)$$

where the values of parameters V_3, V'_3, V''_3 are given in reference 47 for some molecules with two rotating tops of C_{3v} symmetry, then we can obtain the elements of the Hamiltonian matrix and torsional frequencies of the molecule, as above. Note that since Hamiltonian matrix elements, in the case of several internal torsions, can be represented as products of one-dimensional integrals then the problem does not require a calculation of any new integrals, but a multiplication of the well-known one-dimensional ones.

We have calculated the fundamental torsional frequency for propane. The vibration–torsional interaction (both kinematic and potential) was ignored for the reasons mentioned above. The calculation gives for the torsional frequency the value of 260 cm^{-1}, and this is in satisfactory agreement with the experimental value of 223 cm^{-1} for this molecule[48].

References

1. A. I. Pavluchko and L. A. Gribov. *Opt. and Spectr.*, **58**, 1247 (1985) (in Russian).
2. H. H. Nielsen. *Rev. Mod. Phys.*, **23**, 90 (1951).
3. H. H. Nielsen, in *Handbuch der Physik* (ed. S. Flugge), Vol. 38, Springer Verlag, Berlin (1959).
4. R. C. Herman and W. H. Shaffer. *J. Chem. Phys.*, **16**, 453 (1948).
5. M. P. Barchewitz. *Spectroscopie Infrarouge*, Vol. I, Gauthier-Villars, Paris, (1961).
6. S. Califano. *Vibrational States*, Ch. 9, John Wiley, Chichester/New York (1976).
7. L. A. Gribov. *Opt. and Spectr.*, **31**, 842 (1971) (in Russian).
8. I. Suzuki. *Bull. Chem. Soc. Japan*, **44**, 3277 (1971).
9. G. D. Carney and C. W. Kern. *Int. J. Quant. Chem. Symp.*, **9**, 317 (1975).
10. R. J. Whitehead and N. C. Handy. *J. Molec. Spectr.*, **55**, 356 (1975); **59**, 459 (1976).
11. I. Suzuki. *Bull. Chem. Soc. Japan*, **48**, 3565 (1975); **52**, 1606 (1979).
12. G. D. Carney and R. N. Porter. *J. Chem. Phys.*, **65**, 3547 (1976).
13. G. D. Carney, S. R. Langhoff and L. A. Curtiss. *J. Chem. Phys.*, **66**, 3724 (1977).
14. A. I. Pavluchko and L. A. Gribov. *J. Appl. Spectr.*, **33**, 904 (1980) (in Russian).
15. L. A. Gribov, A. I. Pavluchko and G. F. Lozenko. *Opt. and Spectr.*, **53**, 812 (1982) (in Russian).
16. L. A. Gribov, A. N. Pavluchko and G. F. Lozenko. *J. Appl. Spectr.*, **36**, 87, 274 (1982) (in Russian).
17. A. I. Pavluchko, G. F. Lozenko and L. A. Gribov. *Optica and Spectr.*, **50**, 450 (1981) (in Russian).

18. A. I. Pavluchko and L. A. Gribov. *Opt. and Spectr.*, **54**, 644 (1983) (in Russian).
19. L. A. Gribov and A. I. Pavluchko. *Opt. and Spectr.*, **56**, 1015 (1984) (in Russian).
20. A. I. Pavluchko, G. F. Lozenko and L. A. Gribov. *Opt. and Spectr.*, **58**, 1175 (1985) (in Russian).
21. A. I. Pavluchko and L. A. Gribov. *Opt. and Spectr.*, **60**, 491 (1986) (in Russian).
22. A. R. O. Hoy, I. M. Mills and G. Strey. *Mol. Phys.*, **24** 1265 (1972).
23. I. Suzuki, M. A. Pariseaw and J. Overend. *J. Chem. Phys.*, **44**, 3561 (1966).
24. J. Strey and I. M. Mills. *Mol. Phys.*, **26**, 129 (1973).
25. M. A. Pariseaw, I. Suzuki and J. Overend. *J. Chem. Phys.*, **23**, 2335 (1965).
26. K. Kuchitsu and Y. Morino. *Bull. Chem. Soc. Japan*, **38**, 805 (1965).
27. I. M. Mills. *Theor. Chem.*, **1**, 110 (1974).
28. I. Suzuki. *J. Molec. Spectr.*, **25**, 479 (1968).
29. G. F. Smith, Jr and J. Overend. *J. Chem. Phys.*, **53**, 1157 (1971).
30. V. K. Wang and J. Overend. *Spectr. Acta*, **32A**, 1043 (1976).
31. A. Schmiedekamp, C. W. Bock and P. George. *J. Molec. Struct.*, **67**, 107 (1980).
32. U. Wanlgren, J. Pacansky and P. S. Bagus. *J. Chem. Phys.*, **63**, 2874 (1975).
33. P. Botschwina. *Chem. Phys.*, **68**, 41 (1982).
34. P. Botschwina. *Mol. Phys.*, **47**, 241 (1982).
35. L. A. Gribov, T. T. Merzlyak and I. S. Perelygin. *J. Molec. Struct.*, **67**, 1 (1980).
36. W. J. Orville-Thomas (ed.), *Internal Rotation in Molecules*, John Wiley, London/New York (1974).
37. D. G. Lister, J. N. Macdonald and N. L. Owen. *Internal Rotation and Inversion*, Academic Press, London (1978).
38. L. A. Gribov, M. R. Rasovsky. *J. Molec. Struct.* (*Theochem.*), **122**, 15 (1985).
39. J. D. Lewis *et al. J. Molec. Struct.*, **12**, 427 (1972).
40. T. Sundius. *J. Molec. Spectr.*, **75**, 368 (1979).
41. R. L. Redington, W. B. Olson and P. C. Cross. *J. Chem. Phys.*, **36**, 1311 (1962).
42. P. A. Giguere and I. D. Liu, *J. Chem. Phys.*, **20**, 136 (1952).
43. J. S. Ziomek and M. D. Zeidler, *J. Mol. Spectrosc.*, **11**, 163 (1963).
44. L. A. Gribov and M. R. Rasovsky, *J. Appl. Spectrosc.* (*USSR*), **39**, 278 (1983).
45. G. Winnewisser, M. Winnewisser and W. Gordy, *J. Chem. Phys.*, **49** 3465 (1968).
46. A. P. Gray and R. C. Lord, *J. Chem. Phys.*, **26**, 690 (1957).
47. L. Radom and J. A. Pople, *J. Am. Chem. Soc.*, **92**, 4786 (1970).
48. J. N. Gavles and W. T. King, *Spectrochim. Acta*, **21**, 543 (1965).

Chapter 11

The Theory of Intensity in the Infra-red Spectra of Polyatomic Molecules

11.1 Expressions of the Dipole Transitions for Matrix Elements

According to the basic tenets of the quantum theory of radiation, the greatest contribution to the intensity of the absorption, and emission bands is made by dipole transitions. The probability of a dipole transition for a spontaneous emission, an absorption or a forced emission is defined by the square of the modulus of the dipole moment's matrix element.

If we take the wave function of the molecule to be $\Psi = \Psi_e(\vec{r}, Q)\Psi_v(Q)$ and the dipole moment operator to be

$$\vec{\mu} = \sum_\alpha Z_\alpha \vec{R}_\alpha - \sum_i \vec{r}_i \tag{11.1}$$

where $\sum_\alpha Z_\alpha \vec{R}_\alpha$ is the nuclear component of the dipole moment and $\sum_i \vec{r}_i$ is the electronic component, we then obtain the following expression for the matrix element:

$$\vec{\mu}^{\,\prime,\prime\prime} = \int \Psi_e''(\vec{r}, Q)\Psi_v''(Q)\left(\sum_\alpha Z_\alpha \vec{R}_\alpha - \sum_i \vec{r}_i\right)\Psi_e'(\vec{r}, Q)\Psi_v'(Q)\,\mathrm{d}V\,\mathrm{d}Q \tag{11.2}$$

If the transition takes place between the vibrational energy levels of a single electronic state, then $\Psi_e'' = \Psi_e' = \Psi_e$. Since the dipole moment operator is not differential, the expression

$$\int \Psi_e(\vec{r}, Q)\Psi_v''(Q)\left(\sum_\alpha Z_\alpha \vec{R}_\alpha - \sum_i \vec{r}_i\right)\Psi_e(\vec{r}, Q)\Psi_v'(Q)\,\mathrm{d}V\,\mathrm{d}Q$$

can be rewritten in the form:

$$\int \rho(\vec{r}, Q)\Psi_v''\left(\sum_\alpha Z_\alpha \vec{R}_\alpha - \sum_i \vec{r}_i\right)\Psi_v'(Q)\,\mathrm{d}V\,\mathrm{d}Q \tag{11.3}$$

Here $\rho(\vec{r}, Q)$ is the electronic density in the molecule and is a function of the vibrational co-ordinates, a particular case being normal co-ordinates. When a vibration takes place about the equilibrium position, the electron density will change and its value can be written as a series in vibrational co-ordinates:

$$\rho(\vec{r},Q)=\rho(\vec{r},0)+\sum_n\left(\frac{\partial\rho}{\partial Q_n}\right)_0 Q_n+\tfrac{1}{2}\sum_{n,m}\left(\frac{\partial^2\rho}{\partial Q_n\partial Q_m}\right)_0 Q_nQ_m+\cdots$$

By taking into account that $\sum_i\int\rho(\vec{r},0)\vec{r}_i\,\mathrm{d}V=\text{const}$ and $\int\Psi''_v\Psi'_v\,\mathrm{d}Q=0$ and integrating with respect to the electron and nuclear co-ordinates, we find that

$$\vec{\mu}^{\prime,\prime\prime}=\int\Psi''_v\Psi'_v\left(\sum_\alpha Z_\alpha\vec{R}_\alpha\right)\mathrm{d}Q-\sum_n\int\left(\frac{\partial\rho}{\partial Q_n}\right)_0\left(\sum_i\vec{r}_i\right)\Psi''_vQ_n\Psi'_v\,\mathrm{d}V\,\mathrm{d}Q$$
$$-\tfrac{1}{2}\sum_{n,m}\int\left(\frac{\partial^2\rho}{\partial Q_n\partial Q_m}\right)_0\left(\sum_i\vec{r}_i\right)\Psi''_v\Psi'_vQ_nQ_m\,\mathrm{d}V\,\mathrm{d}Q+\cdots$$

Expanding the nuclear dipole moment operator into a series in vibrational co-ordinates using the formula

$$\sum_\alpha Z_\alpha\vec{R}_\alpha=\sum_\alpha Z_\alpha\vec{R}_\alpha^{(0)}+\sum_{n,\alpha}Z_\alpha\left(\frac{\partial\vec{R}_\alpha}{\partial Q_n}\right)_0 Q_n$$
$$+\tfrac{1}{2}\sum_{n,m,\alpha}Z_\alpha\left(\frac{\partial^2\vec{R}_\alpha}{\partial Q_n\partial Q_m}\right)_0 Q_nQ_m+\cdots \tag{11.4}$$

we obtain the following from the $\vec{\mu}^{\prime,\prime\prime}$ matrix element:

$$\vec{\mu}^{\prime,\prime\prime}=\sum_{\alpha,n}Z_\alpha\left(\frac{\partial\vec{R}_\alpha}{\partial Q_n}\right)_0\left(\int\Psi''_vQ_n\Psi'_v\,\mathrm{d}Q\right)$$
$$+\tfrac{1}{2}\sum_{\alpha,n,m}Z_\alpha\left(\frac{\partial^2\vec{R}_\alpha}{\partial Q_n\partial Q_m}\right)_0\left(\int\Psi''_vQ_nQ_m\Psi'_v\,\mathrm{d}Q\right)$$
$$-\sum_{n,i}\left(\int\left(\frac{\partial\rho}{\partial Q_n}\right)_0\vec{r}_i\,\mathrm{d}V\right)\left(\int\Psi''_vQ_n\Psi'_v\right)\mathrm{d}Q$$
$$-\tfrac{1}{2}\sum_{i,n,m}\left(\int\left(\frac{\partial^2\rho}{\partial Q_n\partial Q_m}\right)_0\vec{r}_i\,\mathrm{d}V\right)\left(\int\Psi''_vQ_nQ_m\Psi'_v\,\mathrm{d}Q\right) \tag{11.5}$$

We took into account here that, since the $(\partial\rho/\partial Q_n)_0$ and $(\partial^2\rho/\partial Q_n\partial Q_m)_0$ derivatives were taken for equilibrium values of the molecule's geometrical parameters, they will depend only on the electronic co-ordinates. As a result, electronic and nuclear co-ordinates can be separated.

Noting further that $\Psi_v=\prod_n\psi_n(Q_n)$ in a harmonic approximation, the following relations hold:

$$\int\Psi''_vQ_n\Psi'_v\mathrm{d}Q=\int\psi''_n(Q_n)Q_n\psi'_n(Q_n)\,\mathrm{d}Q_n\prod_{m\neq n}\psi''_m(Q_m)\psi'_m(Q_m)\,\mathrm{d}Q_m \tag{11.6}$$

$$\int\Psi''_vQ_nQ_m\Psi'_v\,\mathrm{d}Q=\int\psi''_n(Q_n)Q_n\psi'_n(Q_n)\,\mathrm{d}Q$$
$$\times\int\psi''_m(Q_m)Q_m\psi'_m(Q_m)\,\mathrm{d}Q\prod_{\substack{k\neq n\neq m\\ k\neq m}}\psi''_k(Q_n)\psi'_n(Q_k)\,\mathrm{d}Q_k \tag{11.7}$$

$$\int \Psi_v'' Q_n^2 \Psi_v' \, dQ = \int \psi_n''(Q_n) Q_n^2 \psi_n'(Q_n) \, dQ_n \prod_{m \neq n} \psi_m''(Q_m) \psi_m'(Q_m) \, dQ_m \qquad (11.8)$$

In view of the fact that the vibrational eigenfunctions are orthogonal to the same normal co-ordinate for different states, then integral (11.6) is non-zero only when $\psi_m''(Q_m) = \psi_m'(Q_m)$ for all $m \neq n$, i.e. the state of the mth oscillator does not change. For integral (11.7) to be non-zero, $\psi_k''(Q_k) = \psi_k'(Q_k)$ must be true for all $k \neq n$, m, i.e. the states of the normal oscillators with indices other than n and m do not change. Again, integral (11.8) is non-zero if $\psi_m''(Q_m) = \psi_m'(Q_m)$ for $m \neq n$.

When there are transitions with a change in one of the quantum numbers v_n (i.e. a vibrational quantum number) from 0 to 1 or 2 on the basis of the matrix elements for the co-ordinate of a harmonic oscillator, which are equal to

$$\int \psi_n'' Q_n \psi_n' \, dQ_n = \begin{cases} 0, & v_n'' \neq v_n' \pm 1 \\ (v_n' + 1)^{1/2} Q_{on}, & v_n'' = v_n' + 1 \\ (v_n')^{1/2} Q_{on}, & v_n'' = v_n' - 1 \end{cases}$$

$$\int \psi_n'' Q_n^2 \psi_n' \, dQ_n = \begin{cases} 0, & v_n'' \neq v_n' \pm 2 \\ v_n'^{1/2} (v_n' + 1)^{1/2} Q_{on}^2, & v_n'' = v_n' - 2 \\ (v_n' + 1)^{1/2} (v_n' + 2)^{1/2} Q_{on}^2, & v_n'' = v_n' + 2 \end{cases}$$

$$Q_{on} = \text{const} \frac{1}{\sqrt{v_n}}$$

we obtain the following expression for the transitions:

$$\vec{\mu}^{0,1} = \sum_\alpha Z_\alpha \left(\frac{\partial \vec{R}_\alpha}{\partial Q_n} \right)_0 Q_{on} - \left[\frac{\partial}{\partial Q_n} \left(\int \sum_i \rho \vec{r}_i \, dV \right) \right]_0 Q_{on} \qquad (11.9)$$

These transitions correspond to the fundamental and, as a rule, the most intense absorption bands. The intensity of the bands is thus determined by the first derivatives of $\sum_\alpha Z_\alpha \vec{R}_\alpha$ and $\sum_i \rho \vec{r}_i \, dV$ with respect to normal co-ordinates.

We obtain the following for transitions with a change from 0 to 2 in one quantum number v_n:

$$\vec{\mu}^{0,2} = \sum_\alpha Z_\alpha \left(\frac{\partial^2 \vec{R}}{\partial Q_n^2} \right)_0 Q_{on}^2 - \left[\frac{\partial}{\partial Q_n} \left(\int \sum_i \rho \vec{r}_i \, dV \right) \right]_0 Q_{on}^2 \qquad (11.10)$$

These transitions correspond to the first harmonics. It is clear that these transitions are possible (i.e. the matrix elements are non-zero) only if the second derivatives of $\sum_\alpha Z_\alpha \vec{R}_\alpha$ and $\int \rho \vec{r}_i \, dV$ (which are, respectively, the nuclear dipole moment and the average electronic dipole moment for a given electronic state) are not zero. If these second derivatives and the higher-order ones are non-zero, then we can speak of the appearance of electro-optical anharmonics.

For transitions from the ground state to a level in which two quantum numbers v_n *and* v_m equal unity (these transitions combination correspond to combination frequencies), we obtain the following:

$$\vec{\mu}^{0,01} = \left[\frac{\partial^2}{\partial Q_n \partial Q_n} \left(\sum_\alpha Z_\alpha \vec{R}_\alpha - \sum_i \int \rho \vec{r}_i \, \mathrm{d}V \right) \right] Q_{on} Q_{om} \tag{11.11}$$

It is not too difficult to obtain the appropriate expressions for the matrix elements corresponding to the transitions between any arbitrary initial and final vibrational quantum numbers. It is clear from expressions (11.9)–(11.11) that being able to calculate the appropriate matrix element (and consequently, the transition intensity) depends on whether it is possible to calculate the corresponding derivatives of the average electronic dipole moment of the molecule (calculating the derivatives of the nuclear components is not difficult), or on whether in fact the derivative of the electronic density with respect to normal co-ordinates can be calculated.

Before continuing with our presentation we note that in a single determinant approximation the following equation is needed:

$$\int \rho \left(\sum_i \vec{r}_i \right) \mathrm{d}V = \sum_i \int \rho_i \vec{r}_i \, \mathrm{d}V$$

where the summation is over all occupied electronic states. By transforming it to the LCAO form we obtain

$$\sum_i \int \rho_i \vec{r}_i \, \mathrm{d}V = \sum_{\alpha,\beta,i} C_\alpha^{(i)} C_\beta^{(i)} \sigma_i \int \chi_\alpha \chi_\beta \vec{r} \, \mathrm{d}V = Sp(PM) \tag{11.12}$$

where P is the density matrix and M is the square-symmetric matrix whose elements are $\int \chi_\alpha \chi_\beta \vec{r} \, \mathrm{d}V$ integrals, which can be called the partial matrix elements of the molecule's dipole moment.

By using equation (11.12) and assuming that the atomic orbitals are unchanged, it is possible to rewrite equations (11.9)–(11.11) into the form:

$$\vec{\mu}^{0,1} = \tilde{Z} \left(\frac{\partial \vec{R}}{\partial Q_n} \right)_0 Q_{on} - Sp \left\{ \left(\frac{\partial P}{\partial Q_n} \right)_0 M \right\} Q_{on} \tag{11.13}$$

$$\vec{\mu}^{0,2} = \tilde{Z} \left(\frac{\partial^2 \vec{R}}{\partial Q_n^2} \right)_0 Q_{on}^2 - Sp \left\{ \left(\frac{\partial^2 P}{\partial Q_n^2} \right)_0 M \right\} Q_{on}^2 \tag{11.14}$$

$$\vec{\mu}^{01,01} = \tilde{Z} \left(\frac{\partial^2 \vec{R}}{\partial Q_n \partial Q_m} \right)_0 Q_{no} Q_{mo} - Sp \left\{ \left(\frac{\partial^2 P}{\partial Q_n \partial Q_m} \right) M \right\} Q_{on} Q_{om} \tag{11.15}$$

Here $\tilde{Z}$ is the row matrix containing the charges on the atoms and $(\partial \vec{R}/\partial Q_n)_0$ is the column matrix containing $(\partial \vec{R}_\alpha/\partial Q_n)_0$ derivatives. The method of calculating $(\partial R_\alpha/\partial Q_n)$ was given in Section 8.8. For small vibrations the $(\partial^2 R/\partial Q_n^2)_0$ and $(\partial^2 \vec{R}/\partial Q_n \partial Q_m)_0$ derivatives vanish. The intensity of the bands and the combination frequencies are then only dependent on the electronic terms.

It is clear that the derivative in an appropriate row of the density matrix can be calculated if a method of calculating the derivatives of the same matrix, with respect to the natural vibration co-ordinates, can be developed.

In contrast to normal co-ordinates, natural vibrational ones have the property of localization, and only describe deformations of separately isolated structural elements in the molecule. Since the distribution of electronic density in the molecule is also localized and the distribution in each region of the molecule is basically determined by the local environment, changes should only be expected in the elements of the density matrix for those atoms that are either included in a given structural element or are directly adjacent to it, when there are deformations in separate parts of the molecule. If the electron density is additive for some structural groupings, then the appropriate derivatives of the density matrix with respect to that part of the matrix which corresponds to the atoms of the given grouping must be constant for a series of molecules that contain that structural grouping. This makes it possible for us later to construct a theory of band intensity in infra-red spectra on a semi-empirical basis, including a parameterization selected in some definite way. We are also freed of the necessity of first solving the quantum-mechanical problems of calculating the density matrix and its various derivatives.

To conclude this section, we note that if a problem is solved in the anharmonic approximation, then, as we showed in Chapter 10, the anharmonical vibration eigenfunctions can, in every case for the lower levels, be represented in the form of a linear combination of harmonical vibration eigenfunctions. The expression for a matrix element of a dipole transition between two vibration states thus acquires the form

$$\vec{\mu}^{\prime,\prime\prime} = \sum_{\substack{k,m \\ k \neq m}} c_k^{\prime\prime} c_m^{\prime} \int \Psi_k \vec{\mu} \Psi_m \, \mathrm{d}Q \tag{11.16}$$

where $c_k^{\prime\prime}$ and $c_m^{\prime}$ are coefficients in the expansion of the eigenfunctions of two combining vibrational states in harmonic base functions. We have taken into account here that the diagonal matrix elements, $\int \Psi_k \vec{\mu} \Psi_k \, \mathrm{d}Q$, approximately equal $\vec{\mu}_0$ if the square of the component of the molecule's dipole moment in normal co-ordinates is neglected. In the same way, by expanding $\vec{\mu}$ into a series in normal co-ordinates we find that

$$\vec{\mu} = \vec{\mu}_0 + \sum_n \left(\frac{\partial \vec{\mu}}{\partial Q_n} \right)_0 Q_n + \cdots$$

Substituting this expression into the integral yields $\vec{\mu}_0$ for the first term. The second term disappears because the average values of the normal co-ordinates appear in the vibrational states, and they are zero for harmonic vibrations. As a result, the sum of the diagonal terms in the expression for the matrix element as a whole for anharmonic vibrations takes the form

$$\sum_k c_k^{\prime\prime} c_k^{\prime} \int \Psi_k^2 \vec{\mu} \, \mathrm{d}Q \approx \vec{\mu}_0 \sum_k c_k^{\prime\prime} c_k^{\prime} = 0$$

because the $c_k^{\prime\prime}$ and $c_k^{\prime}$ coefficients form an orthogonal system.

If we designate the matrix element of the appropriate transition for harmonic vibrations by $\vec{\mu}_{km}^{(0)}$, then we obtain

$$\vec{\mu}'{}'' = \sum_{\substack{k,m \\ k \neq m}} c_k'' c_m' \vec{\mu}_{km}^{(0)}$$

In the simplest case of Fermi resonance we can have a situation in which two close levels almost blend together in the harmonic approximation but have different values in the anharmonic one. If the eigenfunctions for the two levels in the harmonic approximation have the form $\Psi_1^{(0)}$ and $\Psi_2^{(0)}$ then in the anharmonic one we obtain

$$\Psi_1 = c_{11}\Psi_1^{(0)} + c_{12}\Psi_2^{(0)}$$
$$\Psi_2 = c_{21}\Psi_1^{(0)} + c_{22}\Psi_2^{(0)}$$

Assuming for simplicity that the original level has the eigenfunction $\Psi_0^{(0)}$, we obtain the following expressions for the matrix elements of the transition to the Fermi doublet:

$$\vec{\mu}^{0,1} = C_{11}\vec{\mu}_{01}^{(0)} + C_{12}\vec{\mu}_{02}^{(0)}$$
$$\vec{\mu}^{0,2} = C_{21}\vec{\mu}_{01}^{(0)} + C_{22}\vec{\mu}_{02}^{(0)}$$

It is possible that $|\vec{\mu}_{02}^{(0)}| \ll |\vec{\mu}_{01}^{(0)}|$, in which case the total probability of a transition from the original level to the doublet level is

$$J \sim (\vec{\mu}^{0,1})^2 + (\vec{\mu}^{0,2})^2 = (C_{11}^2 + C_{12}^2)(\vec{\mu}_{01}^{(0)})^2 \tag{11.17}$$

The C_{ij} coefficients obey the following relation: $C_{11}^2 + C_{12}^2 = 1$. We conclude from this that the combined intensity of the Fermi doublet is proportional to the square of the matrix element of the more intense transition.

When estimating the anharmonicity it is necessary to redistribute the intensity between the components of the doublet, but the total combined intensity is unchanged. This important property allows us to adapt the theory of intensity to the approximations of harmonic vibrations in many cases and also for the anharmonic variant.

11.2 The Valence-optical Scheme

We established in Section 11.1 that calculating the probability of a transition between the energy levels of a polyatomic molecule—which determines the intensity of the corresponding absorption bands—reduces to finding various kinds of derivatives of the molecule's dipole moment with respect to the normal co-ordinates that characterize the vibrational motion of the system. We also noted that more intense absorption bands are observed for transitions with a change of one of the vibration quantum numbers to unity (fundamental vibrations). The corresponding probability of a transition is related to the first derivative of the dipole moment with respect to a normal co-ordinate. That is why we started our presentation of the theory of intensity in infra-red spectra of polyatomic molecules with the determination of these derivatives. In principle, the derivative can be calculated by the methods presented in Section 11.1 and is wholly based on the adiabatic approximation and LCAO–MO theory supplemented by the superposition of configurations. However, the accuracy of these calculations is not satisfactory at present and the calculations are cumbersome. This is because the changes of a molecule's dipole moment during vibrations are minute (of the order of 0.01 D), which require precise calculations to estimate them. We repeat, though the calculations are, in principle, possible, they are not as yet widespread.

Thus the only way of calculating the intensity in the spectra of polyatomic molecules that is at present realistic is to construct a semi-empirical theory. We noted earlier that *ab initio* and semi-empirical estimates of a polyatomic molecule's spectrum are not mutually exclusive but are complementary. When investigating small molecules, straightforward quantum-mechanical estimates would apparently be preferable. However, it is more advantageous when dealing with complicated molecules to use the property of the conservation of the characteristics of the electron cloud which can be transformed to construct a theory on a semi-empirical basis by the introduction of several parameters. Such an approach, as experience has demonstrated, can make the formidable calculations of vibrational spectra possible, with an acceptable accuracy. This approach has the undoubted advantage of speed of computation, and can operate with large-scale systems up to and including polymers and crystals.

We will now dwell on a few of the principle problems that are related to the construction of a semi-empirical theory of intensity in the infra-red spectra of polyatomic molecules.

Above all, we note that the first derivative of the dipole moment with respect to a normal co-ordinate may be found from experimental data on the intensity (and polarization) of the fundamental absorption bands in an infra-red spectrum. However, the $\vec{\mu}$ and Q_i characterize the molecule's state as a whole. In order to see how a molecule's internal structure and its intramolecular processes can be judged from the intensities of the molecule's vibrations it is necessary to return to a description of the behaviour of the separate parts of the molecule. First, it is clear that a normal co-ordinate may be replaced by those that characterize the position of the molecule's atoms or structural elements for a given vibration. These co-ordinates could be the natural vibrational ones, which were the original choice when the problem concerning a polyatomic molecule's vibrations was first posed. These co-ordinates are related to normal ones: thus $q = L_q Q$.

By using the rule for differentiating complex functions, we obtain

$$\left(\frac{\partial \vec{\mu}}{\partial Q_i}\right)_0 = \left(\frac{\partial \vec{\mu}}{\partial q_1}\right)_0 l_{q1i} + \left(\frac{\partial \vec{\mu}}{\partial q_2}\right)_0 l_{q2i} + \cdots + \left(\frac{\partial \vec{\mu}}{\partial q_n}\right)_0 l_{qni}$$

where $(l_q)_{ki}$ $(k = 1, 2, \ldots, n)$ are the elements of the ith column of the matrix L_q. We have already indicated that we call the set of these elements the form of the vibration, and it is written as a column matrix $l_q^{(i)}$.

In matrix form the above equation is

$$\left(\frac{\partial \vec{\mu}}{\partial Q_i}\right)_0 = \left(\widetilde{\frac{\partial \vec{\mu}}{\partial q}}\right)_0 l_q^{(i)} \tag{11.18}$$

The $(\partial\vec{\mu}/\partial q_k)_0$ elements can be taken to be parameters that characterize the intensities of the vibrational bands in the molecule's infra-red spectrum. Clearly, they will have the same character for all normal vibrations. The expressions for the $(\partial\vec{\mu}/\partial Q_i)_0$ derivative differ from each other solely by the vibrational form $l_q^{(i)}$. A knowledge of the $(\partial\vec{\mu}/\partial q_k)_0$ parameter enables us to judge how the dipole moment changes with alterations in the equilibrium configuration of separate parts of the molecule, i.e. it characterizes to some extent the components of the molecule's structure. However, these derivatives have a whole series of deficiencies. They are dependent on the dipole moment of the molecule as a whole and so cannot be considered as being characteristic of separate structural elements, and change on going from one molecule to another, even if both molecules have the same structural groups. In the general case of arbitrary vibrations $(\partial\vec{\mu}/\partial q_k)_0$ must change even for a series of isotropically substituted compounds.

All these considerations require a molecular dipole moment to be expressed in such a form that it is related to separate elements of the system . In order to do this it is necessary that this form has the following properties. Since a molecular dipole moment must be characterized by several parameters that are evaluated from the experimentally measured values of $(\partial\vec{\mu}/\partial Q_i)_0$, it is desirable that a system of equations, linear with respect to the parameters being introduced, be obtained when the molecular dipole moment is substituted for in a formula such as equation (11.18) for all the molecule's vibrations. Further, the parameters introduced must not depend on isotopic substitution, i.e. the influence of the molecule's mechanical properties is completely excluded. In this respect, it will be possible to take advantage of data for isotopically substituted molecules, which occurs, for example, when solving mechanical problems concerning the frequencies and forms of vibrations in polyatomic molecules as supplementary equations in order to calculate the parameters. Only in this way will the parameters being used be wholly determined by the properties of the molecule's electronic density.

It is also necessary that the form selected for representing the molecular dipole moment should correspond to the basic ideas of structural chemistry and have a fairly universal character, whilst the parameters being used should be close in value for similar molecules. All this is only possible when the parameters are very closely related to the properties of the appropriate elements.

This semi-empirical approach, which has become widespread for the understanding of polyatomic molecular spectra, does not in any way run counter to the direct quantum estimation of $(\partial\vec{\mu}/\partial Q_k)_0$-type derivatives. Rather it is a natural extension and development of the theory as applied to large systems. Consequently, the problem is to select the requisite parameters as simply and naturally as possible.

There are now several semi-empirical theories for the band intensities in polyatomic molecules' infra-red spectra, and these are presented in references 1 and 2, where they can be compared. The most thoroughly developed and practically tested theory is the valency-optical one, and it is reasonably presented in references 3–6. There is now no doubt that the electro-optical parameters (eop) required by the theory can be found for individual bonds and atomic groupings, and that the parameters possess the important property of transferability. The valency-optical theory has been extended to polymers and crystals[7] and used to predict estimates for band intensities in these systems[8]. The theory can also be used to analyse band intensities for internal rotation[9].

Let us consider the determination of the electro-optical parameters using consistent quantum estimations, and discuss the overall problem of rationally constructing a semi-empirical theory for infra-red band intensities.

11.2.1 *Expression for the dipole moment*

It is known that a matrix element of an optical dipole transition in the bounds by one electron state can be given as

$$\mu^{\prime,\prime\prime}=\sum_{\alpha,n}\left[Z_\alpha\left(\frac{\partial\vec{R}_\alpha}{\partial Q_n}\right)_0-\left(\frac{\partial\vec{\mu}_e}{\partial Q_n}\right)_0\right]\langle v'|Q_n|v'\rangle$$
$$+\tfrac{1}{2}\sum_{\alpha,n,m}\left[Z_\alpha\left(\frac{\partial^2\vec{R}_\alpha}{\partial Q_n\partial Q_m}\right)_0-\left(\frac{\partial^2\vec{\mu}_e}{\partial Q_n\partial Q_m}\right)_0\right]\langle v''|Q_nQ_m|v'\rangle+\cdots \tag{11.19}$$

Here Z_α is the nuclear charge on atom α, $\vec{R}_\alpha$ is the atom's radius vector in the external co-ordinate system, Q_n is a normal co-ordinate, $\vec{\mu}_e=\sum_i\int\rho\vec{r}_i\,\mathrm{d}V$ is the electronic constituent vector of the dipole moment, ρ is the electronic density and $\vec{r}_i$ is a radius vector of an electron. The summation is done over all electrons and the integration over the entire molecular volume.

We shall rewrite the expression for $\vec{\mu}_e$ in more detail by assuming that the electronic function can be represented as a linear combination of Slater determinants, constructed from molecular orbitals that are themselves in the form of a linear combination of atomic orbitals.

The coefficients for determinants will be designated as $a_k(a_l)$; those for the atomic orbitals, $\varphi_\alpha(\varphi_\beta)$, in the molecular orbitals of linearly combined atomic orbitals will be designated $c_\alpha(c_\beta)$.

We will mark the molecular orbitals by the subscripts f and g. By taking into account that $\rho=\Psi_e^2$ and substituting the expression for ρ into the formula for $\vec{\mu}_e$, we obtain a very general case:

$$\vec{\mu}_e=\sum_{k,f,\alpha}a_k^2c_{\alpha kf}^2\int\varphi_\alpha^2\vec{r}\,\mathrm{d}V+\sum_{\substack{k,f,\alpha,\beta\\ \alpha\neq\beta}}a_k^2c_{\alpha kf}c_{\beta kf}\int\varphi_\alpha\varphi_\beta\vec{r}\,\mathrm{d}V$$
$$+\sum_{\substack{k,f,\alpha,lg\\ k\neq l\\ f\neq g}}a_ka_lc_{\alpha kf}c_{\alpha lg}\int\varphi_\alpha^2\vec{r}\,\mathrm{d}V+\sum_{\substack{k,l,\alpha,\beta,f,g\\ k\neq l\\ f\neq g\\ \alpha\neq\beta}}a_ka_lc_{\alpha kf}c_{\beta lg}\int\varphi_\alpha\varphi_\beta\vec{r}\,\mathrm{d}V \tag{11.20}$$

The summation in equation (11.20) is over all the indices of the determinants, all atomic orbitals and all occupied orbitals in each determinant (allowing for the occupation numbers), and in the last two terms over every combination of determinant that is distinct from another by not more than one molecular orbital.

We introduce the notation

$$q_\alpha = \sum_{k,f} a_k^2 c_{\alpha k f}^2 + \sum_{\substack{k,l,f,g \\ k \neq l \\ f \neq g}} a_k a_l c_{\alpha k f} c_{\alpha l g} \tag{11.21}$$

$$P_{\substack{\alpha\beta \\ \alpha \neq \beta}} = \sum_{k,f} a_k^2 c_{\alpha k f} c_{\beta k f} + \sum_{\substack{k,l,f,g \\ k \neq l \\ f \neq g}} a_k a_l c_{\alpha k f} c_{\beta l g} \tag{11.22}$$

and so rewrite (11.20) as

$$\vec{\mu}_e = \sum_\alpha q_\alpha \int \varphi_\alpha^2 \vec{r}\, \mathrm{d}V + \sum_{\alpha \neq \beta} P_{\alpha\beta} \int \varphi_\alpha \varphi_\beta \vec{r}\, \mathrm{d}V \tag{11.23}$$

It should be remembered that a summation over α (β) must include all atomic orbitals, and that there may be several of them at one centre. From now on we shall, without limiting the generality of the discussion, consider first, that all atomic orbitals are centred on the atomic nucleus, second, that they are such that their squares are symmetric with respect to an inversion centre that coincides with the centre of the atomic nucleus (this condition is fulfilled by all the usually required atomic orbitals), and third, that they are normalized. Given these conditions, we obtain the following for integrals containing squares of an atomic orbital:

$$\int \varphi_\alpha^2 \vec{r}\, \mathrm{d}V = e\vec{R}_\alpha \tag{11.24}$$

where e is the unit charge and $\vec{R}_\alpha$ is the nucleus's radius vector. For integrals containing the products of atomic orbitals for different atoms, we obtain

$$\int \varphi_\alpha \varphi_\beta \vec{r}\, \mathrm{d}V = S_{\alpha\beta} \vec{R}_{\alpha\beta} \tag{11.25}$$

where $S_{\alpha\beta}$ (the 'bond charge') is numerically equal to the overlap integrals of the atomic orbitals φ_α and φ_β and $\vec{R}_{\alpha\beta}$ is the radius vector of the 'centre of gravity' of the charge distribution defined by the product $\varphi_\alpha\varphi_\beta$. We note here that, while it is uncommon, it may be found that $S_{\alpha\beta} = 0$. In this case, the expression given above is no longer correct, and we must simply calculate the dipole moment's transition vector for the φ_α and φ_β atomic orbitals.

Some integrals with the product ($\varphi_\alpha\varphi_\beta$) are encountered for which the atomic orbitals belong to the same centre. These represent matrix elements of dipole transitons between the atomic orbitals of one atom, and can also be calculated in an explicit form.

In order to make a distinction, in this case we use the subscripts α and α', and so emphasize that the atomic orbitals belong to the same centre. We can now write out the full expression for a molecule's dipole moment, i.e.

$$\begin{aligned}\vec{\mu} = \vec{\mu}_n - \vec{\mu}_e &= \sum_{\alpha} Z_\alpha \vec{R}_\alpha - \sum_{\alpha} e\left(\sum_{\alpha'} q_{\alpha'}\right)\vec{R}_\alpha \\ &\quad - \sum_{\substack{\alpha\beta \\ \alpha\neq\beta}} P_{\alpha\beta} S_{\alpha\beta} \vec{R}_{\alpha\beta} - \sum P_{\alpha\alpha'} \vec{\mu}'_{\alpha\alpha'} - \sum P'_{\alpha\beta} \vec{\mu}'_{\alpha\beta} \\ &= \sum_{\alpha} (Z_\alpha - Z_{e\alpha})\vec{R}_\alpha - \sum_{\alpha\neq\beta} P_{\alpha\beta} S_{\alpha\beta} \vec{R}_{\alpha\beta} - \sum P_{\alpha\alpha'} \vec{\mu}'_{\alpha\alpha'} - \sum P'_{\alpha\beta} \vec{\mu}'_{\alpha\beta}\end{aligned} \tag{11.26}$$

Here $Z_{e\alpha}$ is the complete electronic charge on the atom. According to (11.9), the $(\partial \vec{\mu}/\partial Q_i)_0$ and other derivatives can be used to calculate the intensities of the corresponding transitions between vibrational energy levels. It is worth noting that the first two terms in equation (11.26) are purely classical, and express the system's dipole moment as a sum of point charges distributed around the molecule multiplied to their external co-ordinate radius vectors. The third and fourth terms do not have a classical analogue and appear only in a general quantum treatment. We can, without any particular loss of accuracy, also assume from now on that all two-centre integrals, i.e. $\int \varphi_\alpha \varphi_\beta \vec{r}\,\mathrm{d}V$, and their corresponding terms in equation (11.26) are zero if the α and β atoms do not lie on the same bond. We will also consider that when calculating derivatives we use natural internal vibration co-ordinates (or co-ordinates linearly related to natural ones) and not Cartesian ones, which are considerably less useful for studying molecular vibrations. Since all the quantities in equation (11.26) are calculated strictly, we can find the requisite derivatives either by differentiating analytically or by calculating them numerically from data about $\vec{\mu}$ for a sufficient number of points, corresponding to different deformations of the molecule around its equilibrium position.

11.2.2 Constructing a semi-empirical theory

A semi-empirical theory can be constructed on the basis of equation (11.26)[10–15], and we shall initially limit ourselves to the purely classical part of that equation. Here, a molecule's dipole moment is a sum of point charges multiplied by their radius vectors. The charges are distributed within the molecule's space and associated with its individual structural elements. Since it is known that the charge distribution of a given group is mainly determined by its constituent atoms and is only slightly changed, even by placing a polar substitute outside it, it is natural to associate any parameters introduced into a semi-empirical theory with the different charge terms included in equation (11.26).

There are several possibilities. The first is that by neglecting the moments associated with the $\vec{\mu}'_{\alpha\alpha'}$ and $\vec{\mu}'_{\alpha\beta}$ transitions and neglecting the second term in equation (11.26), $\vec{\mu}$ can be given as

$$\vec{\mu} = \sum_{\alpha} Z'_\alpha \vec{R}_\alpha \tag{11.27}$$

where Z'_α is a theoretical parameter. This leads to simple formulae for $(\partial \vec{\mu}/\partial Q_n)_0$ and other derivatives. However, this approach suffers from two defects. First, a quantum calculation of Z'_α and its derivatives with respect to natural co-ordinates and their substitution into the expression for $(\partial \vec{\mu}/\partial Q)_0$ will certainly not lead to a result equivalent to the direct quantum calculation of the derivative. Second, since it is the internuclear charge distribution itself that is most mobile when the molecule is vibrating, the losses of the corresponding terms mean that the parametric theory thus obtained does not give any opportunity for achieving the accuracy required during the calculation.

It is possible to try to improve the parametric theory—for example, by adding bond charges as additional parameters, i.e. $P'_{\alpha\beta} = \sum P^{(k)}_{\alpha\beta} S^{(k)}_{\alpha\beta}$ (the summation being over all pairs of atomic orbitals centered on a given bond). This would undoubtedly significantly improve the theory but it greatly increases the number of parameters, and this is extremely undesirable for a semi-empirical approach. It is also inconvenient when drawing general conclusions and constructing an intensity theory for polymers, etc. to use radius vectors attached to an external co-ordinate system. However, this deficiency can be partially avoided if we take advantage of the fact that the end of the $\vec{R}_{\alpha\beta}$ radius vectors (the α and β referring to different atoms) will lie on the straight line connecting the α and β atoms or lie close to it. This makes it possible to represent $\vec{R}_{\alpha\beta}$ as $\vec{R}_{\alpha\beta} = \vec{R}_{\alpha\beta} + (\vec{R}_\alpha - \vec{R}_\beta)(l_{\alpha\beta}/J_{\alpha\beta})$ where $l_{\alpha\beta}$ is the distance of the end of the $\vec{R}_{\alpha\beta}$ vector from the β atom (the atom being taken as the origin of the measurement) and $J_{\alpha\beta}$ is the bond length. Substituting this expression into equation (11.26) enables us to eliminate the $\vec{R}_{\alpha\beta}$ vectors and include the 'bond charges' as part of the empirical parameter Z'_α. Unfortunately, we are unable to do this with the third and fourth terms, which contain transition moments. Hence it is apparently impossible to transform equation (11.26) into one such as (11.27), i.e. to obtain in the final analysis an accurate quantum formula for the Z'_α parameter.

We now return to the conventional representation of the molecule's dipole moment $\vec{\mu}$ in valency-optical theory, i.e.

$$\vec{\mu} = \sum_{\alpha,\beta} \mu_{\alpha\beta} \vec{e}_{\alpha\beta}$$

where $\mu_{\alpha\beta}$ is a theoretical parameter and $e_{\alpha\beta}$ is the orientating vector of the bond. The summation is taken over all chemical bonds. Even in reference 3 attention is drawn to the fact that the physical significance of the $\mu_{\alpha\beta}$ parameters can be quite complex, and that in the general case they may not be considered to be characteristic of individual bonds. All the same, the experience gained so far in obtaining estimates has shown that they are clearly related to individual atomic groupings, which explains their transferability. This ability only occurs when the parameters are functions of the charge distribution within separate parts of the molecule. In order to understand the physical significance of the $\mu_{\alpha\beta}$ parameters and at the same time to indicate a way of a consistent quantum calculation, we proceed as follows.

Initially, we limit ourselves to the terms of equation (11.26) that correspond to the classical expression of the dipole moment as a sum of point charge moments. We choose one the molecule's atoms as the origin for the co-ordinate system. It is more convenient to select one of the central atoms. In the presence of symmetry it is useful to select the atom so that it lies on the main symmetry element. Given a centre of inversion, we could also put the origin at the centre of the bond (if one exists) which contains the centre. If the molecule is neutral then the magnitude of the dipole moment will be independent of the position of the origin. Furthermore, we introduce a set of orientating bond vectors, aligning them to take due account of the molecule's symmetry. In order to comply fully with symmetry requirements when there is a single bond containing a centre of inversion, we can introduce two oppositely directed vectors, one along the bond and the other away from the centre, both originating at the midpoint of the bond.

The conditions indicated, however, are not obligatory, and only ensure that the discussion will be easier, since the final result does not depend on the choice of co-ordinate origin. Further, we note that the disposition of any point lying on one of the molecule's bonds can be represented as a linear combination of orientating bonds with coefficients (positive or negaitive, depending on the direction of the bond unit vector) which are equal to the bond length or to the fraction the point lies along it if that point is between the two atoms of the bond. We illustrate this procedure for an ethylene molecule (for two different origins), assuming that the end of the vector $\vec{R}_{\alpha\beta}$ is placed on the bond at some distance $l_{\alpha\beta}$ from the start of the bond. We then have (Figure 11.1)

$$\vec{R}_1 = J_{\mathrm{CH}}\vec{e}_{51}$$
$$\vec{R}_2 = J_{\mathrm{CH}}\vec{e}_{52}$$
$$\vec{R}_3 = J_{\mathrm{CC}}\vec{e}_{56} + J_{\mathrm{CH}}\vec{e}_{63}$$
$$\vec{R}_4 = J_{\mathrm{CC}}\vec{e}_{56} + J_{\mathrm{CH}}\vec{e}_{64}$$
$$\vec{R}_5 = 0$$
$$\vec{R}_6 = J_{\mathrm{CC}}\vec{e}_{56}$$
$$\vec{R}_{51} = l_{\mathrm{CH}}\vec{e}_{51} \qquad \vec{R}_{63} = J_{\mathrm{CC}}\vec{e}_{56} + l_{\mathrm{CH}}\vec{e}_{63}$$
$$\vec{R}_{52} = l_{\mathrm{CH}}\vec{e}_{52} \qquad \vec{R}_{64} = J_{\mathrm{CC}}\vec{e}_{56} + l_{\mathrm{CH}}\vec{e}_{64}$$
$$\vec{R}_{56} = \tfrac{1}{2}J_{\mathrm{CC}}\vec{e}_{56}$$

(Symmetry considerations yield $J_{51} = J_{52} = J_{63} = J_{64} = J_{\mathrm{CH}}$.)

By substituting these equations into equation (11.26), reducing similar terms with bond-orientating vectors and taking into account the symmetry in charge distribution, we obtain the following expressions for the coefficients of the orientating vectors (i.e. the $\mu_{\alpha\beta}$ parameters):

$$\mu_{51} = (Z_{e\mathrm{H}} - Z_{\mathrm{H}})J_{\mathrm{CH}} + P_{\mathrm{CH}}S_{\mathrm{CH}}l_{\mathrm{CH}} = \mu_{52} = \mu_{63} = \mu_{64}$$

$$\mu_{56} = J_{\mathrm{CC}}\left[2(Z_{e\mathrm{H}} - Z_{\mathrm{H}}) + (Z_{e\mathrm{C}} - Z_{\mathrm{C}}) + P_{\mathrm{CC}}\frac{S_{\mathrm{CC}}}{2} + 2P_{\mathrm{CH}}S_{\mathrm{CH}}\right] = 0$$

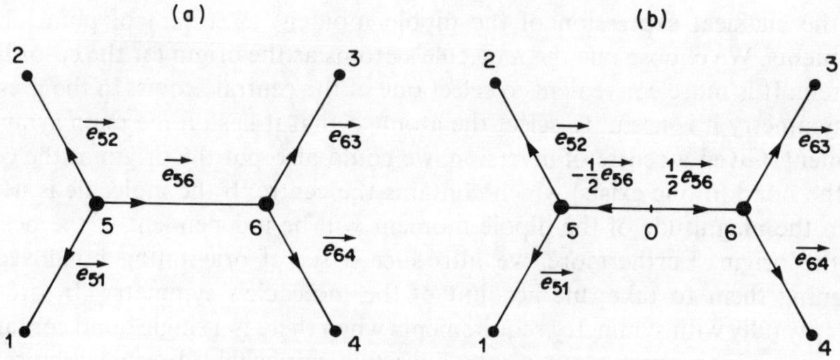

Figure 11.1. In case (a) the origin is placed on atom 5; in case (b) the origin is placed at the centre of symmetry

The last equation is a consequence of the sum of all the charges being zero, and hence the sum of the charges of the right-hand half of the molecule must equal the charges on the left-hand one.

For the second possible origin (case (b) in Figure 11.1) we obtain the expressions

$$\vec{R}_1 = -\tfrac{1}{2}J_{CC}\vec{e}_{56} + J_{CH}\vec{e}_{51}$$
$$\vec{R}_5 = -\tfrac{1}{2}J_{CC}\vec{e}_{56}$$

and so on. Hence we find, as before, that

$$\mu_{51} = \mu_{52} = \mu_{63} = \mu_{64} = (Z_{eH} - Z_H)J_{CH} + P_{CH}S_{CH}l_{CH}$$
$$\mu_{56} = [-\tfrac{1}{2}(Z_{eC} - Z_C) - 2(Z_{eH} - Z_H) - 2P_{CH}S_{CH}l_{CH} + \tfrac{1}{2}(Z_{eC} - Z_C) + 2(Z_{eH} - Z_H) + 2P_{CH}S_{CH}l_{CH}]J_{CC} = 0$$

By carrying out similar manipulations for an arbitrary molecule, taking into account that the orientating vectors of end bonds will be repeated in the expressions for $\vec{R}_\alpha$ and $\vec{R}_{\alpha\beta}$ as many times as the charges $(Z_{e\alpha} - Z_\alpha)$ and $(P_{\alpha\beta}S_{\alpha\beta})$ are 'spread' from the starting atom to the end atom of these bonds, we find that the $\mu_{\alpha\beta}$ parameters for end bonds must be

$$\mu_{\alpha\beta} = (Z_{e\beta} - Z_\beta)J_{\alpha\beta} + \sum_{\substack{\alpha,\beta \\ \alpha\neq\beta}} P_{\alpha\beta}S_{\alpha\beta}l_{\alpha\beta} = \sum_m \mu_{\alpha\beta m} \tag{11.28}$$

The α and β indices refer to the atoms forming the bond. We assume that the orientating vector $\vec{e}$ goes from the α atom to the β one and that the summation in the second term is taken over all combinations of atomic orbital pairs that encompass α and β.

The physical sense of $\mu_{\alpha\beta}$ for the end bonds is that it is the algebraic sum of the absolute values of the dipole moments $\mu_{\alpha\beta m}$ of all the charges (the number of charges is m) spread among the bonds with respect to their origins. If the orientating vector is not pointing towards the outer atom of the bond but is pointed the other way, then the sign of $\mu_{\alpha\beta}$ must be reversed. By differentiating with respect to natural co-ordinates we obtain the appropriate derivatives of these charges. Equation (11.28), which is very simple and clear, can also be used to calculate $\mu_{\alpha\beta}$ (and all its requisite derivatives) via a direct quantum-chemical approach.

We can conclude from the formula that if a bond's charge distribution and the variation of the distribution due to a change in adjacent natural co-ordinates depend only weakly on the bond's environment, then both the parameter $\mu_{\alpha\beta}$ and its derivatives will be unchanged over a series of molecules, i.e. they will be transferable. Clearly (and this is both important and convenient for constructing a semi-empirical theory), $\mu_{\alpha\beta}$ parameters naturally include all the charge distributed in the space of a given bond. Thus they are much more powerful than the Z'_α parameters we considered above.

We return now to internal bonds. In order to obtain rules for constructing the $\mu_{\alpha\beta}$ parameters for such bonds we should consider how the position of the charges with the vectors $\vec{R}$ and $\vec{R}_{\alpha\beta}$ are calculated for the general case. Having selected one of the molecule's atoms (or the centre of inversion) as the co-ordinate origin, we must then choose a series of paths along the bonds such that every atom in the molecule is reached, each new path starting from the origin. Given these movements along these peculiar trees (graphs), the terminal bonds will only be travelled along once and the internal bonds will be travelled along several times in various trees (graphs). We shall consider the propylene molecule by way of example (see Figure 11.2).

We establish the origin at the vertex of the angle formed by bonds 1 and 2. Then we must construct linear combinations of the following orientating bond vectors (series of movements, i.e. trees or graphs) in order to give the dispositions of the atoms lying at the ends of the 3, 4, 5, 6, 7 and 8 bonds: 3; 3–4; 3–5; 3–5–6; 3–5–7; 3–5–8. In these trees bond vector 3 is encountered six times, bond vector 5 four times and bond vectors 4, 6, 7 and 8 once (they are end bonds). Substituting the corresponding linear combinations (in this case they have a

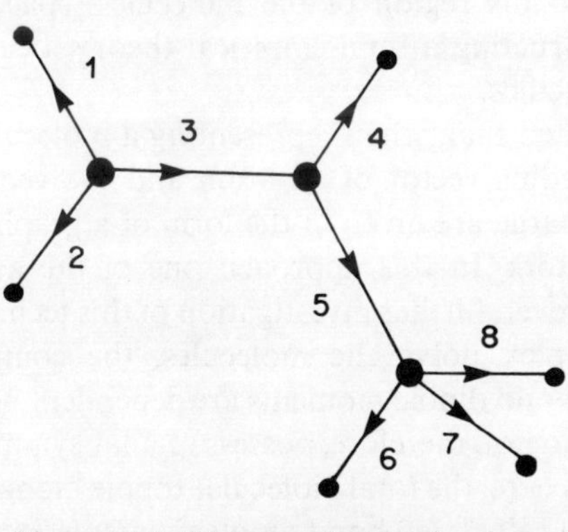

Figure 11.2. The propylene molecule and the bond-orientating vectors described in the text and their numeration

simple form

$$\vec{R}_\alpha = \sum_{\alpha,\beta} J_{\alpha\beta}\vec{e}_{\alpha\beta}$$

where the summation is over all the bond numbers in the graph) into equation (11.26) after reducing similar terms, we find that bond vector 3, which is multiplied by its own bond length and the charge on the graph's end atom, will be repeated as many times as it appears in the various graphs. The same is true for the bond charges (the $\vec{R}_{\alpha\beta}$ vectors). This then gives us a simple rule for compiling $\mu_{\alpha\beta}$ parameters for internal bonds. Accordingly, the parameter must be:

$$\mu_{\alpha\beta} = \sum_k P^{(k)}_{\alpha\beta} S^{(k)}_{\alpha\beta} l^{(k)}_{\alpha\beta} + (Z_{e\beta} - Z_\beta)J_{\alpha\beta} + \sum_\gamma (Z_{e\gamma} - Z_\gamma)J_{\alpha\beta}$$

$$+ \sum_m P^{(m)}_{\gamma\delta} S^{(m)}_{\gamma\delta} J_{\alpha\beta} = \sum_m \mu_{\alpha\beta m} + J_{\alpha\beta} Z_{\text{tot}} \quad (\gamma, \delta \neq \alpha, \beta) \qquad (11.29)$$

Here, $\mu_{\alpha\beta m}$ are the dipole moments of all charges distributed along the bond and calculated with respect to its origin. Z_{tot} is the total charge of that part of a molecule that is located beyond the bond.

The formula for $\mu_{\alpha\beta}$ is once again very simple. At first it seems that the $\mu_{\alpha\beta}$ for internal bonds will not be localized and transferable since they are integral parameters absorbing charge from many bonds, this being especially true in large systems. This, however, is not the case, because the larger a fragment of molecule is, the smaller, as a rule, will be its summed charge and the closer it becomes an electrically neutral system. Thus the role of the third and fourth terms of equation (11.29) must be smaller, the larger the number of atoms and bonds there are for a given bond in the direction of the development of the appropriate graphs. Consequently, the localization and transferability properties which we know the $\mu_{\alpha\beta}$ parameters and their derivatives with respect to natural co-ordinates passes must occur even for internal bonds. It is worth noting even here that the perimeters for both $\mu_{\alpha\beta}$ and internal bonds accumulate a large group of charges in this region of the molecule's space, and thus are very convenient for constructing a semi-empirical theory. Let us consider now the general type of molecules.

It has been suggested that, when representing a molecular dipole moment as $\vec{\mu} = \sum_k \mu_k \vec{e}_k$, each radius vector of an atom and the vectors of the 'centre of gravity' of a bond charge are given in the form of a graph whose edges are the molecular bond vectors. In this approach one of the atoms is taken as the reference point. However, further investigation of this technique has shown that, in the case of complex polycyclic molecules, the computed charge-related contributions to the bond dipole moments are dependent upon the chosen origin of co-ordinates and do not, therefore, possess regular symmetry properties. With the method proposed here, the total molecular dipole moment, obtained through the summation of the calculated bond dipole moments, is invariant with respect to the origin of the co-ordinate system.

This communication gives an outline of a general procedure for computing bond dipole moments within an arbitrary polyatomic molecule, which is free of the above weak points. Consideration is given here only to the charge-related contribution to the total dipole moment, the reason being that it is only this contribution that engenders the problem of going from the representation of the molecular dipole moment as a sum (in the simplest case) of atomic dipole moments to representation as the sum of bond dipole moments.

The first step is to introduce directional vectors for every bond within a molecule, taking into account the absolute symmetry of the system; the bond charges are considered as being positioned on the bonds *per se*, which is a reasonable assumption. At this point it is necessary to introduce the notion of a graph matrix, G, in which the number of rows is equal to the total number of charges within the molecule. Each row in the G matrix corresponds to a given charge (or merely to a given atom, should the bond charges be ignored), comprising non-zero elements which are consistent with the bond numbers included in a graph passing from the reference point to a given atom or bond charge. These non-zero elements are taken as $+1$ when, in passing along the graph, the bond is covered completely in the direction of its unit vector and as -1 when, in passing along the graph, the entire bond is encompassed in the direction opposite to that of the bond unit vector.

When a bond-charge radius vector is included a situation can arise where the bond is incompletely covered in passing along the graph. In this case, a fractional number is entered in the appropriate column of the graph matrix, and is assigned a plus or a minus sign accordingly. The fractional number is equal to the ratio of the bond origin to bond charge distance, to the total bond length. A column matrix of the radius vectors of all the molecular charges can then be represented in the form of the matrix product $\vec{R} = GS\vec{e}$, where S is a bond-length diagonal matrix and $\vec{e}$ is a column matrix of the directional bond vectors.

In the general case we must take into account that it is typical of cyclic and polycyclic molecules to have a variety of graphs, rather than a single one, coming to any one atom or bond charge. Take as an example the naphthalene molecule. If one of the atoms on the middle C—C bond is taken as the reference point, then three graphs arrive at the other atom of the bond, i.e. one along the bond and one graph around each of the rings. In this case, in order to specify a radius vector for the atom a graph must be plotted which is the arithmetic mean of all the graphs coming from the reference point to that atom or bond charge. Given that the matrix G is designed to comprise all the graphs coming from the reference to a given point, then to obtain a matrix comprising only arithmetic mean graphs the initial matrix G is multiplied from the left by the C matrix. It is the action of the latter matrix which results in the summation with the respective weight factors of all the graphs coming to the same centres. The pattern of the C matrix is self-evident.

The row $\tilde{\mu}$ of the bond dipole moments can be expressed as a row matrix $\tilde{q}$ of all the charges, i.e. $\tilde{\mu} = \tilde{q}CGS$. This relationship enables the charge-related contributions to the bond dipole moments to be unambiguously calculated from the distribution of charge within a molecule (the charge distribution may be established by using quantum computations).

It was found, however, that the above formula suffers from several deficiencies. First, in the case of polycyclic molecules individual elements within the row-matrix $\tilde{\mu}$ (bond dipole moments) are no longer invariant with respect to the reference point. Second, the symmetry features of the bond dipole moments for polycyclic molecules are not observed. For example, the dipole moment of the middle C—C bond in the napthalene molecule is found to be non-zero if the reference point is placed at one of the atoms in the bond. The symmetry can be obtained by placing the reference point at the centre of the middle bond in this molecule, but such contrivance is not convenient, and there are cases where the reference point cannot be aligned with any of the points within the bond or with the constituent atom in the molecule to ensure correct symmetry prediction.

To eliminate the above drawbacks (i.e. to obtain bond dipole moments invariant with respect to the reference point and a regular symmetry) it is suggested that the notion of the averaged dipole moments is introduced, and this is given by

$$\tilde{\mu} = \frac{1}{N}\tilde{q}\sum_{n=1}^{N} C_n G_n S \tag{11.30}$$

This relationship implies that the reference point is successively registered with all the atoms in the molecule (the number of atoms being equal to N) to give the summation of the graphs originating from all the reference points. In so doing, the meaning of the dipole moments of the terminal and skeletal bonds within branched molecules remains unchanged. As far as ring-type structures are concerned (for example, in the case of a six-membered cycle where atoms are numbered in a clockwise pattern) the formula becomes

$$\mu_1 = \tfrac{1}{12}[5(Q_2 - Q_1) + 3(Q_3 - Q_6) + (Q_4 - Q_5)]$$

where μ_1 is a dipole moment of the first bond passing from atom 1 to atom 2 and Q_i represents the total combined charges of the ring atoms and atom groups connected to this bond.

For a similar case of a five-membered ring,

$$\mu_1 = \tfrac{1}{5}[2(Q_2 - Q_1) + (Q_3 - Q_5)]$$

It can be seen that the dipole moments of the ring bonds are dependent upon the distribution of charge over the entire ring, but the more important role is played by the charges corresponding to the chosen bond.

Obviously, after the parameters $\mu_{\alpha\beta} = \mu_k$ have been introduced it becomes easy to obtain their derivatives with respect to natural vibrational co-ordinates. Thus

$$\frac{\partial\tilde{\mu}}{\partial\eta} = \frac{1}{N}\frac{\partial\tilde{q}}{\partial\eta}\sum_{n=1}^{N} C_n G_n S + \frac{\tilde{q}}{N}\sum_{n=1}^{N} C_n G_n \frac{\partial S}{\partial\eta} \tag{11.31}$$

Here, η stands for the vibrational co-ordinate that has been selected.

We now return to the last two terms of equation (11.26), which contain the dipole moments of the electronic interatom transitions and the those between the atomic orbitals of the atoms in a given bond if the corresponding $S_{\alpha\beta}$ integral is zero for these pairs of atomic orbitals. These moments are vectors whose *directions and magnitudes can be found by calculating the integrals* $\vec{\mu}'_{\alpha\beta} = \int \varphi_\alpha \varphi_\beta \vec{r} \cdot \mathrm{d}V$ and $\vec{\mu}'_{\alpha\alpha'} = \int \varphi_\alpha \varphi_\beta \vec{r}\, \mathrm{d}V$ for given orientations of the atomic orbitals with respect to the orientating bond vectors. It is natural to associate the origin of the $\vec{\mu}'_{\alpha\alpha'}$ vectors with the centres of the atoms that have the indices α. The origin of the $\vec{\mu}'_{\alpha\beta}$ vectors are related to the atoms of the bond from which the $\vec{\mu}'_{\alpha\beta}$ vector forms an acute angle with an extension of the straight line connecting the α and β atoms.

The $\vec{\mu}'_{\alpha\alpha'}$ and $\vec{\mu}'_{\alpha\beta}$ vectors cannot be included without resorting to artificial constructions in the parametric scheme based on the 'charge' concept of the dipole moment of the whole system. However, this can be done quite easily by relying on the representation of $\vec{\mu}$ as

$$\vec{\mu} = \sum_{\alpha,\beta} \mu_{\alpha\beta} \vec{e}_{\alpha\beta}$$

where the system of $\vec{e}_{\alpha\beta}$ unit vectors itself gives a natural local co-ordinate system for each atomic grouping. Three cases are possible: a linear molecule or portion of the molecule; a planar molecule or grouping; or a non-planar grouping or molecule. We can conclude from symmetry considerations that the $\vec{\mu}'_{\alpha\alpha'}$ and $\vec{\mu}'_{\alpha\beta}$ vectors must be collinear with the orientating bond vectors for all these portions in the linear case, and hence they must add to or subtract from (depending on whether their directions are the same as or opposite to the directions of the $\vec{e}_{\alpha\beta}$ bond vectors) the full value of the corresponding electro-optical parameters.

In the second case $\vec{\mu}'_{\alpha\alpha'}$, and $\vec{\mu}'_{\alpha\beta}$ must lie in the plane of the molecule or that of the structural element, and they can be uniquely decomposed into pairs of bond-orientating vectors that intersect at a given centre. These pairs must be chosen so that the different vectors $\vec{\mu}'_{\alpha\alpha'}$ and $\vec{\mu}'_{\alpha\beta}$ lie between the two $\vec{e}_{\alpha\beta}$ vectors taken as the reference pair. The electro-optical parameters $\mu_{\alpha\beta}$ of the reference pair of bonds will in this case be increased (or decreased, depending on the sign of the projection) by the value of the corresponding projection of the dipole moment of the transition.

If a given structural grouping is three-dimensional then it will always be possible to choose three orientating vectors intersecting a given atom such that the transition moment vectors can always be expressed uniquely in terms of the orientating vectors. This enables us to include all the terms in equation (11.26), which gives an accurate representation of the molecule's dipole moment in terms of the scheme expressed by

$$\vec{\mu} = \sum_{\alpha,\beta} \mu_{\alpha\beta} \vec{e}_{\alpha\beta}$$

It is possible to retain the same form of the dipole moment, i.e. the sum $\vec{\mu} = \sum_k \mu_k \vec{e}_k$, for characterizing any molecule, but the physical meaning of the parameter μ will vary, depending on the properties of a particular molecule.

When the molecule's dipole moment in the form of

$$\vec{\mu} = \sum_{\alpha,\beta} \mu_{\alpha\beta} \vec{e}_{\alpha\beta}$$

is substituted in equation (11.19), derivatives of $\mu_{\alpha\beta}$ with respect to the selected natural vibration co-ordinates appear together with the $\mu_{\alpha\beta}$ terms themselves. These derivates and the $\mu_{\alpha\beta}$ form the system of empirical, 'electro-optical' parameters of the molecule*. These parameters meet all the necessary requirements with respect to the form representing the molecule's dipole moment, i.e. they enter the following system of linear equation as unknowns:

$$\left(\frac{\partial \vec{\mu}}{\partial Q_i}\right)_0 = \vec{F}\left(\mu_k^{(0)}, \left(\frac{\partial \mu_k}{\partial q_n}\right)_0, l_{qin}\right) \quad (k = \alpha, \beta)$$

They are independent of the mechanical properties of the molecule, and only reflect the character of the molecule's electronic cloud, etc. In the special case when a molecule has no cycles and has properties of strict additivity, a bond's dipole moment is only dependent on the extension of its bond and is independent of the co-ordinates that define the vibrations of the other atoms and structural elements. As a result, the derivatives of bond dipole moments with respect to all co-ordinates, other than the co-ordinates of the extension of the bonds, vanish. The remaining parameters correspond in meaning to the analogous quantities for biatomic molecules. The bond interactions should be estimated by introducing all the derivatives of the bond dipole moments (or some of them) with respect to the 'foreign' vibration co-ordinates. This situation, in addition to the requirements we have already mentioned, makes the proposed scheme very convenient, obvious and universal.

It is obvious that all our comments concerning physical significance as characteristics of separate parts of the molecule apply to the $(\partial\mu_k/\partial q_n)_0$ parameters as much as they do to the μ_k. We must pay careful attention to this situation when attempting to use derived values of μ_k and $(\partial\mu_k/\partial q_n)_0$ (a derivative with respect to the extension of the same bond) for characteristics of a specific type of chemical bond. A clear understanding is important, because the derived values of μ_k and its derivatives are frequently interpreted as inadmissible characteristics of bonds.

We emphasize the principal importance of this once again, since the method of measuring intensities in infra-red spectra is an important method of deriving the value of μ_k—the 'bond dipole moment'—and these data are widely used in a variety of applications.

We now summarize our concepts of additivity in a molecule. We shall call the set of natural vibrational co-ordinates that describe the vibrations of a group as if it were a free molecule the internal natural vibrational co-ordinates of the group (or, more simply, the group's internal co-ordinates), and will call

*In the general case $\mu_{\alpha\beta}$ and $\vec{e}_{\alpha\beta}$ are continuous functions of the vibrational co-ordinates and can be differentiated.

all the others the external co-ordinates. If two or more groups can be recognized within a molecule, then we shall also use the term 'intermediate'. We shall call a molecule 'partially additive' if its dipole moment is equal to the vector sum of those of two (or more) groups, and if, moreover, the absolute values of the μ_k parameters in each group are only functions of the internal co-ordinates of the group and are independent of all the other natural external vibrational co-ordinates.

Then, in partially additive molecules,

$$\vec{\mu} = \sum_k \mu_k^{\mathrm{I}}(q_{\mathrm{int}}^{\mathrm{I}})\vec{e}_k^{\,\mathrm{I}}(q_{\mathrm{int}}^{\mathrm{I}}, q_{\mathrm{interm}}, q_{\mathrm{int}}^{\mathrm{II}}) + \sum_{k'} \mu_{k'}^{\mathrm{II}}(q_{\mathrm{int}}^{\mathrm{II}})\vec{e}_{k'}^{\,\mathrm{II}}(q_{\mathrm{int}}^{\mathrm{I}}, q_{\mathrm{interm}}, q_{\mathrm{int}}^{\mathrm{II}})$$

where $q_{\mathrm{int}}^{\mathrm{I}}$, $q_{\mathrm{int}}^{\mathrm{II}}$ are the sets of internal co-ordinates of the first and second groups and $q_{\mathrm{intermediate}}$ is the set of intermediate co-ordinates and the derivatives of the μ_k of the first and second groups with respect to their internal co-ordinates are non-zero. In the extreme case of a fully additive molecule the μ_k are functions only of the bond extensions, and the $(\partial\mu_k/\partial q_n)_0$ derivatives alone are non-zero. The orientating vectors of the bonds in both of these cases are functions of all the molecule's vibrational co-ordinates.

Whether it is possible to retain constant values for these parameters over a series of molecules depends upon the influence of the environment on μ_k. This is important when the intensities and polarizations in the spectra of some molecules are precalculated on the basis of the electro-optical parameters of similar compounds. If a bond in a molecule had the properties of additivity and this additivity were retained over the series of molecules being considered, then it would be possible to expect that the parameters which correspond to it (μ_k and $(\partial\mu_k/\partial q_n)_0$), and which in this case are inherent characteristics of the bond, would have closely similar values for all the molecules in the series.

Note that in the CH_3 group, for example, the bonds do not have the property of additivity but all the structural elements do (the case of partial additivity of a molecule). We should expect therefore that the set of electro-optical parameters characterizing the molecule will be unchanged (within certain limits) over a series of molecules containing this structural element.

This argument leads to the following important finding. If the general deductions from the theory presented (attributes of intensity characterization, theory of overtone and combination bands, etc.) are to have a general character and be true for any class of molecules, then the possible numerical calculations must be limited to those classes of compounds which contain sufficiently complete repeating elements. In other words, numerical calculations are mainly restricted to organic compounds.

Concrete calculations require a rational choice of a working system of vibrational co-ordinates. The choice of system must be determined not only by the convenience of calculation but mainly by the physical sense of the corresponding derivatives. It follows from a consideration of a polyatomic molecule as a set of bonds which are similar in kind to diatomic molecules (and which have their own 'dipole moments') that a choice of co-ordinate system in which changes of bond lengths will be considered is essential. The derivatives thus obtained of the bond dipole moments with respect to the extension of the bonds will correspond (above all, in terms of their content) to the similar derivatives for diatomic molecules, especially when we are dealing with completely additive molecules. The most widespread systems of co-ordinates in the theory of the vibrations of polyatomic molecules—the valence force field and central force-field co-ordinate systems—fulfil these conditions. In the first system, changes of the lengths of the valence bonds and in the valence angles are the vibrational co-ordinates, while in the second, changes in the bond lengths and in the interatomic distances between non-bonded atoms are the co-ordinates. The valence force-field potential is the more convenient, completely justifying itself when studying mechanical problems concerning the frequencies and forms of the vibrations of polyatomic molecules. The application of this co-ordinate system for investigating the frequencies and forms of vibrations and intensities is very convenient, because it allows us to use the results of solutions of mechanical problems directly in the theory of the intensities in infra-red spectra.

Similarly, a theory for the line intensities in Raman spectra can be constructed by introducing empirical parameters, in this case derivatives of the molecule's polarizability tensor with respect to normal co-ordinates. In the additive approximation the parameters used have the sense of components of bond polarizability tensors and their derivatives with respect to natural vibration co-ordinates. It is possible to derive a formula very close in structure to that for $(\partial\vec{\mu}/\partial Q_i)_0$ and to obtain a number of deductions that are, in the main, analogous to those derived below for the intensities in infra-red spectra. However, the application of such a theory is significantly more difficult, because the number of empirical parameters introduced are three times those for infra-red spectra.

11.3 General Formula for $(\partial\vec{\mu}/Q_i)_0$

We start from the expression of the dipole moment of a polyatomic molecule as the sum $\vec{\mu} = \sum_k \mu_k \vec{e}_k$. We shall retain the traditional term 'bond dipole moment' for the parameter μ_k, even though its physical meaning is, as we have shown, much more complex. We will designate the set of bond-lengthening co-ordinates by q and the set of any other co-ordinates (changes in the valence angles, non-planar co-ordinates, changes in the interatomic distances of bound atoms, etc.) by γ.

Differentiating the dipole moment with respect to a normal coordinate yields

$$\left(\frac{\partial\vec{\mu}}{\partial Q_i}\right)_0 = \sum_m \frac{\partial}{\partial q_m}\left(\sum_k \mu_k \vec{e}_k\right)(l_q)_{mi} + \sum_{m'} \frac{\partial}{\partial \gamma_{m'}}(\sum \mu_k \vec{e}_k)(l_\gamma)_{m'i} \tag{11.32}$$

where $(l_q)_{mi}$, $(l_\gamma)_{m'i}$ are the elements of the ith column of the matrix L_q. Completing the differentiations and going over to the simpler and more convenient matrix forms, we obtain

$$\left(\frac{\partial\vec{\mu}}{\partial Q_i}\right)_0 = \left\{\tilde{\vec{e}}\left[\frac{\partial\mu}{\partial q}, \frac{\partial\mu}{\partial\gamma}\right] + \tilde{\mu}\left[\frac{\partial\vec{e}}{\partial q}, \frac{\partial\vec{e}}{\partial\gamma}\right]\right\}\begin{bmatrix} l_q^{(i)} \\ l_\gamma^{(i)} \end{bmatrix} \tag{11.33}$$

We take $\tilde{\vec{e}} = [\vec{e}_1, \ldots, \vec{e}_s]$ to be the row matrix in which all the directional vectors of the molecule's bonds are placed in order, $\tilde{\mu}$ to be the row matrix with elements μ_k and

$$\left[\frac{\partial\mu}{\partial q}, \frac{\partial\mu}{\partial\gamma}\right]$$

to be the rectangular matrix containing the derivatives of the bond dipole moments with respect to vibration co-ordinates. The latter has the form

$$\left[\frac{\partial\mu}{\partial q}, \frac{\partial\mu}{\partial\gamma}\right] = \begin{bmatrix} \left(\frac{\partial\mu_1}{\partial q_1}\right)_0 & \cdots & \left(\frac{\partial\mu_1}{\partial q_s}\right)_0 \left(\frac{\partial\mu_1}{\partial\gamma_1}\right)_0 & \cdots & \left(\frac{\partial\mu_1}{\partial\gamma_r}\right)_0 \\ \cdots & \cdots & \cdots & \cdots & \cdots \\ \left(\frac{\partial\mu_s}{\partial q_1}\right)_0 & \cdots & \left(\frac{\partial\mu_s}{\partial q_s}\right)_0 \left(\frac{\partial\mu_s}{\partial\gamma_1}\right)_0 & \cdots & \left(\frac{\partial\mu_s}{\partial\gamma_r}\right) \end{bmatrix} \quad (11.34)$$

In future and to simplify writing formulae, we shall write the $(\partial\mu_k/\partial q_m)_0$ and $(\partial\mu_k/\partial\gamma_{m'})_0$ derivatives as $\partial\mu_k/\partial q_m$ and $\partial\mu_k/\partial\gamma_{m'}$. Returning to the symbols, $[\partial\vec{e}/\partial q, \partial\vec{e}/\partial\gamma]$ is a rectangular matrix, similar to that above, containing the derivatives of $\vec{e}_k$—the orientating vectors of the bonds—with respect to the vibrational co-ordinates, and finally

$$\begin{bmatrix} l_q^{(i)} \\ l_\gamma^{(i)} \end{bmatrix}$$

is the form of the vibration, i.e. a column matrix containing, in order, all the relative amplitudes of the vibrational co-ordinates corresponding to the given normal vibration (the ith column of L_q).

The matrix $[\partial\mu/\partial q, \partial\mu/\partial\gamma]$ characterizes the change in the absolute values of the bond dipole moments when the molecule's equilibrium configuration is disturbed, whilst $[\partial\vec{e}/\partial q, \partial\vec{e}/\partial\gamma]$ characterizes the change in their orientations caused by rotations of the bonds during vibrations. We call the first term 'valence' and the second, 'deformational'. In the general case of non-additive molecules, the absolute values of μ_k are assumed to be functions of all the vibrational co-ordinates. The elements of the $[\partial\mu/\partial q, \partial\mu/\partial\gamma]$ and the $\tilde{\mu}$ matrices represent the system of electro-optical parameters associated with the molecule. A knowledge of these parameters and the geometry of the molecule, as well as of a solution of the mechanical problems, will enable us unequivocally to calculate the size and direction of $(\partial\vec{\mu}/\partial Q_i)_0$, i.e. to determine the intensity and polarization of a given absorption band.

When the molecule is completely additive all the elements in the $[\partial\mu/\partial q, \partial\mu/\partial\gamma]$ matrix, apart from the diagonal ones in the $[\partial\mu/\partial q]$ submatrix, are assumed to be zero. This variant is often called the zero approximation of the valence optical theory. In the general case the interactions between the bonds are calculated by the introduction of all the derivatives of μ_k or some of them. In future, we will call this operation the choice of the molecule's electric field, because the values of the derivatives allow us to establish the degree to which the atomic groupings interact in the molecule to establish the existence of closed formations, etc.

Consequently, the choice of the initial form of the electric field must be made with the help of a variety of data of this type. The method used to determine electro-optical parameters on the basis of experimental data on the intensities and polarizations of the absorption band in infra-red spectra will be considered in detail in Chapter 13.

We showed in Chapter 8 that the matrix $[\partial\vec{e}/\partial q, \partial\vec{e}/\partial\gamma]$ or simply $[\partial\vec{e}/\partial q]$ is:

$$\left[\frac{\partial\vec{e}}{\partial q}\right] = S^{-1}(\Delta E\tilde{\vec{B}}\theta - \vec{E}'), \quad \theta = T^{-1} \tag{11.35}$$

or

$$\left[\frac{\partial\vec{e}}{\partial Q}\right] = S^{-1}(\Delta E\tilde{\vec{B}}L_p - \vec{E}'L_q) \tag{11.36}$$

In future, we write the matrix $\vec{E}'$ as $[\vec{E}, 0]$. The matrix $(\Delta E\tilde{\vec{B}}\theta - [\vec{E}, 0])$ will be rectangular, with the number of rows equal to that of the bonds in the molecule and of the columns equal to the number of independent vibrational co-ordinates.

Substituting this expression into equation (11.33), we finally arrive at the following relation for the derivative of a molecule's dipole moment with respect to a normal co-ordinate:

$$\left(\frac{\partial\vec{\mu}}{\partial Q_i}\right)_0 = \left\{\tilde{\vec{e}}\left[\frac{\partial\vec{\mu}}{\partial q}, \frac{\partial\vec{\mu}}{\partial\gamma}\right] + \tilde{\mu}S^{-1}(\Delta E\tilde{\vec{B}}\theta - [\vec{E}, 0])\right\}\begin{bmatrix} l_q^{(i)} \\ l_\gamma^{(i)} \end{bmatrix} \tag{11.37}$$

or

$$\left(\frac{\partial\vec{\mu}}{\partial Q_i}\right)_0 = \tilde{\vec{e}}\left[\frac{\partial\mu}{\partial q}\right]l_q^{(i)} - \tilde{\mu}S^{-1}\Delta E\tilde{\vec{B}}l_p^{(i)} - \tilde{\mu}S^{-1}[\vec{E}, 0]l_q^{(i)} \tag{11.38}$$

The expression in the braces in equation (11.37) remains unchanged in both the additive and non-additive approximations for all normal vibrations of the molecule, only the forms of the vibrations being changed. The $(\partial\vec{\mu}/\partial Q_i)_0$ derivative is clearly expressed through the mechanical and electro-optical parameters of the molecule and the frequency and form of the vibration and the co-ordinate vectors $\vec{e}_k, \vec{f}_{ij}, \vec{h}_k$, etc. For numerical calculations it is convenient to introduce a three-dimensional system of co-ordinates from the outset, and to determine the projections of all the unit vectors on the selected axes. The expression in the braces is therefore written in a scalar form, and all the calculations can be computed easily because they reduce to the multiplication and addition of a series of simply structured matrices.

The formula we have derived for $(\partial\vec{\mu}/\partial Q_i)_0$ enables us to find, using the rules for differentiating matrices, the partial derivatives of $(\partial\vec{\mu}/\partial Q_i)_0$ with respect to any parameters of a molecule. This may prove to be very useful for investigating the changes in spectra when the system is weakly perturbed (for instance, by intermolecular interactions). When the partial derivatives with respect to the geometric parameters are calculated, second and first derivatives of μ_k with respect to natural vibrational co-ordinates appear in the expression for the derivative of $(\partial\vec{\mu}/\partial Q_i)_0$ when the matrices $[\partial\mu/\partial q, \partial\mu/\partial\gamma]$ and $\tilde{\mu}$ are differentiated. These values can be found by studying the intensities and polarizations of the absorption bands due to the overtone and combination bands.

The general formula we have derived for $(\partial\vec{\mu}/\partial Q_i)_0$ (equation (11.37)) is of fundamental importance, and can be used to obtain a number of important results dealt with in the following sections.

We shall mention a few of the more general deductions. First, it is clear that the band intensity must be strongly dependent on the form of the vibration. It is known that, in a whole series of cases, vibrations are observed, the forms of which change quite considerably while their frequencies remain almost constant. If only the frequencies of the vibrations are measured, such changes of their forms, which are often related to various kinds of intra- and intermolecular interactions, may remain unnoticed, hence measurements of the intensities are very useful. In general, the intensity and polarization of a given absorption band depends upon a number of molecular parameters that is larger than that of frequencies. Thus the fact that the intensities are more sensitive than the frequencies to various kinds of alterations in a molecule's structure becomes understandable.

Very often we have to deal with quite a strict distinction between valence vibrations (changes mainly in the bond lengths) and deformational vibrations (changes in the valence angles). Frequently, we can make deductions about the derivative of a bond's dipole moment with respect to its elongation from the intensities of the bond-stretching vibrations, and deductions about the dipole moment itself from the intensities of bond-deformation vibrations. It is understandable that this cannot be done in the general case. Even for purely stretching vibrations, in the additive approximation the bond dipole moment enters the expression for $(\partial\vec{\mu}/\partial Q_i)_0$. This is because, in general, rotations of the bonds also occur in these vibrations (see the example below of calculations for a symmetric and non-linear molecule XY_2). On the other hand, in purely deformational vibrations there may be significant contributions made to the band intensities by derivatives of μ_k with respect to angle co-ordinates, which in non-additive molecules may become quite large in addition to the contributions made by μ_k.

Before discussing topics related to the construction of a semi-empirical theory of the intensities in infra-red spectra we rewrite the expression for $(\partial\vec{\mu}/\partial Q_i)_0$ as

$$\left(\frac{\partial\vec{\mu}}{\partial Q_i}\right)_0 = \sum_{\gamma}\left(\frac{\partial\vec{\mu}}{\partial q_j}\right)_0 l_{qij} \tag{11.39}$$

where

$$\left(\frac{\partial\vec{\mu}}{\partial q_j}\right)_0 = \tilde{\vec{e}}\frac{\partial\mu^{(j)}}{\partial q} + \tilde{\mu}S^{-1}(\Delta E\tilde{\vec{B}}\theta - [\vec{E}, 0])^{(j)} \tag{11.40}$$

and j is the index of the corresponding column. The dependence of $(\partial\vec{\mu}/\partial q_j)_0$ on the mass of the atoms is concealed in the derivatives $(\partial\vec{e}_k/\partial q_j)_0$, which are elements of the matrix

$$S^{-1}(\Delta E\tilde{\vec{B}}\theta - [\vec{E},0])$$

In order to deduce the appropriate term in a clear form we will take advantage of the following argument.

Consider some system of connected material points, the position of each is given by a radius vector in some stationary Cartesian co-ordinates framework. We deform our system arbitrarily—it may be uniquely described by a set of selected natural vibrational co-ordinates. When the system is deformed the material points are displaced from their initial positions by a value $\delta\vec{r}'_n$. Obviously, there is no unambiguous relationship between the changes in the system's geometrical parameters and the displacements $\delta\vec{r}'_n$ of the particles which generate the system unless we are given additional conditions limiting arbitrariness. However, it is clear that the set of the displacements of the particles from their initial positions cannot be distinguished from another any better than by a translation of the whole deformed system and a rotation of it about an axis, given any additional conditions and the same deformation of our molecule. In this way, displacements of the atoms from their equilibrium positions during vibrations (which satisfy the conditions of momentum and moment of momentum conservation) may be expressed as the result of three motions. These are (1) a displacement of the atoms when the molecule is deformed without taking into account the conservation of momentum and moment of momentum (the appropriate matrix for transforming from natural co-ordinates to displacement co-ordinates, which we will denote as $\vec{B}''$, can be chosen in many different ways), (2) a translation and (3) a rotation of the deformed system. The last two operations must clearly by chosen from considerations of the compensation that arise from the arbitrary displacements $\delta\vec{r}' = \vec{B}''q$ of the momentum and moment of momentum of the whole molecule.

As already mentioned, the choice of the matrix $\vec{B}''$ is, to a large extent, arbitrary. In order to be specific, we will choose it in the following way. We will consider that when a bond is stretched, one of its atoms is fixed while the other moves a distance $\delta\vec{r}'_n = q_n\vec{e}_n$. The other atoms are either stationary or move parallel to the bond's second atom. For angular or non-planar co-ordinates we fix, respectively, the atom at the vertex of the angle or those lying at the edge of the dihedral angle. We will take the displacements of the remaining atoms as being equal to the vector products between the radius vectors of the atoms with respect to the axis of rotations, which is perpendicular to the plane of the valence angle (parallel to the edge of the dihedral angle), multiplied, respectively, by plus or minus $\frac{1}{2}\gamma$ (γ is the change in the angular or non-planar co-ordinate).

These conditions make it possible to express the $\vec{B}''$ matrix in an explicit form for any molecule. The choice of the $\vec{B}''$ matrix has no effect on the following discussion.

It is not difficult to see that $\vec{B}''$ (like the $\vec{B}'$, which is normally used when calculating the displacements of the atoms from their equilibrium positions during vibrations) satisfies the equation $\vec{B}\cdot\vec{B}''=I$. When the atoms are displaced from their starting positions by $\delta\vec{r}'=\vec{B}''q$ the molecule's centre of mass is displaced by

$$\delta\vec{R}=m_\Sigma^{-1}\sum_n m_n\delta\vec{r}'_n=m_\Sigma^{-1}\tilde{m}\vec{B}''q$$

If now the atoms are displaced from their equilibrium positions then $\delta\vec{r}''=(I-m_\Sigma^{-1}[1]M)\vec{B}''\cdot q$, and the displacements will already be obeying the condition that the momentum of the centre of mass is conserved. Here [1] is the square unit matrix, m_Σ is the mass of the molecule and M is the diagonal matrix of the masses of the molecule's atoms.

When the atoms are displaced by $\delta\vec{r}''_n$, a moment is induced in the system equal to $\delta\vec{N}=\sum_n\rho_n\times m_n\delta\vec{r}''_n$ or, in a somewhat different form,

$$\delta\vec{N}=\sum_n m_n((\rho))_n^+\,\delta\vec{r}''_n=((\tilde{\rho}))^+M\delta\vec{r}'' \tag{11.41}$$

where $((\widetilde{\rho}))^+$ is the row matrix whose elements are the anti-symmetric tensors

$$((\rho))^+=\left(\left(\begin{matrix}0 & -\rho_{nz} & \rho_{ny}\\ \rho_{nz} & 0 & -\rho_{nx}\\ -\rho_{ny} & \rho_{nx} & 0\end{matrix}\right)\right) \tag{11.42}$$

and ρ_n are the atoms' equilibrium radius vectors, it being implicit that all the arguments are for small displacements of the atoms from their equilibrium positions.

On the other hand, by ignoring changes in the molecule's moment of inertia during the vibrations (small second-order variables) it is possible to write $\delta\vec{N}=((I_0))\vec{\omega}\neq 0$, where $\vec{\omega}$ is the twist, corresponding to moment $\delta\vec{N}$, of the deformed molecule around the axis that passes through the centre of mass and $((I_0))$ is the system's moment-of-inertia tensor.

By comparing these two expressions we obtain

$$\vec{\omega}=((I_0))^{-1}((\rho))^+M\delta\vec{r}''=((I_0))^{-1}((\widetilde{\rho}))^+M(I-m_\Sigma^{-1}[1]M)\vec{B}''q \tag{11.43}$$

This enables us to write the formula for $(\partial\vec{\mu}/\partial Q_i)_0$ in the following form:

$$\begin{aligned}\left(\frac{\partial\bar{\vec{\mu}}}{\partial Q_i}\right)_0 &=\left\{\tilde{\vec{e}}\left[\frac{\partial\vec{\mu}}{\partial q}\right]+\tilde{\mu}S^{-1}(\Delta\vec{B}''-[\vec{E},0])\right.\\ &\quad\left.-((\mu_0))((I_0))^{-1}((\widetilde{\rho}))^+M(I-m_\Sigma^{-1}[1]M)\vec{B}''\right\}l_q^{(i)}\\ &=\left\{\tilde{\vec{e}}\left[\frac{\partial\vec{\mu}}{\partial q}\right]+\tilde{\mu}S^{-1}[0,\Delta\vec{B}''_\gamma]\right.\\ &\quad\left.-((\mu_0))((I_0))^{-1}((\widetilde{\rho}))^+M(I-m_\Sigma^{-1}[1]M)\vec{B}''\right\}l_q^{(i)}\end{aligned}$$

where $[0,\Delta\vec{B}''_\gamma]$ is the matrix $\Delta\vec{B}''$ in which the columns corresponding to the co-ordinates of bond extensions are set to zero and $((\mu_0))$ is the anti-symmetric tensor for the molecule's equilibrium dipole moment, given by

$$((\mu_0)) = \begin{pmatrix} 0 & \mu_{0z} & -\mu_{0y} \\ -\mu_{0z} & 0 & \mu_{0z} \\ \mu_{0y} & -\mu_{0x} & 0 \end{pmatrix} \tag{11.44}$$

Even though this formula is not like equation (11.37) in form in fact they differ only by an insignificant modification that is related to the neglected dependency of the molecule's inertia tensor on the natural co-ordinates.

We will use $(\partial\vec{\mu}/\partial q_j)'_0$ to denote the following combination of the parameters $(\partial\mu_k/\partial q_j)_0$ and $\mu_k(0)$:

$$\tilde{e}\left(\frac{\partial\mu}{\partial q}\right)^{(j)} + \tilde{\mu}S^{-1}[\Delta, \Delta\vec{B}''_{\gamma}]^{(j)}$$

The derivative $(\partial\vec{\mu}/\partial q_j)'_0$ characterizes the change in a molecule's dipole moment when the bonds are stretched without the molecule being turned or the valence or dihedral angles being changed when the atoms are displaced following the rules for choosing matrix B''. First, the derivative is clearly independent of the mass of the atoms and depends solely on the molecule's electron shell. Second, it has a fairly simple physical meaning.

Clearly, only the derivative $(\partial\vec{\mu}/\partial q_j)'_0$ can be derived from the inverse spectral problem (see below). Superficially, this is why it is so tempting to use it is an electro-optical parameter, all the more so because it is unnecessary to have a concrete representation of a molecule's dipole moment. Actually, it is possible to construct an empirical theory with the help of this parameter, which has proved to be transferable for a number of additive homologous molecules. The theory can predict the intensities in the spectra of some molecules on the basis of the experimental values of intensities in those of related molecules. The parameter $(\partial\vec{\mu}/\partial q_j)'_0$ acts as a carrier of the additive properties of the molecules, enabling us to separate out the influence of the common dipole moment of the related molecules and the forms of the vibrations.

However, a disadvantage is that the transferability properties will only be possessed by the derivative $(\partial\vec{\mu}/\partial q_j)'_0$ with respect to the co-ordinates of the end additive groups. The derivatives with respect to internal co-ordinates do not have similar transferability, even in additive molecules. Second, the value of any theory must not only be due to its simplicity of calculation for a series of a specific class of compounds but also it must be possible to make deductions of a general character from the theory for a large number of different cases. Thus, for example, it is scarcely possible to solve some principal topics such as questions about the characterizability of vibrations, the dependence of absorption band intensities on the number of similar groups in a molecule and the distribution of intensity in the spectra of polymers and finite length periodic molecules without introducing a representational form of the molecule's dipole moment that is related to the properties of different parts of the molecule. It is also very important when studying complex systems to be able to use a theory based on a simplified model of the molecule being considered. This is normal practice for any method of studying molecules. We have this ability when using the valence optical scheme, but it is absent when $(\partial\vec{\mu}/\partial q_j)'_0$ parameters, or similar ones, are introduced.

These considerations compel us to prefer a semi-empirical theory of infra-red band intensities based on the valence-optical scheme. The complications that arise because, in a number of cases, practical calculations necessitate the use of combinations of the parameters $(\partial\mu_k/\partial q_j)_0$ and $\mu_k(0)$, which have, in fact, the same meaning as $(\partial\vec{\mu}/\partial q_j)'_0$, are not of prime importance. In the final analysis they can be denoted by a single symbol. The expression of this combination may itself prove useful, because it helps to clarify the structure of the $(\partial\vec{\mu}/\partial q_i)'_0$ parameter, and in some cases predicts its behaviour in a number of molecules.

11.4 The Symmetry of Molecules and the Selection Rules with Respect to Symmetry

If a molecule has symmetry elements, then calculating its frequencies and forms of vibration may be considerably easier by transforming to symmetry co-ordinates. The problem of determining $(\partial\vec{\mu}/\partial Q_i)_0$ can also be simplified if we take advantage of these co-ordinates.

Let us express the natural vibrational co-ordinates in terms of those of a certain type of symmetry s, this being done with the help of a matrix $C^{(s)}$ such that

$$l_q^{(i)} = C^{(s)} l_{qs}^{(i)}$$

The formula for displacements of atoms from their equilibrium positions takes the following form for symmetry co-ordinates:

$$\delta\vec{r}^i = E\tilde{\vec{B}}^{(s)}\theta^{(s)} l_{qs}^{(i)} Q_{0i}$$

where $\vec{B}^{(s)}$ is the matrix which relates the symmetry co-ordinates to the displacements of the atoms during vibrations. This is obtained by premultiplying the matrix $\vec{B}$ by that of the transformation from the original vibration co-ordinates to symmetry ones. The matrix $\theta^{(s)}$ is also written in terms of symmetry co-ordinates. Taking this into account and substituting the original co-ordinates for symmetry ones, we obtain, after post-multiplying the expression in the braces in relation (11.37) by the matrix $C^{(s)}$:

$$\left(\frac{\partial\vec{\mu}}{\partial Q_i}\right)_0 = \left\{\tilde{\vec{e}}\left[\frac{\partial\mu}{\partial q}, \frac{\partial\mu}{\partial\gamma}\right]C^{(s)} + \tilde{\mu}S^{-1}(\Delta E\tilde{\vec{B}}^{(s)}\theta^{(s)} - [\vec{E}C^{(s)}, 0])\right\}\begin{bmatrix} l_{qs} \\ l_{\gamma s}\end{bmatrix}^{(i)} \tag{11.45}$$

The product $[\partial\mu/\partial q, \partial\mu/\partial\gamma]C^{(s)}$ may be replaced by a single matrix containing the derivatives of the bond dipole moments with respect to the co-ordinates of the given symmetry type. We will designate this matrix $[(\partial\mu/\partial q^{(s)})(\partial\mu/\partial\gamma^{(s)})]$. We also note that the rows of matrix $[\partial\mu/\partial q^{(s)}, \partial\mu/\partial\gamma^{(s)}]$, corresponding to equivalent bonds, will differ from each other by the products defining the relative amplitudes of the natural co-ordinates of the extensions of these bonds during vibrations of the given symmetry type. The μ_k (and S_k^{-1}) elements for equivalent bonds are equal to each other, while the vectors $\delta\vec{e}_k$ will be transformed during symmetry operations into each other with coefficients equal to the relative symmetry coefficients (if $q \neq 0$) for a given group of bond-extension co-ordinates; this should be clear from equation (11.35). By taking this into account it is clearly possible to write equation (11.45) in the form

$$\left(\frac{\partial\vec{\mu}}{\partial Q_i}\right)_0 = \left\{\tilde{\vec{e}}^{(s)}\left[\frac{\partial\mu}{\partial q^{(s)}}, \frac{\partial\mu}{\partial\gamma^{(s)}}\right]_{\text{equ}} + \tilde{\mu}_{\text{equ}} S_{\text{equ}}^{-1}(\Delta_{\text{equ}} E\tilde{\vec{B}}^{(s)}\theta^{(s)} - [\vec{E}_{\text{equ}}^{(s)}, 0])\right\}\begin{bmatrix} l_{qs} \\ l_{\gamma s} \end{bmatrix}^{(i)} \tag{11.46}$$

Here, $\tilde{\vec{e}}^{(s)}$ is a row of unnormalized orientating vectors of the appropriate symmetry. (We introduced the concept of an orientating vector of symmetry in Chapter 5.) Let us just recall that what we call an orientating vector of symmetry is one which is obtained by adding of orientating unit vectors of equivalent bonds and the symmetry coefficients of the natural co-ordinates of the extensions of these bonds in the given vibrational type. Together with these orientating vectors of symmetry, constructed from the symmetry coefficients of the orthogonal matrix C (normalized coefficients), we will also use in the theory of intensity symmetry vectors constructed from the unnormalized symmetry coefficients. These vectors differ from those introduced above only by their lengths, and we call them 'unnormalized symmetry vectors' to make this distinction. Obviously, the original unit-orientating vectors for equivalent bonds must be chosen in such a way that for any allowed symmetry operation the vectors will be transformed into each other without change. The number of orientating symmetry vectors will equal that of the groups of equivalent bond-extension co-ordinates.

The matrix $[\partial\mu/\partial q^{(s)}, \partial\mu/\partial\gamma^{(s)}]_{\text{equ}}$ in equation (11.45) is obtained from $[\partial\mu/\partial q^{(s)}, \partial\mu/\partial\gamma^{(s)}]$ (the derivatives of the bond dipole moments with respect to the symmetry co-ordinates) by deleting all the rows corresponding to equivalent bonds except for the first row that is retained.

$\tilde{\mu}_{\text{equ}}$ and S_{equ}^{-1} are matrices that are obtained by deleting elements from the matrices $\tilde{\mu}$ and S^{-1}, respectively, such elements being all those that correspond to equivalent bonds, except for the first element from each group of equivalent bonds.

Δ_{equ} is the rectangular matrix with a number of rows equal to that of the groups of equivalent bonds, and with a number of columns equal to that of the atoms in the molecule. This matrix is obtained as the result of adding the row matrices Δ, each of which corresponds to a group of equivalent bonds, term by term. $\tilde{\vec{B}}^{(s)}$ is the transpose of the matrix of the transformation from atom displacements to symmetry co-ordinates and $\vec{E}_{\text{equ}}^{(s)}$ is the diagonal matrix of the normalized orientating vectors of symmetry. Finally,

$$\begin{bmatrix} l_{qs} \\ l_{\gamma s} \end{bmatrix}^{(i)}$$

is the form of the vibration in terms of symmetry co-ordinates.

If equivalents bonds coincide with one of the molecule's symmetry elements and the vibrations are not symmetric with respect to this element, then the extension co-ordinates must be zero. Thus it follows that all the rows in the matrix $[\partial\mu/\partial q^{(s)}, \partial\mu/\partial\gamma^{(s)}]$ that correspond to bonds with zero extension co-ordinates are discarded. This is equivalent to the formal introduction of zero-orientating vectors of symmetry with zero coefficients in front of the orientating vectors of equivalent bonds. The matrix $\vec{E}^{(s)}_{\text{equ}}$ in this case is conveniently written as a rectangular one, whose number of rows equals that of the groups of equivalent bonds and whose numbers of columns equals that of the bond-extension symmetry co-ordinates that are non-zero. A row of $\vec{E}^{(s)}_{\text{equ}}$, corresponding to a group of equivalent bonds whose bond lengths remain constant, will clearly consist entirely of zeros.

The formula we have obtained makes it possible for us to simplify substantially the general expression for the first derivatives of a molecules' dipole moment with respect to normal co-ordinates when the molecule has a definite symmetry. It is convenient here for all the calculations to be made by using the derivatives of the bond dipole moments with respect to the symmetry co-ordinates, and transferring to the derivatives with respect to the natural co-ordinates—i.e. the molecule's electro-optical parameters—only at the last stage.

Moreover, the formula allows us to make a number of general deductions. First, it is clear that the valency terms in the expressions for $(\partial\vec{\mu}/\partial Q_i)_0$ in vibrations during which the extensions of all the bonds are zero (for example, non-planar vibrations of planar molecules and deformational vibrations of linear molecules) will be absent, because all their components will be zero. The deformation terms drop out in vibrations which do not induce twists in the bonds (for example, valency vibrations in linear molecules). Thus non-planar and deformation vibrations are polarized perpendicular to the plane or axis of the molecules, while valency vibrations are polarized along the molecule's axis.

For several symmetry types of vibrations the components of the valency terms and the vectors of the twists of the bonds may be the same and non-zero, but changes in the dipole moment, induced by disturbances of the equilibrium configuration of one part of a molecule, will be completely compensated for by corresponding movements of another part of the molecule symmetric to it. In this case, the valency and deformation terms will simultaneously and identically fall to zero for all the vibrations of the given symmetry type.

Absorption bands from vibrations possessing this property will not appear in the infra-red spectrum, i.e. they are prohibited due to symmetry. The valency terms in this case will be zero when all the non-normalized orientating vectors of symmetry are zero for the given vibration type. By compiling these vectors we can establish immediately whether a vibration of a given type will be active in absorption. For example, the orientating vectors of the first and second bonds in a linear symmetric molecule of the type XY_2 are opposed to each other. Thus the appropriate unique element of the column matrix $\tilde{\vec{e}}^{(s)}$ (the vector $(\vec{e}_1 + \vec{e}_2)$) will be zero for symmetric vibrations in which the bond-extension co-ordinates are transformed without a change in sign. Bands from vibrations of this type will thus be absent in the absorption spectrum. The symmetry vectors for non-linear XY_2-type molecules are not equal to zero in symmetric vibrations, and so the corresponding bands do appear in the spectrum. Note that if only one bond is included in a group of equivalent bonds, then the bond's extension co-ordinate (and its corresponding symmetry vector) is non-zero only during fully symmetric vibrations of the system. The activity or inactivity of these vibrations in an infra-red spectrum is, as a rule, quite apparent. For example, the dipole moment of a bond having a centre of symmetry is zero and clearly does not change during fully symmetric vibrations. Thus the fully symmetric vibrations of those molecules are non-active in their infra-red spectra.

It follows from this that if a molecule possesses a group of equivalent bonds consisting solely of one bond, then the elements of the column matrix $[\partial\mu/\partial q]$, which correspond to this bond's extension co-ordinate, and the elements of the row, corresponding to the derivatives of the bond's dipole moment with respect to the vibration co-ordinates, will not enter the expression for $(\partial\vec{\mu}/\partial Q_i)_0$ for this molecule or for isotopically substituted ones of another symmetry. Thus these elements can be set equal to zero. At the same time, the vector sum, obtained by multiplying the row $\tilde{\vec{e}}$ by the column of the single bond's extension co-ordinate, must be zero, because the derivatives of the dipole moment of the bonds of the symmetry group with respect to the single bond's extension co-ordinate are the same for equivalent bonds due to the symmetry of the molecule's electron cloud. The direction of $\vec{e}_k$ is also given by taking the molecule's symmetry into account. The elements of the row of derivatives of the solitary bond's dipole moment with respect to the vibrational co-ordinates must also be zero, otherwise a non-zero term would appear for fully symmetric vibrations.

Consequently it is essential, when dealing with molecules without a dipole and having a solitary bond, to discard in the $[\partial\mu/\partial q]$ matrix the row and column of the bond's dipole moment and its extension co-ordinates, while in the row matrix $\tilde{\vec{e}}$ the orientating vectors of the bond must be discarded. It is necessary to act in the same way as regards $\tilde{\mu}$ and $[\partial\vec{e}/\partial q]$, since $\mu_k = 0$ for such a bond.

Vibrations in which bond extensions are absent will be active in infra-red spectra if the symmetric vectors formed by the linear combination of the co-ordinate vectors of the appropriate natural vibrational co-ordinates (angular co-ordinates, non-planar co-ordinate, etc.) are zero.

Since the symmetry vectors for different symmetry types of vibrations are orthogonal, it is impossible to have more than three types of fundamental vibrations of different symmetries active in an infra-red spectrum. In fact only vibrations with non-zero symmetry vectors can be active in an infra-red spectrum, and there can only be three or less such vectors. Since triply degenerate vibrations are described by three orthogonal symmetry vectors (and doubly generate vibrations by two), then if a triply degenerate vibration is active in the infra-red spectrum all the remaining vibrations must be prohibited. Similarly, if a doubly degenerate vibration is active, then only one other (necessarily non-degenerate) type of vibration can also be active. The rigorous selection rules for all molecules of a given point group can be established using the rules presented in many texts. However, by using the above examples the activity of a given symmetry-type of vibration can be quickly defined for a particular molecule.

All the vibrations of a given type of symmetry will have the same polarization for the majority of symmetry groups. It is thus sufficient to know the direction of any non-zero orientating vector of symmetry in order to obtain the polarization. It follows, therefore, that fully symmetric vibrations will, if they are active in an infra-red spectrum, be polarized along the molecule's major axis of symmetry.

11.5 Additional Relationships between the Co-ordinates

The set of molecular electro-optical parameters must, in general, include derivatives of the bond dipole moments with respect to changes in the lengths of all the bonds and in the valency angles (or the separations between bounded atoms). Several of the corresponding vibrational co-ordinates may be found to be dependent. These must be excluded during any calculations, otherwise it will not be possible, in particular, to calculate the matrix θ by the method of inverting the matrix of kinematic coefficients. The exclusion of dependent co-ordinates is done as follows. Without limiting generality, we can consider the exclusion of angle co-ordinates.

Since all the displacements of the atoms leading to changes in the molecule's equilibrium configuration are completely described by the independent internal vibrational co-ordinates, then when there are additional relationships between the co-ordinates they must be taken into account before the inversion of T. That is, the row and column corresponding to the excluded co-ordinate must be discarded from this matrix and the rows of the dependent co-ordinates removed from the matrix $\vec{B}$.

It is necessary to add to every column matrix $(\partial\mu/\partial q)$ in the valency term the excluded columns with the coefficients in front of the independent co-ordinates in the equation that expresses an excluded dependent co-ordinate in terms of the independent ones, the addition being done term by term. Then the columns corresponding to the dependent co-ordinate can be discarded.

Suppose, further, that there is a relationship

$$q_n = \sum_{i}^{n-1} a_i q_i$$

between the co-ordinates, the a_i being constant coefficients. Excluding the dependent co-ordinates q_n may be interpreted as a transformation from the co-ordinates q to those $q'_k = q_k (k = 1, \ldots, n-1)$ with the transformation matrix:

$$A = \begin{bmatrix} 1 & 0 & \cdots & 0 \\ 0 & 1 & \cdots & 0 \\ \cdots & \cdots & \cdots & \cdots \\ 0 & 0 & \cdots & 1 \\ a_1 & a_2 & \cdots & a_{n-1} \end{bmatrix} \tag{11.47}$$

The matrix $[\partial\mu/\partial q']$ in the new system of co-ordinates may be obtained by post-multiplying the original matrix $[\partial\mu/\partial q]$ by the co-ordinate transformation matrix, i.e.

$$\begin{bmatrix} \dfrac{\partial\mu_1}{\partial q_1} & \cdots & \dfrac{\partial\mu_1}{\partial q_{n-1}} & \dfrac{\partial\mu_1}{\partial q_n} \\ \cdots & \cdots & \cdots & \cdots \\ \dfrac{\partial\mu_k}{\partial q_1} & \cdots & \dfrac{\partial\mu_k}{\partial q_{n-1}} & \dfrac{\partial\mu_k}{\partial q_n} \end{bmatrix} \begin{bmatrix} 1 & 0 & \cdots & 0 \\ 0 & 1 & \cdots & 0 \\ \cdots & \cdots & \cdots & \cdots \\ 0 & 0 & \cdots & 1 \\ a_1 & a_2 & \cdots & a_{n-1} \end{bmatrix} \tag{11.48}$$

Carrying out the multiplication, we obtain

$$\left[\frac{\partial\mu}{\partial q'}\right] = \begin{bmatrix} \dfrac{\partial\mu_1}{\partial q_1} + a_1 \dfrac{\partial\mu_1}{\partial q_n} \cdots \dfrac{\partial\mu_1}{\partial q_{n-1}} + a_{n-1} \dfrac{\partial\mu_1}{\partial q_{n-1}} \\ \dfrac{\partial\mu_k}{\partial q_1} + a_1 \dfrac{\partial\mu_k}{\partial q_n} \cdots \dfrac{\partial\mu_k}{\partial q_{n-1}} + a_{n-1} \dfrac{\partial\mu_1}{\partial q_{n-1}} \end{bmatrix} \tag{11.49}$$

Consequently, equation (11.37) will retain its earlier form, but the matrix $[\partial\mu/\partial q]$ and the deformation terms will be transformed in the way indicated, while the form of the vibrations will only contain the independent co-ordinates. In most cases, the dependent co-ordinates are more conveniently excluded after the transformation to symmetry ones.

11.6 Examples of Compiling General Formulae for $(\partial\vec{\mu}/\partial Q_i)_0$

We will now provide several examples illustrating the derivation of the formulae for the first derivative of a molecular dipole moment with respect to a normal co-ordinate.

11.6.1 A non-linear symmetric molecule XY_2

Figure 11.3 shows the system of unit vectors used. The two Y atoms are numbered 1 and 2 and the X atom is 0. Vibrations of the molecule are described by the bond-extension co-ordinates q_1 and q_2 and the co-ordinate of the change in angle, γ. The molecule has C_{2v} symmetry and has an axis of symmetry. Two type A vibrations are possible, both symmetric with respect to all the symmetry elements, and one type B vibration, anti-symmetric with respect to the axis of symmetry.

The original vibrational co-ordinates can be expressed in terms of the symmetry ones via the matrix C thus:

$$\begin{bmatrix} q_1 \\ q_2 \\ \gamma \end{bmatrix} = \begin{bmatrix} \dfrac{1}{\sqrt{2}} & 0 & \dfrac{1}{\sqrt{2}} \\ \dfrac{1}{\sqrt{2}} & 0 & -\dfrac{1}{\sqrt{2}} \\ 0 & 1 & 0 \end{bmatrix} \begin{bmatrix} q^{\mathrm{A}} \\ \gamma^{\mathrm{A}} \\ q^{\mathrm{B}} \end{bmatrix} = C \begin{bmatrix} q^{(s)} \\ \gamma^{(s)} \end{bmatrix} \tag{11.50}$$

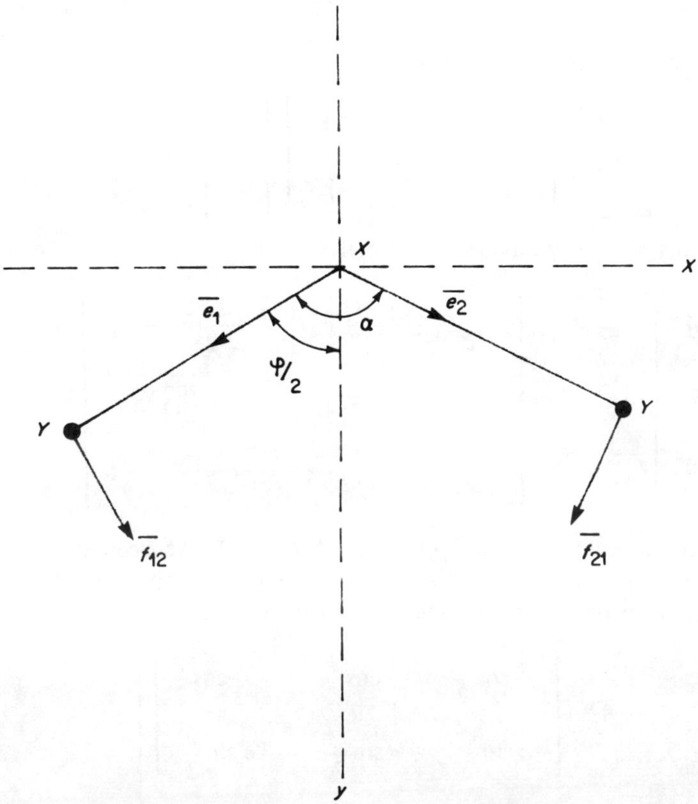

Figure 11.3. The system of unit vectors for a symmetric non-linear triatomic molecule

Relation (11.46) with these symmetry co-ordinates has the form:

Type A

$$\left(\frac{\partial \vec{\mu}}{\partial Q_i}\right)_0 = \left\{(\vec{e}_1 + \vec{e}_2)\left[\frac{\partial \mu_1}{\partial q^{\mathrm{A}}}, \frac{\partial \mu_1}{\partial \gamma^{\mathrm{A}}}\right] + \mu_1 S_1^{-1}\left([\varepsilon_1, \varepsilon_2, -2\varepsilon_0]\vec{B}^{\mathrm{A}}\theta^{\mathrm{A}} - \left[\frac{1}{\sqrt{2}}(\vec{e}_1 + \vec{e}_2), 0\right]\right)\right\}\begin{bmatrix} l_{q^{\mathrm{A}_i}} \\ l_{\gamma^{\mathrm{A}_i}} \end{bmatrix} \quad i = 1, 2 \tag{11.51}$$

Type B

$$\left(\frac{\partial \vec{\mu}}{\partial Q_3}\right)_0 = \left\{(\vec{e}_1 - \vec{e}_2)\frac{\partial \mu_1}{\partial q^{\mathrm{B}}} + \mu_1 J_1^{-1}\left([\varepsilon_1, \varepsilon_2, -2\varepsilon_0]\tilde{\tilde{B}}^{\mathrm{B}}\theta^{\mathrm{B}} - \frac{1}{\sqrt{2}}(\vec{e}_1 - \vec{e}_2)\right)\right\} l_{q^{\mathrm{B}}} \tag{11.52}$$

We can infer from the disposition of the orientating vectors of symmetry that the type A vibrations will be polarized along the axis of second-order symmetry. While the type B vibration is polarized perpendicular to it. We introduce X–Y co-ordinates and place them in the directions of the polarization. This makes it possible for us to go over to a scalar form of expression. We will not compile the matrix $\vec{B}$, for which we need to express the q_1, q_2 and γ co-ordinates in terms of atom displacements (velocities) thus:

$$\begin{bmatrix} \dot{q}_1 \\ \dot{q}_2 \\ \dot{\gamma} \end{bmatrix} = \begin{bmatrix} \vec{e}_1 & 0 & -\vec{e}_1 \\ 0 & \vec{e}_2 & -\vec{e}_2 \\ -\sigma\vec{f}_{12} & -\sigma\vec{f}_{21} & \vec{F}_{12} \end{bmatrix} \begin{bmatrix} \dot{\vec{r}}_1 \\ \dot{\vec{r}}_2 \\ \dot{\vec{r}}_0 \end{bmatrix} = \vec{B}\dot{\vec{r}} \tag{11.53}$$

Transforming to symmetry co-ordinates, we have

$$\begin{bmatrix} \dot{q}^{\mathrm{A}} \\ \dot{\gamma}^{\mathrm{A}} \\ \dot{q}^{\mathrm{B}} \end{bmatrix} = \tilde{C}\vec{B}\dot{\vec{r}} = \begin{bmatrix} \frac{1}{\sqrt{2}}\vec{e}_1 & \frac{1}{\sqrt{2}}\vec{e}_2 & -\frac{1}{\sqrt{2}}(\vec{e}_1 + \vec{e}_2) \\ -\sigma\vec{f}_{12} & -\sigma\vec{f}_{21} & \vec{F}_{12} \\ \frac{1}{\sqrt{2}}\vec{e}_1 & -\frac{1}{\sqrt{2}}\vec{e}_2 & -\frac{1}{\sqrt{2}}(\vec{e}_1 - \vec{e}_2) \end{bmatrix} \dot{\vec{r}} \tag{11.54}$$

In order to calculate $(\partial\vec{\mu}/\partial Q_i)_0^{\mathrm{A}}$, type A ($i = 1, 2$), we need only know the projection of the matrix $\vec{\mathbf{B}}^{\mathrm{A}}$ on the y axis, and to calculate $(\partial\vec{\mu}/\partial Q_3)_0^{\mathrm{B}}$ we need the projection of $\vec{\mathbf{B}}$ on the x axis. These matrices will have the form:

$$\left.\begin{aligned} \vec{B}_y^{\mathrm{A}} &= \begin{bmatrix} \frac{1}{\sqrt{2}}\cos\frac{\varphi}{2} & \frac{1}{\sqrt{2}}\cos\frac{\varphi}{2} & -\frac{2}{\sqrt{2}}\cos\frac{\varphi}{2} \\ -\sigma\sin\frac{\varphi}{2} & -\sigma\sin\frac{\varphi}{2} & 2\sigma\sin\frac{\varphi}{2} \end{bmatrix} \\ \vec{B}_x^{\mathrm{B}} &= \begin{bmatrix} \frac{1}{\sqrt{2}}\sin\frac{\varphi}{2} & \frac{1}{\sqrt{2}}\sin\frac{\varphi}{2} & -\frac{2}{\sqrt{2}}\sin\frac{\varphi}{2} \end{bmatrix} \end{aligned}\right\} \tag{11.55}$$

The terms $2\cos(\varphi/2)$ and $2\sin(\varphi/2)$ are the projections of the respective symmetry vectors.

The matrices θ^{A} and θ^{B} can in this case be found in a general form by inverting the matrices $\mathbf{T}^{\mathrm{A}}$ and $\mathbf{T}^{\mathrm{B}}$, which are complied in the usual way:

$$\theta^{\mathrm{A}} = \begin{bmatrix} \frac{2\sigma^2[\varepsilon + \varepsilon_0(1 - \cos\varphi)]}{2\sigma^2(\varepsilon^2 + 2\varepsilon\varepsilon_0)} & \frac{\sqrt{2}\varepsilon_0\sigma\sin\varphi}{2\sigma^2(\varepsilon^2 + 2\varepsilon\varepsilon_0)} \\ \frac{\sqrt{2}\varepsilon_0\sigma\sin\varphi}{2\sigma^2(\varepsilon^2 + 2\varepsilon\varepsilon_0)} & \frac{[\varepsilon + \varepsilon_0(1 + \cos\varphi)]}{2\sigma^2(\varepsilon^2 + 2\varepsilon\varepsilon_0)} \end{bmatrix} \tag{11.56}$$

$$\theta^{\mathrm{B}} = [\varepsilon + \varepsilon_0(1 - \cos\varphi)]^{-1}$$

Here, $\varepsilon_1 = \varepsilon_2 = \varepsilon = m_0/m_y$ and the dimensionless reciprocals of the bond lengths $\sigma_1 = \sigma_2 = \sigma = (S_0/S_{xy})$.

Having substituted all the necessary quantities, derived in this way, into equations (11.51) and (11.52), we obtain the final equations (11.57) and (11.58) for the derivatives of the molecule's dipole moment with respect to the normal co-ordinates. These clearly express these derivatives in terms of the electro-optical and mechanical parameters of the molecule and of the form of the vibration in symmetry co-ordinates:

$$\left|\left(\frac{\partial\vec{\mu}}{\partial Q_i}\right)_0^{\mathrm{A}}\right| = \sqrt{2}\left(\frac{\partial\mu}{\partial q}+\frac{\partial\mu}{\partial q'}\right)\cos\frac{\varphi}{2}\,l_{q^{\mathrm{A}_i}} + \left(2\frac{\partial\mu}{\partial\gamma}\cos\frac{\varphi}{2}-\mu S^{-1}\sin\frac{\varphi}{2}\right)l_{\gamma^{\mathrm{A}_i}},\quad i=1,2 \tag{11.57}$$

$$\left|\left(\frac{\partial\vec{\mu}}{\partial Q_3}\right)_0^{\mathrm{B}}\right| = \sqrt{2}\left[\frac{\partial\mu}{\partial q}-\frac{\partial\mu}{\partial q'}+\mu S^{-1}\frac{2\varepsilon_0\cos^2(\varphi/2)}{\varepsilon+2\varepsilon_0\sin^2(\varphi/2)}\right]\sin\frac{\varphi}{2}\,l_{q^{\mathrm{B}_3}} \tag{11.58}$$

where $(\partial\mu/\partial q)$ and $(\partial\mu/\partial q')$ are the derivatives of a bond's dipole moment with respect to its own extension and to its neighbour's extension, respectively.

We transform from derivatives of bond dipole moments with respect to symmetry co-ordinates to the derivatives with respect to the original natural ones:

$$\frac{\partial\mu_1}{\partial q^{\mathrm{A}}}=\frac{1}{\sqrt{2}}\left(\frac{\partial\mu_1}{\partial q_1}+\frac{\partial\mu_1}{\partial q_2}\right)=\frac{1}{\sqrt{2}}\left(\frac{\partial\mu}{\partial q}+\frac{\partial\mu}{\partial q'}\right)$$

$$\frac{\partial\mu_1}{\partial q^{\mathrm{B}}}=\frac{1}{\sqrt{2}}\left(\frac{\partial\mu}{\partial q}-\frac{\partial\mu}{\partial q'}\right)$$

It is interesting to note that, despite the fact that during anti-symmetric vibrations the angle between the bonds does not change and the γ co-ordinate is zero, there is a term containing a bond dipole moment in the final expression for $(\partial\vec{\mu}/\partial Q_3)_0^{\mathrm{B}}$. This term arises from the requirement that the moment of momentum has to be conserved for a given vibration when a molecule rocks somewhat in the xy plane without changing the angle between its bonds.

11.6.2 An ethylene molecule

The ethylene molecule belongs to the V_h-point group and has the following symmetry types of vibration:

$$\Gamma = 3A_g + A_u + 2B_{1g} + B_{1u} + 2B_{2u} + B_{2g} + 2B_{3u}$$

The co-ordinate system is shown in Figure 11.4.

All the CH bonds are equivalent. Since the molecule has an axis of symmetry, the fully symmetric A type and the B_{1g}- and B_{2g}-type vibrations are inactive in the infra-red spectrum.

The C=C bond-extension co-ordinate does not take part in the B_{1g}-, B_{2u}- and B_{3u}-type vibrations. Only one element, which corresponds to the orientating vector of symmetry of the CH bonds' orientating vectors, remains in the matrix of non-normalized orientating vectors of symmetry. This symmetry vector is equal to $(\vec{e}_1+\vec{e}_2-\vec{e}_3-\vec{e}_4)$ for B_{3u} to $(\vec{e}_1-\vec{e}_2+\vec{e}_3-\vec{e}_4)$ for B_{2u}, and the vector is $(\vec{e}_1-\vec{e}_2-\vec{e}_3+\vec{e}_4)=0$ for B_{1g}. For this reason, a B_{1g}-type vibrational band does not appear in the absorption spectrum. It can be easily inferred that during B_{2g}-type and torsional vibrations of a CH_2 group around the C=C bond a

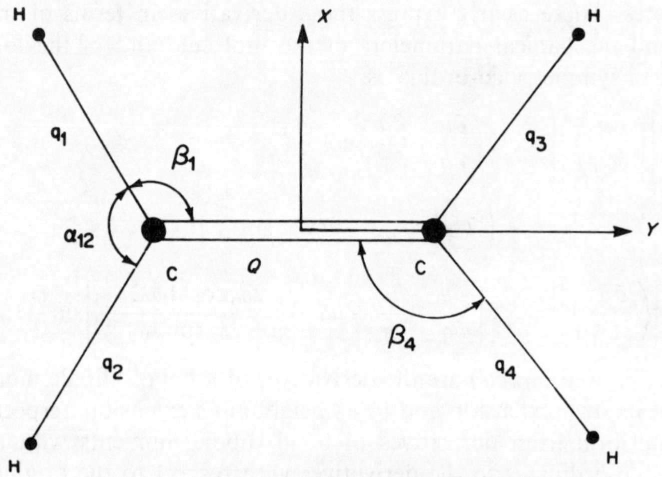

Figure 11.4. Co-ordinate system for the ethylene molecule

change in the molecular dipole moment caused by a movement of one of the CH_2 groups out of the plane of the molecule or a torsional movement of a group is compensated for by an opposite movement of the other CH_2 group. These vibrations are also inactive in the infra-red spectrum. Thus the only active vibrations are the B_{1u}-(non-planar) B_{2u}- and B_{3u}-type (planar) vibrations.

For the planar vibrations equation (11.46) can be written as follows. For type B_{2u} vibrations,

$$\left(\frac{\partial \vec{\mu}}{\partial Q_i}\right)_0 = \Bigg\{ (\vec{e}_1 - \vec{e}_2 + \vec{e}_3 - \vec{e}_4)\left[\frac{\partial \mu_1}{\partial q^{(s)}}, \frac{\partial \mu_1}{\partial \beta^{(s)}}\right] + \mu_1 S_1^{-1}[\varepsilon_H, \varepsilon_H, \varepsilon_H, \varepsilon_H, -2\varepsilon_C, -2\varepsilon_C]\tilde{\tilde{\vec{B}}}^{(s)}\theta^{(s)} - \tfrac{1}{2}[(\vec{e}_1 - \vec{e}_2 + \vec{e}_3 - \vec{e}_4), 0]\Bigg\}\begin{bmatrix} l_{q^{(s)i}} \\ l_{\beta^{(s)i}} \end{bmatrix}, \quad S = B_{2u},\ i = 1, 2 \tag{11.59}$$

and for the B_{3u} type,

$$\left(\frac{\partial \vec{\mu}}{\partial Q_i}\right)_0 = \Bigg\{ (\vec{e}_1 + \vec{e}_2 - \vec{e}_3 - \vec{e}_4)\left[\frac{\partial \mu_1}{\partial q^{(s)}}, \frac{\partial \mu_1}{\partial \beta^{(s)}}\right] + \mu_1 S_1^{-1}([\varepsilon_H, \varepsilon_H, \varepsilon_H, \varepsilon_H, -2\varepsilon_C, -2\varepsilon_C]\tilde{\tilde{\vec{B}}}^{(s)}\theta^{(s)} - \tfrac{1}{2}[(\vec{e}_1 + \vec{e}_2 - \vec{e}_3 - \vec{e}_4), 0])\Bigg\}\begin{bmatrix} l_{q^{(s)i}} \\ l_{\beta^{(s)i}} \end{bmatrix}, \quad S = B_{3u},\quad i = 3, 4 \tag{11.60}$$

Consequently, the B_{3u}-type vibrations are polarized along the C=C bond, while the B_{2u} type are polarized perpendicularly to it. Accordingly, we introduce a right-angled x, y axis. The matrix $\vec{B}$, which relates the vibrational co-ordinates to the displacements (velocities) of the molecule's atoms, has the form

$$\begin{bmatrix} \dot{q}_1 \\ \dot{q}_2 \\ \dot{q}_3 \\ \dot{q}_4 \\ \dot{\beta}_1 \\ \dot{\beta}_2 \\ \dot{\beta}_3 \\ \dot{\beta}_4 \end{bmatrix} = \begin{bmatrix} \vec{e}_1 & 0 & 0 & 0 & -\vec{e}_1 & 0 \\ 0 & \vec{e}_2 & 0 & 0 & -\vec{e}_2 & 0 \\ 0 & 0 & \vec{e}_3 & 0 & 0 & -\vec{e}_3 \\ 0 & 0 & 0 & \vec{e}_4 & 0 & -\vec{e}_4 \\ -\sigma\vec{f}_{1Q} & 0 & 0 & 0 & \vec{F}_{1Q} & -\sigma_Q\vec{f}_{Q1} \\ 0 & -\sigma\vec{f}_{2Q} & 0 & 0 & \vec{F}_{2Q} & -\sigma_Q\vec{f}_{Q2} \\ 0 & 0 & -\sigma\vec{f}_{3Q} & 0 & -\sigma_Q\vec{f}_{Q3} & \vec{F}_{3Q} \\ 0 & 0 & 0 & -\sigma\vec{f}_{4Q} & -\sigma_Q\vec{f}_{Q4} & \vec{F}_{4Q} \end{bmatrix} \begin{bmatrix} \dot{\vec{r}}_1 \\ \dot{\vec{r}}_2 \\ \dot{\vec{r}}_3 \\ \dot{\vec{r}}_4 \\ \dot{\vec{r}}_5 \\ \dot{\vec{r}}_6 \end{bmatrix} \tag{11.61}$$

Here, the H atoms are numbered in the same order as the q_i co-ordinates and the C atoms by 5 and 6. Transforming to symmetry co-ordinates and premultplying $\vec{B}$ by the matrix for the transformation from the original vibration co-ordinates to the symmetry ones, we find the $\vec{B}$ matrices to be for type B_{2u}:

$$\frac{1}{2}\begin{bmatrix} \vec{e}_1 & -\vec{e}_2 & \vec{e}_3 & -\vec{e}_4 & -(\vec{e}_1-\vec{e}_2) & -(\vec{e}_3-\vec{e}_4) \\ -\sigma\vec{f}_{1Q} & \sigma\vec{f}_{2Q} & -\sigma\vec{f}_{3Q} & \sigma\vec{f}_{4Q} & \sigma(\vec{f}_{1Q}-\vec{f}_{2Q}) & \sigma(\vec{f}_{3Q}-\vec{f}_{4Q}) \end{bmatrix} \tag{11.62}$$

and for type B_{3u}:

$$\frac{1}{2}\begin{bmatrix} \vec{e}_1 & \vec{e}_2 & -\vec{e}_3 & -\vec{e}_4 & -(\vec{e}_1+\vec{e}_2) & (\vec{e}_3+\vec{e}_4) \\ -\sigma\vec{f}_{1Q} & -\sigma\vec{f}_{2Q} & \sigma\vec{f}_{3Q} & \sigma\vec{f}_{4Q} & \sigma(\vec{f}_{1Q}+\vec{f}_{2Q}) & -\sigma(\vec{f}_{3Q}+\vec{f}_{4Q}) \end{bmatrix} \tag{11.63}$$

The $\theta^{B_{2u}}$ matrix is

$$\theta^{B_{2u}} = \begin{bmatrix} \dfrac{(\varepsilon_H+\varepsilon_C+\varepsilon_C\cos\varphi)\sigma^2}{(\varepsilon_H^2+2\varepsilon_H\varepsilon_C)\sigma^2} & \dfrac{\varepsilon_C\sigma\sin\varphi}{(\varepsilon_H^2+2\varepsilon_H\varepsilon_C)\sigma^2} \\ \dfrac{\varepsilon_C\sigma\sin\varphi}{(\varepsilon_H^2+2\varepsilon_H\varepsilon_C)\sigma^2} & \dfrac{\varepsilon_H+\varepsilon_C-\varepsilon_C\cos\varphi}{(\varepsilon_H^2+2\varepsilon_H\varepsilon_C)\sigma^2} \end{bmatrix} \tag{11.64}$$

where $\varphi = \angle HCH$

It is also easy to obtain the expression for the $\theta^{B_{3u}}$ matrix. Substituting these matrices into equations (11.59) and (11.60) and carrying out the multiplication, we obtain

$$\left|\left(\frac{\partial\vec{\mu}}{\partial Q_i}\right)_0\right| = 4\left[\frac{\partial\mu_1}{\partial q^{(s)}}\sin\frac{\varphi}{2}l_{q^{(s)}i} + \frac{\partial\mu_1}{\partial\beta^{(s)}}\sin\frac{\varphi}{2}l_{\beta^{(s)}i}\right] - 2\mu S_{CH}^{-1}\cos\frac{\varphi}{2}l_{\beta^{(s)}i}$$
$$i = 1, 2; \quad S = B_{2u} \tag{11.65}$$

we obtain the following relation for the B_{3u}-type vibrations in exactly the same way:

$$\left|\left(\frac{\partial\vec{\mu}}{\partial Q_i}\right)_0\right| = 4\left[\frac{\partial\mu_1}{\partial q^{(s)}}\cos\frac{\varphi}{2}l_{q^{(s)}i} + \frac{\partial\mu_1}{\partial\beta^{(s)}}\cos\frac{\varphi}{2}l_{\beta^{(s)}i}\right] + 2\mu S_{CH}^{-1}\sin\frac{\varphi}{2}l_{\beta^{(s)}i}$$
$$i = 3, 4; \quad S = B_{3u} \tag{11.66}$$

we have taken into account here that the vector $(\partial\vec{\mu}/\partial Q_i)_0$ is directed along the x axis for the B_{2u}-type vibrations and the analogous vector for the B_{3u} type is directed along the y axis.

After taking into account the relationship between the symmetry co-ordinates and the original vibrational ones, and having made the natural assumption that the derivatives of the bond dipole moments respect to the co-ordinates of opposite bonds and angles are zero, we go over to the derivatives with respect to the original co-ordinates. These will be the molecule's electro-optical parameters, i.e. for type B_{2u}:

$$\frac{\partial\mu_1}{\partial q^{(s)}}=\frac{1}{2}\left(\frac{\partial\mu_1}{\partial q_1}-\frac{\partial\mu_1}{\partial q_2}\right);\quad \frac{\partial\mu_1}{\partial\beta^{(s)}}=\frac{1}{2}\left(\frac{\partial\mu_1}{\partial\beta_1}-\frac{\partial\mu_1}{\partial\beta_2}\right)$$

and for type B_{3u}:

$$\frac{\partial\mu_1}{\partial q^{(s)}}=\frac{1}{2}\left(\frac{\partial\mu_1}{\partial q_1}+\frac{\partial\mu_1}{\partial q_2}\right);\quad \frac{\partial\mu_1}{\partial\beta^{(s)}}=\frac{1}{2}\left(\frac{\partial\mu_1}{\partial\beta_1}+\frac{\partial\mu_1}{\partial\beta_2}\right)$$

Consequently, the final formulae can be written as:

$$\left|\left(\frac{\partial\vec{\mu}}{\partial Q_i}\right)_0\right|=2\left[\left(\frac{\partial\mu}{\partial q}-\frac{\partial\mu}{\partial q'}\right)\sin\frac{\varphi}{2}l_{q^{(s)}i}+\left(\frac{\partial\mu}{\partial\beta}-\frac{\partial\mu}{\partial\beta'}\right)\sin\frac{\varphi}{2}l_{\beta^{(s)}i}-\mu S_{\mathrm{CH}}^{-1}\cos\frac{\varphi}{2}l_{\beta^{(s)}i}\right]\quad i=1,2;\quad S=B_{2u} \tag{11.67}$$

$$\left|\left(\frac{\partial\vec{\mu}}{\partial Q_i}\right)_0\right|=2\left[\left(\frac{\partial\mu}{\partial q}+\frac{\partial\mu}{\partial q'}\right)\cos\frac{\varphi}{2}l_{q^{(s)}i}+\left(\frac{\partial\mu}{\partial\beta}+\frac{\partial\mu}{\partial\beta'}\right)\cos\frac{\varphi}{2}l_{\beta^{(s)}i}+\mu S_{\mathrm{CH}}^{-1}\sin\frac{\varphi}{2}l_{\beta^{(s)}i}\right]\quad i=3,4;\quad S=B_{3u}, \tag{11.68}$$

where $\partial\mu/\partial q$ and $\partial\mu/\partial\beta$ are the derivatives of the CH bond dipole moment, μ, with respect to adjacent co-ordinates and $\partial\mu/\partial q'$ and $\partial\mu/\partial\beta'$ are the derivatives with respect to opposite ones.

The valency term is absent in non-planar vibrations of the ethylene molecule, and the vibrations are polarized along an axis perpendicular to the plane of the molecule. The non-planar vibrations are described by the co-ordinates ρ_1 and ρ_2. These are the co-ordinates of the movement of the CH_2 group out of the plane of the molecule. The velocities of the original co-ordinates are related to the velocities of displacements of the appropriate atoms in the following way:

$$\begin{bmatrix}\dot{\rho}_1\\ \dot{\rho}_2\end{bmatrix}=\begin{bmatrix}-\sigma\vec{h} & -\sigma\vec{h} & 0 & 0 & (\sigma_Q+2\sigma)\vec{h} & -\sigma_Q\vec{h}\\ 0 & 0 & -\sigma\vec{h} & -\sigma\vec{h} & -\sigma_Q\vec{h} & (\sigma_Q+2\sigma)\vec{h}\end{bmatrix}\dot{\vec{r}} \tag{11.69}$$

where $\vec{h}$ is a unit vector perpendicular to the plane of the molecule. For symmetry co-ordinates we have

$$\dot{\vec{\rho}}^{(s)} = \frac{1}{\sqrt{2}}[-\sigma\vec{h}, -\sigma\vec{h}, -\sigma\vec{h}, -\sigma\vec{h}, +2\sigma\vec{h}, +2\sigma\vec{h}]\dot{\vec{r}} \qquad (11.70)$$

Since the projections of all the vectors of the twists of the CH bonds into the direction of polarization are similar, we obtain

$$\left|\left(\frac{\partial\vec{\mu}}{\partial Q_s}\right)_0\right| = 2\sqrt{2}\mu S_{\mathrm{CH}}^{-1}\left([\varepsilon_{\mathrm{H}}, -\varepsilon_{\mathrm{C}}]\begin{bmatrix}-\sigma\\ 2\sigma\end{bmatrix}\theta^{(s)} l_{\rho^{(s)}}\right) = -\sqrt{2}\mu S_{\mathrm{CH}}^{-1} l_{\rho^{(s)}}$$

$$S = B_{1u}; \quad \theta^{(s)} = [2\sigma(\varepsilon_{\mathrm{H}} + 2\varepsilon_{\mathrm{C}})]^{-1} \qquad (11.71)$$

One property related to the calculation of $(\partial\vec{\mu}/\partial Q_i)_0$ deserves comment. In several cases the bond twists can be found directly without any computation. For instance, ethylene's C═C bond does not rotate in space during planar vibrations. Hence, the rotation of the C—H bond is simply equal , in absolute terms, to the change in the HCC angle during the vibrations, i.e. it is equal to the co-ordinate $\beta = \frac{1}{2}\beta^{(s)}$. For non-planar vibrations it is again possible to infer from the constancy of the direction of the C═C bond in space that

$$|d\vec{e}_{\mathrm{CH}}| = S_{\mathrm{CH}}^{-1}\rho\cos\frac{\varphi}{2} = \frac{\rho}{2S_{\mathrm{CH}}}$$

The formulae we have obtained enable us to make a number of general conclusions. In most cases, especially in organic compounds containing bonds with light atoms at one end, we have to deal with vibrations that are quite characteristic in form, i.e. vibrations in which only some of the vibrational co-ordinates are non-zero. Then it is only necessary to know about the geometrical structure and distribution of mass (elements of the $[\partial\vec{e}/\partial q, \partial\vec{e}/\partial\gamma]$ are placed for this in a numbered way) to obtain formulae such as (11.68). The compilation is never difficult, even for complicated molecules. This makes it possible in a whole series of cases for us to separate out the main parameters on which the intensity of a given band is dependent. In other words, we can make a crude categorization of intensities, something that is necessary, in principle, for any attempts to use the intensity method to study molecular structure. If, moreover, a series of electro-optical parameters is known, then an approximate quantitative estimate of the band intensity becomes possible.

On the other hand, it is frequently necessary to estimate the electro-optical parameters via the observable intensities of the bands. This is particularly true of the bond dipole moments and their derivatives with respect to the extensions of the same bonds. In this case, an analysis of the general formulae for $(\partial\vec{\mu}/\partial Q_i)_0$ allows us to separate out those vibrations in which the parameters that are of interest are of decisive importance. For example, if we need to find the dipole moment of a CH bond in a molecule such as ethylene then it is plain that we must study the band intensities of the non-planar vibrations. The CH bond dipole moment enters the expression for $(\partial\vec{\mu}/\partial Q_i)_0$ for the planar vibrations in combination with $\partial\mu/\partial\beta$, and cannot be found from the measurement of the band intensity of one deformational vibration. Thus a knowledge of the general formulae for the first derivatives of a molecular dipole moment with respect to normal co-ordinates prevents many errors.

We will now comment on one essential corrollary. We have the following relations for the band intensities of the valency vibrations of ethylene:

$$\left|\left(\frac{\partial\vec{\mu}}{\partial Q_i}\right)_0^{B_{2u}}\right| = 2\left(\frac{\partial\vec{\mu}}{\partial q} - \frac{\partial\vec{\mu}}{\partial q'}\right)\sin\frac{\varphi}{2} l_{q^{(s)}i}$$

$$\left|\left(\frac{\partial\vec{\mu}}{\partial Q_i}\right)_0^{B_{3u}}\right| = 2\left(\frac{\partial\vec{\mu}}{\partial q} + \frac{\partial\vec{\mu}}{\partial q'}\right)\cos\frac{\varphi}{2} l_{q^{(s)}i}$$

Consequently, by comparing the experimental intensities of appropriate absorption bands we can estimate $\partial\mu/\partial q'$, and it is clear from a cursory argument that the normalized forms of both valency vibrations in the CH_2 group may be considered as approximately equal. Since $\partial\mu/\partial q'$ characterizes interactions between bonds it is possible to obtain very important information on the structure and properties of groups using a simple experimental procedure.

Consequently, it is clear that the derivation of the general formulae can be found very useful for choosing the absorption bands in the molecule's spectrum, a knowledge of the intensities of which allows us to solve problems concerning the structure of a compound.

11.7 Relationships between the Intensities of Isotopically Substituted Molecules

The intensity of a fundamental absorption band is proportional to the square of the derivative $(\partial\vec{\mu}/\partial Q_k)_0$. We can write the derivative as

$$\left(\frac{\partial\vec{\mu}}{\partial Q_k}\right)_0 = \frac{\partial\tilde{\vec{\mu}}^{(j)}}{\partial q} l_q^{(k)}$$

where the jth element of row of $\partial\vec{\mu}/\partial q$ is

$$\frac{\partial\vec{\mu}}{\partial q_j} = \tilde{\vec{e}}\frac{\partial\mu^{(j)}}{\partial q} + \tilde{\mu}\frac{\partial\vec{e}^{(j)}}{\partial q}$$

and $l_q^{(k)}$ is the form of the kth vibration. Thus

$$\left(\frac{\partial\vec{\mu}}{\partial Q_k}\right)_0^2 = \tilde{l}_q^{(k)}\frac{\partial\vec{\mu}}{\partial q}\frac{\partial\tilde{\vec{\mu}}}{\partial q} l_q^{(k)} \tag{11.72}$$

The product of the column $\partial\vec{\mu}/\partial q$ and the row $\partial\tilde{\vec{\mu}}/\partial q$ forms a square-symmetric matrix M, whose elements are $(M)_{ij} = \partial\vec{\mu}/\partial q_i \cdot \partial\vec{\mu}/\partial q_s$.

Neglecting the constant the sum of the intensities of all the fundamental absorption bands of a given molecule is equal to

$$\sum_k J_k = \sum_k \tilde{l}_q^{(k)} M l_q^{(k)}$$

This expression can be written in the form

$$\sum_k J_k = Sp\,\tilde{L}_q M L_q$$

Now, taking advantage of the rule about cyclically permutating the members of a matrix product under a trace sign, we find that

$$\sum_k J_k = SpML_q\tilde{L}_q = SpMT_p \tag{11.73}$$

Consequently, if there is a series of isotopic molecules with indices α, for each of which $\sum_\alpha a_\alpha T_p^{(\alpha)} = 0$, then these molecules must also fulfil the relation $\sum_\alpha a_\alpha MT^{(\alpha)} = 0$, provided that the elements of the matrix M do not change as a result of an isotopic substitution. It is the clear that the same molecules will also obey the relation

$$\sum_{\alpha,k} a_\alpha J_k^{(\alpha)} = 0 \tag{11.74}$$

This rule was first obtained by Decius[16], and is called the 'sum rule' for sums of the intensities of isotopic molecules. It is obeyed in every case for which the rule of the sums of the squares of frequencies operates.

Crawford[17] introduced another useful relation. From his conclusions we find that:

$$\sum_k \frac{1}{\lambda_k}\left(\frac{\partial\vec{\mu}}{\partial Q_k}\right)_0^2 = \sum_k \frac{J_k}{\lambda_k} = \sum_k \frac{1}{\lambda_k}\tilde{l}_q^{(k)} M l_q^{(k)}$$

$$= Sp\Lambda^{-1}\tilde{L}_q M L_q = SpML_q\Lambda^{-1}\tilde{L}_q = SpMW \tag{11.75}$$

where W is the inverse of the matrix of force constants. Consequently, if the elements $(M)_{ij}$ do not depend on isotopic substitution, then

$$\sum_k \frac{J_k}{\lambda_k}$$

will also be independent of isotopic substitution. This property enables us to establish a relationship between the intensities of any two pairs of isotopic molecules that is analogous to the product rule for frequencies.

The $(M)_{ij}$ elements will clearly not be changed during isotopic substitutions if the deformational term in the expression for $(\partial\vec{\mu}/\partial Q_k)_0$ does not change. During vibrations of a molecule, the atomic displacements depend on the form of the vibration and the conservation of the moment of momentum for the molecule as a whole. Thus twists of the bonds must be composed of two quantities. Moreover, the first term will be wholly determined by the forms of the vibration, whilst the second will be determined by the vibration of the molecule as a whole. It is clear that only the second component will depend on isotopic substitution. Thus the $(M)_{ij}$ terms will not change for isotopic substitutions if we do not have to allow for a contribution to $(\partial\vec{\mu}/\partial Q_k)_0$ due to any twists of the molecule as a whole.

This can be done when (1) the molecular dipole moment in the equilibrium state is zero and (2) when the moment is non-zero but a twist of the whole molecule in space is either absent or can be neglected for vibrations of a given symmetry type. It should be noted that for several types of vibrations, including rotations of one group of atoms relative to another in the molecule, the $(M)_{ij}$ elements may not be invariant with respect to isotopic molecules even in the absence of twists of the molecule in space.

Usually the matrix $[\partial\vec{e}/\partial q]$ varies very much less than the forms of the vibrations during isotopic substitutions. Thus the type of relationships considered here must be very nearly satisfied for any molecule.

The rules we have derived for the intensities of isotopic molecules are useful for estimating the intensities for bands not experimentally measured. Such estimates are also very helpful in checking the accuracy of values obtained theoretically.

11.8 The Intensity Theory of Overtone and Combination Bands

11.8.1 General formulation and formula of $(\partial^2\vec{\mu}/\partial Q_i\partial Q_j)_0$

We showed in Section 11.1 that, besides fundamentals, transitions between levels during which one quantum number changes by more than unity (overtones) are also possible in a polyatomic molecule, as are transitions with changes in several quantum numbers (i.e. combination bands). Overtones and combination transitions give rise to additional absorption bands in a molecules's spectrum. Frequently, these bands lie in parts of the spectrum that are more accessible for investigation than the fundamental absorption bands. The continuously growing application of vibrational spectroscopy to analytical problems and the solution of a wide diversity of problems concerning the structures and properties of molecules and interatomic effects requires that the maximum use be made of the information included in the experimentally measured data. This makes the study of overtone and combination bands both important and promising. True, the difficulties of accounting for the various factors that determine the positions of the corresponding bands in the spectra and, in particular, their intensities, also increase. The fact that overtones and combinations are observed experimentally often leads to errors in interpreting the spectra.

Usually, the first overtones and the summed combinations where transitions from the ground state to a level with two quantum numbers equalling unity are the only ones encountered in practice. The intensities of these bands are determined, in the harmonic approximation, by the squares of terms such as

$$\left(\frac{\partial\vec{\mu}}{\partial Q_i^2}\right)_0 \sqrt{(2)}Q_{0i}^2$$

and

$$\left(\frac{\partial^2\vec{\mu}}{\partial Q_i\partial Q_j}\right)_0 Q_{0i}Q_{0j}$$

respectively, whilst the polarizations are determined by the directions of these vectors. In this way the task reduces to calculating the second derivative of a molecular dipole moment with respect to normal co-ordinates.

In the general calculation of the matrix elements for probabilities of higher-order transitions it is necessary to take into account both the mechanical and electro-optical anharmonicity. The simplest method of considering the former is to obtain the anharmonic wavefunctions by using the variational harmonic functions as a basis. Such a method, suitable for large and small molecules, has been developed in Chapter 10.

If the anharmonic wavefunction is assumed to be of the form $\Psi = \sum_v c_v \Psi_v^{(0)}$, where $\Psi_v^{(0)}$ are the harmonic basis functions and the molecular dipole moment is expanded into a series of normal co-ordinates of the harmonic approximation, then the expression for the probability of the dipole transition can be written in the form

$$\vec{\mu}^{\prime,\prime\prime} = \sum_n \left(\frac{\partial \vec{\mu}}{\partial Q_n}\right)_0 \tilde{C}'[\langle Q_n \rangle]C'' + \tfrac{1}{2}\sum_{k,n}\left(\frac{\partial^2 \vec{\mu}}{\partial Q_k \partial Q_n}\right)_0 \tilde{C}'[\langle Q_k Q_n \rangle]C'' + \cdots$$

where $[\langle Q_n \rangle]$ and $[\langle Q_k Q_n \rangle]$ denote the square-symmetric matrices of the values of the matrix elements of the normal co-ordinates and their products for all the combinations of the basis harmonic eigenfunctions. Calculation of these functions is straightforward. The symbols $\tilde{C}'$ and C'' denote the row and column matrices, respectively, composed of the c_v coefficients for the combined vibrational states.

We will again proceed from the representation of a molecular dipole moment in the form of a vector sum and take advantage of the formula for the first derivative of the moment with respect to a normal co-ordinate, writing it as:

$$\frac{\partial \vec{\mu}}{\partial Q_i} = \tilde{l}_q^{(i)}\left\{\left[\frac{\partial \tilde{\mu}}{\partial q}\right]\vec{e} + \left[\frac{\partial \tilde{\vec{e}}}{\partial q}\right]\mu\right\} \tag{11.76}$$

where $\tilde{l}_q^{(i)}$ is the row of matrix $\tilde{L}_q$. We differentiate this expression with respect to a normal co-ordinate, allowing for the fact that $\tilde{l}_q^{(i)}$ is a constant and that the derivatives of some functions with respect to a normal co-ordinate are, when transformed to natural co-ordinates, equal to the row matrices of the derivatives of the function with respect to the natural co-ordinates multiplied by the column vibration form. Hence we set

$$\begin{aligned}\left(\frac{\partial^2 \vec{\mu}}{\partial Q_i \partial Q_j}\right)_0 = \tilde{l}_q^{(i)}\Bigg\{&\frac{\partial}{\partial q}\left[\frac{\partial \tilde{\mu}}{\partial q}\right] l_q^{(j)}\vec{e} + \left[\frac{\partial \tilde{\mu}}{\partial q}\right]\left(\frac{\partial \vec{e}^{(1)}}{\partial q} l_{q1j} + \cdots + \frac{\partial \vec{e}^{(n)}}{\partial q} l_{qnj}\right)\\ &+ \left[\frac{\partial \tilde{\vec{e}}}{\partial q}\right]\left(\frac{\partial \mu^{(1)}}{\partial q} l_{q1j} + \cdots + \frac{\partial \mu^{(n)}}{\partial q} l_{qnj}\right)\\ &+ \frac{\partial}{\partial q}\left[\frac{\partial \tilde{\vec{e}}}{\partial q}\right] l_q^{(j)}\mu\Bigg\} = \tilde{l}_q^{(i)}\left\{\frac{\partial}{\partial q}\left[\frac{\partial \tilde{\mu}}{\partial q}\right] l_q^{(j)}\vec{e}\right\}\\ &+ \tilde{l}_q^{(i)}\left\{\left[\frac{\partial \tilde{\mu}}{\partial q}\right]\left[\frac{\partial \vec{e}}{\partial q}\right] + \left[\frac{\partial \tilde{\vec{e}}}{\partial q}\right]\left[\frac{\partial \mu}{\partial q}\right]\right\} l_q^{(j)}\\ &+ \tilde{l}_q^{(i)}\left\{\frac{\partial}{\partial q}\left[\frac{\partial \tilde{\vec{e}}}{\partial q}\right] l_q^{(j)}\mu\right\} l_q^{(j)}\end{aligned} \tag{11.77}$$

It is easy to show that

$$\left\{\frac{\partial}{\partial q}\left[\frac{\partial \mu}{\partial q}\right]\right\} l_q^{(j)} \vec{e}$$

can be expressed in the form of a product of simply structured matrices, which is very useful for practical purposes. Thus:

$$\left[\left\{\frac{\partial^2 \mu_m}{\partial q_k \partial q_e}\right\}_m\right] \vec{\mathscr{E}} l_q^{(j)} \tag{11.78}$$

where

$$\left[\left\{\frac{\partial^2 \mu_{m'}}{\partial q_k \partial q_l}\right\}_m\right] = \left[\begin{array}{ccc:ccc} \left(\dfrac{\partial^2 \mu_1}{\partial q_1^2}\right)_0 & \cdots & \left(\dfrac{\partial^2 \mu_1}{\partial q_1 \partial q_n}\right)_0 & \left(\dfrac{\partial^2 \mu_m}{\partial q_1^2}\right)_0 & \cdots & \left(\dfrac{\partial^2 \mu_m}{\partial q_1 \partial q_n}\right)_0 \\ \hdashline \left(\dfrac{\partial^2 \mu_1}{\partial q_n \partial q_1}\right)_0 & \cdots & \left(\dfrac{\partial^2 \mu_1}{\partial q_n^2}\right)_0 & \left(\dfrac{\partial^2 \mu_m}{\partial q_n \partial q_1}\right)_0 & \cdots & \left(\dfrac{\partial^2 \mu_m}{\partial^2 q_n}\right)_0 \end{array}\right] \tag{11.79}$$

and

$$\vec{\mathscr{E}} = \begin{bmatrix} \vec{e}_1 & 0 & \cdots & 0 \\ 0 & \vec{e}_1 & \cdots & 0 \\ & & \cdots & \\ 0 & & \cdots & \vec{e}_1 \\ & & \cdots & \\ \vec{e}_m & & \cdots & 0 \\ & & \cdots & \\ 0 & & \cdots & \vec{e}_m \end{bmatrix} \begin{matrix} \left.\begin{matrix} \\ \\ \\ \\ \end{matrix}\right\} n \\ \\ \left.\begin{matrix} \\ \\ \\ \end{matrix}\right\} n \end{matrix} \Bigg\} m$$

As a result, by substituting an obvious expression for $[\partial \vec{e}/\partial q]$ we obtain

$$\begin{aligned} \left(\frac{\partial^2 \vec{\mu}}{\partial Q_i \partial Q_j}\right)_0 = \tilde{l}_q^{(i)} \Bigg\{ & \left[\left[\left\{\frac{\partial^2 \vec{\mu}_{m'}}{\partial q_k \partial q_l}\right\}_m\right] \vec{\mathscr{E}} + \left[\frac{\widetilde{\partial \mu}}{\partial q}\right] S^{-1}(\Delta E \vec{\tilde{B}} \theta - [\vec{E}, 0])\right. \\ & + \left(\theta \vec{B} \tilde{\Delta} E - \begin{bmatrix} \vec{E} \\ 0 \end{bmatrix}\right) S^{-1} \left[\frac{\partial \mu}{\partial q}\right] + \left[\left\{\frac{\partial^2 \vec{e}_{m'}}{\partial q_k \partial q_l}\right\}_m\right] M \Bigg\} l_q^{(j)} \end{aligned} \tag{11.80}$$

The matrices $[\{\partial^2 \vec{e}_{m'}/\partial q_k \partial q_l\}_m]$ and M have the same structures as equations (11.78) and (11.79), but with μ and $\vec{e}$ interchanged.

The method of calculating the vectors $(\partial^2 \vec{e}_m/\partial q_k \partial q_l)_0$ has been discussed previously (see Chapter 8), but will be briefly repeated here. The matrix of the partial derivatives of the unit bond vectors is given by

$$\left[\frac{\partial \vec{e}}{\partial q}\right] = S^{-1}(\Delta E \vec{\tilde{B}} \theta - \vec{E}') \tag{11.81}$$

$$\vec{E}' = [\vec{E}, 0]$$

where $[\partial\vec{e}/\partial q]$ is a matrix of partial derivatives of unit bond vectors with respect to the natural co-ordinates and θ is the matrix of kinetic energy in the co-ordinate representation (inverse to the kinematic one). Note that θ must also be calculated for the equilibrium state.

It follows from equation (11.81) that the first derivatives of the unit bond vectors are expressed in terms of the co-ordinate unit vectors $\vec{e}_k, \vec{f}_{kk}$, etc. which appear in the formulae for the elements of the matrices $\vec{B}$ and $\vec{E}'$.

Differentiation of the matrix of the second derivatives $[\partial^2\vec{e}/\partial q_k\partial q_l]$ with respect to a certain co-ordinate with the subscript l yields

$$\left[\frac{\partial^2\vec{e}_k}{\partial q_k\partial q_l}\right]=S^{-1}\left(\Delta E\frac{\partial\vec{B}}{\partial q_l}\theta+\Delta E\tilde{\vec{B}}\frac{\partial\Theta}{\partial q}-\frac{\partial\vec{E}'}{\partial q_l}\right)+\left(\frac{\partial}{\partial q_l}S^{-1}\right)(\Delta E\tilde{\vec{B}}\theta-\vec{E}')$$

When differentiating matrix $\vec{B}$, as well as obtaining the first derivatives of the vectors $\vec{e}_k$ according to the known rules, the vectors $\vec{f}_{kk'}$ and others must also be differentiated with respect to angular co-ordinates. For non-linear regions of a molecule the derivatives of the vectors $\vec{f}_{kk'}$ can be calculated in the simplest way by differentiating the expression

$$\vec{f}_{kk'}=\frac{1}{\sin(\vec{e}_k,\vec{e}_{k'})}(\vec{e}_k\times\vec{e}_{k'})\times\vec{e}_k$$

In calculating the derivatives of the vectors $\vec{f}_{kk'}$ for the linear regions the method used in valence-optical theory for the Raman spectral intensities is applied. Thus (see Chapter 15)

$$\frac{\partial\vec{f}_{kk'}}{\partial q_l}=-\left(\vec{f}_{kk'}\frac{\partial\vec{e}_k}{\partial q_l}\right)\vec{e}_k$$

The unit vectors $\vec{h}$ and $\vec{g}$ appear with non-planar co-ordinates and play the same role in dihedral angles as do the unit vectors $\vec{f}_{kk''}$ in the treatment of angular co-ordinates. Therefore the calculations of the above derivatives with respect to the internal co-ordinates can be performed in exactly the same way.

Having outlined the methods for calculating the derivatives of the elements of the matrix $\vec{B}$ with respect to the internal co-ordinates, it is significant that all the first derivatives are calculated in terms of the values of the co-ordinate unit vectors. This permits us to obtain the second derivatives $(\partial^2\vec{e}_m/\partial q_k\partial q_l)_0$, which are also expressed in terms of the unit vectors of the internal co-ordinates. Since the corresponding expressions are cumbersome they are not given in detail here, but a description of the calculation procedure follows.

Based on the same considerations, the derivatives $(\partial^3\vec{e}_m/\partial q_k\partial q_l\partial q_n)_0$ may be calculated and the quantities $(\partial^3\vec{\mu}/\partial Q_k\partial Q_n\partial Q_m)_0$ thus obtained. Formulae for the higher-order derivatives of the molecular dipole moment with respect to the normal co-ordinates cannot be obtained in simple matrix form. However, this in no way restricts us from formulating the corresponding algorithm as a computer program.

For simplicity, we will consider below everything at the harmonic (mechanical) approximation. Hence we neglect the terms containing second derivatives of the unit-orientating vectors of the bonds.

The product

$$\left[\left\{\frac{\partial^2 \mu_{m'}}{\partial q_k \partial q_l}\right\}_m\right]\vec{\mathscr{E}}$$

and the sum

$$\left(\left[\frac{\widetilde{\partial \mu}}{\partial q}\right]\left[\frac{\partial \vec{e}}{\partial q}\right]+\left[\frac{\widetilde{\partial \vec{e}}}{\partial q}\right]\left[\frac{\partial \mu}{\partial q}\right]\right)$$

are symmetric matrices. The first term in equation (11.80) characterizes the non-linear dependence of the parameter μ_m on changes in the equilibrium configuration of the molecule, the configuration being described by a collection of natural co-ordinates. This is why the elements of the matrix $[\{\partial^2\mu_{m'}/\partial q_k \partial q_l\}_m]$ are called the electro-optical anharmonic parameters, and we shall introduce them, as we did with the bond dipole moment and their first derivatives with respect to natural co-ordinates, as empirical parameters that describe a molecule's electron cloud. Obviously, all the comments we made as regards physical sense when we introduced μ_m and $(\partial\mu_m/\partial q_k)_0$ are extended to the new parameters. The second term in equation (11.80) is called the harmonic term. The formula we have obtained, which expresses in an obvious way the derivative that defines the probability of the transitions forming combinations and overtones (the formula being in terms of all the mechanical and electro-optical quantities of the molecule, the frequencies and forms of the normal vibrations in the usual system of vibrational co-ordinates and co-ordinate vectors), enables us to draw a number of interesting and general conclusions.

It is clear, above all, that the overtones and combination bands will appear in an infra-red spectrum in the absence of both electro-optical and mechanical anharmonicity. The appropriate harmonic term equation (11.80) may have quite a significant value, because it includes parameters which determine the intensity of fundamental vibrations, even in the formula for the first derivative of the dipole moment with respect to a normal co-ordinate.

Thus the appearance of overtone and combination bands in spectra cannot be considered as proof of a significant influence of anharmonicity. The elements of the $[\partial\vec{e}/\partial q]$ matrix are not zero in general, even for purely valency vibrations, which is why even in the additive approximation the harmonic terms may not vanish. In cases of departure from additivity, when the non-diagonal derivatives in the $[\partial\mu/\partial q]$ matrix play a significant role, the harmonic terms may grow even faster. It is not impossible that it is really this factor that explains the widely known fact that in the spectra of a number of molecules the overtone and combination bands are more intense than some of the fundamental absorption bands. For diatomic molecules $\partial\vec{e}/\partial q = 0$, and only the term characterizing electro-optical anharmonicity remains. Apparently, the mechanical transference of this result to polyatomic molecules has led to the widely held opinion concerning the necessity of using anharmonicity to explain the appearance of overtone and combination bands in their spectra.

In the harmonic vibration approximation the intensities of the overtones and combination bands do not depend on the bond dipole moments; they are determined instead only by the bond dipole moments' first and second derivatives with respect to natural vibration co-ordinates.

In the harmonic term the changes in the absolute values of the bond dipole moment due to a normal vibration are multiplied by the rotations in the respective bonds during the vibration. This is why in the general case a weak overtone or combination band may correspond to a strong fundamental band, and vice versa. This should be borne in mind, particularly when interpreting the origin of bands in spectra, as many workers mistakenly consider that strong fundamental bands must correspond to strong overtone or combination bands.

The anharmonic term is very important for overtones and combinations of purely bond-stretching vibrations, while the harmonic term is important for overtones and combinations of purely deformational vibrations. The latter is especially true in non-additive molecules. Both terms are large in combination vibrations, both valency and deformational. This lends credence to the belief that the first and second derivatives of the bond dipole moment with respect to angle co-ordinates are comparable in size in non-additive molecules to the derivatives with respect to the bond-extension co-ordinates.

This is why it would seem that we can expect the intensities of the overtones and combination bands to be, respectively, large for valency–valency vibrations, for deformational–deformational vibrations and for deformational–valency vibrations.

The intensities of these bands are much more dependent on the forms of the vibrations than are those of the fundamentals. It is easy to see that the third derivatives of a molecule's dipole moment with respect to normal co-ordinates will be non-zero only in the presence of electro-optical or mechanical anharmonicity. This is why a sudden decrease in intensity must be observed for the appropriate bands. Actually, such absorption bands are encountered very infrequently, as we have already indicated.

Equation (11.80) provides a simple and rapid method of calculating any second derivative of a molecular dipole moment with respect to a normal co-ordinate if all the necessary quantities are known. Hence, as was the case when analysing fundamental vibrations, a matrix multiplier that is constant for all vibrations can be separated out. Being able to use directly the results of estimates for fundamental frequencies—the matrices containing the derivatives of bond dipole moments with respect to natural co-ordinates and the analogous derivatives of the bond directional vector—makes the calculations very much easier.

Unfortunately, we usually do not have any information about the values of the elements of the matrix

$$\left[\left\{\frac{\partial^2 \mu_{m'}}{\partial q_l \partial q_k}\right\}_m\right]$$

The few investigations carried out have been concerned only with diatomic molecules or with the localized vibration of a single bond in a complex molecule. These results cannot be simply transferred to polyatomic molecules, especially non-additive ones. Hence the determination of anharmonic electro-optical parameters using experimentally measured intensities and polarizations of overtone and combination bands is vital both for the advancement of our theoretical understanding and for practical applications. We can consider, in this respect, that the frequencies and forms of normal vibrations, the element of the matrices $[\partial\mu/\partial q]$ and $[\partial\vec{e}/\partial q]$ and the size and direction of the second derivative $(\partial^2\vec{\mu}/\partial Q_i\partial Q_j)_0$ are all known (although for this to be true of the second derivative all the signs of the $(\partial^2\vec{\mu}/\partial Q_i^2)_0$ derivatives must be the same). By substituting these data into equation (11.79) we obtain a system of vector equations in terms of the unknown elements of the matrix

$$\left[\left\{\frac{\partial^2\mu_{m'}}{\partial q_k\partial q_l}\right\}_m\right]$$

The number of equations will equal that of the overtone and combination frequencies of a molecule. Since the latter yields a number of combinations of two out of $3N-6$ (N being the number of atoms in the molecule), the number of equations is $(3N-6)+\frac{1}{2}(3N-6)\cdot(3N-7)=\frac{1}{2}(3N-6)(3N-5)$, which is equivalent to a system of $3\cdot\frac{1}{2}(3N-6)(3N-5)$ independent scalar heterogeneous equations. The number of unknowns for a non-linear molecule is clearly equal to $\frac{1}{2}(N-1)(3N-6)(3N-5)$. Consequently we have a situation similar to that for the electro-optical parameters μ_m and $\partial\mu_m/\partial q_k$ in that the number of equations for one molecule is less than the number of unknowns once $N>4$. The magnitudes but not the absolute signs of the anharmonic electro-optical parameters can be calculated, though the relative signs of these parameters and the $\partial\mu_m/\partial q_k$ can be determined.

If a molecule has symmetry elements, then the number of different unknowns is reduced, although at the same time that of equations is also decreased. This is because some of the transitions can be prohibited by selection rules on the grounds of symmetry, or the directions of the polarization are determined by vibration symmetry or it may be found that the anharmonic terms may be zero by definition, as happens for the non-planar vibrations of planar molecules. It is not difficult to find the number of unknowns and independent equations for each specific case.

When using data for isotopic molecules it is useful to take into account that relationships between the $(\partial^2\vec{\mu}/\partial Q_i^2)_0$ derivatives and the powers of the derivatives for the overtones of isotopically substituted molecules when there is an absence or near-absence in normal vibrations of a given symmetry type of rotations of the molecule as a whole (excluding, perhaps, some types of rotational vibrations) must be the same as the relationships between the squares of the first derivative of a molecule's dipole moment with respect to normal co-ordinates. In particular, the following relation must hold for isotopically substituted molecules:

$$\sum_k \left(\frac{\partial^2 \vec{\mu}}{\partial Q_k^2}\right)_0 (XYZ) + \sum_k \left(\frac{\partial^2 \vec{\mu}}{\partial Q_k^2}\right)_0 (X'YZ')$$

$$= \sum_k \left(\frac{\partial^2 \vec{\mu}}{\partial Q_k^2}\right)_0 (X'YZ) + \sum_k \left(\frac{\partial^2 \vec{\mu}}{\partial Q_k^2}\right)_0 (XYZ') \tag{11.82}$$

$$\sum_k \frac{1}{\lambda_k}\left(\frac{\partial^2 \vec{\mu}}{\partial Q_k^2}\right) = \text{const.}$$

Furthermore,

$$\left(\frac{\partial^2 \vec{\mu}}{\partial Q_k^2}\right)_0 = \widetilde{l_q^{(k)}}\vec{A}\, l_q^{(k)}$$

i.e. it like $(\partial\vec{\mu}/\partial Q_k)_0^2$, is a quadratic form with a matrix

$$\vec{A} = \left[\left\{\frac{\partial^2 \mu_{m'}}{\partial q_k \partial q_l}\right\}_m\right]\vec{\mathscr{E}} + \left[\widetilde{\frac{\partial \mu}{\partial q}}\right]\left[\frac{\partial \vec{e}}{\partial q}\right] + \left[\widetilde{\frac{\partial \vec{e}}{\partial q}}\right]\left[\frac{\partial \mu}{\partial q}\right] \tag{11.83}$$

whose elements are independent of isotopic substitution if we can neglect all rotations of the molecules as a whole in space.

Equations (11.82) and (11.83) may be used to calculate the appropriate $(\partial^2\vec{\mu}/\partial Q_k^2)_0$ derivatives for the vibrations of particular isotopically substituted molecules if they are known experimentally for other molecules.

The elements of the matrix

$$\left[\left\{\frac{\partial^2 \mu_{m'}}{\partial q_k \partial q_l}\right\}_m\right]$$

will not be invariant with respect to a partial change of co-ordinates if the molecule is neither fully nor partially additive. Let us choose a group in the molecule and number it I. The remaining parts of the molecule will be considered to be the second group (II). We divide all the molecule's co-ordinates into three types, i.e. co-ordinates internal to the first group, internal to the second group and relating the movement of the first group with respect to the second.

The matrix

$$\left[\left\{\frac{\partial^2 \mu_{m'}}{\partial q_k \partial q_l}\right\}_m\right]$$

must then be written in the form

$$\begin{bmatrix} \dfrac{\partial^2 \mu^{\mathrm{I}}}{\partial q^{\mathrm{I}^2}} & 0 & 0 & 0 & 0 & 0 \\ 0 & 0 & 0 & 0 & 0 & 0 \\ 0 & 0 & 0 & 0 & 0 & \dfrac{\partial^2 \mu^{\mathrm{II}}}{\partial q^{\mathrm{II}^2}} \end{bmatrix} \tag{11.84}$$

It is easy to verify that the submatrix $\partial^2\mu^{\mathrm{I}}/\partial q^{\mathrm{I}^2}$ does not change for such a transformation of variables when the internal co-ordinates of the first group do not change, while the second group's co-ordinates and the intermediate ones are substituted by their linear combinations.

If the molecule has some symmetry then this enables us to simplify the compilation of the general formulae for the second derivatives of a molecule's dipole moment with respect to normal co-ordinates. We transform equation (11.80) to go over to symmetry co-ordinates. For a harmonic term this transformation entails a transition from matrices containing bond dipole moment derivatives with respect to natural vibration co-ordinates to matrices containing derivatives with respect to symmetry ones. These matrices are, respectively:

$$\left[\frac{\partial\mu}{\partial q^{(s)}}\right]=\left[\frac{\partial\mu}{\partial q}\right]C^{(s)} \tag{11.85}$$

and

$$\left[\frac{\partial\vec{e}}{\partial q^{(s)}}\right]=\left[\frac{\partial\vec{e}}{\partial q^{(s)}}\right]C^{(s)}=S^{-1}(\Delta E\tilde{\vec{B}}^{(s)}\theta^{(s)}-[\vec{E}C^{(s)},0])$$

where $C^{(s)}$ is the matrix containing the symmetry coefficients for the transformation from the type s symmetry co-ordinates to the natural ones. When calculating the harmonic term we should also take into account the relationships between the row of matrices $[\partial\mu/\partial q^{(s)}]$ and $[\partial\vec{e}/\partial q^{(s)}]$ for equivalent bonds in order to simplify the mathematical operations.

We shall now investigate the anharmonic term. When the term is premultiplied by the transposed matrix for the transformation from the symmetry co-ordinates to the original ones, the rows corresponding to equivalent co-ordinates with coefficients equal to the symmetry ones must be added in the matrix

$$\left[\left\{\frac{\partial^2\mu_{m'}}{\partial q_k\partial q_l}\right\}_m\right]$$

Post-multiplying the matrix

$$\left[\left\{\frac{\partial^2\mu_{m'}}{\partial q_k\partial q_l}\right\}_m\right]\vec{\mathscr{E}}$$

by the matrix for the transformation from symmetry co-ordinates to the original ones is equivalent to adding the normalized symmetry coefficients to the columns of the submatrices containing the second derivatives for every bond in the molecule and equivalent to a transition in the $\vec{\mathscr{E}}$ matrix of the corresponding directional vectors, whose rank is equal to the number of symmetry co-ordinates of a given type, to square submatrices. As a result, the anharmonic term can be written in the form

$$\begin{bmatrix} \dfrac{\partial^2\mu_1}{\partial q_1^{(s_1)}\partial q_1^{(s_2)}} & \cdots & \dfrac{\partial^2\mu_1}{\partial q_1^{(s_1)}\partial q_n^{(s_2)}} & \cdots & \dfrac{\partial^2\mu_m}{\partial q_1^{(s_1)}\partial q_1^{(s_2)}} & \cdots & \dfrac{\partial^2\mu_m}{\partial q_1^{(s_1)}\partial q_n^{(s_2)}} \\ \cdots & \cdots & \cdots & \cdots & \cdots & \cdots & \cdots \\ \dfrac{\partial^2\mu_1}{\partial q_k^{(s_1)}\partial q_1^{(s_2)}} & \cdots & \dfrac{\partial^2\mu_1}{\partial q_k^{(s_1)}\partial q_n^{(s_2)}} & \cdots & \dfrac{\partial^2\mu_m}{\partial q_k^{(s_1)}\partial q_1^{(s_2)}} & \cdots & \dfrac{\partial^2\mu_m}{\partial q_k^{(s_1)}\partial q_n^{(s_2)}} \end{bmatrix}$$

$$\times \overbrace{\begin{bmatrix} \vec{e}_1 & \cdots & 0 \\ \cdots & \cdots & \cdots \\ 0 & \cdots & \vec{e}_1 \\ \cdots & \cdots & \cdots \\ \vec{e}_m & \cdots & 0 \\ \cdots & \cdots & \cdots \\ 0 & \cdots & \vec{e}_m \end{bmatrix}}^{n} \left.\begin{matrix} \}n \\ \\ \}n \end{matrix}\right\} m \tag{11.86}$$

where m is the number of bonds; k is the number of s_1-type symmetry co-ordinates; n is the number of s_2-type symmetry co-ordinates; and s_1 and s_2 are superscripts indicating the first and second normal vibrations. Since the number of s_1- and s_2-type symmetry co-ordinates may be different, the submatrix for the second derivatives of the bond dipole moments with respect to symmetry co-ordinates and the matrix

$$\left(\left[\frac{\partial\widetilde{\mu}}{\partial q^{(s_1)}}\right]\left[\frac{\partial\vec{e}}{\partial q^{(s_2)}}\right]+\left[\frac{\partial\widetilde{\vec{e}}}{\partial q^{(s_1)}}\right]\left[\frac{\partial\mu}{\partial q^{(s_2)}}\right]\right)$$

will generally be rectangular.

It is easy to show that the submatrix containing the second derivatives of bond dipole moments for all the equivalent bonds can be obtained by multiplying the submatrix of a bond, assumed to be the first, by the square of the unnormalized symmetry coefficient (or by the product of the unnormalized symmetry coefficients for the combination frequencies formed by normal vibrations of different types of symmetry), the coefficient determining the ratio of the amplitude of the natural co-ordinate of a given bond's extension to the first bond in a normal vibration of the appropriate type.

Therefore, in order to simplify the anharmonic term further we need retain only one of the submatrices containing the second derivatives of μ_k with respect to symmetry co-ordinates out of the set of equivalent bonds. Meanwhile, we can, in matrix $\vec{\mathscr{E}}$, add, line by line, the submatrices containing the directional vectors of equivalent bonds to their coefficients which are equal to the squares (or products) of the unnormalized symmetry coefficients for the natural coordinates of the extensions of the appropriate bonds. We designate the matrices we obtain this way by the subscript equ (the matrix elements of $\vec{\mathscr{E}}_{\text{equ}}$ we will call the symmetry vectors of the overtone and combination frequencies). Finally, we write:

$$\left(\frac{\partial^2\vec{\mu}}{\partial Q_i\partial Q_j}\right)_0 = \tilde{l}^{(i)}_{q^{(s_1)}}\left[\left[\left\{\frac{\partial^2\mu_{m'}}{\partial q_k^{(s_1)}\partial q_l^{(s_2)}}\right\}_m\right]_{\text{equ}}\vec{\mathscr{E}}_{\text{equ}} + \left[\frac{\widetilde{\partial\mu}}{\partial q^{(s_1)}}\right]\left[\frac{\partial\vec{e}}{\partial q^{(s_2)}}\right] + \left[\frac{\widetilde{\partial\vec{e}}}{\partial q^{(s_1)}}\right]\left[\frac{\partial\mu}{\partial q^{(s_2)}}\right]\right] l^{(j)}_{q^{(s_2)}} \tag{11.87}$$

If the normal vibration for overtones or for combinations belongs to symmetry type such that the extension co-ordinate of a bond is equal to zero by definition, then, by analogy with the case of the first derivative with respect to a normal co-ordinate, the submatrices corresponding to this bond ought to be excluded from the

$$\left[\left\{\frac{\partial^2\mu_{m'}}{\partial q_k^{(s_1)}\partial q_l^{(s_2)}}\right\}_m\right]_{\text{equ}}$$

and $\vec{\mathscr{E}}_{\text{equ}}$ matrices.

This is why, for instance, the overtones and combinations of non-planar vibrations of planar molecules and the deformational vibrations of linear molecules are not dependent on the anharmonic term and are polarized perpendicularly to the plane and axis of the molecule. This property may be used to determine from measurements of the intensities and polarizations of appropriate absorption bands those electro-optical parameters of planar and linear molecules that cannot be found from a study of the fundamental vibrations of this type because the valency term is missing in the expression for the band intensity. The derivatives of bond dipole moments with respect to natural non-planar co-ordinate (shifts of the bond out of the plane of the molecule, changes in the dihedral angle, etc.) are such electro-optical parameters. By contrast, the harmonic term (rotations of the bonds here are zero by definition) disappears for overtones and combination bands for a linear vibration.

As is the case for fundamental vibrations, it may be found that the $(\partial^2\vec{\mu}/\partial Q_i\partial Q_j)_0$ derivatives are zero by definition for some types of symmetry. The corresponding transition will thus be prohibited, in agreement with the selection rules, on grounds of symmetry.

The activity or inactivity of a given vibration in an infra-red spectrum when none of the bond lengths change may be easily established by considering a general form of the harmonic term on the basis of the symmetry vectors for appropriate co-ordinates of the overtone and combination vibrations.

In all other cases the anharmonic and harmonic terms will vanish simultaneously. This is because the activity or inactivity of a vibration in an infra-red spectrum depends on the symmetries of the molecule and of the vibration, and are not related to size of the parameters of the molecule's electron cloud or the particular form of the vibration. Hence in order to find whether an overtone or combination vibration, of a particular symmetry type, is active in an infra-red spectrum it is sufficient to compile the matrix $\vec{\mathscr{E}}_{\text{equ}}$. The direction of the elements—the vectors of this matrix—will show the direction of the absorption bands' polarizations (if the vectors are all collinear). If they are zero then this will indicate that transitions between the levels are absent. Here, the $[\partial^2\mu_1/\partial q_1\partial q_k]$

submatrices of particular bonds in molecules that are highly symmetric and are without dipoles must be zero, as must the elements of the remaining submatrices which correspond to the derivatives with respect to extensions of certain bonds. We have here an analogy with the properties of the $[\partial\mu/\partial q]$ matrix.

When these electro-optical parameters, their corresponding vibrational co-ordinates being dependent, are used, an exclusion of a co-ordinate from an anharmonic term is analogous to that of a dependent co-ordinate from the force constants matrix, whilst an exclusion from the harmonic term entails one governed by the rules presented in Section 11.5. The row and column corresponding to the excluded coordinate should thus be deleted from all the submatrices in $\vec{\mathscr{E}}$ which contain the directional vector of the bond.

11.8.2 The example of the second derivative calculation

We now provide examples of deriving formulae for the second derivatives of a molecular dipole moment with respect to normal co-ordinates. We shall consider the non-linear symmetric molecule XY_2. Following the general formula (11.87), we obtain the following for the A-type overtone and combination normal vibrations:

$$\left(\frac{\partial^2\vec{\mu}}{\partial Q_i \partial Q_j}\right)_0^A = [l_{q^{A_i}}, l_{\gamma i}]\left[\begin{bmatrix} \dfrac{\partial^2\mu_1}{\partial q^{A^2}} & \dfrac{\partial^2\mu_1}{\partial q^A\partial\gamma} \\ \dfrac{\partial^2\mu_1}{\partial\gamma\partial q^A} & \dfrac{\partial^2\mu_1}{\partial\gamma^2}\end{bmatrix}\begin{bmatrix}(\vec{e}_1+\vec{e}_2) & 0 \\ 0 & (\vec{e}_1+\vec{e}_2)\end{bmatrix}\right.$$

$$\times\begin{bmatrix} \dfrac{\partial\mu_1}{\partial q^A} & \dfrac{\partial\mu_2}{\partial q^A} \\ \dfrac{\partial\mu_1}{\partial\gamma} & \dfrac{\partial\mu_2}{\partial\gamma}\end{bmatrix}\begin{bmatrix} \dfrac{\partial\vec{e}_1}{\partial q^A} & \dfrac{\partial\vec{e}_1}{\partial\gamma} \\ \dfrac{\partial\vec{e}_2}{\partial q^A} & \dfrac{\partial\vec{e}_2}{\partial\gamma}\end{bmatrix}$$

$$\left. + \begin{bmatrix} \dfrac{\partial\vec{e}_1}{\partial q^A} & \dfrac{\partial\vec{e}_2}{\partial q^A} \\ \dfrac{\partial\vec{e}_1}{\partial\gamma} & \dfrac{\partial\vec{e}_2}{\partial\gamma}\end{bmatrix}\begin{bmatrix} \dfrac{\partial\mu_1}{\partial q^A} & \dfrac{\partial\mu_1}{\partial\gamma} \\ \dfrac{\partial\mu_2}{\partial q^A} & \dfrac{\partial\mu_2}{\partial\gamma}\end{bmatrix}\right]\begin{bmatrix} l_{q^{A_j}} \\ l_{\gamma j}\end{bmatrix}$$

By taking account that $(\partial\vec{e}_1/\partial q^A) = (\partial\vec{e}_2/\partial q^A) = 0$, through $(\partial\mu_1/\partial q^A) = (\partial\mu_2/\partial q^A)$ and $(\partial\mu_1/\partial\gamma) = (\partial\mu_2/\partial\gamma)$ (see the estimates of the intensities of the fundamental vibrations in Section 11.6), and by multiplying the matrices we obtain

$$\left(\frac{\partial^2\vec{\mu}}{\partial Q_i \partial Q_j}\right)_0 = \left(\frac{\partial^2\mu}{\partial q^{A^2}} l_{q^{A_i}} l_{q^{A_j}} + 2\frac{\partial^2\mu}{\partial q^A\partial\gamma} l_{q^{A_i}} l_{\gamma j} + \frac{\partial^2\mu}{\partial\gamma^2} l_{\gamma i} l_{\gamma j}\right)$$
$$\times(\vec{e}_1+\vec{e}_2) + 2\left(\frac{\partial\mu}{\partial q^A} l_{q^{A_i}} + \frac{\partial\mu}{\partial\gamma} l_{\gamma i}\right)\left(\frac{\partial\vec{e}_1}{\partial\gamma} + \frac{\partial\vec{e}_2}{\partial\gamma}\right) l_{\gamma j} \quad i,j = 1,2 \tag{11.88}$$

The length of the vector

$$\left(\frac{\partial \vec{e}_1}{\partial \gamma} + \frac{\partial \vec{e}_2}{\partial \gamma}\right) l_{\gamma j}$$

is equal to twice the projection of the rotation vector of bond I (or II) onto the direction of the polarization (the y axis) in a normal vibration of a given symmetry type, i.e.

$$\left|\left(\frac{\partial \vec{e}_1}{\partial \gamma} + \frac{\partial \vec{e}_2}{\partial \gamma}\right) l_{\gamma j}\right| = s^{-1} l_{\gamma j} \sin \frac{\varphi}{2}$$

Moreover, $(e_{1y} + e_{2y}) = 2\cos \varphi/2$. Thus in terms of symmetry co-ordinates we have

$$\left|\left(\frac{\partial^2 \vec{\mu}}{\partial Q_i \partial Q_j}\right)_0^A\right| = 2\left(\frac{\partial^2 \mu}{\partial q^{A^2}} l_{q^{A_i}} l_{q^{A_j}} + 2\frac{\partial^2 \mu}{\partial q^A \partial \gamma} l_{q^{A_i}} l_{\gamma j} + \frac{\partial^2 \mu}{\partial \gamma^2} l_{\gamma i} l_{\gamma j}\right) \cos \frac{\varphi}{2}$$
$$+ \frac{2}{s}\left(\frac{\partial \mu}{\partial q^A} l_{q^{A_i}} l_{\gamma j} + \frac{\partial \mu}{\partial \gamma} l_{\gamma i} l_{\gamma j}\right) \sin \frac{\varphi}{2}, \quad i, j = 1, 2 \qquad (11.89)$$

We complete the transformation to the derivatives with respect to the natural vibrational co-ordinates, which represent the electro-optical and anharmonic electro-optical parameters, and arrive at

$$\frac{\partial \mu}{\partial q^A} = \frac{1}{\sqrt{2}}\left(\frac{\partial \mu}{\partial q} + \frac{\partial \mu}{\partial q'}\right), \quad \frac{\partial^2 \mu}{\partial q^{A^2}} = \frac{1}{2}\left(\frac{\partial^2 \mu}{\partial q^2} + 2\frac{\partial^2 \mu}{\partial q \partial q'} + \frac{\partial^2 \mu}{\partial q'^2}\right)$$

$$\frac{\partial}{\partial \gamma}\left(\frac{\partial \mu}{\partial q^A}\right) = \frac{1}{\sqrt{2}}\left(\frac{\partial^2 \mu}{\partial q \partial \gamma} + \frac{\partial^2 \mu}{\partial q' \partial \gamma}\right)$$

By substituting these expressions into equation (11.89) we finally obtain

$$\left|\left(\frac{\partial^2 \vec{\mu}}{\partial Q_i \partial Q_j}\right)_0^A\right| = 2\left\{\frac{1}{2}\left(\frac{\partial^2 \mu}{\partial q^2} + 2\frac{\partial^2 \mu}{\partial q \partial q'} + \frac{\partial^2 \mu}{\partial q'^2}\right) l_{q^{A_i}} l_{q^{A_j}}\right.$$
$$\left. + \frac{1}{\sqrt{2}}\left(\frac{\partial^2 \mu}{\partial q \partial \gamma} + \frac{\partial^2 \mu}{\partial q' \partial \gamma}\right) l_{q^{A_i}} l_{\gamma j} + \frac{\partial^2 \mu}{\partial \gamma^2} l_{\gamma i} l_{\gamma j}\right\} \cos \frac{\varphi}{2}$$
$$+ \frac{2}{s}\left\{\frac{1}{\sqrt{2}}\left(\frac{\partial \mu}{\partial q} + \frac{\partial \mu}{\partial q'}\right) l_{q^{A_i}} l_{\gamma j} + \frac{\partial \mu}{\partial \gamma} l_{\gamma i} l_{\gamma j}\right\} \sin \frac{\varphi}{2}, \quad i, j = 1, 2 \qquad (11.90)$$

Equation (11.87) for overtones of B_2-type vibrations in symmetry co-ordinates takes the form:

$$\left(\frac{\partial^2 \vec{\mu}}{\partial Q_3^2}\right)_0^B = l_q B_3 \left(\frac{\partial^2 \mu_1}{\partial q^{B^2}}(\vec{e}_1 + \vec{e}_2) + \left[\frac{\partial \mu_1}{\partial q^B}, \frac{\partial \mu_2}{\partial q^B}\right]\begin{bmatrix}\dfrac{\partial \vec{e}_1}{\partial q^B} \\ \dfrac{\partial \vec{e}_2}{\partial q^B}\end{bmatrix} + \left[\frac{\partial \vec{e}_1}{\partial q^B}, \frac{\partial \vec{e}_2}{\partial q^B}\right]\begin{bmatrix}\dfrac{\partial \mu_1}{\partial q^B} \\ \dfrac{\partial \mu_2}{\partial q^B}\end{bmatrix}\right) l_{q^{B_3}}$$
$$= l_{q^{B_3}}^2 \left\{\frac{\partial^2 \mu}{\partial q^{B^2}}(\vec{e}_1 + \vec{e}_2) + 2\frac{\partial \mu}{\partial q^B}\left(\frac{\partial \vec{e}_1}{\partial q^B} - \frac{\partial \vec{e}_2}{\partial q^B}\right)\right\} \qquad (11.91)$$

We also have the relations

$$\left(\frac{\partial \vec{e}_1}{\partial q^B} - \frac{\partial \vec{e}_2}{\partial q^B}\right) = [1, -1]\begin{bmatrix} \dfrac{\partial \vec{e}_1}{\partial q^B} \\ \dfrac{\partial \vec{e}_2}{\partial q^B} \end{bmatrix}$$

$$= \frac{1}{s}\left\{[\varepsilon, -\varepsilon, 0]\begin{bmatrix} \dfrac{1}{\sqrt{2}}\vec{e}_1 \\ -\dfrac{1}{\sqrt{2}}\vec{e}_2 \\ -\dfrac{1}{\sqrt{2}}(\vec{e}_1 - \vec{e}_2) \end{bmatrix}\theta^B - \frac{1}{\sqrt{2}}(\vec{e}_1 + \vec{e}_2)\right\} \tag{11.92}$$

$$= \frac{1}{s\sqrt{2}}(\vec{e}_1 + \vec{e}_2)(\varepsilon\theta^B - 1) = \frac{-1}{s\sqrt{2}}\cdot\frac{\varepsilon_0\cdot 2\sin^2(\varphi/2)}{\varepsilon + 2\varepsilon_0\sin^2(\varphi/2)}(\vec{e}_1 + \vec{e}_2)$$

$$\theta^B = [\varepsilon + \varepsilon_0(1 - \cos\varphi)]^{-1}$$

The first and second derivatives of a bond's dipole moment with respect to a symmetry co-ordinate are

$$\frac{\partial \mu}{\partial q^B} = \frac{1}{\sqrt{2}}\left(\frac{\partial \mu}{\partial q} - \frac{\partial \mu}{\partial q'}\right), \quad \frac{\partial^2 \mu}{\partial (q^B)^2} = \frac{1}{2}\left(\frac{\partial^2 \mu}{\partial q^2} - 2\frac{\partial^2 \mu}{\partial q \partial q'} + \frac{\partial^2 \mu}{\partial q'^2}\right)$$

Substituting these in equation (11.91) we obtain,

$$\left|\left(\frac{\partial^2 \vec{\mu}}{\partial Q_3^2}\right)_0^B\right| = \left\{\left(\frac{\partial^2 \mu}{\partial q^2} - 2\frac{\partial^2 \mu}{\partial q \partial q'} + \frac{\partial^2 \mu}{\partial q'^2}\right)\cos\frac{\varphi}{2}\right.$$
$$\left. - 2\left(\frac{\partial \mu}{\partial q} - \frac{\partial \mu}{\partial q'}\right)\frac{1}{s}\frac{2\varepsilon_0\sin^2(\varphi/2)\cos(\varphi/2)}{\varepsilon + 2\varepsilon_0\sin^2(\varphi/2)}\right\} l_{q^{B_3}}^2 \tag{11.93}$$

For combinations with a mixed symmetry type we find that

$$\left(\frac{\partial^2 \vec{\mu}}{\partial Q_i^A \partial Q_3^B}\right)_0 = [l_{q^{A_i}}, l_{\gamma i}]\left\{\begin{bmatrix} \dfrac{\partial^2 \mu}{\partial q^A \partial q^B} \\ \dfrac{\partial^2 \mu}{\partial \gamma \partial q^B} \end{bmatrix}(\vec{e}_1 - \vec{e}_2)\right.$$
$$\left. + \begin{bmatrix} \dfrac{\partial \mu_1}{\partial q^A} & \dfrac{\partial \mu_2}{\partial q^A} \\ \dfrac{\partial \mu_1}{\partial \gamma} & \dfrac{\partial \mu_2}{\partial \gamma} \end{bmatrix}\begin{bmatrix} \dfrac{\partial \vec{e}_1}{\partial q^B} \\ \dfrac{\partial \vec{e}_2}{\partial q^B} \end{bmatrix} + \begin{bmatrix} \dfrac{\partial \vec{e}_1}{\partial q^A} & \dfrac{\partial \vec{e}_2}{\partial q^A} \\ \dfrac{\partial \vec{e}_1}{\partial \gamma} & \dfrac{\partial \vec{e}_2}{\partial \gamma} \end{bmatrix}\begin{bmatrix} \dfrac{\partial \mu_1}{\partial q^B} \\ \dfrac{\partial \mu_2}{\partial q^B} \end{bmatrix}\right\} l_{q^{B_3}} \tag{11.94}$$

By taking into account the results of the preceding manipulations and that the vector

$$\left(\frac{\partial \vec{e}_1}{\partial q^B} + \frac{\partial \vec{e}_2}{\partial q^B}\right)$$

is directed along the x axis and its length is

$$\frac{\sqrt{2}}{s}\cdot\frac{2\varepsilon_0\cos^2(\varphi/2)\sin(\varphi/2)}{\varepsilon+2\varepsilon_0\sin(\varphi/2)}$$

and

$$\frac{\partial\vec{e}_1}{\partial q^A}=\frac{\partial\vec{e}_2}{\partial q^A}=0 \quad\text{and}\quad \frac{\partial\mu_1}{\partial q^B}=-\frac{\partial\mu_2}{\partial q^B}$$

it is possible to transform to the derivatives with respect to natural co-ordinates, i.e.

$$\frac{\partial^2\mu}{\partial q^A\partial q^B}=\frac{\partial}{\partial q^A}\left\{\frac{1}{\sqrt{2}}\left(\frac{\partial\mu}{\partial q}-\frac{\partial\mu}{\partial q'}\right)\right\}=\frac{1}{2}\left(\frac{\partial^2\mu}{\partial q^2}-\frac{\partial^2\mu}{\partial q'^2}\right)$$

$$\frac{\partial^2\mu}{\partial\gamma\partial q^B}=\frac{1}{\sqrt{2}}\left(\frac{\partial^2\mu}{\partial q\partial\gamma}-\frac{\partial^2\mu}{\partial q'\partial\gamma}\right)$$

For this point it is easily found that

$$\begin{aligned}\left|\left(\frac{\partial^2\vec{\mu}}{\partial Q_i^A\partial Q_3^B}\right)\right|&=\left\{\frac{1}{2}\left(\frac{\partial^2\mu}{\partial q^2}-\frac{\partial^2\mu}{\partial q'^2}\right)l_{q^{Ai}}l_{q^{B_3}}\right.\\&\quad\left.+\frac{1}{\sqrt{2}}\left(\frac{\partial^2\mu}{\partial q\partial\gamma}-\frac{\partial^2\mu}{\partial q'\partial\gamma}\right)l_{\gamma i}l_{q^{B_3}}\right\}2\sin\frac{\varphi}{2}\\&\quad-\frac{\sqrt{2}}{s}\left\{\frac{1}{\sqrt{2}}\left(\frac{\partial\mu}{\partial q}+\frac{\partial\mu}{\partial q'}\right)l_{q^{Ai}}l_{q^{B_3}}+\frac{\partial\mu}{\partial\gamma}l_{\gamma i}l_{q^{B_3}}\right\}\\&\quad\times\frac{2\varepsilon_0\cos^2(\varphi/2)\sin(\varphi/2)}{\varepsilon+2\varepsilon_0\sin^2(\varphi/2)}-\frac{1}{s\sqrt{2}}\left(\frac{\partial\mu}{\partial q}-\frac{\partial\mu}{\partial q'}\right)l_{\gamma i}l_{q^{B_3}}\cos\frac{\varphi}{2}\qquad i=1,2\end{aligned}\tag{11.95}$$

11.9 The Valence-optical Theory of Band Intensities in Infra-red Spectra of Molecules with Internal Rotations

We noted in Chapter 10 that one of the most important types of internal motions in a polyatomic molecule together with the vibration of nuclei close to their equilibrium positions is the large-amplitude motion of one part of a molecule with respect to the others. Internal rotations are an example of this kind of motion. In Chapter 10 we presented the variational technique for estimating this type of molecular motion and ascertaining the corresponding energy levels by a direct numerical diagonalization of the Hamiltonian of a molecule with internal rotation. However, the intensities of the lines and bands in the spectra of these molecules is no less interesting. It is clear from elementary physical consideration that the total molecular dipole moment during the rotation of some of its fragments with respect to the others will be a periodic function of the natural co-ordinate that describes the internal rotation. Clearly, this will be reflected in the intensities of the spectral lines and bands in the general case. A more detailed explanation of the nature of the appropriate transitions and the demonstration of a method of estimating their properties are thus of considerable interest.

We will consider a general quantitative theory of intensities for molecules with internal rotations based on the valence optical scheme.

We choose the dihedral angle χ of the rotation of one of the twisting parts of the molecule round another as the co-ordinate that describes the internal rotation. In the general case the complete molecular dipole moment vector will be represented by a function of the normal co-ordinates as well as that of the internal rotation, i.e. $\vec{\mu} = \vec{\mu}(Q, \chi)$. In order to simplify the treatment we will only consider the case with just one rotational degree of freedom. Note that the internal rotation is not included in the set of normal vibrations that are described by the normal coordinate Q.

Without limiting generality, we can expand the expression for the instantaneous value of the molecular dipole moment into a series (around an arbitrarily selected origin, but note that, for co-ordinates which describe the vibrations proper this origin is the equilibrium position) with respect to all the co-ordinates the moment depends upon. Thus we obtain

$$\begin{aligned}\vec{\mu}(Q,\chi) = \vec{\mu}(0) + \sum_i \left(\frac{\partial \vec{\mu}}{\partial Q_i}\right)_0 Q_i + \tfrac{1}{2}\sum_{i,j}\left(\frac{\partial^2 \vec{\mu}}{\partial Q_i \partial Q_j}\right)_0 Q_i Q_j + \cdots \\ + \sum_i \left(\frac{\partial^2 \vec{\mu}}{\partial Q_i \partial \chi}\right)_0 Q_i \Delta\chi + \cdots + \left(\frac{\partial \vec{\mu}}{\partial \chi}\right)_0 \Delta\chi + \frac{1}{2}\left(\frac{\partial^2 \vec{\mu}}{\partial \chi^2}\right)_0 (\Delta\chi)^2 + \cdots\end{aligned} \tag{11.96}$$

Equation (11.96) can be written schematically as

$$\vec{\mu}(Q,\chi) = \vec{\mu}(0) + \vec{\mu}_Q(Q) + \vec{\mu}_{Q\chi}(Q,\chi) + \vec{\mu}_\chi(\chi) \tag{11.97}$$

where $\vec{\mu}_Q(Q)$ is the 'vibrational' part of the dipole moment and $\vec{\mu}_{Q\chi}(Q,\chi)$ and $\vec{\mu}_\chi(\chi)$ are periodic functions of the torsional co-ordinate χ with a period of 2π. This enables us to group the terms $\vec{\mu}_{Q\chi}$ and $\vec{\mu}_\chi$ together and to write them as a Fourier series in a projection onto the x axis of a co-ordinate system which is immobile with respect to the molecule:

$$[\vec{\mu}_1(Q,\chi)]_x = \sum_k \{a_k(Q)\cos k\chi + b_k(Q)\sin k\chi\} \tag{11.98}$$

where $\vec{\mu}_1 = \vec{\mu}_{Q\chi} + \vec{\mu}_\chi$ and the Fourier coefficients must be polynomials with respect to the normal vibrational co-ordinates Q, i.e.

$$a_k(Q) = a_k^0 + \sum_i a_k^i Q_i + \sum_{i,j} a_k^{ij} Q_i Q_j + \cdots \tag{11.99}$$

and

$$b_k(Q) = b_k^0 + \sum_i b_k^i Q_i + \sum_{i,j} b_k^{ij} Q_i Q_j + \cdots \tag{11.100}$$

Equating the coefficients of the same degree of co-ordinates Q and χ in equations (11.96) and (11.98), we obtain correlations that connect derivatives of various order of the X component of a dipole moment in these co-ordinates with the coefficients a and b from equations (11.99) and (11.100):

$$\sum_k a_k^0 = \sum_k a_k^i = \sum_k a_k^{ij} = \cdots = 0$$

Hence for pure 'vibrational' derivatives we have, for the mixed derivatives,

$$\left(\frac{\partial^2 \mu_x}{\partial Q_i \partial \chi}\right)_0 = 2! \sum_k k b_k^i$$

$$\left(\frac{\partial^3 \mu_x}{\partial Q_i \partial Q_j \partial \chi}\right)_0 = 3! \sum_k k b_k^{ij}$$

$$\left(\frac{\partial^3 \mu_x}{\partial Q_i \partial \chi^2}\right)_0 = -\frac{3!}{2!} \sum_k k^2 a_k^i$$

$$\left(\frac{\partial^4 \mu_x}{\partial Q_i \partial Q_j \partial \chi^2}\right)_0 = -\frac{4!}{2!} \sum_k k^2 a_k^{ij}$$

$$\left(\frac{\partial^4 \mu_x}{\partial Q_i \partial \chi^3}\right)_0 = -\frac{4!}{3!} \sum_k k^3 b_k^i$$

$$\left(\frac{\partial^5 \mu_x}{\partial Q_i \partial Q_j \partial \chi^3}\right)_0 = -\frac{5!}{3!} \sum_k k^3 b_k^{ij} \tag{11.101a}$$

while for the 'angle' derivates we obtains

$$\left(\frac{\partial \mu_x}{\partial \chi}\right)_0 = \sum_k k b_k^0$$

$$\left(\frac{\partial^2 \mu_x}{\partial \chi^2}\right)_0 = -\sum k^2 a_k^0$$

$$\left(\frac{\partial^3 \mu_x}{\partial \chi^3}\right)_0 = -\sum_k k^3 b_k^0 \tag{11.101b}$$

Numerical values for all the coefficients a and b can be obtained, therefore, from the equations written here provided sufficient values are known for the electro-optical parameters of the molecule.

Let us consider a problem involving infra-red active transitions that may take place between the torsional states of a molecule. In order to deal with this let us write the expressions for the various matrix elements of Fourier component of the dipole moment $\vec{\mu}_1(Q, \chi)$ (see equation (11.98)), i.e.

$$\langle 0|\cos k\chi|0\rangle = 0 \tag{11.102}$$

$$\langle 0|\cos k\chi|n+\rangle = \begin{cases} \dfrac{1}{\sqrt{2}} & n = \pm k \\ 0 & n \neq \pm k \end{cases} \tag{11.103}$$

$$\langle 0|\cos k\chi|n-\rangle = 0 \tag{11.104}$$

$$\langle m+|\cos k\chi|n-\rangle = 0 \tag{11.105}$$

$$\langle m+|\cos k\chi|n+\rangle=\begin{cases}\frac{1}{2} & m\pm n=\pm k\\ 0 & m\pm n\neq\pm k\end{cases} \tag{11.106}$$

$$\langle m-|\cos k\chi|n-\rangle=\begin{cases}-\frac{1}{2} & m+n=\pm k\\ \frac{1}{2} & m-n=\pm k\\ 0 & m\pm n\neq\pm k\end{cases} \tag{11.107}$$

$$\langle 0|\sin k\chi|0\rangle=0 \tag{11.108}$$

$$\langle 0|\sin k\chi|n+\rangle=0 \tag{11.109}$$

$$\langle 0|\sin k\chi|n-\rangle=\begin{cases}\dfrac{1}{\sqrt{2}} & n=k\\ -\dfrac{1}{\sqrt{2}} & n=-k\\ 0 & n\neq\pm k\end{cases} \tag{11.110}$$

$$\langle m+|\sin k\chi|n-\rangle=\begin{cases}\frac{1}{2} & m=\pm(n-k)\\ -\frac{1}{2} & m=\pm(n+k)\\ 0 & m\neq\pm(n\pm k)\end{cases} \tag{11.111}$$

$$\langle m+|\sin k\chi|n+\rangle=0 \tag{11.112}$$

$$\langle m-|\sin k\chi|n-\rangle=0 \tag{11.113}$$

In these relations the integral $\langle m+|\cos k\chi|n-\rangle$ represents

$$\int_0^{2\pi}\phi_m^{(1)}(\chi)\cos k\chi\,\phi_n^{(2)}(\chi)\,\mathrm{d}\chi$$

We see, therefore, that in the infra-red spectra of molecules with one internal rotation only those dipole transitions are allowed, which satisfy the selection rules (11.103), (11.106), (11.107), (11.110) and (11.111). The transitions with any other changes of a torsional quantum number are prohibited.

It follows from this that a theoretical analysis of the band intensities associated with all possible transitions must, strictly speaking, precede an assignment of experimental frequencies in the spectra of molecules with an internal rotation.

Considering a set of quantum numbers $(v_1, v_2, \ldots, l)$ as a whole, we can introduce the concepts of 'fundamental frequency', 'overtones' and 'combination frequencies' for the case of a pure vibrational problem for a polyatomic molecule. We obtain so far the following classification of the transitions with regard for an internal rotation:

(1) $v_i''=v_i'+1, v_j''=v_j'$ for all $j\neq i$; $l''=l'$

This represents a fundamental vibrational transition (for the ith normal vibration); $l=$ quantum number for internal rotation;

(2) $v_i''=v_i'$ for all i; $l''=l'\pm 1$

representing a fundamental torsional transition;

(3) $v_i'' = v_i' + k, k > 1, v_j'' = v_j'$ for all $j \neq i; l'' = l'$

representing the kth vibrational overtone (for the ith normal vibration);

(4) $v_i'' = v_i'$ for all $i; l'' = l' \pm k, k > 1$; i.e. the kth torsional overtone;

(5) $v_i'' > v_i', v_j'' > v_j', v_k'' = v_k'$ for all $k \neq j \neq i; l'' = l'$

representing a vibrational composite frequency (for the ith and jth normal vibrations);

(6) $v_i'' > v_i'; l'' = l' \pm k$; corresponding to a vibration–torsional composite frequency.

For the calculation of intensities of transitions for internal rotation let us use a valence-optical scheme. We assume, as usual, that

$$\vec{\mu} = \sum_k \mu_k \vec{e}_k$$

where μ_k are some 'effective' dipole moments that are ascribed to valence bonds with the directional unitary vectors $\vec{e}_k$. Note that, since the torsional energy is not very large, we can conclude, with a good degree of precision, that no essential deformation of an electron shell occurs because of the internal rotation. Hence, the dipole moment of bonds $\vec{\mu}_k$ may be considered to be independent of the torsional co-ordinate χ. Therefore, in the expression of the total dipole moment of the molecule only the directional vectors of bonds $\vec{e}_k$ are functions of χ, and this dependence evidently has a periodic character. Hence the total intensity of a torsional band will be determined by the dipole moment of the transition, the expression of the x component of which is

$$\vec{\mu}_x(\chi) = \sum_k \mu_k [\vec{e}_k(\chi)]_x = \sum_{k,n} \mu_k (c_{nk} \cos n\chi + d_{nk} \sin n\chi) \qquad (11.114)$$

The values of matrix elements for dipole transitions have been given above. It remains only to show how the Fourier coefficients c_{nk} and d_{nk} are calculated.

The algorithm is as follows. A molecular model is created in which for the vibrational co-ordinates very large values for the corresponding force constants are introduced. These large values are invariant during all subsequent iterations. Their role, as we shall see, is to 'fix' separate parts of the molecule, and for this it is important for them to be large enough, but the actual numerical values are unimportant. The potential for the torsional co-ordinate is taken to be oscillatory, but its force constants are much smaller. If there are other internal rotations they are also 'fixed' by the introduction of parabolic potentials with a large degree of curvature.

With this model we then have to solve a typical vibrational problem. All vibrational frequencies for this problem will divide naturally into two groups: one of very high frequencies arising from vibrations for all degrees of freedom except the torsional mode being investigated and one of low-frequency vibration.

The model is so constructed that the latter will have a form in which only one torsional co-ordinate will change significantly. We can, then, obtain the vibrational displacements of atoms from their equilibrium positions for this torsional vibration by considering a small increase in the respective normal co-ordinate. We shall use

$$\delta\vec{r} = E\tilde{\vec{B}}L_p\delta Q$$

where $\delta\vec{r}$ is a column matrix of Cartesian displacements; E is the matrix of inverse masses; $\vec{B}$ is a vector matrix which connects Cartesian displacements of atoms with their 'natural' co-ordinates; L_p is a matrix which defines the forms taken by the vibrations (in the momentum representation); and δQ is a column matrix of atomic displacements in normal co-ordinates.

This formula defines the new disposition of every atom in the displaced position which will be, first, of such a type that all remaining parts of the molecule will move as a whole. Second during this small rotation of one part of the molecule around another the impulse and the angular momentum will be invariant.

The operation is then repeated, near a new starting geometrical state of the molecule, etc. until a complete cycle of the internal rotation is performed.

Note that only this method allows the realization of atomic movements such as vector sums of 'microdisplacements' at every step of the interactions where only some designated co-ordinates change, while those at 'rest' remain practically invariant, i.e. the respective elements of the molecule move as a whole. Meanwhile the omission of the 'fixed' degrees of freedom corresponds to a molecular model with 'torn-off' sections. These parts, as a whole, do not carry along the atomic motions when the co-ordinates not related to the designated structural elements of the molecule change. If this occurred it would lead to deformations in all parts of the molecule, and it is these effects which must be avoided. The method described above can be programmed easily.

Now, if the succession of Cartesian displacements, $\Delta_i\vec{r}$, is known it is possible to find the vectors $\Delta_i\vec{e}_k$ of bond rotations. We can then construct point by point the dependences of projections of $\vec{e}_k$ onto the axes of an external (laboratory) co-ordinate system upon the χ co-ordinate, and so obtain the Fourier coefficients for the operator $[\vec{\mu}(\chi)]_x$. Hence the matrix elements of the corresponding dipole transitions can be calculated.

Note that this way of obtaining large atomic displacements as a set of small ones may also be used if we wish to calculate point by point, with the subsequent interpolation of the dependency of $\vec{e}_k$ vectors upon both internal rotational and normal vibrational co-ordinates. The latter property may make unnecessary the analytical calculation of the corresponding derivatives in those cases where it is necessary to know the values of such derivatives for the solution of a problem.

It appears, therefore, that for a calculation of intensities of torsional and vibration–torsional bands occurring in the microwave and infra-red spectra of molecules with an internal rotation it is possible to use the same type of electro-optical parameters which have been widely used for the calculation of vibrational band intensities. This feature is important and valuable in practice. Additional information about general questions of infra-red intensity theory can be found in references 18–22.

11.10 The Valence-optical Theory of Intensities in the Infra-red Spectra of Molecular Ions

Let now us consider the problem of constructing a valence-optical scheme in the case of molecular ions. A solution of this problem is of great interest in connection with the investigation of the ion-molecular beam spectra. The existence of charged entities is also typical of many solutions.

When deriving a formula for a matrix element of a purely dipole transition we should take into account that, irrespective of whether the system under consideration as a whole carries no charge (a molecule) or is a charged entity (a molecular ion), intensity of the fundamental absorption band is proportional to the square of a derivative of the system's dipole moment $|\vec{\mu}|$ taken with respect to an appropriate normal co-ordinate. Note that, though for a charged system the dipole moment $\vec{\mu}$ is not invariant with respect to the selected origin of co-ordinates, a derivative $(\partial\vec{\mu}/\partial Q_i)_0$ with respect to a normal co-ordinate does possess such invariance.

What is more, the same formula for a dipole moment[11.26] applies equally both to the ground state of a molecule and to an ion when employing an adiabatic approximation and representing the electronic wavefunction in the CI-MO-LCAO form.

In order not to complicate further discussion, let us employ a CNDO approximation for different atoms. In this case we can write

$$\vec{\mu} = \sum_{\alpha} Z_\alpha \vec{R}_\alpha + \sum_{\alpha} P_{\alpha\alpha'} \cdot \vec{\mu}'_{\alpha\alpha'} \tag{11.115}$$

where Z_α is a total atomic charge.

Now let us take advantage of the approach set out in detail in Section 11.2, converting to the representation of $\vec{\mu}$ in the form containing bond directional vectors μ_k. Having arbitrarily selected the origin of co-ordinates, let us draw a radius vector $\vec{R}$ from the origin to an atom taken as the first one. Then for the radius vectors of all the rest of the atoms we can write

$$\vec{R}_\alpha = \vec{R}_1 + \sum_k \sigma_{k\alpha}^{(1)} \cdot S_k \cdot \vec{e}_k$$

where $\sigma_{k\alpha}^{(1)} = \pm 1$, depending upon the direction of passing a respective bond when moving along a given graph.

A graph matrix enables us to write an expression for the dipole moment of the entire system as follows:

$$\vec{\mu} = \vec{R}_1 \left(\sum_\alpha Z_\alpha \right) + \tilde{Z} G_1 S \vec{e} + \sum_\alpha P_{\alpha\alpha'} \vec{\mu}_{\alpha\alpha'} \tag{11.116}$$

Here, as previously, we have used matrix designations: $\tilde{Z}$ is a row-matrix of all the atomic charges; G_1 is a matrix of all the graphs coming from the first atom to all the 'rest' ones (if more than one graph arrives at an atom, an averaged graph is used as a row in a respective matrix), S is a diagonal matrix of bond lengths; and $\vec{e}$ is a column-matrix of the bond directional vectors.

Clearly, the value of $\vec{\mu}$ will not change if atom number 2 is taken as an initial one instead of atom number 1, writing as follows:

$$\vec{\mu} = R_2\left(\sum_\alpha Z_\alpha\right) + \tilde{Z}G_2 S\vec{e} + \sum_\alpha P_{\alpha\alpha'}\vec{\mu}'_{\alpha\alpha'} \tag{11.117}$$

Now, we add together (as was done in Section 11.2 in order to achieve correct symmetry) equalities of types (11.116) and (11.117) written sequentially for all the atomic numbers, and divide the result by the number of atoms N, thus obtaining

$$\vec{\mu} = \frac{Z}{N}\tilde{1}\vec{R} + \frac{\tilde{Z}}{N}\sum_\alpha G_\alpha S\vec{e} + \sum_\alpha P_{\alpha\alpha'}\vec{\mu}'_{\alpha\alpha'}$$

where $Z = \sum_\alpha Z_\alpha$ is a total charge of the system, $\tilde{I}$ is the row matrix consisting entirely of ones and $\vec{R}$ is the column matrix of vectors $\vec{R}_\alpha$. By further expanding the vectors $\vec{\mu}'_{\alpha\alpha'}$ for each of the atoms in terms of the local co-ordinate axes formed by the vectors $\vec{e}_k$ around each atom we finally obtain

$$\vec{\mu} = \frac{Z}{N}\tilde{1}\vec{R} + \sum_k \mu_k\vec{e}_k = \frac{Z}{N}\tilde{1}\vec{R} + \tilde{\mu}\vec{e} \tag{11.118}$$

It can easily be seen that here the electro-optical parameters μ_k have the same sense as in the case of neutral molecules. For $Z = 0$ we arrive at an ordinary expression consistent with the valence-optical pattern.

Representation of the dipole moment as given by equation (11.118) is not the only possible one. Indeed, if we multiply equalities as given by equations (11.116) and (11.117) by m_α, where m_α is the mass of atom with subscript α, and then add them together and divide the result by a total mass of the molecule M, then the term ZR_c will be separated in the expression for the dipole moment, where $\vec{R}_c$ is the radius vector of the molecular centre of mass. This term disappears, when changing over to the derivative $(\partial\vec{\mu}/\partial Q_i)_0$, as the molecular vibrations do not cause shifting of the centre of mass. As a result, all the formulae of the valence-optical schemes remain true for ions but electro-optical parameters become dependent of the m_α/M ratio, which is inconvenient in a parametric theory, as parameters are changing upon conversion from one isotope-substituted modification to another, with the electron shell being the same. This is why it is worth utilizing expression (11.118), explicitly retaining the total charge of the system.

Having done that, we obtain for a derivative of the molecular dipole moment with respect to the normal co-ordinate:

$$\left(\frac{\partial\vec{\mu}}{\partial Q_i}\right)_0 = \vec{e}\left[\frac{\partial\mu}{\partial q} - \left(\delta\frac{\mu}{J}\right)_{\!,}\frac{\partial\mu}{\partial\gamma}\right] l_q^{(i)} + \left(\frac{Z}{N}\tilde{1} + \tilde{\mu}S^{-1}\Delta\right)E\tilde{\vec{B}}l_p^{(i)} \tag{11.119}$$

This formula is distinguished from that previously derived only in the presence of an additional term $(Z/N)\tilde{1}E\vec{B}l_p^{(i)}$. We have written the formula as applied to the case where we can find dependent co-ordinates among the natural ones. The following designations are used here: $[\delta(\mu/J)]$ is a diagonal matrix of the ratios between the bond dipole moments and bond lengths, q are the bond-extension co-ordinates and γ denotes all the rest co-ordinates.

It can be seen that the entire computational scheme of valence-optional theory is preserved when converting to molecular ions, wherein the contribution made by the term containing Z is markedly dependent upon vibration modes, being greatly reduced in local normal vibrations. The role played by this term will also reduce in the spectra of overtones and combination frequencies.

References

1. W. B. Person and G. Zerbi (eds). *Vibrational Intensities in Infrared and Raman Spectroscopy*, Elsevier, Amsterdam (1982).
2. A. Rupprecht. *J. Molecular Spectroscopy*, **89**, 356–86 (1981).
3. L. A. Gribov. *Intensity Theory for Infrared Spectra of Polyatomic Molecules*, Cons. Bureau, NY (1964).
4. M. V. Volkenstein, L. A. Gribov, M. A. Elyashevich and B. I. Stepanov. *Vibrations of Molecules*, Nauka, Moscow (1972) (in Russian).
5. L. M. Sverdlov, M. A. Kovner and E. P. Krainov. *Vibrational Spectra of Polyatomic Molecules*, Halsted Press, London (1974).
6. M. Gussoni, S. Abbate and G. Zerbis. In *Vibrational Spectroscopy, Modern Trends*, A. J. Barnes and W. J. Orville-Thomas (eds), Elsevier, Amsterdam (1977).
7. L. A. Gribov. *The Theory of Infrared Spectra of Polymers*, Nauka, Moscow (1977) (in Russian).
8. L. A. Gribov, B. A. Dementiev and A. T. Todorovsky. *Interpretated Vibrational Spectra of Alkanes, Alkenes and Benzene Derivatives*, Nauka, Moscow (1986).
9. L. A. Gribov and M. R. Rasovsky, *J. Molec. Struct.*, (*Theochem.*), **122**, 15 (1985).
10. L. A. Gribov. *J. Molec. Struct.* (*Theochem.*), **110**, 61 (1984).
11. L. A. Gribov and S. V. Kotov. *J. Molec. Struct.* (*Theochem.*), **136**, 391 (1986).
12. L. A. Gribov, *J. Molec. Struct.*, **141**, 127 (1986).
13. L. A. Gribov and S. B. Kotov. *J. Struct. Chem.* **27**, 13 (1986) (in Russian).
14. S. V. Kotov, L. A. Gribov and V. L. Stepanian. *J. Struct. Chem.*, **27**, 19 (1986) (in Russian).
15. S. V. Kotov and L. A. Gribov. *J. Appl. Spectr.*, **45**, 443 (1986) (in Russian).
16. J. C. Decius. *J. Chem. Phys.*, **20**, 1039 (1952).
17. B. L. Crawford. *J. Chem. Phys.*, **20**, 977 (1952).
18. M. Gussoni. *J. Molec. Struct.*, **141**, 63 (1986).
19. P. Jona, M. Gussoni and G. Zerbi. *J. Molec. Struct.*, **95**, 15 (1982).
20. M. Gussoni, C. Castiglioni and G. Zerbi. *Chem. Phys. Lett.*, **95**(6), 483 (1983).
21. M. Gussoni. *J. Appl. Spectr.*, **42**, 265 (1985) (in Russian).
22. W. J. Orville-Thomas, S. Suzuki and G. Riley. In *Vibrational Spectroscopy, Modern Trends*, A. J. Barnes and W. J. Orville-Thomas (eds) Elsevier, Amsterdam (1977).

Chapter 12

Characteristic Vibrations of Polyatomic Molecules

12.1 General Principles

The concept of characteristic vibrations is one of the most important in the theory of polyatomic molecular vibrations[1]. The existence of vibrations that can be characterized by frequency has, in many respects, determined the spread of vibrational spectroscopy in research in spectrochemistry. Hence it is very important to know under what conditions the appearance of characteristic lines and bands can be expected in infra-red and Raman spectra. However, not every change in the characteristic frequencies associated with the vibrations of a group can be used as a criterion that there has been a change in the group's structure. We shall see below that frequency shifts may occur when the same group occurs in a number of molecules, even if the group itself is completely unchanged. It is therefore very clear that, without an understanding of all the reasons that lead to an observed frequency change in the position of a band in a spectrum, we cannot make a reliable judgement about the character of the change in a molecule.

Let us consider a particular group of atoms forming part of a molecule. Since each normal vibration has a characteristic frequency and form, the vibration is naturally classified by its form and frequency attributes. We have already met this classification when considering valency and deformational vibrations. The form of a vibration has a decisive influence on the intensity of its absorption band in the infra-red spectrum or on that of its line in the Raman spectra of a series of molecules that contain the given group. The different intensities are due to the various forms of the vibrations.

We say a vibration is characteristic (with respect to frequency) for a given group if one of a molecule's normal vibrations coincides with respect to frequency with one of the normal vibrations of the group when it is acting as a free molecule. If one of the molecule's normal vibrations coincides in form with some of the group's natural ones, the vibration is said to be characteristic with respect to form for the group. If both conditions are satisfied, the vibration is then said to be characteristic with respect to both form and freqency, or to be fully characteristic. This division, even though it is to some extent tenuous, completely justifies itself in practice, since the different kinds of characterizability are not identical in concept, and vibrations that may be characteristic with respect to frequency may not be so with respect to form, and vice versa.

These properties of a normal vibration may be retained over a series of molecules containing the same group. We can then speak of the stable characterizability with respect to form or frequency (or form *and* frequency) for that group. The existence of a vibration with stable characterizability is dependent on whether its absorption band (infra-red) or line (Raman) appears in the same region for the series of molecules containing the group. However, the existence of such vibrations allows tables of frequency–structure correlations to be compiled. These can be used for quantitative analysis to confirm the presence of certain groups and structural elements.

Let us consider a series of molecules or radicals containing the same atomic group. It is obvious that the appearance of characteristic vibrations for this group can be expected only if its geometric and force parameters do not change. These groups can be called 'additive'. Our task is to explain whether the additivity of a group is a sufficient condition in order to explain the appearance of characteristic normal vibrations or whether it is only a necessary one.

We shall designate the set of natural vibration co-ordinates that describe a group's natural vibrations as if the group were a free molecule by q^{I} and the remaining co-ordinates of the molecule containing the group by q^{II}. Using the results in Chapter 9, we can confirm that a group contained in a molecule will be additive, i.e. in particular the force constants belonging to the interactions between the co-ordinates of the q^{I} group will depend on the properties of the group if, for any choice of q^{I} co-ordinate, a set of q^{II} co-ordinates can be found which together with the q^{I} ones form a complete system of vibration co-ordinates for the molecule such that the force constant matrix has the quasi-diagonal form:

$$\begin{bmatrix} U_{11} & 0 \\ 0 & U_{22} \end{bmatrix}$$

Here U_{11} is the force-constant submatrix for the q^{I} co-ordinate set and U_{22} is the analogous submatrix for the q^{II} co-ordinates. If, moreover, the geometrical parameters of the group considered do not change over the series of molecules in question, then, as shown in Chapter 8, the submatrix T_{11} must also be retained for the q^{I} co-ordinates in the general matrix

$$T = \begin{bmatrix} T_{11} & T_{12} \\ T_{21} & T_{22} \end{bmatrix}$$

of the molecule's kinematic coefficients.

The principle here is that if the force constant matrix in natural co-ordinates is diagonal or quasi-diagonal for a certain choice of co-ordinates and interatomic forces, then the kinematic coefficient matrix in natural co-ordinates will never be diagonal and cannot be quasi-diagonalized, even for additive groups. In reality, the fact that the non-diagonal elements are zero in the force-constant matrix is entirely related to the character of the interatomic forces in the model chosen for them, and is not dependent on whether structural elements in the molecules (changes in which are taken as the natural vibration co-ordinates) share common atoms or not.

The situation is the reverse for the kinematic coefficient matrix. All those non-diagonal kinematic coefficients for which the corresponding interacting natural vibrations have even one common atom are non-zero. Thus the kinematic coefficient matrix also may not be quasi-diagonal in natural co-ordinates, since this would mean that one group was completely separated from the remaining parts of the molecule.

A consequence of this property of kinematic coefficient matrices (which, from now on, will be called the kinematic relation between groups in a molecule) is that in the great majority of cases we can only speak of the approximate characterizability of a molecule's normal vibrations for different groups. Kinematic relations between different groups in a molecule also means that characteristic vibrations may be absent even if a molecule has additive groups.

Additive groups were introduced above such that, for their internal co-ordinates, there is no force relationship with the molecule's remaining ones. The U matrix is then quasi-diagonal, while the group's force constants depend only on its properties.

The force constants are determined first, by the character of the molecule's electron cloud and its variations during vibrations. At Chapter 9 we showed that, in general, even the simplest model of a polyatomic molecule's electronic functions does not result in a quasi-diagonal form for the force-constant matrix with the exception of a linear non-conjugate system. Even then, it must be an approximation in which the repulsions of nuclei not belonging to a given bond are not accounted for. The quasi-diagonal form may only be approximate even when the electron clouds of separate groups are significantly closed. Thus, clearly, the particular properties of the force-constant matrix which lead, given certain supplementary conditions (see below), to the appearance of characteristic vibrations may arise only in the absence of conjugation between a given group and the rest of the molecule.

We will, however, use the concept of additivity, both in this chapter and when analysing intensities in infra-red spectra, bearing in mind that it is a useful (if not a particularly accurate) approximation that reflects, to some degree, the properties of a non-conjugated system's electron cloud. It enables us to reach a series of conclusions on the basic conditions for the appearance of characteristic vibrations, and we shall now cover this topic in more detail.

Let us assume that a molecule has an additive group and its kinematic coefficient and force-constant matrices in terms of the q^{I} and the q^{II} set of co-ordinates have the following respective forms:

$$T = \begin{bmatrix} T_{11} & T_{12} \\ T_{21} & T_{22} \end{bmatrix} \quad U = \begin{bmatrix} U_{11} & 0 \\ 0 & U_{22} \end{bmatrix} \tag{12.1}$$

We can compile and solve secular equations for the T_{11} and U_{11} and the T_{22}

and U_{22} submatrices, i.e.

$$(T_{11}U_{11} - \lambda^{\mathrm{I}}\mathrm{I})q^{\mathrm{I}} = 0 \tag{12.2}$$

$$(T_{22}U_{22} - \lambda^{\mathrm{II}}\mathrm{I})q^{\mathrm{II}} = 0 \tag{12.3}$$

As a result, we find new sets of co-ordinates Q^{I} and Q^{II} which are associated with the old ones by the relationships

$$q^{\mathrm{I}} = L_{q1}Q^{\mathrm{I}}$$

and

$$q^{\mathrm{II}} = L_{q2}Q^{\mathrm{II}}$$

and for which the T_{11} and T_{22} matrices are reduced to identity ones and the U_{11} and U_{22} matrices are diagonalized to Λ^{I} and Λ^{II}.

The Q^{I} co-ordinates and elements of the Λ^{I} matrix are clearly the normal co-ordinates and squares of the frequencies of the natural vibrations the additive group would have were it a free molecule, but with the geometrical parameters and force constants it has as part of the molecule.

We change the variables thus:

$$\begin{bmatrix} q^{\mathrm{I}} \\ q^{\mathrm{II}} \end{bmatrix} = \begin{bmatrix} L_{q1} & 0 \\ 0 & L_{q2} \end{bmatrix} \begin{bmatrix} Q^{\mathrm{I}} \\ Q^{\mathrm{II}} \end{bmatrix} \tag{12.4}$$

In Q^{I} and Q^{II} co-ordinates the T and U matrices take the form

$$\begin{bmatrix} \mathrm{I} & T'_{12} \\ T'_{21} & \mathrm{I} \end{bmatrix} \quad \text{and} \quad \begin{bmatrix} \Lambda^{\mathrm{I}} & 0 \\ 0 & \Lambda^{\mathrm{II}} \end{bmatrix} \tag{12.5}$$

while the secular equation can be written as

$$\left(\begin{bmatrix} \Lambda^{\mathrm{I}} & T'_{12}\Lambda^{\mathrm{II}} \\ T'_{21}\Lambda^{\mathrm{I}} & \Lambda^{\mathrm{II}} \end{bmatrix} - \lambda \mathrm{I} \right) \begin{bmatrix} Q^{\mathrm{I}} \\ Q^{\mathrm{II}} \end{bmatrix} = (D - \lambda \mathrm{I}) \begin{bmatrix} Q^{\mathrm{I}} \\ Q^{\mathrm{II}} \end{bmatrix} = 0 \tag{12.6}$$

In the open form the D matrix is

$$D = \left[\begin{array}{cccc:cccc} \lambda_1^{\mathrm{I}} & 0 & \cdots & 0 & d_{11} & d_{12} & \cdots & d_{1\mathrm{m}} \\ 0 & \lambda_2^{\mathrm{I}} & \cdots & 0 & d_{21} & d_{22} & \cdots & d_{2\mathrm{m}} \\ & & & & & & & \\ 0 & 0 & \cdots & \lambda_{\mathrm{k}}^{\mathrm{I}} & d_{\mathrm{k}1} & d_{\mathrm{k}2} & \cdots & d_{\mathrm{km}} \\ \hdashline & & & & \lambda_1^{\mathrm{II}} & 0 & \cdots & 0 \\ & & D_{21} & & 0 & \lambda_2^{\mathrm{II}} & \cdots & 0 \\ & & & & \cdots & \cdots & \cdots & \cdots \\ & & & & 0 & 0 & \cdots & \lambda_{\mathrm{m}}^{\mathrm{II}} \end{array} \right]. \tag{12.7}$$

We see immediately that in order for a molecule's first normal vibration to be characteristic with respect to form and frequency for the additive group (i.e. for the appropriate eigenvalue to be equal to λ_1^{I}) while the form has the value

$$Q_1^{\mathrm{I}} = 1, \quad Q_i^{\mathrm{I}} = 0 \quad (i \neq 1), \quad Q^{\mathrm{II}} = 0$$

where i is the index of a normal co-ordinate of the additive group's vibrations, it is necessary and sufficient that all the elements $d_{11}, d_{12}, \ldots, d_{1m}$ of the first row of the D_{12} submatrix and the first column of D_{21} be zero.

Even if only one of the $d_{1i'}$ elements is non-zero, then the normal vibration will not have frequency ν_1^{I} (equal to $(\lambda_1^{\mathrm{I}})^{1/2}$), while the form of the vibration will also include non-zero constituents for the Q^{II} co-ordinates. The square of such a vibration's frequency and its form may be found (and this can easily be concluded from the form of the D matrix) by solving the equation

$$\left(\begin{bmatrix} \lambda_1^{\mathrm{I}} & d_{1i'} \\ d_{i'1} & \lambda_{i'}^{\mathrm{II}} \end{bmatrix} - \lambda \mathrm{I}\right)\begin{bmatrix} Q_1^{\mathrm{I}} \\ Q_{i'}^{\mathrm{II}} \end{bmatrix} = 0 \qquad (12.8)$$

This is the equation which describes the vibration of two oscillators (the normal vibration of the additive group and the rest of the molecule) with the given bond coefficients.

Hence in order to clarify under what conditions different types of characteristic vibrations will appear for an additive group it is sufficient to consider how one of the group's natural vibrations interacts with one of those of the rest of the molecule. If the d_{ij} elements are large for all the natural vibrations of the additive group (the Q^{I} co-ordinates), then not even one of the normal vibrations of a molecule will coincide with respect to form or frequency with any one of the group's natural vibrations. Hence not one of them will be characteristic of the group. If this situation remains true over a series of molecules, then the vibrations of these molecules will change with respect to both frequency and form, due solely to the kinematic relation between the additive group and the rest of the molecule. This is true even if the structure of the group itself is not changed and it continues to be additive. Thus shifts in any of the bands in a spectrum or changes in their intensities (the latter depend very strongly on the form of the vibration) cannot, in general, be considered to be a sufficient sign of disturbances in the structure of the group in a molecule under consideration. On the other hand, invariability of the frequencies and, in particular, intensities of vibrational lines and bands in a large series of molecules sharing some grouping testifies to the additivity of the grouping.

Obviously, in principle, it may be true that the kinematic coefficients of the bond between two groupings in a molecule in the series being studied may have very close values in a series of molecules, even if they are non-zero. It is then possible that frequencies will be repeated, even though the vibrations are not characteristic with respect to frequency and form for the group. However, this must be exceptional, and it can always be detected using an adequately large number of similar molecules with different substituents.

Shifts due to kinematic relations are especially large for groups in the middle of a molecule, but even for terminal groups they may be as much as $100\,\mathrm{cm}^{-1}$ or more. This is one of the difficulties encountered when investigating polyatomic molecules by vibrational spectroscopy. A reasonably unambiguous conclusion about the changes in the structures of individual groups for different kinds of substitutions outside these groups can be made by measuring the shifts in the absorption bands or Raman spectra lines after the kinematic factors have been taken into account, and these factors almost always lead to different kinds of model calculations having to be carried out.

Note also that, other things being equal, the influence of the kinematic factors is clearly reduced if the number of atoms common to the additive group and the rest of the molecule is decreased. This is why characteristic vibrations are most often encountered for groups connected to the molecule via only one atom.

The effect of the remainder of the molecule on a selected grouping can be considered as a perturbation. We know from the general theory of vibrations that those of a system are not affected if the disturbance is at the modes of the vibrations or acts perpendicular to the displacements of the system's particles during its natural vibrations. This may occur within molecules in two ways.

First, the atoms shared by the group and the rest of the molecule may not move at all during the group's natural vibrations. This is often encountered in polyatomic groups, and is why normal vibrations localized within them are often found in molecules containing polyatomic groups. Second, the atoms shared by the group and the rest of the molecule may move in one direction during the group's natural vibrations, and then in another, perpendicular to the first, during the natural vibrations of the rest of the molecule.

Kinematic relations are absent in these two cases, and vibrations characteristic with respect to form and frequency are observed for the group. In every other case (except for one that is considered below) the group's vibrations will be perturbed by the effect of the rest of the molecule.

In order to make the physics of the situation clearer, let us consider the vibrations of two connected oscillators. We will also assume that their coordinates have been normalized so that the kinematic coefficient and force-constant matrices have the following respective forms:

$$T = \begin{bmatrix} 1 & \tau \\ \tau & 1 \end{bmatrix}, \quad U = \begin{bmatrix} \lambda_1^{(0)} & 0 \\ 0 & \lambda_2^{(0)} \end{bmatrix} \tag{12.9}$$

Here, $\lambda_1^{(0)}$ and $\lambda_2^{(0)}$ are the squares of the frequencies of the respective natural vibrations of the two oscillators, and τ is the normalized kinematic non-diagonal coefficient. The D matrix of the dynamic coefficients is written in this case as

$$D = \begin{bmatrix} \lambda_1^{(0)} & \lambda_2^{(0)}\tau \\ \lambda_1^{(0)}\tau & \lambda_2^{(0)} \end{bmatrix} \tag{12.10}$$

We shall now consider several limiting cases. Let us assume $\lambda_1^{(0)} = \lambda_2^{(0)}$. Then, by solving the secular equation with the D matrix, we find that, given $\tau \neq 0$, two normal vibrations arise as a result of the kinematic interactions between the pair of oscillators. The squares of their frequencies are $\lambda_{1,2} = \lambda_1^{(0)} \pm \lambda_1^{(0)}\tau$ and their forms (unnormalized) are

$$\begin{bmatrix} 1 \\ 1 \end{bmatrix} \text{ and } \begin{bmatrix} 1 \\ -1 \end{bmatrix}.$$

Thus the interactions between the two natural vibrations that each have the same frequency lead to the frequencies splitting and a 'smearing' of the amplitudes. This phenomenon is called 'resonance'. In order for this to occur it is essential that the frequencies of any pair of natural vibrations of the additive group and the rest of the molecule must coincide, and that the displacements of the shared atoms, through which the additive group is connected to the rest of the molecule, are not mutually orthogonal for that pair of vibrations.

In the presence of resonance the vibration of an additive group spreads throughout the molecule and is not characteristic with respect to either form or frequency for the group. Resonance is often observed in molecular skeletal vibrations, an hence these are virtually always uncharacteristic for individual bonds.

Let us assume now that $\lambda_1^{(0)}$ and $\lambda_2^{(0)}$ are different, and that $\lambda_1^{(0)} \gg \lambda_2^{(0)}$. Hence, $\tau\lambda_1^{(0)} \gg \tau\lambda_2^{(0)}$, and under certain conditions $\tau\lambda_2^{(0)}$ can be neglected. The D matrix can then be written in the simple triangular form:

$$D = \begin{bmatrix} \lambda_1^{(0)} & 0 \\ \lambda_1^{(0)}\tau & \lambda_2^{(0)} \end{bmatrix} \tag{12.11}$$

The eigenvalues of such a triangular matrix are equal to the diagonal elements of the matrix and the eigenvectors are thus, respectively,

$$\begin{bmatrix} 1 \\ \dfrac{\tau\lambda_1^{(0)}}{\lambda_1^{(0)} - \lambda_2^{(0)}} \end{bmatrix} \text{ and } \begin{bmatrix} 0 \\ 1 \end{bmatrix} \tag{12.12}$$

Therefore in this case we obtained two normal vibrations which have frequencies that are the same as the natural ones of the two interacting oscillators. The lower frequency is characteristic with respect to form for the second oscillator while the upper one also extends to the second oscillator with the coefficient

$$\tau\lambda_1^{(0)}/\lambda_1^{(0)} - \lambda_2^{(0)}$$

The result we have just obtained can be directly generalized to a multidimensional case for which the D matrix has the following block structure:

$$D = \begin{bmatrix} \Lambda^{\mathrm{I}} & T'_{12}\Lambda^{\mathrm{II}} \\ T'_{21}\Lambda^{\mathrm{I}} & \Lambda^{\mathrm{II}} \end{bmatrix} \tag{12.13}$$

If the non-diagonal blocks of this matrix are such that one of them can be neglected, i.e. the D matrix can be replaced by the triangular matrix

$$D' = \begin{bmatrix} \Lambda^{\mathrm{I}} & 0 \\ T'_{21}\Lambda^{\mathrm{I}} & \Lambda^{\mathrm{II}} \end{bmatrix} \tag{12.14}$$

then, by using the general properties of the eigenvalues and eigenvectors of triangular matrices, we can state that the vibrations of the group selected from the molecule will be characteristic with respect to frequency and form if its natural frequencies are considerably smaller than those of the rest of the molecule, and will be characteristic with respect to frequency, but not form, if its natural frequencies are significantly greater than those of the rest of the molecule.

Because of this property, vibrations which are characteristic with respect to frequency appear for bonds such as C≡C, N≡N and C≡N, which lie, as a rule, in the neighbourhood of single bonds whose natural frequencies are significantly different.

An additional point must be made here: we have shown that the appearance of characteristic vibrations when kinematic relations are present is due to the non-symmetry of the *D* matrix (the product of two symmetric matrices) when, in general, one of its blocks can be neglected. Generally speaking, it is not essential for the condition about the significant difference between the natural frequencies of the interacting parts of the molecule to be introduced in order to neglect a block. However, frequently the block can only be neglected in real molecules using this condition.

We provide a simple example (the HCN molecule) which demonstrates conveniently the interaction between the vibrations of two groups as discussed above. The force constant for the interaction between the extension co-ordinate of the C—H bond and that of the C≡N bond is not very large, and hence at a certain level of approximation the two bonds can be considered to be additive. The frequency of the natural vibration of the C—H bond is significantly higher than that of the C≡N one. Moreover, during vibrations of the H atom the C atom does not move significantly, since the latter is much heavier than the former. Thus we ought to expect vibrations to appear that are characterizable with respect to form or frequency for the C—H bond. In fact, the calculated form of the vibration is

$$\begin{bmatrix} 1 \\ -0.136 \end{bmatrix}$$

(the first co-ordinate is with respect to the C—H bond extension).

When hydrogen is replaced by deuterium the vibration of the frequency of the C—D bond is approximately equal to the natural frequency of the C≡N bond. As a result, resonance vibrations occur, and the frequency of the C≡N bond, for example, is $2089\,\text{cm}^{-1}$ in the HCN molecule but is shifted more than $100\,\text{cm}^{-1}$ (to $1921\,\text{cm}^{-1}$), while the form becomes

$$\begin{bmatrix} 1.13 \\ 1.0 \end{bmatrix}$$

The vibrations have been 'smeared'.

Resonance and changes in the form of vibrations, in general, are very frequent in deuteration, and this is why, as a rule, deuteration is not only accompanied by shifts in spectral absorption bands but also by abrupt changes in the corresponding intensities.

We can make a number of deductions from the considerations above about the conditions under which vibrations characteristic for additive groups appear.

These conclusions are only qualitative, but they indicate that in many cases it is possible to make estimates of the main features of many of a molecule's vibrations quite quickly, without any calculations or quantitative estimates of the contributions of all the different factors.

(1) Those natural vibrations of an additive group during which the atoms common to it and the rest of the molecule either do not move at all or move orthogonally to the direction they move in during any of the natural vibrations of the rest of the molecule may be characteristic with respect to form or frequency for the additive group. A special case of this condition is when the atoms of the additive group that are shared with the rest of the molecule are significantly heavier than all the group's other atoms.

(2) If some of the natural frequencies of the additive group are the same as some of those of the rest of the molecule and the kinematic relations between the relevant pair of vibrations is non-zero (the shared atoms are mobile during the vibrations and their displacements during the two natural vibrations are not orthogonal), then vibrational resonance occurs. This leads to absorption bands and Raman spectra lines, whose frequencies and intensities depend upon the entire molecule, even for additive groups.

(3) If some of the frequencies of an additive group's natural vibrations are significantly greater than the natural vibration frequencies interacting with them and belonging to the rest of the molecule, then vibrations may be observed which are characteristic with respect to frequency (although not to form) for the additive group. The position of these vibrations within the spectrum will depend on the properties of the additive group, although their intensities will be determined by those of the whole molecule. When substitutions are made outside the group such that the group's additivity is not altered, the frequencies of the vibrations may remain unchanged, although their intensity will undoubtedly be changed.

(4) If some of the frequencies of an additive group's natural vibrations are much lower than the natural ones interacting with them and belonging to the rest of the molecule, then, once again, vibrations characteristic with respect both to form and frequency may be observed.

Thus we see that the appearance of different kinds of vibrations characteristic for additive groups is associated with the properties of both the groups themselves and of their environments, particularly the vibrations of the rest of the molecule. This is the reason for the very frequent alterations of the characteristic nature of vibrations of additive groups in some members of a homologous series that is being studied, even though the vibrations may be characteristic for other compounds in the same series. Thus in order for characteristic frequencies to remain unchanged in a series of molecules it is necessary that the conditions we formulated above are not altered in any of the molecules.

We have considered above a number of limiting cases. However, it has been found that infra-red absorption bands or Raman spectra lines may be shifted or their intensities changed due solely to purely 'mechanical' reasons, even if the properties of the groups themselves remain unchanged.

If a molecule has characteristic vibrations for separate groups, then estimates of their frequencies and forms may be significantly simplified by using the approach described at the beginning of this chapter. This is to transform the equations for a molecule's vibrations by changing to new co-ordinates which are normal ones for the free vibrations of the molecule's separate groups. Using the form of the D_{12} and D_{21} submatrices we can immediately identify those natural vibrations of the parts of the molecule that will be characteristic with respect to form and frequency. In many cases approximate values for the molecule's remaining vibrational frequencies and forms can then be obtained by solving a set of second-order equations such as

$$\left(\begin{bmatrix} 1 & d_{ij} \\ 0 & 1 \end{bmatrix} - \lambda \mathrm{I}\right)\begin{bmatrix} Q_i \\ Q_j \end{bmatrix} = 0$$

where d_{ij} is the largest element of the D_{12} submatrix. In some cases we can use various modifications of the perturbation method.

When a molecule has vibrations that are nearly characteristic it is most helpful to simplify the calculations of the vibrational frequencies and forms, especially for molecules with many atoms and containing relatively closed groupings (for example, aromatic rings, sections of conjugate chains, etc.). For instance, it is frequently very useful to separate out the diagonal submatrices of the relatively closed atomic groups in the force constant and kinematic coefficient matrices in terms of the molecule's natural co-ordinates. As a result, we can, as a first approximation, break down the overall problem into a series of smaller ones. The frequencies and the matrices for the transformations to normal co-ordinates can then be determined for these partial problems. The matrices for the force constants and kinematic relationships between the normal co-ordinates of different atomic groupings can then be easily determined from the known non-diagonal submatrices of the initial force constant and kinematic coefficient matrices, given in terms of natural co-ordinates, and from the matrices for the transformations to the normal co-ordinates that were compiled for each problem of the initial approximation. The fragmentary method[12] which is widely used for rapid estimates was constructed on this basis.

12.2 The Theory of Characteristic Intensities and Polarizations in Infra-red Spectra

It was shown above that certain chemical groups have characteristic frequencies and forms of its vibrations which are relatively little changed when these groups are transferred from one molecule to another. This property of the vibrations of polyatomic molecules is a significant reason why spectral methods have spread so widely into science and industry.

It has therefore become necessary to develop a method of calculating the intensities and polarizations of characteristic vibrations, and to explain the conditions under which infra-red absorption bands that are characteristic with respect to intensity and polarizability occur. This question is of interest both from the point of view of calculating a molecule's electro-optical parameters from infra-red spectral data and of obtaining additional criteria by which the absorption bands observed in a spectrum can be assigned to the molecule's respective normal vibrations. In particular, it is possible to use the fact that valency and deformational vibrations can be distinguished to obtain approximate formulae for the derivative $(\partial\vec{\mu}/\partial Q_i)_0$.

An investigation into characterizability with respect to polarization is, moreover, crucial for measurements whose aim is to establish the arrangement of bonds and atomic groups within a molecule according to the orientating vector of the changes in the dipole moment in a given normal vibration.

We select a chemically characteristic group of atoms within a molecule and describe its vibrations by n' internal natural co-ordinates out of the molecule's total of n co-ordinates. Let us consider the limiting case of vibration, i.e. one that is fully characteristic for this group of n' co-ordinates. By definition, this means that the form and frequency of one of the molecule's vibrations exactly match those of one of the normal vibrations the group would undergo if it were a free molecule.

We turn now to an analysis of the intensities and polarizations of a completely characteristic vibration. The expression for $(\partial\vec{\mu}/\partial Q_i)_0$ may in this case be written as

$$\left(\frac{\partial\vec{\mu}}{\partial Q_i}\right)_0 = \left\{ [\tilde{\vec{e}}^{\,\mathrm{I}}, \tilde{\vec{e}}^{\,\mathrm{II}}] \begin{bmatrix} \dfrac{\partial\mu^{\mathrm{I}}}{\partial q^{\mathrm{I}}} & \dfrac{\partial\mu^{\mathrm{I}}}{\partial q^{\mathrm{II}}} \\ \dfrac{\partial\mu^{\mathrm{II}}}{\partial q^{\mathrm{I}}} & \dfrac{\partial\mu^{\mathrm{II}}}{\partial q^{\mathrm{II}}} \end{bmatrix} + [\tilde{\mu}^{\mathrm{I}}, \tilde{\mu}^{\mathrm{II}}] \begin{bmatrix} (S^{\mathrm{I}})^{-1} & 0 \\ 0 & (S^{\mathrm{II}})^{-1} \end{bmatrix} \right.$$
$$\left. \times \left(\begin{bmatrix} \Delta E^{\mathrm{I}} \\ \Delta E^{\mathrm{II}} \end{bmatrix} [\tilde{\vec{B}}^{\mathrm{I}}, \tilde{\vec{B}}^{\mathrm{II}}] \begin{bmatrix} \Theta_{11} & \Theta_{12} \\ \Theta_{21} & \Theta_{22} \end{bmatrix} - \begin{bmatrix} \vec{E}^{\mathrm{I}} & 0 \\ 0 & \vec{E}^{\mathrm{II}} \end{bmatrix} \right) \right\} \begin{bmatrix} l_{q^{\mathrm{I}}}^{(i)} \\ 0 \end{bmatrix} \tag{12.15}$$

Here the superscripts I and II refer to the sets of variables that refer, respectively, to the selected group and to the rest of the molecule. For example, $\tilde{\vec{e}}^{\,\mathrm{I}}$ and $\tilde{\vec{e}}^{\,\mathrm{II}}$ and $\tilde{\mu}^{\mathrm{I}}$ and $\tilde{\mu}^{\mathrm{II}}$ are, respectively, the submatrices of the row-matrices $\tilde{\vec{e}}$ and $\tilde{\mu}$; $(\partial\mu^i/\partial q^k)$ are rectangular submatrices of the matrix of the derivatives of the bond dipole moment's of the first and second groups with respect to the vibration co-ordinates; $\tilde{\vec{B}}^{\mathrm{I}}$ and $\tilde{\vec{B}}^{\mathrm{II}}$ are rectangular submatrices of the $\tilde{\vec{B}}$ matrix and both have N (number of atoms in the molecule) rows but, respectively, n' and $(n-n')$ column; and

$$\begin{bmatrix} l_{q^{\mathrm{I}}}^{(i)} \\ 0 \end{bmatrix}$$

is a column of the normalized amplitudes of the natural vibrational co-ordinates of either sort. The assumption of full characterizability for the vibration means that the amplitudes of the group II co-ordinates are zero. The Θ matrix is the inverse of

$$T = \begin{bmatrix} T_{11} & T_{12} \\ T_{21} & T_{22} \end{bmatrix}$$

from which $\Theta_{11} = T_{11}^{-1} + \Theta'_{11}$, where

$$\Theta'_{11} = T_{11}^{-1} T_{12}(T_{22} - T_{21} T_{11}^{-1} T_{12})^{-1} T_{21} T_{11}^{-1}$$

By allowing for this and completing the multiplications in equation (12.15) we obtain

$$\left(\frac{\partial \vec{\mu}}{\partial Q^i}\right)_0 = \underline{\left\{ \tilde{\vec{e}}^{\text{I}} \left[\frac{\partial \mu^{\text{I}}}{\partial q^{\text{I}}} \right] + \tilde{\mu}^{\text{I}} (S^{\text{I}})^{-1} (\Delta E^{\text{I}} \tilde{\vec{B}}^{\text{I}} T_{11}^{-1} - \vec{E}^{\text{I}}) \right\} l_{q^{\text{I}}}^{(i)}}$$

$$+ \underline{\underline{\tilde{\vec{e}}^{\text{II}} \left[\frac{\partial \mu^{\text{II}}}{\partial q^{\text{I}}} \right] l_{q^{\text{I}}}^{(i)} + \tilde{\mu}^{\text{I}} (S^{\text{I}})^{-1} (\Delta E^{\text{I}} \tilde{\vec{B}}^{\text{I}} \Theta'_{11}) l_{q^{\text{I}}}^{(i)} + \tilde{\mu}^{\text{I}} (S^{\text{I}})^{-1} (\Delta E^{\text{I}} \tilde{\vec{B}}^{\text{II}} \Theta_{21}) l_{q^{\text{I}}}^{(i)}}}$$

$$+ \underline{\underline{\mu^{\text{II}} (S^{\text{II}})^{-1} \Delta E^{\text{II}} (\tilde{\vec{B}}^{\text{I}} \Theta_{11} + \tilde{\vec{B}}^{\text{II}} \Theta_{21}) l_{q^{\text{I}}}^{(i)}}} = \left(\frac{\partial \mu^{\text{I}}}{\partial Q_i^{\text{I}}}\right)_0 + \left(\frac{\partial \mu^{\text{II}}}{\partial Q_i^{\text{I}}}\right)_0 \quad (12.16)$$

where

$$\Theta_{21} = -(T_{22} - T_{21} T_{11}^{-1} T_{12})^{-1} T_{21} T_{11}^{-1}$$

Hence the expression for $(\partial \vec{\mu}/\partial Q_i)_0$ can be separated into two terms, $(\partial \vec{\mu}^{\text{I}}/\partial Q_i^{\text{I}})_0$ and $(\partial \vec{\mu}^{\text{II}}/\partial Q_i^{\text{I}})_0$. In equation (12.16) they are underlines once and twice, respectively. One of these depends solely on the elements of the selected group and yields the size of the vector $(\partial \vec{\mu}/\partial Q_i)_0$ when the group I atoms vibrate as if they were part of a free molecule. The second term defines the electrical and mechanical interactions of group I with the rest of the molecule.

The physical meaning of the constituents of $(\partial \vec{\mu}^{\text{II}}/\partial Q_i^{\text{I}})_0$ are as follows. The first term describes the size of the change in the bond dipole moments of group II during changes in the equilibrium configuration of group I. Its contribution depends on the degree of overlap between the electron clouds of the two groups and is reduced as the additivity of group I increases. The second term appears because the terms in the Θ_{11} submatrix are not invariant, and may be changed by alterations in group II even if all the elements for group I remain the same. The closer the Θ matrix is to being quasi-diagonal, the smaller this term will be. The third and fourth terms are corrections, respectively, to rotations in space of bonds in the first and second groups. These rotations may be subject to the requirement that the centre of mass and moment of momentum be conserved for the molecule as a whole when a vibration changes the equilibrium bond lengths and angles (this is usually called the 'form' of the vibration) in the selected group of atoms. If group I is not an end then rotations of the bonds in the rest of the molecule may not be zero, even when there is no noticeable effect.

The representation of $(\partial \vec{\mu}/\partial Q_i)_0$ as two components enables us to study the characterizability of vibrations with respect to polarization and intensity, and to see whether this is stable when a group of atoms 'moves' from one molecule to another. The vibrations for which characterizability is stable are those that arouse the greatest interest in physics. It is very clear that the phenomenon may only occur when the influence of the environment of the chemical group is small.

We call a molecule's vibration 'fully characteristic' with respect to polarization and intensity for a given atomic group if $(\partial \vec{\mu}^{\mathrm{II}}/\partial Q_i^{\mathrm{I}})_0 = 0$. It is easy to show that, in general, this can only be true when the group being considered is completely separated from the rest of the molecule. We can thus really only speak of 'approximate' characterizability $(\partial \vec{\mu}/\partial Q_i)_0$. The smaller the size of $(\partial \vec{\mu}^{\mathrm{II}}/\partial Q_i^{\mathrm{I}})_0$ is relative to $(\partial \vec{\mu}^{\mathrm{I}}/\partial Q_i^{\mathrm{I}})_0$, the more characteristic the vibration will be with respect to intensity and polarization. In this sense, $(\partial \vec{\mu}^{\mathrm{II}}/\partial Q_i^{\mathrm{I}})_0$ can be thought of as a measure of non-characterizability.

In all probability, it will be impossible to demonstrate any general conditions that are necessary or sufficient for full characterizability with respect to polarization and intensity. All the $(\partial \vec{\mu}^{\mathrm{II}}/\partial Q_i^{\mathrm{I}})_0$ terms are a series of (matrix) multiplicants, and so the product may be zero even if all the terms are non-zero. Nonetheless, we can name a series of attributes, which do have a physical reality, in the presence of which $(\partial \vec{\mu}^{\mathrm{II}}/\partial Q_i^{\mathrm{I}})_0$ can be neglected. These necessary conditions for vibrations to be characteristic with respect to polarization and intensity for a given group are:

(1) The existence of full characterizability with respect to form and frequency;
(2) Weak interaction between the electron clouds of groups I and II, i.e. the elements of the $[\partial \mu^{\mathrm{II}}/\partial q^{\mathrm{I}}]$ submatrix must be negligibly small compared with those of $[\partial \mu^{\mathrm{I}}/\partial q^{\mathrm{I}}]$;
(3) Groups I and II must share as few common atoms as possible; and
(4) There must be no rotation in any of the bonds in group II during vibrations of group I.

In order for the full characterizability with respect to polarization and intensity of a group's vibration to be conserved from one molecule to another while its full characterizability with respect to form and frequency (and some other conditions) are maintained the value of the group's dipole moment must remain unchanged. This means that $[\partial \mu^{\mathrm{I}}/\partial q^{\mathrm{II}}]$ must be zero. In other words, we have arrived at a property of partial additivity for the molecule. Consequently, the experimental discovery of absorption bands in infra-red spectra that are characteristic with respect to polarizability and intensity and of the stability of this characteristic nature is one of the weighter proofs of the isolation of the electron cloud of the corresponding atomic group.

It may happen that some conservation of intensity exists for a group over a series of molecules, even though characterizability with respect to form is partially altered due to the large size of the group's electro-optical parameters with respect to the rest of the molecule. However, this must be a rare exception.

Stability of characterizability with respect to form and frequency is usually associated with the chemical forces which are shown by the small non-diagonal elements in the U matrix. Since the potential and electrical fields within a molecule must, generally speaking, be interrelated, the small size of the relevant non-diagonal elements in the matrix of the derivatives of the bond dipole moments with respect to vibrational co-ordinates will be accompanied by small analogous elements in the force constant matrix. This enables us to conclude that vibrations characteristic with respect to intensity will, in the great majority of cases, be characteristic with respect to both frequency and form, while those characteristic with respect to polarization will also be accompanied by a characterizability with respect to form. It is natural to classify the absorption bands in a polyatomic molecule's vibrational spectrum as those characteristic with respect to (1) intensity and polarization, (2) polarization only and (3) frequency. These different types can be investigated experimentally. The highest kind of characterizability is that with respect to intensity and polarization, since it requires a whole series of conditions to be met for it to exist, and is therefore only rarely encountered. All things being equal, end groupings have the greatest probability of possessing this property.

Note that characterization with respect to frequency is not a necessary condition for characterizability of $(\partial\vec{\mu}/\partial Q_i)_0$ with respect to direction (polarization). However, vibrations that are characteristic with respect to frequency and form in fact most often correspond to stable characteristic frequencies. Hence special stipulations concerning the conditions for characterizability with respect to polarization are not vitally important.

12.2.1 Bond vibrations

Let us assume that a given normal vibration of a molecule is fully characteristic for one bond (bond number 1) and, furthermore, that the bond is not part of a ring structure and its vibration does not induce rotations in the other bonds in the molecule. This would be true if a straight line passes along the bond and through the centres of mass of the adjacent groups, or if the bond is a terminal one and the mass of one of the bond's atoms is significantly less than that of the rest of the molecule. Clearly, we then obtain

$$\left(\frac{\partial\vec{\mu}}{\partial Q_i}\right)_0 = \vec{e}\,\frac{\partial\mu_1}{\partial q_1}\,l_{q_{1i}} + \sum_{n\neq 1}\vec{e}_n\,\frac{\partial\mu_n}{\partial q_1}\,l_{q_{1i}} \tag{12.17}$$

If $(\partial\mu/\partial q_1) \gg (\partial\mu_n/\partial q_1)$ then

$$\sum_{n\neq 1}\vec{e}_n\,\frac{\partial\mu_n}{\partial q_1}\,l_{q_{1i}}$$

can be neglected, and the vibration will be fully characteristic with respect to polarization and intensity. This, in all probability, is the case for vibrations of OH, CH, NH and similar bonds, in which it is known that their vibrational bands have fairly stable and significant intensities.

In the case of symmetric bonds such as C=C, C—C and N=N (whose vibrations are often characteristic with respect to frequency) the inherent dipole moment of such a bond, taken as if it were a free group, and the derivative

$(\partial\mu_0^{(1)}/\partial q_1)$ will be zero. Consequently, $(\partial\vec{\mu}/\partial Q_i)_0$ is determined wholly by the bonds' environment, and is not characteristic.

It is thus clear why attempts to make judgements on these groups in specific compounds using the observable intensities of their vibrations are ineffectual.

12.2.2 *Vibrations of XY_2 and XY_3 groups*

Let us assume that a molecule contains the non-linear symmetric atomic groupings XY_2 and XY_3, and that their normal vibrations are fully characteristic with respect to both frequency and form. Let us also assume, moreover, that these groups are connected with the rest of the molecule only through the X atoms, and that the electron clouds of the groups being considered overlap the cloud of the rest of the molecule only by a small amount. These properties are possessed in particular by the groups CH_2 and CH_3. It is clear from the general conditions for characterizability that the symmetric vibrations of these groups (especially if a group's axis of symmetry passes through the molecule's centre of mass) are the most characteristic with respect to intensity and polarization.

Anti-symmetric and degenerate vibrations are accompanied by displacement of the X atom and rotations of the group itself such that other parts of the molecule must rotate. Strictly speaking, the level of characterizability is, for this reason, probably less for these vibrations. We can point to the expressions for $(\partial\vec{\mu}/\partial Q_i)_0$ by way of example for the symmetric and antisymmetric valency vibrations of the XY_2 group when it is free and when it is a part of the ethylene molecule. For the valency vibrations symmetric with respect to the C_2 axis we have, for XY_2,

$$\left|\left(\frac{\partial\vec{\mu}}{\partial Q_i}\right)_0\right| = \sqrt{2}\left(\frac{\partial\mu_1}{\partial q_1}+\frac{\partial\mu_2}{\partial q_1}\right)\cos\frac{\varphi}{2}\,l_{q^{(s)i}} \tag{12.18}$$

and for ethylene,

$$\left|\left(\frac{\partial\vec{\mu}}{\partial Q_i}\right)_0\right| = \sqrt{2}\left(\frac{\partial\mu_1}{\partial q_1}+\frac{\partial\mu_2}{\partial q_1}\right)\cos\frac{\varphi}{2}\,l_{q^{(s)i}} \tag{12.19}$$

For the anti-symmetric vibrations we obtain for XY_2

$$\left|\left(\frac{\partial\vec{\mu}}{\partial Q_i}\right)_0\right| = \sqrt{2}\left[\left(\frac{\partial\mu_1}{\partial q_1}-\frac{\partial\mu_2}{\partial q_1}\right)+\mu S^{-1}\frac{2\varepsilon_0\cos(\varphi/2)}{\varepsilon+2\varepsilon_0\sin^2(\varphi/2)}\right]\sin\frac{\varphi}{2}\,l_{q^{(s)i}} \tag{12.20}$$

and for ethylene,

$$\left|\left(\frac{\partial\vec{\mu}}{\partial Q_i}\right)_0\right| = \sqrt{2}\left(\frac{\partial\mu_1}{\partial q_1}-\frac{\partial\mu_2}{\partial q_1}\right)\sin\frac{\varphi}{2}\,l_{q^{(s)i}} \tag{12.21}$$

It is clear from these expressions that even if the normalized amplitudes of the vibrational co-ordinates and the XY_2 group's electro-optical parameters are conserved when the group is transferred to any ethylene-type molecule, then the $(\partial\vec{\mu}/\partial Q_i)_0$ derivatives corresponding to the vibrations of a free XY_2 group and those of a bound group for anti-symmetric vibrations will change in value. Thus these vibrations will not be characteristic with respect to intensity (the polarization in this case being determined by the symmetry of the molecule). By contrast, the symmetric vibrations are characteristic with respect to intensity.

The opposite is true for Y_2XZ and Y_3XZ groups, for if the XZ bond coincides with the axis of symmetry of the XY_2 and XY_3 groups, and if the derivatives of μ_{XZ} with respect to the co-ordinates of the XY_2 and XY_3 groups are non-zero, then the antisymmetric vibrations of these groups will have a great probability of retaining their intensities over a range of molecules, provided that the derivatives are large enough in value. This is because the polarization vectors for these vibrations are directed perpendicularly to the XZ bond, and the corresponding μ_{XZ} derivatives, which, in general, must change more than the electro-optical parameters of the XY_2 and XY_3 group (the group is bonded to the rest of the molecule through the XZ bond) do not, in the main, contribute to the expression for the derivative of the dipole moment with respect to the normal co-ordinate. In addition, the symmetric vibrations of these groups are rarely characteristic with respect to form, especially if the masses of all the atoms are approximately the same. The same situation occurs, for example, for vibrations of the CNO_2 group.

We have shown that characteristic intensities can be encountered only relatively rarely, and this explains the virtual absence of experimental correlations on this question. In fact, virtually no investigations to establish the characterizability of some groups with respect to intensity or polarization have been carried out.

In future, the development of the relevant theory will make it possible to select consciously the objects for study. This, in turn, must accelerate the search for functional groups whose vibrations possess characterizability with respect to intensity and polarization. This will be critically important for the development of applied spectral analysis.

The difference of interpretation of the frequencies and intensities (polarizations) is particularly noticeable in the example of vibrations characterizable with respect to form. If the frequency and form of such a vibration depend solely on the properties of group I, then the intensity of the same vibration may, to a significant degree, be related to the environment. Moreover, if one of the group II atoms is substituted for, then this would require a change in the intensity of a vibration that is characterizable with respect to frequency and form for group I. Such an alteration of 'constant' intensity may be attributed to a disruption in the structure of group I or to intramolecular effects, but in fact this is not what is observed. We must emphasize that the interpretation of intensities may not coincide with the attribution of the vibrational frequencies, even when there is stable characterizability with respect to form and frequency. The interpretations of intensity and frequency will only coincide for vibrations that are characterizable with respect to intensity, i.e. only when an absorption band's intensity and frequency are completely determined by the parameters of the given group.

12.3 The Intensities of Infra-red Absorption Bands of a Polyatomic Molecule that Contains Several Identical Groups

A topic that is related to characteristic intensities and polarization is the dependence of the intensities of the fundamental vibrations on the number of identical groups in a molecule. This problem is very significant for the development of quantitative methods for analysing structures and groups using vibrational absorption spectra.

We shall consider a molecule composed of an unchanging skeleton and n identical atomic groups connected to it. The intensity of an absorption band corresponding to a normal vibration of the molecule will be determined both by vibration of the skeleton and by those of the groups. In addition, when the number of equivalent groups is raised, the molecule's electric and mechanical parameters, the vibrational forms and the frequencies may undergo noticeable changes. Thus in the general case the intensities of the corresponding absorption bands may depend on the number of equivalent groups in an extremely complex and irregular way.

The task can, however, be simplified if vibrations that are characteristic with respect to form for equivalent groups are studied using the assumption that the equilibrium geometrical parameters of the skeleton do not change for the given normal vibration. We shall designate all the quantities (groups of internal natural vibration co-ordinates, the submatrices of the matrices $\vec{B}, \Theta$, $[\partial\mu/\partial q]$, etc.) that refer to the n separate groups by the superscripts $1, 2, \ldots, n$, while the quantities for the skeleton are given the superscript 0. We shall also assume that the $[\partial\mu^{(k)}/\partial q^{(m)}]$ and the U_{km} submatrices (k and $m = 1, 2, \ldots, n$) are zero, i.e. the force and electric fields of the groups being considered do not overlap.

Experimentally it is found that these assumptions normally hold, since they depend on the mutual independence of the equivalent groups. This kind of group independence should not be confused with the independence that freely vibrating molecules possess.

Bearing these factors in mind, we can write the expression for $(\partial\vec{\mu}/\partial Q_i)_0$ as

$$\left(\frac{\partial\vec{\mu}}{\partial Q_i}\right)_0 = \left\{[\tilde{\vec{e}}^{(1)}, \ldots, \tilde{\vec{e}}^{(n)}, \tilde{\vec{e}}^{(0)}] \begin{bmatrix} \dfrac{\partial\mu^{(1)}}{\partial q^{(1)}} & \cdots & 0 & \dfrac{\partial\mu^{(1)}}{\partial q^{(0)}} \\ 0 & \cdots & \dfrac{\partial\mu^{(n)}}{\partial q^{(n)}} & \dfrac{\partial\mu^{(n)}}{\partial q^{(0)}} \\ \dfrac{\partial\mu^{(0)}}{\partial q^{(1)}} & \cdots & \dfrac{\partial\mu^{(0)}}{\partial q^{(n)}} & \dfrac{\partial\mu^{(0)}}{\partial q^{(0)}} \end{bmatrix} + [\tilde{\mu}^{(1)}, \ldots, \tilde{\mu}^{(n)}, \tilde{\mu}^{(0)}]\right.$$

$$\times \begin{bmatrix} (S^{(1)})^{-1} & \cdots & 0 & 0 \\ \cdots & \cdots & \cdots & \cdots \\ 0 & \cdots & (S^{(n)})^{-1} & 0 \\ 0 & \cdots & 0 & (S^{(0)})^{-1} \end{bmatrix}$$

$$\times\left(\begin{bmatrix}\Delta E^{(1)}\\ \cdots\\ \Delta E^{(n)}\\ \Delta E^{(0)}\end{bmatrix}[\tilde{\vec{B}}^{(1)},\ldots,\tilde{\vec{B}}^{(n)},\tilde{\vec{B}}^{(0)}]\begin{bmatrix}\Theta_{11} & \cdots & \Theta_{1n} & \Theta_{10}\\ \cdots & \cdots & \cdots & \cdots\\ \Theta_{n1} & \cdots & \Theta_{nn} & \Theta_{n0}\\ \Theta_{01} & \cdots & \Theta_{0n} & \Theta_{00}\end{bmatrix}\right.$$

$$\left.-\begin{bmatrix}\vec{E}^{(1)} & \cdots & 0 & 0\\ \cdots & \cdots & \cdots & \cdots\\ 0 & \cdots & \vec{E}^{(n)} & 0\\ 0 & \cdots & 0 & \vec{E}^{(0)}\end{bmatrix}\right)\Bigg\}\begin{bmatrix}l^{(i)}_{q^{\mathrm{I}}}\\ \vdots\\ l^{(i)}_{q^{(n)}}\\ 0\end{bmatrix}\frac{v_i}{\sqrt{\left(\sum_k^n \tilde{l}^{(i)}_{q(k)}U_{kk}l^{(i)}_{q(k)}\right)}}=\frac{v_i}{\sqrt{\left(\sum_k^n \tilde{l}^{(i)}_{q(k)}U_{kk}l^{(i)}_{q(k)}\right)}}$$

$$\times\left\{\sum_k^n\left\{\vec{e}^{(k)}\left[\frac{\partial\mu^{(k)}}{\partial q^{(k)}}\right]+\tilde{\mu}^{(k)}(S^{(k)^{-1}})(\Delta E^{(k)}\tilde{\vec{B}}^{(k)}\theta_{kk}-E^{(k)})\right\}l^{(i)}_{q(k)}\right.$$

$$\left.+\text{interaction terms}\right\} \tag{12.22}$$

Here, Θ_{kk} is the submatrix containing the elements of the kth equivalent group and is a part of the Θ matrix. The interaction terms combine all the corrections to the molecule's $(\partial\vec{\mu}/\partial Q_i)_0$, the correction appearing because of the chemical bond each equivalent group has with the molecule's skeleton. That is, the elements of the $[\partial\mu^{(k)}/\partial q^{(k)}]$ submatrix are not zero and the Θ_{kk} submatrix is not invariant with respect to molecules with different numbers of groups. The skeleton can rotate as a whole, given that the vibrations occurs within the functional groups being considered. The corrections also arise due to those to bond rotations of the groups that are needed in order to conserve the centre of mass and angular momentum of the whole molecule during vibrations of a given form.*

When the interaction terms can be neglected, equation (12.22) takes the form

$$\left(\frac{\partial\vec{\mu}}{\partial Q_i}\right)_0=\frac{v_i}{\sqrt{\left(\sum_k^n \tilde{l}^{(i)}_{q(k)}U_{kk}l^{(i)}_{q(k)}\right)}}\sum_k^n(\vec{F}_k l^{(i)}_{q(k)}) \tag{12.23}$$

where

$$\vec{F}_k=\tilde{\vec{e}}^{(k)}\left[\frac{\partial\mu^{(k)}}{\partial q^{(k)}}\right]+\tilde{\mu}^{(k)}S^{(k)^{-1}}(\Delta E^{(k)}\tilde{\vec{B}}^{(k)}\theta_{kk}-\vec{E}^{(k)})$$

However, even then, $(\partial\vec{\mu}/\partial Q_i)_0$ depends on the number of identical groups in the molecule in a very complex way (even when $\vec{F}_k$ and U_{kk} are conserved), because both the vibrational frequency and form $l^{(i)}_{q(k)}$ may change from one molecule to another. Nevertheless, equation (12.23) can be used for worthwhile calculations.

The simplest case is when the $l^{(i)}_{q(k)}$ forms differ from each other by a constant

*A normalization factor is taken out and the vibrational form is assumed to be unnormalized. The $\vec{E}^{(k)}$ matrix, like the $E^{(0)}$ one contains null columns for deformational co-ordinates.

factor, i.e.

$$l^{(i)}_{q(k)} = \eta_{ki} l^{(i)}_{q(1)}$$

from which we obtain

$$\left(\frac{\partial \vec{\mu}}{\partial Q_i}\right)_0 = \frac{1}{\sqrt{(\tilde{l}^{(i)}_{q(1)} U_{kk} l^{(i)}_{q(1)})}} \cdot \frac{\nu_i}{\sqrt{\left(\sum_k^n \eta^2_{ki}\right)}} \sum_k \eta_{ki} \vec{F}_k l^{(i)}_{q(1)} \tag{12.24}$$

We shall represent the vector $\vec{F}_k l^{(i)}_{q(1)}$ as $A_{ki}\vec{h}_k$, where $\vec{h}_k$ is a unit-orientating vector and A_{ki} is the absolute value of the $\vec{F}_k l^{(i)}_{q(1)}$ vector. Since we are dealing with groups that have the same structure, we have $U_{kk} = U_{11}$ and $A_{ki} = A_{1i}$ for all k.

We shall assume, in addition, that the $\vec{F}$ and U matrices and the $l^{(i)}_{q(k)}$ form remain the same over the series of molecules being studied, and that the interaction terms remain small all the time. In other words, we shall be dealing with stable characteristics in the $\vec{F}_k l^{(i)}_{q(k)}$ vectors for the equivalent groups. The intensity of a normal vibration then depends in a simple way on the number of identical groups there are in the molecule, i.e.

$$\mathscr{I}_i(n) = \text{const}\, \frac{\nu_i^2(n)}{\sum_k^n \eta^2_{ki}} \left(\sum_k^n \eta_{ki} \vec{h}_{ki}\right)^2 \tag{12.25}$$

A special case is when all the $\eta_k = 1$, and then $\vec{h}_{ki}$ are parallel and the $\mathscr{I}_i(n) \sim \nu_i^2(n) \cdot n$. On the other hand, if it can be experimentally established that a band's intensity follows equation (12.25) over a series of molecules, then this is an impressive argument that the vibrational form and electro-optical parameters of the atomic groups in question should be conserved.

Note that for degenerate vibrations

$$\mathscr{I} \sim \sum_m \left(\frac{\partial \vec{\mu}}{\partial Q_m}\right)_0^2$$

where the summation runs over all the degenerate components. The dependence of each component on the number of groups may initially be found from equations (12.23)–(12.25).

These considerations indicate that, in the general case, band intensities do depend upon the number of identical groups in the molecule in a complex and irregular way. If the vibrations of the groups have stable characterizability of the $\vec{F}_k l^{(i)}_{q(k)}$ vectors and the interaction terms are small, then the intensities of the relevant bands may depend on the number of groups quite simply, given that the mutual arrangement of the groups and the values of the η_{ki} are known.

The above refers to bands corresponding to one vibrational mode for each independent group. When different vibrations are superimposed, the dependency (on group number) of the intensities of the complicated bands that are obtained will be quite different, and, given the large number of superpositions in real cases, the dependency on the number of identical groups may become linear. It is easy to see that the sum of the intensities of symmetric and anti-symmetric vibrations of two symmetric independent groups is twice as large as the intensity of the vibrations of one group, independent of the angle between the vectors $\vec{h}_1$ and $\vec{h}_2$. The sum of the intensities of the symmetric and degenerate vibrations of three symmetric groups, also independently of the $\vec{h}_i$ directions, will be tripled. It may happen that only one group vibrates in each of the normal vibrations corresponding to those of the equivalent groups, i.e. the forms are:

$$\begin{aligned} \nu_1 &\to \{1, 0, \ldots 0\} \\ \nu_2 &\to \{0, 1, \ldots 0\} \\ &\ldots\ldots\ldots\ldots\ldots\ldots \\ \nu_n &\to \{0, 0, \ldots 1\} \end{aligned}$$

This may occur if the groups are far apart and separated by heavy atoms. In such cases, $\nu_1 = \nu_2 = \cdots = \nu_n$ is usually the rule, and an absorption band with an intensity that grows linearly as the number of identical groups in the molecule increases will be observed.

Vibrations whose frequencies are changed very dramatically by alterations in the number of groups must be used with care, because this behaviour may be related to changes in the force fields and the electro-optical parameters in the equivalent groups themselves.

In some cases these simplified formulae may help to give a fairly accurate explanation of the variation in intensity over a series of molecules having different numbers of identical groups, and this kind of work does not require any complicated manipulations.

12.4 Characteristic Intensities and Polarizations for Primary Overtones and Combination Bands

In Section 12.3 we considered the intensities and polarizations of fundamental absorption bands and how these intensities depend on the number of identical groups in polyatomic molecules. The analogous investigations for overtones and combination bands are no less important both to establish general patterns that will facilitate the interpretation of spectra, the study of topics associated with the appearance of closed atomic groupings and with quantitative group and structural analysis.

We select from a molecule a group of atoms whose vibrations are described by the system of n_1 internal vibration co-ordinates that are a subset of all the molecule's natural vibrational ones. The set of group co-ordinates we designate q^{I} and the group's bond dipole moments and orientating vectors μ^{I} and $\vec{e}^{\,\mathrm{I}}$, respectively. The corresponding set of quantities that describe the rest of the molecule are designated q^{II}, μ^{II} and $\vec{e}^{\,\mathrm{II}}$, respectively.

Rewriting the expression for the $(\partial^2\vec{\mu}/\partial Q_i \partial Q_j)_0$ derivatives, we obtain

$$\left(\frac{\partial^2\vec{\mu}}{\partial Q_i \partial Q_j}\right)_0 = [\tilde{l}^{(i)}_{q^{\mathrm{I}}}\tilde{l}^{(i)}_{q^{\mathrm{II}}}]$$

$$\times\left\{\begin{bmatrix} \dfrac{\partial^2\mu^{\mathrm{I}}}{\partial q^{\mathrm{I}^2}} & \dfrac{\partial^2\mu^{\mathrm{I}}}{\partial q^{\mathrm{I}}\partial q^{\mathrm{II}}} & \dfrac{\partial^2\mu^{\mathrm{II}}}{\partial q^{\mathrm{I}^2}} & \dfrac{\partial^2\mu^{\mathrm{II}}}{\partial q^{\mathrm{I}}\partial q^{\mathrm{II}}} \\ \dfrac{\partial^2\mu^{\mathrm{I}}}{\partial q^{\mathrm{I}}\partial q^{\mathrm{II}}} & \dfrac{\partial^2\mu^{\mathrm{I}}}{\partial q^{\mathrm{II}^2}} & \dfrac{\partial^2\mu^{\mathrm{II}}}{\partial q^{\mathrm{II}}\partial q^{\mathrm{I}}} & \dfrac{\partial\mu^{\mathrm{II}}}{\partial q^{\mathrm{II}^2}} \end{bmatrix}\begin{bmatrix} \vec{\mathscr{E}}^{\mathrm{I}} & 0 \\ 0 & \vec{\mathscr{E}}^{\mathrm{I}} \\ \vec{\mathscr{E}}^{\mathrm{II}} & 0 \\ 0 & \vec{\mathscr{E}}^{\mathrm{II}} \end{bmatrix}\right.$$

$$+\begin{bmatrix} \dfrac{\partial\tilde{\mu}^{\mathrm{I}}}{\partial q^{\mathrm{I}}} & \dfrac{\partial\tilde{\mu}^{\mathrm{II}}}{\partial q^{\mathrm{I}}} \\ \dfrac{\partial\mu^{\mathrm{I}}}{\partial q^{\mathrm{II}}} & \dfrac{\partial\tilde{\mu}^{\mathrm{II}}}{\partial q^{\mathrm{II}}} \end{bmatrix}\begin{bmatrix} \dfrac{\partial\vec{e}^{\mathrm{I}}}{\partial q^{\mathrm{I}}} & \dfrac{\partial\vec{e}^{\mathrm{I}}}{\partial q^{\mathrm{II}}} \\ \dfrac{\partial\vec{e}^{\mathrm{II}}}{\partial q^{\mathrm{I}}} & \dfrac{\partial\vec{e}^{\mathrm{II}}}{\partial q^{\mathrm{II}}} \end{bmatrix}$$

$$\left.+\begin{bmatrix} \dfrac{\partial\tilde{\vec{e}}^{\mathrm{I}}}{\partial q^{\mathrm{I}}} & \dfrac{\partial\tilde{\vec{e}}^{\mathrm{II}}}{\partial q^{\mathrm{I}}} \\ \dfrac{\partial\tilde{\vec{e}}^{\mathrm{I}}}{\partial q^{\mathrm{II}}} & \dfrac{\partial\tilde{\vec{e}}^{\mathrm{II}}}{\partial q^{\mathrm{II}}} \end{bmatrix}\begin{bmatrix} \dfrac{\partial\mu^{\mathrm{I}}}{\partial q^{\mathrm{I}}} & \dfrac{\partial\mu^{\mathrm{I}}}{\partial q^{\mathrm{II}}} \\ \dfrac{\partial\mu^{\mathrm{II}}}{\partial q^{\mathrm{I}}} & \dfrac{\partial\mu^{\mathrm{II}}}{\partial q^{\mathrm{II}}} \end{bmatrix}\right\}\begin{bmatrix} l^{(j)}_{q^{\mathrm{I}}} \\ l^{(j)}_{q^{\mathrm{II}}} \end{bmatrix} \tag{12.26}$$

As in the case of fundamental vibrations, it is clear that one of the conditions required for the $(\partial^2\vec{\mu}/\partial Q_i\partial Q_j)_0$ derivative to be independent of the environment of group I (i.e. for the appropriate absorption bands to be characteristic with respect to intensity and polarization) must be that the *i*th and *j*th normal vibrations are characteristic with respect to form and frequency for the q^{I} co-ordinate set. Thus the row and column of the normalized forms will take the form

$$[\tilde{l}^{(i)}_{q^{\mathrm{I}}}, 0] \quad \text{and} \quad \begin{bmatrix} l^{(j)}_{q^{\mathrm{I}}} \\ 0 \end{bmatrix}$$

and we obtain

$$\left(\frac{\partial^2\vec{\mu}}{\partial Q_i\partial Q_j}\right)_0 = \tilde{l}^{(i)}_{q^{\mathrm{I}}}\left(\frac{\partial^2\mu^{\mathrm{I}}}{\partial q^{\mathrm{I}^2}}\mathscr{E}^{\mathrm{I}} + \frac{\partial^2\mu^{\mathrm{II}}}{\partial q^{\mathrm{I}^2}}\mathscr{E}^{\mathrm{II}}\right)l^{(j)}_{q^{\mathrm{I}}}$$

$$+\,\tilde{l}^{(i)}_{q^{\mathrm{I}}}\left(\frac{\partial\tilde{\mu}^{\mathrm{I}}}{\partial q^{\mathrm{I}}}\frac{\partial\vec{e}^{\mathrm{I}}}{\partial q^{\mathrm{I}}} + \frac{\partial\tilde{\mu}^{\mathrm{II}}}{\partial q^{\mathrm{I}}}\frac{\partial\vec{e}^{\mathrm{II}}}{\partial q^{\mathrm{I}}} + \frac{\partial\tilde{\vec{e}}^{\mathrm{I}}}{\partial q^{\mathrm{I}}}\frac{\partial\mu^{\mathrm{I}}}{\partial q^{\mathrm{I}}} + \frac{\partial\tilde{\vec{e}}^{\mathrm{II}}}{\partial q^{\mathrm{I}}}\frac{\partial\mu^{\mathrm{II}}}{\partial q^{\mathrm{I}}}\right)l^{(j)}_{q^{\mathrm{I}}} \tag{12.27}$$

We can conclude from equation (12.27) that the relevant derivative of the molecule's dipole moment with respect to the normal co-ordinates will not depend on the environment of group I when the molecule's dipole moment is made up by adding the dipole moments of groups I and II, and also when the two groups share as few common atoms as possible (i.e. the T matrix is close to being quasi-diagonal). The first condition means that the elements of the $(\partial^2\mu^{\mathrm{II}}/\partial q^{12})$ and $(\partial\mu^{\mathrm{II}}/\partial q^{\mathrm{I}})$ submatrices must be zero, while the second follows from the requirement that the $(\partial\vec{e}^{\mathrm{I}}/\partial q^{\mathrm{I}})$ submatrix is independent of group II. As a result, we can formulate the following criteria for overtones and combination bands to be characteristic with respect to intensity and polarization.

A primary overtone of a normal vibration will be characteristic with respect to intensity if:

(1) The normal vibration is fully characteristic with respect to form and frequency for the set of co-ordinates that describe the changes in the equilibrium configuration of the group of atoms in the molecule;
(2) The molecule's dipole moment is the sum of the moments of the group and the rest of the molecule (i.e. the molecule is partially additive);
(3) The group shares as few atoms as possible in common with the rest of the molecule.

The $(\partial^2\vec{\mu}/\partial Q_i\partial Q_j)_0$ derivative will not depend on the environment of group I, even when the overall dipole moment is not the sum of the two parts of the molecule, so long as there is no electro-optical anharmonicity and there are no rotations of the bonds in group II during changes in the equilibrium configuration of group I. However, these criteria are too general.

It can easily be shown that these conditions are the same as those needed for fundamental bands to be characteristic with respect to intensity. However, there is in this case the additional limitation on the absence of rotations of the bonds in group II during vibrations within group I.

Thus the requirements for overtones to be characterizable with respect to intensity are less severe. Consequently, if a fundamental vibration of the molecule is characterizable with respect to intensity then the corresponding primary overtone must also be characteristic. By contrast, it does not follow from the characterizability of the primary overtone with respect to intensity that the fundamental band is also characteristic. In any case, it is apparent that overtones which are characteristic with respect to intensity must be encountered no less frequently than fundamental bands which are thus characterizable. This conclusion is very significant in practical terms.

One comment must be made at this stage. All arguments used have assumed that the molecules containing group I strictly satisfy the conditions for characterizability with respect to intensity and polarization for a given vibration. However, it should be emphasized that, even given substantial disturbances to the forms of the vibrations, the intensities of the overtones and combinations in some molecules undergo larger changes than those for fundamentals, because the overtones depend on vibrational form to the fourth degree while the fundamentals only depend on the form quadratically. Therefore our conclusion concerning the relationship between the characterization with respect to intensity of the fundamental frequencies and those of the overtones is only true when all the characterizability conditions are strictly fulfilled by the molecule or range of molecules. Should one of these conditions with respect to form and its retention over the series of molecule be changed, then this will be most strongly reflected by the overtones.

A combination band arising from two normal vibrations will be characteristic with respect to intensity when both vibrations are themselves fully characteristic with respect to form and frequency for the group's set of vibrational co-ordinates and both the second and third conditions for overtone intensity characterization are satisfied. The intensity of a combination band formed by two normal vibrations, each of which is characteristic with respect to intensity for the fundamental tone, must also be characterizable. The inverse is not, in general, true. The $(\partial^2\vec{\mu}/\partial Q_i\partial Q_j)_0$ derivative for combinations formed by two normal vibrations such that the first is fully characterizable with respect to form and frequency for group I whilst the second is the same for group II, can be found using equation (12.26) by substituting forms such as $[\vec{l}^{(i)}_{q^{\mathrm{I}}}, 0]$ and $[0, l^{(j)}_{q^{\mathrm{II}}}]$. It is easy to show that the derivative will be zero if the molecule is partially additive and there are no rotations of bonds in group I during vibrations within group II and vice versa. It follows in particular that the intensity of the combination absorption formed by two vibrations, each of which is characterizable with respect to intensity for a fundamental transition for the two groups, will be zero, even if the band is allowed by the selection rules with respect to symmetry.

12.5 The Dependence of the Overtone and Combination Band Intensities on the Number of Identical Groups in the Molecule

We will now discuss how overtone and combination band intensities depend on the number of identical groups possessed by the molecule, and shall consider a molecule to be composed of an unchanging skeleton with n identical groups attached. We found for fundamentals that the intensities of some of the separate absorption bands, which correspond solely to selected groups, depend on the number of groups in a relatively simple and regular way only when the vibrations of each equivalent group satisfy the conditions for characterizability with respect to intensity and polarization, and when the vibrational forms of each group differ only by a constant factor. That is, the dependence was relatively simple only when the interactions between the groups themselves, or between the groups and the skeleton, were ignored, and only the mutual arrangement of the groups and the relative phases and amplitudes of their vibrations needed to be considered. The situation is the same for overtones and combinations.

The molecular dipole moment can be represented as an additive sum, i.e.

$$\vec{\mu} = \vec{\mu}_1 + \vec{\mu}_2 + \cdots + \vec{\mu}_n + \vec{\mu}_0$$

The forms of the normal vibrations are assumed to be such that the skeleton's equilibrium configuration does not change, i.e.

$$\vec{l}^{(i)}_q = [\vec{l}^{(i)}_{q(1)}, \vec{l}^{(i)}_{q(2)}, \ldots \vec{l}^{(i)}_{q(n)}, 0]$$

and

$$\vec{l}^{(j)}_q = [\vec{l}^{(j)}_{q(1)}, \vec{l}^{(j)}_{q(2)}, \ldots \vec{l}^{(j)}_{q(n)}, 0]$$

Moreover,

$$\vec{l}^{(i)}_{q(k)} = \eta_{ki}\vec{l}^{(i)}_{q(1)}$$

and

$$\vec{l}^{(j)}_{q(k)} = \eta_{kj}\vec{l}^{(j)}_{q(1)}$$

where η_{ki} and η_{kj} are scalar multipliers that describe the relative phases and amplitudes of the vibrations of each equivalent group. The superscripts $1, 2, \ldots, n$ refer to the equivalent groups, while the 0 subscript refers to the skeleton.

Using the general formula for $(\partial^2\mu/\partial Q_i \partial Q_j)_0$ and assuming that all

$$\left[\frac{\partial \vec{e}^{(k)}}{\partial q^{(m)}}\right] l^{(j)}_{q^{(m)}} = 0 \quad (k, m = 1, 2, \ldots n; k \neq m)$$

i.e. that no band rotation in an equivalent group leads to rotations in any other, and having separated out the normalizing factors in an explicit form, we obtain

$$\begin{aligned}\left(\frac{\partial^2 \vec{\mu}}{\partial Q_i \partial Q_j}\right)_0 &= N_i N_j \sum_k^n \tilde{l}^{(i)}_{q^{(k)}} \left[\frac{\partial^2 \mu^{(k)}}{\partial q^{(k)2}} \vec{\mathscr{E}}^{(k)} + \frac{\partial \tilde{\mu}^{(k)}}{\partial q^{(k)}} \frac{\partial \vec{e}^{(k)}}{\partial q^{(k)}} + \frac{\partial \tilde{\vec{e}}^{(k)}}{\partial q^{(k)}} \frac{\partial \mu^{(k)}}{\partial q^{(k)}}\right] l^{(j)}_{q^{(k)}} \\ &= N_i N_j \sum_k^n \eta_{ki} \eta_{kj} \tilde{l}^{(i)}_{q^{(1)}} \left[\frac{\partial^2 \mu^{(k)}}{\partial q^{(k)2}} \vec{\mathscr{E}}^{(k)} + \frac{\partial \tilde{\mu}^{(k)}}{\partial q^{(k)}} \frac{\partial \vec{e}^{(k)}}{\partial q^{(k)}} + \frac{\partial \tilde{\vec{e}}^{(k)}}{\partial q^{(k)}} \frac{\partial \mu^{(k)}}{\partial q^{(k)}}\right] l^{(j)}_{q^{(1)}}\end{aligned} \tag{12.28}$$

Assuming that the $(\partial^2\mu^{(k)}/\partial q^{(k)2})$ and $(\partial\mu^{(k)}/\partial q^{(k)})$ submatrices are all equal because the groups are equivalent, and that $\vec{\mathscr{E}}^{(k)}$ and $\partial\vec{e}^{(k)}/\partial q^{(k)}$ differ only in direction in space, the vectors have the form:

$$\tilde{l}^{(i)}_{q^{(1)}} \left[\frac{\partial^2 \mu^{(k)}}{\partial q^{(k)2}} \vec{\mathscr{E}}^{(k)} + \cdots\right] l^{(j)}_{q^{(1)}} = |\vec{A}| \vec{h}^{(k)}_{ij}$$

The unit vectors $\vec{h}_{ij}$ are given by the groups' vibrational forms and their spatial distribution.

Assuming further that the electro-optical parameters and mechanical characteristics of the groups and their unnormalized vibrational forms are unchanged over a series of molecules that have differing numbers of groups, and by allowing for the normalizing factors we obtain:

$$\left(\frac{\partial^2 \vec{\mu}}{\partial Q_i \partial Q_j}\right)_0 = \text{const} \frac{v_i(n) v_j(n) \left(\sum_k^n \eta_{ki} \eta_{kj} \vec{h}^{(k)}_{ij}\right)}{\left[\sqrt{\left(\sum_k^n \eta_{ki}^2\right)}\right]\left[\sqrt{\left(\sum_k^n \eta_{kj}^2\right)}\right]} \tag{12.29}$$

This relation expresses the dependence of the derivative of the molecule's dipole moment with respect to normal co-ordinates on the number of identical groups in the molecule.

Note that $i = j$ for overtones, and hence equation (12.29) has much in common with the formula obtained previously for the first derivative of the dipole moment with respect to a normal co-ordinate. It is interesting to note that, when all the $\eta_{ki} = \eta_{kj} = 1$ and the $\vec{h}^{(k)}_{ij}$ vectors are parallel, the second derivatives do not depend on the number of groups (if the v_i and v_j do not change). At the same time in the analogous special case, the first derivative is proportional to $\sqrt{n}$, while the intensity of the corresponding band is proportional to the number of equivalent groups.

As is the case for fundamental vibrations, the formula that has been derived refers to bands arising from one vibration from each identical group present in the molecule. When several vibrations are superimposed, the dependence of the intensity of the summed bands in quite different.

12.6 Characteristic Electronic Transitions

In conclusion, let us discover under what conditions the characteristic (local) electronic transitions should be expected to appear for individual structural groupings (known as chromophores). Obviously, the effect can happen only in those conditions where at least one of the associated MOs is localized within the limits of a given structural grouping, and the properties of the orbital and the respective energy values are only slightly dependent upon the environment of the grouping.

To obtain an insight into the problem, let us assume that two groupings have been distinguished within a molecule, and let us denote these by indices I and II. Admittedly, we can always proceed in such a way as to make all the atoms of the molecule and the respective AOs belong to either grouping I or grouping II. A stricter requirement (though not always a principal one) is that based on the occupancy of electron shells of groupings I and II. In the initial basis of AOs a matrix of the overlap integrals must have the following form:

$$S = \begin{bmatrix} S_{11} & S_{12} \\ S_{21} & S_{22} \end{bmatrix}$$

where S_{II} and S_{22} are the diagonal submatrices of the overlap integrals for the AOs in groupings I and II, respectively, and $S_{12} = \tilde{S}_{21}$ is a submatrix of the overlap integrals for the AOs in the neighbouring atomic groupings.

Now let us consider a new basis of localized MOs constructed separately from the AOs of groups I and II. This basis can be chosen so as to ensure that the submatrices S_{11} and S_{22} are capable of being reduced to a diagonal form.

In this new basis, overlap integrals for the MOs for the neighbouring groups will be given as

$$S_{ij}^{\mathrm{I,II}} = \sum_{k_1,k_2}^{N_1,N_2} C_{k_1}^{\mathrm{I}_i} C_{k_2}^{\mathrm{II}_j} \int \chi_{k_1}^{\mathrm{I}} \chi_{k_2}^{\mathrm{II}} \, \mathrm{d}v$$

Here, $C_{k_1}^{\mathrm{I}_i}$ and $C_{k_2}^{\mathrm{II}_j}$ are the coefficients of AOs in a linear combination of atomic orbitals (LCAO), which conform to the AOs in the first and second groups (and MOs with indices *i* and *j*), N_1 and N_2 are the number of AOs in groups I and II. Since orthogonal transformations reduce the submatrices S_{11} and S_{22} to a diagonal form, we obtain

$$\sum_{k_1}^{N_1} (C_{k_1}^{\mathrm{I}_i})^2 = \sum_{k_2}^{N_2} (C_{k_2}^{\mathrm{II}_j})^2 = 1$$

The following alternatives can occur:

(1) The localized MO includes all the atoms in a given grouping;
(2) The localized MO includes only those atoms which are not located at a bond connecting groups I and II; or
(3) The MOs are localized on atoms pertaining to the bonds connecting groups I and II and the bonds adjacent to these groups.

In the first case, due to the requirement

$$\sum_{k_1}^{N_1} (C_{k_1}^{\mathrm{I}_i})^2 = 1$$

it can be stated that

$$|C_{k_1}^{\mathrm{I}_1}| \sim \frac{1}{\sqrt{N_1}}$$

Then for two MOs of type (1) of groups I and II, the integral $S^{\mathrm{I,II}}$ will diminish in proportion to $1/\sqrt{N_1 N_2}$ as the number of atoms in the groups increases, if, in addition to that, the two groupings are 'intertwined' through a small number of atoms (in the limit, they are connected by a single bond).

For the MO overlap integrals of type (2) $S^{\mathrm{I,II}} = 0$, because the centres of the AOs are spaced apart, all the integrals $\int \chi_{k_1}^{\mathrm{I}} \chi_{k_2}^{\mathrm{II}}\, dv$ with non-zero products of coefficients in LCAO are equal to zero.

The MO overlap integrals of type (3) can have a substantial magnitude and the overlap integrals for combinations of MOs of types (1) and (2) must also be equal to zero. The overlap integrals for MOs of types (1) and (2) must diminish in proportion to $1/\sqrt{N_1}$, integrals for combinations of MOs of types (2) and (3) are also equal to zero.

Incidentally, it can be concluded that in those cases where a molecule comprises fairly large atomic groupings connected via a single bond, the matrix S' is produced in passing to the localized orthogonalized orbitals, wherein only a few of the elements in the submatrix S'_{12} are found to differ substantially from zero. It should be borne in mind that the probability of a MO occurring for case (3) is rather small. It is not difficult to see that at a further stage of construction of MOs for the entire molecule these non-zero elements of the submatrix S'_{12} will play a smaller role than the orders of the submatrices S_{11} and S_{12} (i.e. the size of the fragments of the molecules is much more important).

Thus, the so-called CNDO approximation should be the more precise for the groupings of MOs localized within different groups of atoms. The larger the size of these groupings, the smaller the number of bonds connecting these groupings and the lower the influence of these bonds.

Confining ourselves to the Hartree–Fock approximation in constructing complete MOs let us assume that the Hartree–Fock problems have been solved separately for groupings I and II. As a result, we obtain sets of MOs localized in groups I and II.

In combining these groupings into a molecule the states conforming to the solutions of the Hartree–Fock problems for separate groupings will perturb each other. This perturbation will be seen in the impact of one group upon the electronic cloud of another, and, in particular, on the attraction of the localized electron clouds of one group for the positive centres of the other. This will lead to the appearance of additional terms in the respective diagonal submatrices of a single-electron portion of the entire Hartree–Fock matrix for the molecule as a whole.

The impact of the electron–electron repulsion of electronic clouds of the two groups I and II will be reflected in the non-diagonal submatrices of the Hartree–Fock matrix. The respective exchange two-electron terms and resonance single-electron integrals will be small for the same reasons as those applicable to the similar terms in the matrix for the overlap integrals.

It should further be noted that if the molecular fragments I and II are almost electrically neutral, then the electron–nuclear attraction of electrons in one group for the nuclei in the other will be almost completely offset by electron–electron repulsion.

These considerations lead to the conclusion that the Hartree–Fock matrix for the entire molecule will have, in the conditions under consideration, an almost quasi-diagonal form, with a small number of elements appreciably different from zero in a non-diagonal matrix of interaction of groups I and II.

As a result, it will be found that the positions of a number of the energy levels are dependent only upon the properties of 'their' groups, and only some of them will interact with each other. If, for the sake of simplicity, we take into account only the interaction between pairs of levels, then we obtain for the energy of a molecule

$$E_{1,2} = \frac{E^{\mathrm{I}} + E^{\mathrm{II}}}{2} \pm \tfrac{1}{2}\sqrt{[(E^{\mathrm{I}} - E^{\mathrm{II}})^2 + 4h_{12}^2]}$$

where E^{I} and E^{II} are the energy levels in the combined first and second groupings of the molecule and h_{12} is a matrix element of interaction of these levels.

The respective eigenfunctions can be presented as

$$\Psi_1 = \cos\varphi \cdot \Psi^{\mathrm{I}} + \sin\varphi \cdot \Psi^{\mathrm{II}}$$
$$\Psi_2 = -\sin\varphi \cdot \Psi^{\mathrm{I}} + \cos\varphi \Psi^{\mathrm{II}}$$

Here, Ψ^{I} and Ψ^{II} are also the eigenfunctions of the combined groupings. For $E^{\mathrm{I}} = E^{\mathrm{II}}$, resonance will take place and the eigenfunctions become $\Psi_{1,2} = 1/\sqrt{2}(\Psi^{\mathrm{I}} + \Psi^{\mathrm{II}})$, i.e. no localization of the groupings remains on one combination. It is only natural that in this case we cannot expect the appearance of the characteristic (localized) transitions in any of the groupings, even if the coefficient h_{12} is small. Another extreme situation is that where $E^{\mathrm{I}} \gg E^{\mathrm{II}}$. This can happen if the effective charges of atoms in one of the groupings differ substantially from those in the other. Then, in the course of interaction of the energy levels, the relative displacement of a level having a higher energy will be smaller than that of the level with lower energy, so that we can accept the occurrence of a transition which is almost 'characteristic' in terms of frequency. A transition can also become characteristic in terms of intensity if the 'admixing' of the function Ψ^{II} of the neighbouring group is small. Clearly, this depends on the difference between the levels E^{I} and E^{II} as well as on the value of the interaction coefficient h_{12} and, in addition, if the matrix element of the transition is local for the second group and small in magnitude.

To summarize, we can conclude that the occurrence of purely electronic (Franck–Condon) transitions which are characteristic (local) for a given grouping is likely in cases where this grouping is connected to the rest of the molecule only through a single bond, provided that the length of the bond is either much longer than those of the adjacent ones in the grouping under consideration (this results in small coefficient h_{12}) or if the bond comprises very different atoms.

The conditions giving rise to a resonance invariably result in the occurrence of non-localized transitions throughout the entire molecule.

References

1. M. V. Volkenshtein, L. A. Gribov, M. A. Elyashevich and B. I. Stepanov. *Vibrations of Molecules*, 2nd edn, Nauka, Moscow (1972) (in Russian).
2. L. A. Gribov and V. A. Dementiev. *Methods and Algorithms for Calculations in the Theory of Vibrational Spectra of Molecules*, Nauka, Moscow (1981) (in Russian).

Chapter 13

Solutions of Inverse Spectral Problems

13.1 General Aspects

From a general mathematical point of view, the identification and examination of the structures of polyatomic molecules by spectral methods is achieved by solving the so-called 'inverse spectral problem'. To make their solution possible, it is necessary to set up models of molecules with different degrees of complexity and to know how to construct their spectra and compare them with experiment. Models can be either discrete (for example, a set of structural formulae) or continuous (space molecular models, depending on geometrical parameters functionally). By varying the models systematically and comparing their spectra with experimental ones, coincidence of theoretical and experimental spectra can be achieved. Typical molecular models used in spectroscopy are shown in Table 13.1.

The simplest model of the molecule is the Brutto formula, as in this model only the number and types of atoms are indicated. This model is a very indefinite one: for example, for the Brutto formula C_6H_6 there are 217 isomers (topological ones), which are constructed in accordance with the classical valency rules.

The second model consists of the Brutto formula plus some indications of atomic groups. The mathematical expression of this model includes not only number and types of atoms but also adjacency matrices for individual fragments. This model is more detailed and possesses a greater degree of definition. Usually only some topological isomers exist for such a model.

The third model is the structural formula of the molecule. The mathematical expression of this model includes the indication of charges, number of atoms and the total adjacency matrix.

More complicated is the solid three-dimensional model of the structure. The mathematical expression includes atomic charges, adjacency matrix, bond lengths and valence bond angles.

The models mentioned above are so-called 'discrete' ones. Additional information and the use of so-called 'functional' models are necessary to obtain more detailed descriptions for molecules.

The spectra corresponding to these molecular models are listed in Table 13.2, and it is important to note that each model of a molecule has its own corresponding type of spectral display.

Table 13.1. Hierarchy of the models of molecules

Hierarchy of models	Mathematical representation
Discrete models	
(1) Brutto formula	Number and types of atoms
(2) CH_2, CO, R: only some groups are indicated	Charges of atoms, adjacency matrices for atomic groups: $\begin{array}{c\|ccc} & C & HI & H2 \\ \hline C & 0 & I & I \\ HI & I & 0 & 0 \\ H2 & I & 0 & 0 \end{array}$
(3) Structural formulae 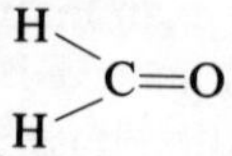	Full adjacency matrix: (plane graph) $\begin{array}{c\|cccc} & H1 & H2 & C & O \\ \hline H1 & 0 & 0 & I & 0 \\ H2 & 0 & 0 & I & 0 \\ C & I & I & 0 & 2 \\ O & 0 & 0 & 2 & 0 \end{array}$
(4) A solid three-dimensional model	Charges of atoms, adjacency matrix, values for bond lengths and valence angles
Functional models	
(5) Soft harmonic three-dimensional model	M_a: mass of atoms, adjacency matrix; S_i, A_{ij}: bond lengths and valence angles, force constants
(6) Soft anharmonic three-dimensional model	M_a, S_i, A_{ij} quantities and potential functions' parameters
(7) Soft (harmonic or anharmonic) three-dimensional model and charge (Q, P) distribution	Set of parameters: M_a, S_i, A_{ij}, Q_n, P_{nm} (and their distortions), parameters of potential function (V)
(8) Soft harmonic or anharmonic three-dimensional models for the ground and excited electronic states	Sets of parameters: M_a, S_i, A_{ij}, Q_n, P_{nm}, V for respective states

There are two types of problems in the theory of polyatomic spectra. The first is the direct spectral problem (i.e. the simulation of the spectrum) corresponding to the model. The second is the inverse spectral problem, the solution of which provides quantitative information on the molecular parameters for the models.

Iterative methods are used to solve inverse spectral problems. The first step involves the simplest model whose theoretical spectrum is derived and to compare it with experiment. This makes clear whether the model should be changed or not. If necessary, it is made more accurate, and this process is repeated until the theoretical and experimental spectra coincide with sufficient

Table 13.2. Models and spectra

Type of model	Corresponding spectra
(1) Brutto formula	Atom emission or absorption spectra
(2) Atomic group indication	Some regions of emission or absorption and (in the best case) a qualitative indication of intensity (strong, weak, medium ...)
(3) Structural formula	Detailed spectra in the form of emission or absorption regions
(4) Solid three-dimensional model	Pure electronic spectrum and rotational spectrum (only position of lines given)
(5) Soft harmonic model	Vibrational spectrum, rotational–vibrational spectrum and interactions
(6) Soft anharmonic model	Vibrational spectrum of high orders
(7) Soft model with charge distribution	Vibrational, vibrational–rotational and rotational spectra (position and intensities of bands and lines given)
(8) Soft model for the ground and excited electronic states	Electronic–vibrational (electronic band contours) spectrum

accuracy. The model of the next degree of complexity is then constructed and the whole procedure repeated.

It is clear that, the more complicated the model used, the greater is the quantity of input information necessary to proceed. As a consequence, more valuable information about the molecular structure can be obtained when the corresponding inverse problems are solved.

The practical solution of these problems should be based on a definite set of initial axioms kept in databanks. These define a group of problems which can be solved. These are important questions, and will be discussed in detail.

For the simplest molecular model the first set of axioms which are necessary to solve inverse problems consist of the system of well-known spectra–structure correlations. Note that the possibility of these correlations is based on the existence of relatively closed atomic groups. Although this condition is not enough (see Chapter 12), since there exists a set of additional constraints depending on vibrational properties this limits the appearance of characteristic spectral bands. Even if there are groups with characteristic features, the establishment of spectra–structure correlations requires the preliminary knowledge of a large number of spectra containing these groups. We can therefore say that the first set of axioms are completely definite and lead to useful indications of the structural elements possessed by the system. This does not, of course, mean that the results are unimportants. As in geometry, where some initial postulates lead to a large number of theorems, the combination of groups examined in our case can lead to the construction of a large number of different molecules which fit the spectra.

The second group of axioms are valency rules, and these are general enough to construct molecular models in terms of structural formulae. In practice, when using these rules we can construct the structural formulae of highly exotic molecules the existence of which can be proved by experience. However, these rules also have limits, and cannot be realized in molecules and ions deficient in electrons. Axioms based on unfavourable combinations of atoms are very useful for reducing the number of possible structures at one of the stages of generating molecular models.

The next group of axioms combine vector atomic models and geometrical models of separate structural elements. Structural formulae of molecular models and the corresponding theoretical spectra can be constructed on the basis of such a set of initial axioms. The appearance of absorption lines or other spectral features should be expected to be given with adequate accuracy. The comparison of these spectra with experiment allows us to reject incorrect structures and to identify an approximately correct structural formula or space model.

A more complete set of axioms is necessary to describe more accurate molecular models. Values of force constant and electro-optical parameters, atom–atom potential constants and parameters for quantum calculations, etc. are needed. At this level of approximation we obtain theoretical spectra which correspond, for example, to possible vibrational frequencies or, in the general case, complete spectral curves, which can be compared with experiment.

The most reliable method of obtaining the sets of axioms is as follows. The first step is to define the minimum set of different molecule fragments, which can then be combined to yield the classes of molecules examined earlier. For this purpose, spectral and structural databanks can be used. Further analysis of these fragments based on statistical examination and solution of corresponding inverse problems can give the whole set of axioms needed. Preliminary analysis shows that the number of such groups will not exceed 10^3, which makes the problem not only complete enough but one that can be solved. There are no axioms concerning the unfavourable atomic combinations in this scheme: since they are not included amongst the molecules taken for analysis it is impossible to find them using these methods.

Inverse problems (which, of late, are attracting increasing attention) are those which serve to establish characteristics of molecules on the basis of measured spectra. A typical inverse problem is the determination of force constants from observed vibrational spectra. Inverse problems, as a rule, fall into the class of the so-called 'incompletely defined problems'; they all have one common feature, i.e. that unless some special *a priori* constraints are imposed they do not have a unique solution. Consequently, the solution of inverse problems in many cases of practical importance becomes rather cumbersome, and involves a certain degree of subjectivity which distorts the final results. In the theory of vibrational spectra inverse problems are usually solved to find atomic groups, force constants, anharmonicity factors and electro-optical parameters. On the basis of solutions to these problems, parameters have been derived which serve as tools for calculating spectra of complicated systems such as polymers and molecular crystals.

Identification of a molecule from its experimental spectra can be achieved by comparing sets of spectral characteristics with those available in the databanks. Essentially, this is a search by the 'fingerprint' technique. Generally speaking, the software of data-retrieval systems can also be used for solving other problems (for example, determination of the presence of certain structural components in a molecule, the classification of the system, etc.).

The capacity of data-retrieval systems is directly proportional to the number of compounds whose spectral characteristics are stored in the bank. The largest systems store spectral data for about 10^5 molecules. This may seem a very large number, but it has to be remembered that this is about the amount of new organic compounds synthesized in the world annually. It becomes impossible to accumulate spectral information in a bank of any size as soon as it appears. Moreover, data on new compounds presenting the most interest for identification are often not programmed into the computer. Accordingly, although data-retrieval systems claim to be multipurpose, they will inevitably fall behind progress in synthetic chemistry. Because of the complexity of data accumulation in the databank and the necessity of standardizing them, these data do not actually meet modern spectral experimental requirements. Some spectral databanks are based on infra-red spectra obtained in the range up to 500–700 cm^{-1}. Meanwhile, the range of modern infra-red spectrometers has been considerably increased. The absorption bands most advantageous analytically lie in the low-frequency range.

The application of laser Raman spectroscopy, low-temperature techniques and high-resolution methods has considerably increased the data available. All this can hinder the application of spectral data obtained by the new techniques to the identification of polyatomic molecules by data retrieval. This is a major disadvantage, which would be very difficult to eliminate by improvements in the databank.

As a rule, any data-retrieval system contains not only much useful information but also a large amount of extraneous data. It is well known, for example, that, given information on the spectra and structure of some molecules, the spectra of other molecules can be established. Thus, mathematically, only a limited number of spectra are independent. The availability of the dependent information makes the system irrationally complicated. For an accurate spectral identification of a molecule by the fingerprint method, experimental and standard spectra should be compared, strictly speaking, not at the level of single features but at that of the total sum of their features. However, a detailed description of spectra stored in the system makes it still more complicated.

There are other shortcomings, but those outlined above seem adequate to support the conclusion that specialized data-retrieval systems offer little potential for improvement, and have various inherent disadvantages. Thus their improvement cannot be regarded as a guideline for the completely automated identification of polyatomic molecules by means of spectral data. This does not imply that any further development of data-retrieval systems is ineffective. Databanks in such systems are useful as a source of primary information for building up more sophisticated artificial intelligence systems. More immediately, specialized data-retrieval systems can be very helpful when designed for analysing particular industrial products or identifying the pollution status of natural and waste waters, atmospheres, etc.

Identification of compounds in initial data-processing systems following the artificial intelligence pattern is also achieved by comparing experimental molecular spectra with computed ones. However, unlike data-retrieval systems, such comparative spectra are not originally stored in the computer but are generated by following certain rules in the process of solution. In this process, automatically formulated hypotheses on the structure of the molecule studied are involved. These hypotheses are continuously updated, and there is no need to store a large amount of data in the bank. It is sufficient to 'memorize' relatively few basic axioms. Great complexity of the algorithms for a computer is not crucial in designing such systems. Artificial intelligence systems, by their very nature, are free of the shortcomings inherent in data-retrieval systems. They can be multipurpose and flexible, and can be readily updated by using any new knowledge from the theory of molecular structure and spectra and quantum chemistry. This enables artificial intelligence systems to be subject to continuous and theoretically unlimited improvement.

Fairly sophisticated artificial intelligence systems are now available for molecular spectroscopy. With such systems, sufficiently reliable identification of isomeric structures can be achieved even when a computed spectrum is not identical but only relatively close to the experimental spectrum because of restrictions in different theories.

In the first stage, the computer must isolate a set of 'probable' structures, based on a number of features. In the second, the sequential generation of more accurate spectra and their comparison with experimental spectra gradually excludes false structures from the initial set until an unambiguous result is achieved. Hence, any system of artificial intelligence must have two large blocks: one for building hypothetical structures and the other for generating spectra at different levels of accuracy. Moreover, there must be a feedback for comparing theoretical and experimental spectra. Mathematically, the first block formulates a set of initial hypotheses, whereas the second and the feedback test the validity of these hypotheses. The algorithms of the first block must be capable of structural group analysis and computation of molecular structures at the adjacency matrix level and of spatial models for the 'probable' structures at varying levels of accuracy.

The algorithms of the second block, using the data on molecular structures, should be able to generate features which include complete spectral curves for these structures. For sequential estimation of the validity of the initial hypotheses it is advisable to perform a primary filtering of hypotheses by comparing only very roughly calculated theoretical spectra with experimental ones. It is possible, for example, to build a reasonably representative vibrational spectrum of a structure by analysing the closed additive groupings comprising it and the corresponding spectra–structure correlations. As experience has shown, this simple procedure can sometimes reduce the number of possible structures by half, or even more. Further testing of the 'probable' structures requires more accurate calculations of spectra. These are possible only when detailed spatial characteristics are available. Thus a certain hierarchy of algorithm complexity is obtained, corresponding to the sequence of steps undertaken by the system to yield the final result and prove its unambiguity. This proof becomes possible because of the great differences in molecular spectra revealed by detailed analysis.

Since the spectra of different molecular models are formed by the computation system itself there is no problem in generating not only spectra of components but also those of their mixtures. This opens up the possibility of applying this method to the analysis of mixtures.

One more important aspect should be stressed. By one cycle in the artificial intelligence system it is possible not only to identify a molecule but also to gain detailed information on its geometrical and electronic structure. The results can be applied to the solution of different problems of a wider scope than those initially formulated. Components of the artificial intelligence system can be applied efficiently to the solution not only of analytical problems but also of those arising in various branches of the chemistry and physics of polyatomic molecules.

13.2 Symbolic Logic Methods for Spectrochemical Investigations

13.2.1 Problem of structural group analysis of polyatomic molecules according to infra-red spectra[1,2]

Before proceeding to the statements of the problem, some basic relevant concepts will be briefly recapitulated. We define a proposition as a complete sentence about which it may be definitely said that its content is true or false. A simple proposition is designated by a letter, and is called an elementary proposition. For example, an elementary proposition ω may be read as: 'In the spectrum the ω frequency has been detected'.

Propositions formed by combining elementary propositions are called combined propositions or Boolean functions. For the construction of Boolean functions the following main logical operations are used:

(1) The negation of the proposition ω corresponds to the particle 'no' and is represented by $\bar{\omega}$ ('There is no ω frequency in the spectrum').
(2) The conjunction (logical multiplication) of the propositions ω_1 and ω_2 corresponds to the conjunction 'and' and is designated by $\omega_1.\omega_2$ ('There are the ω_1 and ω_2 frequencies in the spectrum').
(3) The disjunction (logical addition) of ω_1 and ω_2 corresponds to the non-excluding conjunction 'or' and is designated by $\omega_1 + \omega_2$ ('Either the ω_1 or the ω_2 frequency or both together are present in the spectrum').

(4) The excluding disjunction of ω_1 and ω_2 corresponds to the excluding disjunction 'or' and is represented by $\omega_1 {}^{\vee} \omega_2$ ('Either the ω_1 or the ω_2 frequency is observed in the spectrum but not both simultaneously').
(5) The proposition $A \rightarrow \omega$ is an implication: 'If A, then ω', 'A implies ω' ('If the molecule contains the structural group A, then the frequency ω is observed in the spectrum').
(6) The equivalence of propositions A and ω means that they always have the same truth value. It is designated by $A = \omega$ ('The frequency ω is observed in the spectrum if and only if the molecule contains the structural group A').

Complex propositions that are always true irrespective of the truth value of the elementary propositions of which they are constituted, are called 'tautologies', and are designated by I: for instance, $\omega + \bar{\omega} = I$ ('There is an ω frequency or there is no ω frequency in the spectrum'). The negation of I, namely $\bar{I}$, is the universally false proposition and is designated by 0: for instance, $\omega . \bar{\omega} = 0$ ('There is the ω frequency in the spectrum and there is no ω frequency in the spectrum').

The following list contains some of the most important formulae of Boolean algebra:

$$\begin{array}{lll} A + A = A & 0.A = 0 & (A \rightarrow B) = \bar{A} + B \\ A.A = A & A.\bar{A} = 0 & A {}^{\vee} B = \bar{A}.B + A.\bar{B} \\ A + I = I & A + \bar{A} = \mathrm{I} & A + B.C = (A + B).(A + C) \\ A.I = A & \overline{A.B} = \bar{A} + \bar{B} & A + A.B = A \\ 0 + A = A & \overline{A + B} = \bar{A}.\bar{B} & A + \bar{A}.B = A + B \end{array} \tag{13.1}$$

Now let us pass directly to the statement of the problem. First, we note that the connection of some functional group (of the structural element) A_i with the characteristic frequency ranges indicated in the correlation tables may be represented in terms of mathematical logic by

$$A_i \rightarrow \{\Delta\omega_1 . \Delta\omega_2 \cdots \Delta\omega_j \cdots \Delta\omega_{q(A_i)}\} \tag{13.2}$$

where $q(A_i)$ is the number of regions in which the characteristic vibrations of the A_i group are displayed.

Because of the intersection of sets of characteristic frequencies corresponding to various functional groups, the inverse implication

$$\{\Delta\omega_1 . \Delta\omega_2 \cdots \Delta\omega_j \cdots \Delta\omega_{q(A_i)}\} \rightarrow A_i$$

does not necessarily hold.

However, the arbitrary frequency ω_j may always be compared with a set of structural elements A_i, and this correspondence may be expressed by the implication

$$\omega_j \rightarrow (A_1 + A_2 + \cdots + A_{nj})$$

where nj is the number of functional groups of which the frequency ω_j may be considered characteristic.

Now let us assume that the infra-red spectrum of a certain specimen is known and that data concerning the relations between infra-red absorption bands and structural elements of molecules are available. All possible combinations of functional groups, whose presence in the specimen conforms with the spectrum, have to be established. Let $\tilde{\omega}=\{\omega_j\}, j=1\div m$ be the set of spectral signs (for instance, frequencies) observed in the infra-red spectrum of the specimen, and $\mathscr{A}=\{A_i\}, i=1\div N$, be the set of structural elements (functional groups) having at least one characteristic frequency among the elements of the set $\tilde{\omega}$.*

$\bar{\omega}=\{\omega_j\},\ j=m+1\div M$, is the set of frequencies which corresponds to some $A_i \in \mathscr{A}$, the frequencies of which are not observed in the spectrum, and $\omega=\tilde{\omega}+\bar{\omega}=\{\omega_j\}, j=1\div M$, is the set of all the frequencies considered during the solution of the problem.

Let us write the logical connections existing between elements of sets $\mathscr{A}$ and ω as a Boolean function $T(\mathscr{A},\omega)$† and describe the infra-red spectrum as a Boolean function $R(\omega)$. It is obvious that the result of a logical comparison of the functions $T(\mathscr{A},\omega)$ and $R(\omega)$ will be expressed by the Boolean function $f(\mathscr{A})$.

Thus the problem of the structural-group analysis of complex molecules based on infra-red spectra amounts to the calculation of the $f(\mathscr{A})$ function from the logical equation

$$T(\mathscr{A},\omega)\rightarrow\{R(\omega)\rightarrow f(\mathscr{A})\} \tag{13.3}$$

This equation is the most general formulation of the problem of qualitative spectral analysis in terms of the logical system described, and a solution of this equation always exists.

Let us show by a simple example how the $f(\mathscr{A})$ function may be calculated in practice.

Algorithm for calculation of $f(\mathscr{A})$ Let there be a product of some chemical reaction in which, according to the conditions of synthesis, the presence of a C=C group is assumed. A structural-group analysis of this sample according to the infra-red spectrum should be carried out after we have verified that the presence of C=C group is confirmed by the spectrum.‡ Let us assume that bands at about 1380 and 1460 cm^{-1} have been observed in the infra-red spectrum of this sample.

It is known that the spectral sign of the double bond C=C is the presence of an absorption band at about 1650 cm^{-1}. However, the absence of a band in the given region does not mean that there is no C=C group in the product

*In the set A, groups may also be included whose presence or absence in the specimen under investigation must be established (The case of analysis for predetermined groups).

†Into the function $T(\mathscr{A},\omega)$ may be introduced all additional information on the object being studied, represented in the form of Boolean functions of $\mathscr{A}$ and ω, as well as the assumed logical dependences between the elements of the set $\mathscr{A}$.

‡We shall deliberately limit ourselves to the consideration of the unreal situation which includes only three groups and three frequencies $(M+N=6)$, as larger values of $(M+N)$ would lead to cumbersome calculations of little further use.

under investigation. Indeed, if this group is present at the centre of symmetry of the molecule, its stretching vibrations are not active in the infra-red spectrum. Taking this fact into consideration, we write the following logical relations:

$$
\begin{aligned}
&(\mathrm{C{=}C}) && \rightarrow (1650\,\mathrm{cm}^{-1}) + \overline{(1650\,\mathrm{cm}^{-1})} = I \\
&(1650\,\mathrm{cm}^{-1}) && \rightarrow (\mathrm{C{=}C}) \\
&(1460\,\mathrm{cm}^{-1}) && \rightarrow (\mathrm{CH_2}) + (\mathrm{CH_3}) \\
&(1380\,\mathrm{cm}^{-1}) && \rightarrow (\mathrm{CH_3}) \\
&(\mathrm{CH_2}) && \rightarrow (1460\,\mathrm{cm}^{-1}) \\
&(\mathrm{CH_3}) && \rightarrow (1460\,\mathrm{cm}^{-1}).(1380\,\mathrm{cm}^{-1})
\end{aligned}
$$

When we introduce the designations $\omega_1 = 1460\,\mathrm{cm}^{-1}$, $\omega_2 = 1380\,\mathrm{cm}^{-1}$, $\omega_3 = 1650\,\mathrm{cm}^{-1}$, $A_1 = \mathrm{CH_2}$, $A_2 = \mathrm{CH_3}$ and $A_3 = \mathrm{C{=}C}$, we shall obtain $\mathscr{A} = \{A_1, A_2, A_3\}$, $\tilde{\omega} = \{\omega_1, \omega_2\}$, $\bar{\omega} = \{\omega_3\}$, $\omega = \{\omega_1, \omega_2, \omega_3\}$, and $R(\omega) = \omega_1.\omega_2.\bar{\omega}_3$, and the function $T(\mathscr{A}, \omega)$ will have the form:

$$
\begin{aligned}
T(\mathscr{A}, \omega) &= (\omega_3 \rightarrow A_3).(\omega_1 \rightarrow A_1 + A_2).(\omega_2 \rightarrow A_2).(A_1 \rightarrow \omega_1).(A_2 \rightarrow \omega_1.\omega_2) \\
&= (\bar{\omega}_3 + A_3).(\bar{\omega}_1 + A_1 + A_2).(\bar{\omega}_2 + A_2).(\bar{A}_1 + \omega_1).(\bar{A}_2 + \omega_1.\omega_2)
\end{aligned}
$$

Let us write down in columns all possible conjunctions (complexes) of the elements $A_i \in \mathscr{A}$ and $\omega_j \in \omega$ ($i = 1 \div 3$, $j = 1 \div 3$), placing the figure one in the row against the corresponding element if the element is present in the complex in an affirmative form and a zero if it is present in a negative form (Table 13.3).

Table 13.3 is a standard logical basis for six logical variables and is constructed according to the principle of the number of zeros or ones standing next to one another in a group of figures, periodically repeated in the lth row equals 2^l ($l = 0 \div M + N - 1$).

It can easily be seen that the number of columns is equal to 2^{M+N}, and that each column, if read upwards, forms a binary number equal to the number of the column which runs from 0 to $2^{M+N} - 1$. To each $A_i \in \mathscr{A}$ and $\omega_j \in \omega$ a binary number is compared according to the corresponding row of the standard basis, this being designated by the symbol # and called the designation number. For example:

$\#A_2 =$
0011 0011 0011 0011 0011 0011 0011 0011 0011 0011 0011 0011
0011 0011 0011 0011

Table 13.3. Standard Logical Basis

A_1	0101	0101	0101	0101	0101	0101	0101	0101	0101	0101	0101	0101	0101	0101	0101	0101
A_2	0011	0011	0011	0011	0011	0011	0011	0011	0011	0011	0011	0011	0011	0011	0011	0011
A_3	0000	1111	0000	1111	0000	1111	0000	1111	0000	1111	0000	1111	0000	1111	0000	1111
ω_1	0000	0000	1111	1111	0000	0000	1111	1111	0000	0000	1111	1111	0000	0000	1111	1111
ω_2	0000	0000	0000	0000	1111	1111	1111	1111	0000	0000	0000	0000	1111	1111	1111	1111
ω_3	0000	0000	0000	0000	0000	0000	0000	0000	1111	1111	1111	1111	1111	1111	1111	1111

Logical operations are interpreted in terms of designation numbers and calculations with them are carried out in conformity with the system of Boolean algebra equations (13.1).

Obviously, in practice by no means all combinations of elements given by the logical basis may be realized. For instance, the conjunction $A_1.A_2.\bar{A}_3.\bar{\omega}_1.\bar{\omega}_2.\bar{\omega}_3$ has no physical meaning, as it describes a situation where, in the spectrum of the compound containing the groups CH_2 and CH_3, the bands at about 1380 and 1460 cm^{-1} are absent. The elimination of 'superfluous' columns from the standard logical basis is carried out in accordance with the function $T(\mathscr{A}, \omega)$. For this, it is necessary to calculate the $\# T(\mathscr{A}, \omega)$ and to delete from Table 13.3 those columns that correspond to zeros in the designation number of this function.

When we carry out the appropriate operations, we obtain

$\# T(\mathscr{A}, \omega) =$
1000 1000 0100 0100 0000 0000 0011 0011 0000 1000 0000 0100
0000 0000 0000 0011

and the constrained logical basis will be written:

A_1	0011	0101	0101
A_2	0000	1111	0011
A_3	0101	0011	1111
ω_1	0011	1111	0111
ω_2	0000	1111	0011
ω_3	0000	0000	1111

Then we calculate $\# R(\omega)$ in terms of the constrained logical basis $\# R(\omega) = 0000$ 1111 0000 and single out those columns that correspond to ones in $\# R(\omega)$.

The complexes of elements of the set $\mathscr{A}$ which are contained in these columns represent a complete set of conjunctions forming the function $f(\mathscr{A})$:

$$f(\mathscr{A}) = \bar{A}_1.A_2.\bar{A}_3 + A_1.A_2.\bar{A}_3 + \bar{A}_1.A_2.A_3 + A_1.A_2.A_3$$

Thus we have managed to establish that the sample under investigation may contain any of four complexes of functional groups and, consequently, the probability of the presence of each of them is equal to 1/4, the presence of all four complexes simultaneously not being excluded.

In the general case, $f(\mathscr{A})$ will always look like a disjunction of r conjunctions, and the diagnosis will consist in a statement about the possibility of the presence in the sample under investigation of r complexes of functional groups, the probability of each of them being $1/r$.

When we make elementary transformations of $f(\mathscr{A})$, we have

$$f(\mathscr{A}) = A_1.A_2.(A_3 + \bar{A}_3) + \bar{A}_1.A_2.(A_3 + \bar{A}_3) = A_2$$

from which it follows that the probability of the presence of the group CH_3 $p(CH_3) = 1$, and $p(CH_2) = p(C{=}C) = 1/2$. The ambiguity in the answer shows that additional investigation is needed. The planning of this investigation is

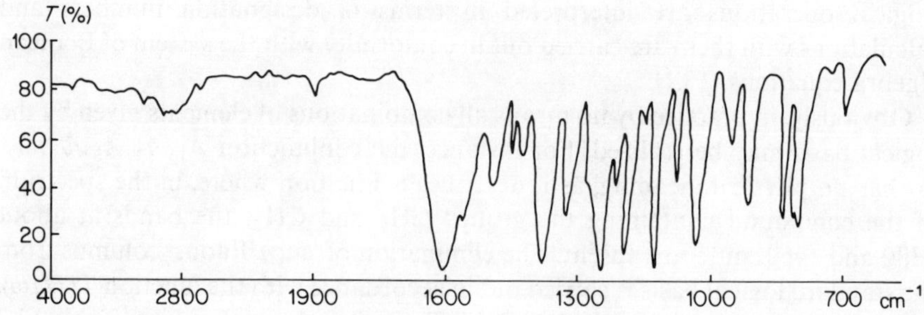

Figure 13.1. Infra-red spectrum of the sample 2-hydroxy-4-methoxyacetophenone

simplified by knowing the function $f(\mathscr{A})$. Apparently, this method may, if necessary, be repeatedly applied at the second stage of analysis, i.e. during the 'building up' of the molecule from functional groups contained in the complex.

Let us illustrate the solution of problems by the following example. Let us assume that it is necessary to determine the structure of a compound with the formula $C_9H_{10}O_3$, the infra-red spectrum of which is represented in Figure 13.1 (KBr disc).

First, as usual, let us single out the bands that are most important for the structural-group analysis, writing them in Table 13.4 (we form the set $\tilde{\omega}$). Comparing each $\omega_j \in \tilde{\omega}$ with functional groups in accordance with the correlation tables[3,4] simultaneously carrying out a rough estimate of the elementary composition (only groups containing C, H, O in different combinations are selected), we can establish the whole complex of elements of the set $\mathscr{A}$ (Table 13.5).

Let us write down the implications given in the correlation tables, replacing the characteristic frequency regions by experimental wave numbers:

$$A_1 \rightarrow \omega_4 . \omega_5 . \omega_{12}$$

$$A_2 \rightarrow \omega_1 . \omega_4 . \omega_5 . \omega_6 . \omega_7 . \omega_{12}$$

$$A_3 \rightarrow \omega_1 . \omega_5 . \omega_6 . \omega_7$$

$$A_4 \rightarrow \omega_2 . (\omega_8 + \omega_9)$$

If the molecule contains CH_2, the band ω_2 and any of the bands ω_8 and ω_9

Table 13.4. Elements of the set

3050 ω_1	2900–3000 ω_2	2835 ω_3	3200–2500 ω_4	1623 ω_5
1580 ω_6	1504 ω_7	1447 ω_8	1442 ω_9	1369 ω_{10}
1332 ω_{11}	1251 ω_{12}	859 ω_{13}	815 ω_{14}	

Table 13.5. Elements of the set $\mathscr{A}$

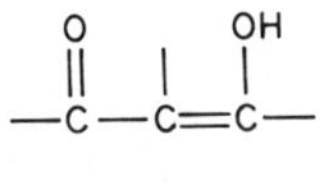 A_1	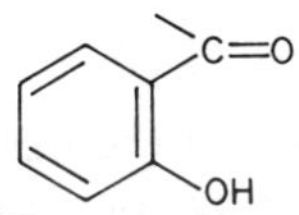 A_2	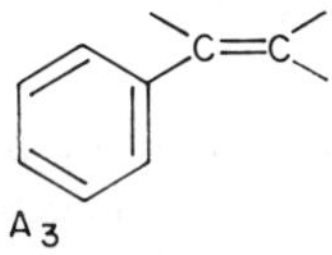 A_3
—CH_2 A_4	—CH_3 A_5	—OCH_3 A_6
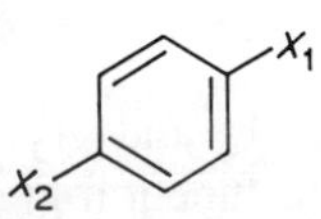 A_7	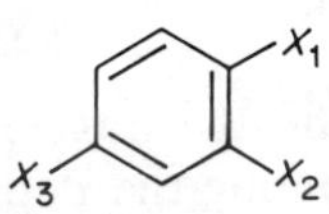 A_8	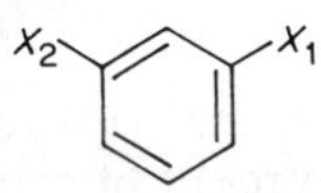 A_9

may be attributed to the vibrations of this group:

$$A_5 \to \omega_2.(\omega_8 + \omega_9).(\omega_{10} + \omega_{11})$$
$$A_6 \to \omega_3.\omega_{12}.(\omega_8 + \omega_9)$$
$$A_7 \to \omega_1.\omega_6.\omega_7.(\omega_{13} + \omega_{14})$$
$$A_8 \to \omega_1.\omega_6.\omega_7.\omega_{13}.\omega_{14}$$
$$A_9 \to \omega_1.\omega_6.\omega_7.\omega_{13}.\omega_{14}.\omega_{15}$$

where ω_{15} is a strong band in the range 680–725 cm^{-1} characterizing meta-substitution but absent in the spectrum.

Then, note that from equation (13.2) it follows

$$(\overline{\Delta\omega_1} + \overline{\Delta\omega_2} + \cdots + \overline{\Delta\omega_j} + \cdots + \overline{\Delta\omega}_{q(A_i)}) \to \bar{A}_i \tag{13.4}$$

Equation (13.4) shows that if just one of the characteristic signs of the group A_i is absent, there is no group A_i in the material. Hence A_9 can be excluded from consideration.

Then, passing to the construction of Boolean functions forming $T(\mathscr{A}, \omega)$, using as much knowledge as is available, we obtain:

(1) $\omega_1 \to A_1 + A_3 + (A_7 {}^{\vee} A_8)$ (A_7 and A_8 cannot be present together (excess of carbon));
(2) $\omega_2 \to A_4 + A_5$;
(3) $\omega_3 \to A_6$;
(4) $\omega_4 \to (A_1 {}^{\vee} A_2)$ (A_1 and A_2 cannot be jointly present because of the excess content of C and O)
(5) $\omega_5 \to A_3 + (A_1 {}^{\vee} A_2) + (A_7 {}^{\vee} A_8)$;

(6) $\omega_6 \rightarrow A_2 + A_3 + (A_7 \vee A_8)$;
(7) $\omega_7 \rightarrow A_2 + A_3 + (A_7 \vee A_8)$;
(8) $\omega_8 \rightarrow A_4 + A_5 + A_6$;
(9) $\omega_9 \rightarrow A_4 + A_5 + A_6$;
(10) $(\omega_{10} + \omega_{11}) \rightarrow A_5$ (if even one of the frequencies ω_{10} or ω_{11} is present in the spectrum, the molecule may contain A_5);
(11) $\omega_{12} \rightarrow (A_1 \vee A_2) + A_6$;
(12) $\omega_{13} \rightarrow (A_1 \vee A_8)$;
(13) $\omega_{14} \rightarrow (A_7 \vee A_8)$;

Let us add two more conditions to the written ones:

(14) $A_2 \rightarrow A_8$;
(15) $A_3 \rightarrow (A_7 \vee A_8)$.

Conditions (14) and (15) relate the fragments A_2 and A_3 to the type of substitution. The result of computing the Boolean function $f(\mathscr{A})$ is shown in Table 13.6. We see from this Table that eight complexes of structural elements satisfy the function $T(\mathscr{A}, \omega)$ and fit the spectrum, partially taking into account the elementary composition.

We can easily see that only the last column, which includes the following structural elements

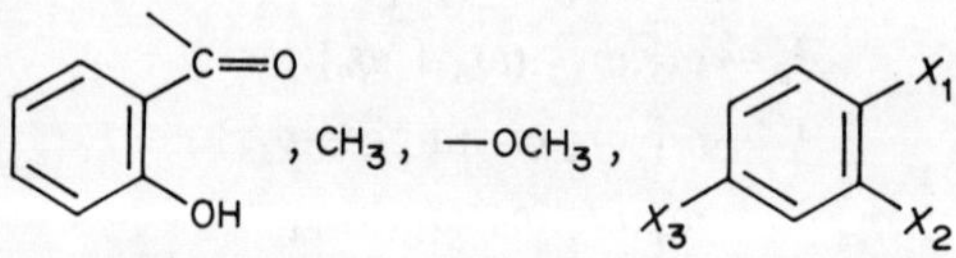

meets the conditions of the problem. Apparently, two structures may be arranged out of these fragments:

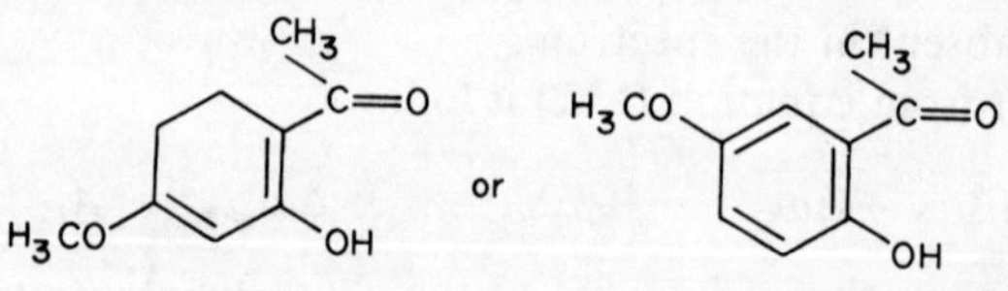

It is not possible to distinguish between them according to the infra-red spectrum.

Table 13.6. Solution of the problem of structural group analysis

A_1	1	1	1	1	0	0	0	0
A_2	0	0	0	0	1	1	1	1
A_3	1	1	0	0	1	1	0	0
A_4	1	0	1	0	1	0	1	0
A_5	1	1	1	1	1	1	1	1
A_6	1	1	1	1	1	1	1	1
A_7	0	0	0	0	0	0	0	0
A_8	1	1	1	1	1	1	1	1

13.2.2 *Establishment of the structure of polyatomic molecules according to infra-red spectra*

As shown previously, the principal logical equation of structural-group analysis of polyatomic molecules according to characteristic frequencies of infra-red spectra has the form

$$T(\mathscr{A}, \omega) \rightarrow [R(\omega) \rightarrow f(\mathscr{A})] \tag{13.3}$$

where $T(\mathscr{A}, \omega)$ is the Boolean function describing all interconnections between structural elements and spectral signs in the given problem, $R(\omega)$ is the Boolean function describing the infra-red spectrum of the compound under investigation and $f(\mathscr{A})$ is the required Boolean function which contains the enumeration of all complexes of structural elements, the presence of which is possible in the given sample.

Here we shall widen somewhat the range of definition of the functions $T(\mathscr{A}, \omega)$ and $R(\omega)$ by assuming that arguments ω_j may generally be not only frequencies of the infra-red spectrum but also other characteristics of the compound under investigation and of separate functional groups (for instance, information about the number of atoms of different types in the empirical formula of the compound, chemical shifts in the NMR spectra, etc.). In contrast to frequencies of infra-red spectra ω_j, we shall designate these characteristics as ω_j^*, and the introduction of this additional information will allow us greater use of the available information.

The preparation of initial data for solution of the problem of determining structures from infra-red spectral data (assuming that the empirical formula of the compound is known) consists of the following stages:

(1) The elements of the set $\tilde{\omega}$ are determined. This set includes the frequencies of the infra-red spectrum together with any of the other important characteristics.

(2) Implications of the type

$$\omega_j \rightarrow (A_{1j} + A_{2j} + \cdots + A_{nj})$$

are written, where A_{ij} are the structural elements of which the ω_j frequency is characteristic. Into the relations of type (2) only those A_{ij} which satisfy the qualitative elementary composition are introduced. Even one important characteristic frequency of the functional A_i group is absent in the spectrum, this group is *a priori* excluded from consideration.

(3) Logical relations are constructed which connect the ω_j with functional groups, and an approximate estimate of the quantitative elementary composition is made. This is attained by constructing formulae which impose a ban on the joint presence of definite structural groups. It suffices to write down only those relations which are easily revealed by comparing the number of atoms of different elements entering the functional groups under consideration with those indicated by the empirical formula.

All this information is codified and introduced into the computer. In the calculation of the function $f(\mathscr{A})$ an additional subprogram may be used which ensures that the result of the logical analysis is compatible with the empirical formula. We indicate this subprogram as PEF.

The principle of the control consists of the following. Let the empirical formula have the form

$$Z^1_{n1}, Z^2_{n2}, \ldots Z^k_{nk}$$

where $Z^1, Z^2, \ldots Z^k$ are varieties of atoms entering the composition of the sample under investigation and $n_1, n_2 \ldots n_k$ are the numbers of atoms of the corresponding elements. Let each of the A_i groups under investigation ($i = 1 \div N$) have as coefficients the set of numbers $\alpha_{1i}, \alpha_{2i}, \ldots \alpha_{ki}$, each of which equals the number of atoms of the elements Z^1, $Z^2, \ldots Z^k$ in the given group. Subprogram PEF selects only such complexes of functional groups for which all inequalities are fulfilled:

$$\frac{\sum_{i=1}^{N} \alpha_{1i} \leqslant n_1}{\sum_{i=1}^{N} \alpha_{ki} \leqslant n_k}$$

Thus the final result of the analysis is the singling out of such complexes of functional groups which, in the total, contain as many or fewer atoms of all types as exist in the empirical formula. In such an approach we succeed in allowing for the possibility of a duplication of some functional group in the structure of the compound under examination.

Let us now pass directly to the consideration of concrete examples. The conditions of the problem have been adopted from Nakanishi's manual[5].

Example 1 The infra-red spectrum of a $C_{10}H_{10}O$ compound is given (Figure 13.2). The NMR spectrum shows one resonance corresponding to the protons of a methyl group. The structure of this compound is to be determined. Using known correlations, let us single out elements of $\tilde{\omega}$ and $\mathscr{A}$ sets (Tables 13.7

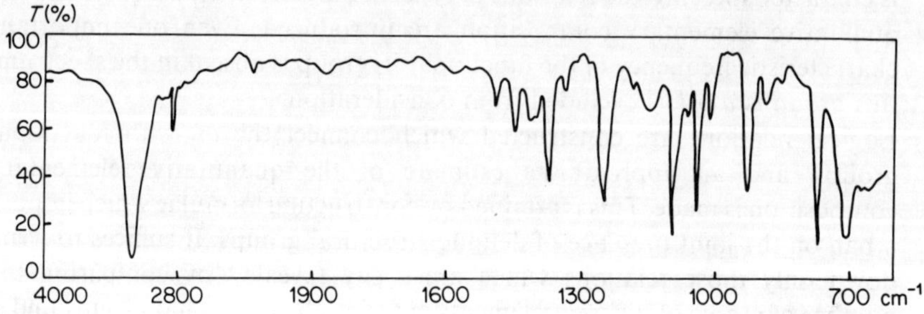

Figure 13.2. Infra-red spectrum of the $C_{10}H_{10}O$ compound (2-phenyl-3-butynol-2)

Table 13.7. Elements of the set $\tilde{\omega}$ (Example 1)

3220	2985	2165	1600	1370
ω_1	ω_2	ω_3	ω_4	ω_5
1230	1095	771	704	650
ω_6	ω_7	ω_8	ω_9	ω_{10}

Table 13.8. Elements of the set $\mathscr{A}$ (Example 1)

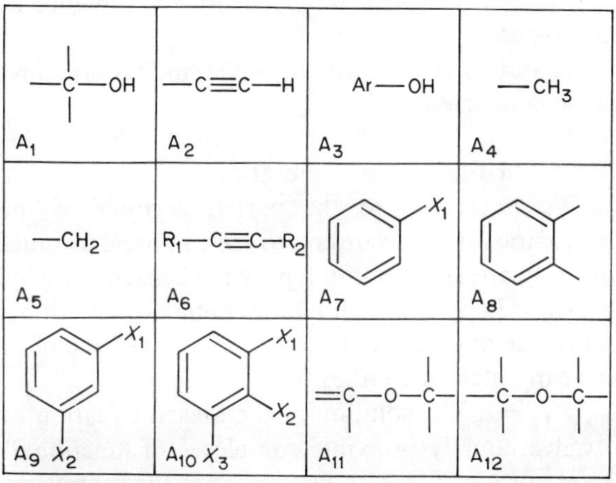

and 13.8) and represent all initial assumptions as logical relations, first taking into account only the qualitative elementary composition.

The absorption at 3220 cm^{-1} may be connected with stretching vibrations of the OH bond of an alcohol or a phenol, with the stretching vibration of ≡C—H or with the presence in the molecule of any combination of these groups which, in a symbolic way, will be written:

(1) $\omega_1 \to A_1 + A_2 + A_3$. Some infra-red absorption bands at 3000 cm^{-1} are associated with the methyl and methylene group or with their joint presence:

(2) $\omega_2 \to A_4 + A_5$. The band at 2165 cm^{-1} may be ascribed to stretching vibrations of C≡C, at least of one of the A_2 and A_6 groups:

(3) $\omega_3 \to A_2 + A_6$. Absorption at 1600 cm^{-1} indicates the presence of the benzene ring. According to relations (8) and (9) below, there is at least one type of substitution of the aromatic ring:

(4) $\omega_4 \to A_7 + A_8 + A_9 + A_{10}$. The $\omega_j \to A_i$ implication requires the obligatory presence of the group A_i in the molecule if ω_j appears in the spectrum (this band is only connected with one group of $\mathscr{A}$). Therefore in cases where there is no certainty that the group A_i has been distinctly revealed, caution requires the introduction of this logical connection in the form

$A_i \rightarrow \omega_j$. However, the presence in infra-red spectrum and the NMR spectrum of bands characteristic of the methyl group gives sufficient ground for writing down the implication:

(5) $\omega_5 \rightarrow A_4$. The band at 1230 cm^{-1} must obviously be connected with the groups A_1, A_3, A_{11} which contain the C—O bond:

(6) $\omega_6 \rightarrow A_1 + A_3 + A_{11}$. The band at 1095 cm^{-1} may be attributed to the —C—O vibrations of alcohols and saturated ethers:

(7) $\omega_7 \rightarrow A_1 + A_{12}$. The maxima at 774 and 704 cm^{-1} most likely belong to out-of-plane vibrations of non-substituted hydrogen atoms of the benzene ring. Let us test the 774 cm^{-1} frequency for the presence of the following substitution types:

(8) $\omega_8 \rightarrow A_7 + A_8 + A_9 + A_{10}$. With $\omega_9 = 704$ cm^{-1} only three substitution types may be connected:

(9) $\omega_9 \rightarrow A_7 + A_9 + A_{10}$. In the 650 cm^{-1} region the absorption picture is very typical of ≡C—H deformation vibrations:

(10) $\omega_{10} \rightarrow A_2$. We now pass to the partial account of the quantitative elementary composition. According to the empirical formula, the molecule can contain only one substitution type of the benzene ring hydrogen atoms:

(11) $A_7 \dot{\ } A_8 \dot{\ } A_9 \dot{\ } A_{10} = I$ ($\dot{\ }$ is the symbol of excluding disjunction). The molecule contains only one oxygen atom and therefore only one of the A_1, A_3, A_{11}, A_{12} groups can enter its composition:

(12) $A_1 \dot{\ } A_3 \dot{\ } A_{11} \dot{\ } A_{12} = I$. The solution of the logical equation of this problem contains twelve equally probable complexes of functional groups. Only the A_1, A_2, A_4 groups prove to be present in the sample with probability equal to 1, but there are no A_3, A_8, A_{11}, A_{12} groups in the compound being studied. The subroutine PEF eliminates nine other complexes and the final solution is therefore represented by the following sets of structural elements:

	A_1	A_2	A_3	A_4	A_5	A_6	A_7	A_8	A_9	A_{10}	A_{11}	A_{12}
(1)	1	1	0	1	1	0	0	0	0	1	0	0
(2)	1	1	0	1	0	0	1	0	0	0	0	0
(3)	1	1	0	1	0	0	0	0	1	0	0	0

Bearing in mind the empirical formula, the following structures may be constructed:

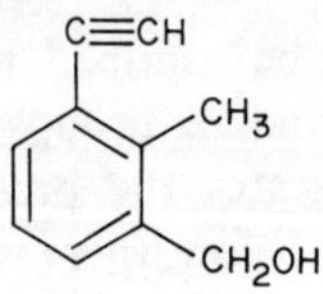

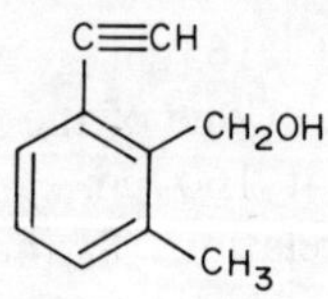

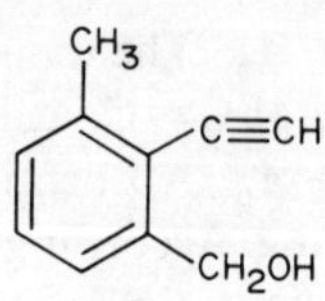

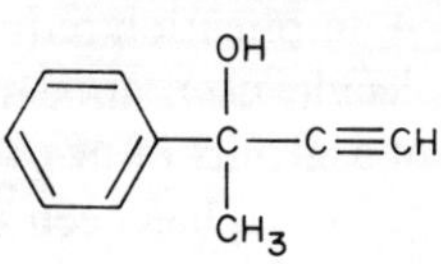

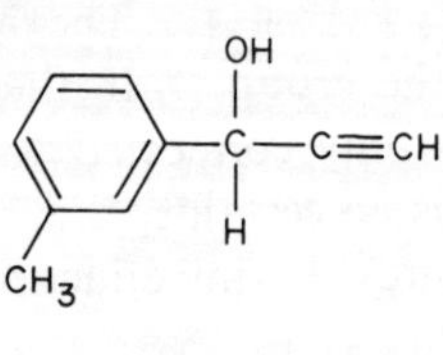

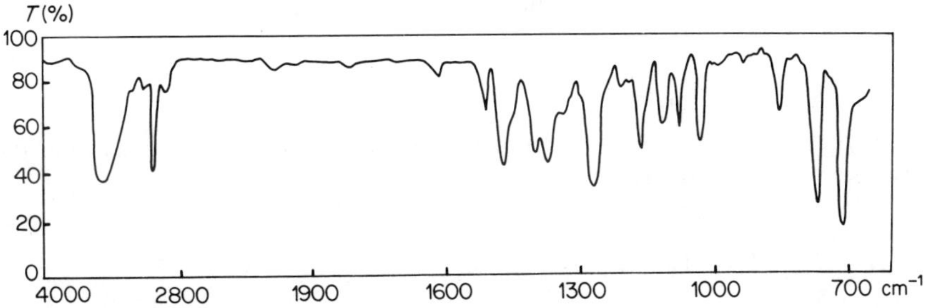

Figure 13.3. Infra-red spectrum of the $C_9H_{12}O_2$ compound (2-hydroperoxycumene)

Each of them is completely consistent with the infra-red spectrum given in Figure 13.2. The positive identification of the molecular structure of the sample (structure 2) is therefore only possible if additional information is provided (for instance, a detailed analysis of the NMR spectrum).

Example 2 The infra-red spectrum of a $C_9H_{12}O_2$ compound is given (Figure 13.3) and its structure has to be determined. Let us begin with the determination of the elements of $\tilde{\omega}$ and $\mathscr{A}$ sets (Tables 13.9 and 13.10) and then, by analogy with the preceding example, write down the initial information

Table 13.9. Elements of the set $\tilde{\omega}$ (Example 2)

3460	3000 1453	1380	1360	1267	835	763	701	C_9
ω_1	ω_2	ω_3	ω_4	ω_5	ω_6	ω_7	ω_8	ω_9^*

Table 13.10. Elements of the set $\mathscr{A}$ (Example 2)

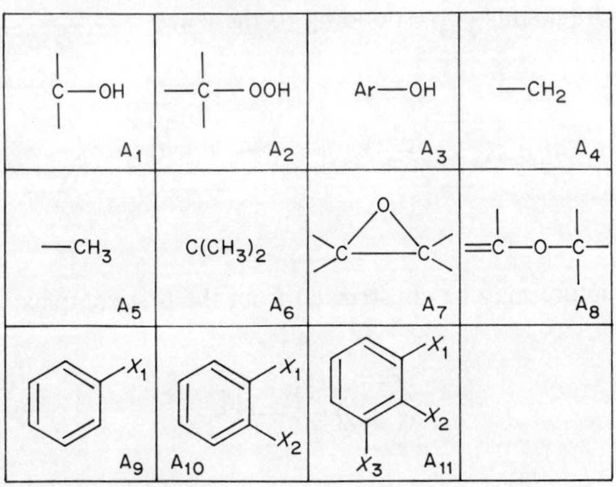

in symbolic form, giving brief explanations where necessary:

(1) $\omega_1 \to A_1 + A_2 + A_3$;
(2) $\omega_2 \to A_4 + A_5 + A_6$.

The presence of the $\omega_3.\omega_4$ doublet gives sufficient ground to assume the presence of the group $C(CH_3)_2$:

(3) $\omega_3.\omega_4 \to A_6$;
(4) $\omega_5 \to A_1 + A_2 + A_3 + A_7 + A_8$;
(5) $\omega_6 \to A_2 + A_7$;
(6) $\omega_7 \to A_9 + A_{10} + A_{11}$;
(7) $\omega_8 \to A_9 + A_{11}$.

As the molecule contains only nine carbon atoms, only one substitution type of hydrogen atom of the benzene ring is possible:

(8) $\omega_9^* \to A_9 {}^{\vee} A_{10} {}^{\vee} A_{11}$.

If the molecule contains the hydroperoxide group, other oxygen-containing groups cannot enter its composition (two oxygen atoms are contained in the molecule):

(9) $A_2 \to \bar{A}_1.\bar{A}_3.\bar{A}_7.\bar{A}_8$.

Under these conditions the table of solutions will contain 56 variants. However, if we introduce the one more obvious requirements of $A_6 {}^{\vee} A_7 = I$ (the presence of the benzene ring is obvious, leaving three carbon atoms to be associated with the other groups), the disjunctive normal form (DNF) of function $f(\mathscr{A})$ will only consist of eight elementary conjunctions.

The subsequent application of subroutine PEF gives two variants:

	A_1	A_2	A_3	A_4	A_5	A_6	A_7	A_8	A_9	A_{10}	A_{11}
(1)	0	1	0	0	0	1	0	0	1	0	0
(2)	0	1	0	0	0	1	0	0	0	0	1

The molecular fragments corresponding to these are:

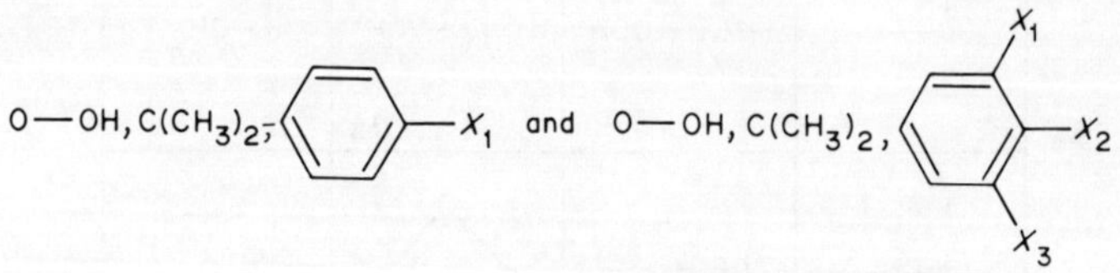

A single structure may be constructed from the first complex:

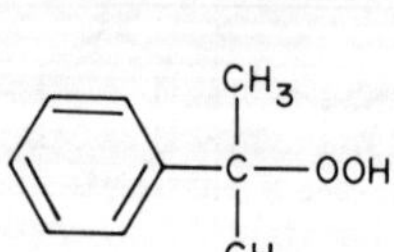

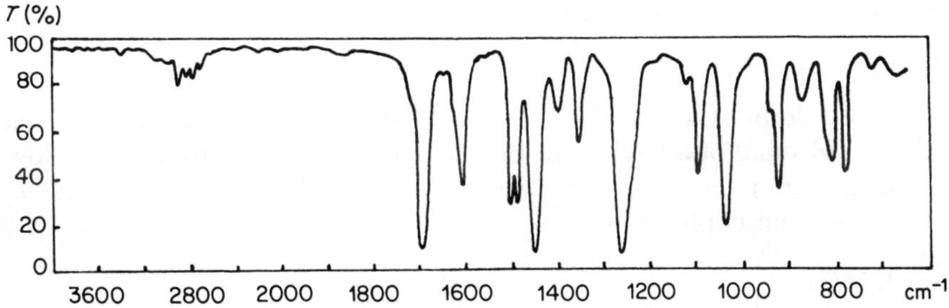

Figure 13.4. Infra-red spectrum of the $C_8H_6O_3$ compound (3,4-methylenedioxybenzaldehyde)

It is not possible to construct from the second complex a molecule which would fulfil the requirements of the empirical formula.

Example 3 The infra-red spectrum of the $C_8H_6O_3$ compound is given (Figure 13.4), and the structure of this compound has to be determined. We put the elements of sets $\tilde{\omega}$ and $\mathscr{A}$ into Tables 13.11 and 13.12 and write down

Table 13.11. Elements of the set $\tilde{\omega}$ (Example 3)

3000	2700	1690	1601	1448	1366	1258	1100	1037	927
ω_1	ω_2	ω_3	ω_4	ω_5	ω_6	ω_7	ω_8	ω_9	ω_{10}
Three bands in the region 900–750 ω_{11}				The molecule contains C_8 ω_{12}^*				The molecule contains H_6 ω_{13}^*	

Table 13.12. Elements of the set $\mathscr{A}$ (Example 3)

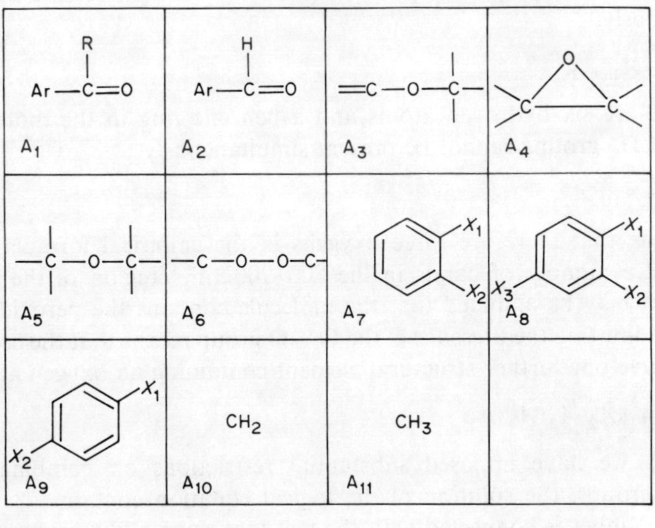

the logical relations in the following way:

(1) $\omega_1 \rightarrow A_{10} + A_{11}$.

The absorption at about 2700 cm^{-1} may be associated with the vibrations of the C—H bond of an aldehyde, but the intensity of this absorption is low. Therefore, caution requires us to introduce the dependence between A_2 and ω_2 in the form of an implication:

(2) $A_2 \rightarrow \omega_2$;
(3) $\omega_3 \rightarrow A_1 + A_2$;
(4) $\omega_4 \rightarrow A_7 + A_8 + A_9$;
(5) $\omega_5 \rightarrow A_7 + A_8 + A_9 + A_{10} + A_{11}$.

If the connection of A_{11} with ω_6 was introduced in the form $\omega_6 \rightarrow A_{11}$ this would mean that the CH_3 group is present with a probability of 1. However, as the molecule contains, in all, six hydrogen atoms and the presence of the benzene ring is obvious, it is reasonable to write the relation between A_{11} and ω_6 by the following implication:

(6) $A_{11} \rightarrow \omega_6$;
(7) $\omega_7 \rightarrow A_1 + A_2 + A_3 + A_4$;
(8) $\omega_8 \rightarrow A_1 + A_5 + A_7 + A_8 + A_9$;
(9) $\omega_9 \rightarrow A_3 + A_7 + A_8 + A_9$;
(10) $\omega_{10} \rightarrow A_2 + A_4$.

Three bands in the 750–900 cm^{-1} region are associated with one of the three substitution types of the benzene ring:

(11) $\omega_{11} \rightarrow A_7 \veebar A_8 \veebar A_9$.

The molecule contains eight carbon atoms. Hence, after accounting for six of these in the ring, the ketone and aldehyde groups cannot be present simultaneously:

(12) $\omega^*_{12} \rightarrow A_1 \veebar A_2$.

As there are six hydrogen atoms and a benzene ring in the molecule, the CH_2 and CH_3 groups cannot be present simultaneously:

(13) $\omega^*_{13} \rightarrow A_{10} \veebar A_{11}$.

If we note that there are three oxygens in the empirical formula and the presence of a number of bands in the 800–900 cm^{-1} region of the infra-red spectrum, it may be assumed that the molecule contains the peroxide group. However, admitting the presence of the O—O group means that the compound can only have one further structural element containing an oxygen atom:

(14) $A_6 \rightarrow A_1 \veebar A_2 \veebar A_3 \veebar A_4 \veebar A_5$.

Although we have imposed substantial restrictions on combinations of functional groups, the solution of the logical equation contains 78 variants. This large number is connected with the fact that some additional constraints are not immediately revealed during the first comparison of the elementary composition of the rivalling groups. For instance, if we note that $A_4 \rightarrow \bar{A}_1 . \bar{A}_2$ and add this condition to the initial set of ratios, we could reduce the number of solutions to 30. However, the use of the subroutine PEF makes the search for additional restrictions unnecessary.

If we confront each structural element with a minimal set of atoms of different types

	A_1	A_2	A_3	A_4	A_5	A_6	A_7	A_8	A_9	A_{10}	A_{11}
C	1	1	0	2	0	0	6	6	6	1	1
H	0	1	0	0	0	0	4	3	4	2	3
O	1	1	1	1	1	2	0	0	0	0	0

we obtain five solutions:

	A_1	A_2	A_3	A_4	A_5	A_6	A_7	A_8	A_9	A_{10}	A_{11}
(1)	0	1	1	0	1	0	0	1	0	1	0
(2)	0	1	1	0	0	0	0	1	0	1	0
(3)	0	1	0	0	1	0	0	1	0	1	0
(4)	0	1	0	0	0	1	0	1	0	1	0
(5)	0	1	0	0	0	0	0	1	0	1	0

We see that the molecule definitely contains aromatic aldehyde, benzene ring substitution in positions 1, 2, 4 and a CH_2 group. The requirements of the empirical formula and of organic structural theory are satisfied by the second and the fourth variants, which give the following structures:

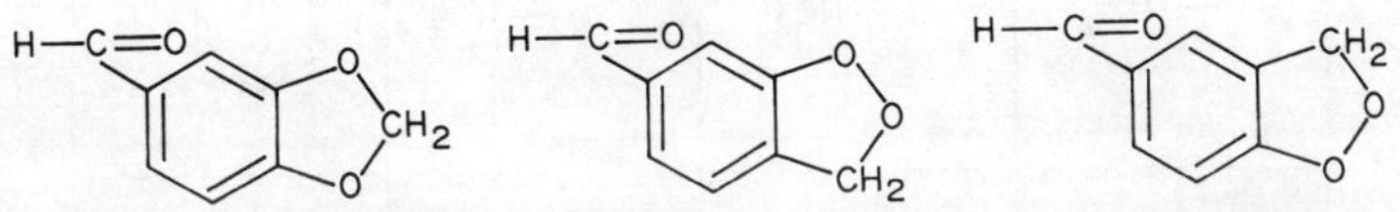

These structures fully agree with the infra-red spectrum.

Example 4 The infra-red spectrum of a $C_9H_5NO_4$ compound is given (Figure 13.5). As seen from Tables 13.13 and 13.14, the set $\tilde{\omega}$ contains 13 characteristics and the set $\mathscr{A}$ consists of 16 rivalling structural elements.

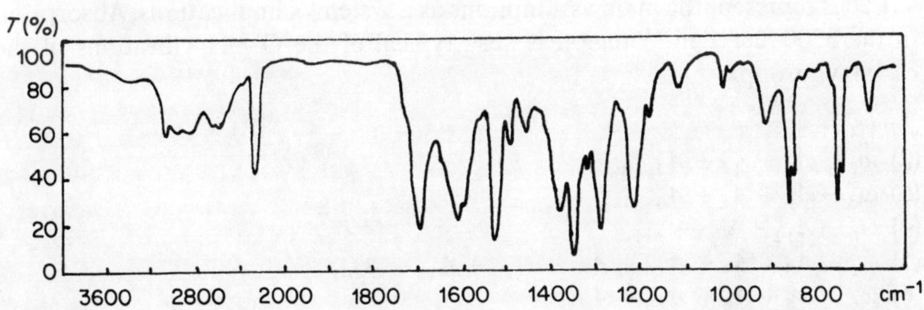

Figure 13.5. Infra-red spectrum of the $C_9H_5NO_4$ compound (p-nitrophenylpropiolic acid)

Table 13.13. Elements of the set $\tilde{\omega}$ (Example 4)

3200–2600	2280	1697	1606	1528	1347	1290	920	863	756
ω_1	ω_2	ω_3	ω_4	ω_5	ω_6	ω_7	ω_8	ω_9	ω_{10}

The molecule contains H_5	The molecule contains N	The molecule contains C_9
ω_{11}^*	ω_{12}^*	ω_{13}^*

Table 13.14. Elements of the set $\mathscr{A}$

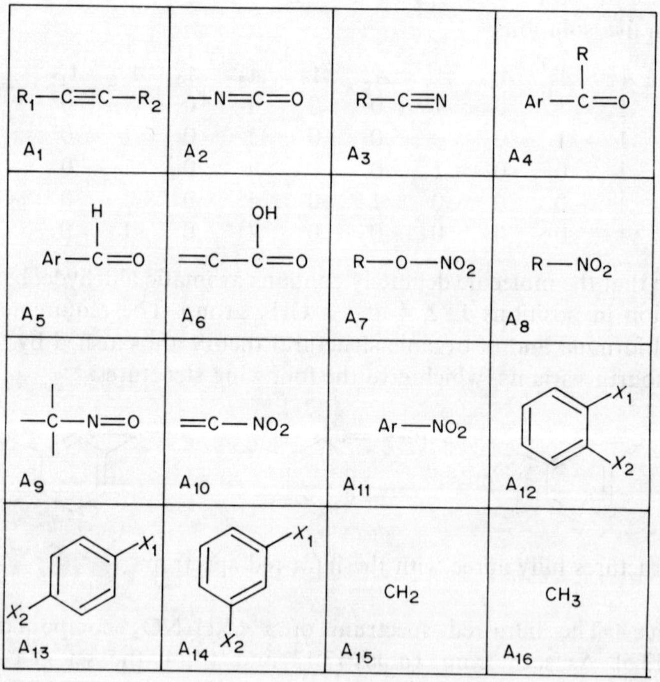

Let us represent the main assumptions as a system of implications. Absorption in the 3200–2600 cm^{-1} region is very typical of the O—H vibrations of the carboxyl group:

(1) $\omega_1 \rightarrow A_6$;
(2) $\omega_2 \rightarrow A_1 + A_2 + A_3$;
(3) $\omega_3 \rightarrow A_4 + A_5 + A_6$;
(4) $\omega_4 \rightarrow A_{12} + A_{13} + A_{14}$;
(5) $\omega_5 \rightarrow A_8 + A_9 + A_{10} + A_{11} + A_{12} + A_{13} + A_{14}$;
(6) $\omega_6 \rightarrow A_5 + A_8 + A_{10} + A_{11}$;
(7) $\omega_7 \rightarrow A_5 + A_6 + A_7$.

The band at $920\,\text{cm}^{-1}$ may be connected with the carboxyl group absorption, but the detection of this band does not prove the presence of COOH as, in the $900\,\text{cm}^{-1}$ region, vibrations of other structural elements also give rise to absorptions. The fact that $\omega_8 = 920\,\text{cm}^{-1}$ is a necessary but not a sufficient sign of the COOH group is also reflected by an implication being equivalent to relation (8): if there is no ω_8 band in the spectrum, the molecule does not contains A_6:

(8) $A_6 \rightarrow \omega_8$;
(9) $\omega_9 \rightarrow A_{11} + A_{13} + A_{14}$;
(10) $\omega_{10} \rightarrow A_{11} + A_{12} + A_{14}$.

Nothing definite may be said about the presence or absence of absorption in the region of stretching C—H vibrations of the methyl and methylene groups, but in the regions which are typical of deformation vibrations of CH_2 and CH_3, a number of bands are observed. Taking this into account, as well as the elementary composition, we conclude that either the molecule contains only one of these groups or both are absent:

(11) $\omega_{11}^* \rightarrow \bar{A}_{15} + \bar{A}_{16}$.

The following restriction is conditioned by the presence of not more than four oxygen atoms:

(12) $A_6 \rightarrow \bar{A}_7$.

The molecule contains one nitrogen atom:

(13) $\omega_{12}^* \rightarrow A_2 \vee A_3 \vee A_7 \vee A_8 \vee A_9 \vee A_{10} \vee A_{11}$.

Only one substitution-type of hydrogen atoms of the benzene ring may exist in the molecule:

(14) $\omega_{13}^* \rightarrow A_{12} \vee A_{13} \vee A_{14}$.

The solution of the logical equation contains 90 variants. Having attributed to each structural element a minimum set of atoms

	A_1	A_2	A_3	A_4	A_5	A_6	A_7	A_8	A_9	A_{10}	A_{11}	A_{12}	A_{13}	A_{14}	A_{15}	A_{16}
C	2	1	1	1	1	1	0	0	0	0	0	6	6	6	1	1
H	0	0	0	0	1	1	0	0	0	0	0	4	4	4	2	3
N	0	1	1	0	0	0	1	1	1	1	1	0	0	0	0	0
O	0	1	0	1	1	2	3	2	1	2	2	0	0	0	0	0

after the use of PEF we obtain five sets:

	A_1	A_2	A_3	A_4	A_5	A_6	A_7	A_8	A_9	A_{10}	A_{11}	A_{12}	A_{13}	A_{14}	A_{15}	A_{16}
(1)	1	0	0	0	0	1	0	0	0	1	0	0	0	1	0	0
(2)	1	0	0	0	0	1	0	0	0	0	1	1	0	0	0	0
(3)	1	0	0	0	0	1	0	0	0	0	1	0	1	0	0	0
(4)	1	0	0	0	0	1	0	0	0	0	1	0	0	1	0	0
(5)	1	0	0	0	0	1	0	1	0	0	0	0	0	1	0	0

Sets (1)–(4) satisfy both the empirical formula and the requirements of structural theory, and correspond to

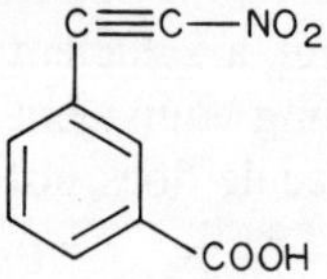

(1) ;

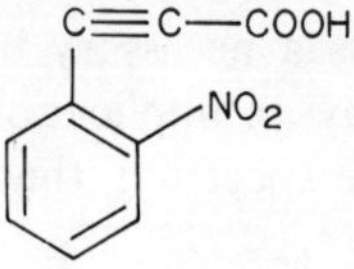

(2) ;

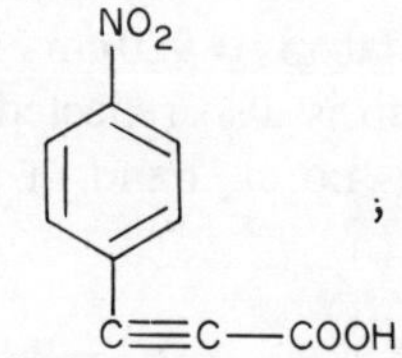

(3) ;

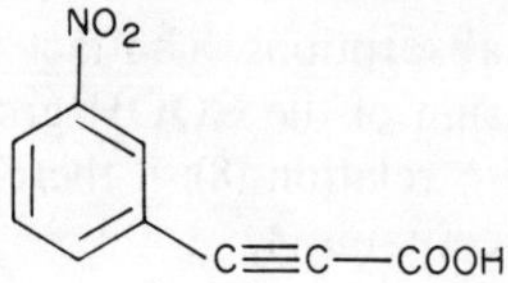

(4)

Generally speaking, structures (1), (3) and (4) are in agreement with the infrared spectrum, but compound (2) may obviously be distinguished in a spectral way, due to the presence of an intramolecular hydrogen bond, the peculiarities of which are displayed during dilution.

We see from the examples examined that, despite the additional restrictions, the number of variants contained in the solution of the logical equation generally proves to be very large (for the last example, up to 90). The addition of new restrictions substantially reduces the number of complexes and lessens the vagueness of the answer. As is known, entropy is the measure of uncertainty of the system. If the system has n equally probable states, the entropy is $H = \log_2 n$. It may easily be seen that the use of the formal-logic method (FLM) allows us to classify logical constraints according to the significance of their information. Indeed, if the number of variants before and after introduction of the constraint $T\ (A, \omega)$ be r_1 and r_2, then we shall find that the amount of information contained in the given constraint is given by the formula $\mathscr{I} = \log(r_1/r_2)$.

The final selection of variants is accomplished by PEF, which, as a rule, singles out a small number of complexes. The construction of all possible structures which satisfy the empirical formula is carried out directly from sets of molecular fragments. Those sets from which it is not possible to construct any realistic molecular structure are rejected. The algorithmization of this final stage of analysis would apparently allow a full automatation of the solution of structural problems.

In conclusion, let us note some evident advantages which the FLM of spectrum interpretation has in comparison with ‘intuitive’ analysis:

(1) During ‘intuitive’ analysis it is extremely difficult (and often practically impossible) for the specialist to bear in mind continuously the conditions which logically relate the functional group to the frequencies, and through them the functional groups with one another. The application of FLM guarantees that no logical connection will be lost during machine operation.
(2) FLM makes visible the whole system of evidence of the object structure studied and allows us to observe easily how a hypothesis may influence the result of analysis.
(3) FLM makes it possible to detect rapidly mistakes in the initial data: if the

$T(\mathscr{A}, \omega)$ function contains contradictory conditions, the machine will yield no single complex of structural elements.

(4) It is generally known that when interpreting the same spectrum different investigators frequently come to different conclusions. FLM makes it possible to estimate the depth and completeness of argument of each specialist and to establish precisely the cause of discrepancy. In particular, it may be found that different investigators proceed from equivalent assumptions but, not being able to take into account all possible variants in 'intuitive' analysis, they arbitrarily single out some of them. In this case, FLM will help to establish the insufficiency of initial information on the object being studied.

(5) The use of FLM not only allows us to include all initial assumptions but also, in a certain sense, to take into consideration the reliability of correlations. Thus, for instance, if $\omega_j \to A_i$ and there are grounds for assuming that the ω_j band occurs in the characteristic region of A_i group by chance, the connection of ω_j with A_i has to be introduced by the implication $A_i \to \omega_j$ (if A_i, then ω_j, but if ω_j, then A_i may or may not be).

(6) Obviously, the reliability of the result of an analysis is wholly determined by that of the initial data. If there are correlation tables precisely reflecting the relationships between structural elements and spectral information, FLM automatically ensures correct results.

(7) As the initial data may include spectral data without specific reference to its physical nature, there are grounds for assuming that the FLM approach will prove to be very useful for joint application of various spectral methods—infra-red, ultra-violet, nuclear magnetic resonance, mass spectra, etc.

13.2.3 Additional constraints

As mentioned, equation (13.3) can give only sets of structural groups which, to some degree of probability, may exist in the molecule under investigation. However this information alone is inadequate for constructing the structural formula of the compound. Moreover, since the solution of equation (13.3) yields information on the presence of one or more functional groups in the molecule but not their number, in general it is possible by repeatedly using these groups to construct an infinite number of molecular structures possessing certain specific spectral characteristics. It is in this sense that the inverse spectral problem may be said to be indeterminate—a situation common to all incompletely defined problems. A unique solution can be derived only be imposing certain additional conditions that are not implied in the postulates already stipulated. We shall now consider these conditions further.

The only additional condition necessary to that a finite set of solutions may be obtained is that the empirical formula of a molecule under identification should be satisfied. This information, in principle, cannot be derived from optical molecular spectroscopy; it must be obtained from an independent experiment. The need for an additional condition arises, in particular, for the following reason. The entire set of frequencies of a molecule may be divided into two groups: characteristic frequencies of functional groups and non-characteristic frequencies. It is known that even those functional groups which are sufficiently closed as regards their chemical structure may nonetheless not exhibit typical spectral characteristics. It is possible that such groups may escape detection at the time of solution of equation (13.3). This fact, and the possibility of multiple repetition of individual fragments without resulting in any change in the characterstic frequencies, prevent the construction of the molecular structural formula only with the help of formal logical analysis. At best, we can recognize only the class of the substance.

The condition that the empirical formula should be satisfied leads to a finite (though sometimes a large) set of possible solutions to our inverse problem. In addition, by virtue of valency rules it has been found that there is an upper limit for the number of all structures consistent with the vibrational frequencies introduced. These structures therefore have to be recognized as equally probable.

Another essential constraint is that the molecular structures so derived should be in compliance with the basic laws of chemistry, and with admissible and inadmissible combinations of atoms, in particular. Examples of inadmissible combinations are

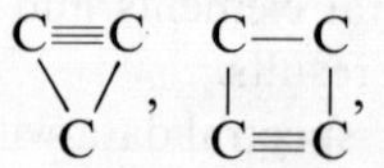 etc.

Thus the inverse spectral problem under investigation should include at least two steps: first, solution of a logical structural-group problem, and second, construction of possible molecular structures consistent with the empirical formula and basic principles of chemical structure. The second step can be carried out by means of the algorithmic graph theory[6].

Construction of structural formulae is a kind of a filter, which limits the arbitrariness in the solution of the inverse problem and there by permits the identification of the compound.

Note here that even if inconsistent situations do not arise in the first step, such contradictions may arise in the second if the constraints mentioned above are imposed. For example, let us assume that a system of initial implications contains the following relations:

$$1715\,\mathrm{cm}^{-1} \rightarrow \left\{ \begin{matrix} \mathrm{O} \\ \| \\ \mathrm{C{-}C{-}C} \end{matrix} \right\}, \quad 1725\,\mathrm{cm}^{-1} \rightarrow \left\{ \begin{matrix} \mathrm{O} \\ \| \\ \mathrm{C{-}C{-}H} \end{matrix} \right\}.$$

and that if we now impose the costraint that the molecule should contain only the oxygen atom, then we have the following implication: $\omega_3 \rightarrow (\bar{A}_1 \cdot A_2) + (A_1 \cdot \bar{A}_2)$, where ω_3 stands for the constraint and

$$A_1 \equiv \left\{ \begin{matrix} \mathrm{O} \\ \| \\ \mathrm{C{-}C{-}C} \end{matrix} \right\}, \quad A_2 \equiv \left\{ \begin{matrix} \mathrm{O} \\ \| \\ \mathrm{C{-}C{-}C} \end{matrix} \right\}$$

On logically multiplying the left- and right-hand parts of these implications we obtain the relation $\omega_1 \cdot \omega_2 \cdot \omega_3 \to 0$, where $\omega_1 = 1715\,\text{cm}^{-1}$ and $\omega_2 = 1725\,\text{cm}^{-1}$. This obviously shows that the spectral features are not consistent with the empirical formula. Thus there is no solution to our problem, and hence we have to introduce additional information to eliminate the inconsistency.

On imposing the condition that the empirical formula should be consistent with chemical laws we find that the number of alternatives is significantly reduced, and only in some particular cases does the solution prove to be unique. This is largely due to the fact that only a relatively small number of functional groups are identified correctly by means of spectra–structure correlations, the groups in general containing only a small number of atoms. In particular, compounds containing several phenyl radicals of different types do not easily yield to identification.

In order to reduce the number of alternative solutions we have to include a third step, i.e. comparison of a hypothetical spectrum with the experimental one. The inclusion of this third step is necessary because, in the course of mathematical synthesis of molecular structures from fragments it is likely that we may obtain new combinations of structural elements that are characterized by their own specific spectral features. These false structures can be eliminated by checking the presence or absence of these spectral characteristic in the experimental spectrum. Frequently, comparison of hypothetical and experimental spectra acts as a filter to screen out the unique solution. In order to automate such a comparison an algorithm has been designed for analysing the molecular structural formulae represented in the form of adjacency matrices[7].

Thus to solve the molecule identification problem we have to impose several additional conditions, each of which is in itself a complicated problem.

13.2.4 *A practical approach to identification of molecules by their spectra*

The general procedure described above for solving the inverse spectral problem has been realized in the STREC programs[7–9] with a system of algorithms. The program can execute all logical and combinatorial operations automatically, and these can be monitored and modified by a specialist.

The empirical formula of a substance, its spectra and additional data (if any) containing the constraints on the presence or absence of any particular fragments in the molecule are fed into the program input. This system includes simultaneous or separate data processing of infra-red and Raman spectra, PMR and NMR C^{13} spectra and, in addition, ultra-violet and mass-spectra of low resolution. The following conclusion can be drawn. The system can solve a great number of various problems step by step using different mathematical formalism, i.e. theory of combinations, mathematical logic, graph theory, methods of general theory of molecular structures and spectra, and quantum chemistry. The system can identify and analyse molecules consisting of 40–50 atoms in the main skeleton. The computer programs are written in FORTRAN and the main part of system is block STREC (structural recognition), use of which allows us to obtain results in the form of structural formulae, which is sufficient in many cases. Block STREC provides structural-group analysis in accordance with spectra–structure correlations (the corresponding library is in the computer memory) and mathematical synthesis of structures from atoms and fragments after structural-group analysis (when the Brutto formula is given) and also compares the approximate spectra of each synthesized structure with the corresponding experimental one.

There are two variants of this system: STREC-1 and STREC-2. The STREC-1 system is described in detail in references 7 and 8 and will not be discussed here. The STREC-2 system[9] is a development of the STREC-1 system and differs from it in using more modern mathematical formalism, a better set of libraries of axioms, etc.

As a result of using the STREC-2 system a set of spectra can be processed simultaneously using appropriately large fragments and operating with molecular structures, containing up to 40–50 atoms in the skeleton. A very important new element is the introduction of a dialogue mode that allows the operator to take part in reaching the final solution and to use his experience and intuition. The large fragments, used as independent discrete units, allows us to represent molecular structure by the combination of additive elements, which form the basis of the concept of characteristic spectral features. For each fragment information about possible connections with hetero-atoms, aromatic rings, multiple bonds and protons bound to the atoms of fragments is indicated. This allows the complete characterization of the properties of fragments and the use of these properties for reducing the number of generating structures at the stage of mathematical synthesis of structural formulae.

The use of large fragments is necessary in constructing special systems to solve problems posed by industrial or academic laboratories. The new method uses frequencies and intensities from experimental spectra, with a five-mark scale and the halfwidth or lines. Chemical shifts, spin–spin coupling constants and multiplicity are obtained from NMR spectra. Thus much detailed information from spectra and molecular structure are included in a system.

When intensities between experimental and simulated spectra are compared, a negative result will be obtained if a strong band is predicted and weak line appears experimentally.

As already mentioned, the possibilities of working in a dialogue mode exists for STREC-2 system. The exchange of information between the operator and the computer can be done with the help of a visual display unit. A special program questions the operator and at the same time gives the expected form of answer. The operator can show what kind of spectrum he would like to deal with, examine the dependence of the result of structural-group analysis on the set of input spectral data, exclude or add separate molecular fragments, take part in the process of building up structural formulae, identify prohibited atom combinations or (on the contrary) the most possible ones. Also, the operator can analyse the process of the solution and inform the computer about maximum bond orders and display the readout mode of the structural formulae.

Constructed in such way, the dialogue mode allows a connection between the operating possibilities of the computer and the experience and intuition of the investigator. This greatly increases the degree of accuracy and the rate of work on artificial intelligence systems.

The STREC-2 system is now operative, and many examples of the complete identification of structures showing its high versatility exist. Examples of structures identified are given in Table 13.15. It is interesting to note that the system described can recognize phosphorganic compounds together with large organic molecules if standard fragments and spectra–structure correlation are included in the library. It is also important that information kept in such a library is basic and, in practice, is not changed until the system is improved.

The molecular identification algorithm described in the previous section is based on discrete spectra–structure correlations and characteristic vibrational frequencies. Such an algorithm may, in principle, not be satisfactory for distinguishing isomers having identical or very closely characteristic frequencies. However, the vibrational spectrum in such cases may be rendered more informative if, in addition to the characteristic frequencies, consideration is given to non-characteristic ones which reflect the relative movement of individual functional groups, and hence essentially depend on their combinations. Such a situation arises precisely in the identification of molecules devoid of characteristic frequencies and consisting, at least partly, of additive structural groups. Since each molecule has its own specific set of vibrational frequencies, a unique solution may be guaranteed for every problem, provided that all the frequencies of the compound are used in the investigation. It is useful at this stage to assess the amount of information contained in both the characteristic and in the non-characteristic spectral features.

Let N be the number of structural isomers having a given empirical formula and n that of structures resulting from the solution of the identification problem. In principle, N may be determined by mathematical synthesis of all the isomers using the algorithm[7]. Then the amount of information from the spectrum is expressed as $I = \log_2 N - \log_2 n$.

In this case, I stands for the amount of information that can be obtained only from spectra–structure correlations by means of the logical–combinatorial analysis described above. Assuming that the molecule is uniquely identified in a detailed analysis of the whole spectrum, we may evaluate the amount of information obtained in solving the inverse problem by means of logical–combinatorial analysis.

Information estimates made on the basis of problems already solved* show that between 45% and 100% of structural information can be derived in such an approach; moreover, the percentage of unique solutions is quite high (see Table 13.16). This suggests that this method is quite effective for solving inverse problems. Information contained in non-characteristic frequencies is, in this case, superfluous. Nonetheless, certain isomeric structures still remain indistinguish-

*Only those problems with such empirical formulae may be considered for which comparatively low computer time is needed to compute N.

Table 13.15. Examples of automatically recognized compounds

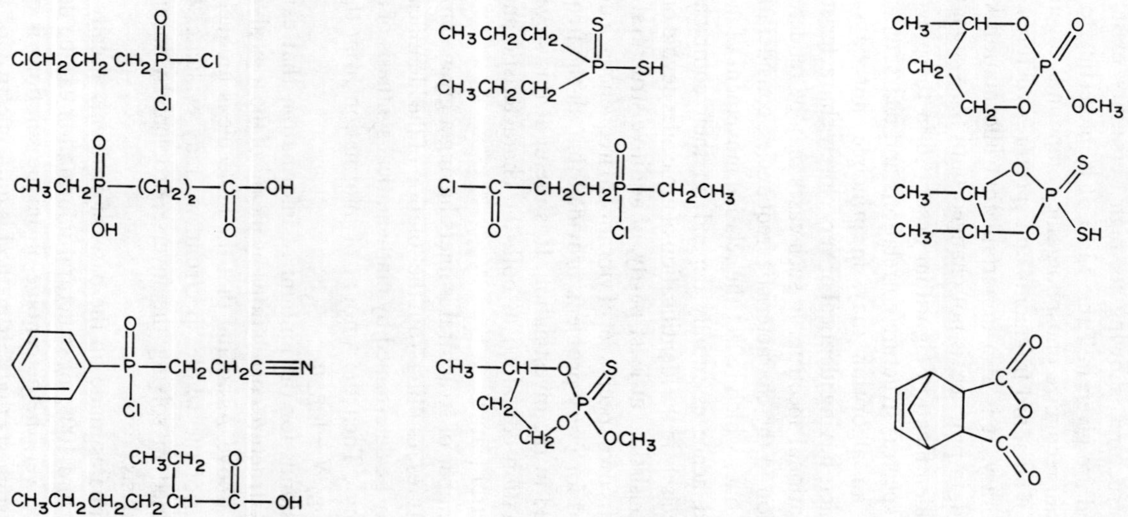

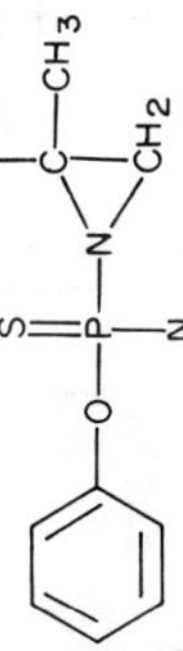

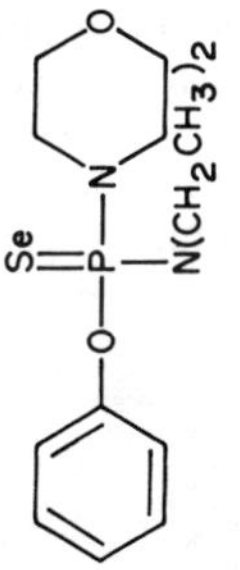

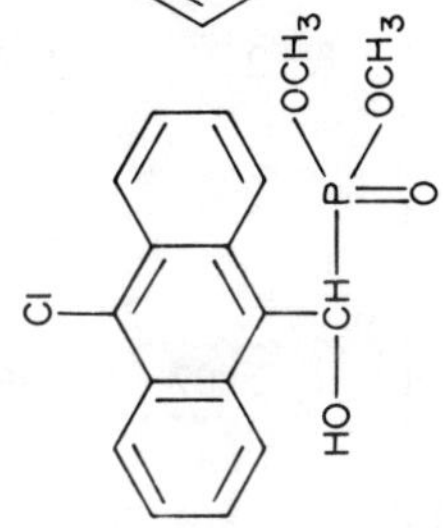

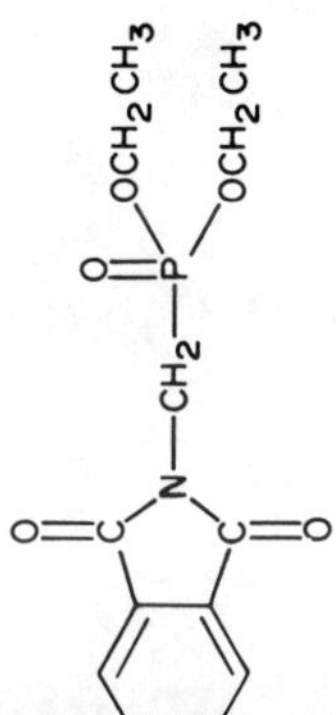

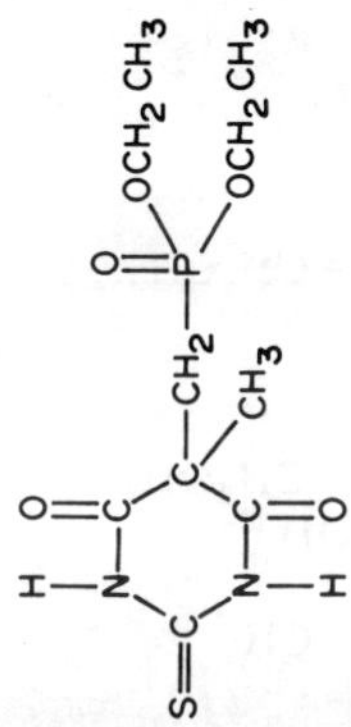

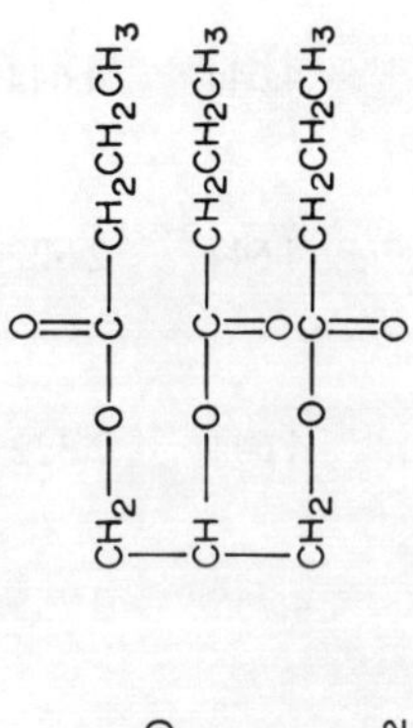

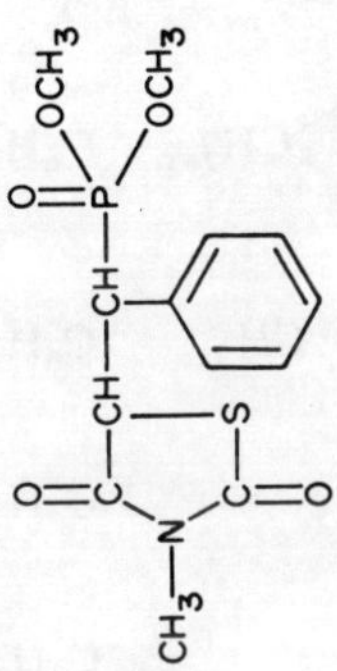

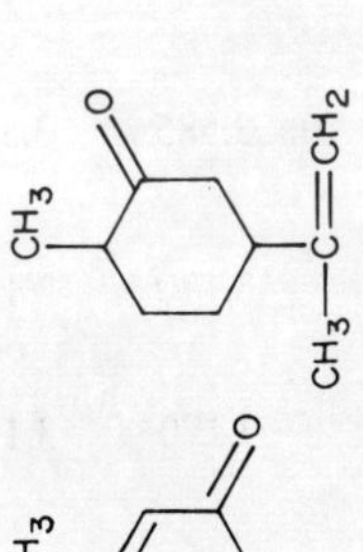

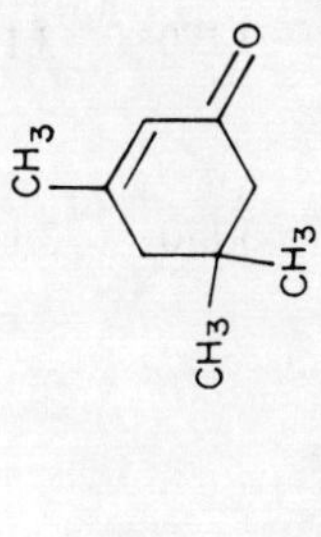

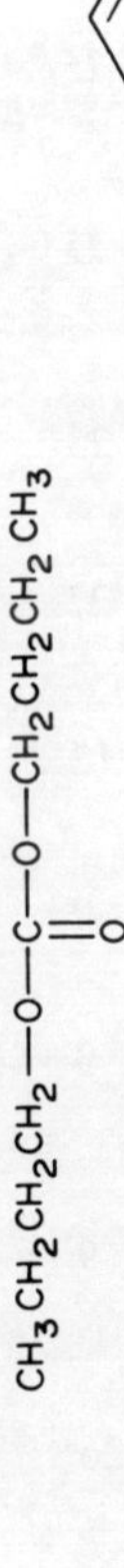

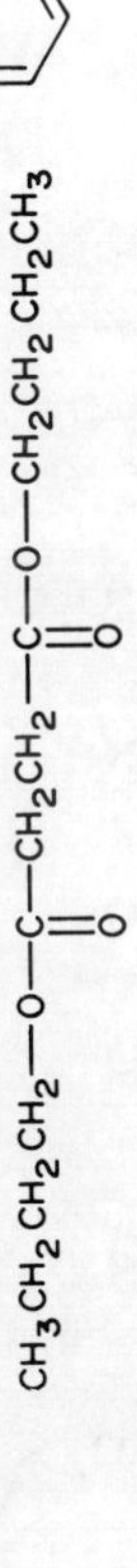

Table 13.16. Evaluation of the amount of information contained in characteristic vibrational frequencies

Structure to be found	Brutto formula	N	n	$I_0 = \log_2 N$	I	(I/I_0) (%)
1	2	3	4	5	6	7
$(CH_3)_3C{-}NH_2$	$C_4H_{11}N$	8	1	3	3	100
$CH_3CH_2NHCH_2CH_3$	$C_4H_{11}N$	8	2	3	2	67
$CH_3CH_2CH_2CH_2NH_2$	$C_4H_{11}N$	8	2	3	2	67
$\begin{smallmatrix}CH_2\\ \vert \\CH_2\end{smallmatrix}{>}NH$	C_2H_5N	4	2	2	2	100
$(CH_3)_3N$	C_3H_9N	4	1	2	2	100
$CH_3{-}\underset{OH}{\underset{\vert}{CH}}{-}CH_2CH_3$	$C_4H_{10}O$	7	1	2.807	2.807	100
$CH_3CH_2CH_2CH_2OH$	$C_4H_{10}O$	7	1	2.807	2.807	100
$(CH_3)_3C{-}OH$	$C_4H_{10}O$	7	1	2.807	2.807	100
$CH_2{=}CH{-}CH_2{-}CH{<}\begin{smallmatrix}CH_3\\CH_3\end{smallmatrix}$	C_6H_{12}	25	1	4.644	4.644	100
$CH_2{<}\begin{smallmatrix}CH_2-CH_2\\CH_2-CH_2\end{smallmatrix}{>}CH_2$	C_6H_{12}	25	5	4.644	2.322	50
$\begin{smallmatrix}CH_2-CH_2\\ \vert \\CH_2-CH_2\end{smallmatrix}{>}CH{-}CH_3$	C_6H_{12}	25	5	4.644	2.322	50
$(CH_3)_3COOH$	$C_4H_{10}O_2$	27[a]	2	4.755	3.755	80
$\begin{smallmatrix}CH_2\\ \vert \\CH_2\end{smallmatrix}{>}N{-}CH_3$	C_3H_7N	12	1	3.585	3.585	100
$CH_2{=}CH{-}CH_2OH$	C_3H_6O	9	1	3.170	3.170	100
$\overset{O}{CH_2{-}CH}{-}CH_3$ (epoxide ring)	C_3H_6O	9	1	3.170	3.170	100
$CH_3{-}\underset{O}{\underset{\Vert}{C}}{-}CH_3$	C_3H_6O	9	1	3.170	3.170	100
$O{<}\begin{smallmatrix}CH_2-O\\CH_2-CH_2\end{smallmatrix}{>}CH{-}CH_3$	$C_5H_{10}O_2$	226[a]	35			

Table 13.16 (*Contd.*)

Structure to be found	Brutto formula	N	n	$I_0 = \log_2 N$	I	(I/I_0) (%)
1	2	3	4	5	6	7
$CH_3CH_2OCH_2CH_2OH$	$C_4H_{10}O_2$	27[a]	6	4.755	2.170	46
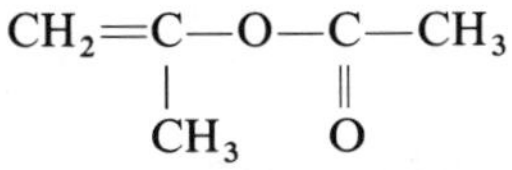	$C_5H_8O_2$	535[a]	7	9.063	6.256	70
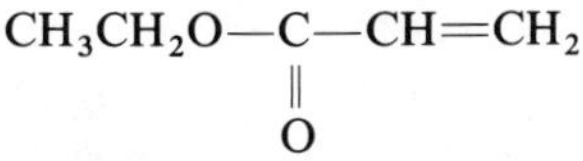	$C_5H_8O_2$	535[a]	4	9.063	7.063	78
$NH_2{-}(CH_2)_2{-}OH$	C_2H_7NO	8	1	3	3	100
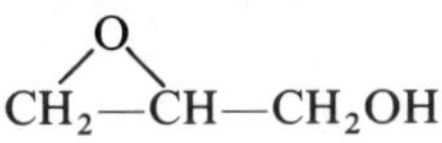	$C_3H_6O_2$	23	1	4.523	4.523	100
$(CH_3)_2N{-}C{\equiv}N$	$C_3H_6N_2$	136	1	7.087	7.087	100
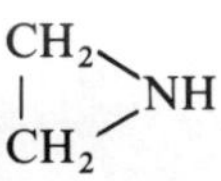	C_2H_5N	4	1	2	2	100
$(CH_3)_3N$	C_3H_9N	4	1	2	2	100
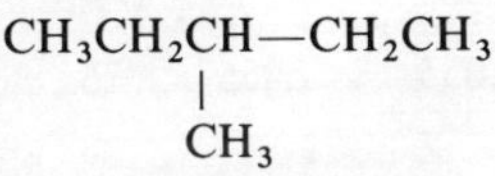	C_6H_{14}	5	2	2.322	1.322	57
$NH_2CH_2CH_2CH_2OH$	C_3H_9NO	21	2	4.392	3.392	77
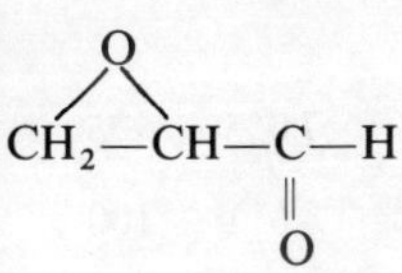	$C_3H_4O_2$	36	1	5.170	5.170	100
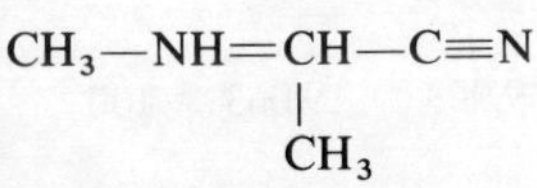	$C_4H_8N_2$	633	4	9.306	7.306	79
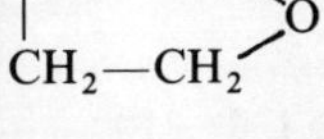	C_4H_8O	26	3	4.70	3.115	66
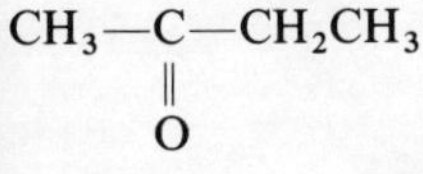	C_4H_8O	26	1	4.7	4.7	100

Table 13.16 (*Contd.*)

Structure to be found	Brutto formula	N	n	$I_0 = \log_2 N$	I	(I/I_0) (%)
1	2	3	4	5	6	7
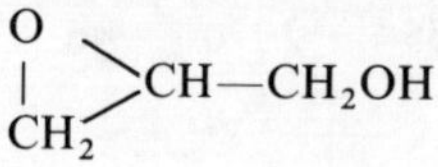	$C_3H_6O_2$	34	1	5.087	4.087	100
$CH_3CH_2C(=O)OH$	$C_3H_6O_2$	34	1	5.087	5.087	100
$CH_2{=}CH{-}CH_2CH_2CH{=}CH_2$	C_6H_{10}	77	2	6.267	5.267	56
$CH_3CH_2NO_2$	$C_2H_5NO_2$	86	1	6.426	6.426	100
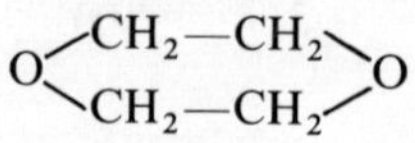	$C_4H_8O_2$	122	8	6.93	3.93	57
$CH_3{-}C(=O){-}O{-}CH_2CH_3$	$C_4H_8O_2$	122	2	6.93	5.93	86
$CH{\equiv}C{-}CH_2CH_2C{\equiv}CH$	C_6H_6	217	1	7.762	7.762	100
$(CH_3)_2CHCH_2C(=O)OH$	$C_5H_{10}O_2$	266[a]	2	8.055	7.055	75
$HOCH_2CH_2CH_2C(=O)CH_3$	$C_5H_{10}O_2$	266[a]	2	8.055	7.055	75
$(CH_3)_3C{-}C(=O)OH$	$C_5H_{10}O_2$	266[a]	2	8.055	7.055	75
$CH_3CH_2CH_2CH_2C(=O)OH$	$C_5H_{10}O_2$	266[a]	2	8.055	7.055	75
$CH_3CH_2CH_2NO_2$	$C_3H_7NO_2$	391	1	8.611	8.611	100
$CH_2{=}CH{-}CH_2{-}O{-}C(=O)CH_3$	$C_5H_8O_2$	535[a]	1	9.063	9.063	100
$CH_2{=}CH{-}CH_2CH_2C(=O)OH$	$C_5H_8O_2$	535[a]	1	9.063	9.063	100
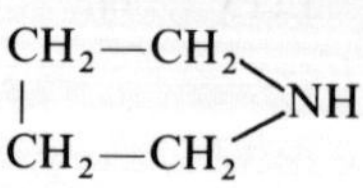	C_4H_9N	35	1	5.130	5.130	100

Table 13.16 (*Contd.*)

Structure to be found	Brutto formula	N	n	$I_0 = \log_2 N$	I	(I/I_0) (%)
1	2	3	4	5	6	7
CH_3—O—CH_2CH_2—C≡N	C_4H_7NO	762	3	9.573	7.989	83
CH_2—CH_2 |　　　　>$CHCH_2OH$ CH_2—O	$C_5H_{10}O_2$	266[a]	10	8.055	4.734	59

[a] The number of relatively stable isomers for a given Brutto formula.

able. In such cases an analysis of the entire set of vibrational frequencies on the basis of appropriate computation methods and a comparison of theoretical and experimental spectra may prove to be promising.

Additional information about problems which have been discussed above can be found in reference 10 and some new ideas have been published in reference 11.

13.3 Mathematical Formulation of the Inverse Spectral Problem for Continuous (Functional) Models

In the last section we presented the general principles for solving the inverse spectral problem for discrete molecular models. We shall now consider the mathematical formulation of the corresponding problem for continuous (or functional) models. Examples of the inverse spectral problem are finding the values of Coulomb and resonance integral matrix elements or the value of the force-constant matrix. If these quantities are, by definition, functions of some parameters, then the values of these parameters must be found. As a basis for the solution, we have a collection of absorption or emission band maxima in the electronic spectra and the frequencies of the absorption bands or emission lines in the infra-red and Raman spectra, the spectra being for a set of isotopic modifications, isomers and closely related members of a homologous series. The assumption made in the above is that the relevant overlap-integral and kinematic-coefficient matrices are known and initial values of the H and U matrices are approximated, so that the number of parameters to be determined, is, under all conditions, less than or equal to that of given experimental values. This condition clearly makes the formulation of the inverse spectral problem mathematically precise and corresponds to sensible physical requirements.

An important situation that makes the inverse spectral problem solvable in principle when dealing with molecular spectra is that the empirical parameters introduced are transferable. These parameters are the Coulomb and resonance integrals and force constants in a series of molecules containing the same structural elements. It follows from the theory of force constants and Coulomb and resonance integrals that these parameters are basically associated with the properties of the given group and its immediate neighbourhood. In other words, only small-scale interactions appear in them for polyatomic molecules. Thus if a structural element and its immediate neighbourhood is retained over a series of molecules then we can expect that the numerical values of the relevant parameters will also be retained. This is observed in practice, and has been established on the basis of many calculations.

The values of the Coulomb and resonance integrals, as well as those of the force constants of individual bonds and groups within a polyatomic molecule, are conserved when the groups are transferred from one molecule to another is a consequence of the well-known fact that there is an essential additivity when a polyatomic molecule is formed, this being particularly true of organic molecules. This additivity is seen in many of the properties of molecules, particularly in the widely known retention of bond lengths, bond dipole moments, valency angles, etc. It may happen that some individual bonds are not additive within a molecule because of the presence of strong conjugation, but whole structural groups are additive. Thus the benzene ring, for example, is incorporated into a large number of molecules almost without change, and the force constants or Coulomb and resonance integrals that were obtained for the benzene molecule can be used, also almost without change, in molecules such as toluene and azobenzene.

Additivity and transformability are basic facts which enable us to reduce considerably the number of unknown parameters in the H and U matrices when solving inverse spectral problems. Hence, every time we consider a molecule we try to rely on the experience of molecules previously studied, and to transfer the individual values of the empirical parameters almost without change from related molecules into the matrices that are to describe the molecule under consideration. As a result, the number of unknowns that remain to be determined by solving the inverse spectral problem can, in the great majority of cases, be reduced to less than (or at least to equal to) the number of observable transitions in the electron or vibrational spectrum of the molecule.

We must note here that when the inverse spectral problem is being discussed for the vibrational spectra of polyatomic molecules we have, as a rule, a large number of observable frequencies in the absorption or Raman vibrational spectra for each molecule, and especially for a series of them containing the same structural elements. There may be as many as several hundred of these frequencies for homologous series. Since the vibrational frequencies depend on the force constants, the masses of the atoms and the molecule's geometry, we may still obtain different vibrational frequencies and different inverse-spectral equations, even given the same set force constants and unknowns requiring inverse-spectral problem solution if, in addition to the basic molecules, we are dealing with its isotopic substitutions and isomers.

All this provides much experimental data to the investigator, which makes the inverse spectral problem unambiguously solvable in principle. The situation is much worse *vis-à-vis* the theory of molecular electronic charge clouds. As a rule, we only observe a small number of the possible electronic transitions. At the best of times, only three to five transitions can be observed and used as experimental reference values for each individual molecule. As is the case with vibrational spectra, the simultaneous study of a large number of related compounds enables us to increase the quality of experimental material. However, unlike an investigation of vibrational spectra we can not use isotopically substituted molecules, because their electronic spectra are identical to the basic molecule, and the same is also often true for isomeric modifications. This makes solution of the inverse spectral problem for polyatomic molecules' electronic spectra significantly more difficult. There are, however, a number of ameliorating factors.

It can be seen from the physical meaning of the Coulomb and resonance integrals and from their estimation formulae (see Chapter 3) that their values are determined by the type of atom in the molecule and the type of chemical bond, i.e. on the fine details of the structural elements that form the molecule. Hence, the number of Coulomb and resonance integrals that differ significantly from one another for organic compounds is found to be less than that of significantly differing force constants, which vary greatly because there is a variety of structural elements. Thus the table of Coulomb and resonance integrals, obtained from semi-empirical calculations of the molecular electronic cloud from a large number of compounds, only contain some 20 or 30 different parameters. At the same time, the known force constants number several hundred. It can be said, therefore, that even though we have much poorer experimental material for solving the inverse spectral problem for electronic spectra than we do for vibrational ones, we do not think that the number of unknowns is significantly greater.

In order to clarify the various features of the solution of the problem we shall conduct our discussion in three stages, starting with a consideration for one molecule, then for molecules and their isotopic modifications and finally for a homologous series of molecules. To make our discussion more practical we shall assume that all the electronic levels are known to us and that all the vibrational frequencies of all the molecules (i.e. the eigenvalues of the appropriate problem) are also known.

A natural system of equations for determining the force constants (or the parameters on which the constants depend) and the respective Coulomb and resonance integrals is that based on the relationships between combinations of the squares of eigenvalues for the respective quadratic forms and the sums of the principal minors of the H and D matrices.

The sums of all the possible principal minors for every square matrix and, in particular, for matrices forming quadratic forms are invariant with respect to changing variables in the form. Since the eigenvalues of the relevant problem appear along the diagonal when the matrix of diagonal form is being diagonalized, we find that the sum of all the possible products of the eigenvalues of the quadratic form, each product being one of the permutations of m factors from n

diagonal elements, is equal to that of all the principal minors of the quadratic form's original nth order matrix. We thus obtain the following simple result for the first-order minors:

$$\sum_{k}^{n} \lambda_k = Tr(\text{or } Sp)A$$

and the following for the second-order minors:

$$\sum_{k,k'}^{n} \lambda_k \lambda_{k'} = \sum_{k,k'} M_k^{(k')}$$

In general, for the mth-order minors

$$\sum_{k} \prod_{m} \lambda_m = \sum_{k} M_k^{(m)} \tag{13.5}$$

Finally, for the nth-order minor, which is simply the full matrix, we obtain

$$\prod_{k}^{n} \lambda_k = \text{Det } A \tag{13.6}$$

Equations (13.5) and (13.6) contain the eigenvalues of the submatrices obtained when diagonalizing the quadratic form (they can be determined experimentally) on their left-hand sides and combinations of their elements on the right-hand sides. They form the natural system of equations for determining the unknown parameters when solving the inverse spectral problem. The unknown force constants and Coulomb or resonance integrals will be included on the right-hand sides, raised to the first power if $m = 1$, the second power if $m = 2$, etc. Finally, in equation (13.6) they are raised to the power of n, n being the order of the matrix $A = TU$ for the vibrational spectra and the $A = S^{-1}H$ matrix for the electronic spectra.

The number of equations for one molecule equals the order of the matrix, n and hence the system can only be used to determine n unknown parameters or less. The number of solutions of the system of non-linear equations will equal the product of the orders of all the equations making up the system. Obviously, it may happen that the solutions obtained may not all be real, and hence some must be disregarded. All the same, the number of possible solutions for all real cases will still remain very large, especially for higher-order matrices. Even if the number of unknowns is lower than the order of the matrix (this can happen by selecting relevant initial approximations of the force constants, Coulomb or resonance integrals), then we may still obtain several possible solutions due to the non-linearity of the system of equations. Not all these solutions may have a real physical sense, nor will they all come within the bounds set by estimates made initially using quantum-mechanical calculations or on the basis of an analysis of the data in the literature concerning the values of the corresponding parameters. Hence some other solutions should be disregarded. This weeding out can be applied from the start of the calculation process if the unknown parameters are only searched for within the range of values the parameters can take. This is always done when looking for actual solutions of the inverse spectral problems. Unfortunately, even under these conditions the inverse spectral problem may not have a unique solution.

We shall now assume that when studying vibrational spectra we can synthesize any isotopic modification of the molecule and measure its vibrational frequencies. Hence the analogous equations for the isotopic modifications must be added to the equation system (13.5 and 13.6), but many dependent equations will be encountered among the additional equations. Thus if only one of the N atoms in the molecule is successively substituted for by their isotopes, then only two of the $\sum_k \lambda_k = Tr D$ equations, one for the basic molecule and the other for one of the isotopically substituted molecule, will be independent. Only three of the $\sum_k \lambda_i \lambda_j = \sum M_k^{(2)}$ equations for three of the isotopic modifications will be independent. The equation $\prod_k \lambda_k = \mathrm{Det}\, D$ is only generally independent for one molecule. As a result, it is clear that if all a molecule's atoms but one were substituted one by one and successively by one of its possible isotopes, then we would obtain N independent first-order equations of the form $\sum_k \lambda_k = Tr\, D$ (here N is the number of atoms in the molecule). These first-order equations all form an equation system linear with respect to the unknown force constants. By solving this system we can find a unique set of force constants if the number of unknowns does not exceed N. No other set of isotopic molecules will give any more new independent first-order equations. Consequently, if the number of elements to be determined in the force-constant matrix exceeds that of the atoms in the molecule it will be impossible to reduce the inverse spectral problem to a system of linear equations and thus obtain a unique solution. However, this is precisely the situation that occurs for the simplest force functions in a molecule in which only the diagonal elements of the force-constant matrix (the number of which is $3N-6$ for an N-atom molecule) are conserved.

If the molecule has symmetry elements, some of its force constants will be the same. However, even in this case the possible set of isotopically substituted molecules for non-linear and non-planar molecules does not enable us to determine uniquely more than N^{equ} different force constants. Here, N^{equ} is the number of different sets of equivalent atoms in the molecule. Remember that independent equations are only obtained for symmetric molecules for one-off substitutions of the whole of a given set of equivalent atoms by the same isotope, while partial substitutions within a set of equivalent atoms do not yield any new independent equations.

We come therefore to our first important conclusion, which is that the inverse spectral problem does not have a unique solution, even for the simplest of diagonal force constant matrices and for the general case of non-linear and non-planar molecules. This is true even if all the possible isotopic modifications of the molecules are utilized and unless the number of unknowns can be reduced to N (or N^{equ} for symmetric molecules) by using some additional physical considerations.

In the case of linear or planar molecules there are additional relationships between combinations of squares of the frequencies of the isotopic molecules. In addition, the whole problem can be solved separately for linear and non-linear vibrations and for planar and non-planar vibrations. Thus, given an XYZ-type linear molecule, for example, the set of squared frequencies for the isotopic modifications X^*YZ and XY^*Z together with those for the unsubstituted molecule enables us to compile a system of linear equations of the $\sum_k \lambda_k = Tr\,D$ type for all these molecules and to find a unique set of three force constants for the matrix associated with the linear vibrations. However, this is more often the exception than the rule.

If more than N (or N^{equ}) unknown force constants need to be determined, then we must include the quadratic

$$\sum_{i,j} \lambda_i \lambda_j = \sum_{i,j} M_{ij}^{(2)}$$

equations in the system alongside the linear $\sum_k \lambda_k = Tr\,D$ equation for the isotopic modifications. All the possible atomic substitutions into the molecules should, in principle, only yield an addition $N(N+1)/2$ independent equations.

In fact, the

$$\sum_{i,j} \lambda_i \lambda_j = \sum_{ij} M_{ij}^{(2)}$$

equations can be written by taking into account that the second-order principal minors are second-order homogeneous functions of the reciprocals of the dimensionless masses of the atoms. Hence we obtain

$$\sum_{i=1}^{N} \alpha_i \varepsilon_i^2 + \sum_{i>j}^{N} \beta_{ij} \varepsilon_i \varepsilon_j = \sum_{k,m} \lambda_k \lambda_m \tag{13.7}$$

The α_i and β_{ij} coefficients here are combinations of the force constants and geometric parameters of the molecule, and are not changed during isotopic substitutions. The number of α_i coefficients is equal to N, while there are, respectively, $(N^2 - N)/2 = (N-1)N/2$ of the β_{ij} coefficients. The overall number of α_i and β_{ij} coefficients is therefore $N(N+1)/2$.

If each of the atoms of a molecule is successively substituted for by an isotope, then another, the $(N+1)$th, independent equation can be obtained in order to determine the α_i and β_{ij} coefficients. Another N equations obtained by substituting each of the atoms by isotopes of another mass can be added to the system. A further substitution of atoms one by one will not lead to any more new equations. However, new equations can be obtained if whole groups of non-equivalent atoms are substituted for by differing isotopes.

If the total number of different isotopes of the molecule obtained in this way is $N(N+1)/2$, then no other isotopic modification will yield more new information about the force constants.

Adding the $N(N+1)/2$ second-order equations to N first-order ones gives us a total of $N(N+3)/2$ independent equations. This enables us to determine $N(N+3)/2$ unknown force constants. However, there are clearly 2^m different solutions to the system, where $m = N(N+3)/2$. Once again, there will not be a unique solution, and there will be considerable arbitrariness. The number of independent equations can be further increased by utilizing the

$\sum \lambda^m \cdots \lambda = \sum M^{(m)}$ type equations for $m \geqslant 3$ for the various isotopic molecules, but since the equations will be third order or higher with respect to the unknown force constants, this will greatly increase the number of different possible solutions to the inverse problem.

If the number of atoms in the molecules is fairly large then the number of possible different solutions to the inverse problem will become so large that it will be intractable unless additional physical considerations are used to limit the permissible solutions to as small a group as possible.

In practice, an investigator has only a limited number of isotopically substituted molecules available. Moreover, essentially different equations are only obtained when compounds containing deuterium or tritium are used. Even so, the proportional shifts in the frequency obtained due to deuteration and the corresponding equations, for solving the inverse problem may not be very different, due to the significant characterization of the internal vibrations of groups with terminal hydrogen atoms.

The situation is consequently very difficult. The way forward is therefore to use related members of a homologous series together with the basic molecule and its isotopically substituted modifications. We shall also assume* that the force constants of the separate bonds, valency angles and atomic groups are either retained unchanged over a series of molecules or are related to one another via known functions. Hence the force constants can be expressed in terms of a set of independent parameters which are common to all the molecules.

In order to determine the force constant or unknown parameters we must compile a system of equations on the basis of equations (13.5) and (13.6), having used the frequencies of a series of molecules that are close in structure and those of the isotopic modifications.

To obtain a unique solution to the inverse spectral problem under these conditions it is necessary and sufficient that the set of $\sum_k \lambda_k = Tr\, D$ equations taken from every molecule forms a system of consistent linear equations, so that its order equals the number of unknown force constants or parameters. It is clear that this may be attained in practice if and only if the number of atoms in the molecule is not very large and, moreover, when the molecule possesses a significant amount of symmetry, so that the secular equations have low degree. These considerations compel us to start determining the force constants from the simplest molecules in a homologous series and to go on from the simpler molecules to the more complex ones. When the force constants are to be determined in large molecules an investigator must already have available data on the force constants of related compounds that he can use to choose a satisfactory initial approximation to the force constant matrix as well as those constants that he will determine more accurately. As a rule, what is then being discussed is the refining of the force constants within certain limits around their initial approximations. Moreover, the choice of both the initial approximation

*This assumption is the major weakness of semi-empirical theories and, to a large extent, limits their use to organic molecules.

and the force constants to be refined, as well as the success of the inverse problem's solution, all depend significantly on the experience and intuition of the investigator. Experience has shown that when a large collection of related molecules is studied sequentially, values for the force constants of the bonds, angles and other structural units can be chosen so as to provide an accuracy of calculation for the vibration frequencies of between 10 and 20 cm^{-1}. This is adequate for the majority of practical problems.

When dealing with the electronic cloud of a molecule we need not be limited to small numbers of atoms in the molecule; indeed, this is an irrelevant condition. We can consider a polyatomic molecule from the start. However, the atoms must be such that similar atoms are encountered for the most part and their bonds are chemically close structurally. The number of different unknowns will then not be too large.

Thus when considering aromatic compounds, for example, we can in the simplest approximation equate the Coulomb integrals to the same value for the whole π-electron system and equate the reasonance integrals for all the C—C bonds. Hence only two unknown parameters are left overall and these can be used to describe the spectra of a large number of different types of aromatic compounds. The inverse spectral problem is clearly easily solvable in this case.

In practice, numerical solution of the inverse spectral problem using equations such as (13.3)–(13.6) is only possible for matrices no higher than third or fourth order. With higher orders the accuracy of solution falls off catastrophically. This is due mostly to the fact that numbers are used in the equations that differ considerably in value.

Another method has now become widespread and is based on the method of least squares. It has become popular recently because it is fully automatic and can be implemented simply on a digital computer. We shall now describe it.

We shall assume that we know an initial approximation of the original matrix. First, the direct spectral problem is solved with a given set of appropriate elements for the basic modifications (if we are dealing with vibrational frequencies) and for a close member of the homologous series. As a result, a set of energy level values are squares of vibration frequency is obtained which we shall call λ_i. These values are then compared with the experimental ones for the electronic energy levels or vibration frequency squares. One of the following may occur:

(1) One of the λ_i^{cal} lies close to one of the λ_i^{exp};
(2) More than one of the λ_i^{cal} lies close to some of the λ_i^{exp}; or
(3) None of the λ_i^{cal} lie close to the λ_i^{exp}.

We shall assume that the number of experimental values λ_i^{exp} is greater than, or at least equal to, the number of parameters to be refined and varied.

Considering the first case, we expand the λ_i^{cal} for all the molecules into a series about the initial approximation and truncate the series to include linear terms only. We thus have

$$\lambda_i^{\text{cal}} = \lambda_i^{(0)} + \sum_k \left(\frac{\partial \lambda_i}{\partial x_k} \right)_0 \Delta x_k \qquad (13.8)$$

Here, x_k is a parameter being varied and Δx_k is the increment to the parameter. If the $\lambda_i^{(0)}$ is close enough to the λ_i^{exp} then the x_k parameter or the Δx_k may clearly be varied until λ_i^{cal} equals the λ_i^{exp}. If the number of λ_i^{exp} equals that of unknown x_k (or Δx_k), then the problem reduces to the solution of a system of linear heterogeneous equations of the form

$$\lambda_i^{\text{exp}} - \lambda_i^{(0)} = \sum_k \left(\frac{\partial \lambda_i}{\partial x_k} \right)_0 \Delta x_k \qquad i = 1, \ldots, n \qquad (13.9)$$

In matrix terms this is

$$J \Delta x = \Delta \lambda \qquad (13.10)$$

Here, Δx is the column matrix of the unknowns, J is the square matrix (Jacobian) containing the $(\partial \lambda_i / \partial x_k)_0$ elements and $\Delta \lambda$, the column matrix containing the elements, is $\Delta \lambda_i = \lambda_i^{\text{exp}} - \lambda_i^{(0)}$. The $(\partial \lambda_i / \partial x_k)_0$ derivative may be found numerically.

If the number of parameters being varied is less than that of λ_i^{exp} (this being the situation in most cases) then the equation system becomes redundant and inconsistent. The latter is due to inaccuracies in the experimental measurements of λ_i^{exp} and to those in the theory itself, incurred by the whole series of assumptions that have been made so far. It is thus impossible to find any linear extrapolations of the increments Δx_k that would satisfy all equation (13.10), even if the λ_i^{cal} values lies close to the λ_i^{exp}. We can then assume that $\lambda_i^{\text{cal}} = \lambda_i^{\text{exp}}$. The problem thus becomes one of finding increments to the unknowns that will best fit all the system's equations. The concept of 'best fit' can be interpreted in different ways, and we shall use the condition applied in the least squares method. The best set of Δx_k is then considered to be the one for which the sum taken over all equations of the squares of the deviations of the left- and right-hand sides of equation (13.10) is minimum. We find from this that the best set of Δx_k values are those that correspond to the minimum of the function

$$\Phi(\Delta x_1, \ldots, \Delta x_m) = \sum_i (\lambda_i^{\text{cal}} - \lambda_i^{\text{exp}})^2$$

and hence satisfy the system of equations

$$\frac{\partial \Phi}{\partial x_k} = \frac{\partial}{\partial x_k} \left[\sum_i (\lambda_i^{\text{exp}} - \lambda_i^{\text{cal}})^2 \right] = 0 \quad k = 1, \ldots, m \qquad (13.11)$$

Expanding equation (13.11) gives

$$\sum_i \left[\Delta \lambda_i - \sum_k \left(\frac{\partial \lambda_i}{\partial x_k} \right)_0 \Delta x_k \right] \left(\frac{\partial \lambda_i}{\partial x_k} \right)_0 = 0 \qquad (13.12)$$

In matrix notation equation (13.12) is written as

$$\tilde{J}J\Delta x = \tilde{J}\Delta\lambda \qquad (13.13)$$

By solving equation (13.13) we obtain the desired increments Δx_k.

As was the case above when the number of equations equals the number of unknowns, the new Coulomb and resonance integral or force constant matrices obtained using these increments may result in values of the new approximations of λ_i^{cal} that lie:

(1) Close enough to the λ_i^{exp} to satisfy the accuracy of solution of the inverse spectral problem (i.e. coincide with the λ_i^{exp});
(2) Closer to the experimental energy levels or squares of vibrational frequency than do the respective initial approximations;
(3) Further away from the λ_i^{exp} than do the initial approximations.

In the first and second cases the procedure should be either stopped or applied again. The results of case (3) indicate that linear extrapolations are not possible, and the approach suggested by Levenberg[12] should be adopted. Formally, this means that equation (13.13) is replaced by

$$(\tilde{J}J + \beta I)\Delta x = \tilde{J}\Delta\lambda \qquad (13.14)$$

where I is the identity matrix and β is a positive number. The introduction of this number enables us to reduce the step for each iteration whilst retaining linear extrapolation. At each iteration in this procedure the search is not for the minimum of the function $\phi(\Delta x) = \sum_i (\lambda_i^{\text{exp}} - \lambda_i^{\text{cal}})^2$ but for those Δx_k which reduce the $\phi(\Delta x)$ function. It can be shown theoretically that, given a sufficiently large β, this method of trial calculation can always reduce the sum of the squares of the differences between the λ_i^{exp} and λ_i^{cal} to below that of the initial approximation. Levenberg's form of least squares always yields a solution, even when the original approximation is poor and when it is not possible to go from the $\lambda_i^{(0)}$ to λ_i^{exp} via a simple linear extrapolation.

All this refers to the situation when only one λ_i^{cal} value lies in the neighbourhood of each λ_i^{exp}, i.e. the initial approximation for the parameters being sought for is reasonably good. If the initial approximation has not been well chosen or if the λ_i^{exp} lie very close together, then a certain indeterminacy arises. It then becomes unclear which λ_i^{exp} value should be identified with which λ_i^{cal}. Thus the indeterminacy is one of principle in the general case and cannot be overcome. We will obtain different solutions of the inverse spectral problem for each different collection of λ_i^{cal} and λ_i^{exp} values. The choice of the proper set can only be made on the basis of the data for the possible ranges that the force constants or Coulomb or resonance integrals must lie within, or by verifying the suitability of the sets of Δx_k, obtained for the various permutations of λ_i^{cal} and λ_i^{exp}, for calculating the spectra of related compounds. The procedure for solving the inverse spectral problem using equations (13.5) and (13.6) is free from this major defficiency. However, in practice, it is only suited to very low-order matrices. In turn, the method of identifying λ_i^{cal} with λ_i^{exp} may be used effectively, even when the vibrational frequencies or electronic energy levels have not all been measured experimentally, whereas all the information on the energetic spectra of the relevant operators must be available in order to compile equations (13.5) and (13.6).

Note also that, even though it is formally permissible for the number of unknown Δx_k to be only slightly less than that of the experimental (measurable) frequencies when applying the least squares method, it may not in fact yield a stable solution. As a result, the addition of new experimental frequencies or energy levels may lead to the solutions of the inverse spectral problem being considerably changed. Hence, in dubious cases it is best to solve the problem for different sets of experimental values and then, by comparing the solutions obtained, to clarify their level of credibility.

When a molecule has symmetry and degeneracy appears, the inverse problem must be solved for each symmetry type separately. When random degeneracy appears, the fact that the $(\partial\lambda_i/\partial x_k)_0$ derivative is indeterminate must be taken into account. It is then necessary to take a sum of the derivatives by allowing for all the components of the degeneracy.

We have another important comment to make here. It has already been demonstrated that for practical solutions of the inverse spectral problem we are not in fact determining the elements of the H or U matrices, when relatively nothing is known about them, but we are only refining them when approximate values are available. Moreover, as a rule, permissible ranges can be obtained for the changes in the desired quantities by relying on physical arguments or on data for similar molecules. It is therefore convenient to supplement the least squares method described with a system of inequalities that limits the possible changes in the parameters being sought in the H or force-constant metrices. Experience has shown that this approach simplifies the solutions of inverse problems and enables the least squares method to be applied, even when dependent co-ordinates are being used.

An important advantage of the least squares method is that it can be used to solve simultaneously the inverse problem for several molecules. It is clear that the system of normal co-ordinates compiled simultaneously for a series of molecules does not change its form. We have only one essential comment to make. As the inverse problem is being solved iteratively using the method described above it is essential that the initially chosen set of λ_i^{cal} and λ_i^{exp} values is justified in order for the best convergence. Since this is difficult to do for almost identical vibrations frequencies (although this is a fairly special case) simply by relying on the values of the λ_i^{cal}, the best approach is to compare the appropriate eigenvectors at each iteration stage.

Another merit of the least squares method is that it is possible to calculate, in addition to the corrections to the parameter x being sought, the mean square error of these corrections δx_k, i.e.

$$\delta x_k = \sqrt{\left[(\tilde{J}J)_{kk}^{-1}\frac{\Phi}{n-m}\right]} \tag{13.15}$$

where Φ is the sum of the squared discrepancies in the λ and $(\tilde{J}J)^{-1}_{kk}$ is a diagonal element of the inverse matrix of the system of normal equations. The δx_k have a defined statistical meaning if the inconsistency of the system of normal equations is due to the errors in the measurements of $\lambda^{(\exp)}$ and these errors are normally distributed. Hence, if the mathematical expectation of the corrections Δx_k is given as $E(\Delta x_k)$, then

$$[\Delta x_k - E(\Delta x_k)]/\delta x_k$$

will have a distribution with $(n - m)$ degrees of freedom, where n is the number of vibration frequencies used in the estimate and m the unknowns being sought.

As a rule, however, the indeterminacy in the system of normal equations is not just due to measurement errors in the experimental energy levels or vibration frequencies but also to the use of approximations. This complicates the interpretation of the statistical sense of the δx_k values and the construction of correct ranges for the selected parameters. However, even in this case the least squares method yields a value for Δx_k with a minimum error and also estimates this error.

In practice, when a least squares treatment is applied to the solution of inverse spectral problems it is inevitable that badly conditioned $(\tilde{J}J)$ matrices are sometimes encountered. The reason for this may be first, that the derivatives (with respect to some of the parameters being varied) of all the frequencies being compared are close to zero. Second, there may be a very much smaller number of frequencies for some parameter on which they depend than for the other parameters. Finally, there may be some dependency between the frequencies being compared. The first two reasons are discovered when large jumps in the increments to the parameters appear from one iteration to another. It is then necessary to reformulate the problem or avoid the jumps using Levenberg's method. In the third case the casting of the problem must be re-examined. These problems can be avoided if the $(\partial\lambda_i/\partial x_k)_0$ derivatives are calculated before starting on the solution and then analysed to ensure that the problem is correctly cast.

The least squares method and the solution to the inverse spectral problem in the variant described need not only be used to determine the elements of the H and U matrices but also to refine the coefficients in the expansions of potential surfaces into a series in vibrational co-ordinates. This can be done if both the frequencies of the basic transitions and those of the corresponding overtone and combination transitions are known and if the variational method of solving the anharmonic problem is used in the form described in Chapter 10. We then obtain a matrix that corresponds to the anharmonic problem, each of whose elements depends linearly on the coefficients of $(\partial^2 V/\partial q_i \partial q_j)_0$, $(\partial^3 V/\partial q_i \partial q_j \partial q_k)_0$, etc. in the series expansion of the potential function in vibrational co-ordinates. The anharmonic energy levels will vary with the coefficients. By selecting the coefficients, using the least squares method in a way similar to that just described, we can make the calculated and experimental energy levels agree, and hence determine the anharmonic constants that characterize the relevant potential surface. Practical calculations have shown this method to be quite effective. Note that the large number of observable energy levels makes it possible to solve the inverse problem for one molecule.

To sum up overall, we can say that if the solution to the direct problem admits a complete formalization, is unique and can be completely programmed into a computer, then that to the inverse spectral problem may also be completely formalized at different stages, even though its logical complexity (especially when the initial approximation, the number of parameters to be varied, their places in the H and U matrices and the collections of λ_i^{cal} and λ_i^{exp} are all being chosen) may require the investigator to participate constantly in the procedure and to control it. Hence, it would not appear to be desirable to translate the whole problem into a computer program, as this would lead to unproductive losses of machine time.

The solution should therefore be under the continuous control of an operator and be conducted interactively. This makes the inverse spectral problem very difficult to solve, because it cannot be completely reduced to a purely calculational problem. However, despite the difficulties, the inverse problem has been solved for a large number of molecules. At present, there is a wide set of empirical values for the Coulomb and resonance integrals as well as for the force constants for separate bonds, angles and atomic groups, all of which are suitable for calculating the spectra of a large number of organic and inorganic molecules.

When the inverse spectral problem is being solved the choice of statistical weights for the experimental values plays an essential role. However, it is suggested that in all cases these be set to unity, and this topic will now be discussed. We do this because we do not have any unambiguous physical criteria at present which enables the weights to be chosen in specific cases.

13.4 The Inverse Electro-optical Problem

The determination of electro-optical parameters from the experimentally measured intensities and polarizations of absorption bands (the reverse electro-optics problem) will now be considered.

Principally, the problem is as follows. Generally speaking, we can find two quantities from our experiments. First, there is the derivative of a molecules's dipole moment with respect to a normal co-ordinate (although the sign is indeterminate, because the intensity of an absorption band is proportional to the square of the derivative). Second, we can find the direction of the vector $(\partial\vec{\mu}/\partial Q_i)_0$, i.e. the polarization. Moreover, the forms of the vibrations can be considered known from a solution of the mechanical problem. We proceed from

$$\left(\frac{\partial\vec{\mu}}{\partial Q_i}\right)_0 = \left\{\tilde{\vec{e}}\left[\frac{\partial\mu}{\partial q}\right] + \tilde{\mu}\left[\frac{\partial\vec{e}}{\partial q}\right]\right\} l_q^{(i)}; \quad \left[\frac{\partial\vec{e}}{\partial q}\right] = S^{-1}(\Delta E\tilde{\vec{B}}\Theta - \vec{E}') \quad (13.16)$$

The elements of the matrices $[\partial\mu/\partial q]$ and $\tilde{\mu}$ form the set of electro-optical parameters we are looking for. The number of derivatives of the bond dipole moments with respect to vibrational co-ordinates is equal, for molecules not containing ring structures, to $(3N-6)(N-1)$, and that of the bond dipole moments is equal to $(N-1)$. Consequently, the total number of unknowns is $(3N-5)(N-1)$, N being the atoms in the molecule in molecules not possessing any symmetry elements. By substituting the appropriate experimental values for $(\partial\vec{\mu}/\partial Q_i)_0$ and the form of the vibrations $l_q^{(i)}$ into equations (13.16) we obtain a vector equation in terms of electro-optical parameters that is equivalent to a system of three independent scalar equations.

In order to transform to scalar equations it is possible to introduce either a single system of rectangular co-ordinates for all the vibrations or a separate one for each vibration in such a way that one axis coincides with the direction of polarization. The choice of the working system of co-ordinates is dictated solely by convenience of calculation. Both sets of three scalar equations thus obtained are essentially equivalent, because we can go from one to the other simply by changing the co-ordinate system.

The original vector equations of equation (13.16) are independent for different normal co-ordinates, hence the corresponding scalar equations will also be independent. By using all the normal vibrations of the molecule we will obtain $3(3N-6)$ scalar equations of the form

$$(\partial\vec{\mu}/\partial Q_i)_{0\alpha}^{\exp} = \sum_{k,n} a_{kn}^{(i)}\left(\frac{\partial\mu_k}{\partial q_n}\right)_0 + \sum_k b_k^{(i)}\mu_k, \quad \alpha = x, y, z \tag{13.17}$$

which will be used to calculate the electro-optical parameters. The coefficients for the unknowns $(\partial\mu_k/\partial q_n)_0$ and μ_k are $a_{kn}^{(i)} = q_n^{(i)}\vec{e}_k$ and $b_k^{(i)} = \sum_n(\partial\vec{e}_k/\partial q_n)_0 q_n^{(i)}$, respectively. They can be calculated unambiguously. It is possible to take advantage of the identity $\vec{\mu} = \sum\mu_k e_k$ as an additional equation, and this is also equivalent to three scalar equations if the direction of the dipole moment's vector direction is known with respect to the molecule. The total number of equations for determining the electro-optical parameters of non-symmetric molecule is thus $3(3N-5)$. When $N \geqslant 5$ the number of unknowns already far exceeds that of the possible equations.

Strictly speaking, the system of equations (13.17) will be consistent only when all the μ_k and $(\partial\mu_k/\partial q_n)_0$ parameters are incorporated, i.e. when all the interactions between the bonds are taken into account. It is more accurate to refrain from putting the equals sign between the experimental value of the $(\partial\vec{\mu}/\partial Q_i)_0$ derivative and an expression for it such as equation (13.16) if some of the parameters are arbitrarily discarded. True, in a number of specific cases some of the interactions can be neglected, and equation (13.17) written with a smaller number of unknowns, but each case must be specifically justified and substantiated. In some papers, μ_k and $(\partial\mu_k/\partial q_n)_0$ have been determined from systems of equations compiled using the additive approximation of the valency optical scheme. In this case, if the molecule does not in reality possess additivity, the system of equations will not be correct and consequently the parameters derived from it will not be accurate.

Sometimes the dipole moments of bonds that are derived from different vibrations (when the experimental values of the derivatives of the molecule's dipole moment with respect to the appropriate normal co-ordinates are substituted) are substantially different. We have here a clear contradiction. Since the idea of an equilibrium dipole moment in the form of the sum $\vec{\mu} = \sum_k \mu_k \vec{e}_k$ was the starting point, the μ_k must also refer to the equilibrium condition, and may not be related to a characteristic of a vibration. The reason for the contradiction is obvious. Usually, chemical bonds are not additive (although a whole atomic group does possess additivity) and the derivatives with respect to angle co-ordinates are large. Equations that allow for this take on a complete form and, as calculations have shown, the contradiction does not arise. An analogous phenomenon occurs when calculating the derivatives of a CH bond's dipole moment with respect to the extension of its own bond length in an ethane molecule using the equations for the valency vibrations of the CH_3 group. The role of the derivative with respect to adjacent bonds here is very significant. Consequently, it is essential to make it a rule to compile the equations (to be used to calculate the electro-optical parameters), taking into account all the elements of the matrix $[\partial\mu/\partial q]$. It is only possible to neglect some of them, by a choice of the form of the electric field, if there is sufficient justification for doing so. The invariability of the parameters introduced for all the vibrations of a given molecule and the various types of its isotopic substitutes may act as a criterion of the validity of the selected model. We must also say that, by using such a general approach, it is automatically decided whether a parameter can be determined individually or whether it will appear in combination with others. For instance, the equation for $(\partial\vec{\mu}/\partial Q_i)_0$ in the methane molecule, and all its deuterium derivatives, does not contain the derivative of a CH bond's dipole moment with respect to changes in its own length. The equation only contains the difference between it and the derivative of a CH bond's dipole moment with respect to the extensions of adjacent bonds. We can understand, therefore, the crude errors that can be found if this situation is not taken into account and an experimental value is used to make conclusions about the CH bond type for a given molecule, or if a parameter is compared with data from other molecules, in which the μ_{CH} derivative (with respect to the extension of its own bond) is actually encountered separately.

Let us assume that a molecule has elements of symmetry. Then the dipole moments of equivalent bonds are equal to one another, while in the matrix $[\partial\mu/\partial q]$ the rows corresponding to equivalent bonds and the columns corresponding to equivalent co-ordinates will contain similar elements, although possibly in different orders. As a result, the number of different electro-optical parameters that can be determined is reduced. However, that of the equations that can be used to calculate them decreases at the same time. In this respect, when there is a measure of symmetry the direction of polarization and that of $\vec{\mu}$ is frequently only defined for the symmetry of the molecule. In this case, the scalar equation for the projections onto the directions perpendicular to either the polarization axis or $\vec{\mu}$ reduces to identities, and so cannot be applied to calculations of the numerical values of the parameters required.*

The same can be said of inactive infra-red vibrations. Moreover, when there is symmetry the electro-optical parameters may enter all the equations in certain repeated combinations, which makes it impossible to calculate them separately. Consequently, the number of different electro-optical parameters, and the independent equations which can be used to determine them, depends on the molecule's configuration. Thus there are eight possible electro-optical parameters for the non-linear molecule XYZ. The number of parameters and equations for a C_{2v}-type molecule XY_2 is reduced to four. A tetratomic non-symmetric molecule will be characterized by 21 parameters, for which 21 independent equations can be compiled. A tetratomic molecule XYZ_2 of the C_{2v} type has 12 different electro-optical parameters that are related by seven equations. In specific cases the search for the number of electro-optical parameters and equations is comparatively simple. Hence we conclude with a few comments.

Since the derivative of a molecule's dipole moment with respect to a normal co-ordinate is

$$\left(\frac{\partial \vec{\mu}}{\partial Q_i}\right)_0 = \sum_j \left(\frac{\partial \vec{\mu}}{\partial q_j}\right)_0 l_{q^i j}$$

it is clear that, generally speaking, we can find only the quantity $(\partial\vec{\mu}/\partial q_j)_0$ from experiments. We showed in Chapter 11 that $(\partial\vec{\mu}/\partial q_j)_0$ can be represented as a sum of the derivative $(\partial\vec{\mu}/\partial q_i)'_0$, which is independent of intermittent rocking of the molecule as a whole during vibrations and a supplementary term that is related to rocking. This last term is abbreviated to $((\mu_0))(\partial\vec{\omega}/\partial q_j)_0$. Hence

$$\left(\frac{\partial \vec{\mu}}{\partial q_j}\right)_0 = \left(\frac{\partial \vec{\mu}}{\partial q_j}\right)'_0 + ((\mu_0))\left(\frac{\partial \vec{\omega}}{\partial q_n}\right)_0 \tag{13.18}$$

If a molecular dipole moment is known or equal to zero, then we can, on the basis of this formula, calculate the parameter $(\partial\vec{\mu}/\partial q_i)'_0$. This parameter, as we noted in Section 11.3, may be represented as a linear combination of the electro-optical parameters μ_k and $(\partial\mu_k/\partial q_n)_0$. As a result, we obtain a system of heterogeneous equations which, when solved, yields a system of electro-optical parameters for the molecule. We also showed, in this respect, that the number of electro-optical parameters will, as a rule, exceed the number of equations possible in principle. However, the number of unknown electro-optical parameters can be reduced by taking advantage of the properties of additivity in molecules and by setting the 'interaction' $(\partial\mu_k/\partial q_n)_0$ derivatives of bonds equal to zero.

Such a simplified description of a molecule's properties is found to be permissible in a whole series of cases, and enables us to calculate the distribution of intensities in the absorption spectra of polyatomic molecules with a reasonable degree of accuracy.

*In such cases when all the orientating vectors of a given symmetry type are collinear, the system of vector equation are replaced by a system of scalar equations.

The solution of the inverse problem of electro-optical parameters has a whole series of common links with that of the inverse problem of the calculation of force constants. Naturally, this is why the use of the spectra of isotopically substituted molecules has been attempted in order to increase the number of equations.

The influence of isotopic substitution is seen in the size of the second term of equation (13.18), since the derivative $(\partial\vec{\omega}/\partial q_j)_0$ clearly changes in the general case during isotopic substitutions.

If a molecule does not have a dipole moment, then the second term in the formula for $(\partial\vec{\mu}/\partial q_j)_0$ will be absent. Clearly, this is because the use of the spectra of isotopically substituted molecules does not, under these conditions, allow us to construct additional independent equations for calculating electro-optical parameters. However, these spectra are found to be useful for averaging and finding the relative signs of the $(\partial\vec{\mu}/\partial Q_i)_0$ derivatives.

If the molecule has a known dipole moment then the contribution of the $((\mu_0))(\partial\vec{\omega}/\partial q_j)_0$ term can be calculated for every isotopic molecule, and in this way the derivative $(\partial\vec{\mu}/\partial q_j)'_0$ can be calculated. Since this derivative is invariant with respect to isotopic substitutions, we again conclude that the spectra of isotopic molecules do not carry new information about a molecule's electro-optical parameters. If, however, the value of $\vec{\mu}$ is not known, then it is clearly possible to determine it from the data of the system of isotopic molecules. It is only in this latter case that it can be said that the spectra of isotopic molecules reveal any additional information about the values of the electro-optical parameters. In this case, there will be no more than three unnecessary independent equations.

Thus it is clear that a solution of the inverse electro-optical problem is even more difficult than solving the inverse problem for force constants.

The use of homologous series and the additive scheme does, nevertheless, allow us to select values for parameters that can predict intensities for quite complex compounds. It should, however, be noted that the calculation of electro-optical parameters suffers from much larger inaccuracies and indeterminacy than is the case for force constants as a result of the above reasons and because of experimental difficulties while measuring intensities.

In the presence of symmetry the matrix $(\partial\mu/\partial q)$ replaces that containing the derivatives with respect to symmetry co-ordinates. These derivatives $(\partial\mu/\partial q^{(s)})$, consisting of a combination of the original electro-optical parameters, will occur in the expression for $(\partial\vec{\mu}/\partial Q_i)_0$. In general, different combinations enter the expression for vibrations of different symmetries. The total number of such combinations for one molecule will be equal to the product of the number of groups of equivalent bonds and the number of symmetry co-ordinates of the infra-red active vibrations. Combinations corresponding to inactive vibrations disappear, even though they are not zero, because the products of $\tilde{\vec{e}}$ and the corresponding column matrices $(\partial\mu/\partial q^{(s)})$ must be zero.

Clearly, the expression for $(\partial\vec{\mu}/\partial Q_i)_0$ in a series of isotopically substituted molecules can always lead to a form containing only $(\partial\mu/\partial q^{(s)})$ (and combinations of parameters) for the symmetry co-ordinates of the active vibrations of any of the molecule's isotopic modifications. In order to do this it is necessary to take linear combinations of the natural co-ordinates, corresponding to the whole set of a selected molecule's symmetry co-ordinates, as the vibration ones. This can always be done using specific linear combinations. Although in this case the vibrational equations will no longer reduce to a quasi-diagonal form for isotopically substituted molecules of other symmetries, when the $\tilde{\vec{e}}\,(\partial\mu/\partial q^{(s)})$ factors for a series of molecules is multiplied by the form of the vibration expressed in terms of the symmetry co-ordinates of one molecule, then only the $(\partial\mu/\partial q^{(c)})$ for the active vibrations of the selected molecule will enter the formula for $(\partial\vec{\mu}/\partial Q_i)_0$. In this respect, when the row matrix containing the bond-orientating vectors $\tilde{\vec{e}}$ is multiplied by the column matrices $(\partial\mu/\partial q^{(s)})$ corresponding to the symmetry co-ordinates of the vibrations that are inactive for the original molecules, we obtain zeros. This procedure enables us to express the $(\partial\mu/\partial Q_i)_0$ for a series of isotopically substituted compounds in terms of the $\partial\mu_k/\partial q_m^{(s)}$ parameters of more symmetric molecules. If active degenerate vibrations exist in the infra-red spectra it is essential to introduce $\partial\mu_k/\partial q_m^{(s)}$ parameters for every pair or triplet of such co-ordinates. The orientations of the vectors of symmetry for these co-ordinates will be mutually perpendicular. This even enables us to express $(\partial\vec{\mu}/\partial Q_i)_0$ for a series of isotopically substituted molecules, given any special orientation of $(\partial\vec{\mu}/\partial Q_i)_0$, even when only degenerate co-ordinates are active in a highly symmetric molecule.

The principal difficulty when calculating electro-optical parameters lies in the indeterminacy of the relative sign of $(\partial\vec{\mu}/\partial Q_i)_0$. This is related to the impossibility of determining the relative phase of a molecule's normal vibrations. Thus we have to calculate the sign of $(\partial\mu/\partial Q_i)_0$ for several different independent combinations. Moreover, the corresponding set of electro-optical parameters has to be obtained. It is clear that combinations obtained from each other by changing the signs in front of all their derivatives simultaneously yield the same values of electro-optical parameters but with opposite signs. It follows that only the absolute value of the electro-optical parameters can be found from solutions of the inverse electro-optical problem. This obviously does not refer to their relative signs.

In future, a choice of the true combination of electro-optical parameters can be made either by using an indirect argument or by substituting the calculated parameters in isotopically substituted molecules and comparing the predicted intensities with experimental values. Homologous series can also be used for this exercise. Note that the accuracy with which solutions of the direct and inverse electro-optical problems can be obtained depends critically on that with which the forms of the vibrations have been obtained, i.e. in the final analysis they depend on the accuracy of the force-constant matrix used.

We wish to comment further on this. Note, primarily, that the signs of the μ_k parameters with respect to one another depend on the accepted directions of the vectors $\vec{e}_k$, remembering that a molecule's dipole moment is represented in the form of a sum $\sum_k \mu_k \vec{e}_k$. If the sign of one $\vec{e}_k$ is reversed, then the sign of μ_k must also be changed, because only then will the whole sum $\vec{\mu} = \sum_k \mu_k \vec{e}_k$ remain constant. This must be done so that the molecular dipole moment is independent of the choice of the direction of $\vec{e}$. The signs of all the derivatives of the dipole moment of this bond with respect to natural co-ordinates will also be changed at the same time. Thus the signs of the set of electro-optical parameters corresponding to different bonds in the molecule are determined not only by the properties of the electronic state but also by the choice of the co-ordinate vector system.

By studying the properties of the force-constant matrix it has been established that when the variables (co-ordinates) are partially substituted all the elements of the matrix are changed. We will explore what happens to the electro-optical parameters. Although μ_k must be independent of the co-ordinate system, this is not the case for the matrix $(\partial\mu/\partial q)$. We will change the co-ordinate system so that the first n co-ordinates remain as before, the remainder being changed via some linear transformation of the old co-ordinates. The relationship between the new and old co-ordinates is of the form

$$\begin{bmatrix} q^{\mathrm{I}} \\ q^{\mathrm{II}} \end{bmatrix} = \begin{bmatrix} \mathrm{I} & 0 \\ A_{21} & A_{22} \end{bmatrix} \begin{bmatrix} (q^{\mathrm{I}})' \\ (q^{\mathrm{II}})' \end{bmatrix} \tag{13.19}$$

where I is the identity matrix. We obtain the matrix $(\partial\mu/\partial q')$ in the new co-ordinate system by post-multiplying the original matrix by

$$\begin{bmatrix} \mathrm{I} & 0 \\ A_{21} & A_{22} \end{bmatrix}$$

As a result, obtain*

$$\begin{bmatrix} \dfrac{\partial\mu^{\mathrm{I}}}{\partial(q^{\mathrm{I}})'} & \dfrac{\partial\mu^{\mathrm{I}}}{\partial(q^{\mathrm{II}})'} \\ \dfrac{\partial\mu^{\mathrm{II}}}{\partial(q^{\mathrm{I}})'} & \dfrac{\partial\mu^{\mathrm{II}}}{\partial(q^{\mathrm{II}})'} \end{bmatrix} = \begin{bmatrix} \dfrac{\partial\mu^{\mathrm{I}}}{\partial q^{\mathrm{I}}} + \dfrac{\partial\mu^{\mathrm{I}}}{\partial q^{\mathrm{II}}} A_{21}; & \dfrac{\partial\mu^{\mathrm{I}}}{\partial q^{\mathrm{II}}} A_{22} \\ \dfrac{\partial\mu^{\mathrm{II}}}{\partial q^{\mathrm{I}}} + \dfrac{\partial\mu^{\mathrm{II}}}{\partial q^{\mathrm{II}}} A_{21}; & \dfrac{\partial\mu^{\mathrm{II}}}{\partial q^{\mathrm{II}}} A_{22} \end{bmatrix} \tag{13.20}$$

Hence, when the co-ordinates are altered in the general case all elements of the matrix $(\partial\mu/\partial q)$ are changed, even those corresponding to derivatives with respect to those co-ordinates which were retained. Consequently, the elements of the matrix containing $(\partial\mu/\partial q)$ terms are not invariant with respect to the system of co-ordinates selected for the calculations. In this sense, the matrix elements have much in common with force constants. However, there is a considerable difference. If a molecule has partial additivity, then, by definition, its dipole moment can be represented as the sum of the dipole moments of two (or more) groups (indexed as 1 and 2) such that the bond dipole moments of

*Here and subsequently in order to abbreviate the submatrices containing $(\partial\mu_k/\partial q_m)_0$ derivatives for the bonds and co-ordinates of some sets we shall write the submatrices as $\partial\mu^{\mathrm{I}}/\partial q^{\mathrm{I}}$, $\partial\mu^{\mathrm{I}}/\partial q^{\mathrm{II}}$, etc., without enclosing them in square brackets.

the first group are functions of the internal co-ordinates of that group and those of the second group are only functions of the internal co-ordinates of that group.

We recall that the internal co-ordinates were the whole set of a molecule's natural co-ordinates which characterize the vibrations of a given group as if it was a free molecule. The set of co-ordinates that describe the moment of one group relative to another we called external co-ordinates. The matrix $\partial\mu/\partial q$ then takes the following form for the case we are considering:

$$\left[\frac{\partial \mu}{\partial q}\right] = \begin{bmatrix} \dfrac{\partial \mu^{\mathrm{I}}}{\partial q^{\mathrm{I}}} & 0 & 0 \\ 0 & 0 & \dfrac{\partial \mu^{\mathrm{II}}}{\partial q^{\mathrm{II}}} \end{bmatrix} \tag{13.21}$$

Here, the middle column corresponds to the external co-ordinates. When the variables are altered we must bear the following in mind. Clearly, the bond-extension co-ordinates cannot be changed. This follows from the form of representing the dipole moment as $\sum_k \mu_k \vec{e}_k$, which means that only the co-ordinates that describe the positions of unconnected bonds can be varied. It is clear that varying the internal co-ordinates of an additive group is only possible via a linear transformation of these co-ordinates themselves. External co-ordinates may be substituted for by a linear combination of all the co-ordinates. Thus a transformation of the co-ordinates can be written as

$$\begin{bmatrix} q^{\mathrm{I}} \\ q^{\mathrm{ext}} \\ q^{\mathrm{II}} \end{bmatrix} = \begin{bmatrix} A_{11} & 0 & 0 \\ A_{21} & A_{22} & A_{23} \\ 0 & 0 & A_{33} \end{bmatrix} \cdot \begin{bmatrix} (q^{\mathrm{I}})' \\ (q^{\mathrm{ext}})' \\ (q^{\mathrm{II}})' \end{bmatrix} \tag{13.22}$$

In the new system we obtain

$$\left[\frac{\partial \mu}{\partial q'}\right] = \begin{bmatrix} \dfrac{\partial \mu^{\mathrm{I}}}{\partial q^{\mathrm{I}}} A_{11} & 0 & 0 \\ 0 & 0 & \dfrac{\partial \mu^{\mathrm{II}}}{\partial q^{\mathrm{II}}} A_{33} \end{bmatrix} \tag{13.23}$$

Consequently, the additive form of the matrix is not disturbed during permissible alternations of the variables, though if the co-ordinates q^{I} and q^{II} are not changed then the elements of the $(\partial\mu/\partial q)$ matrix will remain unchanged. We conclude from this that if a molecule has partial (or full) additivity the elements of the $(\partial\mu/\partial q)$ matrix are invariant with respect to the co-ordinate system.

The dependency of the electro-optical parameters on the choice of co-ordinate vectors and natural coordinates leads to some rules that are important when intensities and polarizations in the infra-red spectra of polyatomic molecules are being analysed:

(1) A comparison of the results of the estimates of the electro-optical parameters obtained by different workers for the same molecule can only be made using the same system of co-ordinates and co-ordinate vectors (or the appropriate

transformations to a common system have been made) or when the molecule has partial or full additivity.

(2) During calculations for a series of similar molecules with repeating groups it is essential that the co-ordinate systems of these groups be retained.

(3) A comparison of the electro-optical parameters of the same group in different molecules, carried out to study its properties, is only permissible when these molecules have additivity properties. This point is also obvious from the physical sense of the electro-optical parameters.

When calculating electro-optical parameters it is most convenient to proceed in two stages. First, an approximate formula for $(\partial\vec{\mu}/\partial Q_i)_0$ is compiled on the basis of the known vibrational forms and orders of magnitude of the electro-optical parameters, and taking into account the contributions of only the main elements. The equations thus obtained are used to obtain approximate values for a number of parameters and to look for the respective signs of $\sqrt{|(\partial\vec{\mu}/\partial Q_i)_0|^2}$. Then these equations are made more accurate and the remaining parameters are determined from the full system of equations. In order to simplify the calculations we recommend that the same sign be retained for the first co-ordinate in the forms of the different vibrations.

A satisfactory set of electro-optical parameters can be obtained using a least squares technique similar to that used for calculating force constants. This method, used within the bounds of the valency optical theory, yields a set of electro-optical parameters that are common to a whole series of isotopic modifications and to members of a homologous series.

Fragments of the compound under study can be studied previously as independent small molecules and their parameters introduced into the initial approximation of the electro-optical parameter system for the larger molecule. The system is now made more accurate so that the sum of the squares of the deviations of the calculated absolute intensities from the experimental intensities is minimized. We will also require that the electro-optical parameters vary within relatively narrow limits (the limits being given previously and selected on the basis of additional physical considerations) as the problem is being solved.

The latter requirement, as experience has shown, is very important, and enables us to eliminate in practice the ambiguity of the solutions that exist in principle for similar non-linear problems. Many of the atomic groups that are included in different organic compounds are almost additive, so their internal parameters only vary between certain limits from one molecule to another. At the same time, the nature of the change in the internal parameters of additive groups is in itself the object of study when a particular series of compounds is being investigated. Clearly, the formulation of the inverse electro-optical problem given above corresponds to the requirements indicated.

The electro-optical parameters of the new bonds that arise when two additive groups are combined, can be determined a new in the same way if they retain relatively wide limits of change. From the algebraic point of view, superimposing restrictions on the limits that parameters can change between is equivalent to increasing the number of independent equations from which the electro-optical parameters are determined.

The technique described above for solving the inverse spectral problem for electro-optical parameters is based on a set of given values for the absolute intensities of individual absorption bands. The technique cannot, however, be applied to complex molecules. First, measuring absolute intensities is difficult, and second, the bands repeatedly overlap, which makes separation into individual components extremely difficult, if not impossible. Meanwhile, it is precisely for complex molecules, particularly polymers and crystals, that estimating the intensities in their spectra is vital. Furthermore, any investigation of the structure and properties of polyatomic molecules via their spectra requires a comparison of their calculated and experimental spectra. As indicated in the Introduction to this book, such a comparison for a complex molecule can only be made if the whole spectral curve is calculated. This is because typically fewer bands are observed in a molecule's spectrum than there are normal vibrations, due to superposition and the presence of low-intensity absorption or emission bands. If the calculation is carried out just to determine the vibrational frequencies, the comparison of the calculated and experimental spectra will be incorrect.

In conclusion, it can be stated that electro-optical parameters, like force constants, are carriers, and retain their value for every molecule that contains given structural elements. This transferability is related to the retention of the properties of the electron cloud for each separate section of the molecules. We refer to this topic in Chapter 9. True, the transferability of the electro-optical parameters is not as good an approximation as that of force constants. However, experience of calculating a very large number of parameters for a variety of molecules has shown that it is adequate when predicting spectra to within a 10–15% error in the intensities. For most practical purposes, this is quite satisfactory.

We can, by relying on this transferability, do the following. Let there be a group of bands which correspond to vibrations in the molecule that are highly characteristic in terms of intensity (we have discussed characterizability in terms of intensity in Chapter 12). The valency vibrations of the CH bond in particular have this property. Now if we take into account that the CH bond, as a rule, has a fair degree of additivity we can take its electro-optical parameters to be the same as those found in small molecules on the basis of an investigation of absolute intensities. This now allows us to calculate the absolute values of the intensities of the absorption bands due to these bonds in a large molecule, and take them to be an internal parameter in terms of which the intensities of the other bands, and their sums, can be evaluated. Thus it is no longer necessary to measure the absolute intensities experimentally; only the relative intensities need be measured.

A further step is not just to use individual values of absolute or relative intensities of bands and their combinations but also the whole of the spectral curve. Furthermore, only at this stage can we allow for the superimposition of

Table 13.17. Absorption band halfwidths (cm^{-1}) for alkanes, alkenes and substituted benzenes

Compound	Spectral interval											
	600 – 730	731 – 830	831 – 915	916 – 975	976 – 1074	1075 – 1120	1121 – 1195	1196 – 1250	1251 – 1300	1301 – 1400	1401 – 1500	2500 – 3500
Alkanes												
Propane	8	—	7	4	10	—	5	—	—	8	10	30
Butane	8	—	—	4	—	—	—	—	10	8	10	30
Pentane	9	—	7	3	10	—	5	40	10	8	10	30
Heptane	8	3	7	3	15	40	5	40	10	8	10	30
Octane	9	3	7	—	12	35	—	40	—	8	10	30
Hexadecane	8	3	7	4	10	35	—	40	10	7	10	30
Isobutane	—	4	12	4	—	—	8	—	—	8	10	30
Neopentane	9	—	7	—	15	—	—	—	8	8	10	30

Compound	400 – 830	831 – 999	1000 – 1060	1061 – 1100	1101 – 1197	1198 – 1300	1301 – 1400	1401 – 1500	1501 – 1700	2800 – 3500
Alkenes										
Propylene	8	15	28	—	8	20	12	14	15	28
Isobutylene	8	14	20	6	—	12	10	18	20	28
Tetramethylethylene	8	14	—	—	8	—	10	15	—	28
Trimethylethylene	8	8	20	4	—	12	11	18	25	28
Cis-butene-2	8	14	20	—	8	20	10	15	20	28
Trans-butene-2	—	16	28	—	—	20	12	14	—	28
But-l-ene	8	14	28	6	8	20	10	14	22	28
3-Methyle-but-l-ene	8	14	20	—	9	20	10	14	25	28

Compound	300 – 600	601 – 700	701 – 800	801 – 890	891 – 950	951 – 980	981 – 1000	1001 – 1020	1021 – 1120	1121 – 1210	1211 – 1490	1491 – 1600	1601 – 1700	2800 – 3500
Methyl derivatives of benzene														
Toluene	4	5	8	—	20	6	18	10	21	6	13	6	5	30
p-Xylene	4	—	10	—	—	6	—	18	21	—	13	5	—	30
o-Xylene	4	—	10	12	20	4	—	10	21	—	13	7	4	30
m-Xylene	4	5	10	20	34	6	18	—	16	6	13	7	5	30
Mesitylene	4	5	—	10	20	—	18	—	17	6	13	—	4	30
1, 2, 4, 5-Tetramethylbenzene	4	5	10	10	—	5	—	10	—	30	14	5	—	30
1, 2, 3, 5-tetramethylbenzene	4	3	—	13	20	6	18	12	22	6	12	5	5	30
1, 2, 3, 4-Tetramethylbenzene	4	5	—	10	—	6	18	12	26	6	13	6	5	30
Pentamethylbenzene	4	—	9	10	—	6	18	10	21	—	14	—	5	30
Hexamethylbenzene	—	—	8	—	—	—	—	10	19	—	14	—	—	30

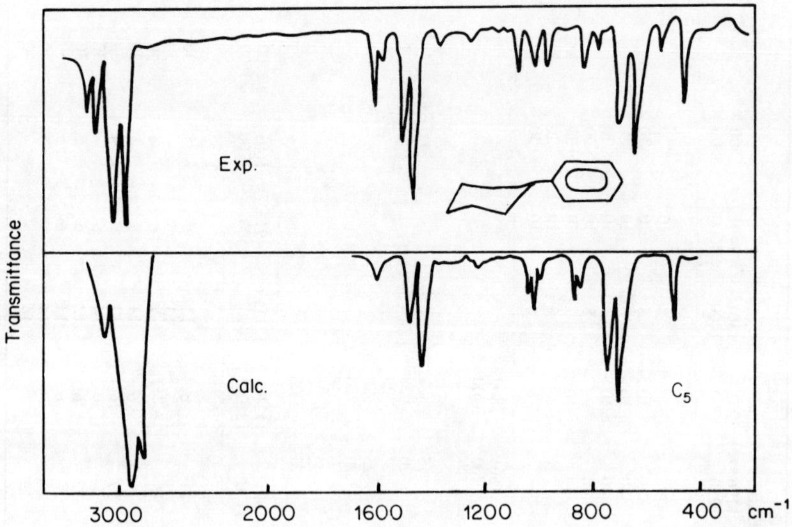
Exp.
Calc.
C_5
Transmittance
3000
2000
1600
1200
800
400
cm^{-1}

(a)

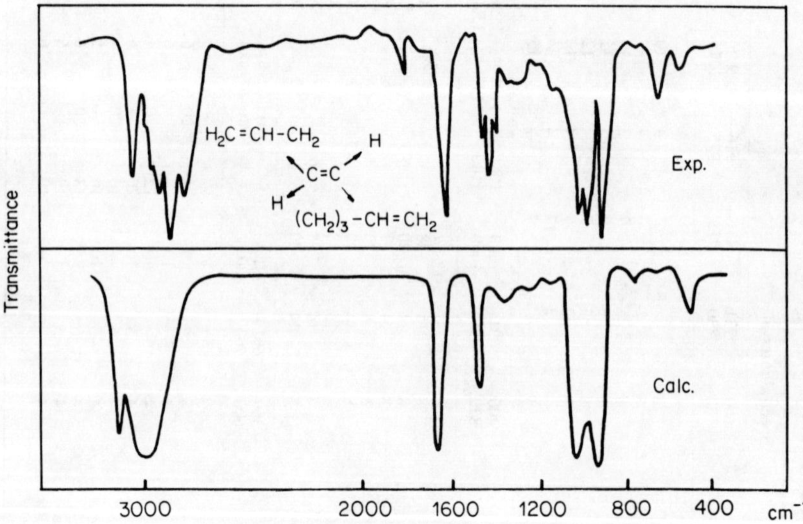
$H_2C=CH-CH_2$
H
C=C
H
$(CH_2)_3-CH=CH_2$
Exp.
Calc.
Transmittance
3000
2000
1600
1200
800
400
cm^{-1}

(b)

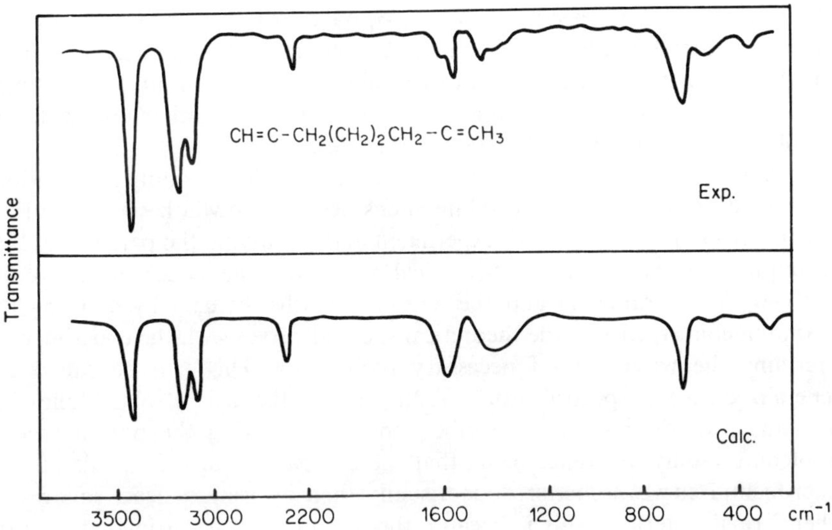

(c)

Figure 13.6. Experiment and calculated infra-red spectra for an organic molecule

absorption bands sufficiently correctly. In order to do this we need to have some values of the half-bandwidths at our disposal, and to adopt some standard model as a contour of the lines and bands. For example, we can take either the Gaussian or Lorentzian curves as the standard contours. By taking into account that the calculations have no pretence towards great accuracy, we can neglect the differences between the two curves. Thus we use the Gaussian representation as a contour of the spectral bands in the examples given here. The full spectrum can thus be represented as a sum of Gaussian bands and written analytically as

$$D(\nu) = \sum_i \frac{I_i}{2.5\sigma_i} \exp\left[-\frac{(\nu - \nu_i)^2}{2\sigma_i^2} \right]$$

where I_i is the integral intensity of an absorption band which is proportional to the square of the $(\partial\vec{\mu}/\partial Q_i)_0$ derivative, σ_i is the half-width of a band and ν_i is the band frequency.

The most convenient curve when comparing with experiments is

$$A(\nu) = 1 - \exp[-\beta D(\nu)]$$

where β is a scale factor. Variation of β corresponds either to changing the path length of the cell or the concentration of the solution.

The quantities I_i and ν_i can be calculated in the way described above. However, we do not have any quantitative theory that would enable us to calculate the value of σ_i *a priori* for an individual absorption band. This means that when calculating the whole spectral curve the σ_i have to be included as additional empirical values and adopted directly from experiment. Fortunately, it is found that these parameters also have transferability properties within known limits for different spectral ranges and for a sequence of similar molecules (a homologous series). This is illustrated in Table 13.17.

We can, by relying on similar tables, attempt to predict not only the positions of the bands in spectra or the absolute intensities but also whole spectral curves. By comparing these curves with experiment and by varying the parameters that are required to construct the theoretical curves (i.e. the force constants), the electro-optical parameters and the half-bandwidths we can, by securing the maximum convergence of the theoretical spectral curves with the experimental, determine the whole set of necessary parameters. This can be called the 'generalized inverse spectral problem'. In practice, the task is solved following the same procedure as was described above for solving the partial inverse problem, the only difference being that the comparison of the calculated and experimental curves is better done visually than by using a least squares or other formal criterion. This is because the accuracy of the estimated spectral curve that can be achieved using such an approach is not very high, although it is sufficient (see examples) for the spectrum to be interpreted and the estimates to be used to identify compounds, etc.

Although the technique that we have described for determining electro-optical parameters on the basis of an internal standard and given experimental spectral curves has now been well worked out and ensures that satisfactory results are obtained for very complex systems, it should, however, be noted that its practical

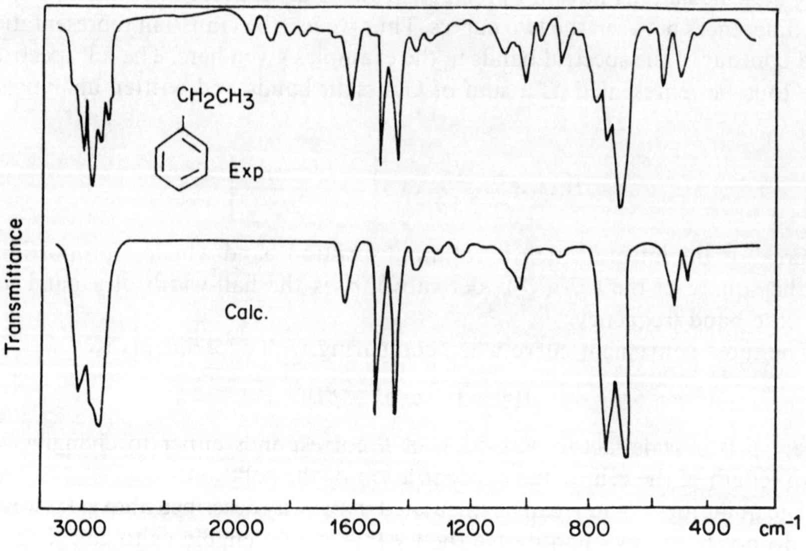

Figure 13.7. Transmission spectra for ethyl benzene: experiment and calculation

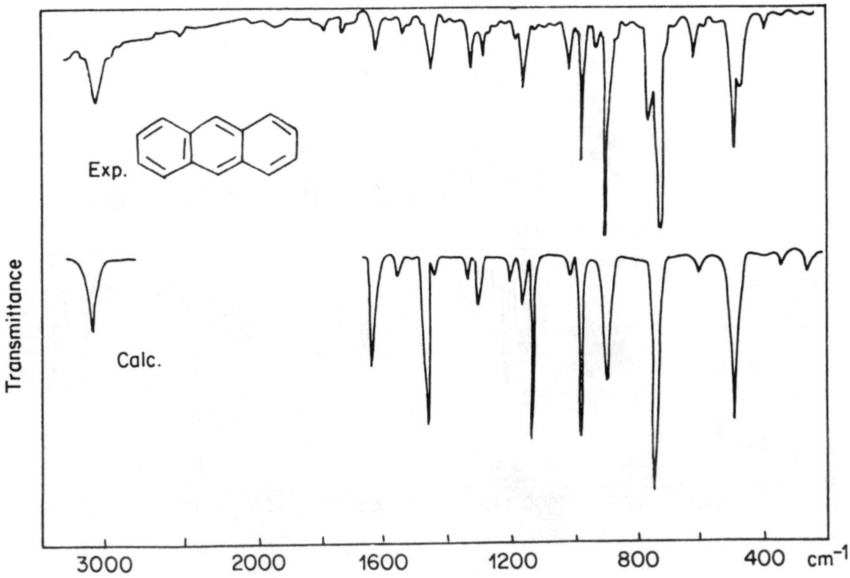

Figure 13.8. Transmission spectra for anthracene: experiment and calculation

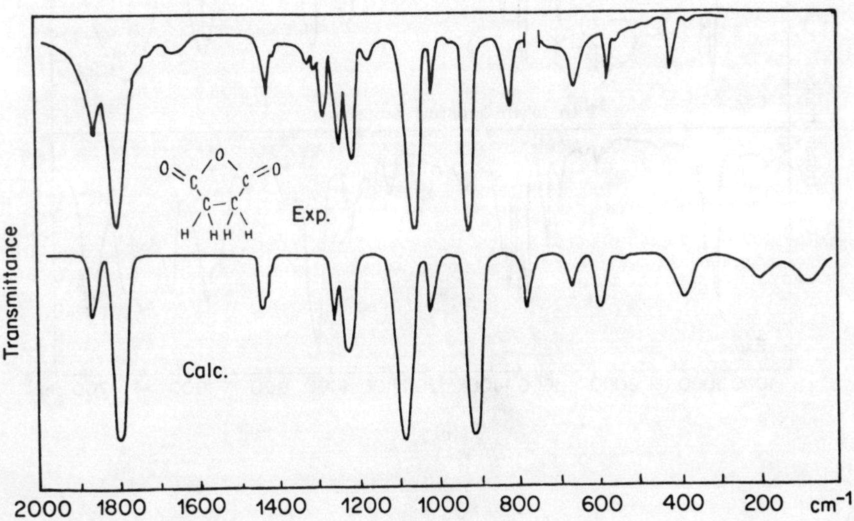

Figure 13.9. Transmission spectra for succinic anhydride: experiment and calculation

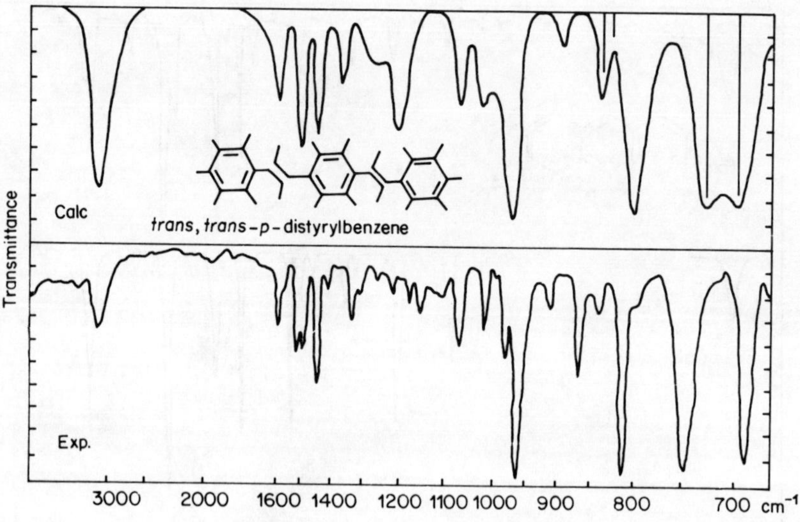
Calc
trans, trans-p-distyrylbenzene
Transmittance
Exp.
3000
2000
1600
1400
1200
1100
1000
900
800
700 cm⁻¹

(a)

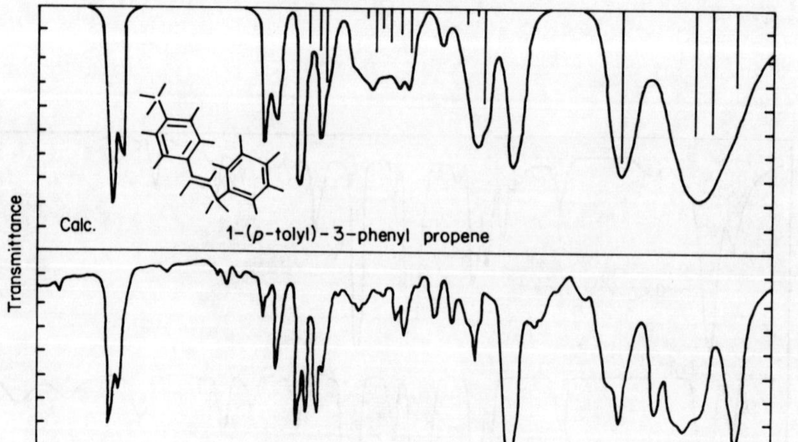
Calc.
1-(p-tolyl)-3-phenyl propene
Transmittance
Exp
4000
3000
2000
1600
1400
1200
1100
1000
900
800
700 cm⁻¹

(b)

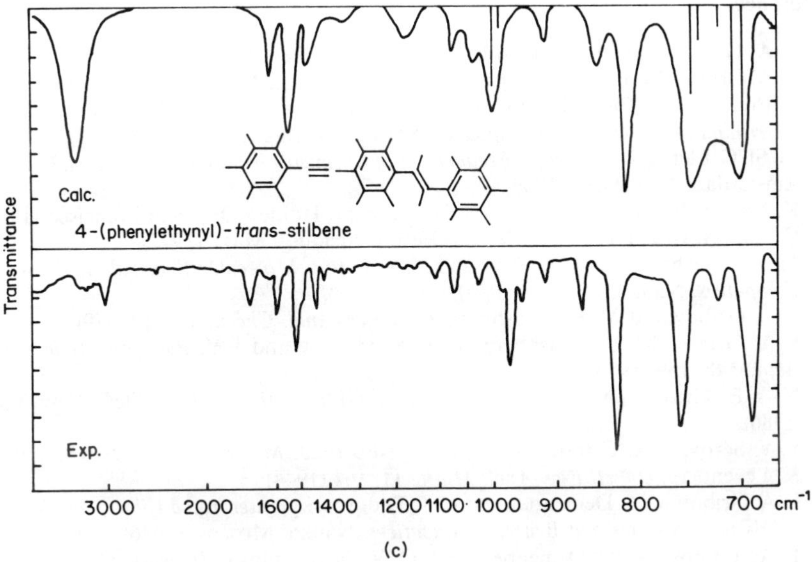

Figure 13.10. Results of simulation infra-red spectra with a special databank for force constants and electro-optical parameters (by O. Novosiolova)

application makes great demands on the investigator, because the whole process leading to the final solution cannot be fully formalized. This inevitably results in the danger of subjectivity. It is for this reason that estimates for whole classes of molecules are more significant than those for individual ones. It is possible in a whole series, by careful securing of the transferability of the parameters, to reject some of those parameters and obtain completely reliable results, despite the complexity of the task as a whole.

It is found that, by using this approach, reasonable values for many of the parameters can be allotted when starting to solve the problem. Consequently the debate has shifted from defining the entire set of parameters, especially the electro-optical ones, for a complex system to defining some of them. Many of the electro-optical parameters can thus be set equal to zero when choosing the initial representation of the matrix of the parameters. This situation is true for unrepeated bonds in symmetric molecule of certain types, to the non-diagonal elements of the $(\partial\mu/\partial q)$ matrix which correspond to the derivatives of a bond dipole moment with respect to those natural co-ordinates not containing the bond's atoms and to some other co-ordinates.

Methods of solving inverse spectral problems and the determination of semi-empirical parameters considered above have enabled us to make calculations of infra-red spectra for a great number of different organic compounds over the last few years, and examples of such calculations are given in Figures 13.6–13.10. Many examples and tables of force constants and electro-optical parameters are given in references 13 and 14.

References

1. L. A. Gribov and M. E. Elyashberg. *J. Molec. Struct.*, **5**, 179 (1970).
2. L. A. Gribov, M. E. Elyashberg and L. A. Moscovkina. *J. Molec. Struct.*, **9**, 357 (1971).
3. E. Pretsch, J. T. Clerc, J. Seibl and W. Simon. *Tables of Spectral Data for Structure Determination of Organic Compounds*, Springer Verlag, Berlin (1983).
4. M. St. C. Flett. *Characteristic Frequencies of Chemical Groups in the Infrared*, Elsevier, Amsterdam/New York (1969).
5. K. Nakanishi. *Infrared Absorption Spectroscopy*, Holden-Day, San Francisco (1962).
6. F. Harary, *Graph Theory*, Addison-Wesley, Reading, Mass. (1972).
7. M. E. Elyashberg, L. A. Gribov and V. V. Serov. *Molecular Spectral Analysis and Computers*, Nauka, Moscow (1980) (in Russian).
8. L. A. Gribov and M. E. Elyashberg. *Crit. Rev. Anal. Chem.*, **8**, 111 (1979).
9. L. A. Gribov, M. E. Elyashberg, V. N. Koldashov and I. V. Pletnjov. *Anal. Chim. Acta.*, **148**, 159 (1983).
10. N. A. B. Gray. *Computer-Assisted Structure Elucidation*, John Wiley, New York (1986).
11. V. V. Serov, L. A. Gribov and M. E. Elyashberg. *J. Molec. Struct.*, **129**, 183 (1985).
12. K. Levenberg. *Quart. Rev. Appl. Math.*, **11**, 164 (1944)
13. L. A. Gribov, V. A. Dementiev and A. T. Todorovsky, *Interpreted Vibrational Spectra of Alkanes, Alkenes and Benzene Derivatives*, Nauka, Moscow (1986) (in Russian).
14. L. A. Gribov, V. A. Dementiev and O. V. Novosiolova, *Interpreted Vibrational Spectra of Hydrocarbons with Isolated and Conjugated Multiple Bonds*, Nauka, Moscow (1987) (in Russian).

Chapter 14

Methods of Calculation of Vibronic Spectra of Polyatomic Molecules

14.1 The Calculation of Matrix Elements

In Chapter 2 we considered the general plan of the calculation of electronic–vibrational (vibronic) spectra of polyatomic molecules. It was pointed out that such spectra can only be calculated on the basis of the adiabatic approximation. For the practical calculation of electronic–vibrational spectra (the vibrational components of electronic spectra) it is necessary to calculate the matrix elements for dipole electronic–vibrational transitions. These matrix elements have two terms (Franck–Condon and Herzberg–Teller), and we must take into account the correspondence between normal co-ordinates for the ground and excited states. Three major problems are encountered:

(1) Determination of the molecular structure in excited electronic states (choice of the geometrical and force-field models of a molecule);
(2) Calculation of the overlap integrals for the vibrational wavefunctions of the combining electronic states which determine the probability of the vibronic transitions; and
(3) Allowance for vibrational interaction.

The determination of the geometry and force-field characteristics of molecules in excited states is a complicated problem which has not yet been resolved. Many papers in the literature have been devoted to the development and use of quantum methods to calculate molecular geometry[1–5]. Calculations performed for numerous molecules have shown that the accuracy of these methods is ± 0.02 Å for bond lengths and $\pm$ several degrees for bond angles, which, unfortunately, is insufficient for determining molecular geometry changes on electronic transition since these changes are of the same order of magnitude. Thus, quantum-chemical methods are of use for qualitative estimations only. Hence, semi-empirical methods are generally used in evaluating the theoretical molecular geometry changes of molecular vibronic spectra. These methods determine the changes for electronic transitions that involve various semi-empirical dependences between the molecular bond lengths and orders;

for example, π orders of C—C bonds[6-9]. Coulson[10] and Coulson and Golebievski[11] were the first to derive such a dependence, and a simplified linear form was suggested by Longuet-Higgins and Salem[12]. Various modifications and clarifications of these expressions have been introduced[13-16]. Popov *et al.*[17-19] obtained similar dependence plots for the C—C bond as well as C—O and C—N bonds. The results of Popov *et al.* seem to be the most accurate, and hold good over a wide range of π orders and bond lengths, since no *a priori* analytical forms were assumed by those authors while constructing the dependences.

The correlations are known to have been obtained for the ground electronic states of the molecules. There are no differences in principle, however, between the ground and excited molecular states (for stable configurations), but the alteration in the electronic structure of a molecule which undergoes changes can be seen in a changed density matrix (particularly in changed bond orders). Hence, there are no grounds for assuming that the bond order–bond length correlation will be considerably changed on transition to excited molecular states. It has been shown[20] that the relationship derived by Longuet-Higgins and Salem is approximately justified, both in the ground and in the excited molecular states. The same point is indirectly supported by the results of reference 21. The evaluation of changes in σ-type bond lengths and valence angles has not yet been resolved.

Let us now consider the determination of potential surface parameters (in terms of the harmonic approximation of the force constants) of molecules in excited states. It is known that quantum methods of calculating force constants are subject to great uncertainties[22-5], whereas semi-empirical calculations based on a solution of the inverse spectra problem are not realizable (with rare exceptions), since experiments do not yield sufficient information about the vibrational frequencies of molecules in excited electronic states.

If the potential surfaces of the combining electronic states are alike, then, due to the similarity of the normal co-ordinate systems, the overlap in integral of the vibrational wave functions is factorized and the problem reduces to the calculation of one-dimensional integrals. Strictly speaking, this is the case with diatomic molecules only. With polyatomic molecules, however, different forms of the potential surfaces would lead to mixing (the Dushinsky effect[26]) and shift of the normal co-ordinate system and, hence, would involve calculation of multi-dimensional overlap integrals. The problem of calculating the overlap integrals has been extensively discussed elsewhere[27-45].

Computational details are available for the approximation that does not consider normal co-ordinate mixing (one-dimensional case)[30,31]. Calculation methods have been developed both for Franck–Condon integrals and also for the so-called Herzberg–Teller integrals which arise when the vibronic interaction is treated within the adiabatic approximation[31] (the Herzberg–Teller effect). More often the vibronic spectra of polyatomic molecules are interpreted using the formulae derived by McCoy and Ross[36] (the Condon approximation) and by Osad'ko[40] (with due reference to vibronic interaction) that assume the absence of normal co-ordinate mixing and low temperatures ($T \approx 0\,K$) (see, for example, references 5–7, 36 and 46–52).

The wide application of the formulae obtained in references 36 and 40 to the analysis of the vibrational progressions in vibronic spectra is mainly accounted for by their simplicity. However, inaccuracies resulting from the neglect of normal co-ordinate mixing have not been estimated, at least in a qualitative way.

As the dependence of the molecular vibration behaviour on the molecular geometry and force-field changes is rather complicated, it is difficult to establish *a priori* the effect of the neglect of normal co-ordinate mixing. It is necessary to carry out direct calculations of molecular vibrations in the combining electronic states followed by calculation of the normal co-ordinate mixing matrix, especially as the theory and techniques of the solution of vibrational problems are well developed[54].

It is widely recognized that, to interpret vibronic spectra, we must allow for normal co-ordinate mixing; this is particularly emphasized in references 41–3. The main problem involved in considering normal co-ordinate mixing arises, first, from the fact that at present no methods have been developed for calculating multidimensional overlap integrals of vibrational wave functions. Expressions for Franck–Condon factors with due reference to normal co-ordinate mixing have been derived by Sharp and Rosenstock[37], but they have considered only transitions from the zero vibrational level of the initial electronic state to the vibronic state with a single normal vibrational excited, apart from some particular formulae for concrete transitions. Lukashin[39] has given the most general expression for Franck–Condon integrals corresponding to transitions from the zero vibrational level of the initial electronic state to the vibronic one with an arbitrary set of vibrational quantum numbers (this case corresponds to low-temperature vibronic spectra, $T \approx 0\,K$). The results of reference[39], however, are hardly applicable to practical calculations and interpretations of vibronic spectra, since the formulae obtained are unwieldy. In addition, another case may be of greater practical concern, i.e. transitions between electronic states in which two or three (not more) normal vibrations are excited, because it is such lines that essentially show up in the vibronic spectra, as well as at non-zero temperatures. Indeed, a great majority of electron spectra reproduced elsewhere have been obtained at room or near-room temperature (see, for example, reference 55, which contains about 25 000 molecular electron spectra), whereas low-temperature spectra are far fewer in number.

Methods for calculating the Herzberg–Teller type of overlap integrals have been examined. Preem[31] confines himself to the case of no mixing of normal co-ordinates. However, as pointed out in references 42 and 56, consideration of the differences between the normal co-ordinates of the combining electronic states is of particular value when studying intensity distributions in vibronic bands in the presence of the Herzberg–Teller effect.

When vibronic interactions are considered within the adiabatic approximation (only this case is treated here) the problem arises of calculating the matrix elements of the vibronic interaction operator-integrals of vibronic interaction.

General methods for calculating quantum-chemical integrals that would provide a solution to the general problem have only recently been developed[57,58].

14.1.1 *Algorithms of matrix element calculation for vibronic transitions*

As a rule, the vibronic spectra (states) of polyatomic molecules are considered using various model representations (approximations) that help to simplify the problem and render it more distinct, with insignificant losses in generality of approach and accuracy of solution. Basic to such a general approach is the adiabatic theory.

In terms of the adiabatic approximation the vibronic wave function is represented in the form

$$\Psi_{tv}(r, Q) = \Psi_t(r, Q)\Psi_v(q) \tag{14.1}$$

where r and Q are the electronic and nuclear co-ordinates, respectively. The parametric dependence of the electronic component of the vibronic wave function (14.1) can be determined in terms of the Herzberg–Teller approach[59-62]:

$$\begin{aligned}\Psi_t(r,Q) = {} & \Psi_t(r,0) + \sum_{u\neq t}\sum_{k}\frac{\langle u|A_k^t|t\rangle}{\varepsilon_t^0 - \varepsilon_u^0}Q_k^t\Psi_u(r,0)\\ & -\tfrac{1}{2}\sum_{u\neq t}\sum_{k,k'}\frac{\langle u|A_k|t\rangle\langle u|A_{k'}|t\rangle}{(\varepsilon_t^0 - \varepsilon_u^0)^2}Q_k^tQ_{k'}^t\Psi_t(r,0)\\ & +\sum_{u\neq t}\sum_{k,k'}\left[-\frac{\langle t|A_k^t|t\rangle\langle u|A_{k'}^t|t\rangle}{(\varepsilon_t^0 - \varepsilon_u^0)^2} + \sum_{u'\neq t}\frac{\langle u|A_k^t|u'\rangle\langle u'|A_{k'}^t|t\rangle}{(\varepsilon_t^0 - \varepsilon_{u'}^0)(\varepsilon_t^0 - \varepsilon_u^0)}\right]\\ & \times Q_k^tQ_{k'}^t\Psi_u(r,0) + \text{terms of higher orders of smallness}\end{aligned} \tag{14.2}$$

where $|t\rangle \equiv \Psi_t(r, 0)$, ε_t^0 are the wave functions and the energy of the tth electronic state corresponding to an equilibrium molecular geometry (the Condon approximation),

$$A_k^t = (\partial U/\partial Q_k^t)_{Q_k^t=0}$$

is the vibronic interaction operator (with due reference only to the linear terms of the expansion of the potential electron-nuclear attraction energy $U(r, Q)$ in the normal co-ordinates Q_k) and the derivative with respect to the normal co-ordinates of the tth electronic state Q_k^t is taken at the point corresponding to the equilibrium geometry of the molecule. Here only non-degenerate electronic states will be considered.

The following expression is produced for the matrix element of the dipole moment of the vibronic transition:

$$\begin{aligned}\langle t, v'|\hat{\vec{\mu}}|s, v\rangle = {} & \langle t|\hat{\vec{\mu}}|s\rangle\langle v'|v\rangle + \sum_{u\neq t}\sum_{k}\frac{\langle t|A_k^t|u\rangle\langle u|\vec{\mu}|s\rangle}{\varepsilon_t^0 - \varepsilon_u^0}\langle v'|Q_k^t|v\rangle\\ & + \sum_{u\neq s}\sum_{k}\frac{\langle t|\vec{\mu}|u\rangle\langle u|A_k^s|s\rangle}{\varepsilon_s^0 - \varepsilon_u^0}\langle v'|Q_k^s|v\rangle + \cdots\end{aligned} \tag{14.3}$$

In the Condon approximation only the first term in expression (14.3) is preserved. Thus in the Herzberg–Teller approximation the vibrational structure of the electron spectrum of a molecule is determined by the overlap integrals of the vibrational wave functions

$$\langle v'|f(Q)|v\rangle \tag{14.4}$$

where $f(Q)=\prod_{i=1}^{M} Q_i^{\alpha_i}$ and $\alpha_i=0,1,2,\ldots,$ and by the vibronic interaction integrals $\langle t|A_k^t|u\rangle$.

It is worth noting that as expression (14.2) takes account of the second-order (quadratic) terms of the perturbation theory, we need not consider the quadratic term in the vibronic interaction operator

$$B_{kk'}^t=\left(\frac{\partial U}{\partial Q_k^t \partial Q_{k'}^t}\right)_{Q_k^t,Q_{k'}^t=0}$$

since this term determines the potential energy of molecular vibrations, $E(q)$ (together with the potential nuclear-repulsion energy), and is implicitly taken into consideration in solving the vibrational problem for a given electronic state.

Methods for calculating the overlap integrals of the vibrational wave functions The overlap integrals of the vibrational wave functions are multidimensional values due to mixing of the normal co-ordinates of the initial and final electronic states[26] (the Dushinsky effect):

$$Q'=IQ+K \tag{14.5}$$

where the normal co-ordinate mixing is governed by the expression

$$I=L_q'^{-1}L_q$$

and the shift vector is $K=L_q'^{-1}\delta$; $\delta=q'-q$ is the change of the equilibrium natural co-ordinates q and q' at electronic excitation; L_q and L_q' are the matrices of the vibrational modes in the initial and final states, respectively.

However, if in calculating molecular vibrations use is made of dependent natural co-ordinates (which is generally the case in formation and solution of the vibrational equations by computer), the matrices L_q and L_q', with the number of rows equal to that of the normal co-ordinates and that of the columns equal to that of the natural co-ordinates, are rectangular. In this case expression (14.5) does not apply to the matrix I.

It is known from molecular vibration theory[53] that the expression $\tilde{L}_pL_q=1$ holds good both in the independent and dependent natural co-ordinates, i.e. the role of L_q^{-1} is performed by the transposed matrix of conjugate impulse transformation $\tilde{L}_p$. Then a general expression is obtained for the mixing matrix which is valid in any basis of vibrational co-ordinates $I=\tilde{L}_p'L_q$. The shift vector K is calculated in a similar way.

Franck–Condon integrals for the $(0,\ldots,0)\rightarrow(v_1',\ldots,v_M')$ *(method A)* The Hermite polynomial generating function method employed in reference 65 to calculate the multidimensional Franck–Condon integrals provides an explicit expression for the overlap integrals not only for vibronic transitions with a single excited normal oscillator[37] but also for two, three, etc. excited normal oscillators, i.e. for the following transition types: $(0,\ldots,0)\rightarrow(v_1',\ldots,v_M')$ and $(v_1,\ldots,v_M)\leftarrow(0,\ldots,0)$, where the primed values represent the vibrational quantum numbers of higher electronic states.

As shown in reference 37, the expression for the Franck–Condon integrals in the implicit form is

$$\sum_{v}\sum_{v'} T^{v}U^{v'}\frac{2^{v}2^{v'}}{v!v'!}I(v;v') = I_0\exp[(T^{+}AT + T^{+}B) + (U^{+}CU + U^{+}D) + U^{+}FT] \tag{14.6}$$

where $A = 2\Gamma'^{1/2}I(I^{+}\Gamma'I+\Gamma)^{-1}I^{+}\Gamma'^{1/2} - E$

$B = -2\Gamma'^{1/2}[I(I^{+}\Gamma'I+\Gamma)^{-1}I^{+}\Gamma' - E]K$

$C = 2\Gamma^{1/2}(I^{+}\Gamma'I+\Gamma)^{-1}\Gamma^{1/2} - E$

$D = -2\Gamma^{1/2}(I^{+}\Gamma'I+\Gamma)^{-1}I^{+}\Gamma'K$

$F = 4\Gamma^{1/2}(I^{+}\Gamma'I+\Gamma)^{-1}I^{+}\Gamma'^{1/2}$

Γ, Γ' are the diagonal matrices of the reduced frequencies $4\pi v_i/h$ and $4\pi v_i'/h$; E is the unit matrix; and I and K are determined from expression (14.5).

If one of the vibrational states is the ground state ($v = 0$), this expression can be greatly simplified, as the cross-term in the power of the exponent ($U^{+}FT$) disappears:

$$\sum_{v'} U^{v'}\frac{2^{v'}}{v'!}I(0;v') = I_0\exp(U^{+}CU + U^{+}D) \tag{14.7}$$

By expanding the exponent on the right-hand side in series in powers of U, an expression is obtained for the integrals $I(0;v')$. Generally, this is rather difficult to perform, but in particular cases of non-zero, two, three, etc. quantum numbers, v', there are no special problems.

Thus for a vibronic transition with excitation of two normal vibrations

$$\sum_{v_i',v_j'} U_i^{v_i'}U_j^{v_j'}\frac{2^{v_i'}2^{v_j'}}{v_i'!v_j'!}I(0;0,\ldots,v_i',\ldots,v_j',\ldots,0) = I_0\exp(U_i c_{ij}U_j + U_i c_{ji}U_j + U_i^2 c_{ii} + U_j^2 c_{jj} + U_i d_i + U_j d_j) \tag{14.8}$$

Using the expression for the generating function of the Hermite polynomials $H_n(z)$:

$$\exp(-\tfrac{1}{2}t^2 + zt) = \sum_{n=0}^{\infty} H_n(z)\frac{t^n}{n!} \tag{14.9}$$

$$H_n(z) = \sum_{p=0}^{[n/2]}(-1)^p\frac{n!}{(n-2p)!}\frac{z^{n-2p}}{2^p p!} \tag{14.10}$$

and taking into account the expression for the integral[65]

$$I(0;0,\ldots,v_k',\ldots,0) = \left(\frac{v_k'!}{2^{v_k'}}\right)^{1/2}\sum_{q=0}^{[v_k'/2]}\frac{c_{kk}^{q}d_k^{v_k'-2q}}{(v_k'-2q)!q!} \tag{14.11}$$

gives the final expression for the integral sought:

$$I(0;0,\ldots,v_i',\ldots,v_j',\ldots,0)=\left(\frac{v_i'!}{2^{v_i'}}\right)^{1/2}\sum_{p=0}^{[v_i'/2]}\frac{c_{ii}^p d_i^{v_i'-2p}}{p!(v_i'-2p)!}$$
$$\times\sum_{r=0}^{(v_i'-2p)} C_r^{v_i'-2p}\left(\frac{v_j'!2^r}{(v_j'-r)!}\right)^{1/2}\left(\frac{c_{ij}}{d_i}\right)^r I(0;0,\ldots,v_j'-r,\ldots,0) \qquad (14.12)$$

where

$$v_i'\leqslant v_j',\quad c_n^k=\frac{k!}{n!(k-n)!}$$

and c_{ij} and d_i are the elements of the matrices C and D.

In a similar way we derive expressions for the Franck–Condon integrals corresponding to an excitation of three or more normal vibrations.

Accurate method for calculating arbitrary overlap integrals of the vibrational wave functions (method B) The overlap integrals of vibrational wave functions (equation (14.4)) can be accurately calculated for the arbitrary combining vibrational states $\Psi_v(Q)$ and $\Psi_{v'}(Q')$. Expression (14.4) can be written as:

$$\langle v'|f(\xi)|v\rangle=N\int\prod_{k=1}^{M}H_{v_k'}(\xi_k')\prod_{i=1}^{M}H_{v_i}(\xi_i)$$
$$\times\exp\left[-\frac{1}{2}\sum_{l=1}^{M}(\xi_l'^2+\xi_l^2)\right]f(\xi)\mathrm{d}\xi_1\cdots\mathrm{d}\xi_M \qquad (14.13)$$

In equation (14.13) we have passed to the normal co-ordinates normalized to zero-point vibration amplitudes $\xi=Q_0^{-1}Q$, $\xi'=Q_0'^{-1}Q'$; Q_0, Q_0' are the diagonal matrices of the zero-point vibrations $Q_{oi}=(\hbar/m_i\omega_i)^{1/2}$, $Q_{oi}'=(\hbar/m_i\omega_i')^{1/2}$; and N is the product of the normalization factors of the wave functions of a harmonic oscillator. It is evident that on a certain unitary transformation $\xi=L\zeta-L\Lambda^{-1}L\tilde{J}\Delta$ and $\xi'=F\zeta+(1-F\Lambda^{-1}\tilde{F})\Delta$, $F=JL$, $J=Q_0'^{-1}IQ_0$, $\Delta=Q_0'^{-1}K$, which would transform the power of the exponent to the quadratic form

$$-\frac{1}{2}\sum_{l=1}^{M}(\xi_l'^2+\xi_l^2)=-\frac{1}{2}\sum_{l=1}^{M}\Lambda_l\zeta_l^2+c \qquad (14.14)$$

Integral (14.3) becomes factorized, each resulting integral being of the type

$$P(w,\lambda)=\int_{-\infty}^{\infty}\chi^w e^{-1/2\lambda\chi^2}\,\mathrm{d}\chi=\begin{cases}(w-1)!!(2\pi/\lambda^{w+1})^{1/2}, & w\text{ even}\\ 0 & w\text{ odd}\end{cases} \qquad (14.15)$$

With due reference to expression (14.10) for the Hermite polynomials and after the appropriate transformations, the polynomial part of the integrand in integral (14.13) becomes

$$K = \prod_{t=1}^{M} H_{v_t'}(\xi_t') \prod_{r=1}^{M} H_{v_r}(\xi_r)\xi_r^{\alpha_r}$$

$$= \prod_{t=1}^{M} \prod_{r=1}^{M} \Bigg\{ \sum_{q_t=0}^{[v_t'/2]} (-1)^{q_t} \frac{v_t'!2^{v_t'}}{4^{q_t}(v_t'-2q_t)!q_t!} \sum_{p_r=0}^{[v_r/2]} (-1)^{p_r} \frac{v_r!2^{v_r}}{4^{p_r}p_r!(v_r-2p_r)!}$$

$$\times \sum_{m_1^t=0}^{(v_t'-2q_t)} \sum_{n_1^r=0}^{(v_r-2p_r+\alpha_r)} b_{t_1}c_{r_1}\zeta_1^{(k_t+\rho_r)} \times \prod_{i=2}^{(M-1)} \left[\sum_{m_i^t=0}^{(m_{i-1}^t)} \sum_{n_i^r=0}^{n_{i-1}^r} b_{ti}c_{ri}\zeta_i^{(w_t^i+\theta_r^i)} \right]$$

$$\times \sum_{m_M^t=0}^{m_{M-1}^t} \sum_{n_M^r=0}^{n_{M-1}^r} b_{tM}c_{rM}\zeta_M^{(w_t^M+\theta_r^M)} \Bigg\} \tag{14.16}$$

where

$$k_t = v_t' - 2q_t - m_1^t, \quad \rho_r = v_r - 2p_r - n_1^r + \alpha_r$$

$$w_t^i = m_{i-1}^t - m_i^t, \quad \theta_r^i = n_{i-1}^r - n_i^r$$

$$b_{t1} = \frac{f_{t1}^{k_t}}{k_t!}, \quad c_{r1} = \frac{l_{r1}^{\rho_r}}{\rho_r!}$$

$$b_{ti} = \frac{f_{ti}^{w_t^i}}{w_t^i!}, \quad c_{ri} = \frac{l_{ri}^{\theta_r^i}}{\theta_r^i!} \quad (M > i \geqslant 2)$$

$$b_{tM} = \frac{f_{tM}^{w_t^M} g_t^{m_M^t}}{m_M^t! w_t^M!}, \quad c_{rM} = \frac{l_{rM}^{\theta_r^M} h_r^{n_M^r}}{\theta_r^M! n_M^r!}$$

f_{ti}, l_{ri}, g_t and h_r are the elements of the matrices F, L, $G = (1 - F\Lambda^{-1}F)\Delta$ and $H = -L\Lambda^{-1}L\tilde{J}\Delta$, respectively.

For integral (14.13) we finally obtain

$$\langle v'|f(\xi)|v\rangle = \pi^{-M/2}e^{-c/2}\left(\prod_{t=1}^{M} \frac{v_t'!v_t!2^{v_t'}2^{v_t}}{Q_{0t}'Q_{0t}}\right)^{1/2}$$

$$\times \sum_{q_1=0}^{[v_1'/2]} \frac{(-1)^{q_1}}{4^{q_1}q_1!} \times \cdots \times \sum_{q_M=0}^{[v_M'/2]} \frac{(-1)^{q_M}}{4^{q_M}q_M!} \sum_{p_1=0}^{[v_1/2]} \frac{(-1)^{p_1}}{4^{p_1}p_1!} \times \cdots$$

$$\times \sum_{p_M=0}^{[v_M/2]} \frac{(-1)^{p_M}}{4^{p_M}p_M!} \sum_{m_1^1=0}^{(v_1'-2q_1)} \cdots \sum_{m_1^M=0}^{(v_M'-2q_M)} \sum_{n_1^1=0}^{(v_1-2p_1+\alpha_1)} \cdots \sum_{n_1^M=0}^{(v_M-2p_M+\alpha_M)}$$

$$\times b_{11} \times \cdots \times b_{M1} \times c_{11} \times \cdots \times c_{M1}\, \Phi(k, \rho, \lambda_{11})$$

$$\times \prod_{i=2}^{M-1} \Bigg(\sum_{m_i^1=0}^{m_{i-1}^1} \cdots \sum_{m_i^M=0}^{m_{i-1}^M} \sum_{n_i^1=0}^{n_{i-1}^1} \cdots \sum_{n_i^M=0}^{n_{i-1}^M} b_{1i} \times \cdots \times b_{Mi}c_{1i} \times \cdots$$

$$\times c_{Mi}\Phi(\omega^i, \theta^i, \lambda_{ii}) \Bigg) \times \sum_{m_M^1=0}^{m_{M-1}^1} \cdots \sum_{m_M^M=0}^{m_{M-1}^M} \sum_{n_M^1=0}^{n_{M-1}^1} \cdots \sum_{n_M^M=0}^{n_{M-1}^M} b_{1M} \times \cdots$$

$$\times b_{MM}c_{1M} \times \cdots \times c_{MM}\Phi(\omega^M, \theta^M, \lambda_{MM}) \tag{14.17}$$

where

$$\Phi(x, y, a) = P(w, a), \quad w = \sum_{i=1}^{M} (x_i + y_i)$$

At first sight, expression (14.17) seems to be rather cumbersome, as it contains a great many sums, the number being equal to $M_1(2M_2 - 1)$, where M_1 is the maximum number of normal vibrations excited in each combining electronic state and M_2 is that of mixing normal co-ordinates (integral dimension). Under actual conditions, however, it is transitions between vibronic states with not more than two excited normal vibrations ($M_1 = 2$) that show up in vibronic spectra. In this case, the number of sums is $4M_2 - 1$, and the method can be applied fairly well to calculate the overlap integrals corresponding to a mixing of as many as ten normal co-ordinates ($M_2 \leqslant 10$), which, it is known from experience, will cover almost all cases met in practice.

Overlap integral calculation with slight variations in the potential surface shapes (method C) If the accurate method for overlap integral calculation is found to be too unwidely it is expedient to consider the techniques for approximate calculation of these integrals for practical purposes. With slight variations in the potential surface shapes, we can resort to an expansion of the vibrational wave function of the final electronic state in the Taylor series with normal co-ordinates of the initial electronic state around the minimum of its potential surface, taking several first expansion terms only into consideration.

Indeed, in large polyatomic molecules with many electrons, potential surfaces corresponding to the combination of lower electronic states should not differ very much from one another, since the geometry of a molecule depends on the combined effect of all the electrons, and hence is not sensitive to variations in the state of one electron. In this case, the normal co-ordinate mixing matrix I is close to unity and the shift vector column K is low. Hence we can write $Q' = Q + \Delta Q$, where $\Delta Q = (I - E)Q + K$ is low.

The vibrational wave function of the final electronic state $\Psi'_{v'}(Q')$ can be expanded in the Taylor series in the vicinity of $\Delta Q = 0$; at low ΔQ values a finite number of expansion terms can be considered:

$$\begin{aligned}\Psi'_{v'}(Q') &= (spQ'_0)^{-1/2}\Psi'_{v'}(\xi') \\ &= \left(\frac{spQ_0}{spQ'_0}\right)^{1/2}\left(\Psi_{v'}(\xi) + [\widetilde{\Delta\xi}][\hat{P}_{\xi}\Psi_{v'}(\xi)] + \frac{1}{2!}[\widetilde{\Delta\xi}][\hat{P}_{\xi n}\hat{P}_{\xi m}\Psi_{v'}(\xi)][\Delta\xi] + \cdots\right)\end{aligned} \tag{14.18}$$

$\hat{P}_{\xi n} = -i\hbar\partial/\partial\xi_n$ is the operator of the impulse conjugate to the co-ordinate ξ_n. $spQ'_0 \equiv \prod_{i=1}^{M} Q_{0i}$.

By substituting the expansion obtained $\Psi'_{v'}(Q')$ into the Franck–Condon integral we have

$$\begin{aligned}I(v;v') &\equiv \int \Psi_v(Q)\Psi'_{v'}(Q')\,\mathrm{d}Q \\ &= \left(\frac{(spQ_0)^3}{spQ'_0}\right)^{1/2}\left\{\prod_{i=1}^{M}\delta_{v_i v'_i} + \frac{1}{(-i\hbar)}\sum_n \Delta_n\left(\int \Psi_v \hat{P}_{\xi n}\Psi_{v'}\,\mathrm{d}\xi\right)\right.\end{aligned}$$

$$+\frac{1}{(-i\hbar)}\sum_{n,m} J_{nm}\left(\int \Psi_v \xi_m \hat{P}_{\xi n} \Psi_{v'}\,\mathrm{d}\xi\right)$$

$$+\frac{1}{2\hbar^2}\sum_{n,m} \Delta_n \Delta_m\left(\int \Psi_v \hat{P}_{\xi n} \hat{P}_{\xi m} \Psi_{v'}\,\mathrm{d}\xi\right)$$

$$+\frac{1}{2\hbar^2}\sum_{n,m,k} J_{nm}\Delta_k\left(\int \Psi_v \xi_m \hat{P}_{\xi k} P_{\xi n} \Psi_{v'}\,\mathrm{d}\xi\right)+\cdots\Bigg\} \tag{14.19}$$

where $J = Q_0'^{-1} I Q_0 - E$, $\Delta = Q_0'^{-1} K$ and δ_{mn} is the Kroneker δ symbol.

The $\int \Psi_v f(\xi, \hat{P}) \Psi_{v'}\,\mathrm{d}\xi$ integrals arising in expression (14.19) are reduced to the form

$$I^{\alpha,\beta}_{v,v'} = \int \Psi_{v_i}(\xi_i)\xi_i^{\alpha}\hat{P}^{\beta}_{\xi i}\psi_{v_i'}(\xi_i)\,\mathrm{d}\xi_i \tag{14.20}$$

The expansion $\Psi'_{v'}(Q')$ can be accomplished at the point $\Delta Q = K$ rather than at $\Delta Q = 0$. This is a reasonable thing to do when the elements of the normal co-ordinate mixing matrix J are small and the elements of the shift vector K are finite. Then we obtain

$$I(v; v) = \left(\frac{(spQ_0)^3}{spQ_0'}\right)^{1/2}\Bigg\{\int \Psi_v(\xi)\Psi_{v'}(\xi+\Delta')\,\mathrm{d}\xi + \frac{1}{(-i\hbar)}\sum_{k,l} J_{kl}\int \Psi_v(\xi)(\xi_l+\Delta_l')\hat{P}_{\xi k}\Psi_{v'}(\xi+\Delta')\,\mathrm{d}\xi+\cdots\Bigg\} \tag{14.21}$$

where $\Delta' = Q_0^{-1} I^{-1} K$ and the problem reduces to calculating integrals of the type

$$I^{k,n}_{v,v'}(\Delta') = \int \Psi_v(\xi)(\xi+\Delta')^k \hat{P}^n_{\xi}\Psi_{v'}(\xi+\Delta')\,\mathrm{d}\xi \tag{14.22}$$

If use is made of the expression for the Hermite polynomial generating function (14.9) we can derive the analytical expression for integrals (14.22):

$$I^{k,n}_{v,v'}(\Delta') = \sum_{m=0}^{[n/2]}\sum_{l=0}^{n-2m} c^n_{ml}\sum_{q=0}^{[k+l/2]}\sum_{r=0}^{k+l-2q}\sum_{u=0}^{r} d^{k+l}_{qru} \times a_{uw}(-1)^{v-u}(\Delta')^{r-u}U_{v-u,v'-w}\left(\frac{\sqrt{2}}{2}\Delta'\right) \tag{14.23}$$

where

$$w = n+k-r-2(m+q)$$

$$c^n_{ml} = (-1)^{m-l}2^{n-l/2-2m}\frac{n!}{l!m!(n-2m-l)!}$$

$$d^p_{qru} = \frac{p!}{q!u!(r-u)!(p-2q-r)!}2^{-r+u-p/2-q}$$

$$U_{ij}(x) = \begin{cases} j!(-1)^j x^{i-j} L_j^{i-j}(x^2), & i \geqslant j \\ i!(-1)^i x^{j-i} L_i^{j-i}(x^2), & i \leqslant j \end{cases}$$

$L^k(z)$ is the Laguerre polynomial

$$a_{uw} = \frac{\sqrt{(v!v'!)}}{(v-u)!(v'-w)!}$$

at $u \leqslant v$ and $w \leqslant v'$, otherwise $a_{uw} = 0$.

For the matrix elements of a harmonic oscillator (equation (14.20)) expression (14.23) is considerably simplified:

$$I^{k,n}_{v,v'} = \sum_{m=0}^{[n/2]} \sum_{l=0}^{n-2m} c^n_{ml} \sum_{q=0}^{[k+l/2]} \sum_{r=0}^{k+l-2q} b^{k+l}_{qr} a_{rw}(v-r)! \qquad (12.24)$$

where

$$b^p_{qr} = \frac{p!}{q!r!(p-2q-r)!} 2^{-q-p/2}$$

Calculation of the Franck–Condon integrals using the approximate formula (14.19) with due reference to the first five terms of the expansion $\Psi'_{v'}(Q')$ together with a comparison with the exact calculation by equations (14.11) and (14.12) have revealed that the former calculations are less sensitive to the magnitude of the elements of the shift matrix Δ compared with those of the matrix characterizing a rotation of axes of the normal co-ordinate system J.

Figure 14.1 is a schematic representation of the range of Δ and J values (shaded) over which the error of two-dimensional Franck–Condon integral calculations is within 5% of the $I(0; 0)$ value. The greater the dimensionality of the Franck–Condon integrals, the less the accuracy of calculations by equation (14.19). For instance, for the matrices J and Δ with elements of the order of 0.2 and 0.3, respectively, the error in the two-dimensional case is under 0.5%, whereas in the three-dimensional case for several integrals at amounts to 2.5% of the $I(0; 0)$ value. Note that at $J = 0$ accuracy is practically independent of the overlap integral dimension.

Calculations of the Franck–Condon integrals have also shown that the overlap integral dimension is less sensitive to the magnitude of non-diagonal elements, and the elements $J_{mn} \leqslant J_{mm}$ can be neglected within an accuracy of

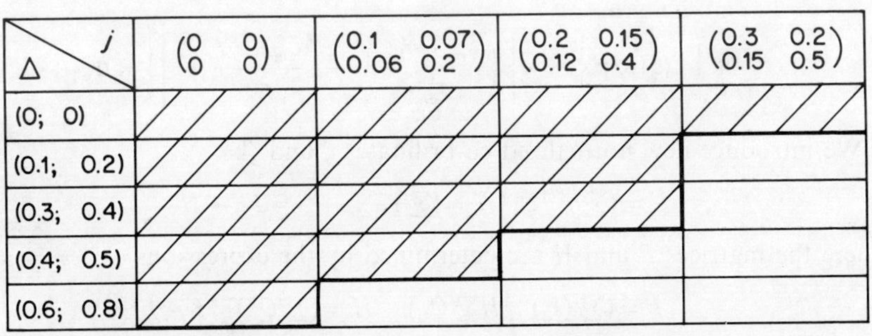

Figure 14.1. Δ and j range (shaded) where the error of two-dimensional Franck–Condon integral calculations are < 5% of the $I(0; 0)$ value

0.01 of the $I(0;0)$ value. This point is very important, since in some cases we can reduce the number of mixing normal co-ordinates and, hence, break the multidimensional Franck–Condon integral into a product of several lower-dimensionality integrals.

The structure of equation (14.19) also permits calculation of Herzberg–Teller-type integrals, which arise from vibronic interaction considerations. In order to calculate integral (14.4) we simply substitute $(\alpha+\alpha_i)$ for α in the corresponding integrals (14.20).

Overlap integral calculation with approximation of the vibrational functions by Gaussian exponents (method D) An alternative approximate method for calculating multidimensional Franck–Condon and Herzberg–Teller integrals lies in approximating the wave function of a harmonic (or anharmonic) oscillator by a linear combination of a finite number of Gaussian exponents. The choice of the Gaussian functions is accounted for by the fact that by a corresponding rotation of the co-ordinate system axes the multidimensional integral of the Gaussian function can be represented as a product of uni-dimensional integrals which are tabulated.

If the vibrational wave functions of the combining electronic states are represented in the form

$$\Psi_v(\xi)=\prod_{m=1}^{M}\psi_{v_m}(\xi_m)\approx\prod_{m=1}^{M}\frac{1}{(Q_{0m})^{1/2}}\sum_{i_m}^{s_m}a_{i_m}^{v_m}\exp\left[-\left(\frac{\xi_m+\delta_{i_m}^{v_m}}{\sigma_{i_m}^{v_m}}\right)^2\right]$$
$$\Psi_{v'}(\xi)=\prod_{k=1}^{M}\psi_{v_k'}(\xi_k')\approx\prod_{k=1}^{M}\frac{1}{(Q_{0k}')^{1/2}}\sum_{j_k}^{s_k}a_{jk}^{v_k'}\exp\left[-\left(\frac{\xi_k'+\delta_{jk}^{v_k'}}{\sigma_{jk}^{v_k'}}\right)^2\right]\tag{14.25}$$

the overlap integrals of the vibrational wave functions of the Herzberg–Teller type are reduced to the overlap integrals of the Gaussian exponents:

$$\begin{aligned}I_{n,l}^{\alpha,\beta}(v;v')&\equiv\int\Psi_{v'}'(\xi')f(\xi)\Psi_v(\xi)\,\mathrm{d}\xi\\&\approx\sum_{i_1,\ldots,i_M}\sum_{j_1,\ldots,j_M}a_{i_1}^{v_1}\cdots a_{i_M}^{v_M}a_{j_1}^{v_1'}\cdots a_{j_M}^{v_M'}\\&\quad\times\int f(\xi)\exp\left\{-\sum_m\left[\left(\frac{\xi_m+\delta_{i_m}^{v_m}}{\sigma_{jm}^{v_m}}\right)^2+\left(\frac{\xi_m'+\delta_{jm}^{v_m'}}{\sigma_{jm}^{v_m'}}\right)^2\right]\right\}\times\mathrm{d}\xi_1\cdots\mathrm{d}\xi_M\end{aligned}\tag{14.26}$$

We introduce new normalized co-ordinates ζ and ζ':

$$\zeta'=(\Sigma^j)^{-1}\xi',\quad\zeta=(\Sigma^i)^{-1}\xi,\quad\zeta'=F\zeta+H$$

where the matrices F and H are determined by the expressions

$$F=(\Sigma^jQ_0')^{-1}I(\Sigma^iQ_0),\quad H=(\Sigma^jQ_0')^{-1}K$$

and Σ^i, Σ^j are the diagonal matrices with the elements $\sigma_{i_m}^{v_m}$ and $\sigma_{jm}^{v_m'}$, respectively. The exponent under the integral in expression (14.26) is transformed as follows:

$$\exp\left\{-\sum_{n}\left[\left(\frac{\xi_m+\delta_{i_m}^{v_m}}{\sigma_{i_m}^{v_m}}\right)^2+\left(\frac{\xi'_m+\delta_{j_m}^{v'_m}}{\sigma_{j_m}^{v'_m}}\right)^2\right]\right\}$$
$$=\exp\left[-(\tilde{\zeta}G^{ij}\zeta+2\tilde{D}^{ij}\zeta+C^{ij})\right]$$
$$=\exp\left[-(\tilde{\eta}\lambda\eta+2\tilde{d}\eta+C^{ij})\right]$$

where $G^{ij}=\tilde{F}F+E$

$$D^{ij}=(\tilde{H}+\tilde{\Delta}^j(\Sigma^j)^{-1})F+(\Sigma^i)^{-1}\Delta^i$$
$$C^{ij}=(\tilde{H}+\tilde{\Delta}^j(\Sigma^j)^{-1})(H+\Delta^j(\Sigma^j)^{-1})+\tilde{\Delta}^i(\Sigma^i)^{-2}\Delta^i$$
$$\Lambda=\tilde{L}G^{ij}L$$

Δ^i, Δ^j are the diagonal matrices with the elements $\delta_{i_m}^{v_m}$ and $\delta_{j_m}^{v'_m}$, respectively, and L is some orthogonal transformation $\zeta=L\eta$ which transforms the matrix G^{ij} to the diagonal form Λ. Then, the integrals in equation (14.25) reduce to unidimensional integrals of type (14.15).

After appropriate calculations a general expression is obtained for integrals (14.26) (in this case, for the sake of simplicity, let us consider $f(\xi)=\xi_n^{\alpha}\xi_l^{\beta}$):

$$I_{n,l}^{\alpha,\beta}(v;v')\approx(sp(Q_0Q'_0))^{-1/2}\sum_{i_1,\ldots,i_M}\sum_{j_1,\ldots,j_M}a_{i_1}^{v_1}\cdot\ldots\cdot a_{i_M}^{v_M}$$
$$\times a_{j_1}^{v'_1}\cdot\ldots\cdot a_{j_M}^{v'_M}\exp(\tilde{d}\Lambda^{-1}d-c^{ij})\times\frac{\text{Det }A^i}{\sqrt{(sp\,\Lambda)}}\cdot N \qquad (14.27)$$

where

$$d=\tilde{L}D^{ij},\quad A^i=\Sigma^iL$$
$$N=\begin{cases}1 & \text{at } \alpha=\beta=0\\ -(\Sigma^iL\Lambda^{-1}d)_n & \text{at } \alpha=1,\ \beta=0\\ (\Sigma^iL\Lambda^{-1}dRd\Lambda^{-1}\tilde{L}\Sigma^i)_{nl} & \text{at } \alpha=\beta=1\end{cases}$$

$R=E+\frac{1}{2}\Lambda d^{-2}$ and E is the unit matrix.

The parameters of the Gaussian functions providing an optimal approximation of the wave functions $\psi_n(x)$ in the basis $s=n+1$ of the Gaussian functions for $n=1-4$ are listed in Table 14.1.

Note that the errors of the Franck–Condon and Herzberg–Teller integral calculations by the last of the above methods depend solely on the accuracy of the approximation concerned, since all the integrals arising from substitution of the vibrational wave functions by linear combinations of the Gaussian exponents are calculated readily and exactly.

The approximation accuracy can be determined from a root-mean-square deviation of the approximate from the exact values. It is evident from Table 14.1 that these errors are within 10^{-2} of the maximum value $\psi_n(x)$. Estimates have revealed that the errors of the Franck–Condon integral calculations by equation (14.27) are under 1% in this case.

Evidently, this method also applies, to the same extent, to the anharmonic case. The algorithm for calculating the overlap integrals of the vibrational wavefunctions is not changed but the parameters of the Gaussian functions must be changed as appropriate.

Table 14.1. Parameters of the Gaussian functions in expansion (14.25), $s=n+1$

n		k					$\dfrac{\sqrt{\left[\sum_{i=1}^{50}(\Delta\psi_n(x_i))^2\right]}}{\sqrt{(50)}\psi_n^{\max}(x)}$
		1	2	3	4	5	
1	a_k^1	−1.550	1.550	0	0	0	2×10^{-3}
	δ_k^1	−0.474	0.474				
	σ_k^1	1.360	1.360				
2	a_k^2	0.866	−1.053	0.866	0	0	6×10^{-3}
	δ_k^2	−1.442	0	1.442			
	σ_k^2	1.131	0.918	1.131			
3	a_k^3	−0.880	1.016	−1.016	0.880		4×10^{-3}
	δ_k^3	−1.832	−0.669	0.669	1.832		
	σ_k^3	1.108	0.835	0.835	1.108		
4	a_k^4	0.801	−0.866	0.738	−0.866	0.801	4×10^{-3}
	δ_k^4	−2.310	−1.114	0	1.114	2.310	
	σ_k^4	1.017	0.716	0.633	0.716	1.017	

Table 14.2. Parameters of the Gaussian functions in expansion (14.22), $\psi_2(x)$ and $\psi_4(x)$, in the minimum basis set ($s=s_{\min}$)

n		k			$\dfrac{\sqrt{\left[\sum_{i=1}^{50}(\Delta\psi_n(x_i))^2\right]}}{\sqrt{(50)}\psi_n^{\max}(x)}$
		1	2	3	
2	a_k^2	3.724	−4.614	0	1×10^{-2}
	δ_k^2	0	0		
	σ_k^2	1.754	1.144		
4	a_k^4	34.767	−73.278	39.328	1×10^{-2}
	δ_k^4	0	0	0	
	σ_k^4	1.677	1.418	1.189	

The basis s of the Gaussian functions, where $s=n+1$, is not minimal. A fairly good approximation of the wave functions of a harmonic oscillator with an even n value can also be obtained on the basis $s=s_{\min}=(n/2)+1$ of the Gaussian functions. Table 14.2 exemplifies the expansion parameters $\psi_2(x)$ and $\psi_4(x)$ for $s=s_{\min}$.

In order to compare the alternative methods for calculating the overlap integrals of the vibrational wave functions, Table 14.3 lists the values of two-dimensional Franck–Condon integrals for some transitions, calculated by the three methods involved—A, C and D. It can be seen that for slight variations in the potential surface shapes (J,Δ small) all three methods yield practically similar values of the integrals (within 0.02). With increasing J and Δ, method C becomes less exact because of poor convergence of the Taylor series (14.9),

Table 14.3. Comparison of the methods for the Franck–Condon integral calculation

J, Δ	Transition		Franck–Condon integral (rel. un.)		
			Method A	Method C	Method D
$J = \begin{pmatrix} 0.1 & 0.07 \\ 0.06 & 0.2 \end{pmatrix}$	(0; 0)	(0; 0)	1	1	1
	(0; 0)	(0; 1)	0.221	0.224	0.221
$\Delta = (0.3 \quad 0.4)$	(0; 0)	(1; 0)	0.177	0.179	0.179
	(0; 0)	(0; 2)	0.160	0.161	0.158
	(0; 0)	(1; 1)	0.095	0.095	0.095
$J = \begin{pmatrix} 0.3 & 0.2 \\ 0.15 & 0.5 \end{pmatrix}$	(0; 0)	(0; 0)	1	1	1
	(0; 0)	(0; 1)	0.154	0.132	0.153
$\Delta = (0.3 \quad 0.4)$	(0; 0)	(1; 0)	0.128	0.149	0.127
	(0; 0)	(0; 2)	0.279	0.364	0.277
	(0; 0)	(2; 0)	0.186	0.225	0.184

whereas the accuracy of method D is independent of J and Δ so that the errors involved in this method do not exceed 1%.

Thus the above methods for calculating the overlap integrals of the vibrational wavefunctions (methods A–D) enable the calculation of both the Franck–Condon and the Herzberg–Teller integrals for arbitrary vibrational levels of the combining electronic states. Hence, it is possible to consider a temperature distribution of molecules over the vibrational sublevels with due reference to both shift and mixing of normal co-ordinates.

It seems expedient now to consider another alternative that has not been discussed so far. This is the problem of molecular vibrations in the excited state and its solution by the variational method in the basis of ground state vibrational wave functions. In this respect the experience gained in the application of the variational method to the anharmonic case[63–6] is useful. It is reported in references 63–6 that, despite all the difficulties involved, the variational method for the anharmonic case is not only practicable but most promising.

When calculating molecular vibrations in excited electronic states it is reasonable to take as the basis functions the vibrational wave functions of the ground electronic state (the wave functions of a harmonic oscillator) from the 'shifted' argument, i.e. with an account of the shift of a minimum of the potential surfaces of the combining electronic states. This permits a considerable reduction in the dimensionality of the basis set. Moreover, it is evidently not necessary to yield diagonalization of the whole Hamiltonian matrix arising in the variational method and, certainly, this is valuable in those cases which have a very large dimensionality (the latter is about an order of magnitude greater than the number of degrees of freedom of a molecule). As generally not more than three to five normal co-ordinates mix in the combining electronic states, this matrix has the form of a quasi-block, the dimensionality of each block being less than several tens. Diagonalization of such a matrix is not difficult.

Such an approach practically eliminates the difficulty of calculating the overlap integrals of the vibrational wave functions, and the problem is reduced to calculating the matrix elements of a harmonic oscillator of type (14.22), for which the general expression (14.23) has been derived.

This approach, as it is, seems therefore to apply to the anharmonic case as well.

Method for calculating vibronic interaction integrals In the theory of vibronic transitions of polyatomic molecules, in order to consider vibronic interactions within the adiabatic approximation based on the Herzberg–Teller approach we should calculate the elements of vibronic interaction of the type $\langle t|A_k^t|u\rangle$ (with restriction to the linear terms of the expansion of the potential electron–nuclear interaction energy in series with normal co-ordinates). In the general case, the problem reduces to calculating the integrals of force and force constants of the nth order.

Here, discussion will be confined to the case of linear terms in the vibronic interaction operator. This will not affect the generality of the considerations, because linear terms make the greatest contributions, and the present considerations can be readily extended to a more general case.

Let us rewrite the vibronic interaction operator in the form

$$A_k^t \equiv \frac{\partial U(r,q)}{\partial Q_k^t} = -\sum_i \sum_m \frac{Z_m e^2}{r_{im}^3} (\vec{r}_{im} \vec{S}_{mk}^t) \tag{14.28}$$

where the indices i and m number the electrons and nuclei, respectively; Z_m is the charge of the mth nuclei in units of electron charge e; $\vec{r}_{im}$ is the radius vector binding the ith electron and the mth nucleus; $\vec{S}_{mk}^t = \partial \vec{r}_m / \partial Q_k^t = \sum_l E_{mm} \vec{B}_{mk} L_{plk}^t$ is the displacement vector of the mth nucleus at the kth normal vibration; E is the diagonal matrix of nuclear inverse masses; $\vec{B}$ is the matrix of transition from the Cartesian co-ordinates of atom displacement to the natural ones; and L_p^t is the impulse transformation matrix[53].

Then, taking into consideration the identity of all the electrons (N), the expression for the vibronic interaction operator will be

$$\langle t|A_k^t|u\rangle = N \sum_m Z_m e^2 \left(\left\langle t \left| \nabla_1 \left(\frac{1}{r_{1m}} \right) \right| u \right\rangle \vec{S}_{mk}^t \right) \tag{14.29}$$

Electronic wave functions in the states t and u are determined by one of the quantum-mechanical methods. If use is made of the MO–LCAO method, with due reference to configuration interaction, the electronic wavefunction acquires the form $\Psi_t = \sum_k c_k^t \Phi_k$, where the configurations Φ_k are constructed of molecular orbitals in the LCAO form $F_k = \sum_\mu a_\mu^k \psi_\mu$, where ψ_μ is the basis of atomic orbitals (AO). The integral $\langle t|\nabla_1(1/r_{1m})|u\rangle$ is transformed as follows:

$$\begin{aligned} \left\langle t \left| \nabla_1 \left(\frac{1}{r_{1m}} \right) \right| u \right\rangle &= \sum_k \sum_l c_k^t c_l^u \left\langle k \left| \nabla_1 \left(\frac{1}{r_{1m}} \right) \right| l \right\rangle \\ &= \frac{1}{N} \sum_k \sum_l c_k^t c_l^u \left\langle F_k \left| \nabla_1 \left(\frac{1}{r_{1m}} \right) \right| G_l \right\rangle \\ &= \frac{1}{N} \sum_k \sum_l c_k^t c_l^u \sum_\mu \sum_\nu a_\mu^F a_\nu^G \left\langle \mu \left| \nabla_1 \left(\frac{1}{r_{1m}} \right) \right| \nu \right\rangle \end{aligned}$$

where the kth and the lth configurations differ by not more than one molecular orbital; otherwise the matrix element $\langle k|\nabla_1(1/r_{1m})|l\rangle$ decreases to zero.

Thus the problem of calculating the vibronic interaction integral (14.29) is reduced to calculation of the integrals over AO$\langle\mu|\nabla_1(1/r_{1m})|\nu\rangle$:

$$\langle t|A_k^t|u\rangle = \sum_m Z_m e^2 \sum_{k,l} c_k^t c_l^u \sum_{\mu,\nu} a_\mu^F a_\nu^G \left(\left\langle \mu \middle| \vec{\nabla}_1\left(\frac{1}{r_{1m}}\right)\middle| \nu\right\rangle \vec{S}_{mk}^t\right) \qquad (14.30)$$

Calculation of the integrals $\langle\mu|\nabla_1(1/r_{1m})|\nu\rangle$ involves certain difficulties due either to operator complexity or to the AO. To solve the problem of calculating the above integrals, approximate methods have been suggested based on replacement of the radial parts of the AO of the Slater type and of the operator r^{-1} by sums of Gaussian exponents[67,68]. The problem of calculating quantum-mechanical integrals has been recently solved completely for the Slater-type AO[57,58]. Calculations with the HLF*-type AO seem to be at a premium if only because all the functions centred at the same nucleus are orthogonal and the Slater-type AO can be represented by a finite linear combination of HLF. These methods will be considered more closely.

Accurate method for calculating vibronic interaction integrals Here, we apply the general method to analytical calculation in the HLF basis of the integrals of the forces corresponding to the first term in the Herzberg–Teller expansion, namely,

$$F = \left\langle \mu \middle| \nabla_1\left(\frac{1}{r_{1m}}\right)\middle| \nu\right\rangle = \int \psi_1(\vec{r}-\vec{R}_1)\psi_2^*(\vec{r}-\vec{R}_2)\frac{\partial}{\partial \vec{R}_c}|\vec{r}-\vec{R}_c|^{-1}d^3\vec{r} \qquad (14.31)$$

where ∇_1 and $\partial/\partial\vec{R}_c$ are gradient operators (designations: $\mu\equiv 1$, $\nu\equiv 2$, $m\equiv C$, $r_1\equiv r$). Let us rewrite equation (14.31) as a derivative of a one-electron integral with respect to $\vec{R}_c$:

$$\frac{\partial Q}{\partial \vec{R}_c} = \frac{\partial}{\partial \vec{R}_c}\int \psi_1(\vec{r}-\vec{R}_1)\psi_2^*(\vec{r}-\vec{R}_2)|\vec{r}-\vec{R}_c|^{-1}\,d^3\vec{r} \qquad (14.32)$$

and let us find an expression for the one-electron integral Q. Substituting into the integral Q instead of the potential $|\vec{r}-\vec{R}_c|^{-1}$, its Fourier transformation and the integral can be written as

$$Q = (2\pi^2)^{-1}\iint \mathrm{d}^3\vec{p}\,d^3\vec{r}\,\vec{p}^{\,-2}\psi_1(\vec{r}-\vec{R}_1)\psi_2^*(\vec{r}-\vec{R}_2)\mathrm{e}^{i\vec{p}(\vec{r}-\vec{R}_c)} \qquad (14.33)$$

Moving the origin of co-ordinates to the point $\vec{R}_1$,

$$Q = (2\pi^2)^{-1}\iint \mathrm{d}^3\vec{p}\,\mathrm{d}^3 r\,\vec{p}^{\,-2}\psi_1(\vec{r})\psi_2^*(\vec{r}-\vec{R}_{21})\mathrm{e}^{i\vec{p}(\vec{r}-\vec{R}_{c1})} \qquad (14.34)$$

*HLF denotes hydrogen-like functions.

Thus the integral Q is an integral of a two-centre form factor:

$$F_{12}(\vec{p})=\int d^3\vec{r}\,\psi_1(\vec{r})\psi_2^*(\vec{r}-\vec{R}_{21})e^{i\vec{p}\vec{r}} \tag{14.35}$$

$$Q=(2\pi^2)^{-1}\int d^3\vec{p}\,e^{-i\vec{p}\vec{R}_{c1}}F_{12}(\vec{p})\vec{p}^{\,-2} \tag{14.36}$$

Let us expand $e^{i\vec{p}\vec{r}}$ in a series in HLFs and 0(4) spherical harmonics on a sphere of radius p_0:

$$e^{i\vec{p}\vec{r}}=4(\pi p_0)^{1/2}\sum_{NLM}\psi_{Np_0,NLM}(p_0\vec{r})(p_0^2+\vec{p}^{\,2})^{-1}\Psi^*_{p_0,NLM}(\vec{p}) \tag{14.37}$$

where $\Psi_{p_0,NLM}(\vec{p})$ is 0(4) harmonic in a momentum space[69]. Then we arrive at the following expansion for the form factor (14.35) in series in 0(4) harmonics:

$$F_{12}(\vec{p})=4(\pi p_0)^{1/2}\sum_{NLM}(p_0^2+\vec{p}^{\,2})^{-1}\Psi^*_{p_0,NLM}(\vec{p})$$
$$\times\int d^3\vec{r}\,\psi_1(\vec{r})\psi_2^*(\vec{r}-\vec{R}_{21})\psi_{Np_0,NLM}(p_0\vec{r}) \tag{14.38}$$

where integrals in the right-hand part of equation (14.38) are expansion coefficients depending on the indices $n_1l_1m_1$, $n_2l_2m_2$, NLM and constants ζ_1, ζ_2, p_0, $\vec{R}_{21}$.

We now introduce a notation for these coefficients:

$$y_1(n_1l_1m_1,n_2l_2m_2;NLM;\zeta_1,\zeta_2,p_0,\vec{R}_{21})$$
$$=\int d^3\vec{r}\,\psi_1(\vec{r})\psi_2^*(\vec{r}-\vec{R}_{21})\psi_{Np_0,NML}(p_0\vec{r}) \tag{14.39}$$

Quantities in this expression are calculated in a closed form because they are reduced to one- and two-centre overlap integrals for functions of HLF types. Computing these coefficients leads to the expansion of equation (14.38).

If the coefficients (14.39) are found, the integral Q (equation (14.36)) is represented by a form of the following analytical expression:

$$Q=(2\pi^2)^{-1}4(\pi p_0)^{1/2}\sum_{NLM}y_1(NLM,\vec{R}_{21})$$
$$\times\int d^3\vec{p}\,e^{-i\vec{p}\vec{R}_{c1}}\vec{p}^{\,-2}(p_0^2+\vec{p}^{\,2})^{-1}\Psi^*_{p_0,NLM}(\vec{p}) \tag{14.40}$$

Integrals on the right-hand side of equation (14.40) can also be calculated in a closed form. Denoting them by $q(NLM,p_0,\vec{R}_{c1})$, integral (14.40) can be rewritten in the form

$$Q(n_1l_1m_1,n_2l_2m_2;\vec{R}_{21},\vec{R}_{c1})$$
$$=(2\pi^2)^{-1}4(\pi p_0)^{1/2}\sum_{NLM}y_1(n_1l_1m_1,n_2l_2m_2;NLM,\zeta_1,\zeta_2,p_0,\vec{R}_{21})$$
$$\times q(NLM,p_0,\vec{R}_{c1}) \tag{14.41}$$

Thus for calculation of gradient Q with respect to vector $\vec{R}_c$ it is sufficient to compute the gradient of the integral q.

We now calculate the quantities y_1 and q. Let us substitute in equation (14.39), instead of function $\psi_2^*(\vec{r}-\vec{R}_{21})$ its Fourier transform:

$$\psi_2^*(\vec{r}-\vec{R}_{21})=(2\pi)^{-3/2}\int e^{-i\vec{k}(\vec{r}-\vec{R}_{21})}\psi_2^*(\vec{k})d^3\vec{k} \tag{14.42}$$

We come to a sixfold integral:

$$y_1=(2\pi)^{-3/2}\iint d^3\vec{r}d^3k\psi_1(\vec{r})e^{-i\vec{k}\vec{r}}\psi_{Np_0,NLM}(p_0\vec{r})e^{i\vec{k}\vec{R}_{21}}\psi_2^*(\vec{k}) \tag{14.43}$$

Let us expand $e^{-i\vec{k}\vec{r}}$ is a series in 0(3) spherical harmonics[72,73]:

$$e^{-i\vec{k}\vec{r}}=\frac{4\pi}{kr}\sum_{l=0}^{\infty}\sum_{m=-l}^{l}(-i)^l j_l(kr)Y^*_{lm}(\theta_{\vec{r}},\varphi_{\vec{r}})Y_{lm}(\theta_{\vec{k}},\varphi_{\vec{k}}) \tag{14.44}$$

Substituting expansion (14.44) into integral (14.43) and integrating over angle variables $(\theta_{\vec{r}},\varphi_{\vec{r}})$ gives

$$y_1=(2\pi)^{-3/2}4\pi\sum_{l=|l_1-L|}^{l_1+L}\sum_{m=-l}^{l}\tilde{C}^{lm}_{l_1m_1,LM}(-i)^l\times\int d^3\vec{k}k^{-1}e^{i\vec{k}\vec{R}_{21}}\psi_2^*(\vec{k})Y_{lm}(\theta_k,\varphi_k)$$
$$\times\int_0^\infty dr\,rR_{n_1l_1}(\zeta_1r)R_{NL}(p_0r)j_l(kr) \tag{14.45}$$

where

$$\tilde{C}^{lm}_{l_1m_1,LM}=\int Y_{l_1m_1}(\theta_{\vec{r}},\varphi_{\vec{r}})Y_{LM}(\theta_{\vec{r}},\varphi_{\vec{r}})Y^*_{lm}(\theta_{\vec{r}},\varphi_{\vec{r}})d\Omega_3 \tag{14.46}$$

To calculate integrals dr in equation (14.45), let us substitute into equation (14.45) expressions of radial functions:

$$\begin{aligned}
i_1&=\int_0^\infty dr\,rR_{n_1l_1}(\zeta_1r)R_{NL}(p_0r)j_l(\vec{k}r)\\
&=C_{n_1l_1}C_{NL}(2\zeta_1)(2p_0)^L\int_0^\infty dr\,r^{l_1+L+1}e^{-(p_0+\zeta_1)r}j_l(\vec{k}r)\\
&\quad\times\sum_{\sigma=0}^{n_1-l_1-1}\sum_{S=0}^{N-L-1}\binom{n_1+l_1}{n_1-l_1-1-\sigma}\binom{N+L}{N-L-1-S}\frac{1}{\sigma!S!}(-2\zeta_1r)^6(-2p_0r)^S\\
&=C_{n_1l_1}C_{NL}(2\zeta_1)^{l_1}(2p_0)^L\\
&\quad\times\sum_{\tau=0}^{n_1+N-l_1-L-2}\sum_{\sigma=0}^{n_1-l_1-1}\binom{n_1+l_1}{n_1-l_1-1-\sigma}\binom{N+L}{N-L-1-\tau+\sigma}\\
&\quad\times\frac{1}{\sigma!(\tau-\sigma)!}\left(\frac{\zeta_1}{p_0}\right)^\sigma(-2p_0)^\tau\int_0^\infty dr\,r^{l_1+L+\tau+1}e^{-(p_0+\zeta_1)r}j_l(kr)
\end{aligned} \tag{14.47}$$

Now the integral in equality (14.47) can be expressed as a table integral[74]. In fact,

$$i_2 = \int_0^\infty \mathrm{d}r\, r^{l_1+L+\tau+1} \mathrm{e}^{-(p_0+\zeta_1)r} j_l(kr) = D_\zeta^{l_1+L-l+\tau+1} \int_0^\infty \mathrm{d}r\, r^l \mathrm{e}^{-\zeta r} j_l(kr)$$
$$= 2^l l!\, k^{l+1} D_\zeta^{l_1+L-l+\tau+1} [k^2+\zeta^2]^{-l-1} \tag{14.48}$$

where $\zeta = p_0 + \zeta_1$, $D_z = \mathrm{d}/\mathrm{d}z$. Taking into consideration an equality

$$D_\zeta^\omega [k^2+\zeta^2]^{-l-1} = \sum_{t=0}^{\omega} \frac{(2\zeta)^{\omega-2t}}{t!} (\omega)_{-2t} D_{\zeta^2}^{\omega-2t} [k^2+\zeta^2]^{-l-1} \tag{14.49}$$

we obtain for the integral i_1 the following:

$$i_1 = C_{n_1 l_1} C_{NL} (2\zeta_1)^{l_1} (2p_0)^L \times \sum^{n_1+N-l_1-L-2} a_\tau(n_1 l_1 NL; \zeta_1, p_0)(-2p_0)^\tau$$
$$\cdot(-1)^{l_1+L-l+\tau+1} 2^l l!\, k^{l+1} \sum_{t=0}^{\omega} \frac{(2\zeta)^{\omega-2t}}{t!} (\omega)_{-2t} D_{\zeta^2}^{\omega-2t} (k^2+\zeta^2)^{-l-1} \tag{14.50}$$

where $\omega = l_1 + L - l + \tau + 1$, $(\omega)_{-2t} = \omega(\omega-1)\cdots(\omega-2t+1)$,

$$a_\tau(n_1 l_1 NL; \zeta_1 p_0) = \sum_{\sigma=0}^{n_1-l_1-1} \binom{n_1+l_1}{n_1-l_1-1-\sigma} \binom{N+L}{N-L-1-\tau+\sigma} \frac{1}{\sigma!(\tau-\sigma)!} \left(\frac{\zeta_1}{p_0}\right)^\sigma$$

using the relation

$$D_{\zeta^2}^{\omega-2t} (k^2+\zeta^2)^{-l-1} = (-1)^{\omega-2t} (l+1)_{\omega-2t-1} (k^2+\zeta^2)^{-l-1-\omega+2t}$$

Thus the integral i_1 has been calculated in a finite form. To calculate y_1 (equation (14.45)) we have to find integrals over d^3k (denoting them by i_2):

$$i_2 = \int \mathrm{d}^3k\, k^{-1} \mathrm{e}^{i\vec{k}R_{21}} \psi_2^*(\vec{k}) k^{l+1} (k^2+\zeta^2)^{-l-1-\omega+2t} Y_{lm}(\theta_{\vec{k}}, \varphi_{\vec{k}}) \tag{14.51}$$

Again, Fourier transform (14.51) is reduced to the Hankel transform over a variable k after integration over angle variables:

$$i_2 = \frac{4\pi}{R_{21}} \sum_{l'=|l_2-l|}^{l_2+l} \sum_{m'=-l'}^{l'} i^{l'} Y_{l'm'}(\theta_R, \varphi_R) \tilde{C}_{lm,l_2m_2}^{l'm'}$$
$$\times \int_0^\infty \mathrm{d}k\, k^{l+1} \tilde{\Pi}_{n_2l_2}^*(k, \zeta_2)(k^2+\zeta^2)^{-l-1-\omega+2t} j_{l'}(kR_{21}) \tag{14.52}$$

where

$$\tilde{\Pi}_{nl}(k) = (\pi/\sqrt{8}) \zeta^{5/2} (\zeta^2+k^2)^{-2} b_{nl} \times \left(\frac{2\zeta k}{(\zeta^2+k^2)}\right)^l C_{n-l-1}^{l+1} \left(\frac{\zeta^2-k^2}{\zeta^2+k^2}\right) \tag{14.53}$$

To calculate Hankel transform (14.52) let us represent the Gegenbauer polynomial C_{n-l-1}^{l+1} as a Gauss hypergeometric function[70]. Then function (14.53) will have a form of a finite sum of inverse powers of a quadratic form $\zeta^2 + k^2$:

$$\tilde{\Pi}_{nl}(k) = ab_{nl}(-1)^{n-l-1}\frac{(2l+2)_{n-l-1}}{(n-l-1)!}(2\zeta)^l k^l$$
$$\times \sum_{n=0}^{n-l-1} \frac{(\alpha)_n(\beta)_n(-1)^{n+l+1}}{(\gamma)_n n!(n+l+1)!}\zeta^{2n} D_{\zeta^2}^{n+l+1}(\zeta^2+k^2)^{-1} \quad (14.54)$$

where $\alpha = -n+l+1$, $\beta = n+l+1$, $\gamma = l+3/2$, D_{ζ^2} is a differentiation operator.

The function $(k^2+\zeta^2)^{-l-1-\omega+2t}$ in equation (14.52) may also be expressed as a derivative of $(k^2+\zeta^2)^{-1}$ with respect to ζ^2. As a result, integral (14.52) can be expressed via a table integral of the type

$$I = \int_0^\infty \mathrm{d}k k^{l+l_2+1}(k^2+\zeta_2^2)^{-1}(k^2+\zeta^2)^{-1} j_{l'}(k, R_{21}) \quad (14.55a)$$

Since $l + l_2 - l'$ is even, equation (14.55) may be rewritten in the form

$$I = \int_0^\infty \mathrm{d}k k^{2u} k^{l'+1}(k^2+\zeta_2^2)^{-1}(k^2+\zeta^2)^{-1} j_{l'}(kR_{21}) \quad (14.55b)$$

Let us decompose the rational function in equation (14.55a) into elementary fractions according to

$$(k^2+\zeta_2^2)^{-1}(k^2+\zeta^2)^{-1} = \frac{1}{\zeta^2-\zeta_2^2}\left(\frac{1}{k^2+\zeta^2} - \frac{1}{k^2+\zeta^2}\right)$$

Equation (14.55a) becomes

$$I = \frac{1}{\zeta^2-\zeta_2^2}(-1)^u \frac{\pi}{2R_{21}^{l'}}[\zeta_2^{2u}\hat{k}_{l'+1/2}(\zeta_2 R_{21}) - \zeta^{2u}\hat{k}_{l'+1/2}(\zeta R_{21})] \quad (14.56)$$

where $\hat{k}_{l+1/2}$ is a reduced Bessel function (RBF)

With the help of the formula for I the final form of equation (14.45) is a finite sum of elementary functions:

$$y_1 = (2\pi)^{-3/2}(4\pi)^2 C_{n_1l_1}C_{NL}(2\zeta_1)^{l_1}(2p_0)^L a_2 b_{n_2l_2}(-1)^{n_2-l_2-1} \times \frac{(2l_2+2)_{n_2-l_2-1}}{(n_2-l_2-1)!}$$
$$\times (2\zeta_2)^{l_2} \sum_{l=|l_1-L|}^{l_1+L} \sum_{m=-l}^{l} \tilde{C}^{lm}_{l_1m_1LM}(-i)^l \times \sum_{\tau=0}^{n_1+N-l_1-L-2} a_\tau(n_1l_1NL;\zeta_1p_0)$$
$$\times (-2p_0)^\tau(-1)^{l_1+L-l+\tau+1}2^l l! \sum_{t=0}^{\omega}\frac{(2\zeta)^{\omega-2t}}{t!}(\omega)_{-2t}(-1)^l\frac{1}{l!} \times \sum_{l'=|l_2-l|}^{l_2+l}$$
$$\times \sum_{m'=-l'}^{l'} i^{l'}\tilde{C}^{l'm'}_{lm,l_2m_2}R_{21}^{-1}Y_{l'm'}(\theta_{\vec{R}},\varphi_{\vec{R}}) \sum_{n=0}^{n_2-l_2-1}\frac{(\alpha)_n(\beta)_n(-1)^{n+l_2+1}}{(\gamma)_n n!(n+l_2+1)!}\zeta_2^{ln}$$
$$\times D_{\zeta^2}^{\omega-2t+l} D_{\zeta_2^2}^{n+l_2+1} I(l, l_2, l', \zeta_2^2, \zeta^2; R_{21}) \quad (14.57)$$

where $\omega = l_1 + L - l + \tau + 1$.

Calculation of derivatives in equation (57) can be done by using the properties of RBFs[57,71], their recurrence relations and decompositions of products of powers and RBF into finite sums of RBF. Here we restrict ourselves to equation (14.57).

It follows from equations (14.56) and (14.57) that an important peculiarity of the integral y_1 is that the sum (14.57) yields a sum of multiple potentials weighted with RBFs, which decrease exponentially with increasing R_{21}. Thus, at large distances R_{21} the exponential decreasing with anisotropic behaviour in a space which is described by spherical harmonics, $Y_{lm}(\theta_{\vec{R}}, \varphi_{\vec{R}})$ in equation (14.57), is a predominating factor.

Now let us consider the integrals q in equation (14.41):

$$q = \int d^3\vec{p}\, e^{-i\vec{p}\vec{R}_{c1}} \vec{p}^{\,-2} (p_0^2 + \vec{p}^{\,2})^{-1} \Psi^*_{p_0,NLM}(\vec{p}) \tag{14.58}$$

Transforming the HLF $\Psi^*_{p_0,NLM}$ to 0(3) representation by

$$\psi_{NLM}(\vec{p}) = a(p_0)(p_0^2 + \vec{p}^{\,2})^{-2} \Psi_{p_0,NLM}(\Omega) \tag{14.59}$$

gives

$$\begin{aligned} q &= a^{-1} \int d^3\vec{p}\, e^{-i\vec{p}\vec{R}_{c1}} \vec{p}^{\,-2} (p_0^2 + \vec{p}^{\,2}) \psi^*_{NLM}(\vec{p}) \\ &= a^{-1} \int d^3\vec{p}\, e^{-i\vec{p}\vec{R}_{c1}} \psi^*_{NLM}(\vec{p}) \\ &\quad + a^{-1} p_0^2 \int d^3\vec{p}\, e^{-i\vec{p}\vec{R}_{c1}} \vec{p}^{\,-2} \psi^*_{NLM}(\vec{p}) \end{aligned} \tag{14.60}$$

The first integral in equality (14.60) is a Fourier image of HLF, defined in a momentum space, i.e.

$$\psi^*_{NLM}(p_0 \vec{R}_{c1}) = (2\pi)^{-3/2} \int d^3\vec{p}\, e^{-i\vec{p}\vec{R}_{c1}} \Psi^*_{p_0,NLM}(\vec{p}) \tag{14.61}$$

The second integral is a Coulombic integral. Expanding the exponential in equation (14.60) in a series in 0(3) spherical harmonics, we integrate over a variable angle, resulting in one integral over dp which can be readily transformed into the table form, as in equation (14.55). The final expression for integral q is

$$\begin{aligned} q &= a^{-1}(2\pi)^{3/2} \psi^*_{NLM}(p_0 \vec{R}_{c1}) + a^{-1} 2\pi^2 i^L R_{c1}^{-L-1} Y_{LM}(\theta_{\vec{R}}, \varphi_{\vec{R}}) a_{NL} b_{NL} (-1)^N \\ &\quad \times \frac{(2L+2)_{N-L-1}}{(N-L-1)!} 2^{-L} \sum_{n=0}^{N-L-1} \frac{(\alpha)_n (\beta)_n (-1)^{n+L+1}}{(\gamma)_n n! (n+L+1)!} \\ &\quad \times p_0^{2n} \left[\frac{(2L)!}{L!} (-1)^n (n+L+1)!\, p_0^{-2(n+L+2)} \right. \\ &\quad \left. + 2^{-(n+L+1)} R_{c1} p_0^{-2n-1} \hat{k}_{n+1/2}(p_0 R_{c1}) \right] \end{aligned} \tag{14.62}$$

It follows from equation (14.62) that integral q consists of a multipole part, described by a solid harmonic of type $R_{c1}^{-L-1} Y_{LM}(\theta_{\vec{R}}, \varphi_{\vec{R}})$ and by an exponentially decreasing part weighted with a polynomial of R_{c1} and spherical harmonic $Y_{LM}(\theta_{\vec{R}}, \varphi_{\vec{R}})$.

To obtain formulae for the force matrix element we need to calculate a gradient of the quantity q. For this purpose, the gradient operator is expressed in cyclic form[72]:

$$\nabla_1 = -\frac{1}{\sqrt{2}}(\nabla_x + i\nabla_y), \quad \nabla_0 = \nabla_z, \quad \nabla_{-1} = \frac{1}{\sqrt{2}}(\nabla_x - i\nabla_y)$$

Since the angle θ, φ dependence in equation (14.62) is separated from the radial one, a differentiation should be carried out in accordance with the following[72]:

$$\nabla_0 R(r) Y_{LM}(\theta, \varphi) = \sqrt{\left[\frac{(L+1)^2 - M^2}{(2L+1)(2L+3)}\right]}\left(\frac{\mathrm{d}R}{\mathrm{d}r} - \frac{L}{r}R\right) Y_{L+1,M}(\theta, \varphi)$$
$$+ \sqrt{\frac{L^2 - M^2}{(2L-1)(2L+1)}}\left(\frac{\mathrm{d}R}{\mathrm{d}r} + \frac{L+1}{r}R\right) Y_{L-1,M}(\theta, \varphi)$$

$$\nabla_{\pm 1} R(r) Y_{LM}(\theta, \varphi)$$
$$= \sqrt{\left[\frac{(L \pm M + 1)(L \pm M + 2)}{2(2L+1)(2L+3)}\right]}\left(\frac{\mathrm{d}R}{\mathrm{d}r} - \frac{L}{r}R\right) Y_{L+1,M+1}(\theta, \varphi)$$
$$- \sqrt{\left[\frac{(L \mp M - 1)(L \mp M)}{2(2L-1)(2L+1)}\right]}\left(\frac{\mathrm{d}R}{\mathrm{d}r} + \frac{(L+1)}{r}R\right) Y_{L-1,M+1}(\theta, \varphi) \qquad (14.63)$$

We may readily calculate HLF and RBF derivatives of r with the help of recurrence relations for the functions. Therefore a differentiation formula takes place for RBF:

$$\frac{\mathrm{d}}{z\mathrm{d}z}\hat{k}_\nu(z) = -\hat{k}_{\nu-1}(z) \qquad (14.64)$$

Finally, the force matrix element (14.31) has been expressed as a series in multipoles and functions of HLF and RBF types with exponential decreasing as $R_c \to \infty$. A convergence of the series follows from the two-centre form factor (14.35) expansion (14.38) in HLF.

At $R_{21} = 0$ the series breaks and the force integral has the form of a finite sum of the elementary functions.

Approximate method for calculating vibronic interaction elements Let us rewrite the integral $\langle \mu | \nabla_1 (1/r_{1m}) | \nu \rangle$ in the form

$$J_{\mu\nu} = \left\langle \mu \middle| \nabla_1 \left(\frac{1}{r_{1m}}\right) \middle| \nu \right\rangle$$
$$= -\int \psi_\mu \frac{1}{r_{1m}} (\nabla_1 \psi_\nu) \mathrm{d}\tau_1 - \int (\nabla_1 \psi_\mu) \frac{1}{r_{1m}} \psi_\nu \mathrm{d}\tau_1 = I_{\mu\nu} + I_{\nu\mu} \qquad (14.65)$$

In order to calculate integrals of the type $I_{\mu\nu}$, we need approximate AO ψ_n and the values r_{1m}^{-1} from sums of the Gaussian exponents[68]. In this case we obtain

$$I_{\mu\nu}^x = -\left(\int \psi_\mu \frac{1}{r_{1m}} \varphi_\nu^{(1)} \mathrm{d}\tau_1 + \int \psi_\mu \frac{1}{r_{1m}} \varphi_\nu^{(2)} \mathrm{d}\tau_1\right) \tag{14.66}$$

where

$$\psi_\mu \approx N_\mu (x_1 - X_\mu)^{l\mu} (y_1 - Y_\mu)^{m\mu} (z_1 - Z_\mu)^{n\mu} \sum_{i=1}^{h_1} \mathrm{d}_i^\mu \exp(-\sigma_i^\mu (\vec{r}_1 - \vec{R}_\mu)^2)$$

$$\varphi_\nu^{(1)} \approx N_\nu l_\nu (x_1 - X_\nu)^{l\nu - 1} (y_1 - Y_\nu)^{m\nu} (z_1 - z_\nu)^{n\nu} \times \sum_{i=1}^{h_1} \mathrm{d}_i^\nu \exp(-\sigma_i^\nu (r_1 - R_\nu)^2), \quad (l_\nu = 0 \rightarrow \varphi_\nu^{(1)} = 0)$$

$$\varphi_\nu^{(2)} \approx -N_\nu 2 (x_1 - X_\nu)^{l\nu + 1} (y_1 - Y_\nu)^{m\nu} (z_1 - Z_\nu)^{n\nu} \times \sum_{i=1}^{h_1} \mathrm{d}_i^\nu \sigma_i^\nu \exp(-\sigma_i^\nu (\vec{r}_1 - \vec{R}_\nu)^2)$$

$$\frac{1}{r_{1m}} \approx \sum_{k=1}^{h_2} b_k \exp(-\sigma_k (\vec{r}_1 - \vec{R}_m)^2)$$

$R_\eta(X_\eta, Y_\eta, Z_\eta)$ is the radius vector of the point around which the AO ψ_n is centred or where the nucleus number η is located; N_ν are the normalization factors determined from the relationship $\langle \mu | \nu \rangle = \delta_{\mu\nu}$

Then the integrals in equation (14.66) are reduced to those having the form

$$\int \psi_\mu \frac{1}{r_{1m}} \psi_\nu \mathrm{d}\tau_1$$

with the corresponding factors and values of the quantum numbers l_ν, m_ν, n_ν. It can be shown that these integrals are equal to

$$\int \psi_\mu \frac{1}{r_{1m}} \psi_\nu \mathrm{d}\tau_1 = N_\mu N_\nu \sum_{i=1}^{h_1} \sum_{j=1}^{h_1} \mathrm{d}_i^\mu \mathrm{d}_j^\nu \sum_{k=1}^{h_2} b_k \times F_{ijk}^x F_{ijk}^y F_{ijk}^z \cdot \exp\{-[\sigma_i^\mu \sigma_j^\nu (\vec{R}_\mu - \vec{R}_\nu)^2 + \sigma_k (\sigma_i^\mu (\vec{R}_\mu - \vec{R}_m)^2 + \sigma_j^\nu (\vec{R}_\nu - \vec{R}_k)^2] \delta_{ijk}^{\mu\nu}\} \tag{14.67}$$

where

$$F_{ijk}^x = \sum_{p=0}^{l_\mu} \sum_{q=0}^{l_\nu} \Gamma\left(\frac{l_\mu + l_\nu - p - q + 1}{2}\right) (\delta_{ijk}^{\mu\nu})^{l\mu + l\nu - p - q + 1/2} \times \binom{l_\mu}{p} \binom{l_\nu}{q} (Q_x - R_{\mu x})^p (Q_x - R_{\nu x})^q$$

$(l_\mu + l_\nu - p - q)$ is even

$$\delta_{ijk}^{\mu\nu} = (\sigma_i^\mu + \sigma_j^\nu + \sigma_k)^{-1}$$
$$Q_x = (\sigma_i^\mu R_{\mu x} + \sigma_j^\nu R_{\nu x} + \sigma_k R_{kx})\delta_{ijk}^{\mu\nu}$$

F_{ijk}^y and F_{ijk}^z are derived from F_{ijk}^x by substituting y, z for x and m, n for l.

For normalization factors

$$N_\eta^{-2} = [\Gamma(l_\eta + 1/2)]^3 \sum_{i=1}^{h_1} \sum_{j=1}^{h_1} \mathrm{d}_i^\eta \mathrm{d}_j^\eta (\sigma_i^\eta + \sigma_j^\eta)^{-3l_\eta - 3/2}$$

The accuracy within which vibronic interaction integrals (14.29) are calculated for the given molecular orbitals t and u by equations (14.66) and (14.67) is governed by the that of the approximation of the Slater orbitals and the functions r^{-1} by Gaussian exponents. The calculations have revealed that the error of calculation for integrals such as equation (14.67) is very low, i.e. about 1%.

14.2 Spectral Distribution Curves of Absorption Coefficients

In this section the vibrational structures of the absorption spectra of butadiene, hexatriene, diphenylhexatriene, toluene, p-xylene and benzonitrile are calculated in terms of the suggested method which is the most general from those available for calculating the vibronic spectra of polyatomic molecules; the extent of the manifestations of normal co-ordinate mixing and vibronic interaction in the vibronic spectra are analysed.

14.2.1 Method for calculating vibrational structure from the electron spectra of polyatomic molecules

As noted earlier, the methods available for calculating the intensity of vibronic transitions suffer from some disadvantages. These are due mainly to the absence of reliable methods for calculating matrix elements. Moreover, a determination of the geometric and field force characteristics of molecules in excited electronic states involves certain difficulties. Although the geometry of molecules in the ground state can be established quite accurately by X-ray diffraction or electron diffraction analyses and the frequencies and modes of vibrations calculated from the infra-red and Raman spectra available, for excited states no such information exists. As pointed out in Section 14.1, the accuracy of quantum methods is not sufficient to calculate the molecular geometry and force constants, and the diffuse character of electron spectra does not provide the information necessary for solving the inverse spectral problem in the excited electronic state.

The method suggested for calculating the vibrational structure of electron spectra of molecules is based on general methods of integral calculation treated in detail in Section 14.1 together with geometric and force models of a molecule in excited electronic states, constructed with the help of semi-empirical dependences between the lengths, force constants and indices of bonds.

Such dependences between the lengths, force constants, and indices have been obtained for the C—H and C—C bonds. In calculating the indices of bonds for the methane, ethane, ethylene, acetylene, diacetylene, butadiene, hexatriene, benzene, *p*-xylene, hydrocyanic acid, methanol, formaldehyde and acetaldehyde molecules, use was made of the CNDO/S quantum-chemical method, with the corresponding experimental values of bond lengths and force constants taken from the literature. These dependences are shown in Figure 14.2. By the bond index between atoms μ and ν is implied the value

$$\xi_{\mu\nu} = \sum_{i\in\mu,\, j\in\nu} p_{ij} S_{ij}$$

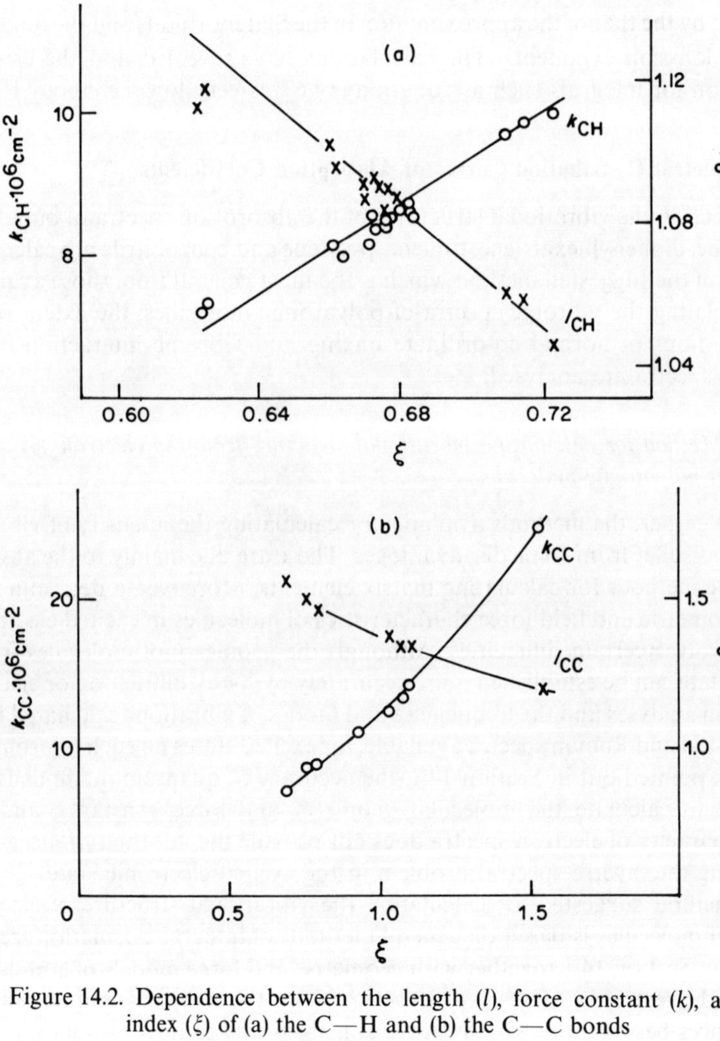

Figure 14.2. Dependence between the length (l), force constant (k), and index (ξ) of (a) the C—H and (b) the C—C bonds

where p_{ij} is the density matrix element and S_{ij} the overlap integral of the ith and jth atomic orbitals. This value characterizes, in a way similar to the bond order of a π bond, the magnitude of the charge on the $\mu\nu$ bond and, hence, reflects the rigidity of this bond. The dependences of Figure 14.2 can be used to estimate the behaviour of both the σ and π bonds. Our calculations have revealed that the results on π bonds obtained by the correlations of Popov *et al.*[17–19] and our dependences (Figure 14.2) are in fairly good agreement, as expected.

It is worth noting that the semi-empirical dependences used have been derived for the ground state of the molecules. No appreciable change, however, is expected on transition to excited states, since the physical difference between the ground and excited states lies in a changed distribution of electron density, which shows up in the density matrix variations and is accounted for in these dependences. This point is supported by the results of reference 20, where a simpler, linear relationship by Longuet-Higgins and Salem[12] between the order and length of the C—C bond is shown to be approximately valid both for the ground and excited electronic states. A similar conclusion follows from the analysis performed in reference 21.

However, a definitive conclusion about the applicability of ground state dependences to excited states and the validity of the models chosen for molecules in excited states can be reached only after a rigorous calculation of the vibrational structure of the electron structure of molecules, followed by a comparison between the results obtained with experiment.

A positive aspect of this approach is that, in calculating the vibrational structure of the electron spectra of a molecule, hardly any information is taken from the spectrum calculated, so this approach can be considered as non-empirical and the calculation concerned is of a predicative nature.

When calculating the vibrational structure of the electron spectrum of a molecule and comparing it with experiment we should take into account the density of spectral range filling by vibrational components. To this end, within our approach every vibrational component is represented by a Gaussian curve, which permits, on a corresponding summation, the construction of a contour for the vibronic band which is directly comparable with experiment. Since, at present, it is impossible to estimate vibronic line halfwidths theoretically, these are treated in our approach as empirical parameters, which are taken to be the same for all the vibrational components of the vibronic band and chosen with the aim of getting the best possible agreement between the calculated contour and the experimental spectral line. Vibrational component halfwidths are the only information obtained from the calculated spectrum. Such a choice of halfwidths does not imply the fitting of theoretical vibronic spectrum to experiment over a wide spectral range.

In order to verify the efficiency of the suggested method and investigate the results of various effects in the vibronic spectra of polyatomic molecules, the vibrational structure of the electron spectra of the butadiene, hexatriene, diphenylhexatriene, *p*-xylene, benzonitrile and toluene molecules have been calculated.

14.2.2 *Calculation of vibrational structure from the electron spectra of molecules*

To calculate the electronic structure and spectra of the molecules concerned, use was made of the semi-empirical CNDO/S method, with due reference to configuration interaction[75]. The predicted energies of electronic transitions are in satisfactory agreement with the observations (the calculation error was under 10%). Based on the calculated matrices of electron density in the combining electronic states and on the correlations derived (Figure 14.2), variations in the lengths and force constants of molecular bonds on transition to the first excited state were determined. For butadiene and hexatriene these variations were as follows: $\Delta l_1 = +0.098$ Å, $\Delta l_2 = -0.06$ Å, $\Delta k_1 = -3.0 \times 10^6\ \mathrm{cm}^{-2}$, $\Delta k_2 = 1.13 \times 10^6\ \mathrm{cm}^{-2}$, and $\Delta l_1 = 0.053$ Å, $\Delta l_2 = -0.04$ Å, $\Delta l_3 = 0.072$ Å, $\Delta k_1 = -1.6 \times 10^6\ \mathrm{cm}^{-2}$, $\Delta k_2 = 1.08 \times 10^6\ \mathrm{cm}^{-2}$, $\Delta k_3 = -2.9 \times 10^6\ \mathrm{cm}^{-2}$, respectively (for numbering of the bonds, see Figure 14.3). The variations in the C—H bond parameters for all the molecules concerned are negligible. This is in accord with the fact that the C—H vibrations do not show up in vibronic spectra.

The vibrations of molecules in the combining electronic states were calculated by the known techniques[54]. The calculated frequencies of molecular vibration differ from the experimental values by no more than 25 cm^{-1}.

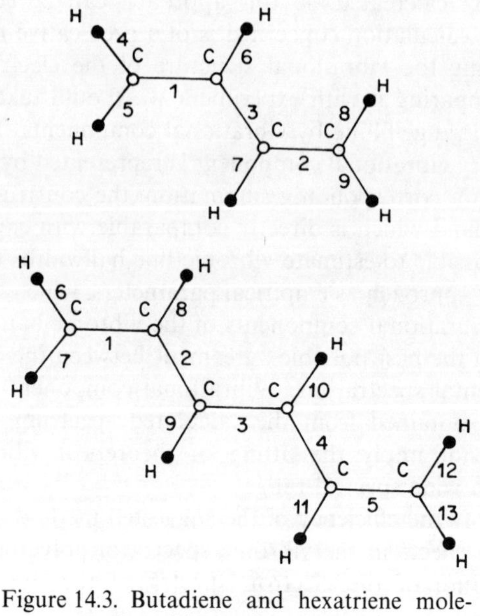

Figure 14.3. Butadiene and hexatriene molecules with numbered bonds

Then, based on the developed methods and calculated vibration frequencies and modes in the combining electronic states, the vibrational structure of the first allowed electronic transition of the molecules in question have been calculated.

The Condon approximation: evaluation of the normal co-ordinate mixing effect. The first-phase calculations were conducted in the Condon approximation, and the results obtained are given in Figures 14.4–14.8. In the case of butadiene, hexatriene, benzonitrile and *p*-xylene, the discrimination level (DEL) was taken to be 0.15, i.e. all the non-diagonal elements of the normal co-ordinate rotation matrix, which were of absolute value under 0.15, were considered to be zero and the mixing between the corresponding normal co-ordinates was not taken into account. With this discrimination level, the calculation error of the vibronic component intensity was less than 0.05 rel. un. (the intensity of the purely electronic transition was assumed to be unity). The corresponding values for diphenylhexatriene were 0.4 and 0.1.

In Figures 14.4–14.8 we can see both the relative intensities of individual vibrational lines (the vertical lines in Figures 14.4–14.6) and the absorption band contours (curves 2) corresponding to the representation of each vibrational line in the form of a Gaussian curve with a halfwidth of 400 cm^{-1} (butadiene), 250 cm^{-1} (benzonitrile) and 200 cm^{-1} (hexatriene, diphenylhexatriene, *p*-xylene).

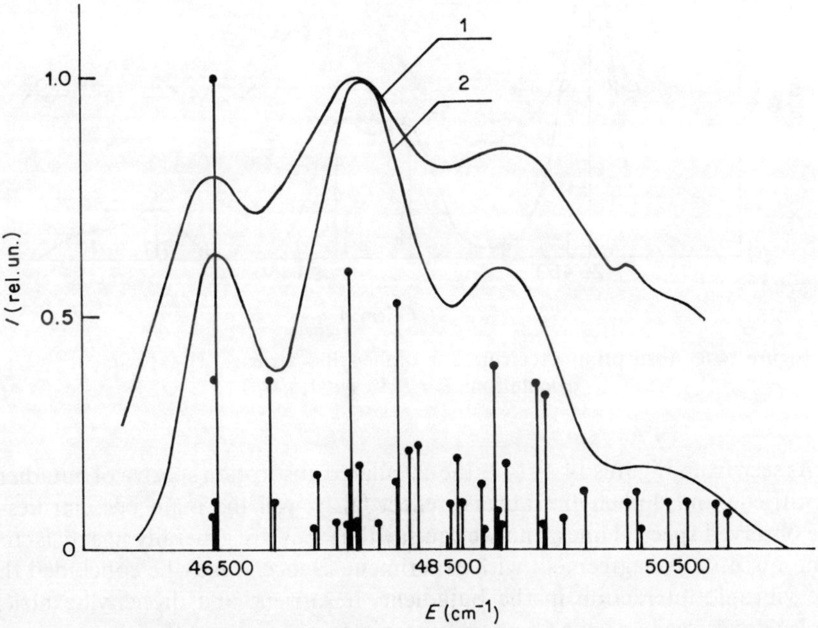

Figure 14.4. Absorption spectrum for butadiene: (1) experiment[77]; (2) calculation; $T = 195$ K, $v \leqslant 2$, $v' \leqslant 3$

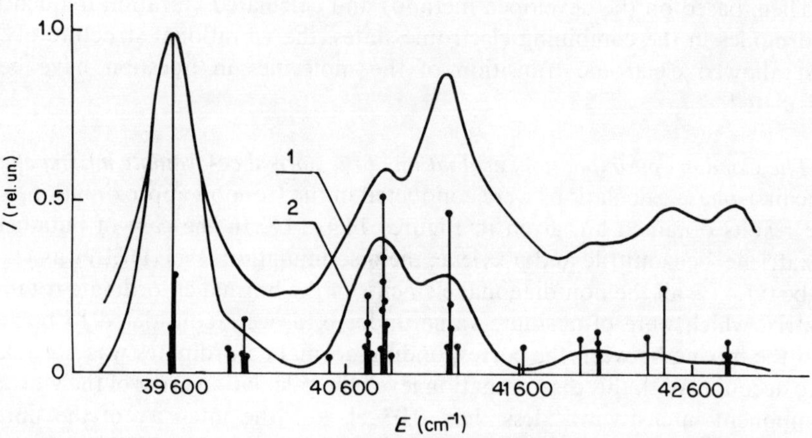

Figure 14.5. Absorption spectrum for hexatriene: (1) experiment[78]; (2) calculation; $T = 300$ K, $v \leqslant 2$, $v' \leqslant 3$

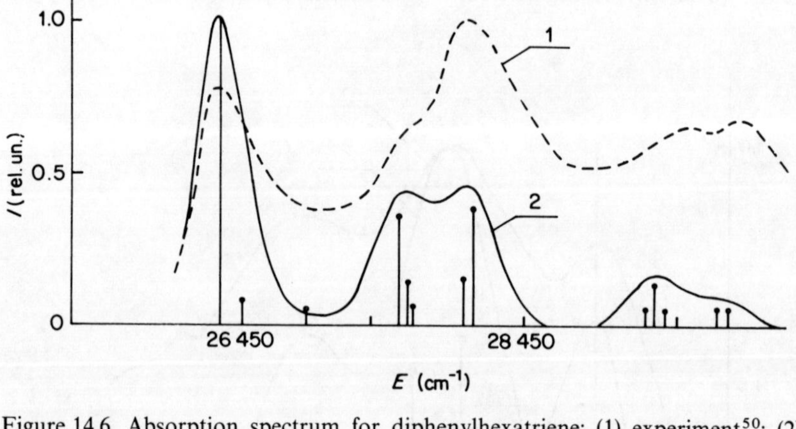

Figure 14.6. Absorption spectrum for diphenylhexatriene: (1) experiment[50]; (2) calculation; $T = 77$ K, $v \leqslant 1$, $v' \leqslant 3$

As seen from Figures 14.4–14.6, the calculated absorption spectra of butadiene, hexatriene and diphenylhexatriene reflect fairly well the main peculiarities of the observed spectral lines, and we can say that they are generally in satisfactory semi-quantitative agreement with experiment. Hence, it may be concluded that the vibronic interaction in the butadiene, hexatriene and diphenylhexatriene molecules is weak and the Condon approximation is quite sufficient to interpret their vibronic spectra.

Comparison of the calculated and observed spectra of butadiene, hexatriene

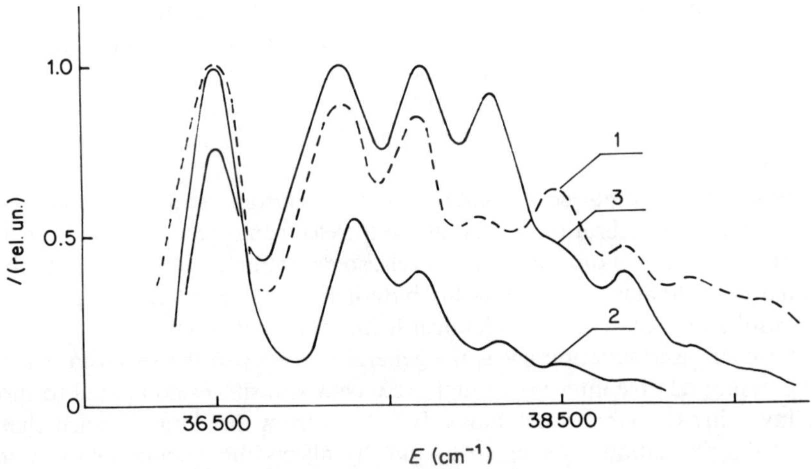

Figure 14.7. Absorption spectrum for *p*-xylene: (1) experiment[55]; (2) calculation—the Condon approximation; (3) the Herzberg–Teller approximation; $T = 300$ K, $v \leqslant 2$, $v' \leqslant 3$

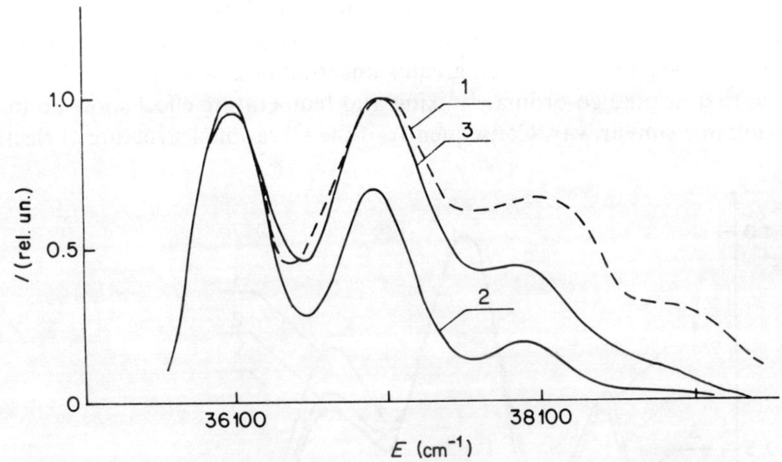

Figure 14.8. Absorption spectrum for benzonitrile: (1) experiment[55]; (2) calculation—the Condon approximation; (3) the Herzberg–Teller approximation; $T = 300$ K, $v \leqslant 2$, $v' \leqslant 3$

and diphenylhexatriene indicates that the chosen models are close to the actual structure of the molecules in the excited electronic states.

The intensity of the high-frequency region of the absorption spectra for *p*-xylene and benzonitrile is too low in the Condon approximation (Figures 14.7 and 14.8, curve 2). This inconsistency is due to a rather strong vibronic interaction, typical of the molecules of the substituted benzenes. This, however, will be treated in greater detail below.

One of the objectives of the present section was to investigate the effect of normal co-ordinate mixing on the vibronic spectra of polyatomic molecules. Such an investigation allows us both to determine how these effects show themselves in the vibronic spectra and to examine cruder methods of calculating the vibrational structure of electron spectra that do not take these effects into account.

The effect of normal co-ordinate mixing on vibronic spectra was analysed by calculating the vibrational structure of the electron spectra of butadiene and hexatriene at various discrimination levels, so determining the number of mixed normal co-ordinates. The results for butadiene are presented in Figure 14.9. The results for hexatriene are identical to those for butadiene.

At higher discrimination levels the general tendency in the spectrum change is increasing relative intensity of high-frequency transitions compared to those not involving the vibrational states. It follows from the data obtained that a rise of discrimination level up to 0.4 hardly affects the calculation accuracy, and the resulting error is $< 10\%$. Any further rise of discrimination level would make the calculations far less accurate. For butadiene, DEL = 0.6 corresponds to absolute neglect of normal co-ordinate mixing, the calculation error amounting to 30% or even higher (as seen from Figure 14.9). Note that in this case the so-called 'frequency effect' is taken into consideration, therefore methods assuming the normal co-ordinate rotation matrix to be unity (for example, in reference 31) are subject to even greater uncertainties.

Note that normal co-ordinate mixing and temperature effect show up in the spectrum in a similar way. Consequently, if the vibrational structure of electron

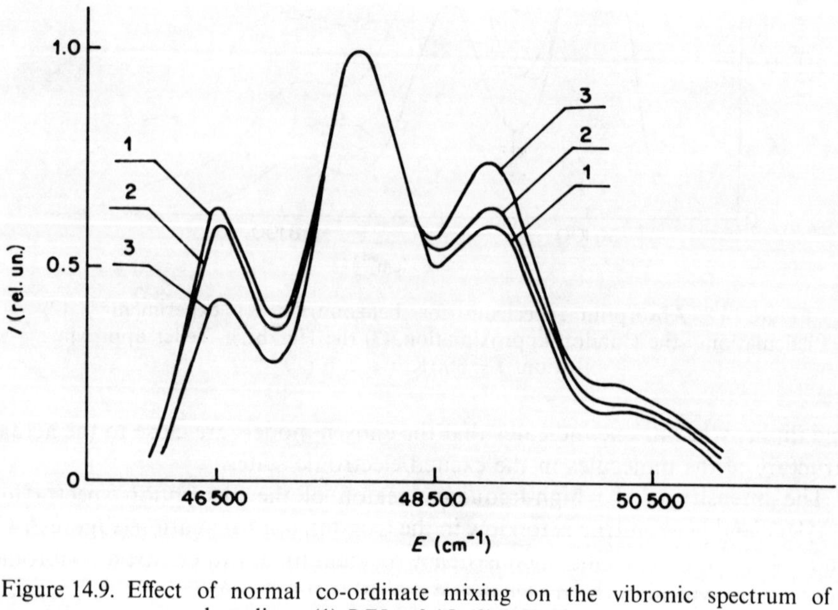

Figure 14.9. Effect of normal co-ordinate mixing on the vibronic spectrum of butadiene: (1) $DEL = 0.15$, (2) 0.40, (3) 0.60

spectra is analysed by methods which take no account of these effects (for example, see references 36 and 40), the results are subject to even greater uncertainties.

In summary, it has been shown that in calculating the vibrational structure of the electron spectra of polyatomic molecules, we should also consider co-ordinate mixing. In some cases, however, particularly those requiring fast calculation of numerous molecules to obtain a qualitative picture of the vibronic spectra of molecules, we may restrict ourselves to cruder approximations.

The Herzberg–Teller effect and convergence of the Herzberg–Teller expansion
In order to study the effect of vibronic interaction on the vibronic spectrum of a molecule the vibrational structure of the first symmetrically allowed electronic transition of the *p*-xylene and benzonitrile absorption spectra have been calculated. The presence of the methyl group in the *p*-xylene molecule and the CN group in benzonitrile reduced the molecular symmetry (compared to benzene) and, hence, removes the electronic state degeneracy. However, the electronic level splitting is slight, so that we expect a significant manifestation of vibronic interaction in the spectra. In addition, the oscillator strengths are much greater for the 0–2, 0–3, and 0–4 transitions than for the 0–1 transition, which is also responsible for the increased contributions of the Herzberg–Teller components to the vibronic spectrum, compared to the Condon treatment.

The elements of vibronic interaction for *p*-xylene and benzonitrile were calculated for all normal vibrations, as we cannot *a priori* evaluate the contribution of each of them, except where the vibronic interaction elements are equal to zero, due to their symmetry.

Table 14.4 lists, as an example, the calculated elements of vibronic interaction for some vibrations of the *p*-xylene molecule in the first excited electronic state

Table 14.4. Elements of vibronic interaction $\langle r|A_k^t|s\rangle$ of *p*-xylene for the first excited electronic state ($t = 1$)

Vibrational frequency (cm^{-1})	$r = 1$, $s = 2$	$r = 1$, $s = 3$	$r = 1$, $s = 4$
307	−0.0132	0	−0.0694
608	0.0478	0	0.2203
979	−0.0073	0	−0.0407
1256	−0.0296	0	−0.1518
1442	−0.0358	0	−0.0183
1597	−0.0558	0	−0.2879
3056	+0.0178	0	+0.0696
1190	0	−0.2392	0
1032	0	−0.2057	0
806	0	−0.1478	0
413	0	−0.1219	0
1691	0	−0.1949	0

$t = 1$, which contributes most of all to the vibronic spectrum. Such values of vibronic interaction elements for vibrations with a frequency of $608\,cm^{-1}$ of the *p*-xylene molecule, say, correspond to the coefficient before the Herzberg–Teller term $\langle v'|Q_k^t|v\rangle$ in the Herzberg–Teller expansion equal to 1.1, if the coefficient before the Condon component is assumed to be unity. Thus, the contribution of the Herzberg–Teller component is rather significant.

The results calculated from the vibrational structure of the electron spectra of *p*-xylene and benzonitrile are shown in Figures 14.7 and 14.8 and tabulated in Tables 14.5–14.7. These data have been obtained with due reference to the first (linear in the normal co-ordinate) terms in the Herzberg–Teller expansion. In the calculations, account has been taken of the vibronic perturbation of the excited electronic state only, since that of the ground electronic state is negligibly small.

In Figures 14.7 and 14.8 we can see the experimental absorption spectra for *p*-xylene and benzonitrile, respectively, at $T = 300\,K$[14] (curves 1) and the calculated spectral curves plotted in the Condon (curves 2) and Herzberg–Teller (curves 3) approximations. In this case, the Condon approximation provides a diffuse spectrum with the vibrational components too low in intensity in the high-frequency spectral region, whereas, as can be easily seen from Figures 14.7 and 14.8, an allowance for Herzberg–Teller components which are just linear in the normal co-ordinates would greatly clarify the calculated spectrum. A vibrational structure appears in the high-frequency region of the absorption spectrum and the observations are in satisfactory semi-quantitative agreement with the predicted spectral curves.

In evaluating the accuracy of the method used and interpreting the molecular vibronic spectrum, a factor of great value is the convergence rate of the Herzberg–Teller expansion. Here, the case of a weak vibronic interaction (the cases of degenerate and quasi-degenerate electronic states require special consideration) is implied.

As the coefficients before the quadratic terms in the normal co-ordinates and the subsequent terms in the Herzberg–Teller expansion are complex, we cannot evaluate *a priori* their orders of magnitude. Hence, to estimate the convergence of the Herzberg–Teller series these coefficients have been calculated for toluene, which permits an estimation of the contribution of the quadratic terms of the Herzberg–Teller expansion to be made. The calculations performed have shown that the quadratic terms do not exceed 0.035 and, as a result, their contribution to the vibronic spectrum is negligible.

The results for the toluene molecule are significant because, despite the high coefficients before the linear terms in the Herzberg–Teller expansion (comparable with those before the Condon component), the subsequent terms are negligibly small.

Such a result seems always to be the case when the contribution to the vibronic spectrum due to vibronic interaction is largely independent of the proximity of electronic states ($\varepsilon_t^0 - \varepsilon_u^0$ is low) and, hence, their coupling, but depends rather on the relationship of the probabilities for electronic transitions

Table 14.5. Frequencies and relative intensities of the vibrational components of the absorption electron spectrum for *p*-xylene ($T = 300$ K)

Assignment and frequency (cm^{-1}) of vibronic transition[a]		Intensity (rel. un.)	Assignment and frequency (cm^{-1}) of vibronic transition[a]		Intensity (rel. un.)
x-components			$2 \times \nu'_{35} + \nu'_{42} - \nu_{42}$	1609	0.297
$\nu'_{36} - \nu_{36}$	−10.3	0.149	$2 \times \nu'_{35} + \nu'_{1} - \nu_{1}$	1610	0.278
$\nu'_{31} - \nu_{31}$	−6.1	0.219	$2 \times \nu'_{35} + \nu'_{19} - \nu_{19}$	1612	0.403
$\nu'_{18} + \nu'_{19} - (\nu_{18} + \nu_{19})$	−6.0	0.130	$2 \times \nu'_{35} + \nu'_{25} - \nu_{25}$	1612	0.604
$\nu'_{18} - \nu_{18}$	−5.6	0.267	$2 \times \nu'_{35}$	1612	1.239
$\nu'_{25} + \nu'_{42} - (\nu_{25} + \nu_{42})$	−3.9	0.130	$\nu'_{34} + \nu'_{35}$	1838	0.164
$\nu'_{19} + \nu'_{42} - (\nu_{19} + \nu_{42})$	−3.7	0.194	$\nu'_{33} + \nu'_{35}$	1996	2.132
$\nu'_{42} - \nu_{42}$	−3.3	0.399	$2 \times \nu'_{35} + \nu'_{36}$	2025	0.461
$\nu'_{1} + \nu'_{19} - (\nu_{1} + \nu_{19})$	−2.8	0.181	$2 \times \nu'_{33} + \nu'_{19} - \nu_{19}$	2380	0.401
$\nu'_{1} - \nu_{1}$	−2.4	0.373	$2 \times \nu'_{33}$	2380	0.823
$\nu'_{19} + \nu'_{25} - (\nu_{19} + \nu_{25})$	−1.0	0.263	$3 \times \nu'_{35}$	2419	0.423
$\nu'_{25} - \nu_{25}$	−0.6	0.542	$\nu'_{33} + 2 \times \nu'_{35}$	2802	1.038
$\nu'_{19} - \nu_{19}$	−0.4	0.810	$2 \times \nu'_{33} + \nu'_{35}$	3189	0.740
	0	1.667	*y*-components		
$\nu'_{36} + \nu'_{42} - \nu_{42}$	409.3	0.228	ν'_{1}	307.4	0.249
$\nu'_{25} + \nu'_{36} - \nu_{25}$	411.9	0.310	$2 \times \nu'_{2} - \nu_{2}$	589.7	0.138
$\nu'_{19} + \nu'_{36} - \nu_{19}$	412.2	0.464	$\nu'_{2} + \nu'_{31} - \nu_{31}$	601.5	0.183
ν'_{36}	412.6	0.953	$\nu'_{2} + \nu'_{18} - \nu_{18}$	602.1	0.224
$\nu'_{35} + \nu'_{36} - \nu_{36}$	795.9	0.171	$\nu'_{2} + \nu'_{42} - \nu_{42}$	604.3	0.334
$\nu'_{35} + \nu'_{31} - \nu_{31}$	800.1	0.250	$\nu'_{1} + \nu'_{2} - \nu_{1}$	605.2	0.312
$\nu'_{35} + \nu'_{18} - \nu_{18}$	800.6	0.305	$\nu'_{2} + \nu'_{25} - \nu_{25}$	607.0	0.453
$\nu'_{35} + \nu'_{42} - \nu_{42}$	802.9	0.456	$\nu'_{2} + \nu'_{19} - \nu_{19}$	607.2	0.677
$\nu'_{35} + \nu'_{1} - \nu_{1}$	803.8	0.425	ν'_{2}	607.6	1.388
$\nu'_{35} + \nu'_{25} - \nu_{25}$	805.5	0.618	$\nu'_{2} + \nu'_{36}$	1020	0.232
$\nu'_{35} + \nu'_{19} - \nu_{19}$	805.8	1.082	ν'_{4}	1256	0.298
ν'_{35}	806.2	2.22	$\nu'_{2} + \nu'_{35}$	1414	1.004
ν'_{34}	1032	0.227	$\nu'_{6} + \nu'_{18} - \nu_{18}$	1591	0.134
$\nu'_{33} + \nu'_{36} - \nu_{36}$	1180	0.176	$\nu'_{6} + \nu'_{42} - \nu_{42}$	1594	0.200
$\nu'_{31} + \nu'_{33} - \nu_{31}$	1184	0.258	$\nu'_{1} + \nu'_{6} - \nu_{1}$	1594	0.187
$\nu'_{18} + \nu'_{33} - \nu_{18}$	1185	0.315	$\nu'_{6} + \nu'_{25} - \nu_{25}$	1596	0.271
$\nu'_{33} + \nu'_{42} - \nu_{42}$	1187	0.412	$\nu'_{6} + \nu'_{19} - \nu_{19}$	1596	0.405
$\nu'_{1} + \nu'_{33} - \nu_{33}$	1188	0.441	ν'_{6}	1597	0.832
$\nu'_{25} + \nu'_{33} - \nu_{25}$	1189	0.638	$\nu'_{2} + \nu'_{33}$	1798	0.645
ν'_{33}	1190	1.96	$\nu'_{6} + \nu'_{36}$	2009	0.139
$\nu'_{19} + \nu'_{33} - \nu_{19}$	1190	0.956	$\nu'_{4} + \nu'_{35}$	2062	0.216
$\nu'_{35} + \nu'_{36}$	1219	0.893	$\nu'_{2} + 2 \times \nu'_{35}$	2220	0.345
$\nu'_{33} + \nu'_{42}$	1603	0.774	$\nu'_{6} + \nu'_{35}$	2403	0.602

[a] Frequencies are counted from that of a purely electronic transition; for the calculated frequencies of vibrations of *p*-xylene see Table 14.7.

Table 14.6. Frequencies and relative intensities of the vibrational components of the absorption electron spectrum for benzonitrile ($T = 300$ K)

Assignment and frequency (cm^{-1}) of vibronic transition[a]		Intensity (rel. un.)	Assignment and frequency (cm^{-1}) of vibronic transition[a])		Intensity (rel. un.)
x-components			ν'_{10}	957.3	0.96
$\nu'_{39} - \nu_{30}$	−14.9	0.05	ν'_{8}	1179.7	0.05
$\nu'_{15} + \nu'_{22} - (\nu_{15} + \nu_{22})$	−10.0	0.07	$\nu'_{7} + \nu'_{22} - \nu_{22}$	1206	0.08
$\nu'_{15} + \nu'_{33} - (\nu_{15} + \nu_{33})$	−9.8	0.06	$\nu'_{7} + \nu'_{33} - \nu_{33}$	1206	0.08
$\nu'_{12} + \nu'_{22} - (\nu_{12} + \nu_{22})$	−9.6	0.04	ν'_{z}	1208	0.18
$\nu'_{15} - \nu_{15}$	−8.4	0.14	$\nu'_{10} + \nu'_{11}$	1690	0.13
$\nu'_{12} - \nu_{12}$	−8.0	0.09	$2 \times \nu'_{10} + \nu'_{22} - \nu_{22}$	1913	0.21
$\nu'_{21} + \nu'_{22} - (\nu_{21} + \nu_{22})$	−7.4	0.07	$2 \times \nu'_{10} + \nu'_{33} - \nu_{33}$	1913	0.19
$\nu'_{21} + \nu'_{33} - (\nu_{21} + \nu_{33})$	−7.2	0.06	$2 \times \nu'_{10}$	1915	0.44
$\nu'_{21} - \nu_{21}$	−5.8	0.15	$\nu'_{7} + \nu'_{10}$	2165	0.16
$\nu'_{20} - \nu_{20}$	−3.6	0.07	$2 \times \nu_{28}$	2448	0.05
$\nu'_{22} + \nu'_{33} - (\nu_{22} + \nu_{33})$	−2.9	0.19	$\nu'_{11} + 2 \times \nu'_{10}$	2647	0.08
$\nu'_{22} - \nu_{22}$	−1.6	0.44	$3 \times \nu'_{10}$	2872	0.13
$\nu'_{33} - \nu_{33}$	−1.3	0.40	*y*-components		
$\nu'_{32} - \nu_{32}$	−1.1	0.08	$\nu'_{31} + \nu'_{22} - \nu_{22}$	598.1	0.05
	0	0.93	$\nu'_{31} + \nu'_{33} - \nu_{33}$	598.4	0.04
$\nu'_{11} + \nu'_{22} - \nu_{22}$	730.6	0.06	ν'_{31}	599.7	0.10
$\nu'_{7} + \nu'_{33} - \nu_{33}$	730.8	0.06	ν'_{28}	1224	0.03
ν'_{11}	732.2	0.13	$\nu'_{25} + \nu'_{22} - \nu_{22}$	1527	0.05
$\nu'_{10} + \nu'_{15} - \nu_{15}$	948.9	0.13	$\nu'_{25} + \nu'_{33} - \nu_{33}$	1527	0.05
$\nu'_{10} + \nu'_{12} - \nu_{12}$	949.3	0.08	ν'_{25}	1529	0.11
$\nu'_{10} + \nu'_{21} - \nu_{21}$	951.5	0.14	$\nu'_{10} + \nu'_{31}$	1557	0.07
$\nu'_{10} + \nu'_{20} - \nu_{20}$	953.7	0.07	$\nu'_{10} + \nu'_{25}$	2486	0.08
$\nu'_{10} + \nu'_{22} - \nu_{22}$	955.7	0.45			
$\nu'_{20} + \nu'_{33} - \nu_{33}$	956.0	0.41			
$\nu'_{10} + \nu'_{32} - \nu_{32}$	956.3	0.07			

[a] The frequencies are counted from that of a purely electronic transition; for the calculated frequencies of benzonitrile see Table 14.7.

$|\langle u|\hat{\vec{\mu}}|s\rangle|/|\langle t|\hat{\vec{\mu}}|s\rangle|$, i.e. the case of prohibited or low-intensity electronic transitions.

The calculations carried out with due reference to vibronic interaction made it possible to estimate the contribution of the Herzberg–Teller terms to the vibronic spectrum of a molecule as well as to construct an explicit electron wave function for the excited state parametrically dependent on the nuclear co-ordinates.

In the case of the toluene molecule, the electron wave function of the first excited state within the Herzberg–Teller approximation has the form:

Table 14.7. Calculated vibrational frequencies for *p*-xylene and benzonitrile in the ground (v) and the first excited (v') electronic states

No.	*p*-Xylene			Benzonitrile		
	Type of symmetry	v(cm^{-1})	v'(cm^{-1})	Type of symmetry	v(cm^{-1})	v'(cm^{-1})
1	B_g	310	307	A_1	3071	3069
2		626	608		3061	3060
3		982	979		3056	3056
4		1268	1256		2265	2265
5		1442	1442		1617	1553
6		1685	1597		1507	1480
7		3057	3056		1225	1208
8		2968	2968		1181	1180
9		851	849		1017	988
10	A_u	1242	1227		991	957
11		1011	976		749	732
12		1345	1345		432	425
13		661	646	A_2	965	962
14		1618	1577		845	843
15		1077	1071		398	390
16		930	925	B_1	982	980
17		1445	1445		807	905
18		379	374		747	741
19		149	148		702	692
20		3061	3060		539	536
21		2894	2894		378	373
22		2967	2967		154	153
23	B_u	1003	974	B_2	3064	3063
24		959	940		3056	3055
25		233	232		1597	1529
26		1312	1211		1442	1421
27		1486	1409		1322	1307
28		1447	1447		1302	1224
29		3065	3063		1160	1143
30		2968	2968		1058	1028
31		421	415		614	600
32		1305	1288		525	524
33	A_g	1223	1190		176	175
34		1045	1032			
35		839	806			
36		423	413			
37		1345	1345			
38		1776	1691			
39		3068	3066			
40		2894	2894			
41		582	577			
42		296	292			
43		1011	1011			
44		1443	1443			
45		1616	1586			
46		2967	2967			

$$\Psi_1(r,Q)=\Psi_1(r,0)+(-0.117Q_{18}^1-0.099Q_{17}^1+0.052Q_{14}^1+0.039Q_{15}^1)$$
$$\Psi_2(r,0)+(0.136Q_6^1+0.112Q_3^1-0.098Q_{18}^1-0.094Q_{16}^1$$
$$-0.073Q_4^1-0.064Q_{17}^1-0.052Q_{12}^1)$$
$$\Psi_3(r,0)+(-0.119Q_{18}^1-0.103Q_{17}^1-0.082Q_6^1+0.073Q_{15}^1-0.068Q_3^1$$
$$+0.055Q_{16}^1)\Psi_4(r,0)$$

where $\Psi_u(r,0)$ is the electron wave function of the uth excited electronic state in the Condon approximation and Q_k^t is the kth normal co-ordinate of the tth electronic state (the electronic states are numbered, as before, from the first excited state upward, the ground state being assumed zero). The expression for the electron wave function contains only the major terms making the greatest contributions. The coefficients before the quadratic terms in the normal co-ordinates are equal in order of magnitude to the squared coefficients before the linear terms, hence they are small. The calculation accuracy of the coefficients before the normal co-ordinates in the electron function expression is governed by the calculation accuracy of the electronic transition energies ε_u^0 and by that of the matrix elements of the vibronic interaction. The calculation errors of the coefficients before the electron wave functions $\Psi_3(r,0)$ and $\Psi_4(r,0)$ are within 15–20%, whereas those of the coefficients before $\Psi_2(r,0)$ may amount to 50% or even higher. This is accounted for by the close proximity of the first and second excited electronic states. As the vibronic spectrum of toluene exhibits components due to vibronic interaction of the first excited state with the third and fourth states, relatively large errors arising from the determination of coefficients via $\Psi_2(r,0)$ hardly affect the calculation accuracy of the toluene vibronic spectrum in the Herzberg–Teller approximation (this is also true for *p*-xylene and benzonitrile).

As pointed out earlier, all the calculations were conducted in the adiabatic approximation. It is known that the closer energetically are the molecular electronic states, the lower is the accuracy of this approximation. It is shown in reference 79 that the Born–Oppenheimer non-adiabatic component of the dipole moment of the vibronic transition is proportional in magnitude to the Herzberg–Teller component, the coefficient of proportionality being equal to the ratio of the energy of a vibration quantum accomplishing the vibronic coupling to the difference of electronic state energies.

The energy difference of the first excited state of *p*-xylene and benzonitrile is about 8000 cm^{-1} (see, for instance, reference 55), which exceeds the vibration quanta energy by a factor of 5–10. Consequently, non-adiabacity in the vibronic spectra of *p*-xylene and benzonitrile must show up to a far lesser extent than do the Herzberg–Teller effects and, as the calculations show (Figures 14.7 and 14.8), an account of the Herzberg–Teller terms leads to fairly good agreement with experiment.

Unfortunately, the experimental magnitude of the second excited state of toluene is unknown, therefore the contribution of the non-adiabatic terms to the vibronic spectrum cannot be estimated. The estimates obtained from the calculated energies of the excited electronic states are not sufficiently accurate, as the calculation error of the electronic state energies is of the order of 2500–3500 cm^{-1} and is comparable with the calculated energy difference ΔE of 2000 cm^{-1}. The much greater discrepancies between the calculations and observations in the case of the toluene molecule, compared with xylene and benzonitrile, are due to the significant non-adiabacity of the electronic states of toluene.

The general methods suggested for calculating the overlap integrals of vibrational wave functions allow the vibrational structure of the electron spectra of polyatomic molecules to be calculated in the most general approximations (one-electron, adiabatic approximations). Our calculations of the vibronic spectra of butadiene, hexatriene, diphenylhexatriene, *p*-xylene and benzonitrile have indicated satisfactory semi-quantitative agreement between the calculated contours of the vibronic bands and the experimental spectral lines. This implies, in particular, that the models chosen for these molecules are adequate for the molecular structures in the excited electronic states. Therefore the parameters of the excited electronic state potential surface can be derived by introducing corrections to those of the ground state potential surface based on the

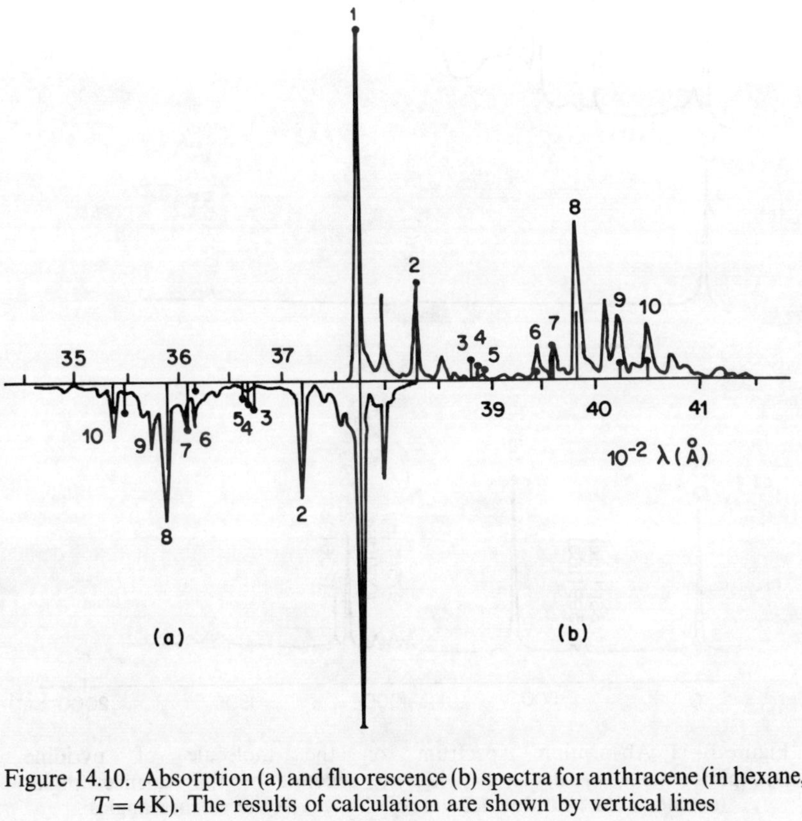

Figure 14.10. Absorption (a) and fluorescence (b) spectra for anthracene (in hexane, $T = 4$ K). The results of calculation are shown by vertical lines

semi-empirical correlations obtained for the indices, lengths and force constants of the bonds.

Comparison of the calculated and experimental spectral curves has revealed the following:

(1) Neglect of normal co-ordinate mixing may lead to an error in the intensity of the vibronic components of the vibronic band as large as 30%.
(2) Though the Franck–Condon approximation is satisfactory in some cases, even in the case of allowed (according to Franck–Condon) transitions the

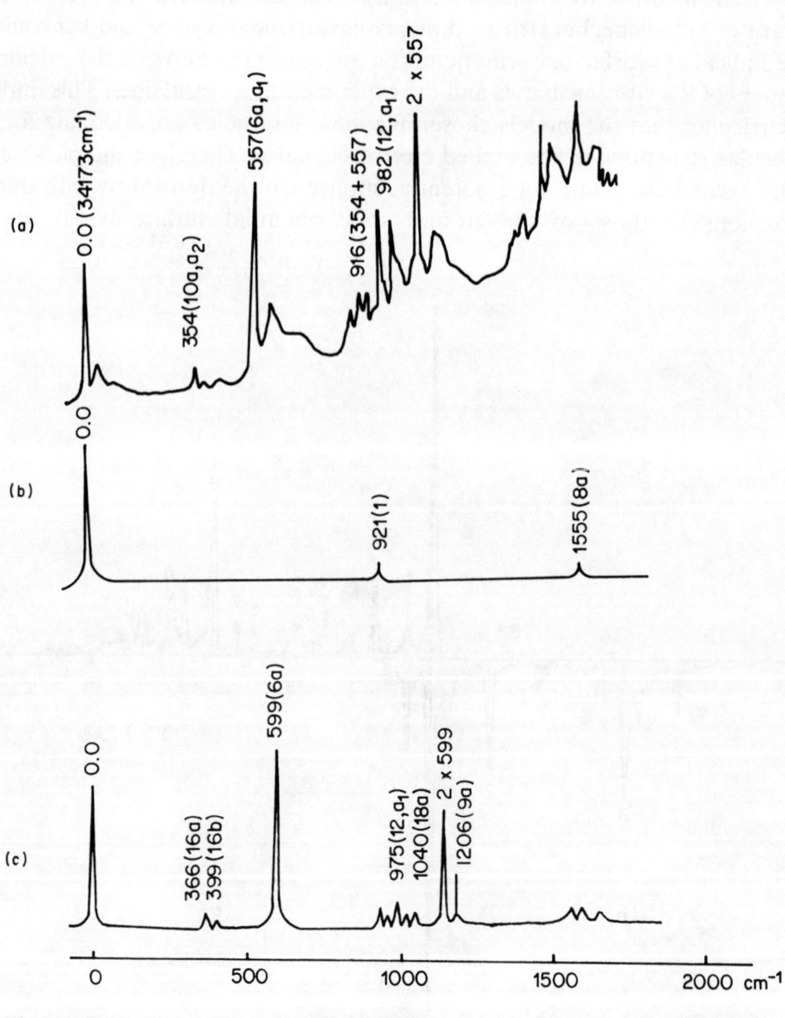

Figure 14.11. Absorption spectrum of the molecule of pyridine; (a) experimental; (b) calculated with no account of the valence angles; (c) calculated with due account of changes in the valence angles

Herzberg–Teller correction may be found to be significant, particularly in the long-wavelength (for absorption) spectral region; this effect is seen very well for *p*-xylene and benzonitrile. A special analysis has shown that, in most cases, in the absence of an accurate value for degeneracy of the electronic states, in order to arrive at satisfactory results we must confine ourselves to the linear terms in the expansion of the electron wave function of the excited state as a power series in normal co-ordinates. This, in turn, implies that the adiabatic approximation still holds good, even when the difference between the electronic state energies is comparable with the vibration quantum magnitude.

(3) An account by the Herzberg–Teller treatment leads to correct changes in the calculated spectrum and, on the whole, the calculated and experimental spectral curves are in better agreement, which is indicative of a sufficient accuracy of the electron wave function derived in the quantum-chemical calculations as well as of the correctness of its utilization in the calculation of the vibrational structure of an electronic transition.

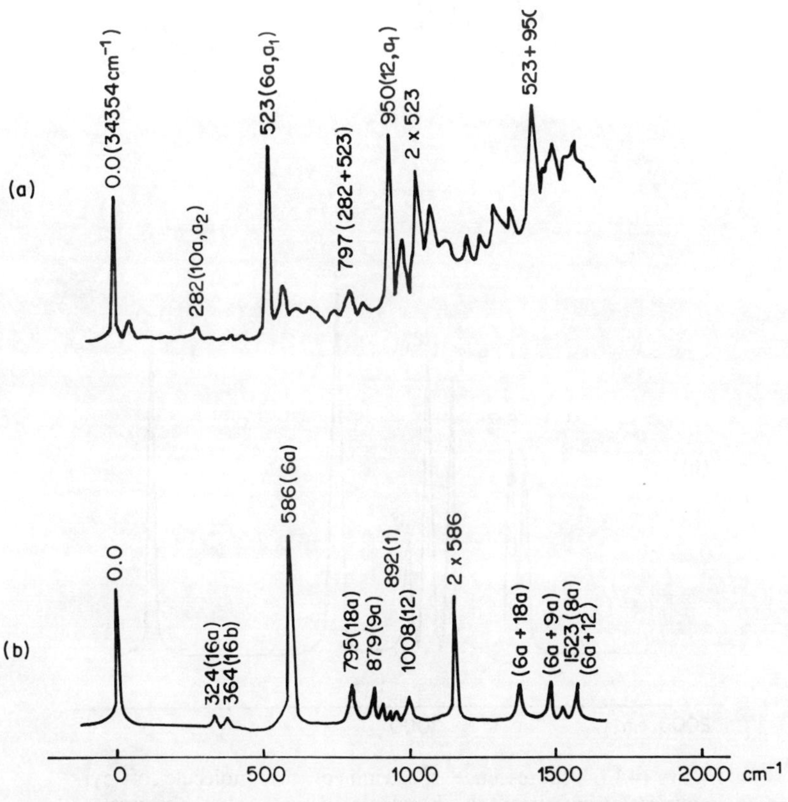

Figure 14.12. Absorption spectrum of the molecule of pyridine-d_5; (a) experimental; (b) calculated with due account of change in the valence angles

Based on the work done[81,82], we can conclude that the method suggested may provide, with due reference to the Franck–Condon and Herzberg–Teller effects and with the accepted estimation of the parameters of the excited state potential surface, theoretical spectra which are in fairly good correlation with observation. This permits the statement to be made that the present theory of vibronic spectra of molecules does not contain rough

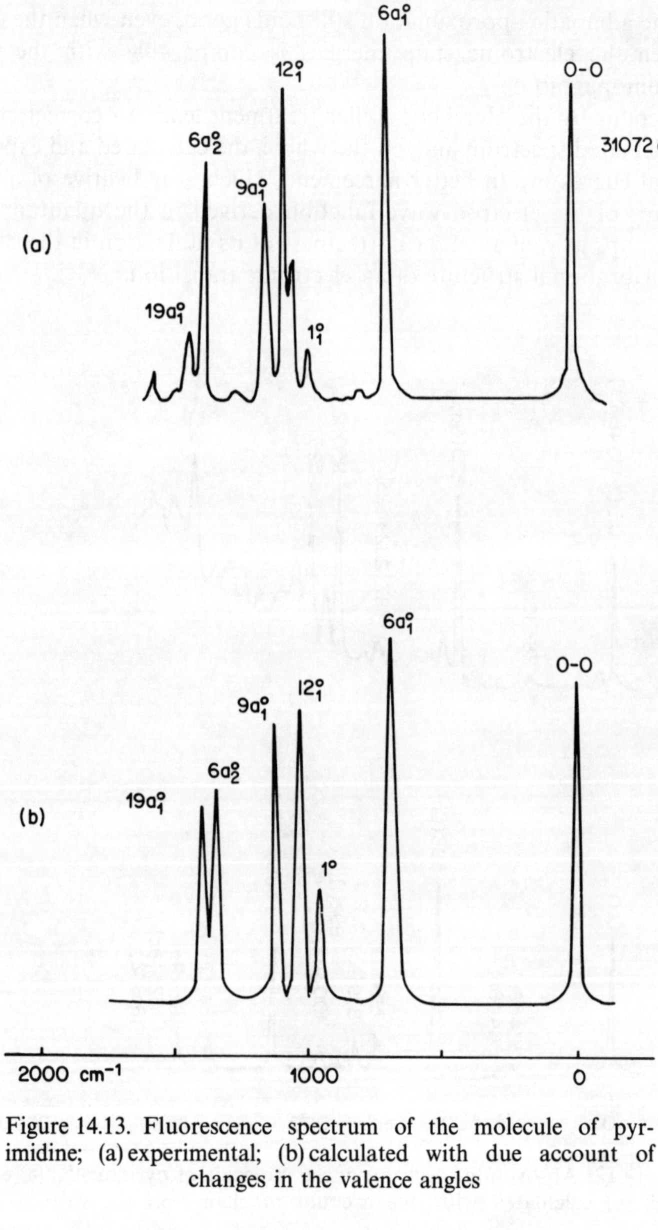

Figure 14.13. Fluorescence spectrum of the molecule of pyrimidine; (a) experimental; (b) calculated with due account of changes in the valence angles

approximations, it is accurate and it may be used to interpret vibronic spectra reliably and clarify the parameters of the excited state potential surface.

The methods of calculation considered above can be used for the construction of the fine structure of electronic bands. However, in this case we must usually take into account not only the distortions of bonds but also of valency angles. From calculation of the corresponding distortions and changes in force constants good results are often obtained using the concept of atomic rehybridization[83]. Examples of such calculation are shown in Figures 14.10–14.13, where experimental and theoretical results are compared.

14.3 Calculation of the Electronic Spectra of Polyatomic Molecules Taking into Account Vibrational Anharmonicity and Internal Rotation

We have studied how to calculate the probability of electronic vibrational transitions at the harmonic approximation in the absence of internal rotation. Now we shall demonstrate how these motions can be taken into account using the methods presented in Chapter 10.

We shall consider that the wave functions describing the internal motion in a polyatomic molecule in combining electronic states were constructed using the variational method, and hence have the form

$$\Psi_{vr}(Q,\chi) = \sum_{w,l} c_{wl}^{vr} \Psi_w(Q)\Phi_l(\chi) \tag{14.68}$$

where $Q = \{Q_i\}$ are the vibrational co-ordinates of the molecule, $\chi = \{\chi_i\}$ is the set of torsional angles that describe the internal rotations in the molecule, $\Psi_w(Q) = \prod_i \psi_{w_i}(Q_i)$, $\Phi_l(\chi) = \prod_i \varphi_{l_i}(\chi_i)$ and $\psi_{w_i}(Q_i)$ are the wave functions obtained in the harmonic approximation and the torsional basis functions are

$$\varphi_{l_i}(\chi) = \begin{cases} N'_{l_i} \sin l_i\chi_i \\ N''_{l_i} \cos l_i\chi_i \end{cases}$$

In the adiabatic approximation the full wave functions describing the electronic and internal nuclei motions in a polyatomic molecule when combining the sth and tth states can be separated into purely electronic components $\Psi_s(Q,\chi,r)$ and $\Psi_t(Q,\chi,r)$, which depend on the internal nuclear co-ordinates using $q = \{Q,\chi\}$ and $q' = \{Q',\chi'\}$ as the parameters, and purely nuclear components $\Psi_{vr}(Q,\chi)$ and $\Psi'_{v'r'}(Q',\chi')$, i.e.

$$\begin{aligned} \Psi_s &= \Psi_s(Q,\chi,r)\Psi_{vr}(Q,\chi) \\ \Psi_t &= \Psi_t(Q',\chi',r)\Psi'_{v'r'}(Q',\chi') \end{aligned} \tag{14.69}$$

Normal co-ordinates in combining electronic states are interrelated by the Dushinsky relations, i.e.

$$Q' = JQ + \Delta \tag{14.70}$$

where J and Δ are matrices of the intersection and shifting of the normal co-ordinates. The co-ordinates characterizing the internal rotations can be assumed to be unaffected by the perturbation $\chi' = \chi$. Indeed, when a part of the molecule is rotated, with the motion being described by a periodic potential, the choice of the initial centre of rotation is not significant. This is because the whole variational function is constructed in the form of a series of harmonic periodic functions. In principle, the problem can also be solved easily for $\chi' = \chi + \delta$. In this case the series in equation (14.69) will be shorter by one, but the procedure for obtaining the appropriate matrix elements will be more complicated.

The component of matrix element of a dipole transition between the t and s electronic states is

$$\mu_{st} = \int \Psi_s^* \mu \Psi_t \mathrm{d}r \mathrm{d}Q = \int \Psi_{vr}^*(Q)\mu_{st}(Q)\Psi'_{v'r'}(Q')\mathrm{d}Q \qquad (14.71)$$

where

$$\mu_{st}(q) = \int \Psi_s^*(Q, r)\mu(r)\Psi_t(Q', r)\mathrm{d}r \qquad (14.72)$$

The matrix element of the dipole moment operator $\mu_{st}(q)$ in equation (14.72) is a slowly changing function of the co-ordinates Q and χ. Moreover, it is clear that the function is periodic with respect to the latter. Since the nuclear vibrations have relatively small amplitudes, $\mu_{st}(q) = \mu_{st}(Q, \chi)$ can be assumed to have the form of a power series in Q around the equilibrium position of $Q = 0$, i.e.

$$\mu_{st}(Q, \chi) = \mu_{st}^{(0)}(\chi) + \sum_k A_k^{st}(\chi)Q_k + \sum_{k,k'} A_{kk'}^{st}(\chi)Q_k Q_{k'} + \cdots \qquad (14.73)$$

where

$$\mu_{st}^{(0)}(\chi) = \mu_{st}(0, \chi), \quad A_k^{st}(\chi) = (\partial\mu_{st}/\partial Q_k)_{Q_k=0}, \quad 2A_{kk'}^{st} = (\partial^2\mu_{st}/\partial Q_k \partial Q_{k'})_{Q_k, Q_{k'}=0}$$

are the periodic functions of the χ co-ordinate set. Equation (14.73) is a notation representing three expressions, one each for the Cartesian components of the $\vec{\mu}_{st}$ vector.

Since the functions $\mu_{st}^{(0)}$, A_k^{st}, etc. are periodic with respect to χ with period 2π, they can be expanded into a Fourier series, i.e.

$$\mu_{st}^{(0)}(\chi) = \mu_{st}^{(0)}(0) + \sum_\lambda (a_\lambda^{st} \sin \lambda\chi + b_\lambda^{st} \cos \lambda\chi)$$

$$A_k^{st}(\chi) = A_k^{st}(0) + \sum_\lambda (a_\lambda^{stk} \sin \lambda\chi + b_\lambda^{stk} \cos \lambda\chi)$$

$$A_{kk'}^{st}(\chi) = A_{kk'}^{st}(0) + \sum_\lambda (a_\lambda^{stkk'} \sin \lambda\chi + b_\lambda^{stkk'} \cos \lambda\chi) \qquad (14.74)$$

where $\lambda = \pm 1, \pm 2$.

Here and later, in order to make the notation compact, it is assumed that only one torsional vibration is present. The solution can easily be generalized to the multidimensional case. In equation (14.74) λ then becomes $\lambda_1, \lambda_2, \ldots, \lambda_n$ and $\lambda\chi = \lambda_1\chi_1 + \cdots + \lambda_n\chi_n$, where n is the number of internal rotations in the molecule.

Consequently, we have for the probability of an electronic transition,

$$\mu_{st} = \mu_{st}^{0}(0)\langle vr|v'r'\rangle + \sum_{\lambda}(a_{\lambda}^{st}\langle vr|\sin\lambda\chi|v'r'\rangle + b_{\lambda}^{st}\langle vr|\cos\lambda\chi|v'r'\rangle)$$
$$+\sum_{k} A_{k}^{st}(0)\langle vr|Q_k|v'r'\rangle + \sum_{k}\Bigg(\sum_{\lambda} a_{\lambda}^{stk}\langle vr|Q_k\sin\lambda\chi|v'r'\rangle$$
$$+ b_{\lambda}^{stk}\langle vr|Q_k\cos\lambda\chi|v'r'\rangle\Bigg) + \sum_{k,k'} A_{kk'}^{st}(0)\langle vr|Q_kQ_{k'}|v'r'\rangle$$
$$+\sum_{k,k'}\Bigg(\sum_{\lambda} a_{\lambda}^{stkk'}\langle vr|Q_kQ_{k'}\sin\lambda\chi|v'r'\rangle + b_{\lambda}^{stkk'}\langle vr|Q_kQ_{k'}\cos\lambda\chi|v'r'\rangle\Bigg) + \cdots \tag{14.75}$$

The first term corresponds to the Franck–Condon approximation, when the purely electronic wave functions can be considered independent of nuclear co-ordinates. The second term depends on the changes in the electronic wave functions when there are internal rotations. The odd terms (third and above) correspond to the usual Herzberg–Teller terms and are characterized by a parametric dependence (because the vibrational amplitudes are small) of the purely electronic wave functions on the vibrational co-ordinates. The remaining terms allow for the dependence of the wave functions on all the integral co-ordinates (vibrational and rotational). By allowing for the vibration–rotation wave functions of equation (14.68) for the combining electronic states we obtain

$$\mu_{st} = \mu_{st}^{(0)}(0)\sum_{l,w,w'} c_{wl}^{vr}c_{w'l}^{v'r'}\langle w|w'\rangle + \sum_{l,l',w,w'} c_{wl}^{vr}c_{w'l'}^{v'r'}\langle w|w'\rangle\sum_{\lambda}(a_{\lambda}^{st}\langle l|\sin\lambda\chi|l'\rangle$$
$$+ b_{\lambda}^{st}\langle l|\cos\lambda\chi|l'\rangle + \sum_{k} A_{k}^{st}(0)\sum_{l,w,w'} c_{wl}^{vr}c_{w'l}^{v}\langle w|Q_k|w'\rangle$$
$$+\sum_{k,\lambda}\sum_{l,l',w,w'} c_{wl}^{vr}c_{w'l'}^{v'r'}\langle w|Q_k|w'\rangle(a_{\lambda}^{stk}\langle l|\sin\lambda\chi|l'\rangle$$
$$+ b_{\lambda}^{stk}\langle l|\cos\lambda\chi|l'\rangle) + \sum_{k,k'} A_{kk'}^{st}(0)\sum_{l,w,w'} c_{wl}^{vr}c_{w'l}^{v'r'}\langle w|Q_kQ_{k'}|w'\rangle$$
$$+\sum_{k,k',\lambda}\sum_{l,l',w,w'} c_{wl}^{vr}c_{w'l'}^{v'r'}\langle w|Q_kQ_{k'}|w'\rangle$$
$$\times(a_{\lambda}^{stkk'}\langle l|\sin\lambda\chi|l'\rangle + b_{\lambda}^{stkk'}\langle l|\cos\lambda\chi|l'\rangle) + \cdots \tag{14.76}$$

where we assumed the orthonormality of the basis functions Φ_l.

In this way the calculation of the electronic spectrum of a molecule with an internal rotational and vibrational anharmonicity reduces to that of the overlap integrals of the vibrational wave functions and those for the internal rotation. Since basis vibrational functions are in fact the wave functions of a harmonic oscillator, the overlap integrals of the vibrational wave functions are similar to the usual Condon and Herzberg integrals, and can be easily found using the appropriate methods of Section 14.1. The overlap integrals of the rotational wave functions can be factorized (for the multidimensional case) and reduce to one-dimensional integrals such as $\langle l_i|\sin\lambda_i\chi_i|l_i'\rangle$ and $\langle l_i|\cos\lambda_i\chi_i|l_i'\rangle$. They are usually calculated thus:

$$\int_0^{2\pi} \Phi_l \cos \lambda\chi \Phi_{l'} \mathrm{d}\chi = \begin{cases} 1/2, & l' = l \pm \lambda \\ 0, & l' \neq l \pm \lambda \end{cases} \tag{14.77}$$

$$\int_0^{2\pi} \Phi_l \sin \lambda\chi \Phi_{l'} \mathrm{d}\chi = \begin{cases} \pm 1/2, & l' = -l \pm \lambda \\ 0, & l' \neq -l \pm \lambda \end{cases} \tag{14.78}$$

These two expressions yield selection rules for the matrix elements. Not only is it important that the integrals for the rotational wave functions are easily calculable, but also most of them vanish due to the selection rules of equations (14.77) and (14.78). Thus allowing for internal rotation does not significantly complicate the calculation programme.

It is easy to see that an account of vibrational anharmonicity when dealing with electronic spectra will also not complicate the calculation unduly if the anharmonic vibrational wave functions for the combining states are constructed by the variational method with a basis of harmonic functions. It is only necessary that the summation be carried out with known coefficients for the integrals calculated in the harmonic approximation.

The description of a molecule's internal rotation requires a calculation of the Fourier coefficients for the matrix elements of the dipole moment operator and its derivatives with respect to the equilibrium vibrational co-ordinates. This, in turn, requires a knowledge of the explicit dependence of the matrix elements on the internal rotational co-ordinates. This may be calculated at significant points by the methods given in Section 14.1 and then fitted to a smooth function to the required accuracy. Whenever the separate internal rotations are independent, in both one- and multidimensional cases, the problem is easy. If there is any dependence between the rotations, then a multidimensional surface has to be approximated, but this very rarely occurs in practice.

Thus the complete structure of the molecule's electronic spectra in the adiabatic approximation accounting for both vibrational anharmonicity and internal rotation can be found by the methods used for the vibrational electronic structure, without having to make any significant changes to the algorithm if the internal motion in the combining electronic states can be described by the variational method with a basis of harmonic vibrational functions and those of a free rotator.

14.4 Determination of Potential Surface Parameters of Excited States for Small Changes in the Latter Compared with the Ground States

It has been mentioned above that the key factor in computing vibronic spectra of polyatomic molecules is to have known values for the parameters of the potential surface in excited electronic states. In order to render these computations feasible the parameters should be expressed in the natural system of co-ordinates. In practice, two extreme cases are most often encountered. In large molecules, where changes in the state of a single electron do not result in the substantial rearrangement of states of all the rest of the electrons, the potential surfaces of excited terms are generally quite similar to those for the ground states. In small molecules, however, these potential surfaces can be very different. Thus in order to establish potential surfaces for excited states in small molecules there is no alternative to carrying out a sequence of calculations similar to those used for the ground state using *ab initio* or good semi-empirical theory. In these calculations the total energy values of the surfaces, or a set of points characterizing the surfaces, are calculated for each of a number of different deformations of the molecule. It is because the molecules are relatively small that such a procedure is feasible in practice.

However, when appreciable changes in the potential surface of combination excited states occur, the above methods of establishing a relationship between the normal co-ordinates (the Dunshinsky relation) is inapplicable. This complication is dealt with in Section 14.6.

Because of small relative changes in the potential surface parameters in large-size molecules, we can apply various semi-empirical correlations of bond order length–force constant to estimate changes in structure. This approach gives fairly good results when applied to certain series of molecules.

It is also useful to be able to carry out similar direct calculations of the potential surface parameters for an excited electronic state on the basis of knowledge of a solution for the electronic problem for a single point within the space of relative nuclear positions. This can be accomplished by expanding the excited-state electronic energy into a series in terms of normal co-ordinates in the neighbourhood of the ground state potential surface minimum[86,87], and enables us to represent the surface in the following quadratic form:

$$E(Q) = E_n^{(0)} + \tilde{a}Q + \tfrac{1}{2}\tilde{Q}AQ \tag{14.79}$$

This is true for relatively small deviations from the molecular equilibrium position in the ground state taken as the origin of the co-ordinate system. In equation (14.79), where $E_n^{(0)}$ is the energy value of the nth electronic state as calculated for a given position of molecular nuclei taken as the origin, $\tilde{a}$ and A are row and symmetric square matrices of the first and second derivatives of E_n with respect to the normal co-ordinates taken at the equilibrium point of the ground state and Q is the vector (column matrix) of the ground state normal co-ordinates.

The normal co-ordinates are all independent, and hence such an expansion can always be made. All that has to be done is to find the derivatives $(\partial E_n/\partial Q_i)_0$ and $(\partial^2 E_n/\partial Q_i \partial Q_j)_0$ in the vicinity of the origin of co-ordinates.

To calculate these derivatives let us resort to the following line of reasoning. The following equalities obviously hold:

$$E_n = \int \Psi_n^* \hat{H}_e \Psi_n \mathrm{d}v \tag{14.80}$$

$$\int \Psi_n^* \Psi_m \mathrm{d}v = \delta_{mn} \tag{14.81}$$

where $\hat{H}_e$ is the Hamiltonian of a purely electronic adiabatic problem and E_n and Ψ_n are solutions.

Differentiating equation (14.80) with respect to the nuclear (normal) coordinate, with due account of equation (14.81) it follows that

$$\int (\partial\Psi_n^*/\partial Q_i)\Psi_n \mathrm{d}v + \int \Psi_n^*(\partial\Psi_n/\partial Q_i)\mathrm{d}v = 0,$$

and hence the Hermitian property of the operator H_e gives[88]

$$\frac{\partial E_n}{\partial Q_i} = \int \Psi_n^* \frac{\partial \hat{H}_e}{\partial Q_i}\Psi_n \mathrm{d}v + \int \frac{\partial \Psi_n^*}{\partial Q_i}\hat{H}_e\Psi_n \mathrm{d}v + \int \Psi_n^* \hat{H}_e \frac{\partial \Psi_n}{\partial Q_i}\mathrm{d}v$$

$$= \int \Psi_n^* \frac{\partial \hat{H}_e}{\partial Q_i}\Psi_n \mathrm{d}v \tag{14.82}$$

Substituting the function $\Psi_n^{(0)}$ established at the point of equilibrium of the ground state into the integral $\int\Psi_n^*(\partial\hat{H}_e/\partial Q_i)\Psi_n\mathrm{d}v$ and processing from equation (14.82) we obtain

$$\left(\frac{\partial E_n}{\partial Q_i}\right)_0 = \int \Psi_n^{(0)*}\frac{\partial \hat{H}_e}{\partial Q_i}\Psi_n^{(0)}\mathrm{d}v \tag{14.83}$$

Thus we have found the derivatives which form the matrix $\tilde{a}$. It is worth noting here that $E_n^{(0)} = \int\Psi_n^{*(0)}\hat{H}_e\Psi_n^{(0)}\mathrm{d}v$, so that it can also be considered as a known value.

It is now necessary to find the second-order derivative $(\partial^2 E_n/\partial Q_i\partial Q_j)_0$, but first note that the set of functions Ψ_m representing the solutions to a purely electronic adiabatic problem form a complete orthonormalized set. In addition, the derivative $\partial\Psi_n/\partial Q_i$ is a certain function in the space of these same variables and can be expanded into a series in terms of Ψ_m. Accordingly, $\partial\Psi_n/\partial Q_i$ may be written as $\partial\Psi_n/\partial Q_i = \sum_m C_{nm}\Psi_m$, where C_{nm} are the real coefficients to be defined.

It can easily be seen that $C_{nn} = 0$. Indeed, bearing in mind that eigenfunctions Ψ_m are orthonormalized, we obtain

$$\int \Psi_n^* \frac{\partial \Psi_n}{\partial Q_i}\mathrm{d}v + \int \Psi_n \frac{\partial \Psi_n^*}{\partial Q_i}\mathrm{d}v = 0 = \sum_m C_{nm}\int \Psi_n^*\Psi_m \mathrm{d}v$$

$$+ \sum_m C_{nm}^* \int \Psi_n\Psi_m^* \mathrm{d}v = 2C_{nn}$$

The rest of the coefficients C_{nm} can be found from the following chain of equalities (with due regard to the Hermitian property of the $\hat{H}_e$ operator):

$$(\hat{H}_e - E_n)\Psi_n = 0; \quad \left(\frac{\partial \hat{H}_e}{\partial Q_i} - \frac{\partial E_n}{\partial Q_i}\right)\Psi_n + (\hat{H}_e - E_n)\frac{\partial \Psi_n}{\partial Q_i} = 0 \tag{14.84}$$

$$\int \Psi_m^* \frac{\partial \hat{H}_e}{\partial Q_i} \Psi_n \mathrm{d}v + \int \Psi_m^* \hat{H}_e \frac{\partial \Psi_n}{\partial Q_i} \mathrm{d}v - E_n \int \Psi_m^* \frac{\partial \Psi_n}{\partial Q_i} \mathrm{d}v$$

$$= \left(\frac{\partial H_e}{\partial Q_i}\right)_{mn} + E_m C_{nm} - E_n C_{nm} = 0 \quad (m \neq n) \tag{14.85}$$

Hence

$$C_{nm} = \left(\frac{\partial \hat{H}_e}{\partial Q_i}\right)_{mn} \Bigg/ E_n - E_m \tag{14.86}$$

Furthermore, taking into account equation (14.82), it follows that

$$\frac{\partial^2 E_n}{\partial Q_i \partial Q_j} = \left(\frac{\partial^2 \hat{H}_e}{\partial Q_i \partial Q_j}\right)_{nn} + \int \left(\Psi_n^* \frac{\partial \hat{H}_e}{\partial Q_i} \frac{\partial \Psi_n}{\partial Q_j} + \frac{\partial \Psi_n^*}{\partial Q_j} \frac{\partial \hat{H}_e}{\partial Q_i} \Psi_n\right) \mathrm{d}v \tag{14.87}$$

so that finally we obtain[89]

$$\left(\frac{\partial^2 E_n}{\partial Q_i \partial Q_j}\right)_0 = \left(\frac{\partial^2 \hat{H}_e}{\partial Q_i \partial Q_j}\right)_{nn}^{(0)} + 2 \sum_{m \neq n} \frac{\left(\frac{\partial \hat{H}_e}{\partial Q_i}\right)_{nm}^{(0)} \left(\frac{\partial \hat{H}_e}{\partial Q_j}\right)_{mn}^{(0)}}{E_n^{(0)} - E_m^{(0)}} \tag{14.88}$$

Note that the form of expression (14.88), which was first derived in reference 85, coincides with that obtained when applying ordinary perturbation theory in the form of the Rayleigh–Schroedinger series. The difference here is that the condition $[(\partial \hat{H}_e/\partial Q_i)_{nm}^{(0)}(\partial \hat{H}_e/\partial Q_j)_{mn}] \ll E_n^{(0)} - E_m^{(0)}$, which is indispensable in applying perturbation theory, is not imposed on when deriving equation (14.88), even though a lack of degeneracy is also necessary in this case. In other words, equation (14.88) is of a more general character.

Substituting the functions $\Psi_n^{(0)}$ into equation (14.88) gives the values of $(\partial^2 E_n/\partial Q_i \partial Q_j)_0$ around the equilibrium position of the ground state, i.e. the elements of the A matrix. A further objective is to find the nth excited state potential surface parameters in terms of the natural dependent co-ordinates; this is done with the aid of equation (14.79).

The geometrical parameters in the natural representation (values of bond lengths, valence angles, etc.) can be easily found from the values of the normal co-ordinates of the ground state, which are consistent with the minimum of the excited state potential surface. This is done with the aid of the relation $\partial E/\partial Q = a + AQ = 0$, which gives $Q_{\min} = -A^{-1}a$. If a vibrational problem for the ground state is solved, then the L_q matrix is known from the relation $q = L_q Q$ between the normal and natural co-ordinates. If, in addition, the equilibrium geometrical parameters s_0 are known, then the geometrical parameters of an excited state at the minimum of the potential surface are given by

$$s_n = s_0 + q_{\min} = s_0 + L_q Q_{\min} = s_0 - L_q A^{-1} a \tag{14.89}$$

From known values of bond lengths, valence angles, etc. for the minimum of an excited electron state potential surface we can then compute the T matrix of kinematic coefficients. Next, the force-constant matrix is found. This is done in terms of the natural co-ordinates for an excited state, assuming the existence of dependent co-ordinates, and taking into account that the natural co-ordinates must be related via the relation

$$q^{(0)} = q^{(n)} + \Delta s \quad (\Delta s = s_n - s_0) \tag{14.90}$$

due to diverse reference points. It should be remembered that, by definition, the natural co-ordinates are the increments of geometrical parameters as collated with their equilibrium values.

Taking the above into account, as well as the fact that $Q^{(0)} = \tilde{L}_p q^{(0)}$ for the electronic ground state in the presence of dependent natural co-ordinates, the variables in equation (14.79) can be interchanged to give

$$E = E_n^{(0)} + \tilde{a}\tilde{L}_p q^{(0)} + \tfrac{1}{2}\tilde{q}^{(0)} L_p A \tilde{L}_p q^{(0)} \tag{14.91}$$

or

$$\begin{aligned} E &= E_n^{(0)} + \tilde{a}L_p(q^{(n)} + \Delta s) + \tfrac{1}{2}(\tilde{q}^{(n)} + \Delta\tilde{s})L_p A \tilde{L}_p(q^{(n)} + \Delta s) \\ &= E_{\min}^{(n)} + \tfrac{1}{2}\tilde{q}^{(n)} L_p A \tilde{L}_p q^{(n)} \end{aligned} \tag{14.92}$$

Here, account is taken of the fact that the term, which is linear with respect to the co-ordinates, equals zero. Hence the force constant matrix being sought is equal to $U^{(n)} = L_p A \tilde{L}_p$. We can find a minimum of the excited state potential surface just as easily:

$$E_{\min}^{(n)} = E_0 + \tfrac{1}{2}\tilde{a}A^{-1}a \tag{14.93}$$

Thus we have found all the values necessary for solving an vibronic problem along the ordinary guidelines. Specific case-study calculations, in which $T_n = T_0$, sometimes gave satisfactory results[86,90,91].

14.5 The Valence-optical Theory of Intensities in the Vibronic Spectra of Polyatomic Molecules

Note that, as a rule, complicated molecules can be divided into relatively large stable fragments. This means that when analysing their properties, semi-empirical approaches can be used. These are based on parameters that are introduced in a special way and which have either been determined by solving inverse spectral problems or by calculating them from first principles. This approach is extremely convenient for constructing special computer libraries, and is capable of analysing large systems, while at the same time radically reducing the expenditure of computer time and simplifying the calculation procedure.

Estimates of small molecules can best be done from first principles (*ab initio*). However, work on complex representatives of homologous series unconditionally requires parametric techniques. This is the situation we now find in the theory of vibrational infra-red molecular spectra, in which, together with *ab initio* estimates of frequencies and band intensities, the parametric valence-optical theory is also widely used. The lack of an analogous theory for vibronic spectra substantially hinders widespread use of these spectra in automatic systems for analysing molecular spectra.

In this section we suggest just such a semi-empirical valence-optical theory[92] for band intensities in the electron–vibrational spectra of polyatomic molecules. It is designed for producing large-scale estimates in particular when using the special computer libraries available for standard complex compound fragments of molecules.

14.5.1 *The Valence-optical Theory of Vibronic Spectra*

As we know, the distribution of intensity over a vibronic absorption or emission band is described in the adiabatic approximation by the value of the matrix elements for dipole transitions of the form:

$$\vec{\mu}^{01}_{v',v''} = \langle 0|\vec{\mu}_e|1\rangle\langle v'|v''\rangle + \sum_j \left(\frac{\partial}{\partial Q_j}\langle 0|\vec{\mu}_e|1\rangle\right)_0 \times \langle v'|Q_j|v''\rangle + \tfrac{1}{2}\sum_{i,j}\left(\frac{\partial^2}{\partial Q_i \partial Q_j}\langle 0|\vec{\mu}_e|1\rangle\right)\langle v'|Q_iQ_j|v''\rangle + \cdots$$

Here, v' and v'' represent the vibrational levels for the combining electronic states, i.e. the ground state (given the superscript 0) and one of the excited states (superscript 1). That we shall only consider transitions from the ground state does not limit the generality of our argument. We shall also not lose too much generality if we also assume that the lower electronic state is described by a single determinantal function, while the excited state is given by a multi-determinant one.

Hence the matrix element of a purely electronic transition $\langle 0|\vec{\mu}_e|1\rangle = \vec{\mu}^{(01)}$ can be given in units of electron charge as

$$\vec{\mu}^{(01)} = \sum_k a_k \int \Phi_k \left(\sum_i \vec{r}_i\right)\Phi_0 \, dV = \sum_k a_k \int \varphi_{kf}\vec{r}\varphi_{og}\, dV = \sum_{k,\alpha} a_k c_{kf\alpha} c_{og\alpha}\int \chi_\alpha^2 \vec{r}\, dV + \sum_{\substack{\alpha,\beta \\ \alpha\neq\beta}} a_k c_{kf\alpha} c_{og\beta}\int \chi_\alpha\chi_\beta \vec{r}\, dV$$

Here, a_k is a coefficient in a linear combination of the determinants that are contained in the electronic function of the excited state:

$$\psi^{(1)} = \sum_k a_k \Phi_k$$

Φ_k is a basis determinantal function; φ_{kf} and φ_{og} are the basis molecular orbitals which have been given the subscripts f and g; $c_{kf\alpha}$ and $c_{og\beta}$ are coefficients in the LCAO–MOs belonging to the α and β atomic orbitals (these may be centred on the same or different atoms); $\vec{r}$ is the current radius vector of the electron; and χ_α and χ_β are atomic orbitals.

We now introduce the notation

$$q_\alpha^{(01)} = \sum_k a_k c_{kf\alpha} c_{og\alpha} \tag{14.94}$$

and

$$p_{\alpha\beta}^{(01)} = \sum_k a_k c_{kf\alpha} c_{og\beta} \tag{14.95}$$

from which we obtain

$$\vec{\mu}^{(01)} = \sum_\alpha q_\alpha^{(01)} \int \chi_\alpha^2 \vec{r}\, \mathrm{d}V + \sum_{\alpha,\beta} p_{\alpha\beta}^{(01)} \int \chi_\alpha \chi_\beta \vec{r}\, \mathrm{d}V$$

assuming that the $\int \chi_\alpha^2 \vec{r}\, \mathrm{d}V$ integrals are numerically equal to $\int \chi_\alpha^2 \vec{r}\, \mathrm{d}V = \vec{R}_\alpha (\vec{R}_\alpha$ is the radius vector of a nucleus) when the square of the normalized atomic orbitals is symmetric with respect to its centre. This condition is satisfied for all the atomic orbitals normally used.

The integrals $\int \chi_\alpha \chi_\beta \vec{r}\, \mathrm{d}V$ is either a product of radius vectors of the distribution centres $(\chi_\alpha \chi_\beta)$ and the overlap integrals (if the latter are non-zero) or has the sense of being the dipole moment of a transition between the χ_α and χ_β atomic orbitals. Thus we obtain

$$\int \chi_\alpha \chi_\beta \vec{r}\, \mathrm{d}V = S_{\alpha\beta} \vec{R}_{\alpha\beta}$$

where $S_{\alpha\beta} \neq 0$ and α and β are the indices of different atoms, and

$$\int \chi_\alpha \chi_\beta \vec{r}\, \mathrm{d}V = \vec{\mu}'_{\alpha\beta}$$

where $S_{\alpha\beta} = 0$ and α and β are the indices of different atoms and

$$\int \chi_\alpha \chi_\beta \vec{r}\, \mathrm{d}V = \int \chi_\alpha \chi_{\alpha'} \vec{r}\, \mathrm{d}V = \vec{\mu}'_{\alpha\alpha'}$$

where α and $\beta = \alpha'$ are the indices of the same atom.

The $\vec{R}_{\alpha\beta}$ radius vector ends between the χ_α and χ_β atomic orbitals, although it need not lie on the straight line connecting their centres. Specific estimates have shown that they are not displaced far from the bond between the atomic orbital centres. Hence we shall in future consider that the end of the $\vec{R}_{\alpha\beta}$ lies on the bond.

Using the above relations we can rewrite the $\vec{\mu}^{(01)}$ integral as

$$\vec{\mu}^{(01)} = \sum_\alpha q_\alpha^{(01)} \vec{R}_\alpha + \sum_{\substack{\alpha,\beta \\ \alpha \neq \beta \\ (S_{\alpha\beta} \neq 0)}} p_{\alpha\beta}^{(01)} S_{\alpha\beta} \vec{R}_{\alpha\beta} + \sum_{\alpha,\alpha'} p_{\alpha\alpha'}^{(01)'} \vec{\mu}'_{\alpha\alpha'} + \sum_{\alpha,\beta} p_{\alpha\beta}^{(01)'} \vec{\mu}'_{\alpha\beta}$$

This integral does not depend on the choice of origin for the co-ordinates, because the Φ_k determinants are constructed of a whole set of mutually orthogonal molecular orbitals for any pair of which the following relation is satisfied:

$$\sum_{\alpha} c_{kf\alpha} c_{og\alpha} + \sum_{\substack{\alpha,\beta \\ \alpha \neq \beta}} c_{kf\alpha} c_{og\beta} S_{\alpha\beta} = 0 \tag{14.96}$$

Future manipulations are more easily done in a system of co-ordinate vectors that are associated with the structural elements of the molecule. This, as we shall see below, leads to the construction of a variant of the valence-optical theory for vibronic spectra.

In order to do this, we follow Section 11.2, in which the physical sense and methods for calculating electro-optical parameters were considered for the usual valence-optical scheme. We introduce a system of directional vectors of the bonds, bearing in mind the symmetry of the molecule's equilibrium configuration. The position of any point on one of the bonds may be represented as a linear combination of the orientating vectors with coefficients (positive or negative, depending on the bond unit vector directions) that equal the length of the bond or the fraction of the bond if the point lies between the centres of the two atoms forming the bond. Thus the radius vector of the ninth atom of propylene molecule (see Figure 14.14) can be written as

$$\vec{R}_9 = -s_{31}\vec{e}_{31} + s_{34}\vec{e}_{34} + s_{46}\vec{e}_{46} + s_{69}\vec{e}_{69}$$

If the origin is taken to be at atom one. Here, s_{31} and $\vec{e}_{31}$, etc. are the vector and length of the relevant bond.

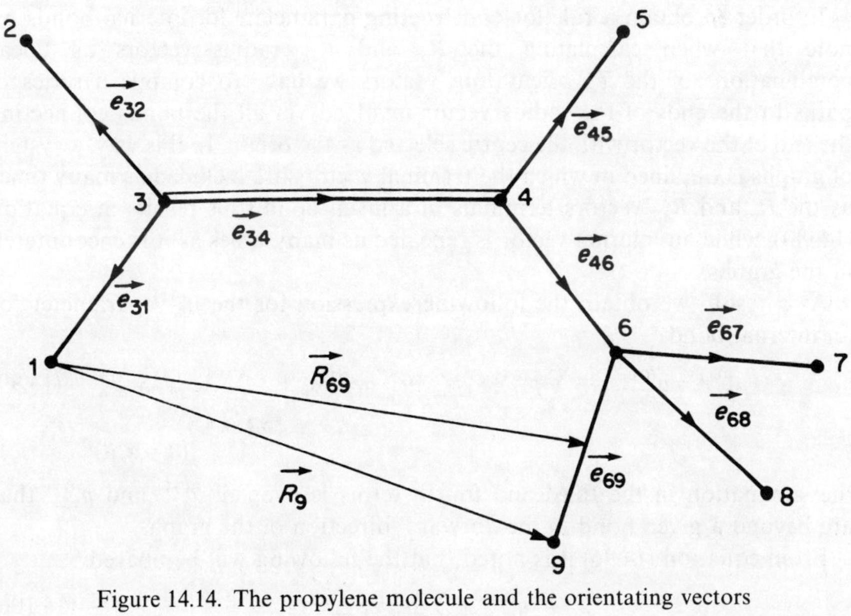

Figure 14.14. The propylene molecule and the orientating vectors

Assuming that its end lies on the bond formed by the 6 and 9 atoms, we obtain the following for the radius vector R_{69}:

$$\vec{R}_{69} = -s_{31}\vec{e}_{31} + s_{34}\vec{e}_{34} + s_{46}\vec{e}_{46} + s'_{69}\vec{e}_{69}$$

where s'_{69} is the distance between the 'centre of gravity' of the distribution $(\chi_\alpha\chi_\beta)$ and the origin of the bond.

By substituting similar expressions for all radius vectors of all the atoms in the molecule and for all the vectors $R_{\alpha\beta}$ (there may be several that are dependent on the number of combinations of χ_α and χ_β for one bond) and by grouping like terms, we obtain

$$\vec{\mu}^{(01)} = \sum_n \mu_n^{(01)}\vec{e}_n + \sum_{\alpha,\alpha'} p_{\alpha\alpha'}^{(01)'}\vec{\mu}'_{\alpha\alpha'} + \sum_{\alpha,\beta} p_{\alpha\beta}^{(01)'}\vec{\mu}'_{\alpha\beta} \qquad n = \alpha\beta \qquad (14.97)$$

Since the orientating vectors of terminal bonds in the expressions for the $\vec{R}_\alpha$ of terminal atoms only appear once (while they appear as many times in the expression for $\vec{R}_{\alpha\beta}$ as this bond has these values), the $\mu_n^{(01)}$ parameter for terminal bonds takes the form

$$\mu_n^{(01)} = q_\beta^{(01)} s_n + \sum_{\substack{\alpha,\beta \\ \alpha\neq\beta}} p_{\alpha\beta}^{(01)} S_{\alpha\beta} s'_{\alpha\beta} \qquad n = \alpha\beta \qquad (14.98)$$

In this expression the α and β indices belong to the atoms forming a bond, and it is assumed that the vector $\vec{e}_n$ points from the α to the β atom and that the summation in the second term is over all combinations of atomic orbital pairs that contain the α and β atoms for which $S_{\alpha\beta} \neq 0$. We also assume that, as is usually done in CNDO methods, the overlap integrals for atoms of different bonds are zero due to the distance between their atomic orbital centres.

In order to obtain a rule for constructing parameters for internal bonds we note that when calculating the $\vec{R}_\alpha$ and $\vec{R}_{\alpha\beta}$ radius vectors as linear combinations of the $\vec{e}_n$ orientating vectors we have to compile a series of paths to the ends of the radius vector involved via all the bonds connecting the end of the vector with the centre selected as the origin. In this way, a system of graphs is obtained in which the terminal vectors are included as many times as the $\vec{R}_\alpha$ and $\vec{R}_{\alpha\beta}$ vectors terminals in a given bond (this results in equation (14.98)), while an internal vector is repeated as many times as it is encountered in the graphs.

As a result, we obtain the following expression for the $\mu_n^{(01)}$ parameter for an internal bond:

$$\mu_n^{(01)} = q_\beta^{(01)} s_n + \sum_{\alpha,\beta} p_{\alpha\beta}^{(01)} S_{\alpha\beta} s'_{\alpha\beta} + \sum_\gamma q_\gamma^{(01)} s_n + \sum_{\substack{\gamma,\delta \\ \gamma,\delta\neq\alpha,\beta}} p_{\gamma\delta}^{(01)} S_{\gamma\delta} s_n \qquad (14.99)$$

$$(n = \alpha, \beta)$$

the summation in the third and fourth terms is over all $q_\gamma^{(01)}$ and $p_{\gamma\delta}^{(01)}$ that are beyond a given bond in the 'forward' direction of the graphs.

From equation (14.96) it is noted that the following will be obeyed:

$$\sum_\alpha q_\alpha^{(01)} + \sum_{\alpha,\beta} p_{\alpha\beta}^{(01)} S_{\alpha\beta} = 0 \qquad (14.100)$$

There are some peculiarities in molecules with cyclic groupings. Here, the graphs must always be constructed so that all the bonds in a cycle are passed through. In symmetric molecules the graphs and their origin must be chosen taking into account the symmetry of the ground state and the disposition of the orientating vectors of the bonds. For additional operations, see Section 11.2.

We turn now to the last two terms of equation (14.97). The relevant quantities are vectors which are called the vectors of the intra-atomic and intrabond dipole transitions. These can be easily found by a direct calculation. It is natural to associate the origin of a $\vec{\mu}'_{\alpha,\alpha'}$ vector with the centre of its atomic orbitals, while that of a $\vec{\mu}'_{\alpha\beta}$ vector ($S_{\alpha\beta} = 0$) is naturally associated with the bond atoms at which the $\vec{\mu}'_{\alpha\beta}$ vector forms an acute angle with the $\vec{e}_n$ vector. If this structural grouping has a three-dimensional structure (this being the case most frequently encountered) then a triplet of orientating bond vectors can be shown such that the momentum vectors of the bond and atomic transitions occur between them. Hence we can expand the vectors of the transition moments uniquely in terms of the triplet. This transition enables us to incorporate the third and fourth terms into equation (14.97) as parts of the first term and write the matrix element $\vec{\mu}^{(01)}$ simply as $\vec{\mu}^{(01)} = \sum_n \mu_n^{(01)} \vec{e}_n$ (using the same notation for the new μ_n^{01} as in equation (14.97). This expression is exactly the same representation of the molecule's dipole moment as a sum of bond dipole moments that is conventional in the usual valence-optical theory.

Note that in linear and planar molecules the $\vec{\mu}'_{\alpha\alpha'}$ and $\vec{\mu}'_{\alpha\beta}$ vectors may be perpendicular to the bond, and hence additional unit vectors perpendicular to the bonds have to be introduced. We shall dwell on this topic, which does not lead to any real difficulty, later.

We now progress to a determination of the value of $\mu_n^{(01)}$, which we shall call the dipole moment of a bond's electron transitions.

The discussion above has shown first, that there is a method of making subsequent quantum-chemical calculations of these quantities in terms of the coefficients a_k, c_{kf}, etc., which are in turn expressed in terms of the 'charge' q_α^{01} and 'bond order' of the electronic transitions, and second, that these parameters are associated with individual structural elements of the molecule. We know that the coefficients belonging to the atomic orbitals in a LCAO–MO wave function have localized properties, which in complicated molecules in particular are seen as very small changes in the coefficients for one part of a molecule when polar elements are substituted into the other.

This properly has been established for the ground states of polyatomic molecules. However, since the electronic states of excited molecules are not very different from the ground states for complicated systems there is every reason to assume that the situation above will also be true for the molecular orbitals in excited states. Hence the $\mu_n^{(01)}$ quantities may also be conserved in certain cases. This is most important, because it means that they can be used as parameters in an appropriate semi-empirical theory. These parameters will be associated with the individual structural groupings of a molecule and, when a grouping is conserved over a series of molecules that have similar structures, the numerical values of the parameters will also be conserved. In other words, we have an analogy with the valence-optical scheme for vibrational transitions. The parameters of the theory will be the values of the $\mu_n^{(01)}$, which are the dipole moments of the transitions for bonds and the $(\partial\mu_n^{(01)}/\partial q_t)_0$ etc., i.e. the first and higher-order derivatives of μ_n^{01} with respect to the natural vibrational co-ordinates.

By analogy with the relevant expression for the valence-optical scheme in the theory of infra-red band intensities[4,5] we can write:

$$\vec{\mu}^{(01)}_{v',v''} = \left(\sum_n \mu_n^{(01)}\vec{e}_n\right)\langle v'|v''\rangle + \sum_j \left\{\tilde{\vec{e}}\frac{\partial\mu^{(01)}}{\partial q} + \tilde{\mu}^{(01)}\frac{\partial\vec{e}}{\partial q}\right\} l_j \langle v'|Q_j|v''\rangle$$
$$+ \tfrac{1}{2}\sum_{j,i}\tilde{l}_j\left\{\frac{\partial^2\mu^{(01)}}{\partial^2 q}\mathscr{E} + \frac{\partial\tilde{\mu}^{(01)}}{\partial q}\frac{\partial\vec{e}}{\partial q} + \frac{\partial\tilde{\vec{e}}}{\partial q}\frac{\partial\mu^{(01)}}{\partial q} + \frac{\partial^2\vec{e}}{\partial q^2}M\right\} l_i\langle v'|Q_jQ_i|v''\rangle \tag{14.101}$$

where

$$\frac{\partial\vec{e}}{\partial q} = S^{-1}(\Delta E\tilde{\vec{B}}\Theta - \vec{E}')$$

Here, $\partial\tilde{\vec{e}}$ is a row-matrix containing the orientating bond vectors of the ground state; $\partial\mu^{(01)}/\partial q$ is a rectangular matrix containing the respective derivatives and is of order $N' \times N$ (N' is the number of bonds and N that of the natural co-ordinates for the molecule); $\tilde{\mu}^{01}$ is a row matrix whose elements are $\mu_n^{(01)}$; S^{-1} is a diagonal matrix containing the reciprocals of the bond lengths; Δ is a matrix which has as many rows as the molecule has bonds (and as many columns as the molecule has atoms) and which contains -1 in an element to indicate the start of a bond and $+1$ for the end of bond, the other elements being zero; $\vec{B}$ is the matrix reflecting the relationship between the rates of change of the natural co-ordinates (generalized velocity) and the Cartesian atom velocities ($\dot{q} = \vec{B}\vec{R}$); E is the matrix containing the reciprocals of the atom masses; θ is the inverse matrix to the kinematic coefficients matrix ($\theta^{-1} = \vec{B}E\tilde{\vec{B}}$); and $\vec{E}'$ is a matrix of order NN' in which the columns corresponding to the numbers of natural co-ordinates of the changes in the bond lengths contain unit vectors, all the other elements being zero.

$\mathscr{E}$ refers to a matrix with the structure (see Section 11.8):

$$\mathscr{E} = \left.\begin{bmatrix} \left.\begin{matrix}\vec{e}_1, 0\cdots 0\\ 0\ \vec{e}_1\cdots 0\\ \cdots\cdots\ \vec{e}_1\end{matrix}\right\} N \\ \cdots\cdots\cdots \\ \left.\begin{matrix}\vec{e}_n\ 0\cdots 0\\ 0\ \vec{e}_n\cdots 0\\ 0\cdots\cdots\ \vec{e}_n\end{matrix}\right\} N\end{bmatrix}\right\} N'$$

M refers to a matrix with the same structure as but containing μ_n^{01} elements instead of $\vec{e}_n$ ones. The expression $\partial^2\mu^{(01)}/\partial q^2$ indicates the matrix

$$\frac{\partial^2 \mu^{(01)}}{\partial q^2} = \left[\begin{array}{cccc:cccc} \dfrac{\partial^2 \mu_1^{(01)}}{\partial q_1^2} & \cdots & \dfrac{\partial^2 \mu_1^{(01)}}{\partial q_1 \partial q_N} & & \dfrac{\partial^2 \mu_{N'}^{(01)}}{\partial q_1^2} & \cdots & \dfrac{\partial^2 \mu_{N'}^{(01)}}{\partial q_1 \partial q_N} \\ \hdashline \dfrac{\partial^2 \mu_1^{(01)}}{\partial q_N \partial q_1} & \cdots & \dfrac{\partial^2 \mu_1^{(01)}}{\partial q_N^2} & & \dfrac{\partial^2 \mu_{N'}^{(01)}}{\partial q_N \partial q_1} & \cdots & \dfrac{\partial^2 \mu_{N'}^{(01)}}{\partial q_N^2} \end{array}\right]$$

All the derivatives are calculated for the equilibrium geometry of the molecule. The method of calculating the elements of the $\partial^2 \vec{e}/\partial q^2$ matrix, which has the same structure as the $\partial^2 \mu^{(01)}/\partial q^2$ one, is given in Section 8.9. The form of the normal-vibrations is described by the column-matrices l_i and l_j, q_t here meaning a natural vibration co-ordinate.

We shall limit ourselves to second-order terms in the expansion in terms of normal co-ordinates. Equation (14.101) contains terms that are analogous in structure to those in the usual valence-optical theory, and this enables us to make a series of deductions by relying directly on an analogy with the latter.

14.5.2 General consequences of the valence-optical theory for vibronic transitions

For simplicity, we shall assume from the beginning that the potential surfaces of the ground and excited electronic states are similar and differ from each other only by shifts along the energy axis. For complicated systems this is not a bad approximation, because many investigations have shown that excited states actually are not very different from the ground states for single-electron transitions.

The vibronic transitions that take place between states without changes in vibrational quantum numbers can be completely separated from those that occur with changes in these numbers if the contribution of the third term in equation (14.101) is neglected. In addition, due to the selection rules, separate matrix elements are obtained for the matrix elements of powers of the harmonic oscillators. These have the form

$$\vec{\mu}_{00}^{(01)} = \left(\sum_n \mu_n^{(01)} \vec{e}_n \right) \tag{14.102}$$

$$\vec{\mu}_{(01)_i}^{(01)} = \left\{ \tilde{\vec{e}} \frac{\partial \mu^{(01)}}{\partial q} + \tilde{\mu}^{(01)} \frac{\partial \vec{e}}{\partial q} \right\} l_i Q_{0i} \tag{14.103}$$

$$\vec{\mu}_{(02)_i}^{(01)} = \frac{\sqrt{2}}{2} \tilde{l}_i \left\{ \frac{\partial^2 \mu^{(01)}}{\partial q^2} \vec{\mathscr{E}} + \frac{\partial \tilde{\mu}^{(01)}}{\partial q} \cdot \frac{\partial \vec{e}}{\partial q} + \frac{\partial \tilde{\vec{e}}}{\partial q} \frac{\partial \mu^{(01)}}{\partial q} + \frac{\partial^2 \vec{e}}{\partial q^2} M \right\} l_i Q_{0i}^2 \tag{14.104}$$

$$\vec{\mu}_{(0,1,1)i,j}^{(01)} = \tfrac{1}{2} \tilde{l}_j \left\{ \frac{\partial^2 \mu^{(01)}}{\partial q^2} \vec{\mathscr{E}} + \frac{\partial \tilde{\mu}^{(01)} \partial \vec{e}}{\partial q\, \partial q} + \frac{\partial \tilde{\vec{e}}}{\partial q} \frac{\partial \mu^{(01)}}{\partial q} + \frac{\partial^2 \vec{e}}{\partial q^2} M \right\} l_i Q_{0i} Q_{0j} \tag{14.105}$$

These elements determine the intensities of purely electronic 0–0 transitions, transitions with the excitation of fundamental vibrations (0–1), transitions with the excitation of overtones (0–2) and combinations (0–1, 1). For simplicity, we shall only discuss absorption spectra, and assume that the system started in the ground electronic and fundamental vibrational states.

We can make some general deductions from these formulae:

(1) Above all, it is clear that because the μ_n^{01} parameters are 'genetically' related to the coefficients of the atomic orbitals in the LCAO–MO's (which for a symmetric molecule have well-defined symmetry), we shall find that the μ_n^{01} parameters for a given electronic transition and which belong to a given set of equivalent bonds will be transformed into each other by a symmetry operation that corresponds to the symmetry of the ground state of the molecule. This will also be true for the parameters $\partial\mu_n^{01}/\partial q_t$, $\partial^2\mu_n^{01}/\partial q_t \partial q_u$, etc. In turn, we can take advantage of the same approaches we used in the usual valence-optical theory in order to allow for symmetry in equations (14.102)–(14.105), that is, we can introduce vectors of symmetry. We need only bear in mind that the μ_n^{01} parameters, as distinct from the usual bond dipole moments μ_n, are not in general equal to each other for the whole set of equivalent bonds of one type. Thus the symmetry vectors in formulae such as equations (14.102)–(14.105), allowing for the symmetry of both the electronic transitions and the vibrations, will not be identical to the usual valence-optical theory. Similar considerations can be used to find selection rules for the components of vibronic bands.

(2) The intensities of the lines corresponding to the 0–1 transitions for purely valency vibrations will only depend on $\partial\mu_n^{01}/\partial q_t$, while the purely deformational line intensities will depend on both the μ_n^{01} and the $\partial\mu_n^{01}/\partial q_t$ parameters (if non-zero for the deformational coordinates).

If the combination of the last two terms in (14.97) is small, then the moment vector of the 0–0 transition will always be in the plane of the molecule if it is planar, or the vector lies along the molecular axis if it is linear. In this case, the component of the dipole vector of a 0–1 transition that is perpendicular to the plane of a planar molecule or to the axis of a linear molecule will only be defined by the μ_n^{01} parameter and will only appear for non-planar, non-linear vibrations. The perpendicular component of μ_{0-1}^{01} is absent for planar vibrations of planar molecules.

The overtones and combinations of valency vibrations will only depend on the $(\partial^2\mu_n^{01}/\partial q_t \partial q_u)_0$ parameters, i.e. on a particular electro-optical anharmonicity. In planar and linear molecules they will be polarized along the plane or axis of the planar or linear molecule.

(3) 0–0 transitions will not change their properties when isotopes are substituted into the molecule. The values of the derivatives of the bond-orientating vectors with respect to natural vibrational co-ordinates for many vibrations of real molecules change little, while the corresponding terms in the purely valency vibrations must be zero. If this condition is fulfilled, albeit approximately, then the sum rule for the line intensities of the 0–1 transitions must also be fulfilled. This means that if we have a particular set of isotopic substitutions in a molecule and we can write the usual sum rule for the squares

of the frequencies for them, then we must have analogous relationships for the sums of the intensities of the respective vibrational components of the electronic band. In these conditions the sum $\sum_i (A_i/\lambda_i)$, where A_i is the intensity of a transition and λ_i is the square of the frequency of a normal vibration and the summation is over all normal vibrations, will not depend on the isotopic substitutions. This conclusion is analogous to that for the intensities in infra-red spectra and is drawn from the same arguments (see Section 11.7). The same relationships must also exist between the $\partial^2 \overline{\mu}^{(01)}/\partial Q_i^2$ coefficients for the overtones of vibronic transitions (see Section 11.7).

(4) An important question concerns the characterizability with respect to intensity for the different kinds of transitions in the vibronic spectra. It is clear that, above all, the 0–0 transition will only be characteristic if the $\mu_n^{(01)}$ parameters are non-zero for some relatively closed grouping. This means that the combining molecular orbitals within a structural element of the molecule must be localized. Note that this condition is not essential for 0–1 transitions. For these components of the vibronic bands to show characterizability it is necessary that:

(a) The normal vibrations be fully characteristic with respect to both frequency and form for the group in question (i.e. only the natural co-ordinates inside the group change during a vibration);
(b) The non-diagonal elements of the $\partial\mu^{(01)}/\partial q$ matrix that correspond to 'interactions' between the μ_n^{01} parameters of the group and the rest of the molecule are close to zero. In effect, this means that the coefficients of the atomic orbitals in the LCAO–MOs, belonging to the atoms of the group, must be changed very little during deformations of the rest of the molecule, and vice versa; and
(c) During vibrations of a given form the bonds not included in the group should rotate very little.

Given these same conditions, characterization with respect to intensity will also be observed for 0–2 transitions. Note that for polyatomic molecules the intensities of the overtones and combinations will be non-zero, even when the electro-optical and mechanical anharmonicity terms $((\partial^2\mu^{01}/\partial q_t \partial q_u)_0$ and $(\partial^2\mu_n^{01}/\partial q_t \partial q_u)_0$, respectively) are absent.

We find that characteristic intensities will be observed for the 0–1, 1 transitions, which result in the appearance of combination bands in the vibronic spectra, when each of the pair of vibrations have characteristic intensities for the fundamental band.

In any case, intensity characterization will be observed if the 0–0 transition is characteristic and the normal vibrations are characteristic with respect to form and frequency. As before, these deductions may be obtained in the same way as was done for the intensities of the absorption bands in infra-red spectra (see Chapter 12). By relying on this analogy it is noted that the solution of the inverse spectral problem for vibronic bands has much in common with vibrational theory, and so it is possible to retain the approaches and software for calculating the $\partial\vec{\mu}^{01}/\partial Q_i$ and $\partial^2\vec{\mu}^{01}/\partial Q_i \partial Q_j$ parameters using computer libraries (the fragmentary method). We also wish to point out that, although the problem is more complicated, it is, in principle, solvable, even when the representation of $\vec{\mu}^{01}$ in the form of a vector sum is insufficient when only the bond-orientating vectors are included, since vectors perpendicular to them must also be included. The derivatives of these vectors with respect to the natural co-ordinates can be calculated without difficulty using the formulae given for the theory of Raman spectra and presented in Chapter 15.

In general, the potential surfaces of the combining electronic states are different. It is obvious, therefore, that these simple general deductions for the intensities of the vibrational components of vibronic bands may not be valid. This is because separate terms cannot be obtained in the general formula for the $\vec{\mu}^{01}$ matrix element. However, the deductions are valid for the coefficients of the matrix elements for vibrational co-ordinates.

To conclude, note that it is apparently possible to construct a valence-optical theory for vibronic transitions for molecules with internal rotations.

14.6 Transitions between very Different States of Polyatomic Molecules

So far, it has been tacitly assumed that when optical transitions occur the electronic states involved do not differ to a large extent from one another. Such an assumption enabled us to employ the Dushinsky relation in its normal or somewhat modified form. However, it is not unlikely that transitions will occur in which the potential surfaces differ considerably from each other in terms of the location of minima and in curvature. This applies not only to small molecules, where such a situation can be rather typical, but also be large ones, where planar regions can become, for example, non-planar. In cases like this, when specifying vibrations within the limits of each of the combination of electronic states it is advisable to use its own system of natural co-ordinates (referred to here as an 'eigensystem'). The normal Dushinsky relation is completely unacceptable in this case.

A characteristic feature of polyatomic molecules is that a large number of different isomers have the same empirical formula. These include inversion and rotational isomers, tautomers and topological isomers (the last are the more numerous). Thus, for example, compounds having the empirical formula C_6H_6 include not only benzene but a further 216 topological isomers. In a very large number of cases topological isomers are so different in their structure and properties that they represent completely different molecules. Hence, from a general point of view topological isomers correspond to different forms of one basic system, and each form corresponds to a particular minimum on a single or different potential surfaces associated with this particular combination of bonded atoms. The minima are fairly deep, being distinctly separated from one another in the multidimensional space of the nuclear co-ordinates.

Under ordinary conditions, and in particular in the course of interaction of light with matter, we can observe only transitions in the areas localized around some of the minima. As a rule, such transitions are accompanied by small changes only in the geometrical structures of the isomeric states. There is a very small possibility of transition from the states in the vicinity of one minimum into those in the vicinity of another, in which a radical rearrangement of the chemical and geometrical structures of a molecule occurs. It is for this reason that these transitions have not been studied in the past.

This does not, however, mean that such transitions are not possible at all, or that they are of no practical interest. Despite their low probability, such transitions are likely to be capable of being initiated by means of the high-density radiation produced by modern lasers. This type of transition can be employed for developing at least two rather important fields. The first is the laser-aided chemical synthesis of complex compounds. The idea is that, having chemically synthesized one of the molecular topological isomers selected on the basis of the simplest procedure of preparation and purification, this isomer is then reconstructed to obtain the required isomer, and this is performed through one or more steps. The sequence of steps is selected to obtain the maximum probability of an optical reconstruction of one isomeric form into another at each of the steps. At present there is a possibility of generating, with the aid of a computer, the structures of all the topological isomers with a given empirical formula and, what is more, with well-defined atomic groupings[93]. This enables us to select a sequence of such steps not in an empirical 'blind' way, which makes no sense because of a great number of variants, but to make it quite deliberately. Such chemical synthesis of compounds may be found to be quite feasible and advantageous in those cases where the purely chemical methods are found to be rather cumbersome since they involve a large number of states.

The second development is to employ the isomeric molecular states for constructing molecular memory elements for a computer. It is well known that to develop such cells would mean a further breakthrough in the field of miniaturization of chips. It is also known that experimental attempts have been made to utilize the molecular isomeric states associated with internal rotation for this purpose. However, this possibility seems to be unpromising, due to the high probability of spontaneous restructuring of one rotamer into another. In this sense, it seems much better to work with the topological isomeric molecular states. Indeed, the transition process *per se* must be made at a very high speed typical of optical transitions in molecules. What is more, both the initial and the final isomeric states in a molecule are long-lived ones, and the possibility of their spontaneous mutual rearrangement is virtually eliminated. In addition, a marked difference in the chemical structure of various topological isomers, as well as the resulting specific features of their spectra, enables us to detect easily one or other states of a molecule, thus reading out information optically.

With these prospects in mind, research in these areas becomes rather attractive. At the same time, it is clear that, because of a great diversity of different variants, it is hardly possible to plan such experiments without an adequate quantitative theoretical investigation.

Accordingly, it is necessary to find a practically useful algorithm for computing the possibilities of optical transitions between arbitrary molecular isomeric states in those cases where such isomeric states correspond to different minima on a single principle potential surface or on differing potential surfaces.

Let us confine ourselves to an adiabatic approximation and dipole transitions. Proceeding from general reasoning, it is clear that such approximations should not lead to any material losses in the problem under investigation, since a precision can be achieved which is comparable with that of similar computations made in the usual type of vibronic problem in the theory of optical spectra of polyatomic molecules.

As a first step, let us assume that the two isomeric states spanned by a dipole optical transition correspond to two substantially different minima on one and the same potential surface. Given that a matrix element for such a transition can be written, as usual, in the following form:

$$\langle\vec{\mu}\rangle^{\prime,\prime\prime} = \int\left\{\int \Psi_e^2(x, Q)\vec{\mu}\,\mathrm{d}x\right\}\varphi_n'\varphi_n''\,\mathrm{d}Q = \int \vec{\mu}(Q)\varphi_n'\varphi_n''\,\mathrm{d}Q \qquad (14.106)$$

It has been taken into account here that the electronic components of the combination adiabatic vibronic functions are not changed upon transition, and that the only nuclear components changed are those obtained in the course of finding a strict solution to the problem of nuclear motion in the presence of a complex multiple-minima potential. Q is used to designate the relative nuclear co-ordinates and x to denote electronic ones. Integration is made over the entire space of possible changes of electronic and nuclear co-ordinates. $\vec{\mu}(Q)$ is the total molecular dipole moment for given current values of the nuclear co-ordinates for a particular nuclear configuration. Electronic and nuclear components of the adiabatic vibronic function are represented by Ψ_e and φ_n, respectively. Now let us take advantage of the fact that we are dealing with two stable isomeric states corresponding to two minima on a multidimensional potential surface separated by a high and wide barrier so that under ordinary conditions there is an extremely low probability of spontaneous (tunnelling) transitions between such isomers.

This means that 'addition' of the vibrational states of a single isomer occurring with the finite motions of nuclei inside a respective potential well to the vibrational states of another isomer should be very small, thus enabling us to write eigenfunctions φ_n' and φ_n'' in the following conventional form:

$$\varphi_n' = \varphi_n^{(1)} + \alpha\varphi_n^{(2)}, \quad \varphi_n'' = \varphi_n^{(2)} + \beta\varphi_n^{(1)} \qquad (14.107)$$

where α and β are small parameters.

Given that the product $\varphi_n'\varphi_n''$ is as in equation (14.106), we obtain

$$\varphi_n'\varphi_n'' = \varphi_n^{(1)}\varphi_n^{(2)} + \alpha\varphi_n^{(2)^2} + \beta\varphi_n^{(1)^2} + \alpha\beta\varphi_n^{(1)}\varphi_n^{(2)} \simeq \varphi_n^{(1)}\varphi_n^{(2)}$$

in which $\varphi_n^{(1)}$ and $\varphi_n^{(2)}$ are ordinary vibrational functions of the two states in combination of the isomers under investigation. It is assumed that these states are specified by multidimensional harmonic vibrational functions.

The product $\varphi_n^{(1)}\varphi_n^{(2)}$ is certainly different from zero only for small variations of the nuclear co-ordinates, and has a maximum in this region. By expanding the function $\vec{\mu}(Q)$ of the dipole moment into a series around this maximum we finally obtain

$$\langle\vec{\mu}\rangle^{\prime,\prime\prime}=\vec{\mu}(0^{\max})\int\varphi_n^{(1)}\varphi_n^{(2)}\,\mathrm{d}Q+\sum_i\left(\frac{\partial\vec{\mu}}{\partial Q_i}\right)_{0^{\max}}\int\varphi_n^{(1)}\varphi_n^{(2)}Q_i\,\mathrm{d}Q \quad (14.108)$$

Computation of $\vec{\mu}(0^{\max})$, $(\partial\vec{\mu}/\partial Q_i)_{0^{\max}}$, etc. can be made by means of quantum-chemical methods for any point within the configurational space of the nuclear co-ordinates. Therefore, the problem is one of selecting a suitable co-ordinate system,* and computing the integrals

$$\int\varphi_n^{(1)}\varphi_n^{(2)}\,\mathrm{d}Q;\quad\int\varphi_n^{(1)}\varphi_n^{(2)}Q_i\,\mathrm{d}Q$$

etc., provided that the expansion is made in terms of a large number of nuclear co-ordinate exponents.

Before dealing with this problem the second case (i.e. a transition between the states of isomers corresponding to different potential surfaces) will be considered.

Retaining designations Ψ'_{e} and Ψ''_{e} we note that in this case only those MOs remain in the integrals which differ from one another in the combination of two electronic states. Expanding Ψ'_{e} and Ψ''_{e} into a series in terms of nuclear states in the vicinity of minima corresponding to the isomers selected (each minimum is assigned to its own potential surface), and assuming that the minima are well separated, the following expression is found for a matrix element:

$$\langle\vec{\mu}\rangle^{\prime,\prime\prime}=\int\left(\Psi'_{\mathrm{e}}(0^{(1)})+\sum_i\left(\frac{\partial\Psi'_{\mathrm{e}}}{\partial Q_i}\right)_{0^{(1)}}Q_i\right)\vec{\mu}\left(\Psi''_{\mathrm{e}}(0^{(2)})+\sum_i\left(\frac{\partial\Psi''_{\mathrm{e}}}{\partial Q_i}\right)_{0^{(2)}}Q_i\right)$$
$$\times\,\varphi_n^{(1)}\varphi_n^{(2)}\,\mathrm{d}x\,\mathrm{d}Q \quad (14.109)$$

It is again obvious that the integrals

$$\int\Psi'_{\mathrm{e}}(0^{(1)})\vec{\mu}\Psi''_{\mathrm{e}}(0^{(2)})\,\mathrm{d}x\ \text{etc.}$$

have to be computed. This can be done by the ordinary methods used in the theory of vibronic spectra of polyatomic molecules, as well as the calculation of overlap integrals $\int\varphi_n^{(1)}\varphi_n^{(2)}\mathrm{d}Q$, etc. of the vibrational functions of the combination of isomeric states. It will be noted here that all the configurations of all the possible molecular states, including the isomeric ones, must be positioned within the real Cartesian space while these co-ordinate systems are materially different for various isomers, since their spatial structure is different. What is more, even the number of the internal degrees of freedom can change

*For a variety of reasons (which cannot be discussed here) the Cartesian system is completely unacceptable for achieving our objectives.

(linear–non-linear conversion of the geometrical configuration), whereas in the ordinary theory of vibronic spectra, systems of natural co-ordinates are virtually indistinguishable, since only a minute shift occurs. All these features greatly complicate the problem of introducing a unified system of co-ordinates for two isomers, as required for computing the integrals. The difficulties arising here can be overcome as follows.

To obtain the desired transformation between the normal co-ordinates for two polyatomic isomers, denoted by indices 1 and 2, we will proceed as follows[94]. Let us assign numerical numbers to all the atoms within a molecule, maintaining this order of atomic numerical designation for any molecular deformations or intramolecular rearrangements. We will introduce as co-ordinates the distance between atoms (having given numerical designations), the angles between the straight lines connecting pairs of atoms forming a triatomic grouping and chosen according to their numbers so that each pair includes one commonly shared atom, the dihedral angles between the planes passing through the triples of atoms, etc. Bearing in mind that the zero values of all such co-ordinates give a situation where all the atoms merge into a single one, certain non-zero values of all these geometrical parametres are taken as a reference point or origin, rather than the zero values. Let us denote these values as S_0, a column matrix. Then the variables (co-ordinates *per se*) are represented by the increments of the chosen geometrical parameters as compared with their initial values. We denote such increments by q, a column matrix. It can be asserted, without loss of generality, that having properly selected a set of the co-ordinates introduced above, we can specify quite unambiguously any relative location of nuclei in a molecule. Obviously, in so doing, no limitations are imposed as to the smallness of increments q or concerning the independence of the geometrical parameters used.

It can be seen that if we relate the choice of bond lengths, bond angles, etc. to the chemical structure of a given isomer, then, taking as an origin of co-ordinates the equilibrium bond lengths, values of the valence angles and similar parameters of this isomer, we would be dealing with an ordinary natural system of vibrational (or, generally, internal) co-ordinates. Let us call such a system of co-ordinates an eigensystem.

On introducing an eigensystem of natural co-ordinates for the first isomer, then clearly the geometrical configuration of the second isomer can be specified with the aid of the same system of co-ordinates. However, in order to avoid excessive increments of co-ordinates in the course of vibrations let us reposition the origin so that it is situated at a point consistent with the equilibrium configuration of the second isomer, making no changes in the meaning of the geometrical parameters utilized for specifying the relative positions of nuclei. This will result in a new system of natural co-ordinates (denoted here as q^{int}), which differs from the eigensystem of natural co-ordinates for the first isomer only by a certain shift. Thus we have:

$$q^{\text{int}} = q^{(1)} + \Delta s_0 \tag{14.110}$$

where Δs is a vector (a column matrix) of the shift in the origin of natural co-ordinates and $\Delta s = s_0^{(1)} - s_0^{(2)}$.

Differentiating equation (14.110) with respect to time yields $\dot{q}^{\text{int}} = \dot{q}^{(1)}$. Below we will employ only linear normal co-ordinates. Then $\dot{q}^{(1)} = L_q^{(1)}\dot{Q}^{(1)}$ and $\dot{q}^{\text{int}} = L_q^{\text{int}}\dot{Q}^{\text{int}}$ from which $\dot{Q}^{\text{int}} = \tilde{L}_p^{\text{int}}\cdot q^{\text{int}}$ and $\dot{Q}^{\text{int}} = \tilde{L}_p^{\text{int}} L_q^{(1)}\dot{Q}^{(1)} (\tilde{L}_p L_q = I)$.

To obtain $L_q^{(1)}$ the vibrational problem for the first isomer is solved in terms of its natural co-ordinate eigensystem; the solution is achieved by standard methods.

To obtain L_p^{int} and further matrices we proceed as follows. The vibrational problem for the second isomer is constructed taking the co-ordinate system q^{int} as the natural co-ordinates for this isomer. The system is then genetically related to the first system. This enables us to construct a matrix of kinematic coefficients in these co-ordinates. Assuming that the force-constant matrix is diagonal in terms of the same co-ordinates, it can be shown that the final result is not dependent upon the choice of the latter.

It is now necessary to write the following chain of self-evident equalities:

$$\dot{Q}^{\text{int}} = \tilde{L}_p^{\text{int}} L_q^{(1)}\dot{Q}^{(1)}; \quad \dot{\vec{r}}^{(2)} = E\tilde{\vec{B}}^{\text{int}} L_p^{\text{int}}\dot{Q}^{\text{int}};$$
$$\dot{\vec{r}}^{(2)} = E\tilde{\vec{B}}^{\text{int}} L_p^{\text{int}}\tilde{L}_p^{\text{int}} L_q^{(1)}\dot{Q}^{(1)}$$

Using the eigensystem of natural co-ordinates $q^{(2)}$, the solution of the vibrational problem for the second isomer is then obtained using standard techniques. Taking into account that $\dot{q}^{(2)} = \vec{B}^{(2)}\dot{\vec{r}}^{(2)}$ gives:

$$\dot{q}^{(2)} = \vec{B}^{(2)} E\tilde{\vec{B}}^{\text{int}} L_p^{\text{int}}\tilde{L}_p^{\text{int}} L_q^{(1)}\dot{Q}^{(1)}$$

and finally we obtain

$$\dot{Q}^{(2)} = \tilde{L}_p^{(2)}\vec{B}^{(2)} E\tilde{\vec{B}}^{\text{int}} L_p^{\text{int}}\tilde{L}_p^{\text{int}} L_q^{(1)}\dot{Q}^{(1)} \tag{14.111}$$

Due to the fact that the unit vectors (matrices $\vec{B}^{(2)}$ and matrix $\vec{B}^{\text{int}}$) are related to the equilibrium state of one and the same isomer, and provided that the vibrations of its atoms around the equilibrium positions are small, all the matrix products must be constant and invariant with respect to the turning and translations of the external (laboratory) system of Cartesian co-ordinates.

The fact that the matrix $A = \tilde{L}_p^{(2)}\vec{B}^{(2)} E\tilde{\vec{B}}^{\text{int}} L_p^{\text{int}}\tilde{L}_p^{\text{int}} L_q^{(1)}$ is constant upon integration over time leads to the relationship $Q^{(2)} = \mathscr{A}Q^{(1)} + b$. In turn, it is clear that the constant $b = \tilde{L}_p^{(2)}\vec{B}^{(2)} E\tilde{\vec{B}}^{\text{int}} L_p^{\text{int}}\tilde{L}_p^{\text{int}}\Delta s_0$. This follows immediately from the fact that if we multiply successively the left-hand side of equation (14.110) first by $\tilde{L}_p^{\text{int}}$, then multiply the result by $(E\tilde{\vec{B}}^{\text{int}} L_p^{\text{int}})$ and then by $\tilde{L}_p^{(2)}\vec{B}^{(2)}$, differentiating the final result with respect to time and taking into account that the matrix products are constant, we obtain expression (14.111). Such a chain of successive multiplications means a transition from $\dot{q}^{\text{int}}$ to $\dot{Q}^{\text{int}}$, then to $\dot{\vec{r}}^{(2)}$ and then to $\dot{q}^{(2)}$, etc.

The relationship

$$Q^{(2)} = \widetilde{L}_p^{(2)} \vec{B}^{(2)} E \widetilde{\vec{B}}^{\text{int}} L_p^{\text{int}} \widetilde{L}_p^{\text{int}} (L_q^{(1)} Q^{(1)} + \Delta s_0) \qquad (14.112)$$

can be called a generalized Dushinsky relation[94].

Knowledge of equation (14.112) enables us to express all the integration elements comprising the vibrational functions in a single system of normal co-ordinates of any of the isomers, thus ensuring the possibility of computing these integrals later using standard methods.

Note that in the first of the cases discussed (isomers associated with one potential surface), in order to construct the co-ordinates used for the expansion of the product $\varphi_n^{(1)}\varphi_n^{(2)}$ around the maximum it is helpful to introduce a dummy model of a vibrating molecule with an equilibrium geometry corresponding to this point.

Following this procedure, the problem of computing the probabilities of optical dipole transitions between topologically isomeric states in a polyatomic molecules is solved completely.

References

1. R. C. Bingham, M. J. S. Dewar and D. H. Lo. *J. Am. Chem. Soc.*, **97**, 1294 (1975).
2. O. Kikuchi and A. Maeda. *Bull. Chem. Soc. Jpn*, **50**, 8, 2179 (1977).
3. P. Zahradnik and J. Leska. *Collect. Czech. Chem. Common.*, **42**, 7, 2060 (1977).
4. D. H. Lo and M. A. Whitehead. *Can. J. Chem.*, **46**, 2041 (1968).
5. A. Warshel and M. Karplus. *J. Am. Chem Soc.*, **94**, 5612 (1972).
6. K. Miller and J. N. Murrell. *Theor. Chim. Acta*, **3**, 231 (1965).
7. K. Miller and J. N. Murrell. *Theor. Chim. Acta*, **7**, 69 (1967).
8. K. Ohno. *Chem. Phys. Lett.*, **53**, 571 (1978).
9. A. V. Likashin, V. I. Permogorov and M. D. Franck-Kamenetsky. *Dokl. Akad. Nauk SSSR*, **183**, 875 (1968).
10. C. A. Coulson. *Proc. Roy. Soc. London, Ser. A*, **169**, 413 (1939).
11. C. A. Coulson and A. Golebievski. *Proc. Roy. Soc.*, **78**, 1314 (1961).
12. H. C. Longuet-Higgins and L. Salem. *Proc. Roy. Soc. London, Ser A*, **251**, 172 (1959).
13. H. J. Bernstein. *Trans. Faraday Soc.*, **57**, 1649 (1961).
14. C. W. Scheer, *J. Chem. Phys.*, **21**, 1582 (1953).
15. A. Julg. Ph. François and R. Mourre. *J. Chim. Phys.*, **59**, 363 (1962).
16. P. C. den Boer-Veenendal and D. H. W. den Boer. *Theor. Chim. Acta*, **6**, 354 (1966).
17. E. M. Popov and G. A. Kogan. *Teor. Eksp. Khim.*, **1**, 295 (1965).
18. E. M. Popov and G. A. Kogan. *Usp. Khim.*, **37**, 256 (1968).
19. E. M. Popov, G. A. Kogan and V. M. Zheltova. *Teor. Eksp. Khim.*, **6**, 14 (1970).
20. A. Julg. *J. Chim. Phys.*, **65**, 541 (1968).
21. Kei Nakachi, Masanori Sukuranage and Hideo Suzuki. *J. Phys. Soc. Jpn*, **43**, 272 (1977).
22. M. Allavena and S. Bratož. *J. Chim. Phys.*, **60**, 1999 (1963).
23. D. M. Bishop and M. Randič. *J. Chim. Phys.*, **44**, 2480 (1960).
24. S. Bratož and M. Allavena. *J. Chim. Phys.*, **37**, 2138 (1962).
25. E. Clementi. *J. Chem. Phys.*, **46**, 4731; 4737 (1967); **49**, 1112 (1968).
26. F. Dushinsky. *Acta Physicochim. USSR*, **7**, 551 (1937).
27. E. Hutchisson. *Phys. Rev.*, **36**, 410 (1930).
28. C. Manneback. *Physica*, **17**, 1001 (1951).
29. M. Wagner. *Z. Naturforcsh., Teil A*, **14**, 81 (1959).

30. F. Ansbaches. *Z. Naturforcsh., Teil A*, **14**, 889 (1959).
31. R. A. Preem. *Trudi IF AN SSSR*, **16**, 57 (1961).
32. W. L. Smith. *J. Phys. (Proc. Phys. Soc.)*, **B2**, 1 (1969).
33. E. V. Doktorov, L. A. Malkin, V. I. Man'ko and A. V. Tyurin. Preprint Phys. Inst., No. 36, Moscow (1974).
34. E. V. Doktorov, I. A. Malkin and V. I. Man'ko. Preprint Phys. Inst., No. 142, Moscow (1974).
35. S. Koide. *Z. Naturforcsh., Teil A*, **15**, 123 (1960).
36. P. McCoy and J. R. Ross. *Aust. J. Chem.*, **15**, 573 (1962).
37. T. E. Sharp and H. N. Rosenstock. *J. Chem. Phys.*, **41**, 3453 (1964).
38. A. V. Lukashin and M. D. Franck-Kamenetsky. *Dokl. Akad. Nauk SSSR*, **188**, 391 (1969).
39. A. V. Lukashin. *Opt. Spektrosk.*, **32**, 661 (1972).
40. I. S. Osad'ko. *Fiz. Tverd. Tela*, **17**, 3180 (1975).
41. I. Kakesaka and K. Kawai. *Bull. Chem. Soc. Jpn*, **48**, 2745 (1975).
42. G. J. Small. *J. Chem. Phys.*, **54**, 3300 (1975).
43. M. V. Priyutov and M. A. Kovner. *Opt. Spektrosk.*, **31**, 699 (1971).
44. E. V. Doktorov, I. A. Malkin and V. I. Man'ko. *J. Mol. Spectrosc.*, **56**, 1 (1975).
45. E. V. Doktorov, I. A. Malkin and V. I. Man'ko. *J. Phys. B*, **9**, 507 (1976).
46. A. A. Terekhov, M. V. Priyutov and M. A. Kovner. *Izv. Akad. Nauk SSSR, Ser. Fiz.*, **34**, 1331 (1970).
47. V. I. Permogorov, L. A. Serdyukova and M. D. Franck-Kamenetsky. *Izv. Akad. Nauk SSSR, Ser. Fiz.*, **32**, 1560 (1968).
48. V. I. Permogorov, L. A. Serdyukova and M. D. Franck-Kamenetsky. *Opt. Spektrosk.*, **22**, 979 (1967).
49. V. I. Permogorov, L. A. Serdyukova and M. D. Franck-Kamenetsky. *Opt. Spektrosk.*, **25**, 77 (1968).
50. A. V. Lukashin, V. I. Permogorov and M. D. Franck-Kamenetsky. *Dokl. Akad. Nauk SSSR*, **183**, 875 (1968).
51. N. S. Strokach, E. A. Gastilovich and D. N. Shigoring. *Opt. Spektrosk.*, **35**, 238 (1973).
52. T. N. Bolotnikova, L. A. Klimova, G. N. Nersesova, D. Yu. Shtrokirkh and P. I. Zhidkin. *Zh. Prikl. Spektrosk.*, **26**, 30 (1978).
53. M. V. Volkenstein, L. A. Gribov, M. A. Elyashevich and B. I. Stepanov. *Molecular Vibrations*, Nauka, Moscow (1972) (in Russian).
54. L. A. Gribov and V. A. Dementyev. *Methods and Algorithms for the Calculations in the Theory of Vibrational Spectra of Polyatomic Molecules*, Nauka, Moscow (1981).
55. *Sadtler Standard Spectra Atlas. UV Spectra*, Sadtler Research Laboratory (1961–75).
56. B. Sharf and B. Honig. *Chem. Phys. Lett.*, **7**, 132 (1970).
57. E. Filter and E. O. Steinborn. *J. Math. Phys.*, **19**, 79 (1978).
58. E. O. Steinborn and E. Filter. *Theor. Chim. Acta*, **52**, 189 (1979).
59. G. Herzberg and E. Teller. *Z. Phys. Chem. Abt. B*, **21**, 410 (1933).
60. E. Renner. *Z. Phys.*, **92**, 172 (1934).
61. H. Sponer and E. Teller. *Rev. Mod. Phys.*, **13**, 75 (1941).
62. H. C. Longuet-Higgins. *UFN*, **83**, 137 (1964).
63. I. Suzuki. *Bull. Chem. Soc. Jpn*, **44**, 3277 (1971); **48**, 1685, 3565 (1975).
64. L. A. Gribov. *Opt. Spektrosk.*, **31**, 842 (1972).
65. A. I. Pavluchco and L. A. Gribov. *J. Appl. Spectr.*, **33**, 904 (1980) (in Russian).
66. L. A. Gribov and N. I. Prokofyeva. *Opt. Spektrosk.*, **44**, 1032 (1978).
67. L. A. Gribov. *Introduction to Molecular Spectroscopy*, Nauka, Moscow (1976) (in Russian).
68. B. K. Novosadov. *Zh. Strukt. Khim.*, **16**, 967 (1975).
69. V. A. Fock. *The Principles of Quantum Mechanics*, Nauka, Moscow (1976) (in Russian).

70. A. Erdeliji, W. Magnus, F. Oberhettinger and F. G. Tricomi. *Higher Transcendental Functions*, vol. I, McGraw-Hill, New York (1951).
71. A. Erdeliji, W. Magnus, F. Oberhettinger and F. G. Tricomi. *Higher Transcendental Functions*, vol II, McGraw-Hill, New York (1953).
72. D. A. Varshalovich, A. N. Moskalev and V. K. Khersonsky. *The Quantum Theory of Angular Momentum*, Nauka, Leningrad (1975).
73. G. Kampe de Feriet *et al. Fonctions de la Physique Mathematique*, Centre National de la Recherche Scientifique, Paris (1957).
74. I. S. Gradstein and I. M. Ryzhik. *Tables of Integrals, Sums, Series and Products*, Fizmatghiz, Moscow (1962).
75. J. Del Bense and H. H. Jaffe. *J. Chem. Phys.*, **48**, 1807, 4050 (1968); **49**, 1221 (1968); **50**, 1126 (1969).
76. D. R. Salahub and C. Sandorfy. *Theor. Chim. Acta*, **20**, 227 (1971); **22**, 325 (1971).
77. R. McDiarmid. *Chem. Phys. Lett.*, **34**, 130 (1975).
78. R. M. Gavin, S. Risemberg and S. A. Rice. *J. Chem. Phys.*, **58**, 3160 (1973).
79. G. Orlandi and W. Siebrand. *J. Chem. Phys.*, **58**, 4513 (1973).
80. V. I. Baranov, G. N. Ten and L. A. Gribov. *J. Molec. Struct. (Theochem)*, **137**, 91 (1986).
81. V. I. Baranov, L. A. Gribov and B. K. Novosadov. *J. Molec. Struct.*, **70**, 1 (1981).
82. V. I. Baranov and L. A. Gribov. *J. Chem. Phys.*, **70**, 31 (1981).
83. V. I. Baranov, G. N. Ten and L. A. Gribov. *J. Molec. Struct. (Theochem)*, **137**, 91 (1986).
84. Y. Mochizuki, K. Kaya and M. Jto. *J. Chem. Phys.*, **65**, 4163 (1976).
85. A. E. Knight, C. M. Lawburgh and C. S. Parmenter. *J. Chem. Phys.*, **63**, 4336 (1975).
86. M. V. Priuytov *et al. Doklady AN USSR*, **232**, 332 (1977) (in Russian).
87. L. A. Gribov and V. I. Baranov. *J. Appl. Spectr.*, **44**, 341 (1985) (in Russian).
88. R. P. Feynman. *Phys. Rev.* **56**, 340 (1939).
89. W. B. Brown. *Proc. Camb. Phil Soc.*, **54**, 251 (1958).
90. M. V. Priuytov *et al. J. Struct. Chem.* **24**, 3 (1983) (in Russian).
91. M. V. Priyutov, Yu. V. Kamensky and I. F. Kovalev. *Opt. and Spectr.*, **57**, 588 (1984) (in Russian).
92. L. A. Gribov. *J. Molec. Struct. (Theochem.)*, **110**, 61 (1984).
93. M. E. Elyashberg, L. A. Gribov and V. V. Serov. *Molecular Spectral Analysis and Computers*, Nauka, Moscow (1980) (in Russian).
94. L. A. Gribov. *Opt. and Spectr.*, **59**, 1337 (1985) (in Russian).

Chapter 15

Theory of Intensities in the Raman Spectra of Polyatomic Molecules

15.1 General Problems and the Construction of the Parametric Theory

To consider the theory of Raman intensities we start from some general considerations. It will be shown that this theory can be founded on a semi-empirical basis, and this enables the spectra of large compounds to be calculated by means of the fragmentary method.

The general formula for the tensor components of Raman scattering viewed as a two-quantum process during which two quanta are absorbed and emitted simultaneously is[1]

$$((\alpha'_{xy}))_{0,0,1} = \frac{1}{h}\sum_{n,v}\left\{\frac{\langle\mu_x\rangle^{n,v}_{0,0}\langle\mu_y\rangle^{0,1}_{n,v}}{\nu^{n,v}_{0,0}-\nu} + \frac{\langle\mu_y\rangle^{n,v}_{0,0}\langle\mu_x\rangle^{0,1}_{n,v}}{\nu^{n,v}_{0,1}+\nu}\right\} \tag{15.1}$$

Summation is made here over all excited (n, v) electronic vibrational states. The indices n and v correspond to the electronic and vibrational states, respectively.

In Raman scattering transition occurs between two (one principal and one of the fundamental) vibrational levels of the main electronic state. This typical case is often realized experimentally. In the tensor notation $((\alpha'))_{0,0,1}$ the first index corresponds to the main electronic state, the second denotes the vibrational state from which the transition occurs and the third is the final vibrational state. The scheme of such transitions is shown in Figure 15.1. $\langle\mu_x\rangle^{n,v}_{0,0}$, etc. denote the components of matrix elements for dipole transitions between two combining electronic-optical states. The frequency of incident light is denoted as ν.

The following considerations are based on the adiabatic approximation. All values in equation (15.1) are numbers and not functions. We must differentiate between the cases of normal (or spontaneous) and of resonance Raman scattering. As a rule, in the first case $\nu < \nu^{n,v}_{0,0}, \nu^{n,v}_{0,1}$. In the second, ν is approximately equal to $\nu^{n,v}_{0,0}$ or, if possible, to the frequencies of the other adjacent electronic–vibrational excited states. Then some of the terms in equation (15.1) are of much greater value than others, and we take only these

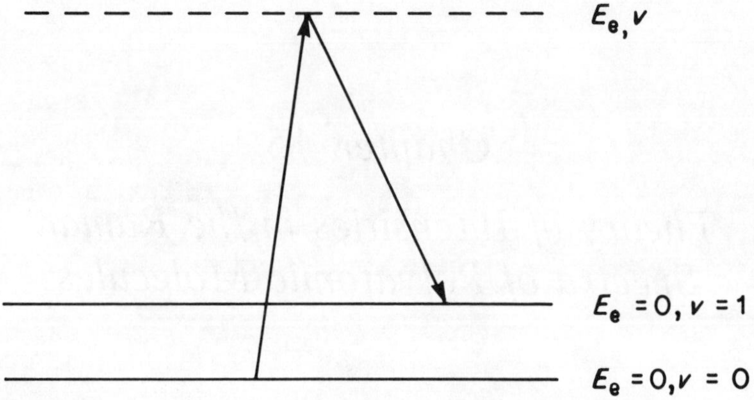

Figure 15.1. The scheme of transitions in the typical case of Raman scattering

terms into consideration. Therefore direct calculations by equation (15.1) becomes possible and there is no need to create a special theory.

Strictly speaking, in non-resonance regions where spontaneous scattering occurs a very large number of terms in equation (15.1) can be significant, and direct calculations become impossible. This forces us to create a theory in order to sum the separate contributions, and the fundamentals of such a theory are considered below[2].

Using a power series expansion it is possible to present the terms in equation (15.1) in the form of infinite sums:

$$(\nu_{0,0}^{n,v} - \nu)^{-1} = \sum_{m=0}^{\infty} \frac{\nu^m}{(\nu_{0,0}^{n,v})^{m+1}}$$

$$(\nu_{0,1}^{n,v} + \nu) = \sum_{m=0}^{\infty} \frac{(-1)^m \nu^m}{(\nu_{0,1}^{n,v})^{m+1}} \tag{15.2}$$

As a result, equation (15.1) has the form:

$$((\alpha'_{xy}))_{0,0,1} = \frac{1}{h} \sum_{n,v,m} \left\{ \frac{\langle \mu_x \rangle_{0,0}^{n,v} \langle \mu_y \rangle_{n,v}^{0,1}}{(\nu_{0,0}^{n,v})^{m+1}} + (-1)^m \frac{\langle \mu_y \rangle_{0,0}^{n,v} \langle \mu_x \rangle_{n,v}^{0,1}}{(\nu_{0,1}^{n,v})^{m+1}} \right\} \nu^m \tag{15.3}$$

This expression for the tensor components $((\alpha_{xy}))_{0,0,1}$ is in the form of a sum of terms, each of which contains a part independent of the incident light and the factor ν^m. These terms will decrease particularly rapidly with the growth of the mth power if $\nu \ll \nu_{0,0}^{n,v}$, $\nu_{0,1}^{n,v}$, i.e. far from resonance. This allows us to restrict the consideration to several terms in equation (15.2). Evidently, there is a term (called Plachek's term) independent from ν and which always exists in the expression for $((\alpha'_{xy}))_{0,0,1}$, becoming predominant in the limit when ν is small.

In the Herzberg–Teller approximation the matrix element $\langle \vec{\mu} \rangle_{0,0}^{n,v}$ is equal to

$$\langle \vec{\mu} \rangle_{0,0}^{n,v} = \langle \vec{\mu} \rangle_0^n \langle v|0 \rangle + \sum_k \left\langle \frac{\partial \vec{\mu}}{\partial Q_k} \right\rangle_0^n \langle v|Q_k|0 \rangle \tag{15.4}$$

if we restrict consideration only to the term linear in Q_k. Therefore, as a result, in equation (15.1) we have the form:

$$\begin{aligned} \langle \mu_x \rangle_{0,0}^{n,v} \langle \mu_y \rangle_{n,v}^{0,1} &= \langle \mu_x \rangle_0^n \langle \mu_y \rangle_0^n \langle v|0 \rangle \langle 1|v \rangle \\ &+ \langle \mu_x \rangle_0^n \sum_k \left\langle \frac{\partial \mu_y}{\partial Q_k} \right\rangle_0^n \langle v|0 \rangle \langle 1|Q_k|v \rangle \\ &+ \langle \mu_y \rangle_0^n \sum_k \left\langle \frac{\partial \mu_x}{\partial Q_k} \right\rangle_0^n \langle 1|v \rangle \langle 0|Q_k|v \rangle + \mathscr{A}(Q_k, Q_l) \end{aligned} \tag{15.5}$$

The term $A(Q_k, Q_l)$ contains the product of matrix elements of normal co-ordinates and will not give any contribution to the intensities associated with principal transitions. Therefore it will be ignored in the following discussion.

For the second-order product $\langle \mu_y \rangle_{0,0}^{n,v} \langle \mu_x \rangle_{n,1}^{0,1}$ we obtain the same expression but with the change of indices x and y.

It is clear that the electronic and vibrational matrix elements are separated. Therefore the summations in equations (15.1) and (15.3) can be made over vibrational states of each excited electronic–vibrational one and then over excited electronic states.

Several simplifications are possible if a number of suppositions are made. First, it is clear that with an increase in the energy of excited electronic states the value of each succeeding member in equation (15.1) over these states will decrease as the frequencies $\nu_{0,0}^{n,v}$, $\nu_{0,1}^{n,v}$ for the corresponding electronic–vibrational transitions are situated in the denominator of that equation.

Generally speaking, this does not mean that highly excited states can be excluded as soon as their summational influence becomes small. Nevertheless, their exclusion in the approximate theory is not likely to lead to a serious falsification of the physical picture.

At the same time, it is well known that comparatively low-energy electronic states (which may be numerous) of polyatomic molecules can be obtained by transition of only one or two electrons to higher levels. Therefore in complicated systems relative changes of distribution of electronic density during pure electronic transition are small. As a result, the parameters of the potential surfaces of the first electronic-excited states do not change much in comparison with the principal state. This has been confirmed by calculation. Experimentally, it is expressed in the fact that, as a rule, in electronic–vibrational spectra the 0–0 transition is the most intense. At the next stage of approximation this allows us to assume that all potential surfaces of electronic-excited states (which really must be taken into account during the summation in equation (15.1)) are very similar, and differ only by a shift on the energy scale.

Certainly this is a very rough approximation, which is justified only by the fact that on the basis of this it is possible to understand the main features of the complete effect.

The next supposition is not so great, and consists of the introduction of a parabolic form for the potentials and the use of the harmonic approximation for the vibrations. We shall assume that the vibrational level with index 1 corresponds to a state for which only one normal vibration with index k is excited; i.e. the state has the set of vibrational quantum numbers $v_{k'=k} = 0, v_k = 1$.

With these assumptions when summing over vibrational states in expressions (15.1) and (15.3) within the limits of the accountable electron states only the terms with the following combinations of vibrational indices would be non-zero;

$$0 \leftarrow\rightarrow 0, 0 \leftarrow\rightarrow 1k, 1k \leftarrow\rightarrow 1k$$

The notation $1k$ is introduced to emphasize the fact that for the excited vibrational level only $v_k = 1$.

Hence equation (15.3) takes the form:

$$((\alpha'_{xy}))_{0,0,1} = \frac{1}{h}\sum_{n,m}\left\{\frac{\langle\mu_x\rangle_0^n\langle(\partial\mu_y/\partial Q_k)\rangle_0^n}{(v_{0,0}^{n,0})^{m+1}} + (-1)^m\frac{\langle\mu_y\rangle_0^n\langle(\partial\mu_x/\partial Q_k)\rangle_0^n}{(v_{0,1}^{n,0})^{m+1}} + \frac{\langle(\partial\mu_x/\partial Q_k)\rangle_0^n\langle\mu_y\rangle_0^n}{(v_{0,0}^{n,1})^{m+1}} + (-1)^m\frac{\langle(\partial\mu_y/\partial Q_k)\rangle_0^n\langle\mu_x\rangle_0^n}{(v_{0,1}^{n,1})^{m+1}}\right\} v^m Q_{ok} \quad (15.6)$$

Further, if we assume that $v_{0,0}^{n,0} \approx v_{0,1}^{n,1}$ and that the energy of the excited vibrational double quantum is much less than that for the first excited electron state, then it follows that $v_{0,1}^{n,0} \approx v_{0,0}^{n,1} \approx v_{0,0}^{n,0} = v_n$.

In this case, as seen from equation (15.6), all terms with odd indices m disappear and we have

$$((\alpha'_{xy}))_{0,0,1} = \frac{2}{h}\sum_{n,m}\frac{v^{2m}}{(v_n)^{2m+1}}\frac{\partial}{\partial Q_k}(\langle\mu_x\rangle_0^n\langle\mu_y\rangle_0^n) \quad 0 \leqslant m \leqslant \infty \quad (15.7)$$

When v_n are constant and the products $\langle\mu_x\rangle_0^n\langle\mu_y\rangle_0^n$ are tensor components then, on introducing tensors $((\alpha_n^{(m)}))$ with components

$$((\alpha_{xyn}^{(m)})) = \langle\mu_x\rangle_0^n\langle\mu_y\rangle_0^n(v_n)^{-(2m+1)} \text{ etc.}$$

equation (15.7) can be rewritten in the form

$$((\alpha'_{xy}))_{0,0,1} = \frac{2}{h}\sum_{n,m} v^{2m}Q_{ok}\frac{\partial}{\partial Q_k}((\alpha_{xyn}^{(m)})) = \frac{2}{h}\sum_n Q_{ok}\frac{\partial}{\partial Q_k}((\alpha_{xyn}^{(0)})) + \frac{2}{h}\sum_{n,m=1} v^{2m}Q_{ok}\frac{\partial}{\partial Q_k}((\alpha_{xyn}^{(m)})) \quad (15.8)$$

At this stage use is made of the valence-optical scheme for the matrix elements of electronic–vibrational transitions.

Then $\langle\mu_x\rangle_0^n$, etc. will be equal $\langle\mu_x\rangle_0^n = \tilde{\mu}_n e_x$, where $\tilde{\mu}_n$ is the row-matrix of dipole moment transitions of bonds and e_x the column-matrix of unit vectors of bonds. It is understood here that for each bond its own orthonormal system of unit vectors is used, one of which is directed along the bond and two others are perpendicular to it.

Then for equation (15.8) we have:

$$
\begin{aligned}
((\alpha'_{xy}))_{0,0,1} &= \frac{2}{h}\sum_{n,m} \nu^{2m} Q_{ok} \frac{\partial}{\partial Q_k}[\tilde{\mu}_n e_x \tilde{e}_y \mu_n (\nu_n)^{-(2m+1)}] \\
&= \frac{2}{h}\sum_{m} \nu^{2m} Q_{ok} \frac{\partial}{\partial Q_k}\{Sp(\tilde{M}[e_x e_y] M \delta\nu_m)\} \\
&= \frac{2}{h}\sum_{m} \nu^{2m} Q_{ok} \frac{\partial}{\partial Q_k}\{Sp(M \delta\nu_m \tilde{M}[e_x e_y]\} \\
&= \frac{2}{h}\sum_{m} \nu^{2m} Q_{ok} \frac{\partial}{\partial Q_k}\{Sp(P_m[e_x e_y])\}
\end{aligned}
\tag{15.9}
$$

Here, $P_m = M\delta\nu_m M$ is the square matrix of order equal to the number of molecular unit vectors, $\delta\nu_m$ is the diagonal matrix of order $n_{\max}$ with the elements $(\nu_n)^{-(2m+1)}$ and M is the rectangular matrix with $n_{\max}$ columns, containing the values of transitional dipole moments of bonds. The matrices P_m may be called the matrices of dynamic polarization.

The square asymmetric matrix $[e_x e_y]$ has the form

$$
[e_x e_y] = \begin{bmatrix} e_{1x}^{(1)} e_{1y}^{(1)}; e_{1x}^{(1)} e_{1y}^{(2)}; \cdots \\ e_{1x}^{(2)} e_{1y}^{(1)}; e_{1x}^{(2)} e_{1y}^{(2)}; \cdots \end{bmatrix}
$$

The subscripts in the notation $e_{1x}^{(1)}$, etc. refer to the bond numbers and the superscript to the unit vectors of these bonds. Directional (unit) bond vectors are marked by (1) and vectors perpendicular to the bonds by (2) and (3).

In equation (15.9) it is important to note that the order of matrices P_m does not depend on $n_{\max}$ at all, and that the electronic states are not taken into account. The dependence upon $n_{\max}$ will be reflected only in the values of the matrix elements. Therefore the elements of matrices P_m can be treated as empirical parameters in constructing the parametric theory.

Now let us consider the case when separate pure electron transitions are localized inside a definite atomic grouping of the molecule. For large molecular fragments far from each other this approach is quite reliable. Combining columns of matrix M in groups so that each group corresponds to the transitions localized inside one fragment a quasi-diagonal matrix whose blocks are rectangular submatrices is obtained. As a result of this, quasi-diagonal matrices P_m of the same structure appear. Under this condition for a set of molecules, containing similar atomic groupings, the elements of such quasi-diagonal matrices P_m can become transferable.

If we denote submatrices of matrices P_m by $P_m^{(r)}$ then equation (15.9) can be written in the form

$$((\alpha'_{xy}))_{0,0,1} = \frac{2}{h}\sum_{m,r} \nu^{2m} Q_{ok} \frac{\partial}{\partial Q_k} \{Sp(P_m^{(r)}[e_x e_y]^{(r)})\} \tag{15.10}$$

where $[e_x e_y]^{(r)}$, being submatrices of the same order as $P_m^{(r)}$, are split out from the general matrix $[e_x e_y]$ along its diagonal.

Since these last submatrices will also contain unit vectors of those molecular fragments, inside which are localized the corresponding electronic transitions, then we can conclude that the tensor components $((\alpha'_{xy}))_{0,0,1}$ would consist of sums of terms containing the characteristics of separate molecular fragments. It is clear that this allows us to construct a parametric theory of line intensities in Raman spectra based on the method of fragmentary calculations by introducing elements of submatrices $P_m^{(r)}$ and their derivative with respect to the natural vibrational co-ordinates as empirical parameters. Unfortunately, the number of parameters is rather large, and it is necessary to introduce further simplifications in order to reduce their number.

The next step can be done if we note that the matrices $P_m = M\delta\nu_m\tilde{M}$ differ from each other only by diagonal matrices $\delta\nu_m$, where $\delta\nu_m = (\delta\nu_0)^{2m+1}$. The matrix $\delta\nu_0$ contains the reciprocal values of frequencies of pure electronic transitions on its diagonal.

In constructing a parametric theory on the basis of elements of the matrix P_m it is possible to change matrix $\delta\nu_0$ for matrix ν_{av}^{-1} by choosing some average value for the frequencies of the electronic transition. Then we obtain new matrices $P_m = \nu_{av}^{-(2m+1)}P'_0$, where $P'_0 = M\tilde{M}$. In this case, equation (15.10) has the form

$$\begin{aligned} ((\alpha'_{xy}))_{0,0,1} &= \frac{2}{h}\sum_{m,r} \frac{\nu^{2m}}{\nu_{av}^{(2m+1)}} Q_{ok} \frac{\partial}{\partial Q_k} \{Sp(P_0^{(r)'}[e_x e_y]^{(r)})\} \\ &= \frac{2}{h}\sum_{m} \frac{\nu^{2m}}{\nu_{av}^{(2m+1)}} Q_{ok} \frac{\partial}{\partial Q_k} \left\{ \sum_r Sp(P_0^{(r)'}[e_x e_y]^{(r)}) \right\} \end{aligned} \tag{15.11}$$

The number of different theoretical parameters are thus greatly reduced, and the factors $\nu^{2m}/\nu_{av}^{(2m+1)}$ which appear rapidly decrease with the growth of m on condition that $\nu < \nu_{cp}$. On the whole, far from resonance when ν is small it is possible to keep in equation (15.11) only one term with $m = 0$ and to ignore the dependence of the tensor $((\alpha'))$ upon the frequency of incident light. This is known as Plachek's approximation.

Hence it can be seen, first, that the tensor defining the line intensities can be presented in the form of a derivative of a dynamic tensor $((\alpha))$ with components

$$((\alpha_{xy})) = \frac{2}{h}\sum_m \nu^{2m} Sp(P_m[e_x e_y])$$

with respect to the corresponding normal co-ordinate multiplied by Q_{ok}. Second, the components of this tensor can be represented approximately in the form

$$((\alpha_{xy})) = \frac{2}{h}\left[\sum_r Sp(P_0^{(r)'}[e_x e_y]^{(r)})\right]\left(\sum_m \frac{\nu^{2m}}{\nu_{av}^{(2m+1)}}\right) \tag{15.12}$$

Therefore calculations of intensities will be reduced to calculation of derivatives of matrices $P_0^{(r)'}$ and $[e_x e_y]^{(r)}$ and spurs of products of matrices. Elements of matrices $P_0^{(r)'}$ and $(\partial/\partial q)P_0^{(r)'}$ (derivatives with respect to natural vibrational co-ordinates) create a set of electro-optical constants introduced into the theory as an empirical parameter. The technique of calculation of derivatives of different bond unit vectors (elements of the matrix $[e_x e_y]$) will be stated below. A unique additional parameter is also introduced. This is the value ν_{av}, which can be ignored in Plachek's approximation since it would be simply some scale factor.

In the case of the most simple parametric theory, known as the valence-optical scheme for Raman line intensities, the matrix P_0 is diagonal. It is not difficult to see that in this case the components of tensor $((\alpha))$ are written as follows:

$$((\alpha_{xy})) = \frac{2}{h}\frac{1}{\nu_{cp}}\left(\sum_{s,g}\alpha_s^{(g)}e_{sx}^{(g)}e_{sy}^{(g)}\right)\sum_{m=0}\left(\frac{\nu}{\nu_{cp}}\right)^{2m} \quad g = 1, 2, 3 \tag{15.13}$$

or

$$((\alpha)) = \left(\sum_{s}\left(\left(\alpha_s\right)\right)\right)\sum_{m=0}\left(\frac{\nu}{\nu_{av}}\right)^{2m}$$

where $((\alpha))$ is the tensor of dynamic bond polarization and s-bond number.

Though the number of empirical parameters ($\alpha_s^{(g)}$ and derivatives $\partial\alpha_s^{(g)}/\partial q$) is increased three times in comparison with the valence-optical theory for vibrational spectra, it is rather convenient not only for general research but also, in many cases, for calculations on specific problems. These questions are considered below.

All that has been said above shows that, although it is possible to create a parametric theory for Raman intensities, this approach has too many assumptions. Therefore in spite of the fact that the parameters introduced can be found by quantum-mechanical calculations, it is clear that the values obtained can differ greatly from those obtained experimentally. In this sense the situation is analogous to that found in semi-empirical theories of electronic spectra. Therefore numerical values of the corresponding parameters can be defined accurately only by means of a solution of the inverse spectral problem.

In conclusion, we return again to the question touched upon above. The transformation to matrices of dynamic polarization is worthwhile only when $\nu \ll \nu_0^1$. In this case, we cannot arbitrarily reduce the number of terms in the sum over electronic states in equation (15.1), etc. since all of them show some summational influence. The introduction of P_m matrices allows us to take into account all this influence as soon as the order of these matrices, for any number of addends in equation (15.1), remains constant. In the same way this method and, consequently, the use of a special parametric theory for Raman intensities is profitable not only for regions far from resonance but also for those in between.

It has already been stated that in the region near to resonance (for instance, $\nu \approx \nu_0^1$), because they are divided by a small quantity, one or more terms in equation (15.1) start to increase abruptly. This, as was noted above, gives an opportunity to perform direct calculations of the tensors of Raman scattering immediately on the basis of equation (15.1) by means of the method developed above for the calculation of electronic–vibrational spectra. In this case there is no need to construct a special parametric theory and, in particular, it is possible to use the parameters, introduced in valence-optical theory for electronic–vibrational spectra. The construction of a special theory and the introduction of new parameters is only essential, therefore, for the description of non-resonance spontaneous Raman spectra.

Far from resonance and in the middle region the dependence of Raman intensities upon the frequencies of the incident light will, according to the assumptions made, depend upon the factor

$$\left[\sum_{m=0}\left(\frac{\nu}{\nu_{av}}\right)^{2m}\right]^2$$

and far from resonance the intensity depends upon the quadratic term

$$\left[\sum_{m=0}\left(\frac{\nu}{\nu_{av}}\right)^{2m}\right]^2 \approx 1+2\left(\frac{\nu}{\nu_{av}}\right)^2$$

Therefore such spectra should be considered as three-dimensional in the co-ordinates of line intensity (vertical axes) and frequency of scattered and incident light. This approach appears to be especially useful for the analysis of multicomponent mixtures.

Finally, it can stated that in obtaining the formula for Plachek's approximation it was shown that in this case the tensor of Raman scattering is found by simple differentiation of the tensor of static molecular polarization. This involved a number of suppositions, some of them rather artificial. However, there is no need to do this, since at all stages of consideration it is possible not to go outside the scope of a more correct approach based on the idea of dynamic polarization.

15.2 General Formulae for the Intensity and Depolarization of a Raman Line

We have shown above that Raman intensities are determined by the tensor $(\partial((\alpha))/\partial Q_i = ((\alpha')))$. Like polarizability, this tensor is symmetric and can be transformed into a diagonal form by appropriate rotations of the co-ordinate axes. The major axis of the tensor $((\alpha'))$ does not, generally speaking, coincide for a vibrating molecule with the major axis of the polarizability tensor $((\alpha))$. Nevertheless, it can always be distinguished. We will derive the expression for the intensity and level of depolarization of the Raman lines for molecules which can freely orientate themselves in a gas or liquid.

We asssume that the light propagates along the ξ axis of a fixed co-ordinate system. Let us suppose, in addition, that the scattered light is observed along a perpendicular axis, η. Since the light wave has a cross-polarization, its field (natural radiation) has components $E_\eta \neq 0$, $E_\zeta \neq 0$ and $E_\xi = 0$. We shall denote the unit vectors of the major axes of the tensor derivative of the polarizability tensor (with respect to a normal co-ordinate) by $\vec{e}_1$, $\vec{e}_2$, $\vec{e}_3$.

The tensor $((\alpha'))$ will have the following form in the system of these unit vectors:

$$\left(\begin{pmatrix} \alpha'_1 & 0 & 0 \\ 0 & \alpha'_2 & 0 \\ 0 & 0 & \alpha'_3 \end{pmatrix}\right)$$

We shall consider a molecule whose tensor $((\alpha'))$ has major axes $\vec{e}_1, \vec{e}_2, \vec{e}_3$ as being arbitrarily orientated with respect to the ξ, η, ζ co-ordinate system, the system being fixed in space.

In accordance with the general rules for transforming a tensor of the second rank when transferring from one orthogonal co-ordinate system to another, the $((\alpha'))$ tensor in the ξ, η, ζ co-ordinate system is written in the form

$$((\alpha')) = \begin{bmatrix} e_{1\xi} & e_{2\xi} & e_{3\xi} \\ e_{1\eta} & e_{2\eta} & e_{3\eta} \\ e_{1\zeta} & e_{2\zeta} & e_{3\zeta} \end{bmatrix} \left(\begin{pmatrix} \alpha'_1 & 0 & 0 \\ 0 & \alpha'_2 & 0 \\ 0 & 0 & \alpha'_3 \end{pmatrix}\right) \begin{bmatrix} e_{1\xi} & e_{1\eta} & e_{1\zeta} \\ e_{2\xi} & e_{2\eta} & e_{2\zeta} \\ e_{3\xi} & e_{3\eta} & e_{3\zeta} \end{bmatrix}$$

$$= \left[\begin{bmatrix} \sum_i^3 \alpha'_i e_{i\xi}^2 & \sum_i^3 \alpha'_i e_{i\xi} e_{i\eta} & \sum_i^3 \alpha'_i e_{i\xi} e_{i\zeta} \\ \sum_1^3 \alpha'_i e_{i\eta} e_{i\xi} & \sum_i^3 \alpha'_i e_{i\eta}^2 & \sum_i^3 \alpha'_i e_{i\eta} e_{i\zeta} \\ \sum_i^3 \alpha'_i e_{i\zeta} e_{i\xi} & \sum_i^3 \alpha'_i e_{i\xi} e_{i\eta} & \sum_i^3 \alpha'_i e_{i\zeta}^2 \end{bmatrix}\right] \tag{15.14}$$

Here, $e_{1\xi}$, $e_{1\eta}$, $e_{1\zeta}$, etc. are the projections of the unit vectors of the co-ordinate system (the major axes) related to the molecule on the ξ, η and ζ axes.

If, as a result of Raman scattering, the molecule undergoes a transition from a state $v_i = 0$ to $v_i = 1$ (the Stokes case), then an electric moment is induced in it with components μ_ξ, μ_η and μ_ζ. These must be related to the electrical vector of the incident wave by the equation $\vec{\mu} = ((\alpha'))\vec{E}$. It follows from this that, by allowing for the form of the tensor in the ξ, η, ζ co-ordinate system and the values of the vector components of the incident wave's electrical field strength,

$$\left.\begin{aligned} \mu_\xi &= \sum_i^3 \alpha'_i (e_{i\xi} e_{i\eta} E_\eta + e_{i\xi} e_{i\zeta} E_\zeta) \\ \mu_\eta &= \sum_i^3 \alpha'_i (e_{i\eta}^2 E_\eta + e_{i\eta} e_{i\zeta} E_\zeta) \\ \mu_\zeta &= \sum_i^3 \alpha'_i (e_{i\zeta} e_{i\eta} E_\eta + e_{i\zeta}^2 E_\zeta) \end{aligned}\right\}$$

Since the observations are made along the η axis, the intensity of the scattered light will be determined by the ξ and ζ components. The intensities of the mutually perpendicularly polarized components of the scattered light will be proportional to μ_ξ^2 and μ_ζ^2, respectively. Averaging over time and noting that

$\overline{E_\xi^2} = \overline{E_\eta^2} = \frac{1}{2}\overline{E^2}$ and $\overline{E_\xi E_\eta} = 0$ (for natural light), it is found that

$$\left.\begin{aligned} J_\xi &\sim \sum_i^3 \alpha_i'^2 (e_{i\xi}^2 e_{i\eta}^2 + e_{i\xi}^2 e_{i\zeta}^2) + 2\sum_{\substack{i\neq j\\ i>j}}^3 \alpha_i'\alpha_j'(e_{i\xi}e_{i\eta}e_{j\xi}e_{j\eta} + e_{i\xi}e_{i\zeta}e_{j\xi}e_{j\zeta}) \\ J_\zeta &\sim \sum_i^3 \alpha_i'^2 (e_{i\zeta}^2 e_{i\eta}^2 + e_{i\zeta}^4) + 2\sum_{\substack{i\neq j\\ i>j}}^3 \alpha_i'\alpha_j'(e_{i\zeta}e_{i\eta}e_{j\zeta}e_{j\eta} + e_{i\zeta}^2 e_{j\zeta}^2) \end{aligned}\right\} \quad (15.15)$$

Since the molecules are freely orientating, the expressions obained must be averaged over all the possible orientations of the $\vec{e}_1$, $\vec{e}_2$, $\vec{e}_3$ unit-vector triplet in the ξ, η, ζ system. This is done by calculating the average of the quantity

$$e_{1\xi}^4 = \cos^4\theta, \quad \theta = \widehat{\vec{e}_1, \xi}$$

for all orientations. By the definition of the average and by using polar co-ordinates, we obtain

$$\overline{\cos^4\theta} = \frac{1}{4\pi}\int_0^\pi \int_0^{2\pi} \cos^4\theta \sin\theta \, d\theta \, d\varphi = \tfrac{1}{5}$$

where θ is the angle between the ξ axis and the $\vec{e}_1$ unit vector of the moving co-ordinate system associated with the molecule and φ is the angle between the projection of $\vec{e}_1$ on the η, ζ plane and the η axis.

Further, since the projections $e_{1\xi}$, $e_{1\eta}$, $e_{1\xi}$, etc. are coefficients in the matrix for the orthogonal co-ordinate transformation, then the following relationships hold:

$$\begin{aligned} e_{i\xi}^2 + e_{i\eta}^2 + e_{i\zeta}^2 &= 1 \qquad i, j = 1, 2, 3 \\ e_{i\xi}e_{j\xi} + e_{i\eta}e_{j\eta} + e_{i\zeta}e_{j\zeta} &= 0 \end{aligned} \quad (15.16)$$

By squaring the first equation and averaging it, we obtain

$$\overline{e_{i\xi}^4} + \overline{e_{i\eta}^4} + \overline{e_{i\zeta}^4} + 2\overline{e_{i\xi}^2 e_{i\eta}^2} + 2\overline{e_{i\eta}^2 e_{i\zeta}^2} + 2\overline{e_{i\xi}^2 e_{i\zeta}^2} = 1$$

Since, by definition, the molecule is freely orientated in space it is clear that

$$\overline{e_{i\xi}^4} = \overline{e_{i\eta}^4} = \overline{e_{i\zeta}^4}$$

and also

$$\overline{e_{i\xi}^2 e_{i\eta}^2} = \overline{e_{i\xi}^2 e_{i\zeta}^2} = \overline{e_{i\eta}^2 e_{i\zeta}^2}$$

Hence by taking this into account we finally find that

$$\overline{e_{i\xi}^2 e_{i\eta}^2} = \tfrac{1}{15}$$

Acting in an analogous way for the second formula in equation (15.16) and having in mind that for free rotations $\overline{e_{i\xi}^2 e_{j\xi}^2} = \overline{e_{i\eta}^2 e_{j\eta}^2}$, etc. and $\overline{e_{i\xi}e_{j\xi}e_{i\eta}e_{j\eta}} = \overline{e_{i\xi}e_{j\xi}e_{i\zeta}e_{j\zeta}}$, etc., it is seen that

$$\overline{e_{i\xi}e_{j\xi}e_{i\eta}e_{j\eta}} = -\tfrac{1}{30}$$

When the numerical values found for the appropriate variables are substituted in the expressions for J_ξ and J_ζ they take the form

$$\left.\begin{aligned} J_\xi &\sim \frac{2}{15}\sum_i^3 \alpha_i'^2 - \tfrac{2}{15}(\alpha_1'\alpha_2' + \alpha_1'\alpha_3' + \alpha_2'\alpha_3') \\ J_\zeta &\sim \tfrac{4}{15}(\alpha_1'^2 + \alpha_2'^2 + \alpha_3'^2) + \tfrac{1}{15}(\alpha_1'\alpha_2' + \alpha_1'\alpha_3' + \alpha_2'\alpha_3') \end{aligned}\right\} \tag{15.17}$$

From this equation it follows that by rewriting the equation in terms of the tensor's invariants the trace is

$$b' = \alpha_1' + \alpha_2' + \alpha_3'$$

and the anisotropy

$$g' = \sqrt{\{\tfrac{1}{2}[(\alpha_1' - \alpha_2')^2 + (\alpha_2' - \alpha_3')^2 + (\alpha_1' - \alpha_3')^2]\}}$$

Hence we finally obtain

$$\left.\begin{aligned} J_\xi &\sim \tfrac{2}{15} g'^2 \\ J_\zeta &\sim \tfrac{1}{45}(5b'^2 + 7g'^2) \end{aligned}\right\} \tag{15.18}$$

It is not difficult to carry out a similar manipulation for other relationships between the components of the incident light and observation conditions. For example, if the incident light is linearly polarized we obtain

$$J_\xi \sim \tfrac{1}{15}(\alpha_1'^2 + \alpha_2'^2 + \alpha_3'^2 - \alpha_1'\alpha_2' - \alpha_1'\alpha_3' - \alpha_2'\alpha_3')$$
$$J_\zeta \sim \tfrac{1}{15}(3\alpha_1'^2 + 3\alpha_2'^2 + 3\alpha_3'^2 + 2\alpha_1'\alpha_2' + 2\alpha_1'\alpha_3' + 2\alpha_2'\alpha_3')$$

or

$$\begin{aligned} J_\xi &\sim \tfrac{1}{15} g'^2 \\ J_\zeta &\sim \tfrac{1}{15}(5b'^2 + 4g'^2) \end{aligned} \tag{15.19}$$

The degree of depolarization $\rho = (J_\xi/J_\zeta)$ will, for natural and linearity polarized light, be equal, respectively, to

$$\rho = \frac{6g'^2}{5b'^2 + 7g'^2}, \quad \rho_p = \frac{3g'^2}{5b'^2 + 4g'^2} \tag{15.20}$$

In the case of a spherically symmetrical molecule for which $\alpha_1' = \alpha_2' = \alpha_3' = \alpha'$ we obtain

$$b'^2 = q\alpha'^2, \quad g' = 0$$

and therefore

$$\rho = 0, \qquad \rho_p = 0$$

The other extreme case is when a molecule is much more polarizable along one axis than it is along the other two (for example, a linear molecule with extensive conjugation). In this case,

$$\alpha_1' = \alpha', \quad \alpha_2' = \alpha_3' = 0$$

and consequently

$$b'^2 = \alpha'^2, \quad g'^2 = \alpha'^2$$

so that

$$\rho = \tfrac{1}{2}, \quad \rho_p = \tfrac{1}{3}$$

Since α_1', α_2' and α_3' can be either positive or negative, then it is possible that $b' = 0$ while $g' \neq 0$. In this case,

$$\rho = \tfrac{6}{7}, \quad \rho_p = \tfrac{3}{4}$$

15.3 The Valence-optical Scheme and the Formulae for Components of a Polarizability Tensor's Derivative with Respect to a Normal Co-ordinate

Some of the general properties of the intensities of Raman lines and the degree of their depolarizations have been considered above, and we did not specify the corresponding tensor to do this. The next problem is to find the links between the intensity and depolarization degree of a Raman line and the structure of the polyatomic molecule and the properties of its normal vibrations. This problem was discussed in Section 15.1. By analogy with infra-red intensity theory, let us choose the representation of a molecule's polarizability tensor for the theory of Raman spectra intensities as $((\alpha)) = \sum_k ((\alpha_k))$, where $((\alpha_k))$ is a tensor which, for an additive bond, has the sense of the tensor of the bond's individual polarizability within the molecule. The major directions of a bond's polarizability tensor are selected so that one of them coincides with the direction of the bond. In what follows, the corresponding unit vector of the tensor will always be designated $\vec{e}_1^{\,(k)}$, where the superscript (k) refers to the number of the bond within the molecule. The two other major directions of the tensor are chosen so that they are perpendicular to the directional vector of the bond, and the unit vectors are designated $\vec{e}_2^{\,(k)}$ and $\vec{e}_3^{\,(k)}$. This choice of the major axes of a bond's tensor makes it identical with a diatomic molecule's polarizability tensor or that of a completely additive bond. This is the basic condition needed so that, at the limit when the molecule is completely additive, its polarizability can be built up from those of its separate bonds.

Thus the general form of a molecule's polarizability tensor takes the form

$$((\alpha)) = \sum_k \left[\begin{bmatrix} \sum_i^3 \alpha_i^{(k)} e_{i\xi}^{(k)2} & \sum_i^3 \alpha_i^{(k)} e_{i\xi}^{(k)} e_{i\eta}^{(k)} & \sum_i^3 \alpha_i^{(k)} e_{i\xi}^{(k)} e_{i\zeta}^{(k)} \\ & \sum_i^3 \alpha_i^{(k)} e_{i\eta}^{(k)2} & \sum_i^3 \alpha_i^{(k)} e_{i\eta}^{(k)} e_{i\zeta}^{(k)} \\ & & \sum_i^3 \alpha_i^{(k)} e_{i\zeta}^{(k)2} \end{bmatrix}\right] \tag{15.21}$$

The $\alpha_i^{(k)}$ coefficients for the bonds represent empirical parameters. Like a μ_k parameter, $\alpha_i^{(k)}$ has a physical meaning as a bond's basic polarizability though it can only be used to characterize a bond for an additive molecule. In non-additive molecules the $\alpha_i^{(k)}$ lose their meaning as characteristics of bonds and become empirical parameters that correspond to the selected representation of a molecule's polarizability tensor, and depend on the properties of both the bond itself and those of its neighbours. In general, they also depend on the environment.

Therefore we retain, as we did with $\vec{\mu}$, the same form of writing a molecule's polarizability tensor, giving the $\alpha_i^{(k)}$ parameters different physical meanings, depending on the properties of the polyatomic molecule. The general conclusions we will make in the future for Raman intensities on the basis of the selected scheme will be true for any class of molecules, both additive and non-additive.

When a molecule's polarizability tensor, as in equation (15.21), is differentiated with respect to a normal co-ordinate, derivatives of $\alpha_i^{(k)}$ with respect to the selected natural vibration co-ordinates appear together with the set of $\alpha_i^{(k)}$ parameters themselves. These derivatives, taken with the $\alpha_i^{(k)}$, form the full system of electro-optical parameters of a molecule for the theory of Raman intensities. These parameters do not depend on the molecule's mechanical properties and only reflect those of the molecular electronic cloud if the scattering that gives rise to a line is such that the wavelength is well removed from the system's own absorption.

For additive molecules all the derivatives of $\alpha_i^{(k)}$ with respect to natural co-ordinates are assumed to be zero, except for the derivative with respect to its own bond-extension co-ordinate. The remaining derivatives will correspond in meaning to the analogous values for diatomic molecules. Account of the interactions between bonds is made by including some or (for completely non-additive molecules) all the derivatives of $\alpha_i^{(k)}$ with respect to the extensions of 'foreign' bonds, and those with respect to changes in the valency angles, etc.

In molecules divisible into additive groupings (i.e. those whose vibrations can be described by a set of natural co-ordinates which are internal to the grouping) all the derivatives of an $\alpha_i^{(k)}$ of a bond in a grouping with respect to all the co-ordinates internal to that grouping are taken to be non-zero.

As we shall see, the concept of additivity carries with it the analogous concept of the additivity of the dipole moment of a group. However, note that, despite all external similarities, the additivity of the tensors of the polarizabilities of groups does not in general follow from the additivity of the dipole moments of the separate groups of a molecule. This is due to the completely different physical meanings of $\vec{\mu}_k$ and $\alpha_i^{(k)}$. Furthermore, we shall see that this has ramifications in the appearance of vibrations characteristic in intensity in infra-red spectra but not characteristic in Raman spectra, and vice versa.

The electro-optical parameters of additive groups must, within certain limits, have the property of transferability from one molecule to another. However, the difficulties of determining experimentally the parameters introduced or their combinations (difficulties caused by the large number of these parameters when compared with that of the μ_k and $\partial\mu_k/\partial q_i$) greatly diminish the computational opportunities open to such a semi-empirical theory. Bearing this in mind, much attention is directed below at obtaining general conclusions.

Let us construct the general formula for the components of the derivative tensor of a polyatomic molecule's polarizability[3]. We start with the expression introduced above for a polyatomic molecule's polarizability tensor. In order to obtain the formula for a component of the $\partial((\alpha))/\partial Q_j$ tensor we shall consider a diagonal element

$$\alpha_{\xi\xi}=\sum_k\sum_i \alpha_i^{(k)}e_{i\xi}^{(k)^2}$$

and a non-diagonal one,

$$\alpha_{\xi\eta}=\sum_k\sum_i \alpha_i^{(k)}e_{i\xi}^{(k)}e_{i\eta}^{(k)}$$

All the remaining elements can then be obtained from the formulae for $\partial\alpha_{\xi\xi}/\partial Q_j$ and $\partial\alpha_{\xi\eta}/\partial Q_j$ simply by changing the indices of the Cartesian axes of the co-ordinates.

In order to simplify the mathematical operations we shall taken into account that

$$\left.\begin{aligned} e_{1\xi}^{(k)^2}+e_{2\xi}^{(k)^2}+e_{3\xi}^{(k)^2}&=1\\ e_{1\xi}^{(k)}e_{1\eta}^{(k)}+e_{2\xi}^{(k)}e_{2\eta}^{(k)}+e_{3\xi}^{(k)}e_{3\eta}^{(k)}&=0.\end{aligned}\right\}$$

Thus the diagonal element $\alpha_{\xi\xi}$ of the $((\alpha))$ tensor can be written as

$$\sum_k[\alpha_1^{(k)}e_{1\xi}^{(k)^2}+\alpha_2^{(k)}e_{2\xi}^{(k)^2}+\alpha_3^{(k)}e_{3\xi}^{(k)^2}]=\sum[\alpha_1^{(k)}+(\alpha_2^{(k)}-\alpha_1^{(k)})e_{2\xi}^{(k)^2}+(\alpha_3^{(k)}-\alpha_1^{(k)})e_{3\xi}^{(k)^2}] \tag{15.22}$$

and the non-diagonal element $\alpha_{\xi\eta}$ as

$$\begin{aligned}&\sum_k(\alpha_1^{(k)}e_{1\xi}^{(k)}e_{1\eta}^{(k)}+\alpha_2^{(k)}e_{2\xi}^{(k)}e_{2\eta}^{(k)}+\alpha_3^{(k)}e_{3\xi}^{(k)}e_{3\eta}^{(k)})\\ &\quad=\sum[(\alpha_2^{(k)}-\alpha_1^{(k)})e_{2\xi}^{(k)}e_{2\eta}^{(k)}+(\alpha_3^{(k)}-\alpha_1^{(k)})e_{3\xi}^{(k)}e_{3\eta}^{(k)}]\end{aligned} \tag{15.23}$$

We can obtain other equally obvious equations if we take the component of the other unit vectors to be independent. For instance, we can write

$$\alpha_{\xi\xi}=\sum_k[\alpha_3^{(k)}+(\alpha_1^{(k)}-\alpha_3^{(k)})e_{1\xi}^{(k)^2}+(\alpha_2^{(k)}-\alpha_3^{(k)})e_{2\xi}^{(k)^2}] \tag{15.24a}$$

and

$$\alpha_{\xi\eta}=\sum_k[(\alpha_1^{(k)}-\alpha_3^{(k)})e_{1\xi}^{(k)}e_{1\eta}^{(k)}+(\alpha_2^{(k)}-\alpha_3^{(k)})e_{2\xi}^{(k)}e_{3\xi}^{(k)}] \tag{15.24b}$$

The first equation can be written in a more convenient form when it is assumed that for most of a molecule's bonds $\alpha_1^{(k)}\neq\alpha_2^{(k)}\neq\alpha_3^{(k)}$ (π bonds). The formulae obtained in this case are more symmetric. The second equation is more convenient when

$$\alpha_1^{(k)}\neq\alpha_2^{(k)}=\alpha_3^{(k)}$$

(σ bonds) for most of a molecule's bonds. In this case, some of the terms in the final formulae obtained below for $\partial\alpha_{\xi\xi}/\partial Q_j$ and $\partial\alpha_{\xi}/\partial Q_j$ vanish. To obtain these quantities the components of a molecule's polarizability tensor are differentiated with respect to a normal co-ordinate, giving

$$\frac{\partial \alpha_{\xi\xi}}{\partial Q_j} = \sum_k \left[\frac{\partial \alpha_1^{(k)}}{\partial Q_j} + \left(\frac{\partial \alpha_2^{(k)}}{\partial Q_j} - \frac{\partial \alpha_1^{(k)}}{\partial Q_j} \right) e_{2\xi}^{(k)^2} + \left(\frac{\partial \alpha_3}{\partial Q_j} - \frac{\partial \alpha_1}{\partial Q_j} \right) e_{3\xi}^{(k)^2} \right.$$
$$\left. + (\alpha_2^{(k)} - \alpha_1^{(k)}) 2 e_{2\xi}^{(k)} \frac{\partial e_{2\xi}^{(k)}}{\partial Q_j} + (\alpha_3^{(k)} - \alpha_1^{(k)}) 2 e_{3\xi}^{(k)} \frac{\partial e_{3\xi}^{(k)}}{\partial Q_j} \right]; \tag{15.25a}$$

$$\frac{\partial \alpha_{\xi\eta}}{\partial Q_j} = \sum_k \left[\left(\frac{\partial \alpha_2^k}{\partial Q_j} - \frac{\partial \alpha_1^{(k)}}{\partial Q_j} \right) e_{2\xi}^{(k)} e_{2\eta}^{(k)} + \left(\frac{\partial \alpha_3^{(k)}}{\partial Q_j} - \frac{\partial \alpha_1^{(k)}}{\partial Q_j} \right) e_{3\xi}^{(k)} e_{3\eta}^{(k)} \right.$$
$$\left. + (\alpha_2^{(k)} - \alpha_1^{(k)}) \left(e_{2\eta}^{(k)} \frac{\partial e_{2\xi}^{(k)}}{\partial Q_j} + e_{2\xi}^{(k)} \frac{\partial e_{2\eta}^{(k)}}{\partial Q_j} \right) + (\alpha_3^{(k)} - \alpha_1^{(k)}) \left(e_{3\eta}^{(k)} \frac{\partial e_{3\xi}^{(k)}}{\partial Q_j} + e_{3\xi}^{(k)} \frac{\partial e_{3\eta}^{(k)}}{\partial Q_j} \right) \right] \tag{15.25b}$$

Here and below we use the first form of writing $\alpha_{\xi\xi}$ and $\alpha_{\xi\eta}$. The formulae for the tensor components using the second form of expression are given below without any derivation. These are completely analogous to the formulae for the tensor components derived using the first way of writing $\alpha_{\xi\xi}$ and $\alpha_{\xi\eta}$.

Using a matrix representation and differentiating with respect to natural co-ordinates equations (15.25a) and (15.25b) take the following forms if there are m bonds and n natural co-ordinates:

$$\frac{\partial \alpha_{\xi\xi}}{\partial Q_j} = \left\{ [1, \ldots, 1] \begin{bmatrix} \dfrac{\partial \alpha_1^{(1)}}{\partial q_1} & \cdots & \dfrac{\partial \alpha_1^{(1)}}{\partial q_n} \\ \cdots & \cdots & \cdots \\ \dfrac{\partial \alpha_1^{(m)}}{\partial q_1} & \cdots & \dfrac{\partial \alpha_1^{(m)}}{\partial q_n} \end{bmatrix} \right.$$
$$+ [e_{2\xi}^{(1)^2}, \ldots, e_{2\xi}^{(m)^2}] \begin{bmatrix} \dfrac{\partial \alpha_2^{(1)}}{\partial q_1} - \dfrac{\partial \alpha_1^{(1)}}{\partial q_1} & \cdots & \dfrac{\partial \alpha_2^{(1)}}{\partial q_n} - \dfrac{\partial \alpha_1^{(1)}}{\partial q_n} \\ \cdots & \cdots & \cdots \\ \dfrac{\partial \alpha_2^{(m)}}{\partial q_1} - \dfrac{\partial \alpha_1^{(m)}}{\partial q_1} & \cdots & \dfrac{\partial \alpha_2^{(m)}}{\partial q_n} - \dfrac{\partial \alpha_1^{(m)}}{\partial q_n} \end{bmatrix}$$
$$+ [e_{3\xi}^{(1)^2}, \ldots, e_{3\xi}^{(m)^2}] \begin{bmatrix} \dfrac{\partial \alpha_3^{(1)}}{\partial q_1} - \dfrac{\partial \alpha_1^{(1)}}{\partial q_1} & \cdots & \dfrac{\partial \alpha_3^{(1)}}{\partial q_n} - \dfrac{\partial \alpha_1^{(1)}}{\partial q_n} \\ \cdots & \cdots & \cdots \\ \dfrac{\partial \alpha_3^{(m)}}{\partial q_1} - \dfrac{\partial \alpha_1^{(m)}}{\partial q_1} & \cdots & \dfrac{\partial \alpha_3^{(m)}}{\partial q_n} - \dfrac{\partial \alpha_1^{(m)}}{\partial q_n} \end{bmatrix}$$
$$+ 2[\alpha_2^{(1)} - \alpha_1^{(1)}, \ldots, \alpha_2^{(m)} - \alpha_1^{(m)}] \begin{bmatrix} e_{2\xi}^{(1)} & \cdots & 0 \\ \cdots & \cdots & \cdots \\ 0 & \cdots & e_{2\xi}^{(m)} \end{bmatrix} \begin{bmatrix} \dfrac{\partial e_{2\xi}^{(1)}}{\partial q_1} & \cdots & \dfrac{\partial e_{2\xi}^{(1)}}{\partial q_n} \\ \cdots & \cdots & \cdots \\ \dfrac{\partial e_{2\xi}^{(m)}}{\partial q_1} & \cdots & \dfrac{\partial e_{2\xi}^{(m)}}{\partial q_n} \end{bmatrix}$$

$$
+\,2[\alpha_3^{(1)}-\alpha_1^{(1)},\ldots,\alpha_3^{(m)}-\alpha_1^{(m)}]\begin{bmatrix} e_{3\xi}^{(1)} & \cdots & 0\\ \cdots & \cdots & \cdots\\ 0 & \cdots & e_{3\xi}^{(m)}\end{bmatrix}\begin{bmatrix}\dfrac{\partial e_{3\xi}^{(1)}}{\partial q_1} & \cdots & \dfrac{\partial e_{3\xi}^{(1)}}{\partial q_n}\\ \cdots & \cdots & \cdots\\ \dfrac{\partial e_{3\xi}^{(m)}}{\partial q_1} & \cdots & \dfrac{\partial e_{3\xi}^{(m)}}{\partial q_n}\end{bmatrix}\Bigg\}\begin{bmatrix}(l_q)_{ij}\\ \vdots\\ (l_q)_{nj}\end{bmatrix}
\tag{15.26a}
$$

$$
\begin{aligned}
\frac{\partial\alpha_{\xi\eta}}{\partial Q_j}=\Bigg\{&[e_{2\xi}^{(1)}e_{2\eta}^{(1)},\ldots,e_{2\xi}^{(m)}e_{2\eta}^{(m)}]\begin{bmatrix}\dfrac{\partial\alpha_2^{(1)}}{\partial q_1}-\dfrac{\partial\alpha_1^{(1)}}{\partial q_1} & \cdots & \dfrac{\partial\alpha_2^{(1)}}{\partial q_n}-\dfrac{\partial\alpha_1^{(1)}}{\partial q_n}\\ \cdots & \cdots & \cdots\\ \dfrac{\partial\alpha_2^{(m)}}{\partial q_1}-\dfrac{\partial\alpha_1^{(m)}}{\partial q_1} & \cdots & \dfrac{\partial\alpha_2^{(m)}}{\partial q_n}-\dfrac{\partial\alpha_2^{(m)}}{\partial q_n}\end{bmatrix}\\
&+[e_{3\xi}^{(1)}e_{3\eta}^{(1)},\ldots,e_{3\xi}^{(m)}e_{3\eta}^{(m)}]\begin{bmatrix}\dfrac{\partial\alpha_3^{(1)}}{\partial q_1}-\dfrac{\partial\alpha_1^{(1)}}{\partial q_1} & \cdots & \dfrac{\partial\alpha_3^{(1)}}{\partial q_n}-\dfrac{\partial\alpha_1^{(1)}}{\partial q_n}\\ \cdots & \cdots & \cdots\\ \dfrac{\partial\alpha_3^{(m)}}{\partial q_1}-\dfrac{\partial\alpha_1^{(m)}}{\partial q_1} & \cdots & \dfrac{\partial\alpha_3^{(m)}}{\partial q_n}-\dfrac{\partial\alpha_1^{(m)}}{\partial q_n}\end{bmatrix}\\
&+[\alpha_2^{(1)}-\alpha_1^{(1)},\ldots,\alpha_2^{(m)}-\alpha_1^{(m)}]\left[\begin{bmatrix} e_{2\eta}^{(1)} & \cdots & 0\\ \cdots & \cdots & \cdots\\ 0 & \cdots & e_{2\eta}^{(m)}\end{bmatrix}\begin{bmatrix}\dfrac{\partial e_{2\xi}^{(1)}}{\partial q_1} & \cdots & \dfrac{\partial e_{2\xi}^{(1)}}{\partial q_n}\\ \cdots & \cdots & \cdots\\ \dfrac{\partial e_{2\xi}^{(m)}}{\partial q_1} & \cdots & \dfrac{\partial e_{2\xi}^{(m)}}{\partial q_n}\end{bmatrix}\right.\\
&\left.+\begin{bmatrix} e_{2\xi}^{(1)} & \cdots & 0\\ \cdots & \cdots & \cdots\\ 0 & \cdots & e_{2\xi}^{(m)}\end{bmatrix}\begin{bmatrix}\dfrac{\partial e_{2\eta}^{(1)}}{\partial q_1} & \cdots & \dfrac{\partial e_{2\eta}^{(1)}}{\partial q_n}\\ \cdots & \cdots & \cdots\\ \dfrac{\partial e_{2\eta}^{(m)}}{\partial q_1} & \cdots & \dfrac{\partial e_{2\eta}^{(m)}}{\partial q_n}\end{bmatrix}\right]\\
&+[\alpha_3^{(1)}-\alpha_1^{(1)},\ldots,\alpha_3^{(m)}-\alpha_1^{(m)}]\left[\begin{bmatrix} e_{3\eta}^{(1)} & \cdots & 0\\ \cdots & \cdots & \cdots\\ 0 & \cdots & e_{3\eta}^{(m)}\end{bmatrix}\begin{bmatrix}\dfrac{\partial e_{3\xi}^{(1)}}{\partial q_1} & \cdots & \dfrac{\partial e_{3\xi}^{(1)}}{\partial q_n}\\ \cdots & \cdots & \cdots\\ \dfrac{\partial e_{3\xi}^{(m)}}{\partial q_1} & \cdots & \dfrac{\partial e_{3\xi}^{(m)}}{\partial q_n}\end{bmatrix}\right.\\
&\left.+\begin{bmatrix} e_{3\xi}^{(1)} & \cdots & 0\\ \cdots & \cdots & \cdots\\ 0 & \cdots & e_{3\xi}^{(m)}\end{bmatrix}\begin{bmatrix}\dfrac{\partial e_{3\eta}^{(1)}}{\partial q_1} & \cdots & \dfrac{\partial e_{3\eta}^{(1)}}{\partial q_n}\\ \cdots & \cdots & \cdots\\ \dfrac{\partial e_{3\eta}^{(m)}}{\partial q_1} & \cdots & \dfrac{\partial e_{3\eta}^{(m)}}{\partial q_n}\end{bmatrix}\right]\Bigg\}\begin{bmatrix}(l_q)_{1j}\\ \vdots\\ (l_q)_{nj}\end{bmatrix}
\end{aligned}
\tag{15.26b}
$$

Using symbolic abbreviations we obtain finally

$$\frac{\partial\alpha_{\xi\xi}}{\partial Q_j}=\left\{\tilde{1}\left[\frac{\partial\alpha_1}{\partial q}\right]+\tilde{e}^2_{2\xi}\left[\frac{\partial\alpha_2}{\partial q}-\frac{\partial\alpha_1}{\partial q}\right]+\tilde{e}^2_{3\xi}\left[\frac{\partial\alpha_3}{\partial q}-\frac{\partial\alpha_1}{\partial q}\right]\right.$$
$$\left.+2[\widetilde{\alpha_2-\alpha_1}]E_{2\xi}\left[\frac{\partial e_{2\xi}}{\partial q}\right]+2[\widetilde{\alpha_3-\alpha_1}]E_{3\xi}\left[\frac{\partial e_{3\xi}}{\partial q}\right]\right\}l^{(j)}_q \tag{15.27a}$$

$$\frac{\partial\alpha_{\xi\eta}}{\partial Q_j}=\left\{\widetilde{e_{2\xi}e_{2\eta}}\left[\frac{\partial\alpha_2}{\partial q}-\frac{\partial\alpha_1}{\partial q}\right]+\widetilde{e_{3\xi}e_{3\eta}}\left[\frac{\partial\alpha_3}{\partial q}-\frac{\partial\alpha_1}{\partial q}\right]\right.$$
$$+[\widetilde{\alpha_2-\alpha_1}]\left(E_{2\eta}\left[\frac{\partial e_{2\xi}}{\partial q}\right]+E_{2\xi}\left[\frac{\partial e_{2\eta}}{\partial q}\right]\right) \tag{15.27b}$$
$$\left.+[\widetilde{\alpha_3-\alpha_1}]\left(E_{3\eta}\left[\frac{\partial e_{3\xi}}{\partial q}\right]+E_{3\xi}\left[\frac{\partial e_{3\eta}}{\partial q}\right]\right)\right\}l^{(j)}_q$$

The structures of the matrices $[\partial\alpha_1/\partial q]$, $[(\partial\alpha_2/\partial q)-(\partial\alpha_1/\partial q)]$, etc. and those of $[\alpha_2-\alpha_1]$ and $[\alpha_3-\alpha_1]$ are similar to the structure of the corresponding matrices in the formula for $\partial\vec{\mu}/\partial Q_j$; $E_{2\xi}$ is a diagonal matrix with elements $e^{(k)}_{2\eta}$, etc.

Before we carry out further manipulations, we will prove some important properties of the trace of the $\partial((\alpha))/\partial Q_j$ tensor.

The expression for the tensor's trace is easily found if we take advantage of the formulae derived above for the tensor components of the polarizability derivative. Doing this, we have

$$b'=\frac{\partial\alpha_{\xi\xi}}{\partial Q_j}+\frac{\partial\alpha_{\eta\eta}}{\partial Q_j}+\frac{\partial\alpha_{\zeta\zeta}}{\partial Q_j}=\left\{\tilde{3}\left[\frac{\partial\alpha_1}{\partial q}\right]+[\widetilde{e^2_{2\xi}+e^2_{2\eta}+e^2_{2\zeta}}]\left[\frac{\partial\alpha_2}{\partial q}-\frac{\partial\alpha_1}{\partial q}\right]\right.$$
$$+[\widetilde{e^2_{3\xi}+e^2_{3\eta}+e^2_{3\zeta}}]\left[\frac{\partial\alpha_3}{\partial q}-\frac{\partial\alpha_1}{\partial q}\right]$$
$$+2[\widetilde{\alpha_2-\alpha_1}]\left(E_{2\xi}\left[\frac{\partial e_{2\xi}}{\partial q}\right]+E_{2\eta}\left[\frac{\partial e_{2\eta}}{\partial q}\right]+E_{2\zeta}\left[\frac{\partial e_{2\zeta}}{\partial q}\right]\right)$$
$$\left.+2[\widetilde{\alpha_3-\alpha_1}]\left(E_{3\xi}\left[\frac{\partial e_{3\xi}}{\partial q}\right]+E_{3\eta}\left[\frac{\partial e_{3\eta}}{\partial q}\right]+E_{3\zeta}\left[\frac{\partial e_{3\zeta}}{\partial q}\right]\right)\right\}l^{(j)}_q$$

Since $e^{(k)2}_{2\xi}+e^{(k)2}_{2\eta}+e^{(k)2}_{2\zeta}=1=e^{(k)2}_{3\xi}+e^{(k)2}_{3\eta}+e^{(k)2}_{3\zeta}$, all the elements in the corresponding row matrices tend to unity. Moreover, we see that

$$E_{2\xi}\left[\frac{\partial e_{2\xi}}{\partial q}\right]+E_{2\eta}\left[\frac{\partial e_{2\eta}}{\partial q}\right]+E_{2\zeta}\left[\frac{\partial e_{2\zeta}}{\partial q}\right]=\left[\vec{e}_2\frac{\partial\vec{e}_2}{\partial q}\right]$$

where $[\vec{e}_2(\partial\vec{e}_2/\partial q)]$ is the rectangular matrix whose elements are equal to the scalar products of the $\vec{e}^{(k)}_2$ vectors and their derivatives with respect to the natural co-ordinates. Since $\partial\vec{e}^{(k)}_2/\partial q_e$ is perpendicular to $\vec{e}^{(k)}_2$, the matrix $[\vec{e}_2(\partial\vec{e}_2/\partial q)]$ is zero. The same is clearly true of the expression

$$\left(E_{3\xi}\left[\frac{\partial e_{3\xi}}{\partial q}\right]+\cdots\right)$$

As a result, and by taking the structures of the matrices

$$\left[\frac{\partial\alpha}{\partial q}\right], \left[\frac{\partial\alpha_2}{\partial q} - \frac{\partial\alpha_1}{\partial q}\right] \quad \text{and} \quad \left[\frac{\partial\alpha_3}{\partial q} - \frac{\partial\alpha_1}{\partial q}\right]$$

into account, we obtain

$$b' = \tilde{1}\left[\frac{\partial\alpha_1}{\partial q} + \frac{\partial\alpha_2}{\partial q} + \frac{\partial\alpha_3}{\partial q}\right] l_q^{(j)} = \tilde{1}\left[\frac{\partial b}{\partial q}\right] l_q^{(j)}$$

where the rectangular matrix $[\partial b/\partial q]$ is structurally similar to $[\partial\alpha_1/\partial q]$, etc. It contains derivatives of the traces of the bond polarizability tensors with respect to natural vibration co-ordinates.

If we use the terminology 'valence' and 'deformational' by analogy with the derivative of a molecule's dipole moment with respect to a normal co-ordinate for the expressions $\partial\alpha_{\xi\xi}/\partial Q_j$ and $\partial\alpha_{\xi\eta}/\partial Q_j$ it can be concluded that the deformational term's trace is zero. Since, in the additive approximation (the zero approximation), all the derivatives of $\partial\alpha_i^{(k)}/\partial q_l$ with respect to angle co-ordinates are zero for purely deformational vibrations, then $\rho = 6/7$ for the degree of polarization when $b' = 0$, as is clear from equation (15.20).

Obviously, this rule is only satisfied approximately for real molecules. However, it can be quite useful either for specifying the frequencies or, when deformational vibrations are characterizable by form, for elucidating whether the additive approximation is applicable to a given molecule.

The trace of the deformational part of the $\partial((\alpha))/\partial Q_j$ tensor is only zero for fundamental vibrations. This condition is never satisfied for overtones or combinations.

We now return to the calculation of the derivatives $\partial\vec{e}_2^{(k)}/\partial q_m$, $\partial\vec{e}_3^{(k)}/\partial q_m$, etc. Here, $\vec{e}_2^{(k)}$ and $\vec{e}_3^{(k)}$ are vectors perpendicular to the directional vector of the $\vec{e}_1^{(k)}$ bond. The derivatives may easily be found in terms of the derivatives of $\vec{e}_1^{(k)}$ with respect to a normal co-ordinate if we neglect the rotations of the ellipsoid of the bond's polarizability around the axis coincident with $\vec{e}_1^{(k)}$. This is quite legimate for σ bonds with axial symmetries ($\alpha_1^{(k)} \neq \alpha_2^{(k)} = \alpha_3^{(k)}$), but, strictly speaking, it is not correct in the general case when all the $\alpha_i^{(k)}$ of a given bond are different and when, during vibrations, the bonds rotate about axes not lying in planes perpendicular to the bond's directional vector. Indeed, it is not enough to know just the positions of the atoms forming a molecule in order to specify the ellipsoid of polarizability unambiguously for a non-axially symmetric bond; we must in fact choose definite orientations of the unit vectors $\vec{e}_2^{(k)}$ and $\vec{e}_3^{(k)}$ perpendicular to the bond's directional vector. The bond vector is instantaneously and fully determined by the positions of the atoms of the bond. The orientation of $\vec{e}_2^{(k)}$ and $\vec{e}_3^{(k)}$ must be given with respect to the molecule's other bonds and atomic groups. This is why even the calculation of the appropriate derivatives with respect to normal and natural co-ordinates must take this situation into account. In general, $\partial\vec{e}_2^{(k)}/\partial q_m$ and $\partial\vec{e}_3^{(k)}/\partial q_m$ will also depend on the derivatives of directional vectors with respect to normal co-ordinates of the neighbouring bonds. Obtaining the corresponding formulae is quite difficult, and so henceforth rotations of the ellipsoid of a bond's polarizability around the axis coincident with the directional vector of the bond will be neglected. This does not limit the generality of the consequent derivations or reduce the practical computability of the semi-empirical theory being developed.

In order to calculate the matrices $[\partial \vec{e}_2/\partial q]$ and $[\partial \vec{e}_3/\partial q]$ in the chosen approximation the increments $\delta\vec{e}_2^{(k)}$ and $\delta\vec{e}_3^{(k)}$ to the vectors $\vec{e}_2^{(k)}$ and $\vec{e}_3^{(k)}$ when the $\vec{e}_1^{(k)}$ bond's directional vector is changed by $\delta\vec{e}_1^{(k)}$ must be found. Let us assume that the change in $\vec{e}_1^{(k)}$ arises when the trio $\vec{e}_1^{(k)}$, $\vec{e}_2^{(k)}$, $\vec{e}_3^{(k)}$ of orthogonal vectors is rotated about the ω axis, which is perpendicular to the plane formed by the original $\vec{e}_1^{(k)}$ vector and its increment $\delta\vec{e}_1^{(k)}$, the rotation being through a small angle $\varphi = |\delta\vec{e}_1^{(k)}|$ in the direction of $\delta\vec{e}_1^{(k)}$ (Figure 15.2). Such a rotation of the three unit vectors must clearly make the $\delta\vec{e}_2^{(k)}$ and $\delta\vec{e}_3^{(k)}$ increments collinear with $\vec{e}_1^{(k)}$; they are equal, in absolute terms, to $\varphi\cos\gamma_1$ and $\varphi\cos\gamma_2$, respectively. Here, γ_1 and γ_2 are the angles between the vectors $\vec{e}_2^{(k)}$ and $\vec{e}_3^{(k)}$ and the unit vector $\vec{g}$, being perpendicular to $\vec{e}_1^{(k)}$ and the axis of rotation and pointed in the same direction as $\delta\vec{e}_1^{(k)}$, However, $\varphi\cos\gamma_1 = |\delta\vec{e}_1^{(k)}|\,|\vec{e}_2^{(k)}|\cos\gamma_1$, and $\varphi\cos\gamma_2 = |\delta\vec{e}_1^{(k)}|\,|\vec{e}_3^{(k)}|\cos\gamma_2$, and are thus the scalar products $(\delta\vec{e}_1^{(k)}\vec{e}_2^{(k)})$ and $(\delta\vec{e}_1^{(k)}\vec{e}_3^{(k)})$, respectively. By allowing for the signs of those scalar products and the directions of the $\delta\vec{e}_2^{(k)}$ and $\delta\vec{e}_3^{(k)}$ vectors (see Figure 15.2) and the one opposite to the rotation, we finally obtain

$$\delta\vec{e}_2^{(k)} = -\vec{e}_1^{(k)}(\vec{e}_2^{(k)}\delta\vec{e}_1^{(k)}), \quad \delta\vec{e}_3^{(k)} = -\vec{e}_1^{(k)}(\vec{e}_3^{(k)}\delta\vec{e}_1^{(k)})$$

It follows from this that because

$$\frac{\partial\vec{e}_2}{\partial Q_j} = -\vec{E}_1\left(\vec{E}_2\frac{\partial\vec{e}_1}{\partial Q_j}\right), \quad \frac{\partial\vec{e}_3}{\partial Q_j} = -\vec{E}_1\left(\vec{E}_3\frac{\partial e_1}{\partial Q_j}\right)$$

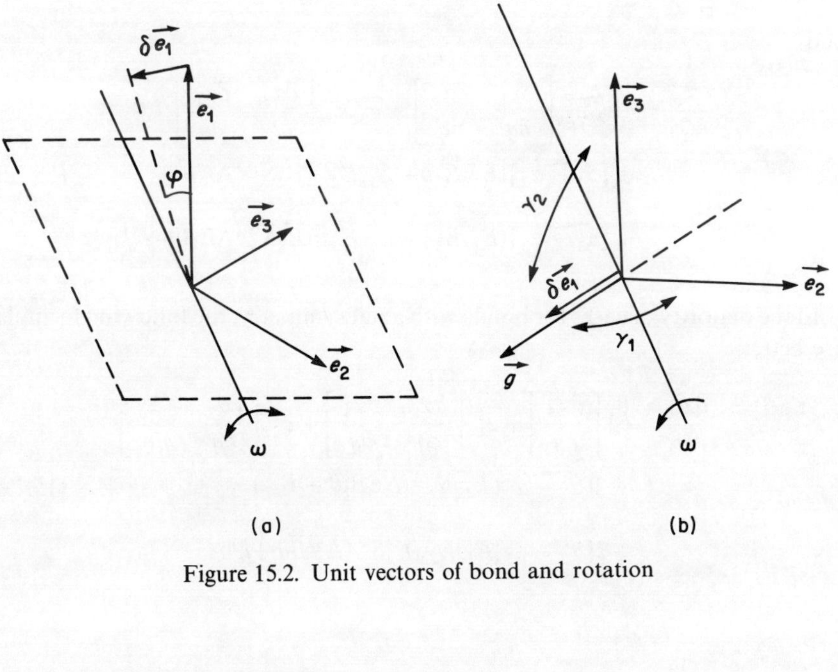

Figure 15.2. Unit vectors of bond and rotation

and since

$$\frac{\partial \vec{e}_1}{\partial Q_j} = S^{-1}(\Delta E \tilde{\vec{B}} \theta - \vec{E}_1) l_q^{(j)}$$

then

$$\frac{\partial \vec{e}_2}{\partial Q_j} = -\vec{E}_1(\vec{E}_2 S^{-1} \Delta E \tilde{\vec{B}} \theta) l_q^{(j)}$$

and

$$\frac{\partial \vec{e}_3}{\partial Q_j} = -\vec{E}_1(\vec{E}_3 S^{-1} \Delta E \tilde{\vec{B}} \theta) l_q^{(j)}$$

due to the orthogonality of the vectors $\vec{e}_i^{(k)}$. Consequently we have

$$\left[\frac{\partial \vec{e}_2}{\partial q}\right] = -\vec{E}_1(\vec{E}_2 S^{-1} \Delta E \tilde{\vec{B}} \theta)$$

$$\left[\frac{\partial \vec{e}_3}{\partial q}\right] = -\vec{E}_1(\vec{E}_3 S^{-1} \Delta E \tilde{\vec{B}} \theta)$$

Substituting these formulae into the expressions for the components of the derivative of the tensor of polarizability with respect to a normal co-ordinate $\partial \alpha_{\xi\xi}/\partial Q_j$ and $\partial \alpha_{\xi\eta}/\partial Q_j$, we obtain the relations

$$\begin{aligned} \frac{\partial \alpha_{\xi\xi}}{\partial Q_j} = \Bigg\{ & \tilde{1}\left[\frac{\partial \alpha_1}{\partial q}\right] + \widetilde{e_{2\xi}^2}\left[\frac{\partial \alpha_2}{\partial q} - \frac{\partial \alpha_1}{\partial q}\right] + \widetilde{e_{3\xi}^2}\left[\frac{\partial \alpha_3}{\partial q} - \frac{\partial \alpha_1}{\partial q}\right] \\ & + 2[\widetilde{\alpha_1 - \alpha_2}] E_{1\xi} E_{2\xi}(\vec{E}_2 S^{-1} \Delta E \tilde{\vec{B}} \theta) \\ & + 2[\widetilde{\alpha_1 - \alpha_3}] E_{1\xi} E_{3\xi}(\vec{E}_3 S^{-1} \Delta E \tilde{\vec{B}} \theta) \Bigg\} l_q^{(j)} \end{aligned} \tag{15.28a}$$

and

$$\begin{aligned} \frac{\partial \alpha_{\xi\eta}}{\partial Q_j} = \Bigg\{ & \widetilde{e_{2\xi} e_{2\eta}}\left[\frac{\partial \alpha_2}{\partial q} - \frac{\partial \alpha_1}{\partial q}\right] + \widetilde{e_{3\xi} e_{3\eta}}\left[\frac{\partial \alpha_3}{\partial q} - \frac{\partial \alpha_1}{\partial q}\right] \\ & + [\widetilde{\alpha_1 - \alpha_2}](E_{1\xi} E_{2\eta} + E_{1n} E_{2\xi})(\vec{E}_2 S^{-1} \Delta E \tilde{\vec{B}} \theta) \\ & + [\widetilde{\alpha_1 - \alpha_3}](E_{1\xi} E_{3\eta} + E_{1\eta} E_{3\xi})(\vec{E}_3 S^{-1} \Delta E \tilde{\vec{B}} \theta) \Bigg\} l_q^{(j)} \end{aligned} \tag{15.28b}$$

In the majority of cases for bonds with axial symmetry, the following formulae are better:

$$\begin{aligned} \frac{\partial \alpha_{\xi\xi}}{\partial Q_j} = \Bigg\{ & \tilde{1}\left[\frac{\partial \alpha_3}{\partial q}\right] + \widetilde{e_{1\xi}^2}\left[\frac{\partial \alpha_1}{\partial q} - \frac{\partial \alpha_3}{\partial q}\right] + \tilde{e}_{2\xi}^2\left[\frac{\partial \alpha_2}{\partial q} - \frac{\partial \alpha_3}{\partial q}\right] \\ & + 2[\widetilde{\alpha_1 - \alpha_3}] E_{1\xi} S^{-1}(\Delta E \tilde{B}_\xi \theta - E_{1\xi}) \\ & + 2[\alpha_2 - \alpha_3] E_{1\xi} E_{2\xi}(\vec{E}_2 S^{-1} \Delta E \tilde{\vec{B}} \theta) \Bigg\} l_q^{(j)} \end{aligned} \tag{15.29a}$$

$$\frac{\partial \alpha_{\xi\eta}}{\partial Q_j}=\left\{\widetilde{e_{1\xi}e_{1\eta}}\left[\frac{\partial \alpha_1}{\partial q}-\frac{\partial \alpha_3}{\partial q}\right]+\widetilde{e_{2\xi}e_{2\eta}}\left[\frac{\partial \alpha_2}{\partial q}-\frac{\partial \alpha_3}{\partial q}\right]\right.$$
$$+[\widetilde{\alpha_3-\alpha_1}](E_{1\xi}S^{-1}(\Delta E\tilde{B}_\eta\theta-E_{1\eta})+E_{1\eta}S^{-1}(\Delta E\tilde{B}_\xi\theta-E_{1\xi}))$$
$$\left.+[\widetilde{\alpha_3-\alpha_2}](E_{2\xi}E_{1\eta}+E_{2\eta}E_{1\xi})(\vec{E}_2S^{-1}\Delta E\tilde{\vec{B}}_\xi\theta)\right\}l_q^{(j)} \qquad (15.29b)$$

These formulae are also obtained if the expressions in equations (15.24a) and (15.24b) are initially used instead.

15.4 Properties of the Intensities and Degree of Polarization of the Lines in the Raman Spectra of Symmetrical Molecules

If a molecule has elements of symmetry then estimating a component of the $\partial((\alpha))/\partial Q$ tensor can be simplified by transforming to symmetry co-ordinates. The derivation of the appropriate formulae is entirely analogous to the determination of the $\partial\vec{\mu}/\partial Q$ derivatives. Hence the details will not be repeated here. We shall only point out that when transforming to symmetry co-ordinates the derivatives with respect to the natural ones in the matrices $[\partial\alpha_1/\partial q]$, $[(\partial\alpha_2/\partial q)-(\partial\alpha_1/\partial q)]$, etc. will be changed into derivatives with respect to the symmetry co-ordinates. Thus by allowing for the structure of these matrices and the identity of these derivatives with respect to natural co-ordinates in the rows which correspond to equivalent bonds, we find, as we did for the $[\partial\mu/\partial q]$ matrix, that having transformed to symmetry co-ordinates, the rows of the matrices $[(\partial\alpha_2/\partial q^{(s)})-(\partial\alpha_1/\partial q^{(s)})]$ and $[(\partial\alpha_3/\partial q^{(s)})-(\partial\alpha_1/\partial q^{(s)})]$ for equivalent bonds are related to each other via the (unnormalized) coefficients of the symmetry of the bond-extension co-ordinates in a vibration of a given type. Since the values of $\alpha_1^{(k)}$, $\alpha_2^{(k)}$ and $\alpha_3^{(k)}$ are, moreover, equal to each other for equivalent bonds, the formulae for $\partial\alpha_{\xi\xi}/\partial Q_j$ and $\partial\alpha_{\xi\eta}/\partial Q_j$ can be written as

$$\frac{\partial \alpha_{\xi\xi}}{\partial Q_j}=\left\{\widetilde{c^{(s)}_{\text{equ}}}\left[\frac{\partial \alpha_1}{\partial q^{(s)}}\right]_{\text{equ}}+\widetilde{(c^{(s)}e_{2\xi}^{(k)2})}_{\text{equ}}\left[\frac{\partial \alpha_2}{\partial q^{(s)}}-\frac{\partial \alpha_1}{\partial q^{(s)}}\right]_{\text{equ}}\right.$$
$$+\widetilde{(c^{(s)}e_{3\xi}^{(k)2})}_{\text{equ}}\left[\frac{\partial \alpha_3}{\partial q^{(s)}}-\frac{\partial \alpha_1}{\partial q^{(s)}}\right]_{\text{equ}}$$
$$+2[\widetilde{\alpha_1-\alpha_2}]_{\text{equ}}(\Sigma_{\text{equ.row}}E_{1\xi}E_{2\xi})(\vec{E}_2S^{-1}\Delta E\tilde{\vec{B}}^{(s)}\theta^{(s)})$$
$$\left.+2[\widetilde{\alpha_1-\alpha_3}]_{\text{equ}}(\Sigma_{\text{equ.row}}E_{1\xi}E_{3\xi})(\vec{E}_3S^{-1}\Delta E\tilde{\vec{B}}^{(s)}\theta^{(s)})\right\}l_{q^{(s)}}^{(j)} \qquad (15.30a)$$

$$\frac{\partial \alpha_{\xi\eta}}{\partial Q_j}=\left\{\widetilde{(c^{(s)}e_{2\xi}^{(k)}e_{2\eta}^{(k)})}_{\text{equ}}\left[\frac{\partial \alpha_2}{\partial q^{(s)}}-\frac{\partial \alpha_1}{\partial q^{(s)}}\right]_{\text{equ}}\right.$$
$$+\widetilde{(c^{(s)}e_{3\xi}^{(k)}e_{3\eta}^{(k)})}_{\text{equ}}\left[\frac{\partial \alpha_3}{\partial q^{(s)}}-\frac{\partial \alpha_1}{\partial q^{(s)}}\right]_{\text{equ}} \qquad (15.30b)$$
$$+[\widetilde{\alpha_1-\alpha_2}]_{\text{equ}}(\Sigma_{\text{equ.row}}[E_{1\xi}E_{2\eta}+E_{1\eta}E_{2\xi}])(\vec{E}_2S^{-1}\Delta E\tilde{\vec{B}}^{(s)}\theta^{(s)})$$
$$\left.+[\widetilde{\alpha_1-\alpha_3}]_{\text{equ}}(\Sigma_{\text{equ.row}}[E_{1\xi}E_{3\eta}+E_{1\eta}E_{3\xi}])(\vec{E}_3S^{-1}\Delta E\tilde{\vec{B}}^{(s)}\theta^{(s)})\right\}l_{q^{(s)}}^{(j)}$$

Here, $[\partial\alpha_1/\partial q^{(s)}]_{\text{equ}}$, $[(\partial\alpha_2/\partial q^{(s)})-(\partial\alpha_1/\partial q^{(s)})]_{\text{equ}}$ and $[(\partial\alpha_3/\partial q^{(s)})-(\partial\alpha_1/\partial q^{(s)})]_{\text{equ}}$ denote the rectangular matrices that are obtained from those containing the derivatives of the $\alpha_i^{(k)}$ parameters with respect to symmetry co-ordinates by deleting all the rows corresponding to each group of equivalent bonds except the one being assumed to be the first in the group. In the row matrices $\widetilde{[\alpha_1-\alpha_2]}_{\text{equ}}$ and $\widetilde{[\alpha_1-\alpha_3]}_{\text{equ}}$, only those parameters that correspond to representatives of each group of equivalent bonds are included. The elements of the row matrices

$$\widetilde{(c^{(s)}e_{2\xi}^{(k)^2})}_{\text{equ}},\ \widetilde{(c^{(s)}e_{2\xi}^{(k)}e_{2\eta}^{(k)})}_{\text{equ}}$$

etc. are obtained by summing the elements of all the original row matrices of each set of equivalent bonds with unnormalized symmetry coefficients. $\sum_{\text{equ.row}}$ designates the addition of the rows of the diagonal matrices $E_{1\xi}E_{2\xi}$, etc. that correspond to each group of equivalent bonds. The superscript, (s), refers to the selected symmetry type.

When using dependent symmetry co-ordinates, account must be taken of the additional relationships in the formulae for the components of the $\partial((\alpha))/\partial Q_j$ tensor in exactly the same way as was done for $\partial\vec{\mu}/\partial Q_j$.

Formulae of this kind are found to be very useful for the study of Raman intensities of symmetric molecules. When using these expressions some simplifications can be made. If the bond-extension co-ordinates are all zero for a given group of equivalent bonds or for one bond, then it is possible to delete all the rows and columns that correspond to those bonds and the co-ordinates of their extensions in all the components of the $\partial\alpha/\partial Q$ tensor in the matrices $[\partial\alpha_1/\partial q^{(s)}]$, $[(\partial\alpha_2/\partial q^{(s)})-(\partial\alpha_1/\partial q^{(s)})]$ and $[(\partial\alpha_3/\partial q^{(s)})-(\partial\alpha_1/\partial q^{(s)})]$. In fact, the valence term in the formulae for $\partial\alpha_{\xi\xi}/\partial Q_j$, $\partial\alpha_{\xi\eta}/\partial Q_j$, etc. is defined by a sum of all the relationships between the matrix elements that have the form $\int\psi_{v''}\delta\alpha^{(k)}(\vec{e}_i^{(k)})^2\psi_{v'}\,\mathrm{d}\rho$, where $\psi_{v'}$ and $\psi_{v''}$ are eigenfunctions in the basic and disturbed states, respectively, and $\delta\alpha^{(k)}(\vec{e}_i^{(k)})^2$ is the valence part of the tensor increment to a bond's polarizability tensor. The unit vectors of this tensor coincide with those of the bond's polarizability tensor. By choosing the ξ, η, ζ axes so that they coincide with the unit vectors of the bond, we obtain the following expressions for the components of the matrix elements being considered:

$$\int\psi_{v''}\delta\alpha_1^{(k)}e_{1\xi}^{(k)2}\psi_{v'}\,\mathrm{d}\rho,\quad \int\psi_{v'}\delta\alpha_2^{(k)}e_{2\eta}^{(k)2}\psi_{v'}\,\mathrm{d}\rho,\quad \int\psi_{v''}\delta\alpha_3^{(k)}e_{3\xi}^{(k)}\psi_{v'}\,\mathrm{d}\rho$$

These integrals will clearly be zero for vibrations during which the bond-extension co-ordinate is zero ($q_k=0$) and which are consequently anti-symmetric with respect to the bond. This means that the appropriate terms are zero in the expressions of the valence term in the formulae for the components of the derivative tensor of polarizability with respect to a normal co-ordinate. Obviously, this result does not change for another choice of Cartesian axes.

It follows that if there are unrepeated bonds (those which have no equivalents during symmetry operations) within the molecule, their electro-optical parameters should only be included in the matrices $[\partial\alpha_1/\partial q^{(s)}]_{\text{equ}}[(\partial\alpha_2/\partial q^{(s)})-(\partial\alpha_1/\partial q^{(s)})]_{\text{equ}}$ and $[(\partial\alpha_3/\partial q^{(s)})-(\partial\alpha_1/\partial q^{(s)})]_{\text{equ}}$ for fully symmetric vibrations.

The valence terms in the formulae for $(\partial\alpha_{\xi\xi}/\partial Q_j)$, $(\partial\alpha_{\xi\eta}/\partial Q_j)$, etc. are completely absent in deformational vibrations of linear molecules and nonplanar vibrations of planar molecules. As can be seen, we have here an exact analogy with the properties of the $[\partial\mu/\partial q^{(s)}]_{\text{equ}}$ matrix in the intensity theory for infra-red spectra.

The properties of the valence term (in deformational and non-planar vibrations in linear and planar molecules, respectively) that we have just mentioned allow us, when solving the inverse spectral problem, to find values for the polarizability of individual bonds.

Since, as mentioned above, the trace of the deformational part of the $\partial((\alpha))/\partial Q_j$ tensor is zero, it follows that $b'=0$ for the deformational and non-planar vibrations in linear and planar molecules, respectively, and consequently $\rho=6/7$.

Overall, because the trace of the tensor, b', has the form $\tilde{1}[\partial b/\partial q]l_q^{(j)}$ and because a transformation to symmetry coordinates allowing for the relationships between the $\partial b^{(k)}/\partial q_{k'}$ derivatives for a given set of equivalent bonds transforms the trace to the form

$$b'=\widetilde{c_{\text{equ}}^{(s)}}\left[\frac{\partial b}{\partial q^{(s)}}\right]_{\text{equ}} l_{q^{(s)}}^{(j)}$$

(where $\widetilde{c_{\text{equ}}^{(s)}}$ is a row matrix whose elements are equal to the sums of the symmetry coefficients for the bond-extension co-ordinates of a given set of bonds that are equivalent in the normal vibration being studied, and the sums are zero for all vibrations except fully symmetric ones), we find that $b'=0$ and $\rho=6/7$ for all vibrations except fully symmetric ones. This rule also directly follows from the properties of the $\int\psi_{v''}b'\psi_{v'}\,\mathrm{d}\rho$ matrix element for fundamental vibrations, because the tensor's trace is invariant with respect to any symmetry operation, and consequently the integral must always be zero for any partially symmetric vibration. Since the rule must be satisfied for any value of a bond's electro-optical parameters, we must also conclude that when $q_k=0$ the parameters of unrepeated bonds must not be included in the matrices

$$\left[\frac{\partial\alpha_1}{\partial q^{(s)}}\right],\left[\frac{\partial\alpha_2}{\partial q^{(s)}}-\frac{\partial\alpha_1}{\partial q^{(s)}}\right] \text{ and } \left[\frac{\partial\alpha_3}{\partial q^{(s)}}-\frac{\partial\alpha_1}{\partial q^{(s)}}\right]$$

When there is a degree of symmetry in the molecule, inactive vibrations will occur in Raman spectra, as they do in infra-red ones. The conclusions reached above about the properties of the $[\partial\alpha_1/\partial q^{(s)}]_{\text{equ}}$ matrix and those of the tensor's trace enable us to confirm that the only vibrations that may be inactive in Raman spectra are the partially symmetric ones.

Overall, the activity or inactivity of vibrations in Raman spectra can be established relatively easily for particular molecules and for vibrations of a given symmetry type. This can be done by studying expressions of the form

$$\sum_k c_{kj}e_{i\xi}^{(k)2},\quad \sum_k c_{kj}e_{i\xi}^{(k)}\frac{\partial e_{i\xi}^{(k)}}{\partial Q_j},\quad \sum_k c_{kj}e_{i\xi}^{(k)}e_{i\eta}^{(k)},\quad \sum_k c_{kj}e_{i\xi}^{(k)}\frac{\partial e_{i\eta}}{\partial Q_j}$$

etc. for every set of equivalent-bond unit vectors and $\partial((\alpha))/\partial Q_j$ tensor components. The vibrations of a given symmetry type will be inactive in the Raman spectrum if all such expressions vanish at the same time.

The relationships between $e_{i\xi}^{(k)}$, $e_{i\eta}^{(k)}$ and $e_{i\zeta}^{(k)}$ for the different k of a set of equivalent bonds and those between the quantities $\partial e_{i\xi}^{(k)}/\partial Q_j$, etc. can be easily established, given the symmetry of the molecule's equilibrium configuration and that of the vibration. Since the activity or inactivity of a vibration with a given symmetry type in a Raman spectrum is solely dependent on the vibration's symmetry and not on its forms or the values of the electro-optical parameters, the deformational and valence parts of the formulae for the components of the $\partial((\alpha))/\partial Q_j$ tensor must vanish at the same time for inactive vibrations. Thus, when not all the bond-extension coordinates are zero it is enough to consider the expressions $\sum_k c_{kj} e_{i\xi}^{(k)2}$, $\sum_k c_{kj} e_{i\xi}^{(k)} e_{i\eta}^{(k)}$, etc. in order to ascertain whether a given type of vibration is active in a Raman spectrum.

It can be proved that no more than six symmetry types of vibrations can be active at the same time (triply degenerate vibrations being counted as three normal vibrations and doubly degenerate ones as two). Remember that in infra-red spectra no more then three symmetry types of vibrations can be active (degenerate vibrations being counted either twice or three times). Hence, if a molecule has even one group of equivalent bonds that contains more than nine bonds, then the vibration must be inactive in both infra-red and Raman spectra.

This is a sufficient thought not a necessary condition for the appearance of inactive vibrations in both spectra. Inactive vibrations may occur for several symmetry types and in molecules that do not possess this property. For instance, A_2-type vibrations are inactive in the infra-red and Raman spectra of a C_{2v}-type molecule.

It is very easy to deduce the alternate exclusion rule with the help of the valence-optical scheme.

15.5 An Example of Calculating the Components of the Derivative Tensor of Polarizability: The H_2O Molecule

We shall now demonstrate the calculation method for the tensor components using the H_2O molecule as an example. Since the tensor's trace and anisotropy do not depend on rotations of the molecule as a whole, we can, for practical purposes, perform all the calculations in a co-ordinate system associated with the molecule.

We choose the directions of the bond unit vectors and those of the x, y, z axes as indicated in Figure 15.3. We shall take the bond-extension co-ordinates (q_1 and q_2) and the angle change co-ordinate (γ) as the natural set.

First, the values of the matrices $\vec{E}_2 S^{-1} \Delta E \tilde{\vec{B}} \theta$ and $\vec{E}_3 S^{-1} \Delta E \tilde{\vec{B}} \theta$ for A- and B-type vibrations are calculated. Since in this case the vectors $\vec{e}_2^{(1)}$ and $\vec{e}_2^{(2)}$ coincide with the co-ordinate unit vectors of the change in the valence angle, i.e. $\vec{f}_{12}$ and $\vec{f}_{21}$, we will henceforth write $\vec{f}_{12}$ and $\vec{f}_{21}$ instead of $\vec{e}_2^{(1)}$ and $\vec{e}_2^{(2)}$. The vectors $\vec{e}_3^{(1)}$ and $\vec{e}_3^{(2)}$ are chosen to be the same for both bonds, and so one symbol only will be used for them. Instead of writing $\vec{e}_1^{(1)}$ and $\vec{e}_1^{(2)}$ we will write $\vec{e}_1$

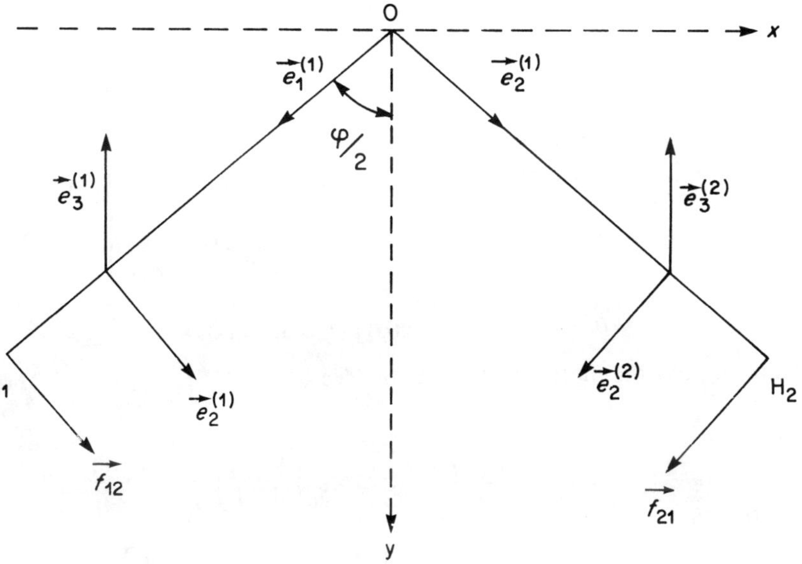

$\vec{e}_3^{(1)} \parallel \vec{e}_3^{(2)} \parallel z \perp x, y$

Figure 15.3. Vectors for bonds of H_2O

and $\vec{e}_2$, respectively. Since atom displacements during vibrations will only occur within the xy plane and the vectors $\vec{e}_3^{(k)} = \vec{n}$ are perpendicular to this plane, the $\vec{E}_3 S^{-1} \Delta E \tilde{\vec{B}} \theta$ scalar matrix product will be zero for both symmetry types. Consequently it only remains to find the expression for $E_2 S^{-1} \Delta E \tilde{\vec{B}} \theta$ which has the following explicit form for A-type vibrations for the H_2O molecule:

$$\frac{1}{S_0}\begin{bmatrix} \vec{f}_{12} & 0 \\ 0 & \vec{f}_{21} \end{bmatrix}\begin{bmatrix} \sigma & 0 \\ 0 & \sigma \end{bmatrix}\begin{bmatrix} \varepsilon & 0 & -\varepsilon_0 \\ 0 & \varepsilon & -\varepsilon_0 \end{bmatrix}\begin{bmatrix} \frac{1}{\sqrt{2}}\vec{e}_1 & -\sigma\vec{f}_{12} \\ \frac{1}{\sqrt{2}}\vec{e}_2 & -\sigma\vec{f}_{21} \\ -\frac{1}{\sqrt{2}}(\vec{e}_1+\vec{e}_2) & \vec{F}_{12} \end{bmatrix}\theta^A$$

where the matrix θ^A is

$$\theta^A = \begin{bmatrix} \dfrac{2\sigma^2[\varepsilon+\varepsilon_0(1+\cos\varphi)]}{2\sigma^2(\varepsilon^2+2\varepsilon\varepsilon_0)} & \dfrac{\sqrt{2}\varepsilon_0\sigma\sin\varphi}{2\sigma^2(\varepsilon^2+2\varepsilon\varepsilon_0)} \\ \dfrac{\sqrt{2}\varepsilon_0\sigma\sin\varphi}{2\sigma^2(\varepsilon^2+2\varepsilon\varepsilon_0)} & \dfrac{[\varepsilon+\varepsilon_0(1+\cos\varphi)]}{2\sigma^2(\varepsilon^2+2\varepsilon\varepsilon_0)} \end{bmatrix}$$

For B-type vibrations we have

$$\frac{1}{S_0}\begin{bmatrix} \vec{f}_{12} & 0 \\ 0 & \vec{f}_{12} \end{bmatrix}\begin{bmatrix} \sigma & 0 \\ 0 & \sigma \end{bmatrix}\begin{bmatrix} \varepsilon & 0 & -\varepsilon_0 \\ 0 & \varepsilon & -\varepsilon_0 \end{bmatrix}\begin{bmatrix} \frac{1}{\sqrt{2}}\vec{e}_1 \\ -\frac{1}{\sqrt{2}}\vec{e}_2 \\ -\frac{1}{\sqrt{2}}(\vec{e}_1-\vec{e}_2) \end{bmatrix}\theta^B$$

where

$$\theta^B=[\varepsilon+\varepsilon_0(1-\cos\varphi)]^{-1},\quad S_0=1.09\text{Å}.$$

Carrying out the matrix multiplications leads to

Type A

$$\vec{E}_2S^{-1}\Delta E\tilde{\tilde{B}}^A\theta^A=-\frac{1}{2S_0}\begin{bmatrix} 0 & 1 \\ 0 & 1 \end{bmatrix}$$

Type B

$$\vec{E}_2S^{-1}\Delta E\tilde{\tilde{B}}^B\theta^B=\frac{\varepsilon_0\sigma\sin\varphi}{S_0\sqrt{(2)}[\varepsilon+\varepsilon_0(1-\cos\varphi)]}\begin{bmatrix} -1 \\ 1 \end{bmatrix}$$

By taking advantage of these expressions we can now write the formulae for the $\partial((\alpha))/\partial Q_j$ tensor components in symmetry co-ordinates. Since we have $e_{1x}=-e_{2x}$, $e_{1y}=e_{2y}$, $(f_{12})_x=-(f_{21})_x$, $(f_{12})_y=(f_{21})_y$ and $e_{1z}=e_{2z}=(f_{12})_z=(f_{21})_z=n_x=n_y=0$, only the diagonal elements of the $\partial((\alpha))/\partial Q_j$ tensor can be non-zero for A-type vibrations, while only the $\partial\alpha_{xy}/\partial Q^B$ component can be non-zero for B-type vibrations.

The formulae for these components will, after substituting in the expressions obtained for $\vec{E}_2S^{-1}\Delta E\tilde{\tilde{B}}^A\theta^S$ and $\vec{E}_2S^{-1}\Delta E\tilde{\tilde{B}}^B\theta^B$, take the form

$$\frac{\partial\alpha_{xx}}{\partial Q_j^A}=\left\{(c_1^A+c_2^A)\left[\frac{1}{\sqrt{2}}\left(\frac{\partial\alpha_1^{(1)}}{\partial q_1}+\frac{\partial\alpha_1^{(1)}}{\partial q_2}\right);\frac{\partial\alpha_1}{\partial\gamma}\right]+(c_1^A(f_{12})_x^2+c_2^A(f_{21})_x^2)\right.$$
$$\cdot\left[\frac{1}{\sqrt{2}}\left(\frac{\partial\alpha_2^{(1)}}{\partial q_1}+\frac{\partial\alpha_2^{(1)}}{\partial q_2}-\frac{\partial\alpha_1^{(1)}}{\partial q_1}-\frac{\partial\alpha_1^{(1)}}{\partial q_2}\right);\frac{\partial\alpha_2^{(1)}}{\partial\gamma}-\frac{\partial\alpha_1^{(1)}}{\partial\gamma}\right]$$
$$\left.-\frac{1}{S_0}(\alpha_1^{(1)}-\alpha_2^{(1)})[e_{1x}(f_{12})_x;\ \ e_{2x}(f_{21})_x]\begin{bmatrix} 0 & 1 \\ 0 & 1 \end{bmatrix}\right\}\begin{bmatrix} l_{q^A} \\ l_\gamma \end{bmatrix}^{(j)}$$

$$\frac{\partial\alpha_{yy}}{\partial Q_j^A}=\left\{(c_1^A+c_2^A)\left[\frac{1}{\sqrt{2}}\left(\frac{\partial\alpha_1^{(1)}}{\partial q_1}+\frac{\partial\alpha_1^{(1)}}{\partial q_2}\right);\frac{\partial\alpha_1}{\partial\gamma}\right]+(c_1^A(f_{12})_y^2+c_2^A(f_{21})_y^2)\right.$$
$$\times\left[\frac{1}{\sqrt{2}}\left(\frac{\partial\alpha_2^{(1)}}{\partial q_1}+\frac{\partial\alpha_2^{(1)}}{\partial q_2}-\frac{\partial\alpha_1^{(1)}}{\partial q_1}-\frac{\partial\alpha_1^{(1)}}{\partial q_2}\right);\frac{\partial\alpha_2^{(1)}}{\partial\gamma}-\frac{\partial\alpha_1^{(1)}}{\partial\gamma}\right]$$
$$\left.-\frac{1}{S_0}(\alpha_1^{(1)}-\alpha_2^{(1)})[e_{1y}(f_{12})_y;\ \ e_{2y}(f_{21})_y]\begin{bmatrix} 0 & 1 \\ 0 & 1 \end{bmatrix}\right\}\begin{bmatrix} l_{q^A} \\ l_\gamma \end{bmatrix}^{(j)}$$

$$\frac{\partial\alpha_{zz}}{\partial Q_j^A}=\left\{(c_1^A+c_2^A)\left[\frac{1}{\sqrt{2}}\left(\frac{\partial\alpha_1^{(1)}}{\partial q_1}+\frac{\partial\alpha_1^{(1)}}{\partial q_2}\right);\frac{\partial\alpha_1}{\partial j}\right]+n_z^2(c_1^A+c_2^A)\right.$$
$$\times\left[\frac{1}{\sqrt{2}}\left(\frac{\partial\alpha_3^{(1)}}{\partial q_1}+\frac{\partial\alpha_3^{(1)}}{\partial q_2}-\frac{\partial\alpha_1^{(1)}}{\partial q_1}-\frac{\partial\alpha_1^{(1)}}{\partial q_2}\right)\right.$$
$$\left.\left.\cdot\left(\frac{\partial\alpha_3^{(1)}}{\partial\gamma}-\frac{\partial\alpha_1^{(1)}}{\partial\gamma}\right)\right]\right\}\left[\frac{l_{q^A}}{l_\gamma}\right]^{(j)},\quad j=1,2$$

$$\frac{\partial\alpha_{xy}}{\partial Q_3^B}=\left\{\frac{1}{\sqrt{2}}(c_1^B(f_{12})_x(f_{12})_y+c_2^B(f_{21})_x(f_{21})_y)\right.$$
$$\times\left(\frac{\partial\alpha_2^{(1)}}{\partial q_1}-\frac{\partial\alpha_2^{(1)}}{\partial q_2}-\frac{\partial\alpha_1^{(1)}}{\partial q_1}+\frac{\partial\alpha_1^{(1)}}{\partial q_2}\right)+(\alpha_1^{(1)}-\alpha_2^{(1)})$$
$$\times[e_{1x}(f_{12})_y+e_{1y}(f_{12})_x;e_{2x}(f_{21})_y+e_{2y}(f_{21})_x]$$
$$\left.\times\frac{\varepsilon_0\sigma\sin\varphi}{S_0\sqrt{2}[\varepsilon+\varepsilon_0(1-\cos\varphi)]}\begin{bmatrix}-1\\1\end{bmatrix}\right\}l_{q^B}.$$

From here and by noting that $e_{1x}=-\sin(\varphi/2)$,

$$e_{1y}=\cos(\varphi/2),\quad (f_{12})_x=\cos(\varphi/2),\quad (f_{12})_y=\sin(\varphi/2)$$

and that $c_1^A=c_2^A=1$ for symmetric vibrations, while $c_1^B=-c_2^B=1$ for anti-symmetric ones, the following relations are finally obtained:

$$\frac{\partial\alpha_{xx}}{\partial Q_j^A}=\sqrt{2}\left(\frac{\partial\alpha_1^{(1)}}{\partial q_1}+\frac{\partial\alpha_1^{(1)}}{\partial q_2}\right)l_{q^{A_j}}$$
$$+\sqrt{2}\cos^2\frac{\varphi}{2}\left(\frac{\partial\alpha_2^{(1)}}{\partial q_1}+\frac{\partial\alpha_2^{(1)}}{\partial q_2}-\frac{\partial\alpha_1^{(1)}}{\partial q_1}-\frac{\partial\alpha_1^{(1)}}{\partial q_2}\right)l_{q^{A_j}}$$
$$+2\frac{\partial\alpha_1^{(1)}}{\partial\gamma}l_{\gamma j}+2\cos^2\frac{\varphi}{2}\left(\frac{\partial\alpha_2^{(1)}}{\partial\gamma}-\frac{\partial\alpha_1^{(1)}}{\partial\gamma}\right)l_{\gamma j}+(\alpha_1^{(1)}-\alpha_2^{(1)})\frac{1}{S_0}\sin\varphi l_{\gamma j}$$

$$\frac{\partial\alpha_{yy}}{\partial Q_j^A}=\sqrt{2}\left(\frac{\partial\alpha_1^{(1)}}{\partial q_1}+\frac{\partial\alpha_1^{(1)}}{\partial q_2}\right)l_{q^{A_j}}$$
$$+\sqrt{2}\sin^2\frac{\varphi}{2}\left(\frac{\partial\alpha_2^{(1)}}{\partial q_1}+\frac{\partial\alpha_2^{(1)}}{\partial q_2}-\frac{\partial\alpha_1^{(1)}}{\partial q_1}-\frac{\partial\alpha_1^{(1)}}{\partial q_2}\right)l_{q^{A_j}}$$
$$+2\frac{\partial\alpha_1^{(1)}}{\partial\gamma}l_{\gamma j}+2\sin^2\frac{\varphi}{2}\left(\frac{\partial\alpha_2^{(1)}}{\partial\gamma}-\frac{\partial\alpha_1^{(1)}}{\partial\gamma}\right)l_{\gamma j}-\frac{1}{S_0}(\alpha_1^{(1)}-\alpha_2^{(1)})\sin\varphi\cdot l_{\gamma j}$$

$$\frac{\partial\alpha_{zz}}{\partial Q_j^A}=\sqrt{2}\left(\frac{\partial\alpha_3^{(1)}}{\partial q_1}+\frac{\partial\alpha_3^{(1)}}{\partial q_2}\right)l_{q^{A_j}}+2\frac{\partial\alpha_3^{(1)}}{\partial\gamma}l_{\gamma j},\quad j=1,2$$

$$\frac{\partial \alpha_{xy}}{\partial Q_3^B} = \left[\frac{1}{\sqrt{2}} \sin \varphi \left(\frac{\partial \alpha_2^{(1)}}{\partial q_1} - \frac{\partial \alpha_2^{(1)}}{\partial q_2} - \frac{\partial \alpha_1^{(1)}}{\partial q_1} + \frac{\partial \alpha_1^{(1)}}{\partial q_2}\right)\right.$$

$$\left. + (\alpha_1^{(1)} - \alpha_2^{(1)}) \frac{\varepsilon_0 \sigma \sin 2\varphi}{S_0 \sqrt{2}[\varepsilon + \varepsilon_0(1 - \cos \varphi)]}\right] l_{q^B}$$

These expressions for $\partial \alpha_{xx}/\partial Q_j^A$ and $\partial \alpha_{yy}/\partial Q_j^A$ can be written in a more compact form as

$$\frac{\partial \alpha_{xx}}{\partial Q_j^A} = \sqrt{2} \sin^2 \frac{\varphi}{2} \left(\frac{\partial \alpha_1^{(1)}}{\partial q_1} + \frac{\partial \alpha_1^{(1)}}{\partial q_2}\right) l_{q^{A_j}} + \sqrt{2} \cos^2 \frac{\varphi}{2} \left(\frac{\partial \alpha_2^{(1)}}{\partial q_1} + \frac{\partial \alpha_2^{(1)}}{\partial q_2}\right) l_{q^{A_j}}$$

$$+ 2\left(\sin^2 \frac{\varphi}{2} \cdot \frac{\partial \alpha_1^{(1)}}{\partial \gamma} + \cos^2 \frac{\varphi}{2} \cdot \frac{\partial \alpha_2^{(1)}}{\partial \gamma}\right) l_{\gamma j} + \frac{1}{S_0} (\alpha_1^{(1)} - \alpha_2^{(1)}) \sin \varphi \cdot l_{\gamma j}$$

$$\frac{\partial \alpha_{yy}}{\partial Q_j^A} = \sqrt{2} \cos^2 \frac{\varphi}{2} \left(\frac{\partial \alpha_1^{(1)}}{\partial q_1} + \frac{\partial \alpha_1^{(1)}}{\partial q_2}\right) l_{q^{A_j}} + \sqrt{2} \sin^2 \frac{\varphi}{2} \left(\frac{\partial \alpha_2^{(1)}}{\partial q_1} + \frac{\partial \alpha_2^{(1)}}{\partial q_2}\right) l_{q^{A_j}}$$

$$+ 2\left(\cos^2 \frac{\varphi}{2} \cdot \frac{\partial \alpha_1^{(1)}}{\partial \gamma} + \sin^2 \frac{\varphi}{2} \cdot \frac{\partial \alpha_2^{(1)}}{\partial \gamma}\right) l_{\gamma j} - \frac{1}{S_0} (\alpha_1^{(1)} - \alpha_2^{(1)}) \sin \varphi \cdot l_{\gamma j}$$

15.6 Solution of the Inverse Spectral Problem in the Theory of Intensities in Raman Spectra

As in the case for infra-red spectra, the $\alpha_i^{(k)}$ and $\partial \alpha_i^{(k)}/\partial q'_k$ parameters are obtained from experimental data on the intensities and depolarization degrees of lines in a Raman spectrum. The latter enable us to find b'^2 and g'^2, which can, in turn, be used to calculate the parameters if the vibrational forms and geometrical structure of the molecule are known.

By extracting the square roots of the experimental values of b'^2 and equating them with the expressions for b' for different vibrations, we can obtain a system of linear equations such as the following:

$$\pm (b'_j)_{\text{equ}} = \sum_k \sum_{k'} \frac{\partial}{\partial q_{k'}} (\alpha_1^{(k)} + \alpha_2^{(k)} + \alpha_3^{(k)}) (l_q)_{k'j}$$

In order to simplify this system we can use the characterizability of the vibrations in form. This enables us to exclude, with a certain degree of approximation, several of the $\partial \alpha^{(k)}/\partial q$ parameters which are multiplied by the $(l_q)_{k'j}$ coefficients whose values are nearly zero.

When the molecule has some symmetry, the number of these equations is sharply reduced because $b' = 0$ for all partially symmetric vibrations. In order to calculate the number of equations, use can be made of the spectra of the deuterosubstituted compounds, the spectra compounds with intermediate symmetries being especially helpful. To do this, all the equations can be reduced to a form containing combinations of $\partial \alpha^{(k)}/\partial q$ parameters, combinations that correspond to derivatives of the $\alpha^{(k)}$ of bonds with respect to the symmetry co-ordinates of fully symmetric vibrations. This reduction is possible if the symmetry co-ordinates of the non-deuterated molecule can, in every case, be used in the description of the vibrational forms.

It is only possible to obtain a system of scalar equations for the $\partial\alpha^{(k)}/\partial q$ and $\alpha^{(k)}$ electro-optical parameters whilst using g'^2 if the molecule is sufficiently symmetric. It is always possible in the special case when one of the non-diagonal components of the $((\alpha'))$ tensor (α'_{xy} or other) becomes non-zero for a certain choice of co-ordinate axes for partially symmetric vibrations. This is the case for B-type vibrations in the H_2O molecule.

When considering the inverse spectral problem in the theory of intensities in infra-red spectra we also encounter the situation that the scalar systems of equations can be obtained if there is symmetry in the molecule. However, in the case of Raman spectra, the probability of obtaining such systems is lower than it is for infra-red spectra. This situation, in addition to the very large number of parameters being used, makes it very difficult to determine them

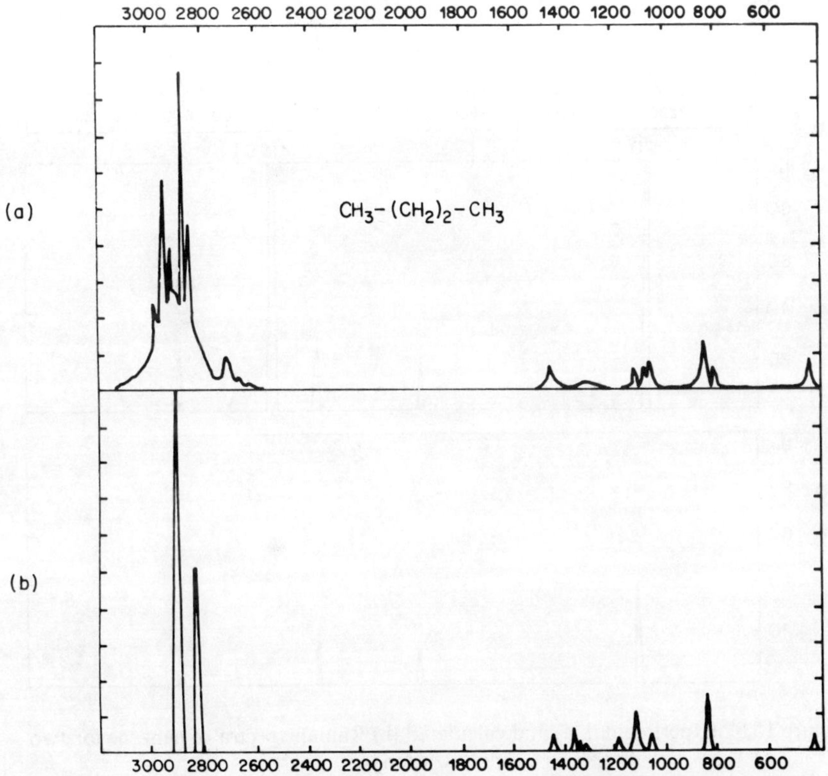

Figure 15.4. Experimental (a) and calculated (b) Raman spectrum of butene

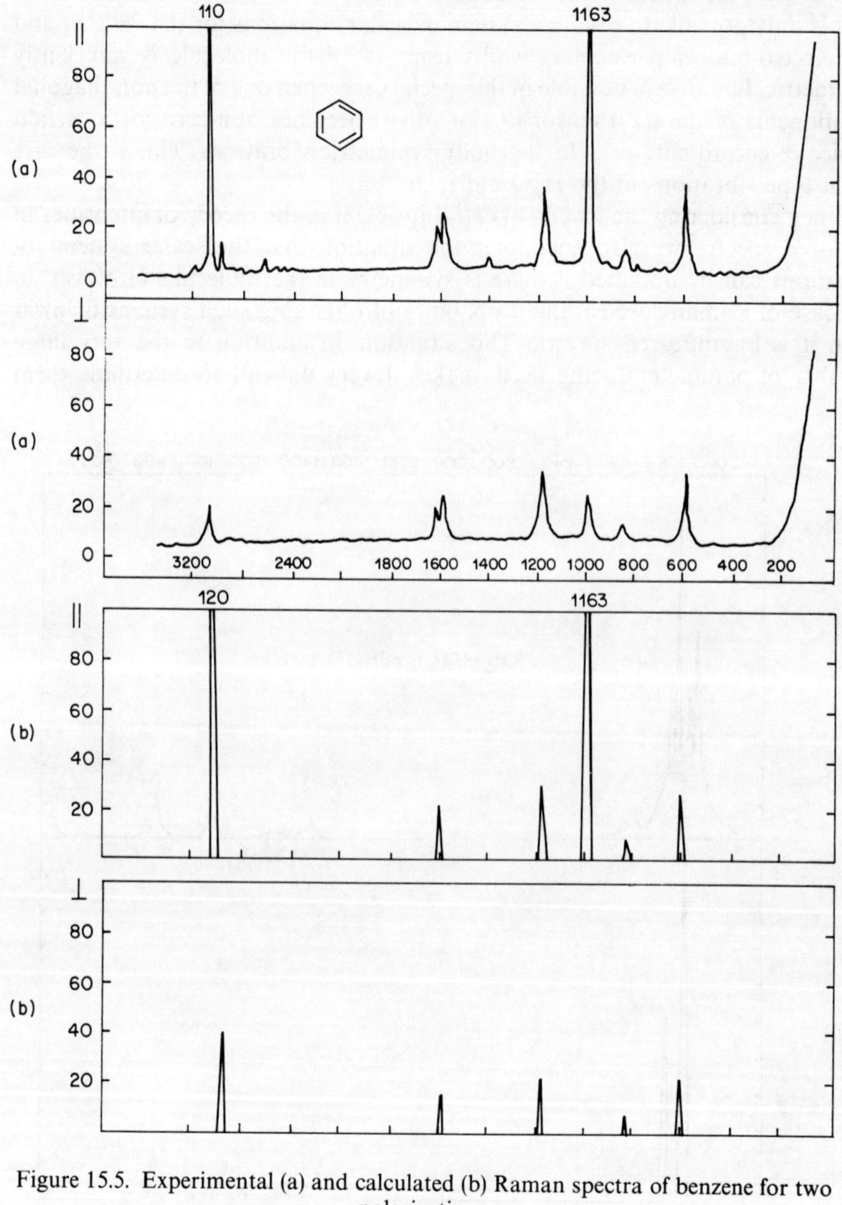

Figure 15.5. Experimental (a) and calculated (b) Raman spectra of benzene for two polarizations

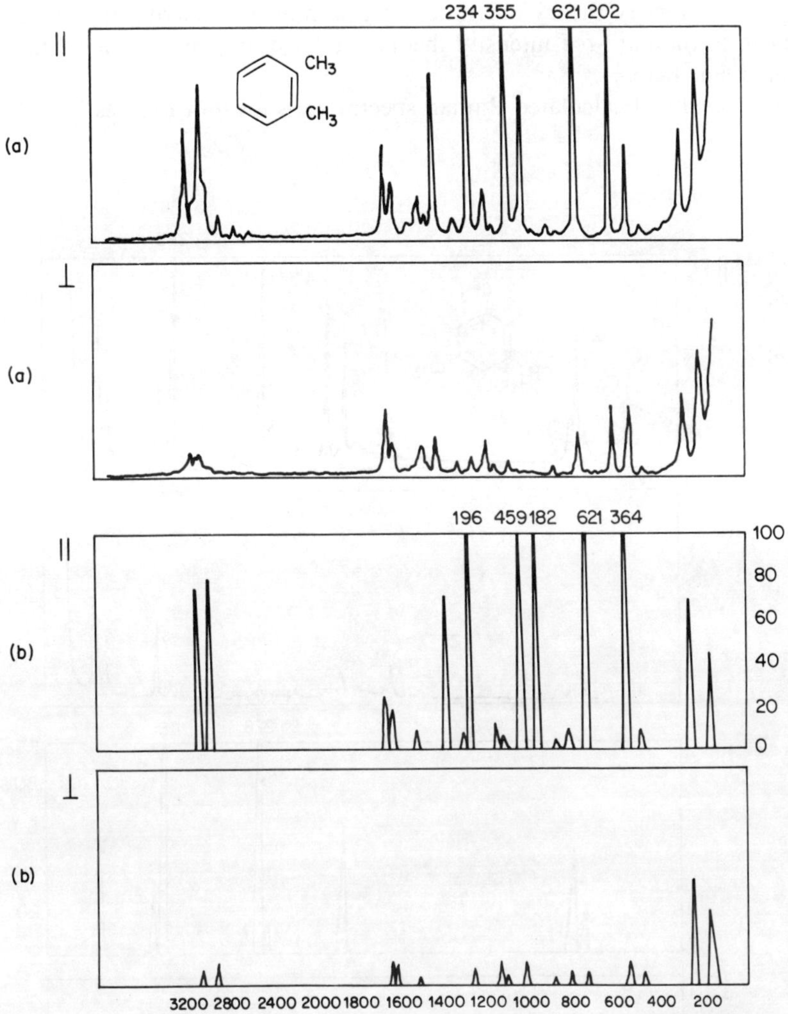

Figure 15.6. Experimental (a) and calculated (b) Raman spectra of o-xylene for two polarizations

even in the simplest cases. This, in turn, radically reduces the computability of the theory being presented. Nevertheless, values have been found for the electro-optical parameters of individual bonds and atom groups in a number of cases, given certain approximations. Since solving the inverse spectral problem in Raman intensity theory has much in common with solving the analogous problem in the infra-red intensity theory, we have only made a small number of comments here.

Some results of calculated Raman spectra are shown in Figures 15.4–15.8.

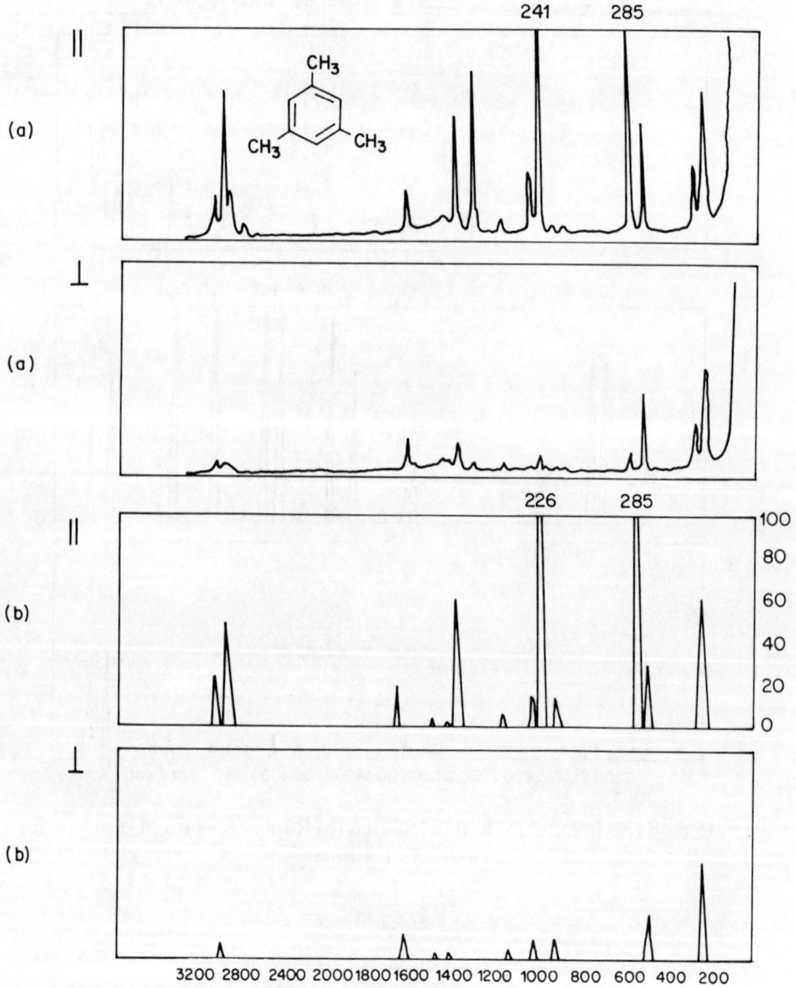

Figure 15.7. Experimental (a) and calculated (b) Raman spectra of mesitelene for two polarizations

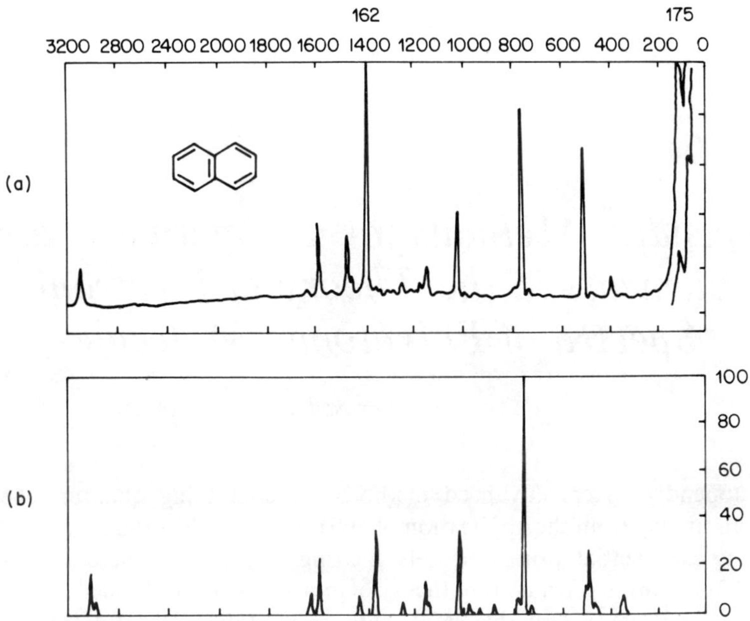

Figure 15.8. Experimental (a) and calculated (b) Raman spectra of naphthalene

References

1. W. Heitler. *The Quantum Theory of Radiation*, 3rd edn., Clarendon Press, Oxford (1954).
2. L. A. Gribov. *J. Appl. Spectr.* **43**, 594 (1985). (in Russian).
3. M. V. Volkenstein, L. A. Gribov, M. A. Elyashevich and B. I. Stepanov. *Vibrations of Molecules*, 2nd edn., Nauka, Main Editorial Board of Physico-Mathematical Literature, Moscow (1972).

Appendix: Methods for Computing Matrix Elements in the Theory of Electronic Spectra of Polyatomic Molecules

B. K. Novosadov

This appendix covers advanced methods for computing quantum-chemical integrals arising from the application of Hartree–Fock–Roothaan theory based on using exponential atomic orbitals. Bearing in mind that these integrals are special functions in the quantum theory of molecules, it was decided not to overburden the main text of this book with computations helpful for a reader interested primarily in the mathematical aspects of molecular quantum theory. At the same time, it has been considered necessary to enable an expert in spectroscopy to make himself familiar in detail with methods for computing matrix elements which occur in the theory of electronic spectra of polyatomic molecules presented in this book. This appendix is not a survey of the problems of computing quantum-chemical integrals but an original analysis by its author.

A.1 Types of Basis Functions in the LCAO–MO Theory

When computing complicated molecular systems using LCAO–MO theory two types of functions are generally considered: exponentially decreasing functions and Gaussian ones. Functions of the first kind are proportional to $\exp(-\zeta r)$, where ζ is a parameter and r the radial coordinate of electron, whilst the second type of function is proportional to $\exp(-\zeta r^2)$. Gaussian functions[1] have been found to have the widest application for practical purposes in making numerical computations of molecules due to their factorizability over the Cartesian electronic co-ordinates. At the same time, they have a substantial disadvantage associated with their asymptotically increasing potential, rather than with the potential vanishing as $r \to \infty$. As a result, these functions are found to be ineffective for solving problems involving interactions between molecules spaced far apart.

In recent years, attention has again been drawn to exponential functions (EF). This can be attributed to the advances in powerful computers and to the necessity of developing banks of quantum-mechanical computations which can be employed in computing complex molecular systems. This problem cannot be adequately solved without resort to EFs which show the correct analytical behaviour on molecular nuclei and for $r \to \infty$. A large variety of EF has been described elsewhere in the literature. Here we will mention only the most important.

(1) *Hydrogen-type orbitals* (HTO) are solutions to the Schroedinger equation for atomic hydrogen:

$$\varphi_n(\vec{r}) = N_{nl} R_{nl}(r) Y_{lm}(\hat{r}) \tag{A.1}$$

where

$$R_{nl}(r) = r^l L^{2l+1}_{n-l-1}(2\xi r)\mathrm{e}^{-\zeta r} \tag{A.2}$$

$$Y_{lm}(\hat{r}) = P_{lm}(\cos\theta)\mathrm{e}^{im\varphi} \tag{A.3}$$

are the spherical functions in R^3 and $P_{lm}(\cos\theta)$ is an associated Legendre polynomial

$$N_{nl} = 2\zeta^{3/2}\{(n-l-1)![n(n+l)]^{-1}\}^{1/2} \tag{A.4}$$

(2) *Slater-type orbitals* (STO) are the asymptotic portion of the HTOs

$$\varphi_s(\vec{r}) = N_{nl}(\zeta r)^{n-1}\mathrm{e}^{-\zeta r} Y_{lm}(\hat{r}) \tag{A.5}$$

where $N_{nl} = (2\zeta)^{n+1/2}[(2n)!]^{-1/2}$ and the numeral n need not be an integer.

(3) *Löwdin–Shull orbitals* (LSO) are like HTOs, the difference being that the upper index in the Laguerre polynomial is increased by 1:[2]

$$\varphi_{\mathrm{LSH}}(\vec{r}) = N_{nl} R_{nl}(r) Y_{lm}(\hat{r}) \tag{A.6}$$

where

$$R_{nl}(r) = (2\zeta r)^l L^{2l+2}_{n-l-1}(2\zeta r)\mathrm{e}^{-\zeta r}$$

and

$$N_{nl} = (2\zeta)^{3/2}[(n-l-1)!/(n+l+1)!]^{1/2} \tag{A.7}$$

(4) *Bessel-type orbitals* (BTO) are associated with Bessel polynomials obtained from reduced Bessel functions (RBF) half-integral indices:

$$\varphi_{\mathrm{B}}(\vec{r}) = N_{nl}\hat{k}_{n-1/2}(\zeta r) Y_{lm}(\hat{r}) \tag{A.8}$$

where

$$\hat{k}_{n-1/2}(z) = (2/\pi)^{1/2} z^{n-1/2} K_{n-1/2}(z) \tag{A.9}$$

is a reduced Bessel function (RBF). Equations (A.8) are given in references 3 and 4. A half-integral-index RBS is a product of the Bessel polynomial and an exponent (see equation (A.59)). An attempt to unify various types of EFs is made in reference 5.

Note, however, that the exponential type functions described above can be combined to form a natural class of solutions to the Schroedinger equation as follows:

$$\left(-\tfrac{1}{2}\Delta - \frac{Z_1}{r} + \frac{Z_2}{r^2}\right)\Psi(\vec{r}) = E\Psi(\vec{r}) \tag{A.10}$$

where Z_1, Z_2, E are parameters.

This section will show that by a proper selection of parameters Z_1, Z_2, E we can obtain the above four types of EFs. Moreover, we will introduce here new types of EFs which can prove helpful in making numerical computations of molecular spectra.

Equation (A.10) can be solved in terms of spherical co-ordinates r, θ, φ to provide the solutions

$$\Psi(\vec{r}) = R_l(r) Y_{lm}(\theta, \varphi) \tag{A.11}$$

where the function $R_l(r)$ satisfies the ordinary differential equation

$$\nabla_r^2 R_l + 2\left(E + \frac{Z_1}{r} - \frac{Z_2}{r^2}\right) R_l - \frac{l(l+1)}{r^2} R_l = 0 \tag{A.12}$$

where

$$\nabla_r^2 = \frac{1}{r^2} \frac{\partial}{\partial r} \left(r^2 \frac{\partial}{\partial r} \right) \tag{A.13}$$

and the spherical function $Y_{lm}(\theta, \varphi)$ satisfies

$$\nabla_{\theta,\varphi}^2 Y_{lm} + l(l+1) Y_{lm} = 0 \tag{A.14}$$

where $l \geqslant |m|$ are integers and $|m| = 0, 1, 2, \ldots, l$.

Introducing the parameter $q = [(l + \frac{1}{2})^2 + 2Z_2]^{1/2} - \frac{1}{2}$ we can write

$$2Z_2 + l(l+1) = q(q+1) \tag{A.15}$$

Parameter q will be real if

$$2Z_2 \geqslant -(l + \tfrac{1}{2})^2 \tag{A.16}$$

The general form of the normalized solutions $R_l(r)$ of equation (A.12) is as follows:

$$R_{kl}(\zeta r) = C_{kl} \mathrm{e}^{-\zeta r} (2\zeta r)^q L_k^{2q+1}(2\zeta r) \tag{A.17}$$

where L_k^{α} is the Laguerre polynomial

$$C_{kl} = \left[\int_0^{\infty} \mathrm{e}^{-2\zeta r} (2\zeta r)^{2q} (L_k^{2q+1}(2\zeta r))^2 \phi(r) \, \mathrm{d}r \right]^{-1/2} \tag{A.18}$$

$\phi(r)$ is a weight function, $k = 0, 1, 2, \ldots$, and

$$\zeta = (-2E)^{1/2} > 0 \tag{A.19}$$

Parameters Z_1, Z_2, ζ are interrelated for solutions (A.17) through the relationship

$$Z_1 \zeta^{-1} = k + q + 1 \tag{A.20}$$

Functions (A.17) are orthogonal and the proof is as follows. Consider two equations (A.12), one for $R_{kl}(\zeta r)$ and the other for $R_{k'l}(\zeta' r)$. If we multiply the first by $R_{k'l'}$, the second by R_{kl}, integrate both relationships with weight r^2 and subtract the second integral from the first, these operations yield the following orthogonality relation:

$$\int_0^\infty R_{k'l}(\zeta' r) R_{kl}(\zeta r)\phi(r)\,\mathrm{d}r = 0$$

where the weight function is

$$\phi(r) = \tfrac{1}{2}(\zeta'^2 - \zeta^2)r^2 + (Z_1' - Z_1)r - Z_2' + Z_2 \tag{A.21}$$

If fixed values are taken for any two parameters we then find eigenvalues for a third parameter from equation (A.20).

Now let us consider three types of solutions (A.17).

(1) $Z_1 = \text{const}$, $Z_2 = \text{const}$. Parameter ζ takes on the following values:

$$\zeta_{kl} = Z_1(k + q + 1)^{-1} \tag{A.22}$$

where $k = 0, 1, 2, \ldots$.

The function $R_{kl}(\zeta_{kl} r)$ is an orthogonal one with weight r^2 according to relation (A.21), where the last two summations become zero. These functions are a generalized representation of HTOs with $Z_2 \neq 0$. For $Z_2 = 0$ we obtain ordinary hydrogen functions with $q = l$. Note that generalized HTOs appear in the course of solving the Dirac equation for a hydrogen atom. Discrete HTOs of this type form an orthogonal system of functions which is, however, incomplete. On being combined with the HTOs of a continuous spectrum ζ this HTO system becomes complete.

(2) $Z_2 = \text{const}$, $\zeta = \text{const}$:

$$Z_1 = \zeta(k + q + 1) \tag{A.23}$$

In this case only the second term remains in the weight function (A.21), and the functions $R_{kl}(\zeta r)$ are orthogonal with weight r. These functions form the Sturm system, i.e. in this case a discrete spectrum Z_1 alone is available.

This type of function appears in solving the problem of motion of an electron in a multi-centre Coulomb nuclear field,[10] where $Z_2 = 0$.

(3) $Z_1 = \text{const}$, $\zeta = \text{const}$. Eigenvalues of Z_2 have the following form:

$$Z_2 = 2^{-1}[(Z_1\zeta^{-1} - k - \tfrac{1}{2})^2 - (l + \tfrac{1}{2})^2] \tag{A.24}$$

These are orthogonal functions with weight 1 in accordance with relation (A.21). Thus we obtain another Sturm series of functions of type (A.17).

(4) Let us consider a general case of functions (A.17). Within the three-dimensional space of parameters Z_1, Z_2, ζ relationship (A.20) defines a set of surfaces, with the trajectories upon these surface determining a collection of solutions (A.17), which are orthogonal with weight function (A.21). Cases 1–3 are consistent with lines upon these surfaces.

Accordingly, EFs of type (A.17) form a wide class of atomic orbitals with different weights of orthogonality, which, however, satisfy differential equations of the same type (A.12). These functions will be referred to as Laguerre atomic orbitals (LO).

The known types of EFs mentioned above can be obtained from equation

(A.17) by selecting values for the parameter q. For example, we obtain:

(1) HTO: $q = l$;
(2) STO: $k = 0, q = n - 1$;
(3) LSO: $q = l + \frac{1}{2}$;
(4) BTO: $Z_1 = 0, 2Z_2 = (n + \frac{1}{2})^2 - (l + \frac{1}{2})^2, q = -n - 1$.

In addition to the above AOs we will mention here other types which can prove helpful in computing various molecules.

(5) Eliminating factor $(2\zeta r)^q$ by taking $q = 0$ (i.e. $2Z_2 = -(l + \frac{1}{2})^2$) we obtain

$$R_k(\zeta r) = C_k e^{-\zeta r} L_k^1(2\zeta r) \tag{A.25}$$

which form an orthogonal system, in which for a fixed parameter ζ we obtain a complete Sturm system of functions which is orthogonal with weight r. A feature of the AOs (equation (A.25)) is that these are independent of the number l.

(6) Note that the STOs are the simplest representatives of the orthogonal systems of functions; i.e. taking $q = n - 1$, where $n = 1, 2, \ldots$, we obtain the following AOs:

$$R_k(\xi r) = C_k e^{-\xi r}(2\xi r)^{n-1} L_k^{2n-1}(2\zeta r) \tag{A.26}$$

Functions (A.25) are a particular case of the above, and these are not dependent upon the number l either: the only condition is that $n > l$.

(7) Let us take $Z_1 = \zeta$. In accordance with equation (A.24), $2Z_2 = (\frac{1}{2} - k)^2 - (l + \frac{1}{2})^2$, so that $q = |\frac{1}{2} - k| - \frac{1}{2}$. Thus we arrive at a system of AOs orthogonal with weight 1:

$$R_k(\zeta r) = C_k e^{-\zeta r}(2\zeta r)^{|1/2-k|-1/2} L_k^{|1-2k|-1}(2\zeta r) \tag{A.27}$$

where $k = 0, 1, 2, \ldots$ for $k > l$.

(8) Let $2Z_2 = -(l + \frac{1}{2})^2 + 4^{-1}(l + 1)^2$, then $q = l/2$ and the AOs can be expressed as

$$R_{kl} = C_{kl} e^{-\zeta r}(2\zeta r)^{l/2} L_k^{l+1}(2\zeta r) \tag{A.28}$$

These functions form a Sturm basis with the weight of orthogonality being equal to r,

The new classes of EFs given in (5)–(8) above have not previously appeared in the literature. Both electronic densities and matrix elements can be expanded in terms of these AOs.

A.2 Fourier Transformation of the Exponential Functions

The Fourier transformation of the EFs are involved in computing matrix elements in LCAO–MO theory. Let us consider equation (A.10), to which a solution is given by equations (A.11) and (A.17), and let us rewrite this solution as follows:

$$\Psi_{klm}(\vec{r}) = C_{kl} e^{-\zeta r}(2\zeta r)^q L_k^{2q+1}(2\zeta r)\, Y_{lm}(\hat{r}) \tag{A.29}$$

Now let us introduce the Fourier transformation of the function (A.29):

$$\bar{\Psi}_{klm}(\vec{p}) = (2\pi)^{-3/2} \int e^{-i\vec{p}\,\vec{r}} \Psi_{klm}(r) d^3 r \tag{A.30}$$

where p is the electronic momentum.

Substituting function (A.29) into integral (A.30) gives

$$\bar{\Psi}_{klm}(\vec{p}) = (2\pi)^{-3/2} C_{kl} \int e^{-\zeta r} (2\zeta r)^q L_k^{2q+1}(2\zeta r) \int_{(\Omega_3)} Y_{lm}(\hat{r}) e^{-i\vec{p}\,\vec{r}} d\Omega_3 r^2 dr \tag{A.31}$$

To take the integral over angles, let us expand the exponent into the Rayleigh series

$$\exp(i\vec{p}\,\vec{r}) = 4\pi (pr)^{-1} \sum_{\lambda\mu} j_\lambda(pr) Y_{\lambda\mu}(\hat{r}) \bar{Y}^*_{\lambda\mu}(\hat{p}) \tag{A.32}$$

where $j_\lambda(pr)$ is the spherical Bessel function and

$$\bar{Y}_{\lambda\mu}(\hat{p}) = (-i)^\lambda Y_{\lambda\mu}(\hat{p}) \tag{A.33}$$

is the modified spherical harmonic.

Taking into account that the spherical functions are orthonormalized, we arrive at a one-dimensional integral for the Fourier transform (A.31):

$$\bar{\Psi}_{klm}(\vec{p}) = (2\pi)^{-3/2} 4\pi p^{-1} C_{kl} \int_0^\infty e^{-\zeta r} (2\zeta r)^q L_k^{2q+1}(2\zeta r) j_l(pr) r \, dr \, \bar{Y}_{lm}(\hat{p}) \tag{A.34}$$

Thus the radial part of function $\bar{\Psi}_{klm}(p)$ is represented in the form of the Hankel–Fourier transformation

$$\Pi_{kl}(p) = (2\pi)^{-1/2} 2 C_{kl} p^{-1} \int_0^\infty e^{-\zeta r} (2\zeta r)^q L_k^{2q+1}(2\zeta r) j_l(pr) r \, dr \tag{A.35}$$

With $q \geqslant l$ being an integer, the integral can be reduced to a linear combination of Sturm HTOs.

It has been shown in reference 6 that, when given in the momentum representation, the HTOs coincide with the four-dimensional spherical harmonics of arguments $\alpha = \arctan(p/\zeta), \theta, \varphi$, thus being capable of being written in a trigonometric form.

A.3 Theorems for the Laguerre Atomic Orbitals

Computation of matrix elements for LOs leads to integrals of the products of these functions, as well as of those dependent upon the difference of the arguments. That is why it seems best to present from the beginning the theorems of 'linearization' of a product of LOs, and the theorem of the addition of the LO arguments, which then enable us to calculate matrix elements in the theory of molecular spectra in a compact form convenient for making numerical computations of the spectra.

A.3.1 Theorem of 'linearization' of a product of LOs in a co-ordinate space

Let us write a product of functions (A.29):

$$\Psi_{klm}(\zeta_1\vec{r})\Psi_{k'l'm'}(\zeta_2\vec{r})$$
$$= C_{kl}(\zeta_1)C_{k'l'}(\zeta_2)\mathrm{e}^{-(\zeta_1+\zeta_2)r}(2\zeta_1 r)^q(2\zeta_2 r)^{q'}L_k^{2q+1}(2\zeta_1 r)L_{k'}^{2q'+1}(2\zeta_2 r)Y_{lm}(\hat{r})Y_{l'm'}(\hat{r}) \quad \text{(A36)}$$

A product of spherical harmonics can be linearized with the aid of the Gaunt coefficients

$$Y_{lm}(\hat{r})Y_{l'm'}(\hat{r}) = \sum_{LM} G_{lm;l'm'}^{LM} Y_{LM}(\hat{r}) \quad \text{(A.37)}$$

where

$$G_{lm;l'm'}^{LM} = \int Y_{lm}(\hat{r})Y_{l'm'}(\hat{r})Y_{LM}^*(\hat{r})\,\mathrm{d}\Omega_3 \quad \text{(A.38)}$$

is a known Gaunt integral (coefficient)[7] related to the Clebsch–Gordan coefficients through the relationship

$$G_{lm;l'm'}^{LM} = \left[\frac{(2l+1)(2l'+1)}{4\pi(2L+1)}\right]^{1/2} C_{lo;l'o}^{LO} C_{lm;l'm'}^{LM} \quad \text{(A.39)}$$

Now let us convert the product of the radial functions to arrive at a linear combination of LOs with parameters $\zeta = \zeta_1 + \zeta_2$ and $Q < q + q'$. To that end, let us express the rational function in the form of a linear combination of the Laguerre polynomials:

$$r^{q+q'-Q}L_k^{2q+1}(2\zeta_1 r)L_{k'}^{2q'+1}(2\zeta_2 r) = \sum_k b_{kq;k'q'}^{KQ} L_K^{2Q+1}(2\zeta r) \quad \text{(A.40)}$$

where $q + q - Q$ is an integer and the coefficient $b_{kq;k'q'}^{KQ}$ is determined by an integral associated with the following radial integral:

$$B_{kl;k'l'}^{KL} = \int_0^\infty R_{kl}(\zeta_1 r)R_{k'l'}(\zeta_2 r)R_{KL}(\zeta r)\phi(r)\,\mathrm{d}r \quad \text{(A.41)}$$

where $\phi(r)$ is the weight function for $R_{KL}(\zeta r)$, parameters q, q', Q and l, l', L are related through relationship (A.15) and $\zeta = \zeta_1 + \zeta_2$.

As a result, we obtain the following theorem of linearization of a product of LOs:

$$\Psi_{klm}(\zeta_1\vec{r})\Psi_{k'l'm'}(\zeta_2\vec{r}) = \sum_K \sum_L v_K B_{kl;k'l'}^{KL} G_{lm;l'm'}^{LM} \Psi_{KLM}(\zeta\vec{r}) \quad \text{(A.42)}$$

where

$$v_K = \left[\int_0^\infty R_{KL}^2(\zeta r)\phi(r)\,\mathrm{d}r\right]^{-1}$$

The limits of summation are determined by the conditions of a triangle for the indices of the Laguerre and Legendre polynomials.

We will now show how to calculate coefficients b. Let us write the theorem of linearization of a product of the rational parts of LOs, which can be inferred from the expansion of equation (A.40) for $\zeta \neq \zeta_1 + \zeta_2$. By further multiplying this equation by the polynomial $r^Q L_K^{2Q+1}(2\zeta r)$, and by integrating it with weight $r^\tau e^{-\zeta r}$ (where $\tau = 0, -1, -2$) we arrive at the following formula for the expansion coefficient:

$$b_{kq;k'q'}^{KQ} \int_0^\infty r^{2Q+\tau} L_K^{2Q+1}(2\zeta r) L_K^{2Q+1}(2\zeta r) e^{-\zeta r}\, dr$$

$$= \int_0^\infty r^{q+q'+Q+\tau} L_k^{2q+1}(2\zeta_1 r) L_{k'}^{2q'+1}(2\zeta_2 r) L_K^{2Q+1}(2\zeta r) e^{-\zeta r}\, dr$$

A.3.2 Addition theorem for LO arguments

Let us write the Fourier transform for the LOs:

$$\Psi_{klm}(\vec{r} - \vec{R}) = (2\pi)^{-3/2} \int e^{i\vec{p}(\vec{r}-\vec{R})} \bar{\Psi}_{klm}(\vec{p}) d^3\vec{p}. \tag{A.43}$$

and expand the exponent $e^{i\vec{p}\vec{r}}$ into a series in terms of the Sturm LOs:

$$e^{-i\vec{p}\vec{r}} = \sum_{k'l'm'} C_{k'l'm'}(\vec{p}) \Psi^*_{k'l'm'}(\zeta \vec{r}) \phi(r) \tag{A.44}$$

where $\phi(r)$ = weight function. On multiplying equation (A.44) by $\Psi_{k'l'm'}(\zeta\vec{r})$ and integrating over $d^3\vec{r}$ with due account of the orthogonality of the LO system, we obtain the following representation for the coefficients:

$$C_{k'l'm'}(\vec{p}) = (2\pi)^{-3/2} v_{k'} \int e^{-i\vec{p}\vec{r}} \Psi_{k'l'm'}(\zeta\vec{r}) d^3\vec{r} \tag{A.45}$$

where

$$v_{k'}^{-1} = (2\pi)^{-3/2} \int \Psi^*_{k'l'm'}(\zeta\vec{r}) \Psi_{klm}(\zeta\vec{r}) \phi(r) d^3\vec{r}$$

The right-hand integral in equation (A.45) is a Fourier transform of LOs, so that it can be rewritten as follows:

$$C_{k'l'm'}(\vec{p}) = v_{k'} \bar{\Psi}_{k'l'm'}(\vec{p})$$

With the aid of this relation, series (A.43) can be finally written as

$$e^{-i\vec{p}\vec{r}} = \sum_{k'l'm'} v_{k'} \Psi_{k'l'm'}(\zeta, \vec{p}) \Psi^*_{k'l'm'}(\zeta\vec{r}) \phi(r) \tag{A.46}$$

It is well known from the literature how to expand the exponent into a series in terms of four-dimensional harmonics[8]:

$$e^{-i\vec{p}\vec{r}} = 4(\pi\zeta)^{1/2} \sum_{NLM} u^*_{NLM}(\zeta\vec{r})(\zeta^2 + p^2)^{-1} \Psi_{NLM}(\Omega_4) \tag{A.47}$$

where the $u^*_{NLM}(\zeta\vec{r})$ HTOs are orthogonal with weight $r^{-1}, \Omega_4 = (\alpha, \theta, \varphi)$.

The following series written in terms of LOs is a generalization of expansion (A.47):

$$e^{-i\vec{p}\vec{r}} = \sum_{KLM} \bar{v}_K \Psi_{KLM}(\zeta, \vec{p}) \Psi^*_{KLM}(\zeta\vec{r}) \bar{\phi}(p) \tag{A.48}$$

where $\bar{\phi}(p)$ is a weight function equal to $(\zeta^2 + p^2)$ in the case of a LO having eigenvalues Z_1.

Substitution of expansion (A.48) into integral (A.43) yields

$$\Psi_{klm}(\vec{r} - \vec{R}) = (2\pi)^{-3/2} \sum_{KLM} \bar{v}_K \Psi^*_{KLM}(\zeta\vec{r}) \int e^{-ipR} \bar{\Psi}_{KLM}(\zeta, \vec{p}) \bar{\Psi}_{klm}(\vec{p}) \bar{\phi}(p) d^3\vec{p} \tag{A.49}$$

When parameter ζ is the same for the functions $\bar{\Psi}_{KLM}$ and $\bar{\Psi}_{klm}$, the integral in the sum in equation (A.49) can be represented in the form of a linear combination of LOs. In the case of HTOs, series (A.49) becomes much more compact:

$$u_{\zeta,nlm}(\vec{r} - \vec{R}) = \sum_{NLM} u_{\zeta,NLM}(\vec{r}) S^{NLM}_{nlm}(-\zeta\vec{R}) \tag{A.50}$$

where the function S is defined by equation (A.49), being a linear combination of HTOs.

It can be seen from equation (A.49) that, to calculate the integrals over momentum, we must resort to a theorem of expansion of a product of LOs within the momentum space.

A.3.3 Theorem of 'linearization' of a product of LOs within the momentum space and operator relationships for HTOs

Consider a product of two LOs within p-space:

$$\bar{\Psi}_{k'l'm'}(\zeta', \vec{p}) \bar{\Psi}_{klm}(\zeta, \vec{p}) \tag{A.51}$$

Of particular interest is a product of HTOs which are proportional to the four-dimensional spherical harmonics. In this case, the theorem of 'linearization' has a particularly simple form, in which the terms of expansion are found to be represented by generalized Gaunt integrals of four-dimensional spherical functions:

$$\Psi_{k'l'm'}(\zeta, \vec{p}) \Psi_{klm}(\zeta, \vec{p}) = \sum_{KLM} \Psi_{KLM}(\zeta, \vec{p}) G^{KLM}_{k'l'm';klm} \tag{A.52}$$

where

$$G^{KLM}_{k'l'm';klm} = (2\pi^2)^{-1} \int \Psi^*_{KLM}(\zeta, \vec{p}) \Psi_{k'l'm'}(\zeta, \vec{p}) \times \Psi_{klm}(\zeta, \vec{p}) d\Omega_4 \quad \text{and} \quad d\Omega_4 = \sin^2\alpha \sin\theta \, d\alpha \, d\theta \, d\varphi \tag{A.53}$$

In reference 8 we can find an explicit form of the Gaunt integral (A.53) expressed through the Clebsch–Gordan coefficients and $9j$-symbols of a special kind:

$$G^{KLM}_{k'l'm';klm} = i^{l'+l-L} [k'kK(2l'+1)(2l+1)]^{1/2} C^{LM}_{l'm';lm} \begin{Bmatrix} \bar{k}' & \bar{k} & \bar{K} \\ \bar{k}' & \bar{k} & \bar{K} \\ l' & l & L \end{Bmatrix} \tag{A.54}$$

where $C^{LM}_{l'm';lm}$ equals CGC,

$$\begin{Bmatrix} \cdots \\ \cdots \end{Bmatrix}$$

is the 9j-symbol and $\bar{k} = (k-1)/2$.

It is obvious that products of LOs with integral indices k', k and the same Slater parameters ζ can be expanded into a final sum of HTOs with the same parameter ζ. In the case of non-integral values of k', k a product of LOs (equation (A.51)) can be expanded into an infinite series in terms of LOs, since in this case the LOs *per se* are no longer represented by rational functions. In the case of $\zeta \neq \zeta'$ and integral k', k, product (A.51) can be expanded into a sum of elementary fractions given in the form of $(\zeta'^2 + p^2)^{-\tau}$ and $(\zeta^2 + p^2)^{-\mu}$. However, these series will comprise fractions with $\tau = \mu = 1$, leading in several cases to diverging matrix elements. The expansion of function (A.51) into elementary fractions is most unsuccessful when the parameters ζ' and ζ are close in value.

Let us consider a HTO written in terms of a Gaussian hypergeometrical function:

$$w_{\zeta,nlm}(\vec{p}) = C_{nl} p^l \bar{Y}_{lm}(\hat{p})(\zeta^2 + p^2)^{-n-1} F\left(-n+l+1, -n+\tfrac{1}{2}; l+\tfrac{3}{2}; -\frac{p^2}{\zeta^2}\right)$$

where

$$C_{nl} = 3^{3/2}\pi^{-1}\zeta^{5/2}(-1)^{n+l+1}2\pi^{1/2}[n(n-l-1)!/(n+l)!]^{1/2} \tag{A.55}$$

Equation (A.55) can be used for obtaining the operator representation of a HTO, and the Fourier transformation of this function enables us to write

$$u_{nlm}(\zeta\vec{r}) = C_{nl}\bar{\mathscr{Y}}_{lm}(i\nabla)F(-n+l+1; -n+\tfrac{1}{2}; l+\tfrac{3}{2}; \Delta/\zeta^2)(\zeta^2 - \Delta)^{-n-1}\delta(\vec{r}) \tag{A.56}$$

where $\delta(\vec{r})$ is the Dirac function.

It can be inferred from equation (A.56) that HTO $u_{nlm}(\zeta\vec{r})$ is a solution to an inhomogeneous differential equation of type

$$(\zeta^2 - \Delta)^{n+1} u_{nlm}(\zeta\vec{r}) = C_{nl}\bar{\mathscr{Y}}_{lm}(i\nabla)F(-n+l+1, -n+\tfrac{1}{2}; l+\tfrac{3}{2}; \Delta/\xi^2)\delta(\vec{r}) \tag{A.57}$$

where the right-hand side of the equation is a derivative of the Dirac δ function.

At the same time, integral

$$\int e^{i\vec{p}\vec{r}}(\zeta^2 + p^2)^{-n-1} d^3\vec{p} = 2^{-n+1}\pi^2(n!)^{-1}\zeta^{-2+1}\hat{k}_{n-1/2}(\zeta r) \tag{A.58}$$

is proportional to the scalar reduced Bessel function (RBF) $\hat{k}_{n-1/2}(z)$, which is defined through

$$\begin{aligned} \hat{k}_{-1/2}(z) &= z^{-1}e^{-z} \\ \hat{k}_{n-1/2}(z) &= z^{-1}e^{-z}\sum_{p=1}^{n}\frac{(2n-p-1)!}{(p-1)!(n-p)!}2^{p-n}z^p \end{aligned} \tag{A.59}$$

where n is a natural number.

It follows from relationship (A.58) that the RBF satisfies the following inhomogeneous differential equation:

$$(\zeta^2 - \Delta)^{n+1}\hat{k}_{n-1/2}(\zeta r) = \pi^{-1/2}2^{n-1/2}n!\zeta^{2n-1}\delta(\vec{r}) \tag{A.60}$$

i.e. function $\hat{k}_{n-1/2}(z)$ is the Green function for the operator $(\zeta^2 - \Delta)^{n+1}$. At $n = 0$, equation (A.60) is transformed into the Helmholtz equation. By applying the differential operator[9,10]

$$\bar{\mathscr{Y}}_{lm}(i\nabla)F(-n+l+1, -n+\tfrac{1}{2}; l+\tfrac{3}{2}; \Delta/\zeta^2)$$

to equation (A.60) we arrive at equation (A.57), thus obtaining the following operator representation of a HTO through the scalar RBF

$$u_{nlm}(\zeta\vec{r}) = \hat{\mathscr{O}}_{nlm}(\zeta, \nabla)\hat{k}_{n-1/2}(\zeta r), \tag{A.61}$$

where operator $\hat{\mathscr{O}}_{nlm}(\zeta, \nabla)$ has the form

$$\hat{\mathscr{O}}_{nlm}(\zeta, \nabla) = K_{nlm}\bar{\mathscr{Y}}_{lm}(i\nabla)F(-n+l+1, -n+\tfrac{1}{2}; l+\tfrac{3}{2}; \Delta/\xi^2) \tag{A.62}$$

and

$$K_{nlm} = \pi^{1/2}2^{-n-1/2}(n!)^{-1}\zeta^{-2n+1}C_{nl}$$

It can be inferred from equation (A.61) that the analytical properties of the HTOs are determined by those for the RBFs, since operator $\hat{\mathscr{O}}_{nlm}$ is a polynomial differential expression in terms of the components of the gradient operator $\nabla_x, \nabla_y, \nabla_z$.

The operator expression (A.61) yields the generalized Dirac function when due account is taken of equation (A.60):

$$u_{nlm}(\zeta r) = \hat{\mathscr{O}}_{nlm}(\zeta, \nabla)\gamma(\xi^2 - \nabla)^{-n-1}\delta(\vec{r}) \tag{A.63}$$

where

$$\gamma = \pi^{-1/2}2^{n+1/2}n!\zeta^{2n-1}$$

An inverse differential operator is an integral one, which can be formally represented in the form of an infinite series of Laplacian powers. Representation of HTOs in the form of equation (A.63) are used in quantum chemistry for computing multi-centre integrals.

A.3.4 Linearization of two-centre distributions over one-centre HTOs

Multi-centre molecular integrals include two-centre electronic distributions and can be represented as the products of HTOs without any loss in generality:

$$f(\vec{r}) = u_{\zeta',\alpha'}(\vec{r})u_{\zeta'',\alpha''}(\vec{r} - \vec{R}) \tag{A.64}$$

Now let us resort to equation (A.50) to expand an HTO of argument $\vec{r} - \vec{R}$ in terms of HTOs of argument $\vec{r}$. Further, let us linearize the product of HTOs to obtain a sum of HTOs. As a result, we will obtain an expansion of function (A.64):

$$f(\vec{r}) = \sum_{\alpha \equiv nlm} n\zeta^{-1}s_\alpha(\alpha', \alpha''; \zeta', \zeta''; -\vec{R})u_{\zeta,\alpha}(\vec{r}) \tag{A.65}$$

where $\zeta = \zeta' + \zeta''$ and

$$s_\alpha(\alpha', \alpha''; \zeta', \zeta''; -\vec{R}) = \sum_{\beta \equiv NLM} G^{lm}_{l'm';LM} B^{nl}_{n'l';NL}(\zeta', \zeta'') S^{\alpha''}_{\beta}(\zeta'', -\vec{R})$$

$$\alpha \equiv nlm, \quad \alpha' \equiv n'l'm', \quad \alpha'' \equiv n''l''m'' \tag{A.66}$$

Series (A.65) can be generalized to cover a multi-centre distribution which is a linear combination of two-centre contributions as follows:

$$f(\vec{r}) = u_{\zeta',\alpha'}(\vec{r}) \sum_{\alpha''_i} C_{\alpha''_i} u_{\zeta''_i,\alpha''_i}(\vec{r} - \vec{R}_i) = \sum_{\alpha} n \sum_{i} \zeta_i^{-1} u_{\zeta_i,\alpha}(\vec{r}) \sum_{\alpha''_i} C_{\alpha''_i} s_\alpha(\alpha', \alpha''_i; \zeta', \zeta''_i; -\vec{R}_i) \tag{A.67}$$

where

$$\zeta_i = \zeta' + \zeta''_i, \quad \alpha''_i \equiv n''_i l''_i m''_i, \quad \alpha = nlm$$

In equation (A.67) coefficients are grouped around HTOs having the same sets of quantum numbers *nlm*, thus enabling us to assess the relative role of the coefficients $s_\alpha(-\vec{R}_i)$ in approximating the multi-centre densities in the Hartree–Fock–Roothaan method.

Now let us derive another expansion of $f(\vec{r})$ in terms of HTOs. Given that

$$f(\vec{r}) = \sum_{\alpha \equiv nlm} f_\alpha(\alpha', \alpha''; \zeta', \zeta'', \zeta; -\vec{R}) u_{\zeta,\alpha}(\vec{r}), \tag{A.68}$$

where parameter $\zeta > 0$ and will be defined below. The coefficient f_α will be found from the condition of orthogonality of HTOs:

$$f_\alpha(\alpha', \alpha''; \zeta', \zeta'', \zeta; -\vec{R}) = \int u_{\zeta',\alpha'}(\vec{r}) u_{\zeta'',\alpha''}(\vec{r} - \vec{R}) u^*_{\zeta,\alpha}(\vec{r}) r^{-1} \mathrm{d}^3\vec{r}$$

$$\times \left[\int u^*_{\zeta,\alpha}(\vec{r}) u_{\zeta,\alpha}(\vec{r}) r^{-1} \mathrm{d}^3\vec{r} \right]^{-1} \tag{A.69}$$

To compute the integrals in equation (A.69) we proceed as follows. For the second integral we obtain

$$\int u^*_{\zeta,nlm}(\vec{r}) u_{\zeta,nlm}(\vec{r}) r^{-1} \mathrm{d}^3\vec{r} = \zeta n^{-1} \tag{A.70}$$

The first integral on the right-hand side of equation (A.69) yields a linear combination of one-electron resonance integrals after having expanded a one-centre product of HTOs according to equation (A.42). Thus we obtain

$$f_\alpha(\alpha', \alpha''; \zeta', \zeta'', \zeta; -\vec{R}) = (-1)^m n \zeta^{-1} (\zeta + \zeta')^{-1} \sum_{\beta \equiv NLM} B^{NL}_{nl;n'l'}(\zeta, \zeta') G^{LM}_{l-m;l'm'} N$$

$$\times \int u_{\zeta+\zeta',\beta}(\vec{r}) u_{\zeta'',\alpha''}(\vec{r} - \vec{R}) r^{-1} \mathrm{d}^3\vec{r} \tag{A.71}$$

The integral on the right-hand side of equation (A.71) is a resonance one-electron integral which can be reduced to the HTO overlap integral.

In the case where $\zeta'' > \zeta'$, assuming $\zeta = \zeta'' - \zeta'$, we can express the integral under consideration through the function $S_{n''l''m''}^{NL-M}(\zeta'', \vec{R})$. Indeed, by applying the Fourier transformation to convert HTOs to the p-representation we arrive at the following integral:

$$J = \int u_{\zeta'',\beta}(\vec{r})u_{\zeta'',\alpha''}(\vec{r} - \vec{R})r^{-1}\mathrm{d}^3\vec{r}$$

$$= (2\zeta''N)^{-1}\int \mathrm{e}^{i\vec{p}\vec{R}}(p^2 + \zeta''^2)w_{\zeta'',\alpha''}(\vec{p})w_{\zeta'',\beta}(-\vec{p})\mathrm{d}^3\vec{p} \qquad \text{(A.72)}$$

where, by replacing the variable $p = \zeta'' \tan(\alpha/2)$, we can convert the above integral to an integral over the surface of a four-dimensional sphere, the latter being proportional to the function S: i.e.

$$J = (-1)^M\zeta''N^{-1}S_{\alpha''}^{NL-M}(\zeta'', \vec{R}) \qquad \text{(A.73)}$$

The final expression for coefficient (A.71) is as follows:

$$f_\alpha(\alpha', \alpha''; \zeta', \zeta'', \zeta; -\vec{R})$$

$$= n\zeta^{-1}\sum_{NLM}(-1)^{M+m}B_{nl;n'l'}^{NL}(\zeta, \zeta')G_{l-m;l'm'}^{LM}S_{\alpha''}^{NL-M}(\zeta', \vec{R}) \qquad \text{(A.74)}$$

where $\zeta = \zeta'' - \zeta'$.

Thus series (A.68) provides for the expansion $f(r)$ in terms of more diffuse HTOs, depending upon the parameter $\zeta = \zeta'' - \zeta'$ as distinct from expansion (A.67), in which the HTO parameter is $\zeta = \zeta' + \zeta''$.

A.35 Differential theorems for HTOs

Special functions in quantum chemistry, such as atomic orbitals and matrix elements of the latter, are often represented in the form of a polynomial series such as the classical orthogonal Legendre, Gegenbauer, Laguerre and Hermite polynomials and the reduced Bessel functions of half-integral index. It is well known that these functions have differential recurrence relations (RR), which can be considered as definitions of the operators for an increase or decrease in the polynomial indices. Because of this, it seems advisable to find a similar RR for basis functions dependent upon three variables.

Atomic orbitals used in quantum chemistry are the product of a solid harmonic by radial polynomials. To obtain a decreasing differential operator for a HTO, use can be made of the Wigner–Biedenharn theorem for four-dimensional harmonics[8,11] for the specific values of the indices:

$$\Psi_{2\tau\sigma}(\Omega_4)\Psi_{nlm}(\Omega_4) = \sum_{\substack{\nu=n\pm1\\ \lambda=l\pm\tau}} C(2\tau\sigma, nlm; \nu\lambda m + \sigma)\Psi_{\nu\lambda m+\sigma}(\Omega_4) \qquad \text{(A.75)}$$

where $\tau = 0, 1$ and $\sigma = 0, \pm 1$.

By introducing the replacement of the Fock variable $p = \zeta \tan(\alpha/2)$ and multiplying equation (A.75) by $a(\zeta^2 + p^2)^{-2}$, where $a = 2^{3/2}\pi^{-1}\zeta^{5/2}$, we obtain the relations for the HTOs in the p-representation.

Now let us take into consideration that

$$\Psi_{200} = -2\frac{\zeta^2 - p^2}{\zeta^2 + p^2}, \quad \Psi_{21\sigma} = -2i\zeta\vec{p}_\sigma/(\zeta^2 + p^2) \tag{A.76}$$

where $\vec{p}_\sigma$ is a covariant cyclic component of vector $\vec{p}$ and $\sigma = -1, 0, 1$[13].

Let us multiply both parts of equation (A.75) by $(\zeta^2 + p^2)$ and make the Fourier transformation of these equations. This will result in the following differential operators for the HTOs in the r-representation:

$$2(\zeta^2 + \Delta)u_{nlm}(\zeta\vec{r}) = (\Delta - \zeta^2) \sum_{\nu = n\pm 1, \nu > l} C(200, nlm; \nu lm)u_{\nu lm}(\zeta\vec{r}) \tag{A.77}$$

$$2\zeta\nabla_\sigma u_{nlm}(\zeta\vec{r}) = (\zeta^2 - \Delta) \sum_{\substack{\nu = n\pm 1 \\ \lambda = l\pm 1, \nu > \lambda}} C(21\sigma, nlm; \nu\lambda m + \sigma)u_{\nu\lambda m+\sigma}(\zeta\vec{r}) \tag{A.78}$$

where ∇_σ is the cyclic gradient operator:

$$\nabla_{\mp 1} = \pm 2^{-1/2}(\nabla_x \mp i\nabla_y), \quad \nabla_0 = \nabla_z \tag{A.79}$$

Hence the RRs (A.77) and (A.78) enable us to decrease (or increase) the HTO indices *nlm*, without disturbing the structure of the basis functions. Recurrence relation (A.78) enables us to reduce the computation of the matrix elements of the gradient operator to the matrix elements of the kinematic energy and to overlap integrals. At the same time, the kinetic energy Laplacian integrals can be reduced to the kinetic energy integrals of the simplest AOs and to overlap integrals. The last section of this appendix deals with recurrence relations for the Wigner–Biedenharm, Clebsch–Gordan and generalized Gaunt coefficients.

A.36 Expansion of atomic orbitals over sphere R^4

When computing quantum-chemical integrals within the momenta space the problem of factorizing the fraction $[p_0^2 + (\vec{p} - \vec{p}\,')^2]^{-k}$ into a series in terms of the functions of momenta $\vec{p}$ and $\vec{p}\,'$ occurs. Here we will give wholly converging series for expanding the above fractions.

Let us consider the fraction

$$f = [p_0^2 + (\vec{p} - \vec{p}\,')^2]^{-k} \tag{A.80}$$

and express the difference $(\Delta p)^2$ through the angles on the sphere R^4:

$$(\vec{p} - \vec{p}\,')^2 = 4\sin^2\frac{\omega}{2}(1 + \cos\alpha)^{-1}(1 + \cos\alpha')^{-1} \tag{A.81}$$

where

$$\cos\omega = \cos\alpha\cos\alpha' + \sin\alpha\sin\alpha'\cos\gamma$$

$$\cos\gamma = \cos\theta\cos\theta' + \sin\theta\sin\theta'\cos(\varphi - \varphi')$$

$$\cos\alpha = \frac{p_0^2 - p^2}{p_0^2 + p^2}, \quad \sin\alpha = \frac{2p_0 p}{p_0^2 + p^2} \tag{A.82}$$

For function (A.80) we obtain

$$f = p_0^{-2k}\cos^{2k}\frac{\alpha}{2}\cos^{2k}\frac{\alpha'}{2}\left(1+\cos^2\frac{\alpha}{2}\cos^2\frac{\alpha'}{2}-\cos^2\frac{\omega}{2}\right)^k \tag{A.83}$$

The denominator of fraction (A.83) is a binomial $(1-x)^k$ for

$$x = \cos^2\frac{\omega}{2}-\cos^2\frac{\alpha}{2}\cos^2\frac{\alpha'}{2}$$

where $|x| \leqslant 1$. Fraction $(1-x)^{-k}$ can be expanded into either a power series or into a Gegenbauer polynomials of arguments x. The series is wholly converging everywhere except perhaps for the point $x = 1$. Because of the numerator of fraction (A.83), integration of the above series does not lead to divergent series and, moreover, the integrated series are wholly convergent. Now let us write these series for the function f:

$$f = p_0^{-2k}\cos^{2k}\frac{\alpha}{2}\cos^{2k}\frac{\alpha'}{2}\sum_{n=0}^{\infty}\binom{-k}{n}\times\left(\cos^2\frac{\omega}{2}-\cos^2\frac{\alpha}{2}\cos^2\frac{\alpha'}{2}\right)^n \tag{A.84}$$

$$f = p_0^{-2k}\cos^{2k}\frac{\alpha}{2}\cos^{2k}\frac{\alpha'}{2}\sum_{n=0}^{\infty}C_n^k\left(\cos^2\frac{\omega}{2}-\cos^2\frac{\alpha}{2}\cos^2\frac{\alpha'}{2}\right) \tag{A.85}$$

where $C_n^k(x)$ is the Gegenbauer polynomial

$$\binom{-k}{n} = k(k+1)\cdots(k+n-1)/n!$$

By truncating series (A.84) and (A.85) we arrive at the approximation of equation (A.83) through the functions on sphere R^4. In contrast with the theorem of summation of HTOs within the p-space, where, on average, the functional series in terms of HTOs are convergent, the above series gives a procedure for constructing wholly convergent series.

Let us now examine an HTO within the p-space, which is dependent upon a difference in momenta:

$$\begin{aligned} w_{\zeta,nlm}(\vec{p}-\vec{p}\,') = {} & C_{nl}\mathscr{Y}_{lm}(p-p')[\zeta^2+(\vec{p}-\vec{p}\,')^2]^{-n-1} \\ & + F(-n+l+1,\,-n+1/2;\,l+3/2;\,-(\vec{p}-\vec{p}\,')^2/\zeta^2) \end{aligned} \tag{A.86}$$

where C_{nl} is defined in equation (A.55).

The use of expansions (A.84) and (A.85) enables us to factorize function (A.86) in terms of the variables $\vec{p}$ and $\vec{p}\,'$. Bearing in mind that the functions on sphere R^4 $\cos^2(\omega/2)$, $\cos^2(\alpha/2)$ and $\cos^2(\alpha'/2)$ can be expressed through the hyperspherical harmonics, equation (A.86) can be expanded into a wholly convergent series in terms of four-dimensional spherical harmonics, and hence, in terms of HTOs. Note that the factor $\mathscr{Y}_{lm}(\vec{p}-\vec{p}\,')F(\ldots;-(\vec{p}-\vec{p}\,')^2/\zeta^2)$ is dependent upon the Cartesian co-ordinates of momentum difference, and hence can be factorized in terms of $\vec{p}, \vec{p}\,'$ in the form of a finite sum of factorized summations[5].

These arguments complete the proof for the theorem of summation of the arguments of HTOs within the p-space, which includes a wholly convergent series of summations factorized in terms of $\vec{p}$ and $\vec{p}\,'$, in which the multipliers of the summations can be presented in the form of linear combinations of HTOs.

A.4 Computation of Matrix Elements in LCAO–MO Theory

We have studied above the properties of atomic orbitals used in making quantum computations of electronic distributions and spectral properties of atoms and molecules, and possessing a detailed knowledge of the analytical properties of AOs would enable us to compute matrix elements and AO integrals in a most compact form.

Let us consider an arbitrary two-electron integral in the HTO basis

$$(AB|CD) = \iint u_{\zeta_A,\alpha_A}(\vec{r}_1 - \vec{A})u_{\zeta_B,\alpha_B}(\vec{r}_1 - \vec{B})r_{12}^{-1} \times u_{\zeta_C,\alpha_C}(\vec{r}_2 - \vec{C})u_{\zeta_D,\alpha_D}(\vec{r}_2 - \vec{D})\mathrm{d}^3\vec{r}_1\,\mathrm{d}^3\vec{r}_2 \tag{A.87}$$

where $\vec{A}$, $\vec{B}$, $\vec{C}$, $\vec{D}$ are the co-ordinates of the centres of AO and $\alpha_A \equiv n_A l_A m_A$; α_B, α_C, α_D are similar.

By applying theorem (A.65) we can write an expansion of integral (A.87) into a double series in terms of two-centre Coulomb integrals:

$$(AB|CD) = \sum_{\alpha_1}\sum_{\alpha_2} n_1\zeta_1^{-1}n_2\zeta_2^{-1} \times s_{\alpha_1}(\alpha_A, \alpha_B; \zeta_A, \zeta_B; \vec{A} - \vec{B})s_{\alpha_2}(\alpha_C, \alpha_D; \zeta_C, \zeta_D; \vec{C} - \vec{D})Q_{\alpha_1,\alpha_2}(\zeta_1, \zeta_2; \vec{A}, \vec{C}) \tag{A.88}$$

where

$$Q_{\alpha_1,\alpha_2}(\zeta_1, \zeta_2; \vec{A}, \vec{C}) = \iint u_{\zeta_1,\alpha_1}(\vec{r}_1 - \vec{A})r_{12}^{-1}u_{\zeta_2,\alpha_2}(\vec{r}_2 - \vec{C})\mathrm{d}^3\vec{r}_1\,\mathrm{d}^3\vec{r}_2 \tag{A.89}$$

and

$$\zeta_1 = \zeta_A + \zeta_B, \quad \zeta_2 = \zeta_C + \zeta_D, \quad \alpha_1 \equiv n_1 l_1 m_1, \quad \alpha_2 \equiv n_2 l_2 m_2$$

Series (A.88) represents the most compact method for computing quantum-chemical integrals with the aid of expansion in terms of two-centre Coulomb integrals. Coefficients s_α and integrals Q_{α_1,α_2} are special functions of parameters which can be expressed through orthogonal polynomials and elementary functions. Recurrence relations are available for these special Q functions which enable us to carry out the computation of the simplest integrals with minimum (or maximum) values of the parameters. These RRs will be discussed below.

The two-centre Coulomb integral Q_{α_1,α_2} can be computed in a finite form as follows. Let us make use of the operator representation of HTOs (equation (A.61)):

$$u_{nlm}(\zeta\vec{r}) = \hat{\mathcal{O}}_{nlm}(\zeta\nabla)\hat{k}_{n-1/2}(\zeta r) \tag{A.90}$$

where $\widehat{\mathcal{O}}_{nlm}$ is given by formula (A.62). Integral (A.89) can now be written as

$$Q_{\alpha_1,\alpha_2}(\zeta_1,\zeta_2;\vec{A},\vec{C})=\widehat{\mathcal{O}}_{(nlm)_1}(-\zeta_1\nabla_1)\widehat{\mathcal{O}}_{(nlm)_2}(-\zeta_2\nabla_C)$$
$$\times\iint\widehat{K}_{n_1-1/2}(\zeta_1|\vec{r}_1-\vec{A}|)r_{12}^{-1}\widehat{k}_{n_2-1/2}(\zeta_2|\vec{r}_2-\vec{C}|)\mathrm{d}^3\vec{r}_1\mathrm{d}^3\vec{r}_2 \tag{A.91}$$

The scalar integral on the right-hand side of equation (A.91) can be computed with the aid of the Helmholtz equation. Note that the following operator representation is true for the integral as a result of relationship (A.60):

$$J=\iint\widehat{k}_{n_1-1/2}(\zeta_1|\vec{r}_1-\vec{A}|)r_{12}^{-1}\widehat{k}_{n_2-1/2}(\zeta_2|\vec{r}_2-\vec{C}|)\mathrm{d}^2\vec{r}_1\mathrm{d}^3\vec{r}_2$$
$$=\pi^{-1}2^{n_1+n_2+1}n_1!n_2!\zeta_1^{2n_1-1}\zeta_2^{2n_2-1}(\zeta_1^2-\Delta_A)^{-n_1-1}(\zeta_2^2-\Delta_C)^{-n_2-1}|\vec{A}-\vec{C}|^{-1} \tag{A.92}$$

Equation (A.92) provides an idea of the general analytical structure of the integral J. The action of the operators upon the function $|\vec{A}-\vec{C}|^{-1}$ can be effected by expanding the binomials into a series in terms of the Laplacians, and taking into account the Helmholtz and Poisson equations.

Now let us consider the following operator procedure for computing integral (A.92) and make use of the inhomogeneous differential equation (A.60):

$$(\zeta^2-\Delta)^{n+1}\widehat{k}_{n-1/2}(\zeta r)=\gamma_{n,\zeta}\delta(\vec{r}) \tag{A.93}$$

where $\gamma_{n,\zeta}=\pi^{-1/2}2^{n+1/2}n!\zeta^{2n-1}$. We apply operator $(\zeta^2-\Delta)^{-n}$ to both parts of equation (A.93), taking into account that

$$(\zeta^2-\Delta)^{-n}\delta(\vec{r})=\gamma_{n-1,\zeta}^{-1}\widehat{k}_{n-1-1/2}(\zeta r) \tag{A.94}$$

Using the procedure for decreasing the RBF indices, we arrive at the following relationship:

$$(\zeta^2-\Delta)\widehat{k}_{n-1/2}(\zeta r)=\gamma_{n,\zeta}(\zeta^2-\Delta)^{-n}\delta(\vec{r})=\gamma_{n-1,\zeta}^{-1}\gamma_{n,\zeta}\widehat{k}_{n-1/2}(\zeta r)\quad n>0$$

from which it can be found that

$$\widehat{k}_{n-1/2}(\zeta r)=\zeta^{-2}[2n\zeta^2\widehat{k}_{n-1-1/2}(\zeta r)+\Delta\widehat{k}_{n-1/2}(\zeta r)] \tag{A.95}$$

Substituting $n-1$ for n in equation (A.95), we obtain

$$\widehat{k}_{n-1-1/2}(\zeta r)=\zeta^{-2}[2(n-1)\zeta^2\widehat{k}_{n-2-1/2}(\zeta r)+\Delta\widehat{k}_{n-1-1/2}(\zeta r)] \tag{A.96}$$

We now substitute equation (A.96) into equation (A.95), making a further recurrence replacement of function $\widehat{k}_{n-2-1/2}(\zeta r)$ by the expression of equation (A.95). Having made n, steps we arrive at the following representation of function $\widehat{k}_{n-1/2}(\zeta r)$:

$$\widehat{k}_{n-1/2}(\zeta r)=\zeta^{-2}\sum_{k=0}^{n}2^k k!C_n^k\Delta\widehat{k}_{n-k-1/2}(\zeta r)+4\pi\zeta^{-3}\delta(\vec{r}) \tag{A.97}$$

The initial RBF is expressed in equation (A.97) through a linear combination of the Laplacians of the RBF and the Dirac δ function, which appears due to equation (A.93) for $n=0$.

If we now substitute relation (A.97) into integral (A.92) we obtain

$$J=\zeta_1^{-2}\sum_{k=0}^{n_1} 2^k k! C_{n_1}^k \iint \Delta\hat{k}_{n_1-k-1/2}(\zeta_1|\vec{r}_1-\vec{A}|)r_{12}^{-1} \times \hat{k}_{n_2-1/2}(\zeta_2|\vec{r}_2-\vec{C}|)\mathrm{d}^3\vec{r}_1\mathrm{d}^3\vec{r}_2 + 4\pi\zeta_1^{-3}\int r_{A2}^{-1}\hat{k}_{n_2-1/2}(\zeta_2|\vec{r}_2-\vec{C}|)\mathrm{d}^3\vec{r}_2 \tag{A.98}$$

where the integral with the δ function has been taken over r_1. To calculate the double integral in equation (A.98) let us transfer the Laplacian from the RBF to the function r_{12}^{-1}. Using to the Poisson equation

$$\Delta_R R^{-1} = -4\pi\delta(\vec{R}) \tag{A.99}$$

we replace the Coulomb-potential Laplacian in the double integral (A.98) with the Dirac δ function, taking the integral over $\vec{r}_1$. We thus obtain

$$J=4\pi\zeta_1^{-2}\sum_{k=1}^{n_1} 2^k K! C_{n_1}^k \int \hat{k}_{n_1-k-1/2}(\zeta_1|\vec{r}_2-\vec{A}|) \times \hat{k}_{n_2-1/2}(\zeta_2|\vec{r}_2-\vec{C}|)\mathrm{d}^3 r_2 + 4\pi\zeta_1^{-3}\int r_{A2}^{-1}\hat{k}_{n_2-1/2}(\zeta_2|\vec{r}_2-\vec{C}|)\mathrm{d}^3\vec{r}_2 \tag{A.100}$$

The integral J is represented in the form of a sum of RBF overlap integrals and the RBF Coulomb integrals. The last integral on the right-hand side of equation (A.100) can be computed by a similar procedure of replacing the RBF with equation (A.97). The result will be

$$\int r_{A2}^{-1}\hat{k}_{n_2-1/2}(\zeta_2|\vec{r}_2-\vec{C}|)\mathrm{d}^3\vec{r}_2 = \zeta_2^{-2}\sum_{k=0}^{n_2} 2^k k! C_{n_2}^k \int r_{A2}^{-1}\Delta\hat{k}_{n_2-k-1/2}(\zeta_2|\vec{r}_2-\vec{C}|)\mathrm{d}^3\vec{r}_2 + 4\pi\zeta_2^{-3}r_{AC}^{-1} \tag{A.101}$$

Integrals on the right-hand side of equation (A.101) are calculated by parts so the Laplacian acts onto function r_{A2}^{-1}, and by applying the Poisson equation (A.99). As a result, integral (A.100) can be written in the form of a sum of contributions of two types: exponential decrease with distance AC (short-range interaction) and the Coulomb decay (long-range interaction):

$$J=4\pi\zeta_1^{-2}\sum_{k=0}^{n_1} 2^k k! C_{n_1}^k \int \hat{k}_{n_1-k-1/2}(\zeta_1|\vec{r}_2-\vec{A}|) \times \hat{k}_{n_2-1/2}(\zeta_2|\vec{r}_2-\vec{C}|)\mathrm{d}^3 r_2 - (4\pi)^2\zeta_1^{-3}\zeta_1^{-3} \times \sum_{k=0}^{n_2} 2^k k! C_{n_2}^k \hat{k}_{n_2-k-1/2}(\zeta_2|\vec{A}-\vec{C}|) + (4\pi)^2(\zeta_1\zeta_2)^{-3}r_{AC}^{-1} \tag{A.102}$$

An expression for the integral J, which would be more symmetrical with respect to parameters ζ_1, ζ_2, can be obtained by substituting series (A.97) for RBFs in the integral (A.92). Having done this, further transformations will give a formula symmetrical with respect to ζ_1, ζ_2:

$$J = (\zeta_1\zeta_2)^{-2}\Bigg[\sum_{k'=0}^{n_1}\sum_{k''=0}^{n_2} 2^{k'+k''}k'!k''!C_{n_2}^{k'}C_{n_2}^{k''}(-4\pi)$$

$$\times \int \hat{k}_{n_1-k'-1/2}(\zeta_1|\vec{r}_2-\vec{A}|)\Delta_2\hat{k}_{n_2-k''-1/2}(\zeta_2|\vec{r}_2-\vec{C}|)\mathrm{d}^3\vec{r}_2$$

$$-(4\pi)^2\zeta_1^{-1}\sum_{k''=0}^{n_2} 2^{k''}k''!C_{n_2}^{k''}\hat{k}_{n_2-k''-1/2}(\zeta_2|\vec{A}-\vec{C}|) \tag{A.103}$$

$$\times (4\pi)^2\zeta_2^{-1}\sum_{k'=0}^{n_2} 2^{k'}k'!C_{n_1}^{k'}\hat{k}_{n_1-k'-1/2}(\zeta_1|\vec{A}-\vec{C}|)$$

$$+(4\pi)^2(\zeta_1\zeta_2)^{-1}|\vec{A}-\vec{C}|^{-1}\Bigg]$$

where the integrals in the double sum are in fact the kinetic energy integrals for RBF AOs.

It is worth noting here that integral J is represented in the form of a sum of short- and long-range terms, and for $AC \gg 1$ there appears the possibility of making simple rough estimates of these integrals through the Coulomb potential of point charges spaced apart at distance AC from each other.

Equation (A.103) does not include the steps of differentiation with respect to parameters ζ_1 and ζ_2. The RBF overlap integrals have been computed in reference 4, and the respective formulae for these are given here without derivation:

$$2^{-n_1-n_2}(n_1!n_2!)^{-1}\int \mathrm{d}^3r\hat{k}_{n_1-1/2}(\alpha|\vec{r}-\vec{A}|)\hat{k}_{n_2-1/2}(\beta|\vec{r}-\vec{C}|)$$

$$=\sum_{t=0}^{n_1} a_{n_2n_1t}(\beta,\alpha)\hat{k}_{t-1/2}(\alpha|\vec{A}-\vec{C}|)2^{-t}(t!)^{-1}$$

$$+\sum_{t=0}^{n_2} a_{n_1n_2t}(\alpha,\beta)\hat{k}_{t-1/2}(\beta|\vec{A}-\vec{C}|)2^{-t}(t!)^{-1} \tag{A.104}$$

$$a_{n_1n_2t}(\alpha,\beta) = (-1)^{n_2+t}4\pi\alpha^{-3}[1-(\beta/\alpha)^2]^{-n_n-1}\times P_{n_2-t}^{(-n_2+t,n_1)}\left(\frac{(\alpha/\beta)^2+1}{(\alpha/\beta)^2-1}\right) \tag{A.105}$$

where $P_N^{(\mu,\nu)}(x)$ is the Jacobian polynomial. For $\alpha=\beta$ we obtain

$$2^{-n_1-n_2}(n_1!n_2!)^{-1}\int \mathrm{d}^3r\hat{k}_{n_1-1/2}(\alpha|\vec{r}-\vec{A}|)\hat{k}_{n_2-1/2}(\alpha|\vec{r}-\vec{C}|)$$

$$=2\pi^{1/2}\alpha^{-3}\hat{k}_{n_1+n_2+1/2}(\alpha|\vec{A}-\vec{C}|) \tag{A.106}$$

Now let us consider the matrix element (A.91). To compute this we determine the action of the operators $\hat{\mathcal{O}}_{nlm}(-\zeta\nabla)$ onto the RBF. Since $\vec{A} \neq \vec{B}$, then, in compliance with the Helmholtz equation,

$$\Delta_A^q \hat{k}_{-1/2}(\zeta|\vec{A}-\vec{C}|) = \zeta^{2q}\hat{k}_{-1/2}(\zeta|\vec{A}-\vec{C}|) \qquad \text{(A.107)}$$

In addition,

$$\bar{\mathscr{Y}}_{lm}(i\nabla_A)\hat{k}_{-1/2}(\zeta|\vec{A}-\vec{C}|) = \zeta^{-1}R_{AC}^{-l-1}Y_{lm}(\vec{R}_{AC})\hat{k}_{l+1/2}(\zeta R_{AC}) \qquad \text{(A.108)}$$

where

$$\bar{\mathscr{Y}}_{lm}(i\nabla_A)R_{AC}^{-1} = R_{AC}^{-l-1}Y_{lm}(\hat{R}_{AC})\hat{k}_{l+1/2}(0) \qquad \text{(A.109)}$$

Then, on taking into account that

$$\nabla_A\hat{k}_{-1/2}(\zeta R_{AC}) = -\nabla_C\hat{k}_{-1/2}(\zeta R_{AC}) \qquad \text{(A.110)}$$

as well as the theorem of expansion for the spherical functions, we obtain the necessary relationship

$$\begin{aligned}&\bar{\mathscr{Y}}_{lm}(-\nabla_A)\bar{\mathscr{A}}_{l'm'}(-i\nabla_C)\hat{k}_{-1/2}(\zeta R_{AC})\\ &\quad = (-1)^l\sum_{\lambda\mu}G_{lm;l'm'}^{\lambda\mu}\zeta^{l+l'-\lambda-1}R_C^{-\lambda-1}Y_{\lambda\mu}(\hat{R}_{AC})\hat{k}_{\lambda+1/2}(\zeta R_{AC})\end{aligned} \qquad \text{(A.111)}$$

where $G_{lm:l'm'}^{\lambda\mu}$ are Gaunt integrals.

A similar relationship for function R_{AC}^{-1} can be derived by multiplying equation (A.111) by ζ, to be followed by taking $\zeta = 0$.

Thus, the action of operators $\hat{\mathcal{O}}_{(nlm)_1}(-\zeta_1\nabla_A)$ and $\hat{\mathcal{O}}_{(nlm)_2}(-\zeta_2\nabla_C)$ on the function $\hat{k}_{-1/2}(\zeta|A-C|)$ can be reduced to the following expression:

$$\begin{aligned}&\hat{\mathcal{O}}_{(nlm)_1}(-\zeta_1\nabla_A)\hat{\mathcal{O}}_{(nlm)_2}(-\zeta_2\nabla_C)\hat{k}_{-1/2}(\zeta|A-C|)\\ &\quad = K_{(nlm)_1}K_{(nlm)_2}F_{(nl)_1}(\zeta^2/\zeta_1^2)F_{(nl)_2}(\zeta^2/\zeta_2^2)\\ &\quad\quad \times(-1)^l\sum_{\lambda\mu}G_{(lm)_1;(lm)_2}^{\lambda\mu}\zeta^{l_1+l_2-\lambda-1}R^{-\lambda-1}Y_{\lambda\mu}(\hat{R}_{AC})\hat{k}_{\lambda+1/2}(\zeta R_{AC})\end{aligned} \qquad \text{(A.112)}$$

where

$$F_{nl}(z) = F(-n+l+1, -n+\tfrac{1}{2}; l+\tfrac{3}{2}; z)$$

A similar formula can be obtained for function R_{AC}^{-1} as follows:

$$\begin{aligned}\hat{\mathcal{O}}_{(nlm)_1}(-\zeta_1\nabla_A)\hat{\mathcal{O}}_{(nlm)_2}(-\zeta_2\nabla_C)R_{AC}^{-1} &= (-1)^l K_{(nlm)_1}K_{(nlm)_2}G_{l_1m_1;l_2m_2}^{l_1l_2m_1m_2}R^{-l_1-l_2-1}\\ &\quad\times Y_{l_1+l_2m_1m_2}(\hat{R}_{AC})\hat{k}_{l_1+l_2+1/2}(0)\end{aligned} \qquad \text{(A.113)}$$

in which the sum over λ, μ is limited to one summation

$$\lambda = l_1 + l_2, \quad \mu = m_1 + m_2$$

Within the integral (A.91) the operators $\hat{\mathcal{O}}$ act upon the integral J (equation (A.103)), where, in addition to the RBF $\hat{k}_{-1/2}$, functions $\hat{k}_{p-1/2}$ are also included and p is an integer. The function $\hat{k}_{p-1/2}$ can be represented by the Rodrigues formula[4] through the weight function $\hat{k}_{1/2}$:

$$\hat{k}_{p-1/2}(z) = (-1)^p z^{2p-1}D_z^p\zeta e^{-z} \qquad \text{(A.114)}$$

Progressing from equation (A.114) we can find a differentiation operator with respect to ζ, which converts the RBF $\hat{k}_{-1/2}(\zeta r)$ into RBF $\hat{k}_{p-1/2}(\zeta r)$:

$$\hat{k}_{p-1/2}(\zeta r)=(-1)^p\zeta^{2p-1}D_\zeta^p\zeta\hat{k}_{-1/2}(\zeta r) \tag{A.115}$$

where $D_\zeta=\zeta^{-1}\mathrm{d}/\mathrm{d}\zeta$.

Substituting relationship (A.115) into equation (A.103), we can express the integral J by the operator technique via RBF $\hat{k}_{-1/2}$, and thus finding the action of operators $\hat{\mathcal{O}}$ upon the integral J, so that all that remains to be done is to carry out differentiation with respect to the Slater parameters ζ_1 and ζ_2.

The Coulomb contribution $|\vec{A}-\vec{C}|^{-1}$ in equation (A.103) can be obtained directly from equation (A.113), which represents a multipole field of order l_1+l_2 at a distance R_{AC} from the origin of the co-ordinates. The multipole's contribution becomes predominant in the Coulomb integral (A.89) as $R_{AC}\to\infty$, determining the long-range interactions within the molecular systems.

The remainder of the contributions within the Coulomb integral (A.89) are substantially of a short-range type, as they decrease in an exponential-multipole way. These contributions define the overlap effects for repulsive electronic clouds. This is true for any AOs of the exponential type, the dependence of which on the radius can be expressed as a product of a polynomial and an exponent.

Computation of one-electron energy integrals can be made in a way similar to that for two-electron ones, the difference being that there is no need to integrate over co-ordinate r_2.

Let us consider an arbitrary three-centre one-electron integral of the Coulomb interaction

$$(AB|r_C^{-1})=\int u_{\zeta_A\alpha_A}(\vec{r}-\vec{A})u_{\zeta_B\alpha_B}(\vec{r}-\vec{B})|\vec{r}-\vec{C}|^{-1}\mathrm{d}^3\vec{r} \tag{A.116}$$

With the aid of theorem (A.65) a product of HTOs can be expanded into a series in terms of HTOs, writing down the expansion of integral (A.116) in terms of the two-centre Coulomb one-electron integrals:

$$(AB|r_C^{-1})=\sum_{\alpha\equiv nlm} n\zeta^{-1}s_\alpha(\alpha_A,\alpha_B;\zeta_A,\zeta_B;\vec{A}-\vec{B})\times Q_\alpha(\zeta;\vec{A},\vec{C}) \tag{A.117}$$

where

$$Q_\alpha(\zeta;\vec{A},\vec{C})=\int u_{\zeta,\alpha}(\vec{r}-\vec{A})|\vec{r}-\vec{C}|^{-1}\mathrm{d}^3\vec{r} \tag{A.118}$$

and

$$\zeta=\zeta_A+\zeta_B$$

Using the operator representation of HTO (equation (A.90)), the integral Q_α can be rewritten in the form

$$Q_\alpha(\zeta;\vec{A},\vec{C})=\hat{\mathcal{O}}_\alpha(-\zeta\nabla_A)\int\hat{k}_{n-1/2}(\zeta|\vec{r}-\vec{A}|)|\vec{r}-\vec{C}|^{-1}\mathrm{d}^3\vec{r} \tag{A.119}$$

Expressing the RBF through the Dirac δ function according to equations

(A.93) and (A.97) and substituting equation (A.97) into integral (A.119) gives

$$Q_\alpha(\zeta; \vec{A}, \vec{C}) = \hat{\mathcal{O}}_\alpha(-\zeta\nabla_A)\Bigg[\zeta^{-2}\sum_{k=0}^{n} 2^k k! C_n^k \int \Delta\hat{k}_{n-k-1/2}(\zeta|\vec{r}-\vec{A}|)|\vec{r}-\vec{C}|^{-1}\mathrm{d}^3\vec{r}$$

$$+4\pi\zeta^{-3}\int \delta(\vec{r}-\vec{A})|\vec{r}-\vec{C}|^{-1}\mathrm{d}^3\vec{r}\Bigg] \tag{A.120}$$

Integrals with the Laplacian are calculated by parts when the derivatives act onto the Coulomb potential followed by further integration with the δ function. Thus we obtain for integral (A.120)

$$Q_\alpha(\zeta; \vec{A}, \vec{C}) = \hat{\mathcal{O}}_\alpha(-\zeta\nabla_A)\Bigg[\zeta^{-2}\sum_{k=0}^{n} 2^k k! C_n^k$$

$$\times(-4\pi)\hat{k}_{n-k-1/2}(\zeta|\vec{C}-\vec{A}|) + 4\pi\zeta^{-3}|\vec{A}-\vec{C}|^{-1}\Bigg] \tag{A.121}$$

where $\zeta = \zeta_A + \zeta_B$.

By combining the operator representation (A.115) for the RBF with relations (A.107)–(A.110) we arrive at the following expression for the Coulomb integral Q_α:

$$Q_\alpha = \Bigg[\zeta^{-2}\sum_{k=0}^{n} 2^k k! C_n^k(-\pi)(-1)^{n-k}\zeta^{2(n-k)-1}$$

$$\times D_\zeta^{n-k} F_{nl}(\zeta^2/\bar{\zeta}^2)\hat{k}_{l+1/2}(\zeta R_{AC}) + 4\pi\zeta^{-3}\hat{k}_{l+1/2}(0)\Bigg] R_{AC}^{-l-1} Y_{lm}(\hat{R}_{AC}) \tag{A.122}$$

where designation F_{nl} is defined in equation (A.112).

Having differentiated with respect to ζ^2 in equation (A.122) we take $\bar{\zeta} = \zeta$ and substitute $F(a, b; c; 1) = \Gamma(c)\ \Gamma(c-a-b)\ \Gamma^{-1}(c-a)\ \Gamma^{-1}\ (c-b)$ for the Gaussian function.

It follows from equation (A.122) that the two-centre one-electron Coulomb integral is proportional to the field of a multipole, in which the proportionality factor includes, along with a constant contribution, the polynomial–exponential contributions in the form of the RBF, which provide for the short-range part of interaction between the electrons and nuclei of a molecule.

Representation of the electronic energy matrix elements in the form of a sum of long- and short-range summations enables us to separate effects associated with the rearrangement of the atomic clouds in forming chemical bonds, and purely electrostatic effects, which predominate at large distances between the nuclei.

The resonance Coulomb one-electron integrals can be expanded in accordance with equation (A.117) into a series in terms of the products of the overlap integrals and two-centre Coulomb one-electron integrals.

However, despite the fact that in a non-relativistic Hamiltonian of a molecule there are no other interactions except for the Coulomb ones, in LCAO–MO theory the energy integrals are conventionally divided into two major types, i.e Coulomb and resonance integrals. In connection with the computations made with reference to one- and two-electron integrals of electronic energy, note here that it would be more advisable to divide the integrals into types judging from their asymptotic behaviour as the interatomic distance increases. Having done that, it would be consistent to group with the Coulomb type those integrals which have a multipole asymptotic behaviour over the maximum of the distances between the centres, while all other integrals have an exponential-multipole asymptotic behaviour, so that their type is consistent with the interaction of overlapping electronic clouds, where the mean length of a cloud is comparable with or greater than the mean distance between the interacting ones.

The exponential-multipole dependence of such integrals upon a distance can be derived from a solution of the Helmholtz equation, i.e. RBF $\hat{k}_{-1/2}(\zeta r)$. Since this solution is also proportional to the Yukawa potential, the integrals of the

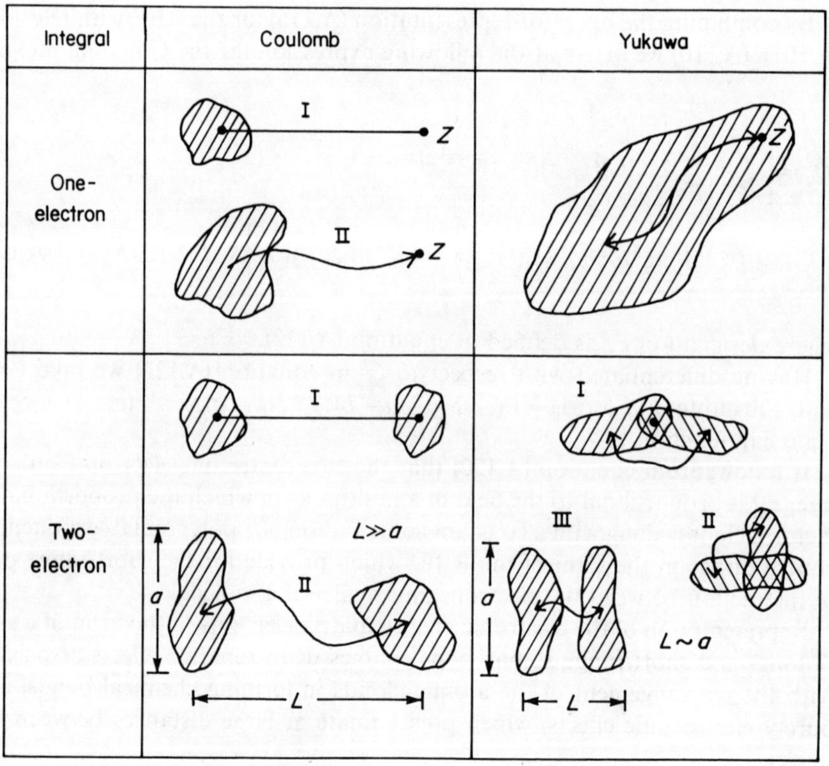

Figure A.1. Types of potential energy integrals occurring in LCAO–MO theory. Shown schematically are the electronic densities under the integral sign and their location in space; Z is a charge of a nucleus. The small circles represent one-centre electronic distributions

exponential-multipole kind can be referred to as Yukawa integrals, to distinguish these from Coulomb integrals. This is sensible, since the word 'resonance' is meaningless in this particular case. Figure A.1 shows the types of Coulomb and Yukawa integrals of one- and two-electron interactions.

Now let us consider the computation of overlap, kinetic-energy and dipole-moment integrals.

A.4.1 HTO overlap integral

This is given by

$$S_{AB} = \int u_{\zeta_A,\alpha_A}(\vec{r} - \vec{A}) u_{\zeta_B,\alpha_B}(\vec{r} - \vec{B}) \mathrm{d}^3\vec{r} \tag{A.123}$$

where $\alpha_A \equiv (nlm)_A$, $\alpha_B \equiv (nlm)_B$

The operator representation of HTO (A.90) enables us to write the overlap integral in the form of the RBF overlap integral:

$$S_{AB} = \hat{\mathcal{O}}_{(nlm)_A}(-\zeta_A \nabla_A)\hat{\mathcal{O}}_{(nlm)_B}(-\zeta_B \nabla_B) \int \hat{k}_{n_A-1/2}(\zeta_A|\vec{r} - \vec{A}|)\hat{k}_{n_B-1/2}(\zeta_B|\vec{r} - \vec{B}|)\mathrm{d}^3\vec{r} \tag{A.124}$$

This integral is given by equation (A.104), from which we obtain the following relation for the HTO overlap integral:

$$\begin{aligned} S_{AB} = {} & 2^{n_A+n_B} n_A! n_B! \hat{\mathcal{O}}_{\alpha_A}(-\zeta_A \nabla_A)\hat{\mathcal{O}}(-\zeta_B \nabla_B) \\ & \times \left[\sum_{t=0}^{n_A} \alpha_{n_B n_A t}(\zeta_B, \zeta_A)\hat{k}_{t-1/2}(\zeta_A R_{AB}) 2^{-t}(t!)^{-1} \right. \\ & \left. + \sum_{t=0}^{n_B} \alpha_{n_A n_B t}(\zeta_A, \zeta_B)\hat{k}_{t-1/2}(\zeta_B R_{AB}) 2^{-t}(t!)^{-1} \right] \end{aligned} \tag{A.125}$$

where coefficients α are given in equation (A.105).

The action of operators $\hat{\mathcal{O}}_\alpha$ upon the RBF was discussed above.

A.4.2 Kinetic energy integral

It is easy to alter equation (A.125) to obtain an expression for the kinetic energy integral:

$$T_{AB} = \int u_{\zeta_A,\alpha_A}(\vec{r} - \vec{A}) \Delta u_{\zeta_B,\alpha_B}(\vec{r} - \vec{B}) \mathrm{d}^3\vec{r} \tag{A.126}$$

Note that

$$\Delta r u_{\zeta_B,\alpha_B}(\vec{r} - \vec{B}) = \Delta_B u_{\zeta_B,\alpha_B}(\vec{r} - \vec{B})$$

so that the kinetic energy integral can be computed by acting with the operator Δ_B on the overlap integral (A.125), i.e.

$$T_{AB} = \Delta_B S_{AB} \tag{A.127}$$

The action of operator Δ_B on the RBF can be easily calculated with the aid of equation (A.107).

It follows from equation (A.125) that the overlap and kinetic energy integrals are dependent in a polynomial–exponential way upon the interatomic distance. They are also proportional, at large distances, to the multipole potential. This results in a rapid decay of these integrals with increasing distance between the centres of the AOs.

A.4.3 Dipole moment integral

This integral is closely associated with the overlap integral, since when the vector of the dipole moment is transformed to the centre of an AO, further expansion of the product of the dipole moment operator and an AO leads to a linear combinations of HTOs.

Without loss in generality, we can consider the following integral

$$\vec{d} = \int (\vec{r} - \vec{A}) u_{\zeta_A, \alpha_A}(\vec{r} - \vec{A}) u_{\zeta_B, \alpha_B}(\vec{r} - \vec{B}) \mathrm{d}^3\vec{r} \tag{A.128}$$

Note that the components of the dipole moment operator $\vec{r} - \vec{A}$ can be considered to be the polynomial part of a HTO $u_{\zeta_A, 21\sigma}(\vec{r} - \vec{A})$, where $\sigma = 0, \pm 1$, and in which the Cartesian components of the moment can be converted to the cyclic components of the vector in accordance with the relations

$$\hat{d}_{\mp 1} = \pm 2^{-1/2}(x \mp iy), \quad \hat{d}_0 = z \tag{A.129}$$

provided that, the components of operator $\hat{d}$ in equation (A.129) coincide with the solid harmonics $\mathscr{Y}_{1\sigma}(r)$.

Now introduce coefficients $\mathscr{E}^{NLM}_{21\sigma, \alpha_A}(\zeta_A)$ which linearize the product of the polynomial parts of HTOs, designated here as $\bar{u}_{\zeta, \alpha}$. Hence we obtain

$$\bar{u}_{\zeta_A, 21\sigma}(\vec{r} - \vec{A}) \bar{u}_{\zeta_A, \alpha_A}(\vec{r} - \vec{A}) = \sum_{NLM} \mathscr{E}^{NLM}_{21\sigma, \alpha_A}(\zeta_A) \bar{u}_{\zeta_A, NLM}(\vec{r} - \vec{A}) \tag{A.130}$$

and

$$\mathscr{E}^{NLM}_{21\sigma, \alpha_A}(\zeta_A) = G^{LM}_{1\sigma; l_A m_A} \bar{B}^{NL}_{21; n_A l_A}(\zeta_A) \tag{A.131}$$

where $G^{LM}_{1\sigma; l_A m_A}$ is the Gaunt coefficients.

Coefficient $\bar{B}^{NL}_{21; n_A l_A}(\zeta_A)$ is defined in equation (A.147), and the limits of summation over NLM are defined when there appear zeros of the Gaunt coefficient and B.

Thus the cyclic components of the dipole moment can be expressed in the form of a sum of the HTO overlap integrals

$$d_\sigma = \sum_{NLM} \mathscr{E}^{NLM}_{21\sigma, \alpha_A} \int u_{\zeta_A, NLM}(\vec{r} - \vec{A}) u_{\zeta_B, \alpha_B}(\vec{r} - \vec{B}) \mathrm{d}^3\vec{r} \tag{A.132}$$

where $\alpha_B \equiv (nlm)_B$, $\sigma = 0, \pm 1$.

Transition from components d_σ in equation (A.132) to the Cartesian components $\vec{d}$ (equation (A.128)) is effected by the inversion of transformation (A.129).

To conclude, let us consider the procedure for computing a two-centre form factor

$$F_{AB}(\vec{k})=\int e^{i\vec{k}\vec{r}}u_{\zeta_A,\alpha_A}(\vec{r}-\vec{A})u_{\zeta_B,\alpha_B}(\vec{r}-\vec{B})d^3\vec{r} \tag{A.133}$$

which specifies the interaction between an electromagnetic wave, characterized by wave vector k, and a two-centre electronic density in the form of a product of HTOs centred on atoms A and B. Expanding the form factor (A.133) into a series in terms of the small radius k, we arrive at a multipole expansion of the electromagnetic interaction of a light wave with the electrons of a molecule.

Resorting to equation (A.65), we can obtain the following expansion of a two-centre form factor into a series in terms of HTOs:

$$F_{AB}(\vec{k})=(2\pi)^{3/2}e^{i\vec{k}\vec{A}}\sum_{\alpha\equiv nlm} n\zeta^{-1}s_\alpha(\alpha_A,\alpha_B;\zeta_A,\zeta_B;\vec{A}\vec{B})w_{\zeta,\alpha}(-\vec{k}) \tag{A.134}$$

where $w_{\zeta,\alpha}(-\vec{k})$ is the momentum representation of a HTO; $\zeta=\zeta_A+\zeta_B$.

In the LCAO–MO method the molecular electronic density can be written in the form of a sum over the paired products of AOs, so that in computing the molecular form factor this will result in the convolution of the density matrix with the matrix of two- and one-centre form factors given by equation (A.133).

Since a HTO $w_{\zeta,\alpha}(-\vec{k})$ is a rational function of momentum $\vec{k}$, its expansion into a series in terms of small radius k presents no difficulties.

It is worth noting that the expansion of the form factor (A.134) in terms of multipoles has already been done, since a HTO is represented in the form of a product of a solid harmonic and a function of a radius. By collecting the HTOs with similar lm we obtain a multipole expansion of the energy of interaction of light with a molecule.

Expansion (A.134) converges, on average, no worse than the expansion of the density under the integral sign in equation (A.133) expressed in terms of the one-centre distributions.

We now present a method for expanding the form factor (A.133) into an absolutely converging series. This equation is equivalent to a convolution of a HTO within the p-space

$$F_{AB}(\vec{k})=e^{i\vec{k}\vec{A}}\int d^3\vec{p}\,\omega_{\zeta_B,\alpha_B}(\vec{p})w_{\zeta_A,\alpha_A}(-\vec{k}-\vec{p})e^{i\vec{p}(\vec{A}-\vec{B})} \tag{A.135}$$

Expanding the function $w_{\zeta_A,\alpha_A}(-\vec{k}-\vec{p})$ into a wholly converging series as described in Section A.3.6 gives an expansion of the form factor as follows:

$$\begin{aligned}F_{AB}(\vec{k})=e^{i\vec{k}\vec{A}}C_{n_Al_A}\zeta_A^{-2n-2}\cos^{2n_A+2}(\alpha_{\vec{k}}/2)\sum_{t=0}^{\infty}\int d^3\vec{p}\,w_{\zeta_B,\alpha_B}(\vec{p})\cos^{2n_A+2}(\alpha_{\vec{p}}/2)\\ \times C_t^{n_A}\left(\cos^2\frac{\omega}{2}-\cos^2\frac{\alpha_k}{2}\cos^2\frac{\alpha_p}{2}\right)\\ \times e^{i\vec{p}\vec{A}\vec{B}}\mathscr{Y}_{l_Am_A}(-\vec{p}-\vec{k})F_{n_Al_A}(-(\vec{p}+\vec{k})^2/\zeta_A^2)\end{aligned} \tag{A.136}$$

where $\cos\alpha_p = (\zeta_A^2 - p^2)(\zeta_A^2 + p^2)^{-1}$, $\cos\alpha_k = (\zeta_A^2 - k^2)(\zeta_A^2 + k^2)^{-1}$, $F_{n_A l_A}(z)$ is the Gauss hypergeometrical polynomial; $\mathscr{Y}_{l_A m_A}(-\vec{p} - \vec{k})$ is a solid harmonic; $\cos\omega = \cos\alpha_k \cos\alpha_p + \sin\alpha_k \sin\alpha_p \cos(-\vec{k}, -\vec{p})$ and $C_t^{n_A}(z)$ is the Gegenbauer polynomial.

The integral in equation (A.136) can be reduced to linear combinations of HTOs as a function of $\vec{k}$, since the variables $\vec{p}$ and $\vec{k}$ can be separated in the polynomial parts of the integrand.

The merit of series (A.136) lies in the fact that it can be used for the purpose of integration of two-electron integrals in the LCAO–MO method, thus providing a wholly converging series for the latter.

Indeed, if we represent the Coulomb potential r_{12}^{-1} in the form of the integral $(2\pi^2)^{-1}\int p^{-2}\exp(i\vec{p}\,\vec{r}_{12})\mathrm{d}^3\vec{p}$, and substitute the latter into equation (A.87), we obtain an integral of a product of two-centre form factors $F_{AB}(\vec{p})\cdot F_{CD}(-\vec{p})$.

This results in the following relation:

$$\iint u_{\zeta_A,\alpha_A}(\vec{r}_1 - \vec{A})u_{\zeta_B,\alpha_B}(\vec{r}_1 - \vec{B})r_{12}^{-1}u_{\zeta_c,\alpha_c}(\vec{r}_2 - \vec{C})u_{\zeta_D,\alpha_D}(\vec{r}_2 - \vec{D})\mathrm{d}^3\vec{r}_1\mathrm{d}^3\vec{r}_2$$

$$= (2\pi^2)^{-1}\int p^{-2}F_{AB}(\vec{p})F_{CD}(-\vec{p})\mathrm{d}^3\vec{p}. \qquad \text{(A.137)}$$

A.5 Recurrence Relations for Matrix Elements and Auxiliary Coefficients

Computation of matrix elements made in Section A.4 results in a series and sums over the integrals which, as will be shown below, are interrelated through recurrence relations. The use of these RRs enables us to facilitate summation of the series, while checking at the same time the computation of the matrix elements. The recurrence properties of the functions under the integral sign emphasize the RRs for matrix elements.

Now let us consider relation (A.50), which includes the functions $S_{nlm}^{n'l'm'}(\vec{R})$. These functions are determined by the following integral of the four-dimensional spherical harmonics:

$$S_{nlm}^{n'l'm'}(\vec{R}) = \iiint \Psi_{nlm}(\alpha,\theta,\varphi)\Psi_{n'l'm'}^{*}(\alpha,\theta,\varphi)\mathrm{e}^{i\vec{p}\vec{R}}\sin^2\alpha\sin\theta\,\mathrm{d}\alpha\,\mathrm{d}\theta\,\mathrm{d}\varphi \qquad \text{(A.138)}$$

On applying the RRs (A.75) to the harmonics (A.138) we obtain

$$\Psi_{2\tau\sigma}(\Omega_4)\Psi_{nlm}(\Omega_4) = \sum_{\substack{\nu = n\pm 1 \\ \lambda = l\pm\tau,\nu>\lambda}} C(2\tau\sigma, nlm; \nu\lambda m+\sigma)\Psi_{\nu\lambda m+\sigma}(\Omega_4) \qquad \text{(A.139)}$$

where $\tau = 0, 1$ and $\sigma = 0, \pm 1$.

Let us multiply RR (A.139) by $\Psi_{n'l'm'}^{*}(\Omega_4)\mathrm{e}^{i\vec{p}\vec{r}}$, integrating over the single sphere in R^4 and associate $\Psi_{2\tau\sigma}(\Omega_4)$ with function $\Psi_{n'l'm'}^{*}(\Omega_4)$ by applying RR

(A.139). This results in the following RRs for functions S:

$$\sum_{\nu',\lambda'} C^*(2\tau-\sigma, n'l'm'; \nu'\lambda'm'-\sigma) S_{nlm}^{\nu'\lambda'm'-\sigma}(\vec{R})$$
$$+(-1)^{\tau+1} \sum_{\nu,\lambda} C(2\tau\sigma, nlm; \nu\lambda m+\sigma) S_{\nu\lambda m+\sigma}^{n'l'm'}(\vec{R}) = 0$$
$$\nu' = n' \pm 1, \quad \lambda' = l' \pm \tau, \quad \nu = n \pm 1, \quad \lambda = l \pm \tau \tag{A.140}$$
$$\nu' > \lambda', \quad \nu > \lambda; \quad \tau = 0, 1; \quad \sigma = 0, \pm 1$$

It can be inferred from RR (A.140) that for $\tau = \sigma = 0$ we obtain a four-term RR; for $\tau = 1$ we obtain an eight-term RR for the functions S.

More complicated RRs can also be obtained by taking the function $\Psi_{3\tau\sigma}(\Omega_4)$ in equation (A.75) and writing a sum similar to equation (A.139) for $\tau = 0, 1, 2$ and $\sigma = 0, \pm 1, \pm 2$.

Using the same procedure as described for the four-dimensional harmonics, we can obtain RRs for the Gaunt integrals of the three-dimensional harmonics:

$$G_{l'm';lm}^{LM} = \int Y_{l'm'}(\theta, \varphi) Y_{lm}(\theta, \varphi) Y_{LM}^*(\theta, \varphi)\, \mathrm{d}\Omega_3 \tag{A.141}$$

Let us consider, for example, the following expansion

$$Y_{1\sigma}(\theta, \varphi) Y_{lm}(\theta, \varphi) = \sum_{L=|l-1|}^{l+1} G_{1\sigma;lm}^{Lm+\sigma}\; Y_{Lm+\sigma}(\theta, \varphi) \tag{A.142}$$

where $\sigma = 0, \pm 1$ and $L \geqslant |m+\sigma|$.

Let us multiply equation (A.142) by $Y_{l'm'}(\theta, \varphi) Y_{LM}^*(\theta, \varphi)$ and integrate over angles θ, φ. Applying RR (A.142) first to the product $Y_{1\sigma} Y_{lm}$ and then to $Y_{1\sigma} Y_{l'm'}$, we obtain the following RRs for the Gaunt integrals:

$$\sum_{\lambda=|l-1|}^{l+1} G_{1\sigma;lm}^{\lambda m+\sigma} G_{l'm';\lambda m+\sigma}^{LM} = \sum_{\lambda=|l'-1|}^{l'+1} G_{1\sigma;l'm'}^{\lambda m'+\sigma} G_{\lambda m'+\sigma;lm}^{LM} \tag{A.143}$$

where the Gaunt integral indices are limited by the conditions of a triangle.

The recurrence relations for matrix elements in LCAO–MO theory can be obtained by using as a general principle a theorem on the linearization of a product of HTOs in the form of relation (A.130). A typical quantum-chemical integral of HTOs can be written as

$$\mathscr{F}(\alpha', \alpha'') = \int u_{\zeta',\alpha'}(\vec{r}-\vec{R}') u_{\zeta'',\alpha''}(\vec{r}-\vec{R}'') f(\vec{r})\, \mathrm{d}^3\vec{r} \tag{A.144}$$

where $f(\vec{r})$ can be either a constant, the Coulomb potential or an integral dependent on $\vec{r}$.

Let us consider the case where $\vec{R}' = \vec{R}''$. We can write the following summation relation for the polynomial parts of the HTOs:

$$\bar{u}_{\zeta',n'l'm'}(\vec{r}) \bar{u}_{\zeta_1,n_1l_1m_1}(\vec{r}) = \sum_{NLM} \mathscr{E}_{n'l'm';n_1l_1m_1}^{NLM}(\zeta', \zeta_1; \zeta) \bar{u}_{\zeta,NLM}(\vec{r}) \tag{A.145}$$

where

$$\bar{u}_{\zeta,nlm}(\vec{r}) = \mathrm{e}^{\zeta r} u_{\zeta,nlm}(\vec{r})$$
$$\mathscr{E}^{NLM}_{n'l'm';n_1l_1m_1}(\zeta', \zeta''; \zeta) = G^{LM}_{l'l';l_1m_1} \bar{B}^{NL}_{n'l';n_1l_1}(\zeta', \zeta_1; \zeta) \tag{A.146}$$

Coefficient B is proportional to integral (A.41), so that it can be computed with the aid of an expansion of the product of the Laquerre polynomials into the series (A.40). As applied to HTOs we can write the following expression for this coefficient:

$$\begin{aligned}\bar{B}^{NL}_{n'l';n_1l_1}(\zeta', \zeta_1; \zeta) &= C_{n'l'} C_{n_1l_1}[C_{NL}\Gamma(N+L+1)]^{-1} \\ &\times \sum_{a=0}^{n'-l'-1} \sum_{n=0}^{n_1-l_1-1} (-1)^{a-b}(a!b!)^{-1}(\zeta'/\zeta)^{l'+a}(\zeta_1/\zeta)^{l_1+b} \\ &\times \binom{n'+l'}{n'-l'-1-a}\binom{n_1+l_1}{n_1-l_1-1-b} \\ &\times (L-l'-l_1-a-b)_{N-L-1}\Gamma(L+l'+l_1+2+a+b)\end{aligned} \tag{A.147}$$

in which the indices NLM can assume the following values:

$$L = |l' - l_1|(2)l' + l_1, \quad N = L + 1(1)n' + n_1 - 1, \quad M = m' + m_1$$

where the index increment is given in parentheses.

Now in relationship (A.145) let us put $\zeta = \zeta'$, $n_1 = 2$, $l_1 = 1$ and $m_1 = \sigma = 0, \pm 1$. Further, let us multiply equation (A.145) by the function $u_{\zeta'',n''l''m''}(\vec{r})\mathrm{e}^{-\zeta' r} f(\vec{r})$, integrating over $\vec{r}$, expanding the product $\bar{u}_{\zeta_1,21\sigma}(\vec{r})u_{\zeta'',n''l''m''}(\vec{r})$ under the integral sign into a series in terms of HTOs. This will result in the following RRs for integrals (A.144):

$$\sum_{N'L'} \mathscr{E}^{N'L'm''+\sigma}_{n''l''m'';21\sigma}(\zeta'', \zeta_1; \zeta'')\mathscr{F}(n'l'm', N'L'm''+\sigma)$$
$$= \sum_{NL} \mathscr{E}^{NLm'+\sigma}_{n'l'm';21\sigma}(\zeta', \zeta_1; \zeta')\mathscr{F}(NLm'+\sigma, n''l''m'') \tag{A.148}$$

where the summation indices take on the following values: $L' = l'' - 1, l'' + 1$; $N' = L' + 1(1)n'' + 1$, $N' > L'$; $L = l' - 1, l' + 1$; $N = L + 1(1)n' + 1$, $N > L$, and the parameter ζ_1 is arbitrary.

To conclude, consider a RR for coefficients B, given in equation (A.147). The simplest RRs can be obtained for non-normalized coefficients B, given by the integral of non-normalized radial functions HTO.

In equation (42′) let us take $Q = L$, $q = l$, $q' = l'$, $K = N - L - 1$, $k = n - l - 1$, $k' = n' - l' - 1$, designating the coefficient thus obtained as $\tilde{B}^{NL}_{n'l';nl}$. Using the four-term RR of type

$$zL^{\alpha+2}_{\nu}(z) - zL^{\alpha+2}_{\nu-1}(z) = (\nu + \alpha + 1)L^{\alpha}_{\nu}(z) - (\nu + 1)L^{\alpha}_{\nu+1}(z) \tag{A.149}$$

for the Laguerre polynomials, which RR is produced by combining the known

RRs[12], we find the required RR for the integral $\tilde{B}$ to be

$$\tilde{B}^{N+1\,L+1}_{n'l';nl} - \tilde{B}^{N\,L+1}_{n'l';nl} = (N+L+1)\tilde{B}^{NL}_{n'l';nl} - (N-L)\tilde{B}^{N+1\,L}_{n'l';nl} \qquad \text{(A.150)}$$

It is clear that, due to the permutation symmetry for pairs of indices NL, $n'l'$ and nl, RR (A.150) can be written for any pair of indices.

References

1. S. F. Boys. *Proc. Roy. Soc.*, **A200**, 542 (1950); **A258**, 402 (1960).
2. H. Shull and P.-O. Lowdin. *J. Chem. Phys.*, **23**, 1362 (1955).
3. I. Shavitt. In *Methods in Computational Physics*, B. Alder, S. Fernbach and M. Rotenberg (eds), Academic Press, New York (1963), p. 16.
4. E. Filter and E. O. Steinborn. *J. Math. Phys.*, **19**, 79–84 (1978).
5. A. V. Niukkanen. *International J. Quant. Chem.*, **25**, 941–55 (1984).
6. V. A. Fock. *Izv. Akad. Nauk SSSR, Ser. Phys.* **2**, 169–79, (1935) (in Russian and in German).
7. J. A. Gaunt. *Phil. Trans. R. Soc. London, Ser. A.*, **228**, 192 (1929) (Appendix).
8. T. Shibuya and C. E. Wulfman. *Proc. Roy. Soc.*, **A286**, 376–89 (1965).
9. B. K. Novosadov. *Int. J. Quant. Chem.*, **24**, 1–18 (1983).
10. L. A. Gribov, V. I. Baranov and B. K. Novosadov. *Methody raschiota electronno-kolebatel'nykh spektrov mnogoatomnykh molekul* (Methods of calculation of electronic–vibrational spectra of polyatomic molecules), Nauka, Moscow (1984), Chs 4 and 6.
11. L. C. Biedenharn. *J. Math. Phys.*, **2**, 433 (1961).
12. A. Erdelyi (ed.). *Higher Transcendental Functions* McGraw-Hill, New York (1953).
13. D. A. Varshalovich, A. N. Moskaliov and V. K. Khersonsky. *Quantum Theory of Angular Momentum*, Nauka Leningrad. (1975) (in Russian).

13. [illegible] Nakamoto, Infrared and Raman Spectra of Inorganic and Coordination Compounds [illegible]

Index